Springer Natural Hazards

More information about this series at http://www.springer.com/series/10179

Sebastiano D'Amico
Editor

Earthquakes and Their Impact on Society

Springer

Editor
Sebastiano D'Amico
Department of Geosciences
University of Malta
Msida
Malta

Springer Natural Hazards
ISBN 978-3-319-34858-2 ISBN 978-3-319-21753-6 (eBook)
DOI 10.1007/978-3-319-21753-6

Springer Cham Heidelberg New York Dordrecht London

Softcover re-print of the Hardcover 1st edition 2016

Printed on acid-free paper

Springer International Publishing AG Switzerland is part of Springer Science+Business Media (www.springer.com)

Preface

The destructive potential of earthquakes depends on their magnitude and the placement of the hypocenter. Worldwide efforts in establishing seismological and geophysical networks has increased considerably and data centers provide information that is essential to identify earthquake locations, understand the physics of the earthquakes and faulting mechanisms, as well as studies on site effects and earthquake recurrence. These types of studies play a key role in mitigating earthquake hazards and planning emergency response. However, recent earthquakes demonstrated the risks to modern societies, especially in terms of health, safety, and economic viability. The study of earthquakes combines science, technology, and expertise in infrastructure and engineering in an effort to minimize human and material losses when their occurrence is inevitable. Although probably the most important, direct shaking effects are not the only hazard associated with earthquakes, other effects such as landslides, liquefaction, and tsunamis have also played important part in destruction produced by earthquakes. Most earthquake-related deaths are caused by the collapse of structures, and the construction practices play a tremendous role in the death toll of an earthquake.

As an example among the others, in the Messina area (southern Italy), more than 120,000 people perished in an earthquake (M = 7.2) that struck the region in 1908, mainly killed due to the easily collapsible structures that dominated both the major cities and villages of the region. A larger earthquake that struck San Francisco three years earlier had killed fewer people (about 700) because building construction practices were less vulnerable and predominantly made by wood. In both areas, these events played a key role in the future of the region. Recently, on January 12, 2010, disaster struck Haiti and its economy, as well as its surrounding nations. Death totals topped 200,000, and reliable industries and structures were destroyed, leaving the nation in a tough situation. It remains clear that the earthquake that has recently stuck Haiti has had a catastrophic impact on Haiti, leaving its economy in shambles. In the last decades, other important examples of catastrophic events are the great Sumatra earthquake (December 26, 2004) and the Japan earthquake (March 11, 2011) associated also with tsunamis of notable energy that caused

Fig. 1 The children drawing presented in this figure has the aim to illustrate the effects of earthquakes on society as seen by children affected by such catastrophic phenomenon

several casualties and damage to both residential buildings and industrial facilities (e.g., Fukushima Daiichi nuclear power plant).

Such catastrophic events take an unimaginable toll on children and their families as demonstrated, for example, by the 2008 Sichuan Province and 2005 Pakistan earthquake. In such disaster management contexts, family members as well as health care and school personnel are the first-line responders and are natural sources of continued social support as children recover. Such kind of natural disasters can really impact on children and consequently on our societies. The children drawing presented in Fig. 1 has the aim to illustrate the effects of earthquakes on society (the word "PRIMA" stands for before, while the word "DOPO" stands for after). The drawing has been realized by one, the children hosted in the Mirandola camp (during the 2012 Emilia seismic sequence, Italy). Such camp was also used by few research teams (I took part in one of them) conducting real-time monitoring and data collection in the epicentral area.

The chapters in this book are devoted to various aspects of earthquake research and analysis of their impact on modern and past societies. In the following a brief outline of each contribution is given.

A Description of Seismicity Based on Non-extensive Statistical Physics: A Review

The manuscript by Vallianatos et al. opens this book tackling the problem of the physics of earthquakes which still has many questions that have not yet been answered since the phenomenon is subjected to many uncertainties. In this chapter, the authors review some fundamental properties of earthquake physics (in the geodynamic laboratory scale) and how these are derived by means of non-extensive statistical physics with the main goal to understand aspects of the underlying physics that lead to the evolution of the earthquake phenomenon.

Recognition of periAdriatic Seismic Zones Most Prone to Next Major Earthquakes: Insights from a Deterministic Approach

Mantovani et al. describe the short-term development of tectonic processes which is closely related to major decoupling earthquakes at plate boundaries. Each of these events triggers a perturbation of the strain field which propagates in the surrounding area and may increase the probability of earthquakes in tectonic zones or may even cause further seismicity when the fault involved is close to failure. This implies that in a given tectonic context, the past distribution of strong shocks can significantly influence the location of the next seismic events. In this work, the authors discuss on how the above interpretation may have influenced the spatiotemporal distribution of major shocks in the central Mediterranean region, with particular regard to the periAdriatic zones. It is argued that the regularity patterns of seismicity so far identified may provide significant information on the possible location of the next strong earthquakes in the Italian peninsula. The results of this analysis suggest that the probability of next strong shocks is higher for the northern periAdriatic zones, with particular regard to the northern Apennines, than in the southern zones (southern Apennines and Calabria). Present knowledge does not allow to give this kind of information for some Italian zones (mainly Sicily, Apulia, and northwestern Italy), for which significant regularity patterns of seismicity have not been recognized yet.

Forecasting Moderate Earthquakes in Northern Algeria and Morocco

In this work, Peláez et al. studied the correlation between locations of $M_W \geq 5.0$ earthquakes and locations of $5.0 > M_W \geq 4.0$ events for northern Algeria and Morocco. A preliminary study shows that it can be observed a relatively good agreement between locations for these two data sets, that is, minor earthquake locations could be used to forecast future places where will happen moderate and moderate to strong earthquakes. They propose a time-independent forecasting model based on the spatially smoothed seismicity rate of $M_W \geq 4.0$ earthquakes. Finally, a time-independent forecasting model is proposed from the computation of $M_W \geq 5.0$ and $M_W \geq 6.0$ earthquake probabilities considering that seismicity follows both a Poisson process and the Gutenberg–Richter magnitude–frequency relationship.

An Earthquake Catalogue (2200 B.C. to 2013) for Seismotectonic and Seismic Hazard Assessment Studies in Egypt

Sawires et al. aim at preparing new and up-to-date unified and Poissonian earthquake catalogue for Egypt including the focal mechanism solution data, so that the earthquake information can be reached from a single source. A catalogue for earthquakes that occurred in Egypt and its vicinity during the period 2200 B.C.–2013 was compiled for achieving a unified magnitude scale. Data were obtained from different sources, local, regional, and international. Earthquake magnitudes are reported in different scales and come from a variety of sources. The initial compiled catalogue comprised a total of 64,613 earthquakes (historical and instrumental events). In addition, a focal mechanism solution database was collected. This database contains 688 fault plane solutions gathered from different published and unpublished sources, covering the time period from 1940 until the end of 2013.

For establishing a common magnitude, namely an equivalent moment magnitude M_W, two new relationships correlating M_S and m_b with M_W were derived.

Probabilistic Seismic Hazard Assessment for Romania

In this study, Vacareanu et al. focus on the probabilistic seismic hazard assessment for Romania using the latest seismicity data and ground motion prediction models. In the first step, an evaluation of the applicability of several ground motion prediction models for seismic sources in Romania is performed on a database which consists of over 300 strong ground motions recorded in Romania, Bulgaria, and Moldova during ten Vrancea subcrustal earthquakes. The testing procedure employed in this study uses several goodness-of-fit measures. A sensitivity analysis for the probabilistic seismic hazard assessment (PSHA) results for four cities in Romania (Bucharest, Focsani, Iasi, and Craiova), located in the forearc of the Carpathian Mountains, is presented. The results provide the basis for defining the computation hypothesis and corresponding weights assigned to the logic tree branches for the PSHA for Romania. The scope is to provide a refined input for the implementation of the seismic action according to the requirements of EN 1998-1.

Practicality of Monitoring Crustal Deformation Processes in Subduction Zones by Seafloor and Inland Networks of Seismological Observations

As well as the 2011 Tohoku earthquake, there will be other megathrust earthquakes around Japan in the future. These earthquakes will surely have a strong effect on various aspects of society, including economic, psychological, infrastructural, and survival problems. In this chapter, Ariyoshi and Kaneda review the progress of studies on monitoring seismic changes in crustal deformation and seismic activity by comparing simulation results for the Tokai and Tonankai earthquakes in order to discuss the validity of modeling of subducting plates and the detectability of preseismic changes based on seafloor and inland observation networks for disaster mitigation. Since leveling change due to shallower VLF swarms is so local as to be incoherent, removal of the moving-averaged data from the data stacked by four nearby observation points in the same node of Dense Oceanfloor Network System for Earthquakes and Tsunamis (DONET) may be useful to detect the short-term local leveling change. This developed oceanfloor observation is expected as short-term forecasting so as to reduce economic loss reasonably.

Neo-deterministic Definition of Seismic and Tsunami Hazard Scenarios for the Territory of Gujarat (India)

Magrin et al. present a neo-deterministic definition of earthquake hazard scenarios. A reliable and comprehensive characterization of expected seismic ground shaking is essential to improve building codes, particularly for the protection of critical infrastructures and for land use planning. So far, one of the major problems in classical methods for seismic hazard assessment consisted in the adequate characterization of the attenuation models, which may be unable to account for the complexity of the medium and of the seismic sources and are often weakly constrained by the available observations.

Geophysical Characterization of Liquefied Terrains Using the Electrical Resistivity and Induced Polarization Methods: The Case of the Emilia Earthquake 2012

Nasser describes liquefaction of a shallow, water-saturated sand and silty sand layer which has resulted in the damage of several buildings as well as of roads and sidewalks during the Emilia–Romagna earthquake (northern Italy, May 20, 2012). In fact, massive surface fracturing, sand upwelling, sand volcanoes, limited blister formation, and lateral spreading liquefaction features took place immediately after the main shock.

Working Strategies for Addressing Microzoning Studies in Urban Areas: Lessons from the 2009 L'Aquila Earthquake

Vessia et al. present working strategies for addressing microzoning studies in urban areas presenting the results through lessons learned from the 2009 L'Aquila earthquake. This is a crucial approach for mitigating risk in prone earthquake areas, and several efforts have been done in the past years by the Department of Civil Protection. This also may help the authorities in estimating economic losses due to the strong earthquakes and comparing the results with losses that occurred in the last decades.

Earthquake-Induced Reactivation of Landslides: Recent Advances and Future Perspectives

In this chapter, Martino presents a nice description of seismic-induced landslide. Earthquake-induced reactivation of landslides is a focus topic in the risk management as severe damages and losses were caused so far from seismically triggered slope failures. Slope stability conditions under seismic action were studied since several decades by pseudostatic approaches as well as by sliding block methods that follow the Newmark approach. These last ones were more recently upgraded by flexible block approaches to provide a more constrained evaluation of earthquake-induced displacements, i.e., by considering the landslide mass resonance during the seismic shaking. Nevertheless, these approaches cannot take into account the very complex interactions between seismic waves and slope pointed out by several case histories that are reported in the literature. Such interactions can be simulated by more sophisticated stress–strain numerical models. Nonetheless, to carry out these simulations, it is necessary to provide very strong constraints to both the geological setting of the slope and the local seismic response. In this regard, a fundamental contribution derives from detailed engineering–geological reconstructions as well as in-site geophysical measurements. Very recent studies experienced theoretical approaches for pointing out the significance of some physical parameters, such as the ratios of characteristic periods related to the seismic wave properties and to the landslide mass geometry, to provide a more exhaustive prevision of earthquake-induced landslide displacements.

Resilience, Vulnerability and Prevention Policies of Territorial Systems in Areas at High Seismic Risk

Teramo et al. present aspects of the seismic vulnerability and resilience of an urban system through a multidimensional and interdisciplinary approach addressed to an effective prevention policy and sustainable territory planning. In particular, in their study, they analyze the actual capability of a territorial system to face the effects of a strong earthquake on the basis of targeted territorial, seismic, health, social, and legal indicators. Specific applications aimed at emergency management and rescue coordination are moreover proposed through the territory monitoring with innovative wireless sensor networks that ensure unexpected effectiveness levels.

Numerical Study of the Seismic Response of a Mid-Rise RC Building Damaged by 2009 Tucacas Earthquake

In this chapter, Vielma et al. present the results of the numerical simulation related to a reinforced concrete framed building who suffered light damage by the 2009 Tucacas Earthquake (6.4 Mw). The building was designed according to the current Venezuelan codes, splitting the structure into three different modules in order to avoid the negative effects of in-plane irregularities and represent a typical mid-rise building located on high seismic-prone areas. In order to improve the original seismic design of the building, a new building was proposed using an innovative energy-based procedure. The set of dynamic analyses was used in order to formulate a new procedure for the determination of fragility curves. Results show that the new procedure is suitable in order to predict the damage state which the building may reach when it is subjected to a strong earthquake, and therefore, it can be really useful for construction practices in order to ensure the life of the population and to reduce the seismic vulnerability.

Analysis of Seismic Vulnerability of Rural Houses in China

Jue et al. present a detailed analysis of seismic vulnerability of rural houses in China. This represents a key point because most of seismic disasters in China have taken place in rural areas. The damage of dwelling houses has been the biggest cause of casualties and asset loss, which causes enormous impact on rural societies and economic systems. In this paper, the seismic vulnerability of counties was assessed with regard to the structural features of rural houses, measured by the number and ratio of potentially damaged houses in the case of seismic disasters.

Finite Element Modelling for Seismic Assessment of Historic Masonry Buildings

Betti et al. discuss on use of the finite element modelling technique for assessment of seismic vulnerability of historic masonry buildings. In particular, two representative case studies are presented: a masonry church and an old residential building. The aim of the chapter is to outline that advanced numerical analyses can provide significant information to understand the actual structural behavior of ancient buildings under seismic loading. A clear understanding of the structural behavior, based on sophisticated tools of structural analysis, can reduce the extent of remedial measures in the restoration of ancient buildings through a reliable strengthening.

Earthquake-Resistant and Thermo-Insulating Infill Panel with Recycled-Plastic Joints

Modern buildings should be endowed with features that allow people to carry out their activities in a sustainable environment and with a high level of safety. The major concern of this work refers to one peculiar aspect that recently is gaining the attention of designers and construction companies: the role of infill panels. There is now an increasing consensus on how these non-structural elements should be carefully designed, both to increase their thermo-insulating capacity and to adapt their earthquake performance to the most recent standards. The solution was found in an innovative constructive system for infill panels, in which the traditional hollow-core masonry blocks are connected, rather than with the traditional mortar layers, through joints made from recycled plastic. The constructive system has the advantage of a rapid assemblage, which reduces the construction times, and allows the insertion of insulating panels that reduce the thermal conductivity to very low levels. Therefore, it will be extremely important for the developments of our modern societies.

Base Isolation and Translation of a Strategic Building Under a Preservation Order

Monti et al. in this chapter deal with a retrofitting intervention on a strategic building. In particular, the seismic retrofitting of a strategic building owned by the Italian Highway Company, built in the 1950s, is presented. The structure is endowed with mushroom-shaped columns with hollow section at the *pilotis* floor, which makes them particularly vulnerable to shear. In order to retrofit the frame, it was decided to design a seismic isolation system at the base of the building, by inserting the isolation devices under the existing columns at the *pilotis* floor. The main benefit of this solution is the possibility to operate exclusively at the ground level, without interrupting the work activities at the upper floors. In general, from the social standpoint, the example presented in this chapter could serve as pilot study that can help in planning the retrofitting of residential building without displacing the occupants to different locations while working on the building.

Lessons from the Wenchuan Earthquake

A strong earthquake like the Wenchuan earthquake on May 12, 2008, in China causes huge losses in human lives and property damages in a short time. Nevertheless, the losses after the main shock are very serious as well, because of the disaster chain, which is often underestimated in its consequences. For this reason, reconstruction projects have been carried out after the earthquake with varying degrees of success. This paper deals with a number of issues that occurred after the Wenchuan earthquake, from which certain lessons can be learned. Inappropriate planning has relocated towns and villages along the active fault zone in seismic risk zones. A safe distance of the buildings from the active fault zones was not respected in all reconstruction projects. The amount of loose source material on the slopes for debris flows after the earthquake was underestimated, leading to ineffective design of hazard control projects. The level of newly designed roads, bridges, and tunnels turned out to be too low due to an underestimation of the rise of the riverbeds in the main valleys. Areas with a high hazard for flooding and debris flows were ignored in the planning of new built-up areas.

Lessons Learned from the Recent Earthquakes in Iran

Although the understanding of the reasons and impacts of natural disasters and models to compute the frequency and severity of disastrous earthquakes has improved, a slight increase can be seen in the earthquake deaths as a percentage of the total global deaths. Iran, like other developing counties, suffers earthquake causalities and economic damages. The residences and other constructions in Iran especially in small villages and towns are generally built without considering seismic design regulations, so they are highly vulnerable. Comparing yearly earthquake death rates among Iran, Japan, and USA during three different periods revealed that while Japan and USA have been reduced their yearly rates, Iran's status has been worsening. Among the principal causes of high death toll and economic losses in Iran, one can refer to building collapse, changes in land use, increases in the concentration of people and capital in high-risk areas, fast and uncontrolled urbanization, the persistence of extensive urban and rural poverty, the depreciation of the region's environment resulting from the mismanagement of natural resources, ineffective public policies, and lagging and misguided investments in infrastructures. Among the important lessons learned from the recent earthquakes in Iran are fundamental earthquake hazard reduction needs to engage national consciousness at all levels of society, public education, solving the problems in the natural disaster preparedness system, and deficiencies in current construction practice.

"The Impact of the Great 1950 Assam Earthquake on the Frontal Regions of the Northeast Himalaya"

In this paper, Devi and Bora discuss the great 1950 Assam earthquake which caused widespread devastation throughout the frontal regions of northeast Himalaya and it is classified as the 10th largest earthquake of the twentieth century. The ground cracked and fissured, bridges and rail lines were destroyed, and riverbeds silted up. Immediately after the shock, several tributaries of Brahmaputra River were blocked by landslips caused by the violent shaking of the earthquake causing drastic flooding afterward. This great earthquake, destructive in Assam and Tibet (China), was an important earthquake event since the introduction of seismological observation stations. This earthquake changed topographical features in the eastern syntaxis and caused havoc in the frontal region of northeast Himalaya, making drastic impact on human civilization.

Archaeoseismology in Sicily: Past Earthquakes and Effects on Ancient Society

This work presents a review of archeological evidence of strong earthquakes occurring in Sicily at the time of Greek and Roman colonization, a period of considerable political, economic, and social instability. In this historical context, the earthquake effects may have been obscured or overlooked to some extent, and consequently, the documentary information on ancient earthquakes, when available, is often sparse and lacking objectivity. The studied cases combine historical and archaeological data together with the evidence of structural damage to archaeological sites. Looking into past, the vocation of archaeoseismology lies in the identification of past seismic events and particularly what the ancient society knew on earthquakes and what kind of seismic effects produced on buildings and sites.

The Earthquakes of Southern Italy from the 18th to the 20th Centuries

In this chapter, Catalani present the earthquakes of southern Italy from the eighteenth to the twentieth centuries. In this period, the high density of population in southern Italy, the characteristics of the territory (mainly mountainous with malarial planes along the coast), and the economic system based on the land had pushed the southern communities to adapt to their territory, finding forms of appropriate use and exploitation of the land itself. Even the building industry had adapted to the territory. However, these people had also learned how to cope with the worst effects of earthquakes. Nevertheless, it is not possible to understand the importance of earthquakes in Italy completely and in particular in the southern regions, if only we consider them as natural phenomena. Earthquakes have surely contributed to the physical alteration of the landscape, just like other natural forces (such as wind, rain, and changes in temperature) together with human activities. However, over the centuries, the earthquake has been marked as a dynamic element of the "cultural landscape." This is because all catastrophic events, therefore even earthquakes, mark the life of communities and become part of their historical memory. Communities build settlements, organize their economy, and establish their social relationships upon which these disastrous events then take effects.

Earthquake and People: The Maltese Experience of the 1908 Messina Earthquake

In this paper, Borg and colleagues describe the perception in the Maltese archipelago of the devastating earthquake that struck southern Italy along the Messina Strait. As a result, the cities of Messina along Sicily's coast and Reggio di Calabria were completely destroyed causing more than 120,000 fatalities and left many without shelter. The 1908 earthquake had a significant impact on buildings and people and local communities which were displaced. The Maltese experience of the Messina 1908 earthquake relied on communication which reached Malta after the event. The chapter discusses information on the building deficiencies and damage, limitations of communication infrastructure during that period, and limits to timely emergency response to support the population and emergency action at the beginning of the twentieth century.

A Web Application Prototype for the Multiscale Modelling of Seismic Input

A Web application prototype is described, aimed at the generation of synthetic seismograms for user-defined earthquake models. The Web application's graphical user interface hides the complexity of the underlying computational engine, which is the outcome of the continuous evolution of sophisticated computer codes, some of which saw the light back in the middle 1980s. With the Web application, even the non-experts can produce ground shaking scenarios at the local or regional scale in very short times, depending on the complexity of the adopted source and medium models, without the need of a deep knowledge of the physics of the earthquake phenomenon. Actually, it may even allow neophytes to get some basic education in the field of seismology and seismic engineering, due to the simplified intuitive experimental approach to the matter. One of the most powerful features made available to the users is indeed the capability of executing quick parametric tests in near real time, to explore the relations between each model's parameter and the resulting ground motion scenario. The synthetic seismograms generated through the Web application can be used by civil engineers for the design of new seismic-resistant structures, or to analyze the performance of the existing ones under seismic load.

Rapid Response to the Earthquake Emergencies in Italy: Temporary Seismic Networks Coordinated Deployments in the Last Five Years

The rapid deployment of a dense temporary network of seismic stations, soon after the occurrence of a damaging earthquake, is an essential action to improve the seismic monitoring and the quality of the studies on the aftershock sequence. Having seismic waves recorded with a dense seismic network can greatly improve the detection of earthquakes and the estimation of hypocenter parameters. Since 1990, the National Institute of Geophysics and Volcanology (Italian: IstitutoNazionale di Geofisica e Vulcanologia (INGV)) manages a portable seismic network structure: an instrumental pool to deploy dense seismic networks for scientific experiments and to monitor aftershocks after the occurrence of damaging earthquakes. Today, this pool includes about 100 seismic stations which, if necessary, can be integrated in the real-time seismic surveillance system of INGV. The real-time data contribute to the monitoring of the seismicity in the epicentral area, and the off-line analysis of the recorded seismograms allows the imaging of the fault system geometry and kinematics. The chapter presents the INGV portable seismic network, its history, and the current coordination projects with other Italian and international institutes. Activities and emergency operations in the last five years with special focus on the “Emilia 2012” and the “Pollino 2011–2014” seismic crises are also discussed.

The Key Role of Eyewitnesses in Rapid Impact Assessment of Global Earthquakes

This chapter presents the strategy and methods implemented at the European Mediterranean Seismological Centre (EMSC) for rapidly collecting in situ observations on earthquake effects from eyewitnesses to reduce uncertainties in rapid impact assessment of global earthquakes. The authors show how Internet and communication technologies are creating new potential for rapid and massive public involvement by both active and passive means underlining the importance of merging results from different methods to improve performance and reliability. They also explore what could be the next technical development phase, by observing that the pervasive use of smartphones changes the way rapid earthquake information is accessed. Finally, they discuss how these approaches not only augment data collection on earthquake phenomenon at little cost but also how they change the way that scientists interface with eyewitnesses and how it pushes us to better understand and respond to the public’s demands and expectations in the immediate aftermath of earthquakes through improved information services.

Real-Time Mapping of Earthquake Perception Areas in the Italian Region from Twitter Streams Analysis

In this chapter, D’Auria and Convertito propose a strategy to retrieve in real time useful information about the area where an earthquake has been perceived and how many people felt it, using data mining of Twitter streams. They show that using a proper normalization of these data allows a quantitative definition of an Earthquake Perception Index based on Twitter posts (EPIT). This index shows a good correlation with ground motion parameters and macroseismic data and hence allows a rapid but realistic mapping of the perception area. The mapping of earthquake perception area is useful to determine how many people have felt it. This is an important issue because even moderate-magnitude earthquakes can affect critical communication infrastructures. However, theoretical estimates of instrumental intensity distribution, derived from ground motion parameters (e.g., ShakeMaps), may be poorly correlated with the actual earthquake perception. Furthermore, the number of people who felt the earthquake depends strongly on the spatial distribution of the population density. In recent years, there has been a growing interest in the data mining of citizen-provided information from social networks, Internet accesses, and Web-based macroseismic surveys aimed at detecting, locating, and characterizing the macroseismic field of moderate and strong earthquakes.

The Easter Sunday 2011 Earthquake Swarm Offshore Malta: Analysis on Felt Reports

This chapter presents the latest efforts done by the Seismic Monitoring and Research Unit (SMRU) at the University of Malta to put in place a "Did you feel an earthquake?" online questionnaire in order to start gathering information from locally felt earthquake-related shaking. In particular, the event occurred on Easter Sunday April 24, 2011, is discussed. During the seismic swarm, the SMRU located 15 earthquakes with magnitudes ranging from ML 1.8 to 4.1 over a period of 4 days. A total of 489 felt reports were submitted through the online questionnaire. The compilation of the data is a first of its kind for the Maltese islands. These reports gave limited qualitative and quantitative information about the shaking experience felt across the islands. Here, the author presents a summary of the reports following the main shock. A maximum intensity value IV on the European Macroseismic Scale was assigned. No structural damage was reported. The data reflect the demographics as well as the different types of buildings found across the archipelago.

Earthquake Readiness and Recovery: An Asia-Pacific Perspective

This chapter discusses lessons learned from the analysis of community responses to earthquakes in New Zealand and Taiwan and how lessons learned from this can inform the development of earthquake readiness strategies in countries whose citizens have to live with high levels of seismic risk. It discusses how individual, social, and societal factors interact to increase resilience and how this can be enacted to reduce risk and enhance response effectiveness. By identifying how personal, community, and cultural characteristics interact to influence earthquake readiness, response, and reduction, this chapter offers insights that can inform the development of the risk communication and community outreach programs required to help answer the call for the development of DRR strategies issued by the ISDR. Following a discussion of the nature of comprehensive earthquake readiness, the chapter addresses the degree to which an earthquake readiness theory can predict readiness in different cultures (New Zealand and Taiwan) and discusses the extent to which people's accounts of their earthquake experiences (in culturally diverse countries) can validate and contribute to developing earthquake readiness theory.

Geoethics, Neogeography and Risk Perception: Myth, Natural and Human Factors in Archaic and Postmodern Society

This chapter combines anthropological–philosophical and geographical–geological research on man's perception of, and reaction to, natural catastrophes such as earthquakes. The first part of the study offers an articulate and cohesive picture of the defense mechanisms man has deployed, since ancient cultures, against this risk, and these are identified with mythical–ritualistic repetition. At critical moments, man develops a series of practical strategies resting on ritual action. The second part of this work is a synthesis of research on the perception of seismic risk in the area of Pollino (southern Italy), where it is been four years that an ongoing earthquake swarm is affecting the area between Calabria and Basilicata. The perception of seismic risk is an important dimension for the schedule. Geoethics can certainly help especially in educating the territory in terms of integrated risk management. In this context, a questionnaire was administered to the students of primary and secondary education and to a sample of adults in some villages affected by the earthquake of Pollino. This study reveals that with the passing of time, the "indisputable certainty" of the earthquake as a form of divine punishment, which dates back to ancient societies, is today falling apart as in particular young people have an understanding of man's responsibility in the causing of natural catastrophes. Improved communications through new information technologies, awareness of the complexity of risk, and the level of preparation would increase the resilience of the territory and allow a more effective planning.

Psychosocial Support to People Affected by the September 5, 2012, Costa Rica Earthquake

This chapter discusses the intervention of the Brigade of psychosocial support of the University of Costa Rica visiting several communities of the area affected by the September 5, 2012, Mw 7,6 Costa Rica earthquake in the weeks following the event. The team was accompanied by members of the Red Sismologica Nacional Preventec and students of the postgraduate program in risk management and emergency response. The interventions included technical talks on earthquakes and tsunamis, incident commands, and care postdisaster. The main fears found in the population were as follows: fear of a future earthquake, of a tsunami, of aftershocks, of the detachment of the Nicoya Peninsula, of emissions of massive poisoning gases, and of the emergence of an underwater volcano. With technical knowledge and psychosocial support, the members of the Brigade took tranquility and calm to people and helped them to return to the joy and normalcy. The work of the interdisciplinary group was extremely successful.

The Lisbon Earthquake in the French Literature

This closing chapter presents the results on how the 1755 great Lisbon earthquake has been impacted the contemporary society and in particular the French literature. This large and catastrophic event caused damages and losses also in Morocco and Algeria; in Europe, it caused considerable damage in Spain, mainly in Madrid and Seville; and the shaking was also widely felt across Europe. This event represents the first earthquake to be studied scientifically for its effects over a large area, and can it be considered as the event which led to the birth of modern seismology and earthquake engineering. It was also widely discussed and dwelt upon by European philosophers, and inspired major developments in theodicy and in the philosophy.

Acknowledgments

I am grateful to all the authors for their close cooperation while preparing their contributions. I also gladly acknowledge all the referees, belonging to a number of research institutions located in European countries, such as Italy, Malta, Moldavia, Portugal, Spain, and Switzerland, and in Australia, India, and USA. Their careful reading and constructive suggestions contribute to the standard of the final versions of each manuscript collected in this book. A special thank goes to Johanna Schwarz, Mannsperger Claudia and all the Editorial staff for their professional assistance and technical support during the entire publishing process.

Sebastiano D'Amico

Contents

A Description of Seismicity Based on Non-extensive Statistical Physics: A Review

Filippos Vallianatos, Georgios Michas and Giorgos Papadakis

1 Introduction

Earthquakes have always been one the most intriguing natural phenomena for mankind. The abruptness of the shaking ground and the devastating consequences for the human environment were always attracting people's fear and wonder. Despite the large amount of effort that has been dedicated in understanding the physical processes that lead to the birth of an earthquake and the significant progress that has been achieved in this field, the prediction of an upcoming earthquake still remains a challenging question (Nature debates 1999).

Concerning the physics of earthquakes, many questions have not yet been answered since the phenomenon is subjected to many uncertainties and degrees of freedom. It is true that we have a good understanding of the propagation of seismic waves through the Earth and that given a large set of seismographic records, we are able to reconstruct a posteriori the history of the fault rupture. However, when we consider the physical processes leading to the initiation of a rupture with a subsequent slip and its growth through a fault system, giving rise to an earthquake, then our knowledge is really limited. Not only the friction law and the rules that govern rupture evolution are largely unknown, but also the role of many other processes such as plasticity, fluid migration, chemical reactions, etc., and the couplings among them, remain unclear (Main et al. 1989, 1992; Sammonds 2005; Sammonds and Ohnaka 1998; Vallianatos et al. 2004).

F. Vallianatos (✉)
Technological Educational Institute of Crete, Laboratory of Geophysics and Seismology, Chania, 73100 Crete, Greece
e-mail: fvallian@chania.teicrete.gr

F. Vallianatos · G. Michas · G. Papadakis
Institute for Risk and Disaster Reduction, University College London, Gower Street, London, WC1E 6BT, UK

S. D'Amico (ed.), *Earthquakes and Their Impact on Society*,
Springer Natural Hazards, DOI 10.1007/978-3-319-21753-6_1

Despite the extreme complexity that characterizes the mechanism of the earthquake generation process, simple phenomenology seems to apply in the collective properties of seismicity. Fault and earthquake populations present scaling relations that seem to be universal in the sense that are appearing in a variety of tectonic environments and scales that vary from the laboratory, to major fault zones and plate boundaries. The best known is the Gutenberg-Richter scaling relation (Gutenberg and Richter 1944) that indicates power-law scaling for the earthquake size distribution. Short and long-term clustering, power-law scaling and scale-invariance have also been exhibited in the temporal evolution of seismicity (Kagan and Jackson 1991). In addition, earthquakes exhibit fractal spatial distribution of epicenters and they occur on fractal-like structure of faults (Turcotte 1997). All these properties provide observational evidence for earthquakes as a nonlinear dynamic process (Kagan 1994).

Due to these properties, concepts such as fractals, multi-fractals, non-linear processes and chaotic dynamical systems are becoming increasingly fundamental for analyzing data and understanding processes in geosciences. In recent years, there is a growing interest on approaching seismicity and other natural hazards, regarding the science of complex systems and the fractal nature of these phenomena (Bak and Tang 1989; Bak et al. 1988; Vallianatos 2009). In the context of critical point phenomena (see Sornette 2004), "self-organized criticality" (SOC) has been proposed by Bak et al. (1987) as a possible driving mechanism that produce the scale-invariant properties of the earthquake populations, such as the G-R scaling relation (see also Bak and Tang 1989; Sornette and Sornette 1989). According to this theory, Earth's crust is in a near critical state that spontaneously organizes into an out-of-equilibrium state to produce earthquakes of fractal size distributions.

Regarding the physics of "many" earthquakes and how this can be derived from first principles, one may wonder:

- How can the collective properties of a set formed by all earthquakes in a given region, be derived?
- How does the structure of seismicity, as formed by all earthquakes, depends on its elementary constituents—the earthquakes? What are these properties?

It may be that these collective properties are largely independent on the physics of the individual earthquakes, in the same way that many of the properties of a gas or a solid do not depend on the constitution of its elementary units. It is natural then to consider that the physics of many earthquakes has to be studied with a different approach than the physics of one earthquake and in this sense we can consider *the use of statistical physics not only appropriate but also necessary to understand the collective properties of earthquakes*. A significant attempt is given in a series of works (Main 1996; Main and Al-Kindy 2002; Rundle et al. 1997, 2003) where classic statistical physics are used to describe seismicity. Then a natural question arises. *What type of statistical physics is appropriate to commonly describe effects from the microscale and crack opening level to the level of large earthquakes and plate tectonics?*

An answer to the previous question could be non-extensive statistical physics (NESP), originally introduced by Tsallis (1988). The latter is strongly supported by the fact that this type of statistical mechanics is the appropriate methodological tool to describe entities with (multi) fractal distributions of their elements and where long-range interactions or intermittency are important, as in fracturing phenomena and earthquakes. NESP is based on a generalization of the classic Boltzmann-Gibbs entropy and has the main advantage that it considers all-length scale correlations among the elements of a system, leading to an asymptotic power-law behavior. So far, NESP has found many applications in nonlinear dynamical systems including earthquakes (Tsallis 2009). In a series of recent publications, it has been shown that the collective properties of the earthquake and fault populations from the laboratory scale (Vallianatos et al. 2011a, 2012a, b, 2013; Vallianatos and Triantis, 2012), to local (Michas et al. 2013), regional (Abe and Suzuki 2003, 2005; Telesca 2010a; Papadakis et al. 2013) and global scale (Vallianatos and Sammonds 2013) can be reproduced rather well using the concept of NESP.

In the present chapter, we review some fundamental properties of earthquake physics (in the geodynamic-laboratory scale) and how these are derived by means of non-extensive statistical physics. The aim is to understand aspects of the underlying physics that lead to the evolution of the earthquake phenomenon. We are focused in a variety of scales, from plate tectonics downscaling to rock fractures and laboratory seismology, to understand better the fundamentals of earthquake occurrence and contribute to the seismic hazard assessment, introducing the new topic of non-extensive statistical seismology.

2 Fundamentals of Non-extensive Statistical Physics

Boltzmann-Gibbs (BG) statistical physics is one of the cornerstones of contemporary physics. It establishes a remarkably useful bridge between the mechanical microscopic laws and macroscopic description using classical thermodynamics. The theory centrally addresses the very special stationary state—denominated thermal equilibrium. This macroscopic state has fundamental importance, since it is the foundation in Boltzmann's famous molecular chaos hypothesis made in 1871. However, BG theory is not universal. It has a limited domain of applicability (Tsallis 2009). Outside this domain, its predictions can be slightly or even strongly inadequate. There was a conflict, among many physicists as well as other scientists, that BG mechanics and standard thermodynamics are always valid and universal. It is certainly fair to say that always valid, in precisely the same sense that Newtonian mechanics is always valid; they indeed are. But again in complete analogy with Newtonian mechanics, we can by no means consider them as universal.

Central in BG statistical physics is the associated entropy that for the discrete states of a system has the form

$$S_{BG} = -k_B \sum_{i=1}^{W} p_i \ln p_i, \quad \text{with} \sum_{i=1}^{W} p_i = 1 \tag{2.1}$$

where S_{BG} is Boltzmann-Gibbs entropy, k_B is Boltzmann's constant, p_i is a set of probabilities and W is the total number of microscopic configurations. One of the main characteristics of S_{BG} is additivity, namely the proportionality to the number of the systems' elements. According to this property, for any two probabilistically independent systems A and B, i.e. if the joint probability satisfies $p_{ij}^{A+B} = p_i^A p_j^B (\forall (i, j))$, S_{BG} satisfies

$$S_{BG}(A + B) = S_{BG}(A) + S_{BG}(B). \tag{2.2}$$

Although BG entropy seems the correct one to be used in a large and important class of physical systems with strongly chaotic dynamics (positive maximal Lyapunov exponent), an important class of weakly chaotic systems (where the maximal Lyapunov exponent vanishes) violates this hypothesis. Additionally, if the effective microscopic interactions and memory are short-ranged (for instance Markovian processes) and the boundary conditions are smooth, then BG statistical mechanics seems to correctly describe nature. On the other hand, if some or all of these restrictions are violated (long-range interactions, non-markovian microscopic memory, multifractal boundary conditions and multifractal structures), then another type of statistical mechanics seems appropriate to describe nature (see for instance Zaslavsky 1999; Tsallis 2001).

Naturally, a question arises: Is it possible to address some of these important, though anomalous in the BG sense, situations with concepts and methods similar to those of BG statistical mechanics? Many theoretical, experimental and observational indications are nowadays available and point towards an affirmative answer. To overcome at least some of these anomalies that seem to violate BG statistical mechanics, non-extensive statistical physics (NESP) was proposed by Tsallis in (1988) that recovers the extensive BG as a particular case. The associated generalized entropic form for the discrete case is

$$S_q = k_B \frac{1 - \sum_{i=1}^{W} p_i^q}{q - 1}, \quad q \in R \text{ with} \sum_{i=1}^{W} p_i = 1 \tag{2.3}$$

where S_q is Tsallis entropy and q is the entropic index that represents a measure of the non-extensivity of a system. S_q recovers S_{BG} in the limit $q \to 1$. Although Tsallis entropy and Boltzmann-Gibbs entropy share a variety of thermodynamical properties like concavity (relevant for the thermodynamical stability of the system), experimental results, extensivity (relevant for having a natural matching with the entropy as introduced in classical thermodynamics), and finiteness of the entropy production per unit time (relevant for a variety of real situations where the system is striving to explore its microscopic phase space in order to ultimately approach some kind of stationary state) (see Tsallis 2009 for the full list of these properties), S_{BG} is additive, whereas S_q ($q \neq 1$) is non-additive. This property is directly related to the definition of S_q in Eq. (2.3). Indeed, for any two probabilistically independent systems A and B, i.e. if the joint probability satisfies $p_{ij}^{A+B} = p_i^A p_j^B (\forall (i, j))$, Tsallis entropy S_q satisfies:

$$\frac{S_q(A+B)}{k} = \frac{S_q(A)}{k} + \frac{S_q(B)}{k} + (1-q)\frac{S_q(A)}{k}\frac{S_q(B)}{k}. \tag{2.4}$$

The origin of non-additivity comes from the last term on the right hand side of this equation and is the fundamental principle of non-extensive statistical physics (Tsallis 2009). The cases $q > 1$, $q = 1$ and $q < 1$ correspond to sub-additivity, additivity and super-additivity respectively.

Tsallis idea of introducing the non-additive entropy S_q was inspired by simple physical principles and the multifractal concept (see Tsallis 2009 for a thorough description). A bias in the probabilities of the different states in a system is introduced by using the entropic index q. Given the fact that generically $0 < p_i < 1$, we have that $p_i^q > p_i$ if $q < 1$ and $p_i^q < p_i$ if $q > 1$. Therefore, $q < 1$ enhances the rare events with probabilities close to zero, whereas $q > 1$ enhances the frequent events, i.e., those whose probabilities are close to unity. Following Tsallis (2009), it is natural to introduce an entropic form passed on p_i^q. The entropic form must be invariant under permutation and the simplest expression which is consistent with this has the form $S_q = F\left(\sum_{i=1}^{W} p_i^q\right)$, where $F(x)$ is a continuous function. The simplest form of $F(x)$ is the linear one, leading to $S_q = C_1 + C_2 \sum_{i=1}^{W} p_i^q$. As any entropy, S_q must be a measure of disorder leading to $C_1 + C_2 = 0$ (Tsallis 2009) and hence $S_q = C_1\left(1 - \sum_{i=1}^{W} p_i^q\right)$. In the limit $q \to 1$ the entropic form S_q approaches the Boltzmann-Gibbs expression and the simplest way for this is when $C_1 = k_B/(q-1)$.

Tsallis entropy is determined by the microscopic dynamics of the system. This point is quite important in practice. If the microscopic dynamics of the system are known, we can determine the corresponding value of entropic index q from the first principles. As it happens, this precise dynamics is most frequently unknown for many natural systems. In this case, a way out that is currently used, is to check the functional forms and then determine the appropriate values of q by fitting. Moreover, there are many complex systems for which one may reasonably argue that they belong to the class that is addressed by non-extensive statistical concepts, but whose microscopic dynamics is inaccessible. For such systems, it appears as a sensible attitude to adopt the mathematical forms that emerge in the theory, e.g. q-exponentials (see below) and then obtain the correct graphs through fitting the corresponding value of q and of similar characteristics.

2.1 Optimizing Tsallis Entropy S_q

Suppose that we have a continuous variable *X* with a probability distribution *p*(*X*). In geophysics, this variable can be for instance seismic moment (M_o), inter-event times (τ) or distances (r) between the successive earthquakes or the length of faults (L) in a given region.

For the probability distribution *p*(*X*) of the continuous variable *X*, Tsallis entropy S_q is given by the integrated formulation as follows:

$$S_q = k_B \frac{1 - \int p^q(X)dX}{q-1}. \tag{2.1.1}$$

where q the entropic index. In the following we set k_B as unity for the sake of simplicity. We require optimizing S_q under the appropriate constraints. The first constraint refers to the normalization condition of *p*(*X*):

$$\int_0^\infty p(X)dX = 1 \tag{2.1.2}$$

The second constraint is the condition about the generalized expectation value (q-expectation value), X_q defined as:

$$X_q = \langle X_q \rangle = \int_0^\infty X P_q(X)dX \tag{2.1.3}$$

where $P_q(X)$ is the escort probability given (Tsallis 2009) as follows:

$$P_q(X) = \frac{p^q(X)}{\int_0^\infty p^q(X)dX} \tag{2.1.4}$$

Using the standard technique of Lagrange multipliers, the following functional is maximized:

$$\Phi(p, a^*, \beta^*) = S_q - a^* \int_0^{x_{\max}} p(X)dX - \beta^* X_q \tag{2.1.5}$$

where α^* and β^* represent the Lagrange multipliers.

Imposing that, $\partial\Phi/\partial p = 0$ we obtain the physical probability:

$$p(X) = \frac{\left[1-(1-q)\beta_q X\right]^{1/1-q}}{Zq} = \frac{\exp_q(-\beta_q X)}{Z_q} \tag{2.1.6}$$

where the q-exponential function is defined as (see Tsallis 2009 and references therein):

$$\exp_q(X) = \begin{cases} [1+(1-\mathrm{q})\mathrm{X}]^{1/(1-\mathrm{q})} & (1+(1-\mathrm{q})\mathrm{X} \geq 0) \\ 0 & (1+(1-\mathrm{q})\mathrm{X} < 0) \end{cases} \tag{2.1.7}$$

whose inverse is the q-logarithmic function: $ln_q(X) = \frac{1}{1-q}(X^{1-q} - 1)$

The denominator of Eq. (2.1.6) is called q-partition function and is defined as:

$$Z_q = \int_0^{x_{\max}} \exp_q(-\beta_q X)dX \tag{2.1.8}$$

where, $\beta_q = \frac{\beta}{c_q+(1-q)\beta Xq}$ and $\mathrm{c_q} = \int_0^{x_{max}} p^q(X)dX$

The q-exponential distribution consists a generalization of the Zipf-Mandelbrot distribution (Mandelbrot 1983), where the standard Zipf-Mandelbrot distribution corresponds to the case $q > 1$ (Abe and Suzuki 2003). In the limit $q \to 1$ the q-exponential and q-logarithmic functions lead to the ordinary exponential and logarithmic functions respectively. If $q > 1$ Eq. (2.1.6) exhibits an asymptotic power-law behavior with slope $-1/(q - 1)$. In contrast, for $0 < q < 1$ a cut-off appears (Abe and Suzuki 2003, 2005).

In non-extensive statistical physics it has been proposed that the quantity to be compared with the observed distribution is not the physical probability $p(X)$ but its associated escort distribution (see Abe and Suzuki 2005; Tsallis 2009; Vallianatos 2009). Following the latter approach, the cumulative distribution function is given by the expression

$$P_{cum}(>X) = \int_{X_{min}}^{\infty} P_q^{esc}(X)dX. \tag{2.1.9}$$

Combining the latter definition with the probability function $p(X)$ we obtain

$$\mathrm{P_{cum}}(>\mathrm{X}) = \exp_\mathrm{q}(-\mathrm{X}/\mathrm{X_o}), \tag{2.1.10}$$

which after simple algebra leads to $\frac{[P_{cum}(>X)]^{1-q}-1}{1-q} = -\frac{X}{X_o}$. The latter equation implies that after estimating the appropriate q, which describes the distribution of the variable X, the $\ln_q(P_{cum}(>X)) = \frac{[P_{cum}(>X)]^{1-q}-1}{1-q} = \left(-\frac{1}{X_o}\right)X$, which express the

q-logarithmic function, is linear with X with slope $-1/X_o$. In Fig. 1 the q-exponential distribution of Eq. (2.1.10) is plotted for various values of the q index.

In cases where $X > X_{min}$, the cumulative distribution of X assumes value 1 for $X = X_{min}$. This implies that the aforementioned equation should be slightly changed (Vallianatos 2013) into the following form, which is more consistent with real observations:

$$P(>X) = \frac{\exp_q(-X/X_o)}{\exp_q(-X_{min}/X_o)}. \tag{2.1.11}$$

We note that in cases where $X_{min} \ll X_o$, the latter introduction does not significantly change the estimated results.

An interesting question that is brought forth in NESP is which distribution we shall compare with the observed distribution. The common approach that is most frequently used is the introduction of the escort probability in the second constraint and the optimization of S_q as described earlier. Other forms have been developed and are described thoroughly in Tsallis (2009). Detailed discussions on this subject can be also found in Wada and Scarfone (2005), Ferri et al. (2005), where it was shown that the different forms related to the second constraint of the expectation value are all equivalent and can be transformed one into the other through simple operations defining q s and X_o s. For instance, if we integrate the physical probability given in Eq. (2.1.6) instead of the escort probability (Eq. 2.1.4), we obtain the cumulative probability:

$$P(>X) = \left[1 - (1-q')\frac{X}{X'_o}\right]^{1/(1-q')}, \tag{2.1.12}$$

where $q' = 1/(2 - q)$ and $X'_o = (2 - q)/X_o$, in relation to q and X_o values in Eq. (2.1.10) (Picoli et al. 2009). If we apply these transformations for q' and X'_o, the following form of the cumulative distribution $P(>X)$ is derived (Michas et al. 2013):

$$P(>X) = \left[1 - (1-q)\frac{X}{X_o}\right]^{\frac{2-q}{1-q}}. \tag{2.1.13}$$

Another type of distributions that are deeply connected to statistical physics is that of the squared variable X^2. In BG statistical physics, the distribution of X^2 corresponds to the well-known Gaussian distribution. If we optimize S_q for X^2, we obtain a generalization of the normal Gaussian that is known as q-Gaussian distribution (see Tsallis 2009) and has the form:

$$p(X) = p_0\left[1 - (1-q)\left(\frac{X}{X_o}\right)^2\right]^{1/(1-q)}. \tag{2.1.14}$$

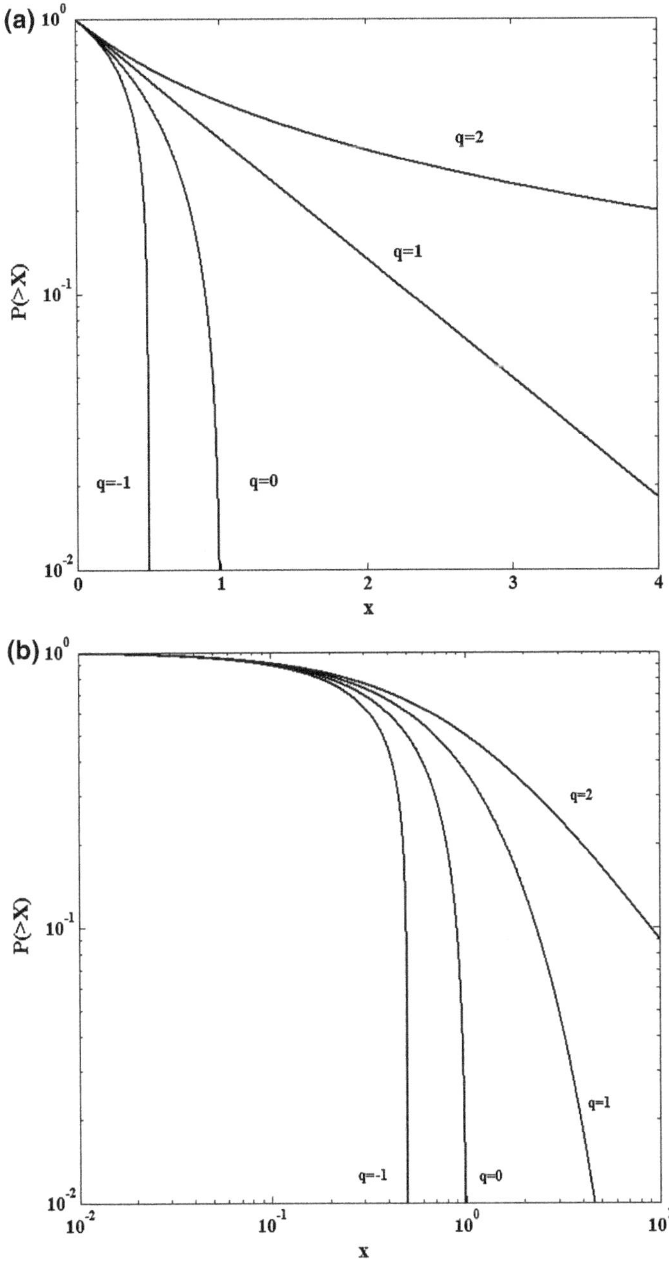

Fig. 1 The q-exponential distribution of Eq. (2.1.10) for various values of q and for $X_0 = 1$ in log-linear (**a**) and log-log scales (**b**). The distribution is convex for $q > 1$ and concave for $q < 1$. For $q < 1$, it has a vertical asymptote at $x = (1 - q)^{-1}$ and for $q > 1$ an asymptotic slope $-1/(q - 1)$. For $q = 1$ the standard exponential distribution is recovered

In the limit $q \to 1$, Eq. (2.1.14) recovers the normal Gaussian distribution. For $q > 1$, the q-Gaussian distribution has power-law tails with slope $-2/(q-1)$, thus enhancing the probability of the extreme values. Typical examples of q-Gaussians are plotted in Fig. 2 for various values of q.

2.2 Cases with Two Slopes

There are various cases in earthquake populations where the observed variable exhibits a distribution with different regions that correspond to different slopes. The most common example is gamma distribution that exhibits a power-law region for small and intermediate values and an exponential tail for greater values of the observed variable. This particular distribution has been used most frequently to model inter-event times (e.g. Corral 2004) and the global earthquake frequency-size distribution (e.g. Kagan 1997). In this latter case, an upper bound or taper in the G-R relation is appearing (Kagan and Jackson 2000) and the G-R relation is modified to include an exponential tail for modeling greater earthquake magnitudes. Estimating this upper bound or the correct distribution that corresponds to earthquake size distribution is of high importance in probabilistic earthquake hazard assessments (see for instance Kagan and Jackson 2013) and fundamental in constraining insurance risk for the largest events (Bell et al. 2013).

In the following we describe how distributions with crossovers and different behavior for large values of the observed variable can be derived in the frame of non-extensive statistical physics, by generalizing the physical probability given in Eq. (2.1.6).

The generalized probability $p(X)$ given by Eq. (2.1.6) can be alternatively obtained by solving the nonlinear differential equation:

$$\frac{dp}{dX} = -\beta_q p_i^q, \tag{2.2.1}$$

where $q \neq 1$; while Boltzmann-Gibbs (BG) formalism is approached in the limit $q \to 1$. We can now further generalize the standard representation of NESP, presented in the previous Sect. (2.1), by considering not only one q index, but a whole distribution of indices (see Tsallis et al. 1999; Tsekouras and Tsallis 2005). In the case where crossover to another type of behavior at larger values of the variable X is observed we can generalize the differential equation Eq. (2.2.1) as:

$$\frac{dp}{dX} = -\beta_r p^r - (\beta_q - \beta_r) p^q. \tag{2.2.2}$$

Equation (2.2.1) is recovered if $r = 0$ or if $r = q$. When $1 \le r < q$, the solution of Eq. (2.2.2) is given by:

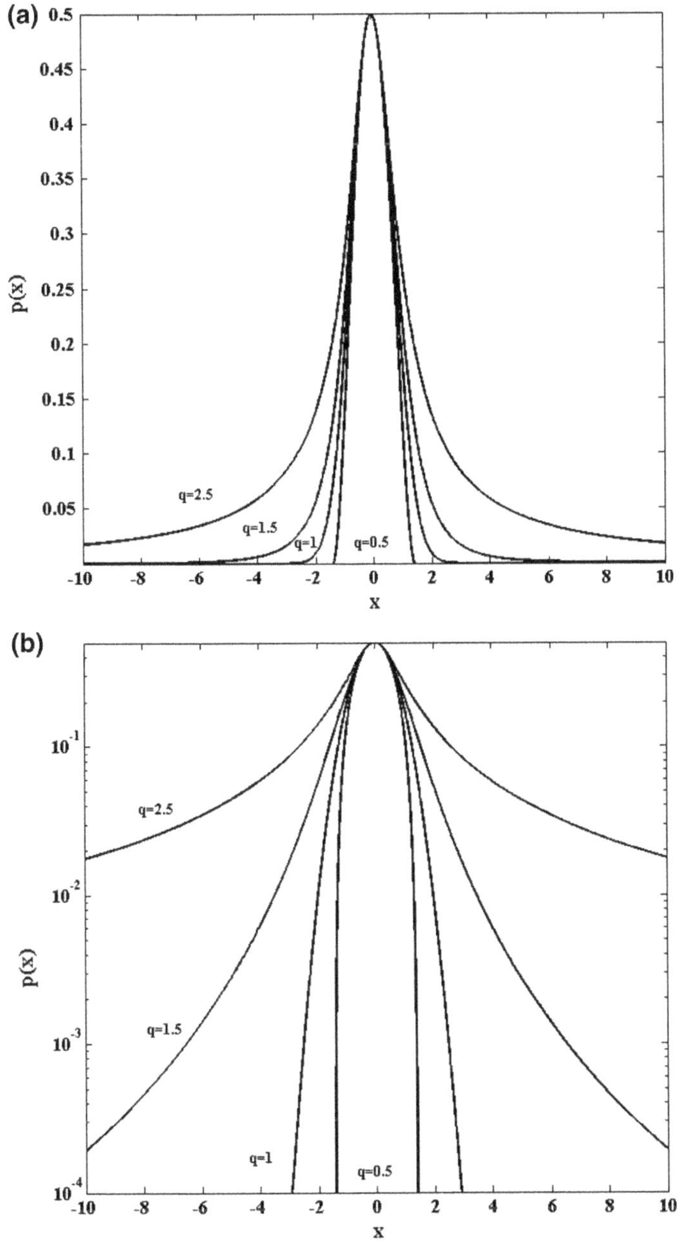

Fig. 2 The q-Gaussian distribution of Eq. (2.1.14) for various values of q and for $p_0 = 0.5$ and $X_0 = 1$ in linear (**a**) and log-linear scales (**b**). For $q = 1$ the normal Gaussian distribution is recovered

$$X = \int_p^1 \frac{dx}{\beta_r x^r + (\beta_q - \beta_r)x^q} = \frac{1}{\beta_r}\int_p^1 dx \left[\frac{1}{x^r} - \frac{\left(\frac{\beta_q}{\beta_r} - 1\right)x^{q-2r}}{1 + (\frac{\beta_q}{\beta_r} - 1)x^{q-r}}\right]$$

and hence

$$X = \frac{1}{\beta_r}\left\{\frac{p^{-(r-1)} - 1}{r-1} - \frac{\frac{\beta_q}{\beta_r} - 1}{1 + q - 2r}\left[H\left(1; q-2r, q-r, \left(\beta_q/\beta_r\right) - 1\right)\right.\right.$$
$$\left.\left. -H\left(p; q-2r, q-r, \left(\beta_q/\beta_r\right) - 1\right)\right]\right\} \tag{2.2.3}$$

with

$$H(\xi; a, b, c) = \xi^{1+\alpha}F\left(\frac{1+a}{b}, 1; \frac{1+a+b}{c}; -\xi^b c\right) \tag{2.2.4}$$

where F is the hypergeometric function.

From the aforementioned solution of Eq. (2.2.2) for $1 \le r < q$ and $\beta_r << \beta_q$ the asymptotic solution defines three regions. The first one is related with very small values of the variable X and

$$p(X) \propto 1 - \beta_q X \quad \text{for } 0 \le X \le X_{c1}, \quad \text{where} \quad X_{c1} = \frac{1}{q-1}\frac{1}{\beta_q} \tag{2.2.5}$$

The second one describes the moderate values and

$$p(X) \propto [(q-1)\beta_q X]^{-1/(q-1)} \text{ for } X_{c1} \le X \le X_{c2}, \quad \text{where} \quad X_{c2} = \frac{[(q-1)\beta_q]^{\frac{r-1}{q-r}}}{[(r-1)\beta_r]^{\frac{q-1}{q-r}}}. \tag{2.2.6}$$

The third region describes the range of large values of X and

$$p(X) \propto [(r-1)\beta_r X]^{-1/(r-1)} \text{ for } \ge X_{c2}. \tag{2.2.7}$$

A special case of the differential Eq. (2.2.2) is when a crossover from anomalous ($q \ne 1$) to normal ($r = 1$) statistical mechanics is appearing for the larger values of the variable X (the truncated G-R relation for instance, described earlier). In this case, the differential equation Eq. (2.2.2) is modified as:

$$\frac{dp}{dX} = -\beta_1 p - (\beta_q - \beta_1)p^q \tag{2.2.8}$$

and includes both the normal and anomalous cases in the first and second term respectively. The solution of Eq. (2.2.8) is given by:

$$p = C\left[1 - \frac{\beta_q}{\beta_1} + \frac{\beta_q}{\beta_1} e^{(q-1)\beta_1 X}\right]^{-1/q - 1} \tag{2.2.9}$$

were C is a normalization factor. For positive β_q and β_1, $p(X)$ decreases monotonically with increasing X (Vallianatos and Sammonds 2010). It can be easily verified that in the case where $\beta_q \gg \beta_1$ Eq. (2.2.9) defines again three regions, according to the value of X. The asymptotic behavior of the probability distributions in these areas, when $r = 1$, is simplified as

$$p(X) \propto 1 - \beta_q X \quad \text{for } 0 \leq X \leq X_{c1} \quad \text{where} \quad X_{c1} = \frac{1}{q-1}\frac{1}{\beta_q} \tag{2.2.10}$$

$$p(X) \propto [(q-1)\beta_q X]^{-1/(q-1)} \quad \text{for } X_{c1} \leq X \leq X_{c2} \quad \text{where} \quad X_{c2} = \frac{1}{[(q-1)\beta_1]} \tag{2.2.11}$$

$$p(X) \propto \left[\frac{\beta_1}{\beta_q}\right]^{-1/(q-1)} e^{-\beta_1 X} \quad \text{for } X \geq X_{c2} \tag{2.2.12}$$

where X_{c1} and X_{c2} are the lower and upper crossover points between the three regions respectively.

3 Applications in Seismicity

As already mentioned in the introduction of this chapter, seismicity and fault systems are among the most relevant paradigms of self-organized criticality (Bak et al. 1987, 2002), representing a complex spatiotemporal phenomenon (e.g. Telesca et al. 2001, 2002, 2003). Despite the complexity that characterizes fracturing and earthquake nucleation phenomena, simple phenomenology seems to apply in their collective properties, where empirical scaling relations are known to describe the statistical properties of the fracture/fault and earthquake populations in a variety of scales. The best known is the Gutenberg-Richter (G-R) scaling relation (Gutenberg and Richter 1944) that expresses fractal power-law dependence in the frequency of earthquakes with energy (seismic moment) E with E:

$$P(E) \sim E^{-B} \tag{3.1}$$

If we consider that the earthquake energy E is related to the magnitude M as $E \sim 10^{1.5M}$ (Kanamori 1978) and for $B = 1 + b/1.5$, the last expression can be alternatively stated as $N(>M) \sim 10^{-bM}$, where $N(>M)$ is the number of earthquakes with magnitude equal or greater than M and b is a constant known as the seismic b-value.

Another well-known scaling relation is the modified Omori formula (Omori 1894; Utsu et al. 1995) where the aftershock production rate $n(t) = dN(t)/dt$ after main earthquakes decays as a power-law with time t:

$$n(t) = K(t + c)^{-p} \tag{3.2}$$

where K and c are constants that are determined from the data and p is the power-law exponent.

Power-laws and fractality have been also found in the space of earthquake locations and laboratory AE (Kagan and Knopoff 1980; Hirata and Imoto 1991) and at the time of their occurrence (Kagan and Jackson 1991; Turcotte 1997).

Crack and fault populations are characterized by scale-invariance so that their length distribution decays as a power-law:

$$N(>L) = AL^{-D} \tag{3.3}$$

where N is the number of faults with length equal or greater than L, A is a constant and D is the scaling exponent (Main 1996; Turcotte 1997).

The properties that are appearing in the earthquake and fault populations, such as those described above are the central subject in the statistical physics approach to seismicity. The necessity of statistical physics in deriving probability distributions for describing seismicity have been highlighted in the early works of Berrill and Davis (1980), Main and Burton (1984), where classic statistical physics and Shannon's information entropy (Shannon 1948) have been used to assess the probability of large earthquakes. Reviews for the classic statistical physics approach in seismicity can be found in Main (1996), Rundle et al. (2003) and Kawamura et al. (2012).

In the following sections we describe how appropriate probability distributions for the description of seismicity and the fault systems can be derived in the frame of NESP by using the maximum entropy principle and how these are applied to fracture and earthquake data, providing various examples for different case studies.

3.1 *Non-extensive Pathways in Earthquake Size Distributions*

3.1.1 The Fragment-Asperity Model

Earthquakes are originating from the deformation and sudden rupture of parts of the earth's brittle crust releasing energy and generating elastic waves that are propagating in the earth's interior. The generated waves are recorded in seismographic stations and properties such as the location and seismic moment of the earthquake are calculated from the waveforms. Primarily, the earthquake generation process is a mechanical phenomenon where stick-slip frictional instability in pre-existing fault zones has a dominant role (e.g. Scholz 1998). Some well-known models, such as the spring-block model (Burridge and Knopoff 1967) and the cellular automaton model (Olami et al. 1992) have been developed to describe the phenomenology of this mechanism. In these models a stick-slip behavior in a set of moving blocks interconnected via elastic springs reproduce well some of the known empirical relations such as the Gutenberg-Richter (G-R) scaling law.

Consisted to the idea of stick-slip frictional instability in faults, Sotolongo-Costa and Posadas (2004) developed the fragment-asperity interaction model to describe earthquake dynamics in a non-extensive context. In this model, the triggering mechanism of an earthquake involves the interaction between the irregular surfaces of the fault planes and the fragments of various sizes and shapes that fill the space between them. When the accumulated stress exceeds a critical value in a particular fault zone, the fault planes are slipping, displacing the fragments and breaking possible asperities that hinder their motion, releasing energy (for an asperity based model for fault dynamics see De Rubeis et al. 1996). Sotolongo-Costa and Posadas considered that the released seismic energy is related to the size of the fragments and by using a non-extensive formalism they established an energy distribution function (EDF) for earthquakes based on the fragments-size distribution. Since the standard Boltzmann-Gibbs formalism cannot account for the presence of scaling in the fragmentation process, NESP seems more adequate to describe the phenomenon. The latter is also supported by the scale-invariant properties of fragments (Krajinovic and Van Mier 2000), the presence of long-range interactions among the fragmented materials (Sotolongo-Costa and Posadas 2004) and laboratory experiments in fracturing processes (Vallianatos et al. 2011, 2012a).

In the following we describe how the fragment-asperity model of Sotolongo-Costa and Posadas (2004), as was later revised by Silva et al. (2006) and Telesca (2012), is derived in the frame of NESP.

In terms of the probability $p(\sigma)$ of finding a fragment of area σ, the maximum Tsallis entropy S_q is expressed as:

$$S_q = k_B \frac{1 - \int p^q(\sigma)\, d\sigma}{q-1}. \tag{3.1.1}$$

The sum of all the possible states in the definition of entropy is here expressed through the integration in all the sizes of the fragments. In what follows we set k_B equal to unity for the sake of simplicity. The probability $p(\sigma)$ is obtained after maximization of S_q under the appropriate two constraints. The first is the normalization of $p(\sigma)$:

$$\int_0^\infty p(\sigma) d\sigma = 1. \tag{3.1.2}$$

The second is the condition about the q-expectation value (Tsallis 2009):

$$\sigma_q = \langle \sigma \rangle_q = \frac{\int_0^\infty \sigma p^q(\sigma) d\sigma}{\int_0^\infty p^q(\sigma) d\sigma}. \tag{3.1.3}$$

This last condition reduces to the definition of the mean value in the limit $q \to 1$.

By using the Lagrange multipliers technique, the functional entropy to be maximized is (Silva et al. 2006):

$$\delta S_q^* = \delta \left(S_q + \alpha \int_0^\infty p(\sigma) d\sigma - \beta \sigma_q \right) = 0, \tag{3.1.4}$$

where α and β are the Lagrange multipliers. After some algebra, the following expression for the fragment size distribution function can be derived (Silva et al. 2006):

$$p(\sigma) = \left[1 - \frac{(1-q)}{(2-q)} (\sigma - \sigma_q) \right]^{1/(1-q)}. \tag{3.1.5}$$

The proportionality between the released relative energy E and the size of the fragments r is now introduced as $E \sim r^3$ (Silva et al. 2006), in accordance to the standard definition of seismic moment scaling with rupture length (Lay and Wallace 1995). The proportionality between the released relative energy E and the three-dimensional size of the fragments r^3 now becomes:

$$\sigma - \sigma_q = \left(\frac{E}{\alpha_E} \right)^{2/3}. \tag{3.1.6}$$

In the last equation, σ scales with r^2 and α_E is the proportionality constant between E and r^3 that has the dimension of volumetric energy density. By using the latter equation, the energy distribution function (EDF) of the earthquakes can be written on the base of the relationship between density functions of correlated stochastic variables (Telesca 2012):

$$p(E) = \frac{1}{\frac{dE}{d\sigma}} p \left[\left(\frac{E}{\alpha_E} \right)^{2/3} + \sigma_q \right] = \frac{d\sigma}{dE} \left[1 - \frac{(1-q)}{(2-q)} \left(\frac{E}{\alpha_E} \right)^{2/3} \right]^{\frac{1}{(1-q)}}, \tag{3.1.7}$$

where the term $d\sigma/dE$ can be obtained by differentiating Eq. (3.1.6):

$$\frac{d\sigma}{dE} = \frac{2}{3}\frac{E^{-\frac{1}{3}}}{\alpha_E^{\frac{2}{3}}}dE. \tag{3.1.8}$$

The EDF now becomes (Silva et al. 2006; Telesca 2012):

$$p(E) = \frac{C_1 E^{-\frac{1}{3}}}{\left[1 + C_2 E^{\frac{2}{3}}\right]^{1/(q-1)}}, \tag{3.1.9}$$

with $C_1 = \frac{2}{3\alpha_E^{\frac{2}{3}}}$ and $C_2 = \frac{(1-q)}{(2-q)\alpha_E^{\frac{2}{3}}}$.

In the latter expression, the probability of the energy is $p(E) = n(E)/N$, where $n(E)$ corresponds to the number of earthquakes with energy E and N is the total number of earthquakes. A more viable expression can now be obtained by introducing the normalized cumulative number of earthquakes given by the integral of Eq. (3.1.9):

$$\frac{N(E > E_{th})}{N} = \int_{E_{th}}^{\infty} p(E)dE, \tag{3.1.10}$$

where $N(E > E_{th})$ is the number of earthquakes with energy E greater than the threshold energy E_{th} and N the total number of earthquakes. Substituting Eq. (3.1.9) in Eq. (3.1.10) the following expression is derived:

$$\frac{N(E > E_{th})}{N} = \left[1 - \left(\frac{1-q_E}{2-q_E}\right)\left(\frac{E}{\alpha_E}\right)\right]^{\frac{2-q_E}{1-q_E}}. \tag{3.1.11}$$

Now the latter expression can be written in terms of the earthquake magnitude M, if we consider that E is related to M as $M = \frac{2}{3}\log(E)$ (Kanamori 1978). Then Eq. (3.1.11) becomes:

$$\frac{N(>M)}{N} = \left[1 - \left(\frac{1-q_E}{2-q_E}\right)\left(\frac{10^M}{\alpha_E^{2/3}}\right)\right]^{\frac{2-q_E}{1-q_E}}. \tag{3.1.12}$$

In real earthquake catalogues the threshold magnitude M_0, i.e. the minimum magnitude M_0 of the catalogue, has to be taken in account and Eq. (3.1.12) should be slightly changed to (Telesca 2012):

$$\frac{N(>M)}{N}=\left[\frac{1-\left(\frac{1-q_E}{2-q_E}\right)\left(\frac{10^M}{\alpha_E^{2/3}}\right)}{1-\left(\frac{1-q_E}{2-q_E}\right)\left(\frac{10^{M_0}}{\alpha_E^{2/3}}\right)}\right]^{\frac{2-q_E}{1-q_E}}. \quad (3.1.13)$$

The fragment-asperity model describes from the first principles the cumulative distribution of the number of earthquakes N greater than a threshold magnitude M, normalized by the total number of earthquakes. The constant a_E expresses the proportionality between the released energy and the fragments of size r, while q_E is the entropic index. This model has been recently applied to various regional earthquake catalogs, covering diverse tectonic regions (Silva et al. 2006; Vilar et al. 2007; Telesca 2010a, b, c, 2011; Michas et al. 2013; Papadakis et al. 2013) and volcano related seismicity (Telesca 2010b; Vallianatos et al. 2013). In comparison to the G-R scaling relation (Eq. 3.1), the fragment-asperity model describes appropriately the energy distribution in a wider range of magnitudes, while for values above some threshold magnitude, the G-R relation can be deduced as a particular case for $b = (2 - q_E)/(q_E - 1)$ (Telesca 2012).

Some relevant paradigms for the application of the fragment—asperity model to earthquake data are given in Figs. 3, 4 and 5. In Fig. 3 the model is applied to the energy distribution function of earthquakes in the West Corinth rift (Greece), according to Eq. (3.1.9). In this case the model describes better than the G-R relation (Eq. 3.1) the observed distribution for the lower earthquake energies, while after some threshold energy the distribution decays as a power-law (Michas et al. 2013). Another example comes from the recent unrest at the Santorini volcanic complex (Vallianatos et al. 2013), where the normalized cumulative magnitude distribution of the volcano seismicity is well described by the model (Eq. 3.1.12) for the value of the entropic index $q_E = 1.39$ (Fig. 4).

The entropic index q_E has been recently used by Papadakis et al. (2013) to geodynamically characterize various seismic zones along the Hellenic Subduction Zone (HSZ). Figure 5 shows the distribution of the relative cumulative number of

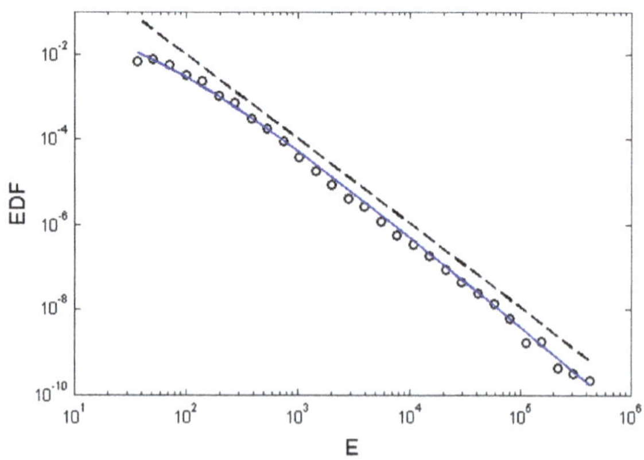

Fig. 3 The energy distribution function (*circles*) and the fitted curve (*solid line*). The *dashed line* represents the G-R relation for $b = 1.51 \pm 0.03$ (Michas et al. 2013)

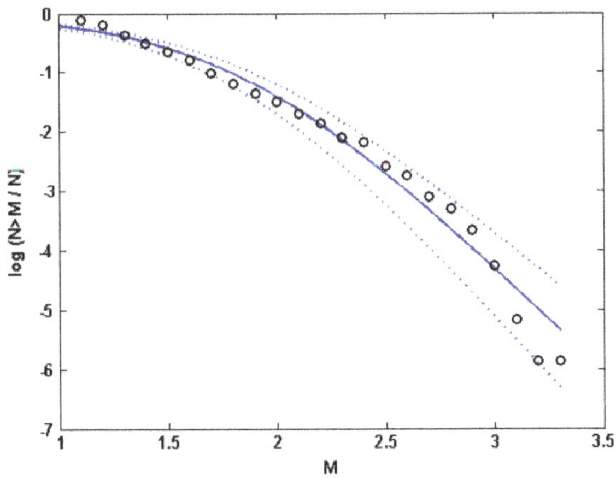

Fig. 4 Normalized cumulative magnitude distribution of the Santorini seismicity (*circles*) and the fitting curve (*solid line*). The values for the best fit regression to the data are $q_E = 1.39 \pm 0.035$ and $a = 286.6 \pm 78$. The 95 % confidence intervals for q_E and a are also plotted (*dashed lines*) (Vallianatos et al. 2013)

Fig. 5 The *black dashed line* indicates the non-extensive fitting curve for the HSZ as a unified system (Papadakis et al. 2013)

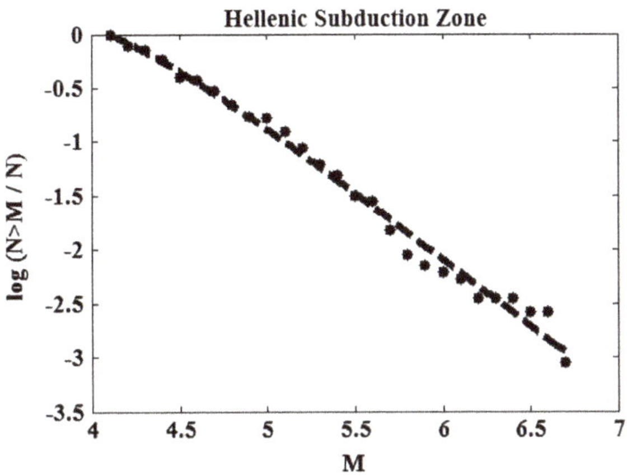

earthquakes as a function of magnitude M for the Hellenic Subduction Zone as a unified system and the fitting according to the model of Eq. (3.1.13), while Fig. 6 shows the variation of the q_E value along the HSZ, where the variation is related to the energy release rate in each seismic zone.

3.1.2 Global Earthquake Size Distribution. The Effect of Mega-Earthquakes

The global earthquake frequency-magnitude distribution is among the long-standing statistical relationships of seismology. Recently, Vallianatos and

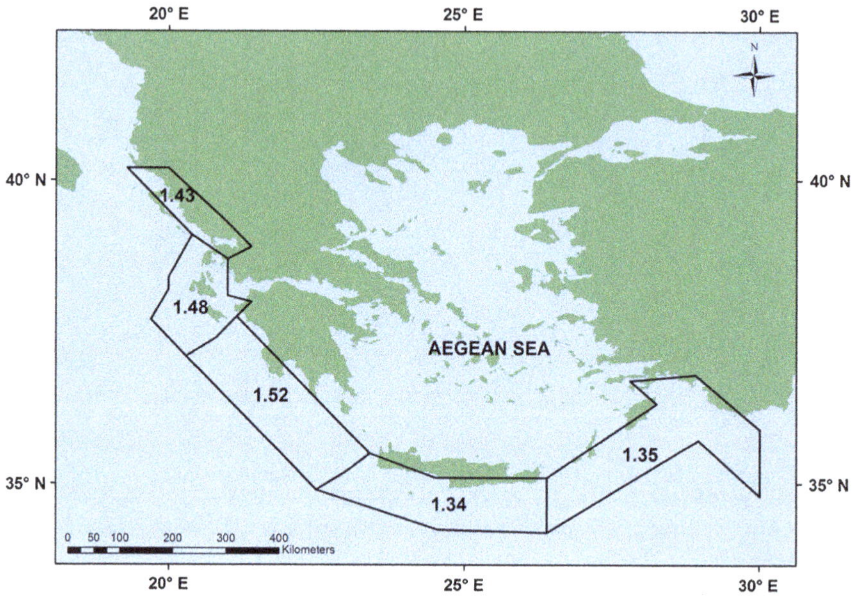

Fig. 6 The variation of the q_E value along the seismic zones of the HSZ (Papadakis et al. 2013)

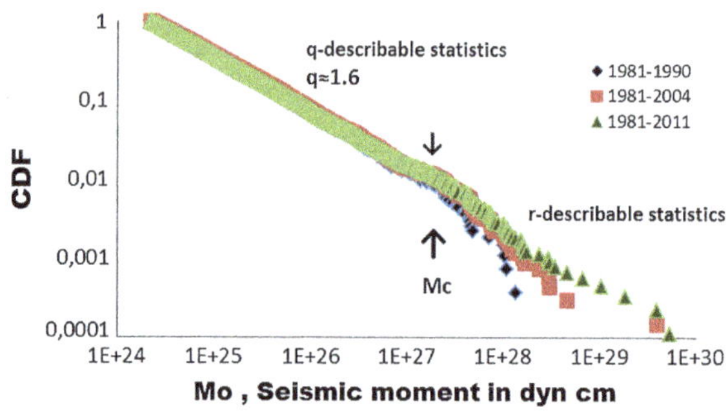

Fig. 7 Distribution of seismicity versus seismic moment for the centroid moment tensor catalogue up to the end of 1990 (before Sumatra mega event, in *black*), the end of December 2004 (after Sumatra, in *red*) and within a week after Honshu mega earthquake (till 17 March 2011, in *green*), for shallow events ($H < 75$ km), with $M_w > 5.5$, since 1 January 1981. This is plotted as a normalized cumulative distribution function (*CDF*) against seismic moment (Vallianatos and Sammonds 2013)

Sammonds (2013) used non-extensive statistical mechanics to characterize the global earthquake frequency–magnitude distribution and to interpret observations from the Sumatran and Honshu earthquakes. Examples of the cumulative distribution function (CDF) from the CMT seismic moment M_o data are given in Fig. 7.

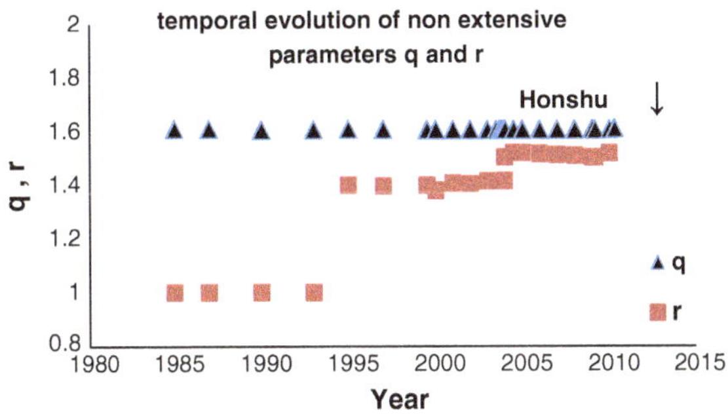

Fig. 8 Temporal evolution of non-extensive parameters q and r extracted from the analyses of seismic moment distribution using the global CMT catalogue. We observe a stable organization in moderate events in contrast to a significant change of r, which supports the concept of the global organization of seismicity before the two recent Sumatra and Honshu mega events (Vallianatos and Sammons 2013)

The aforementioned authors found that global seismicity is described by non-extensive statistical mechanics and that the seismic moment distribution reflects a sub-extensive system, where long-range interactions are important.

Using the cross-over formulation of non-extensive statistical physics (see Sect. 2.2), Vallianatos and Sammonds (2013) conclude that the seismic moment distribution of moderate events yields thermodynamic q-values of $q_E = 1.6$ which seem to be constant for the duration of the Sumatra and Honshu earthquake preparation, while r_M (which describes the seismic moment distribution of great events) varies from 1 that corresponds to an exponential function (Eqs. 2.2.8–2.2.12), to 1.5 and another power-law regime (Eqs. 2.2.2–2.2.7) as we approach the mega events (Fig. 8).

3.1.3 Increments of Earthquake Energies

The probability distribution in the incremental earthquake energies is referred to the probability that an earthquake of energy $S(i)$ will be followed by one with energy $S(i + 1)$ with difference R, expressed as $R = S(i + 1) - S(i)$ ($i = 1, 2, \ldots, N - 1$ where N the total number). Caruso et al. (2007) have calculated this probability for a dissipative Olami–Feder–Christensen model (OFC—Olami et al. 1992) and showed that in the critical regime (small-world lattice) the probability distribution $P(R)$ follows a q-Gaussian, while in the noncritical regime (regular lattice) the distribution $P(R)$ is close to a Gaussian distribution. Then considering the quantity $S = \exp(M)$ as a measure of the energy S of an earthquake of magnitude M, Caruso et al. (2007) showed that the probability distribution $P(R)$ for real earthquakes in Northern California and in the entire world follows a q-Gaussian distribution as well, providing further evidence for self-organized criticality, intermittency and long-range interactions in seismicity.

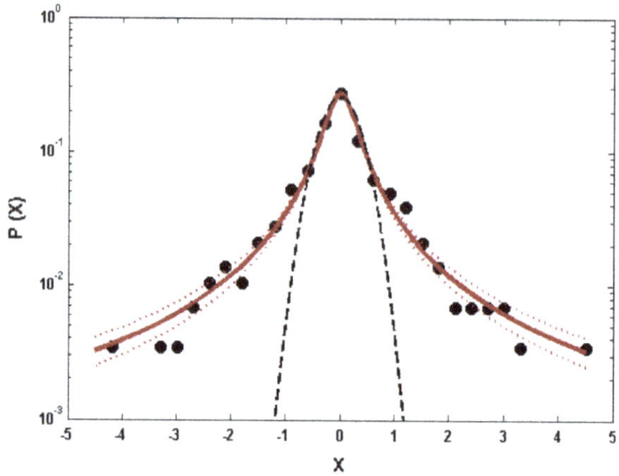

Fig. 9 Probability density function $P(X)$ (*solid circles*) for the 2011–2012 earthquake activity in Santorini volcanic complex on a semi-log plot, where $S = \exp(M)$, $R = S(i + 1) - S(i)$, $X = (R - \langle R \rangle)/\sigma_R$ and $\langle R \rangle$ the mean and σ_R the standard deviation. The *dashed curve* represents the standard Gaussian shape. The data is well fitted by a q-Gaussian curve (*solid line*) for the value of the entropic index $q_R = 2.24 \pm 0.09$ (95 % confidence intervals—*dotted curves*) (Vallianatos et al. 2013)

In addition, Vallianatos et al. (2013) studied the probability distribution of incremental energies in the volcano related seismicity during the 2011–2012 unrest at the Santorini volcanic complex. The probability density function of R, normalized to zero mean and unit variance and subjected to the normalization condition $\int p(R)dR = 1$ exhibits fat tails and can be well described by a q-Gaussian distribution (Eq. 2.1.14) for the value of $q = 2.24 \pm 0.09$ (Fig. 9), indicating non-linear dynamics and self-organized criticality in the observed volcano-related seismicity. Here we provide further evidence by considering the global earthquakes with magnitude $M \geq 7$ that occurred during the period 1900–2012, as these are reported in the latest version of the Centennial earthquake catalog (Engdahl and Villaseñor 2002) (catalog available at http://earthquake.usgs.gov/data/centennial/) and supplemented by the ANSS earthquake catalog (http://www.ncedc.org/anss/) for the period 2007–2012. The probability density of the incremental earthquake energies exhibits fat tails and deviates from the normal Gaussian distribution (Fig. 10). A q-Gaussian distribution with $q = 1.85 \pm 0.1$ can well describe the observed distribution, thus enhancing the probability of large differences in the energies of successive earthquakes in global scale.

3.2 Spatiotemporal Description

Throughout the text we referred to the scale-invariant spatiotemporal properties of seismicity. Considering these properties, Abe and Suzuki proposed that the 3-dimensional hypocentral distances and the time intervals between the successive

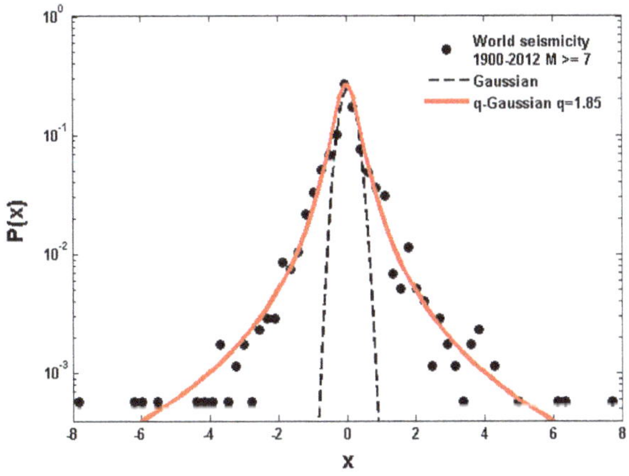

Fig. 10 Probability density function $P(X)$ (*solid circles*) for the 1900–2012 global seismicity with magnitude $M \geq 7$ on a semi-log plot, where $S = \exp(M)$, $R = S(i + 1) - S(i)$, $X = (R - \langle R \rangle)/\sigma_R$ and $\langle R \rangle$ the mean and σ_R the standard deviation. The normalization condition $\int p(R)dR = 1$ applies. The *dashed curve* represents the standard Gaussian shape and the *solid line* the q-Gaussian distribution for the value of $q = 1.85 \pm 0.1$

earthquakes follow a q-exponential distribution (see Sect. 2.1) with $q < 1$ and $q > 1$ respectively and verified their approach for the cumulative distribution of inter-event distances and times of successive earthquakes in California and Japan (Abe and Suzuki 2003, 2005). Since then, this approach has been successfully applied in various studies, covering diverse scales and tectonic regimes (e.g., Darooneh and Dadashinia 2008; Vallianatos et al. 2012a, b; Vallianatos and Sammonds 2013; Papadakis et al. 2013).

Such examples of the non-extensive statistical physics application to the spatiotemporal distributions of earthquakes for regional tectonics and mega-structure geodynamics are given in Figs. 11, 12, 13, 14, 15, 16, 17 and 18. In particular, in Fig. 11 the cumulative inter-event time distribution $P(>\tau)$ for the 1995 Aigion earthquake aftershock sequence is presented that follows a q-exponential distribution for $q_\tau = 1.58$ (Vallianatos et al. 2012b).

Papadakis et al. (2013) estimated the cumulative distribution functions of the inter-event times and distances along the HSZ (Figs. 12 and 13). Figures 14 and 15 show the variation of the calculated q_T and q_D values along the seismic zones of the HSZ. With the exception of seismic zone 4, which is located in the central part of the HSZ and covers the southern area of Crete, the q_T values appear close to each other while the q_D values seem to differ significantly. The latter result possibly reflects the fact that q_T is related with the time evolution of seismicity, which is a long term process in the Hellenic arc, with the highest temporal clustering in the area south of Crete. In addition the q_D variations indicate a different degree of spatial earthquake clustering along the seismic zones.

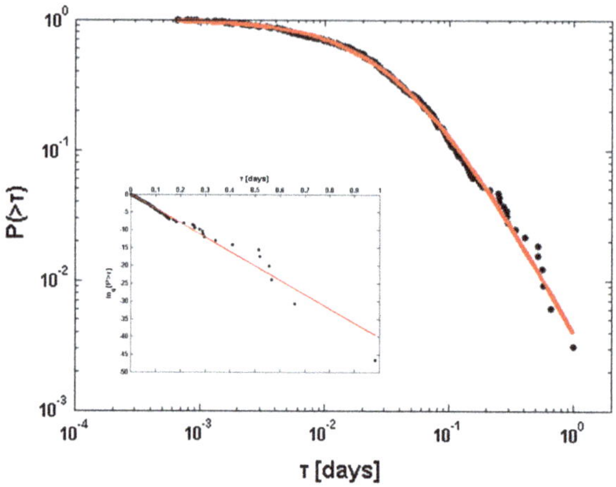

Fig. 11 Log-log plot of P(>τ). The *solid line* represents the q_τ-exponential distribution for the values of $q_\tau = 1.58 \pm 0.02$ and $\tau_0 = 0.025 \pm 0.0003$ days. *Inset* the q_τ-logarithmic distribution $\ln_q(P(>\tau))$, exhibiting a correlation coefficient of $r = -0.9885$. The *straight line* corresponds to the *q*-exponential distribution (Vallianatos et al. 2012b)

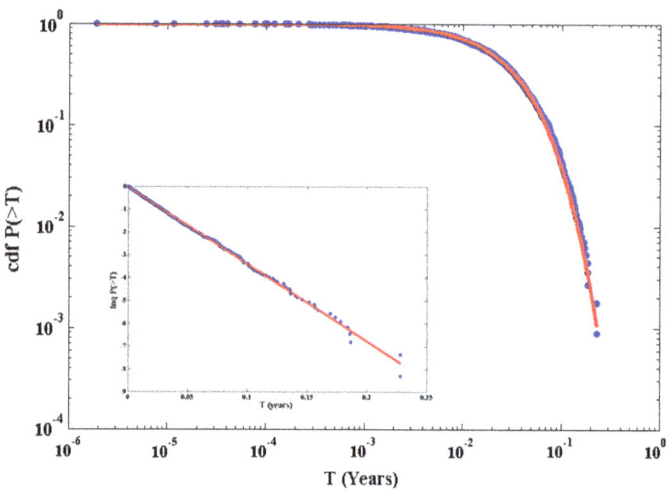

Fig. 12 The log-log plot of the inter-event times cumulative distribution for the HSZ as a unified system. *Inset* the semi-q-log plot of the inter-event times cumulative distribution for the HSZ as a unified system. The *dashed line* represents the q-logarithmic function (Papadakis et al. 2013)

Additionally, Vallianatos and Sammonds (2013), using global shallow seismicity with $M_w > 5.5$ extracted from CMT catalogue, analyzed the inter-event time distribution before and after the Sumatran and Honshu mega earthquakes. Figure 16 shows the inter-event times cumulative probability distribution $P(>\tau)$ and demonstrates that, in spite of the changes observed in non-extensive frequency-seismic

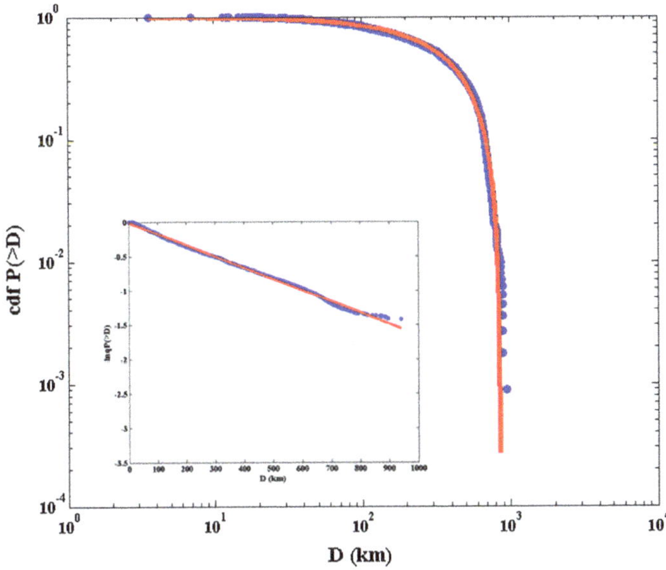

Fig. 13 The log-log plot of the three-dimensional distances cumulative distribution for the HSZ as a unified system. *Inset* The semi-q-log plot of the inter-event distances cumulative distribution for the HSZ as a unified system. The *dashed line* represents the q-logarithmic function (Papadakis et al. 2013)

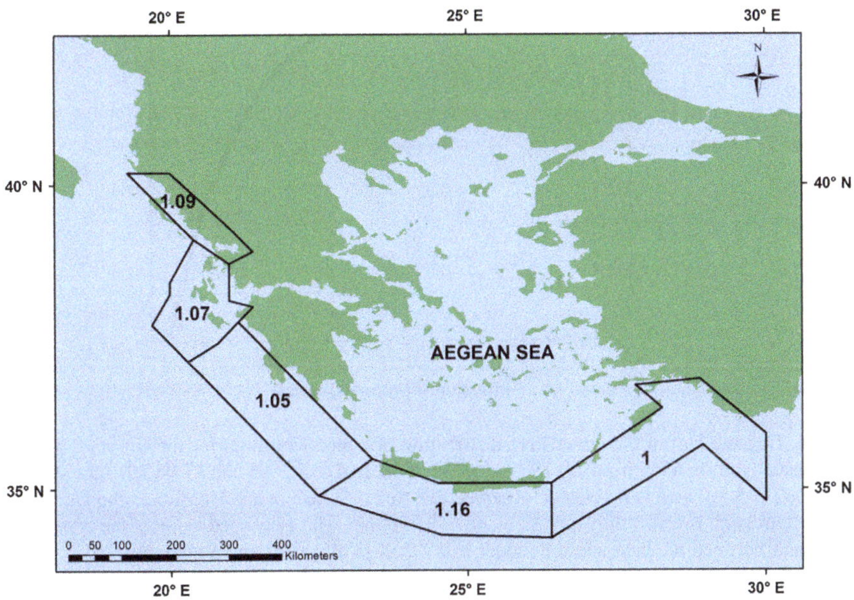

Fig. 14 The q_T variation along the HZS (Papadakis et al. 2013)

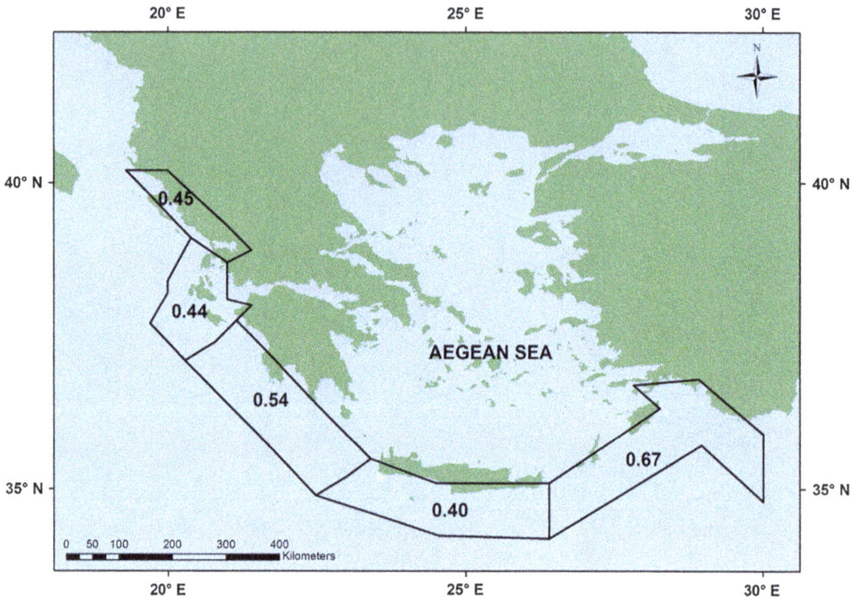

Fig. 15 The q_D variation along the HSZ (Papadakis et al. 2013)

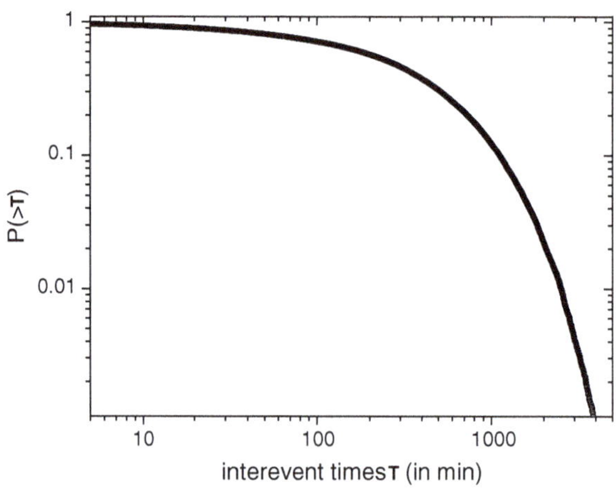

Fig. 16 Log-log plot of the cumulative distribution of inter-event times for the centroid moment tensor catalogue summed from 1.1.1981 to 31.12.2000, to 31.12.2005 and 27.03.2011 for shallow events $M_w > 5.5$. There is no change observed due to the Sumatran and Honshu earthquakes. The best fit regression is realized by $q_\tau = 1.52$ and $\tau_o = 290$ min. The associated value of the correlation coefficient between the data and the model is 0.99354 (Vallianatos and Sammonds 2013)

moment distribution, there is no change in global temporal distribution due to the mega earthquakes. Moreover the inter-event distances cumulative probability distribution *P*(>*D*) (Fig. 17) does not present any change before and after mega-earthquakes.

Along with the cumulative inter-event time distribution, the probability distribution function of the normalized inter-event times of earthquakes in the West Corinth rift has been studied by Michas et al. (2013). In this case, inter-event times τ (in seconds) are scaled to the mean inter-event time $\bar{\tau} = (t_N - t_1)/(N - 1)$ as $\tau' = \tau/\bar{\tau}$. For various threshold magnitudes, the probability distribution function exhibits two-power law regions for short and long inter-event times, indicating the presence of scaling and clustering at both short and long time scales (Fig. 18). This behavior can be well reproduced by a *q* generalized gamma distribution (Queiros 2005) that has the form:

$$p(\tau\prime) = C\tau\prime^{(\gamma-1)}\exp_q(-\tau\prime/\theta), \tag{3.2.1}$$

where C, γ and θ are constants and $e_q(x)$ the q-exponential function (Eq. 2.1.7). In the limit $q \rightarrow 1$, Eq. (3.1) recovers the ordinary gamma distribution. This result indicate that short and intermediate inter-event times, directly related to the production of aftershock sequences, scale with exponent $\gamma - 1$ and long-inter-event times scale with exponent $1/(1 - q)$.

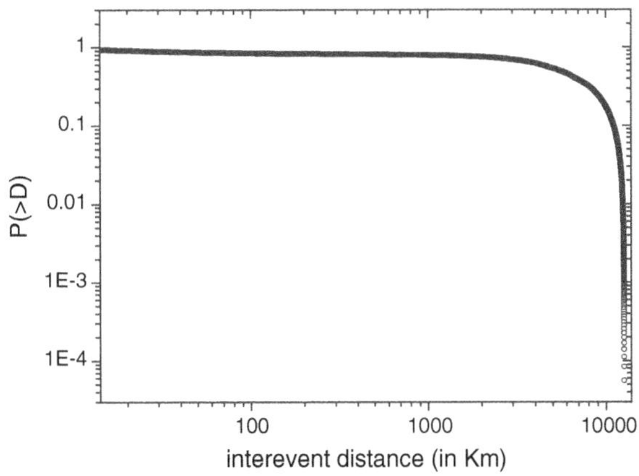

Fig. 17 Log-log plot of the cumulative distribution of inter-event distances for the centroid moment tensor catalogue summed from 1.1.1981 to 31.12.2000, to 31.12.2005 and 27.03.2011 for shallow events $M_w > 5.5$. There is no change observed due to the Sumatran and Honshu earthquakes. The best fit regression is realized by $q_D = 0.29$ and $D_o = 10^4$ km. The associated value of the correlation coefficient between the data and the model is 0.96889 (Vallianatos and Sammonds 2013)

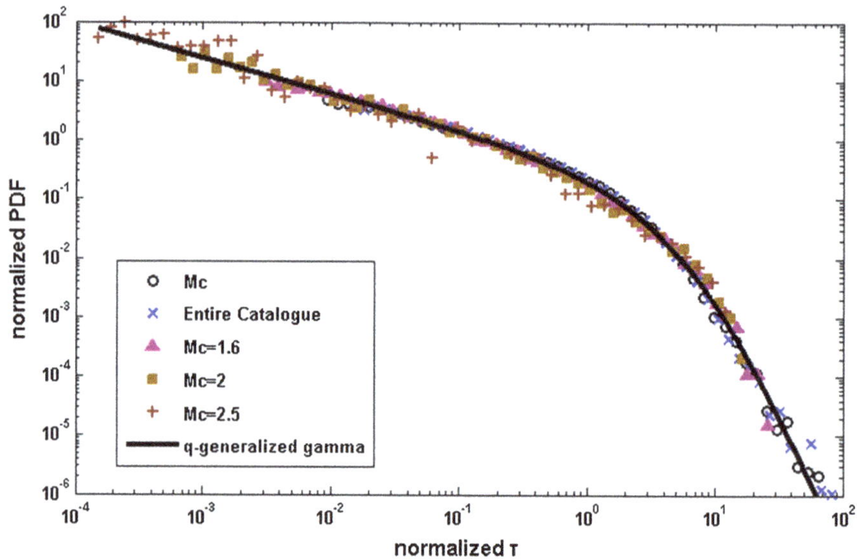

Fig. 18 Normalized probability density $p\ (\tau')$ for the scaled inter-event times τ' for various threshold magnitudes. *Solid line* represents the q-generalized gamma distribution (Eq. 3.2.1) for the values of $C = 0.35$, $\gamma = 0.39$, $\theta = 1.55$ and $q = 1.23$ (Michas et al. 2013)

3.3 *Fault Networks*

Fault systems have been documented on nearly every geologic surface in the solar system (Schultz et al. 2010) and represent a complex scale-invariant network of fractures and faults that is related morphologically and mechanically with the planetary lithosphere deformation and seismicity (Schultz 2003; Knapmeyers et al. 2006). In the last decades, innovative insights into the origin of fault population dynamics have been presented from the point of view of non-equilibrium thermodynamics (Prigogine 1980), fractal geometry (Mandelbrot 1983; Scholz and Mandelbrot 1989), thermodynamics of chaotic systems (Beck and Schlogl 1993) and complexity (Tsallis 2001, 2009).

Vallianatos et al. (2011b) and Vallianatos (2013) used non-extensive statistical physics to explore the distribution of the fault lengths. The aforementioned authors tested the applicability of non-extensive statistical physics in two extreme cases: (a) Crete, in the front of the Hellenic arc and (b) the fault distribution in an extraterrestrial planet, the Mars.

Fault lengths distributions in Central Crete presented by Vallianatos et al. (2011b) (Fig. 19) in the form of log-log plot of the cumulative distribution function (CDF) $P_{cum}(>L)$ of the fault lengths. An analysis of the faults of Central Crete as a single set based on q-exponential distribution leads to $q = 1.16$.

We proceed now to explore using the principles of non-extensive statistical mechanics the fault population statistics derived for an extraterrestrial data set

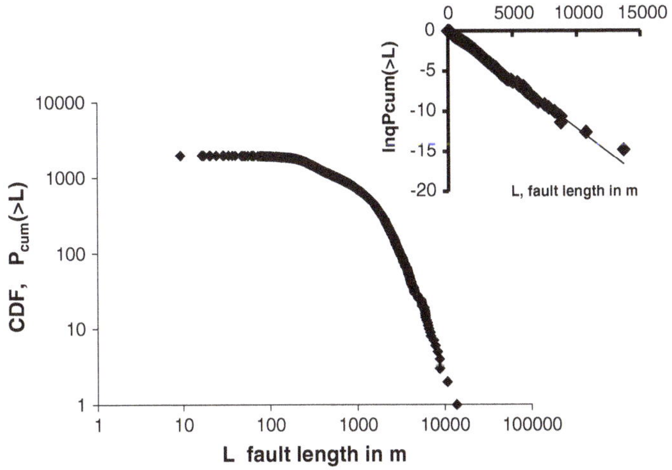

Fig. 19 Unormalized empirical CDF for all fault lengths in Central Crete graben. In the *upper right corner* the semi-q-log plot of the cumulative distribution function CDF of fault lengths for all the examined sets of Central Crete graben is presented (Vallianatos et al. 2011b)

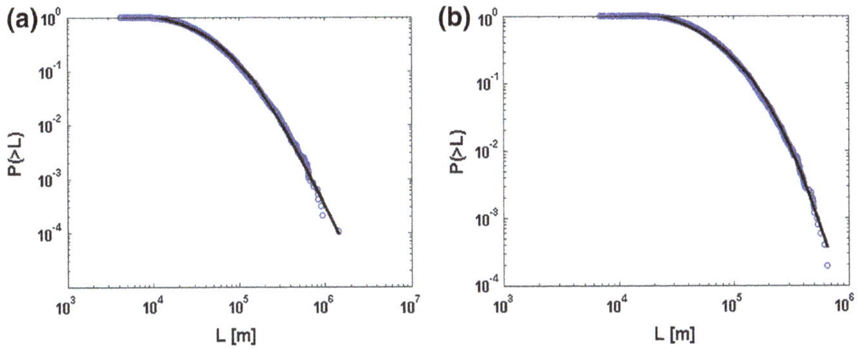

Fig. 20 The normalized cumulative distribution function *P*(>*L*) for Mars **a** compressional and **b** extensional faults. The *black line* is the *q*-exponential fitting for **a** q_c = 1.114 for the trust (compressional) faults and **b** q_e = 1.277 for the normal (extensional) ones (Vallianatos 2013)

selected in a well-studied planet as Mars is. The Valles Marineris Extensional Province on Mars (Lucchitta et al. 1992; Mège and Masson 1996; Schultz 1995, 1997) includes perhaps the largest planetary rift-like structure in the solar system.

The cumulative distribution functions of faults (CDF) *P*(>*L*), are shown as log-log plots in Fig. 20 for the cases of normal and thrust faults. The analysis of fault lengths in Mars indicates that q_e = 1.277 for the extensional (normal) faults, while q_c = 1.114 for the compressional (thrust) faults.

The *q*-values estimated supports the conclusion that the planetary fault system in Mars is a sub-additive one in agreement with a recent result (Vallianatos and

Sammonds 2011) for the Valles Marineris extensional province, Mars and for the regional fault structure in the front of the Hellenic arc (Vallianatos et al. 2011b), with consistency with that observed in local and global seismicity (Vallianatos and Sammonds. 2013; Vallianatos et al. 2013).

3.4 *Plate Tectonics as a Case of Non-extensive Thermodynamics*

In 2003, Bird presented a new global set of present plate boundaries on the Earth (in digital form) and proposed that the distribution of areas of the tectonic plates follows a power law and that this distribution fitted well with the concepts of a few major plates and a hierarchical self-similar organization of blocks at the boundary scale, a fractal plate distribution and a self-organized system.

Vallianatos and Sammonds (2010) applied the concept of non-extensive statistical mechanics to plate tectonics. The aforementioned authors calculated the probability density function for the areas of the tectonic plates. Figure 21 shows the complementary cumulative number $F(>A)$ of plates as a function of area A in steradians, i.e., the number of plates with an area equal to or larger than A. The data are accounted for by the power law, $F(> A) \propto A^{-\mu}$ with μ close to 1/3 except for the three smallest ranks and the largest plates. The results show that three classes (small, intermediate and large) of tectonic plates can be distinguished, which is consistent with the observations of Bird (2003). Vallianatos and Sammonds used the differential equation $dp_i/dA_i = -\beta_1 p_i - (\beta_q - \beta_1)p_i^q$ (see Sect. 2.2 “Cases with two slopes”) in order to further generalize the anomalous equilibrium distribution, in such a way as to have a crossover from anomalous ($q \neq 1$) to normal ($q = 1$) statistical mechanics, while increasing the plate’s area. From a visual inspection of Fig. 21, it might be argued that the deviation from the power law region occurs earlier at the seven largest plates with area more than 1 steradian and belong to a different population than the rest of the plates, indicating that a cross-over exist at $A_{c2} \approx 1$ steradian. Furthermore at the five smallest plates another cross-over exists at $A_{c1} \approx 3 \times 10^{-3}$steradians.

Taking into account that for the intermediate class of tectonic plates the cumulative frequency distribution behaves as a power law with exponent 1/3, the thermodynamic q parameter is calculated equal to $q = 1.75$, which supports the conclusion that the plate tectonics system is a sub-extensive one.

3.5 *Laboratory Seismology*

Recently, the statistical properties of fracture have attracted a wide interest in the statistical physics community (Herrmann and Roux 1990; Chakrabarti and Benguigui 1997). In this context, fracture can be seen as the outcome of the

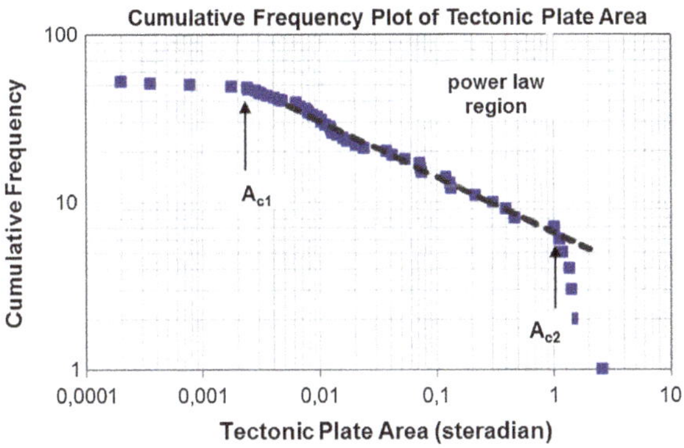

Fig. 21 Complementary cumulative distribution of the areas of tectonic plates compared to the fit with a power law (central long-*dashed line*) with exponent μ = 1/3 (Vallianatos and Sammonds 2010)

irreversible dynamics of a long-range interacting, disordered system. Several experimental observations have revealed that fracture is a complex phenomenon, described by scale invariant laws (Krajcinovic and Van Mier 2000). Examples notably include the acoustic emission (AE) measured prior to fracture and the roughness of the fracture surface (Lei et al. 1992, 2000; and references therein).

Vallianatos et al. (2012a) investigated the statistical physics of fracture in a heterogeneous brittle material (Etna basalt) under triaxial deformation, analyzing the temporal and three dimensional location of moment release of acoustic emissions from micro-fractures that occur before the final fracture.

Using the calculated AE moment, the cumulative distribution function (CDF) $P(>M)$ of the AE scalar moments is shown in Fig. 22a, b presents the log-log plot of the cumulative distribution function $P(>T)$ of the AE inter-event times while Fig. 22c the log-log plot of the cumulative distribution function $P(>D)$ of the AE inter-event Euclidean distances.

The aforementioned authors showed that the scalar moment distribution and the inter-event time distribution of AE, are expressed by the non-extensive statistical mechanics of a sub-additive process with q-values q_M = 1.82 and q_τ = 1.34 respectively supporting the idea of the presence of long-range effects. The inter-event distances described by q-statistics with q_D = 0.65. The above suggests that AEs in Etna's basalt are described by the q-value triplet (q_M, q_τ, q_D) = (1.82, 1.34, 0.65). Furthermore, it should be noticed that the sum of q_τ and q_D indices of the distribution of the inter-event time and distance is $q_\tau + q_D \approx 2$, similar with that observed in regional seismicity data both from Japan and California (Abe and Suzuki 2003, 2005) and verified numerically using the two dimensional Burridge-Knoppoff model (Hasumi 2007, 2009). These results indicate that AEs exhibits a non-extensive spatiotemporal duality similar with that observed with earth seismicity (Abe and Suzuki 2003, 2005; Vallianatos 2009; Vallianatos and Sammonds 2011).

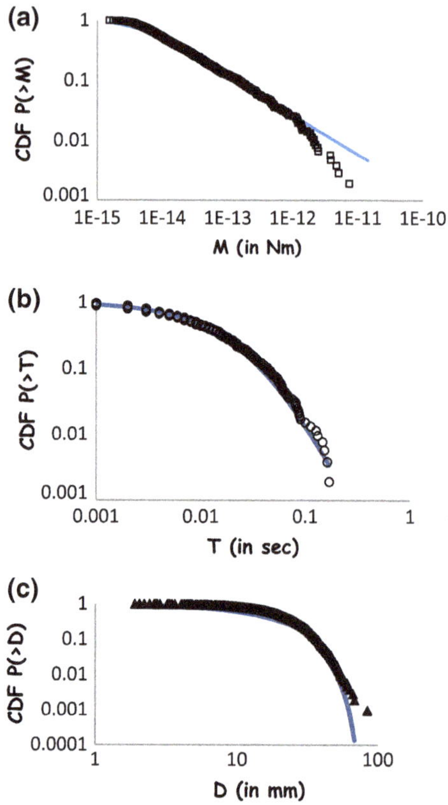

Fig. 22 **a** The cumulative distribution function (*CDF*), *P*(>*M*) of the AEs scalar moment *M*, along with the *q*-exponential fitting curve. **b** The cumulative distribution function *P*(>*T*) of the AEs inter-event time *T*, along with the *q*-exponential fitting curve. **c** The cumulative distribution function *P*(>*D*) of the AEs inter-event Euclidean distance *D*, along with the *q*-exponential fitting curve (Vallianatos and Triantis 2012a)

3.6 Can Non-extensive Statistical Physics Predict Seismicity's Evolution?

Recently, the ideas of non-extensive statistical physics have been used to uncover hidden dynamic features of seismicity before strong events (Papadakis et al. 2014; Vallianatos et al. 2014). These studies examine possible variations of the thermo-statistical parameter q_E before the occurrence of a mainshock. This parameter, which is derived from the fragment asperity model (Sotolongo-Costa and Posadas, 2004), is related to the frequency-magnitude distribution and can be used as an index of the stability of a seismic area. The observed variations are consistent with the evolution of seismicity and seem a very useful tool for the distinction of different dynamical regimes towards a strong earthquake. It should be noticed that this approach has been applied to the strong event of L'Aquila, on April 6, 2009 (M_L = 5.8) (Telesca 2010c). The aforementioned author calculated an increase of the non-extensive parameter in a time interval starting some days before the occurrence of the mainshock.

Vallianatos et al. (2014) applied the concept of non-extensive statistical physics along with the method of natural time analysis (Varotsos et al. 2011) to examine the precursory seismicity of the Mw6.4, October 12, 2013 earthquake in the south-western part of the Hellenic Arc (Fig. 23). Varotsos et al. (2001), proposed the natural time analysis of a complex system, from which we deduce the maximum information from a given time series and we identify the time as we approach towards the occurrence of the mainshock (Varotsos et al. 2011).

Figure 24 presents the temporal evolution of the parameter q_E over increasing (cumulative) time windows. The initial time window has a 100-event width and increases per 1 event over time. The obtained q_E values are associated with the last event included in the window.

The analysis of the frequency-magnitude distribution according to Eq. (3.1.13) reveals that the non-extensive parameter q_E varies during the last period of the earthquake preparatory phase and exhibits a sharp increase a couple of days before the occurrence of the mainshock, indicating an increase in the degree of out-of-equilibrium state before the occurrence of the Mw6.4 earthquake.

Moreover, Papadakis et al. (2014) used the non-extensive formalism to decode the evolution of seismicity towards the January 17, 1995 Kobe earthquake (M = 7.2), in the southwestern part of Japan.

For the detection of possible variations of the non-extensive parameter q_E, the aforementioned authors calculated these variations in different time windows. Figure 25 shows the variations of q_E values over 200-event moving windows (overlapping), having a sliding factor equal to 1. The q_E parameter increases significantly on April 9, 1994 and peaks (q_E = 1.55) as we move towards the 1995

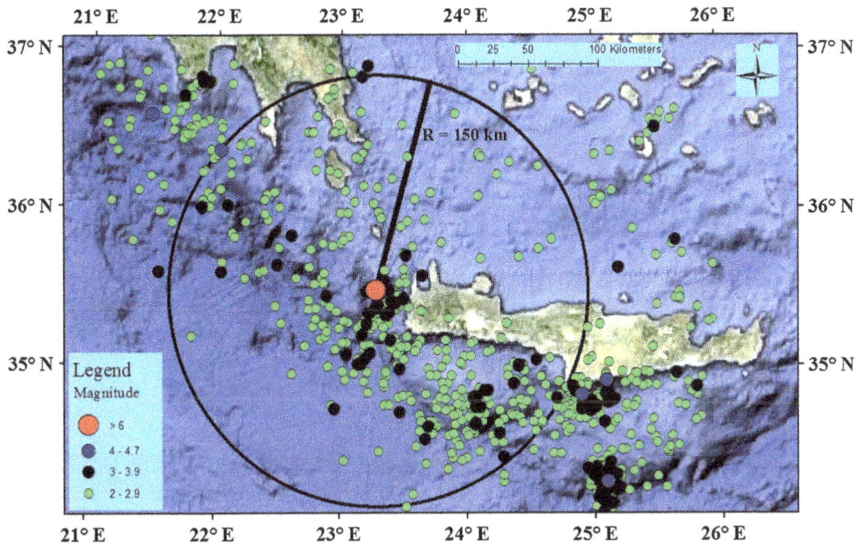

Fig. 23 The observed seismicity in the southwestern segment of the Hellenic Arc during the period 1 July–30 October, 2013

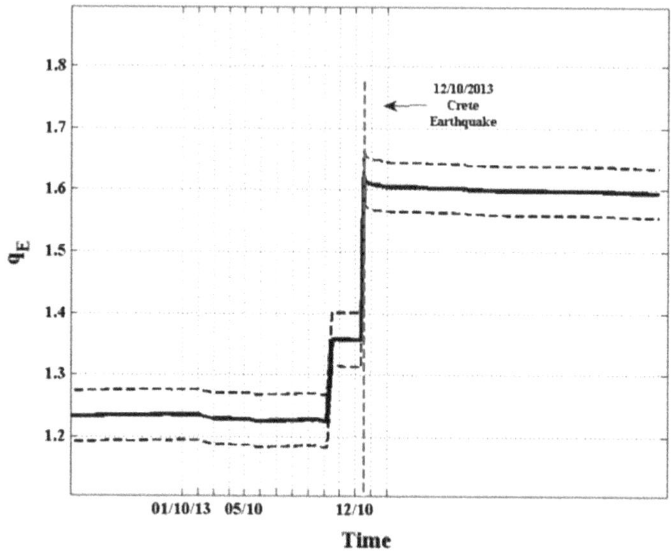

Fig. 24 The temporal evolution of the non-extensive parameter q_E for a circle area with radius R = 150 km around the epicenter of the main event. We note a sharp increase of q_E to the value $q_E \approx 1.36$ a couple of days before the occurrence of the strongest event, while a $q_E \approx 1.6$ is estimated immediately with the Mw6.4 strong event

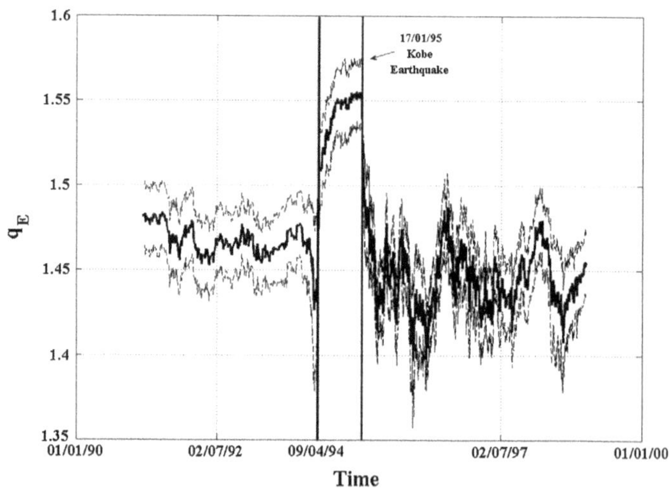

Fig. 25 Time variations of q_E values (*black continuous line*) over 200-event moving windows (*overlapping*), having a sliding factor equal to 1, and the associated standard deviation (*black dashed lines*). On April 9, 1994 the non-extensive parameter increases significantly, indicating the start of a transition phase towards the 1995 Kobe earthquake

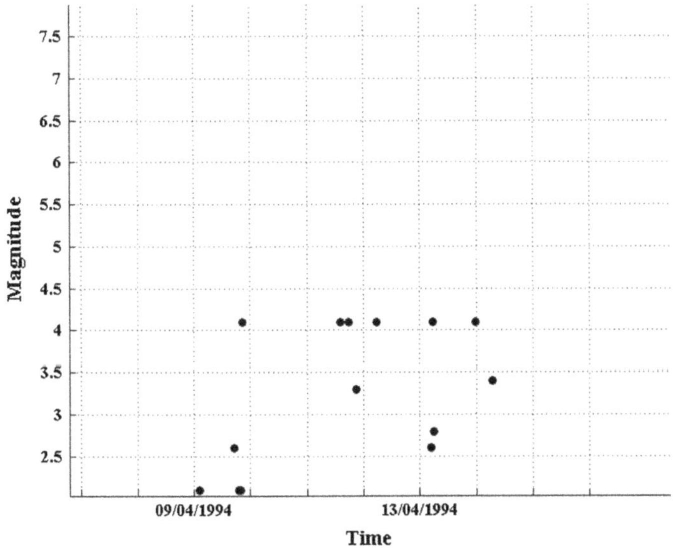

Fig. 26 Time distribution of seismicity, showing the occurrence of six seismic events equal to M = 4.1 between April 9, 1994 and April 13, 1994

Kobe earthquake. After the strong event the non-extensive parameter starts decreasing rapidly.

We detect a significant increase of the non-extensive parameter on April 9, 1994 which coincides with the occurrence of six seismic events equal to M = 4.1 (Fig. 26). The occurrence of these events breaks the magnitude pattern and along with the observed q_E variations indicates a transition phase towards the 1995 Kobe earthquake.

We conclude that the non-extensive statistical physics approach elucidates the physical evolution of a seismic area. Further development of the associated calculated thermostatistical parameter q_E as earthquake precursor improves our ability towards earthquake prediction and becomes beneficial for society and for communities experiencing earthquake hazard worldwide.

4 Quo Vademus?

Many aspects of seismology exhibit complexity. This is an area of research in both the geophysical and statistical physics communities. Although much progress has been made, many questions remain. Relevant areas include scaling laws, temporal and spatial correlations, critical phenomena, and nucleation. Within this complexity, scaling laws are now widely accepted. These include GR frequency–magnitude scaling, Omori's law for the decay of aftershock activity, and Bath's law relating the magnitude of the largest aftershock to the magnitude of the main shock. It can be shown that GR scaling is equivalent to fractal scaling between the number of earthquakes and their rupture area. This scaling is scale invariant, it is robust, but do we understand it? One approach is to directly associate this scaling with the power-law slip-event scaling obtained in slider-block models. But in these models an individual slider block can participate in events of all sizes. This does not seem to be the case for earthquakes on faults; big faults appear to have large earthquakes and small faults small earthquakes. Thus, the GR scaling may be the consequence of a fractal distribution of fault sizes. This in turn can be attributed to scale-invariant fragmentation of the earth's brittle crust in active tectonic regions. The spatio-temporal distribution of seismicity also appears to be universally applicable, but why? A number of explanations have been given on an empirical basis. But the fundamental physics of this spatio-temporal pattern remains controversial.

Models relevant to earthquakes and complexity are at an early stage of development. Slider-block models have certainly played a role but are clearly only weakly related to distributed seismicity. Laboratory seismology also plays a role in understanding the complex behavior of brittle materials. Realistic simulations of distributed seismicity are just beginning to be developed. A major objective of these models is to provide estimates of the seismic hazard.

The study of the non-extensive statistical physics of earthquakes remains wide-open with many significant discoveries to be made. The philosophy is based on the holistic approach to understand the large scale patterns of seismicity. The link between this conceptual approach, based on the successes of statistical physics, and seismology thus remains a highly important domain of research. In particular, statistical seismology needs to evolve into a genuine physically-based statistical physics of earthquakes. In addition, more detailed and rigorous empirical studies of the frequency-size statistics of earthquake seismic moments and how they relate to seismo-tectonic conditions are needed in order to help settle the controversy over the power-law versus the characteristic event regime, and the role of regime-switching and universality. The important debate regarding statistical physics approaches to seismicity would benefit significantly from two points. Firstly, earthquake catalogs contain data uncertainties, biases and subtle incompleteness issues. Investigating their influence on the results of data analyses inspired by statistical physics, increases the relevance of the results. Secondly, the authors should make links with the literature on statistical seismology which deals with similar questions.

The results of the analysis in the cases described previously indicate that the ideas of non-extensive statistical physics can be used to express the non-linear dynamics that control the evolution of the earthquake activity at different scales. The physical models that have been derived using generalized statistical physics (NESP) can successfully describe the statistical properties of the earthquake activity, regarding the magnitude and spatio-temporal scales. These properties, as extracted from first principles, are important for the evolution of the earthquake activity and should be considered in any probabilistic seismic hazard assessment.

Since NESP approach can be evaluated in laboratory scale as well, a future challenging question is to understand how we can scale statistical physics laws in forecasting earthquakes or volcanic eruptions. The laboratory case is important, because it is likely to represent an ideal upper limit for the predictability of time-dependent failure in Earth materials and because many forecasting methodologies intuitively assume a simple scaling from laboratory to field conditions. The global effort to assess the predictability of earthquakes in a rigorous, prospective way has brought the lack of such rigorous evaluations into clearer focus. All forecasting models are subject to the effects of material heterogeneity, measuring error and incomplete data sampling. The key scientific challenge is to understand in a unified way, using NESP principles, the physical mechanisms that drive the evolution of fractures ensembles in laboratory and global scale and how we can use measures of evolution that will forecast the extreme fracture event rigorously and with consistency.

Acknowledgments The work was supported by the THALES Program of the Ministry of Education of Greece and the European Union in the framework of the project entitled "Integrated understanding of Seismicity, using innovative Methodologies of Fracture mechanics along with Earthquake and non-extensive statistical physics—Application to the geodynamic system of the Hellenic Arc. "SEISMO FEAR HELLARC", (MIS 380208).

References

Abe, S., & Suzuki, N. (2003). Law for the distance between successive earthquakes. *Journal of Geophysical Research, 108*(B2), 2113.

Abe, S., & Suzuki, N. (2005). Scale-free statistics of time interval between successive earthquakes. *Physica A, 350*, 588–596.

Bak, P., Christensen, K., Danon, L., & Scanlon, T. (2002). Unified scaling law for earthquakes. *Physical Review Letters, 88*, 178501.

Bak, P., & Tang, C. (1989). Earthquakes as a self-organized critical phenomenon. *Journal of Geophysical Research, 94*, 635–637.

Bak, P., Tang, C., & Wiesenfeld, K. (1987). Self-organized criticality: An explanation of 1/f noise. *Physical Review Letters, 59*, 381–384.

Bak, P., Tang, C., & Wiesenfeld, K. (1988). Self-organized criticality. *Physical Review A, 38*, 364–374.

Beck, C., & Schlogl, F. (1993). *Thermodynamics of chaotic systems: An introduction*. Cambridge: Cambridge University Press.

Bell, A. F., Naylor, M., & Main, I. G. (2013). Convergence of the frequency-size distribution of global earthquakes. *Geophysical Research Letters, 40*, 2585–2589.

Berrill, J. B., & Davis, R. O. (1980). Maximum entropy and the magnitude distribution. *Bulletin of the Seismological Society of America, 70*, 1823–1831.

Bird, P. (2003). An updated digital model of plate boundaries. *Geochemistry, Geophysics, Geosystems, 4*(3), 1027.

Burridge, L., & Knopoff, L. (1967). Model and theoretical seismicity. *Bulletin of the Seismological Society of America, 57*, 341–371.

Caruso, F., Pluchino, A., Latora, V., Vinciguerra, S., & Rapisarda, A. (2007). Analysis of self-organized criticality in the Olami-Feder-Christensen model and in real earthquakes. *Physical Review E, 75*, 055101.

Chakrabarti, B. K., & Benguigui, L. G. (1997). *Statistical physics of fracture and breakdown in disordered systems*. Oxford: Oxford Science Publications.

Corral, A. (2004). Long-term clustering, scaling, and universality in the temporal occurrence of earthquakes. *Physical Review Letters, 92*, 108501.

Darooneh, A. H., & Dadashinia, C. (2008). Analysis of the spatial and temporal distributions between successive earthquakes: Nonextensive statistical mechanics viewpoint. *Physica A, 387*, 3647–3654.

De Rubeis, V., Hallgas, R., Loreto, V., Paladin, G., Pietronero, L., & Tosi, P. (1996). Self-affine asperity model for earthquakes. *Physical Review Letters, 76*, 2599–2602.

Engdahl, E. R., & Villaseñor, A. (2002). Global seismicity: 1900–1999. International Handbook of Earthquake and Engineering Seismology, Part A, Chapter 41, (pp. 665–690). Academic Press, Waltham.

Ferri, G. L., Martínez, S., & Plastino, A. (2005). Equivalence of the four versions of Tsallis's statistics. *Journal of Statistical Mechanics: Theory and Experiment* P04009.

Gutenberg, B., & Richter, C. F. (1944). Frequency of earthquakes in California. *Bulletin of the Seismological Society of America, 34*, 185–188.

Hasumi, T. (2007). Interoccurrence time statistics in the two-dimensional Burridge-Knopoff earthquake model. *Physical Review E, 76*, 026117.

Hasumi, T. (2009). Hypocenter interval statistics between successive earthquakes in the two-dimensional Burridge-Knopoff model. *Physica A, 388*, 477–482.

Herrmann, H. J., & Roux, S. (1990). Modelization of fracture in disordered systems. *Statistical Models for the Fracture of Disordered Media* (pp. 159–188). Elsevier: North-Holland.

Hirata, T., & Imoto, M. (1991). Multifractal analysis of spatial distribution of micro earthquakes in the Kanto region. *Geophysical Journal International, 107*, 155–162.

Kagan, Y. Y. (1994). Observational evidence for earthquakes as a nonlinear dynamic process. *Physica D: Nonlinear Phenomena, 77*, 160–192.

Kagan, Y. Y. (1997). Seismic moment-frequency relation for shallow earthquakes: Regional comparison. *Journal of Geophysical Research, 102*, 2835–2852.

Kagan, Y. Y., & Jackson, D. D. (1991). Long-term earthquake clustering. *Geophysical Journal International, 104*, 117–133.

Kagan, Y. Y., & Jackson, D. D. (2000). Probabilistic forecasting of earthquakes. *Geophysical Journal International, 143*, 438–453.

Kagan, Y. Y., & Jackson, D. D. (2013). Tohoku earthquake: A surprise? *Bulletin of the Seismological Society of America, 103*, 1181–1194.

Kagan, Y. Y., & Knopoff, L. (1980). Spatial distribution of earthquakes: The two point correlation function. *Geophysical Journal Royal Astronomical Society, 62*, 303–320.

Kanamori, H. (1978). Quantification of earthquakes. *Nature, 271*, 411–414.

Kawamura, H., Hatano, T., Kato, N., Biswas, S., & Chakrabarti, B. K. (2012). Statistical physics of fracture, friction and earthquakes. *Review of Modern Physics, 84*, 839–884.

Knapmeyer, M., Oberst, J., Hauber, E., Wahlisch, M., Deuchler, C., & Wagner, R. (2006). Working model for spatial distribution and level of Mars' seismicity. *Journal of Geophysical Research, 111*, E11006.

Krajcinovic, D., & Van Mier, J. G. M. (2000). *Damage and fracture of disordered materials*. New York: Springer.

Lay, T., & Wallace, T. C. (1995). *Modern global seismology*. New York: Academic Press.

Lei, X. L., Kusunose, K., Nishizawa, O., Cho, A., & Satoh, T. (2000). On the spatiotemporal distribution of acoustic emissions in two granitic rocks under triaxial compression: the role of preexisting cracks. *Geophysical Research Letters, 27*, 1997–2000.

Lei, X., Nishizawa, O., Kusunose, K., & Satoh, T. (1992). Fractal structure of the hypocenter distribution and focal mechanism solutions of AE in two granites of different grain size. *Journal of Physics of the Earth, 40*, 617–634.

Lucchitta, B. K., McEwen, S., Clow, G. D., Geissler, P. E., Singer, R. B., Schultz, R. A., & Squyres, S. W. (1992). The canyon system of Mars. In H. H. Kieffer, B. M. Jakosky, C. W. Snyder, & M. S. Matthews (Eds.), *Mars* (pp. 453–492). USA: University of Arizona Press.

Main, I. (1996). Statistical physics, seismogenesis, and seismic hazard. *Reviews of Geophysics, 34*, 433–462.
Main, I. G., & Al-Kindy, F. H. (2002). Entropy, energy, and proximity to criticality in global earthquake populations. *Geophysical Research Letters, 29*(7), 25–1
Main, I. G., & Burton, P. W. (1984). Information theory and the earthquake frequency-magnitude distribution. *Bulletin of the Seismological Society of America, 74*, 1409–1426.
Main, I. G., Meredith, P. G., & Jones, C. (1989). A reinterpration of the precursory seismic b-value anomaly from fracture mechanics. *Geophysical Journal, 96*, 131–138.
Main, I. G., Meredith, P. G., & Sammonds, P. R. (1992). Temporal variations in seismic event rate and b-values from stress corrosion constitutive laws. *Tectonophysics, 211*, 233–246.
Mandelbrot, B. B. (1983). *The fractal geometry of nature*. San Francisco: Freeman.
Mège, D., & Masson, P. (1996). A plume tectonics model for the Tharsis province, Mars. *Planetary and Space Science, 44*, 1499–1546.
Michas, G., Vallianatos, F., & Sammonds, P. (2013). Non-extensivity and long-range correlations in the earthquake activity at the West Corinth rift (Greece). *Nonlinear Processes in Geophysics, 20*, 713–724.
Nature Debates. (1999). Nature debates: Is the reliable prediction of individual earthquakes a realistic scientific goal? Available from http://www.nature.com/nature/debates/.
Olami, Z., Feder, H. J. S., & Christensen, K. (1992). Self-organized criticality in a continuous nonconservative cellular automaton modeling earthquakes. *Physical Review Letters, 68*, 1244–1247.
Omori, F. (1894). On the aftershocks of earthquakes. *Journal of the College of Science, Imperial University of Tokyo 7*, 111–200.
Papadakis, G., Vallianatos, F., & Sammonds, P. (2013). Evidence of nonextensive statistical physics behavior of the Hellenic Subduction Zone seismicity. *Tectonophysics, 608*, 1037–1048.
Papadakis, G., Vallianatos, F., & Sammonds, P. (2014). A nonextensive statistical physics analysis of the 1995 Kobe earthquake, Japan. *Pure and Applied Geophysics* (accepted).
Picoli, S., Mendes, R. S., Malacarne, L. C., & Santos, R. P. B. (2009). q-distributions in complex systems: A brief review. *Brazilian Journal of Physics, 39*, 468–474.
Prigogine, I. (1980). *From being to becoming: Time and complexity in physical systems*. San Francisco: Freeman and Co.
Queirós, S. M. D. (2005). On the emergence of a generalised gamma distribution, application to traded volume in financial markets. *Europhysics Letters, 71*, 339–345.
Rundle, J. B., Gross, S., Klein, W., Ferguson, C., & Turcotte, D. L. (1997). The statistical mechanics of earthquakes. *Tectonophysics, 277*, 147–164.
Rundle, J. B., Turcotte, D. L., Shcherbakov, R., Klein, W., & Sammis, C. (2003). Statistical physics approach to understanding the multiscale dynamics of earthquake fault systems. *Reviews of Geophysics, 41*, 4.
Sammonds, P. (2005). Plasticity goes supercritical. *Nature Materials, 4*, 425–426.
Sammonds, P., & Ohnaka, M. (1998). Evolution of microseismicity during frictional sliding. *Geophysical Research Letters, 25*, 699–702.
Scholz, C. H. (1998). Earthquakes and friction laws. *Nature, 391*, 37–42.
Scholz, C. H., & Mandelbrot, B. B. (1989). *Fractals in geophysics*. Basel: Birkhuser.
Schultz, R. A. (1995). Gradients in extension and strain at Valles Marineris, Mars. *Planet Space Science, 43*, 1561–1566.
Schultz, R. A. (1997). Displacement–length scaling for terrestrial and Martian faults: Implications for Valles Marineris and shallow planetary grabens. *Journal of Geophysical Research, 102*, 12009–12015.
Schultz, R. A. (2003). Seismotectonics of the Amenthes Rupes thrust fault population, Mars. *Geophysical Research Letters, 30*, 1303–1307.
Schultz, R. A., Hauber, E., Kattenhorn, S., Okubo, C., & Watters, T. (2010). Interpretation and analysis of planetary structures. *Journal of Structural Geology, 32*, 855–875.

Shannon, C. E. (1948). A mathematical theory of communication. *Bell System Technical Journal 27*, 379–423, 623–656.

Silva, R., Franca, G. S., Vilar, C. S., & Alcaniz, J. S. (2006). Nonextensive models for earthquakes. *Physical Review E, 73*, 026102.

Sornette, D. (2004). *Critical phenomena in natural sciences, chaos, fractals, self-organization and disorder: Concepts and tools* (2nd ed.). Heidelberg: Springer.

Sornette, A., & Sornette, D. (1989). Self-organized criticality and earthquakes. *Europhysics Letters, 9*, 197–202.

Sotolongo-Costa, O., & Posadas, A. (2004). Fragment-asperity interaction model for earthquakes. *Physical Review Letters, 92*(4), 048501.

Telesca, L. (2010a). Analysis of Italian seismicity by using a nonextensive approach. *Tectonophysics, 494*, 155–162.

Telesca, L. (2010b). Nonextensive analysis of seismic sequences. *Physica A, 389*, 1911–1914.

Telesca, L. (2010c). A non-extensive approach in investigating the seismicity of L'Aquila area (central Italy), struck by the 6 April 2009 earthquake (ML = 5.8). *Terra Nova, 22*, 87–93.

Telesca, L. (2011). Tsallis-based nonextensive analysis of the southern California seismicity. *Entropy, 13*, 1267–1280.

Telesca, L. (2012). Maximum likelihood estimation of the nonextensive parameters of the earthquake cumulative magnitude distribution. *Bulletin of the Seismological Society of America, 102*(2), 886–891.

Telesca, L., Cuomo, V., Lapenna, V., Vallianatos, F., & Drakatos, G. (2001). Analysis of the temporal properties of Greek aftershock sequences. *Tectonophysics, 341*, 163–178.

Telesca, L., Lapenna, V., & Macchiato, M. (2003). Spatial variability of the time-correlated behaviour in Italian seismicity. *Earth and Planetary Science Letters, 212*, 279–290.

Telesca, L., Lapenna, V., & Vallianatos, F. (2002). Monofractal and multifractal approaches in investigating scaling properties in temporal patterns of the 1983–2000 seismicity in the western Corinth graben, Greece. *Physics of the Earth and Planetary Interiors, 131*, 63–79.

Tsallis, C. (1988). Possible generalization of Boltzmann-Gibbs statistics. *Journal of Statistical Physics, 52*, 479–487.

Tsallis, C. (2001). Non extensive statistical mechanics and its applications. In S. Abe, & Y. Okamoto (Eds.), Berlin: Springer.

Tsallis, C. (2009). *Introduction to nonextensive statistical mechanics: Approaching a complex world*. Berlin: Springer.

Tsallis, C., Bemski, G., & Mendes, R. S. (1999). Is re-association of folded proteins a case of non-extensivity? *Physics Letters A, 257*, 93–97.

Tsekouras, G. A., & Tsallis, C. (2005). Generalized entropy arising from a distribution of q indices. *Physical Review E, 71*, 046144.

Turcotte, D. L. (1997). *Fractals and chaos in geology and geophysics* (2nd ed.). Cambridge, UK: Cambridge University Press.

Utsu, T., Ogata, Y., & Matsura, R. S. (1995). The centenary of the Omori formula for a decay law of aftershock activity. *Journal of Physics of the Earth, 43*, 1–33.

Vallianatos, F. (2009). A non-extensive approach to risk assessment. *Natural Hazards and Earth System Sciences, 9*, 211–216.

Vallianatos, F. (2013). On the non-extensivity in Mars geological faults. *Europhysics Letters, 102*, 28006.

Vallianatos, F., Benson, P., Meredith, P., & Sammonds, P. (2012a). Experimental evidence of a non-extensive statistical physics behaviour of fracture in triaxially deformed Etna basalt using acoustic emissions. *Europhysics Letters, 97*, 58002.

Vallianatos, F., Kokinou, E., & Sammonds, P. (2011a). Non extensive statistical physics approach to fault population distribution. A case study from the Southern Hellenic Arc (Central Crete). *Acta Geophysica, 59*, 1–13.

Vallianatos, F., Michas, G., & Papadakis, G. (2014). Non-extensive and natural time analysis of seismicity before the M_w 6.4, 12 Oct 2013 earthquake in the south west segment of the Hellenic arc. (submitted).

Vallianatos, F., Michas, G., Papadakis, G., & Sammonds, P. (2012b). A non-extensive statistical physics view to the spatiotemporal properties of the June 1995, Aigion earthquake (M6.2) aftershock sequence (West Corinth rift, Greece). *Acta Geophysica, 60*, 758–768.

Vallianatos, F., Michas, G., Papadakis, G., & Tzanis, A. (2013). Evidence of non-extensivity in the seismicity observed during the 2011–2012 unrest at the Santorini volcanic complex, Greece. *Natural Hazards and Earth System Sciences, 13*, 177–185.

Vallianatos, F., & Sammonds, P. (2010). Is plate tectonics a case of non-extensive thermodynamics? *Physica A, 389*, 4989–4993.

Vallianatos, F., & Sammonds, P. (2011). A non-extensive statistics of the fault-population of the Valles Marineris extensional province, Mars. *Tectonophysics, 509*, 50–54.

Vallianatos, F., & Sammonds, P. (2013). Evidence of non-extensive statistical physics of the lithospheric instability approaching the 2004 Sumatran-Andaman and 2011 Honsu mega-earthquakes. *Tectonophysics, 590*, 52–58.

Vallianatos, F., & Triantis, D. (2012). Is pressure stimulated current relaxation in amphibolite a case of non-extensivity? *Europhysics Letters, 99*, 18006.

Vallianatos, F., Triantis, D., & Sammonds, P. (2011b). Non-extensivity of the isothermal depolarization relaxation currents in uniaxial compressed rocks. *Europhysics Letters, 94*, 68008.

Vallianatos, F., Triantis, D., Tzanis, A., Anastasiadis, C., & Stavrakas, I. (2004). Electric earthquake precursors: From laboratory results to field observations. *Physics and Chemistry of the Earth, 29*, 339–351.

Varotsos, P. A., Sarlis, N. V., & Skordas, E. S. (2001). Spatio-temporal complexity aspects on the interrelation between seismic electric signals and seismicity. *Practica of Athens Academy, 76*, 294–321.

Varotsos, P. A., Sarlis, N. V., & Skordas, E. S. (2011). *Natural time analysis: The new view of time, precursory seismic electric signals, earthquakes and other complex time series*. Berlin: Springer.

Vilar, C. S., Franca, G. S., Silva, R., & Alcaniz, J. S. (2007). Nonextensivity in geological faults. *Physica A, 377*, 285–290.

Wada, T., & Scarfone, A. M. (2005). Connection between Tsallis' formalisms employing the standard linear average energy and ones employing the normalized q-average energy. *Physics Letters A, 335*, 351–362.

Zaslavsky, G. M. (1999). Chaotic dynamics and the origin of statistical laws. *Physics Today, 52*, 39–45.

Recognition of periAdriatic Seismic Zones Most Prone to Next Major Earthquakes: Insights from a Deterministic Approach

Enzo Mantovani, Marcello Viti, Daniele Babbucci, Caterina Tamburelli, Nicola Cenni, Massimo Baglione and Vittorio D'Intinosante

1 Introduction

The distribution of historical earthquakes indicates that a large portion of the Italian territory could be hit by strong earthquakes (e.g., Guidoboni et al. 2007). Since a large part of the building patrimony in Italy was realized without taking into account adequate antiseismic criteria (e.g., Di Pasquale et al. 2005; Crowley et al. 2009), it would be economically very difficult in the near future to achieve a significant mitigation of seismic risk in such a large zone. This objective could more easily be obtained if reliable information was available about the zones most prone

E. Mantovani (✉) · M. Viti · D. Babbucci · C. Tamburelli
Dipartimento di Scienze Fisiche della Terra e dell'Ambiente, Università di Siena, Siena, Italy
e-mail: enzo.mantovani@unisi.it

M. Viti
e-mail: marcello.viti@unisi.it

D. Babbucci
e-mail: babbucci@unisi.it

C. Tamburelli
e-mail: tamburelli@unisi.it

N. Cenni
Dipartimento di Fisica ed Astronomia, Università di Bologna, Bologna, Italy
e-mail: nicola.cenni@unibo.it

M. Baglione · V. D'Intinosante
Settore Sismica, Regione Toscana, Florence, Italy
e-mail: massimo.baglione@regione.toscana.it

V. D'Intinosante
e-mail: vittorio.dintinosante@regione.toscana.it

S. D'Amico (ed.), *Earthquakes and Their Impact on Society*,
Springer Natural Hazards, DOI 10.1007/978-3-319-21753-6_2

to next strong earthquakes, where the limited resources now available could be concentrated.

An attempt at obtaining such kind of information could be made by taking into account the possible connections between the spatio-temporal distribution of major shocks and the progressive development of tectonic processes in the central Mediterranean region, which are mainly related to the complex short-term kinematics of the Adriatic plate. It is known that each strong shock triggers a perturbation of the strain field that propagates in the surrounding zones (post-seismic relaxation, e.g., Pollitz et al. 2006; Ryder et al. 2007; Ergintav et al. 2009; Ozawa et al. 2011). When the effects of such perturbation reaches other seismic zones they may modify the probability of fault activation or even cause an earthquake when the fault involved is close to seismic failure. The possibility that this phenomenon induces seismicity has been pointed out in a number of papers (e.g., Anderson 1975; Rydelek and Sacks 1990; Pollitz et al. 1998, 2004, 2012; Mikumo et al. 2002; Freed 2005; Freed et al. 2007; Brodsky 2009; Lay et al. 2009; Durand et al. 2010; Luo and Liu 2010; Viti et al. 2012, 2013), which show that the time and place of occurrence of a number of major shocks are compatible with the expected effects of post-seismic relaxation induced by triggering events. In particular, this phenomenon has been recognized for some Italian zones (Southern Apennines and Calabria) whose seismic activity seems to be significantly influenced by major seismic crises in Hellenic and Dinaric zones respectively (Viti et al. 2003; Mantovani et al. 2008, 2010, 2012).

The fact that past seismic activity may affect the spatio-temporal distribution of next shocks in the tectonic context here considered is instead supported by the time pattern of major earthquakes that occurred at the main periAdriatic zones since 1400 A.D. In the next section, we describe the evidence that may support the plausibility of the proposed approach and we discuss on how it may provide insights into the location of next strong earthquakes in the Italian peninsula.

2 Interaction Between Southern Apennine and Southern Dinaric Seismic Sources

The possibility that major earthquakes in the first zone may influence the probability of major shocks in the second zone was first suggested by the fact that the strong earthquake (magnitude M = 7.0) of April 1979 in the Montenegro area of the southern Dinarides was followed on November 1980 by a major event (M = 6.9) in the Irpinia zone of the southern Apennines (Fig. 1). This initial hypothesis has then been reinforced by the knowledge about the seismotectonic setting of the study area (Fig. 1) and the results obtained by the quantification of the post-seismic relaxation triggered by the 1979 Montenegro event, as synthetically described in the following.

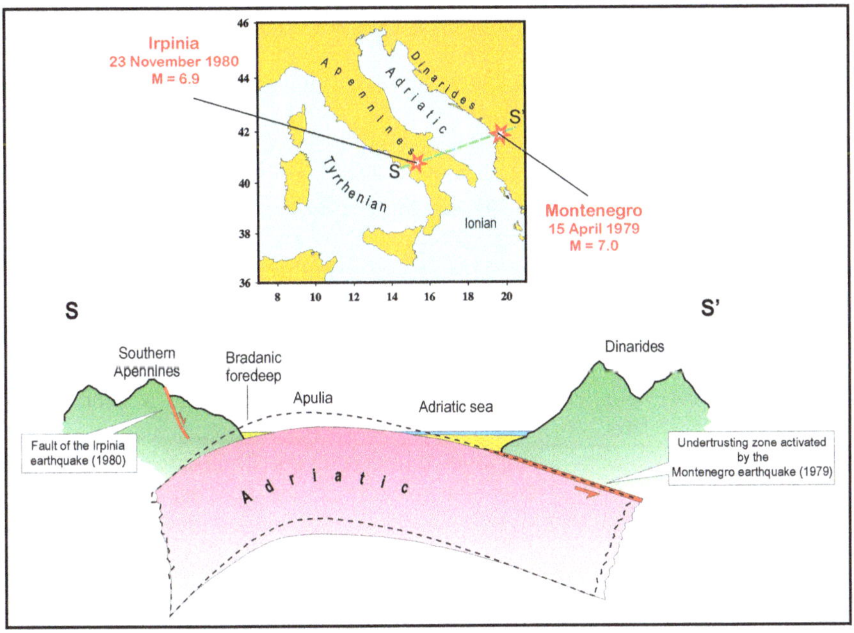

Fig. 1 Structural sketch, through a transversal section in the Southern Adriatic, which points out the vertical flexure of the Adriatic lithosphere overthrusted by the Dinaric belt, on one side, and by the Apennine belt on the other side (e.g., Moretti and Royden 1988; De Alteriis 1995). The vertical scale is exaggerated in order to make more evident the possible effect of a seismic slip (*red arrow*) between the Adriatic lithosphere and the Dinaric belt. The *dashed line* indicates the presumed profile of the Adriatic lithosphere before the Montenegro seismic slip. The epicentres of the 1979 (M = 7.0) Montenegro and 1980 (M = 6.9) Irpinia earthquakes and the trace of the section (green line) are shown in the map

The occurrence of seismic slip at a thrust fault beneath the Southern Dinarides, such as the one that developed with the 1979 Montenegro event (estimated to be 1–2 m, e.g. Benetatos and Kiratzi 2006), implies a roughly NE ward motion of the adjacent part of the Adria plate, which causes a reduction of the vertical flexure in the southern Adriatic domain. As sketched in Fig. 1, such process is expected to induce extensional strain in the Southern Apennines, which may favor the activation of the belt parallel normal faults recognized in that zone (e.g., Ascione et al. 2007).

This hypothesis is confirmed by the results of numerical modelling of the strain perturbation that was presumably induced in the Irpinia zone by the 1979 Montenegro event (Viti et al. 2003; Mantovani et al. 2010, 2012). In particular, by the fact that the strain rate induced by the Montenegro earthquake is expected to reach its maximum amplitude in the Southern Apennines about 1–2 years after the triggering event, i.e. a delay fairly consistent with the time interval that elapsed between the April 1979 Montenegro and November 1980 Irpinia shocks. The possible relationship between stress/strain rate increase and triggering of seismic activity has been pointed out in several works (e.g., Pollitz et al. 1998; Toda et al. 2002; Viti et al. 2003, 2012, 2013).

The possibility that the interaction between Southern Dinaric and Southern Apennine seismic sources is not an isolated phenomenon is suggested by the comparison of the series of major shocks that have occurred in such zones in the last two centuries (Fig. 2). From the list given in this figure, it is possible to note that in the period considered all the shocks with M ≥ 6.0 in the Southern Apennines have been preceded within few years (less than 5) by one or more earthquakes with M ≥ 6 in the Southern Dinarides. Since the probability that such a regular correspondence merely occurs by chance is very small (Mantovani et al. 2010, 2012), it is plausible to suppose that the observed interrelation results from a tectonic connection between the two zones.

The above correspondence does not change significantly if weaker shocks (M ≥ 5.5) are considered, since only one of the 15 Southern Apennine events was not preceded by equivalent earthquakes in the Southern Dinarides. This evidence may imply that a fault in the first zone can hardly activate without the contribution of post-seismic perturbation triggered by one or more major shocks in the other zone.

The fact that this significant time correlation can be recognized for the most recent, complete and reliable part of the seismic catalogue gives good reasons to hope that this phenomenon may provide a tool for recognizing the periods when the probability of strong shocks in Southern Apennines may undergo a significant increase. In this view, the fact that no earthquakes have occurred in the Southern Dinarides since 1996 would imply that at present the probability of major shocks in

Southern Apennines (M ≥ 5.5)	Southern Dinarides (M > 5.5)
1998 (5.6)	**1996 (6.0)**
1990 (5.8, 5.5)	1988 (5.7), 1982 (5.7)
1980 (6.9)	**1979 (7.0**, 5.7, **6.4)**
1962 (6.1, 5.7)	**1962 (6.2), 1959** (5.6, **6.2**, 5.9), 1958 (5.6)
	1935 (5.6), 1934 (5.6)
1930 (6.6)	**1927 (6.0)**, 1926 (5.8, 5.8), 1922 (5.6), 1921 (5.6), 1920 (5.6, 5.6)
1910 (5.7)	**1907 (6.2**, 5.7), **1906 (6.4), 1905 (6.0, 6.6)**
	1896 (6.2, 5.9), **1895 (6.2)**, 1894 (5.9)
1885 (5.5)	**1876** (5.6, **6.3)**
	1870 (6.5), 1869 (6.2, 6.0, 5.6)
1857 (7.0)	**1855 (6.8, 6.6)**, 1853 (5.7)
1853 (5.6), **1851 (6.4)**	**1852 (6.2), 1851 (6.0, 6.6, 6.2, 6.3)**, 1850 (5.6), **1848 (6.4), 1843 (6.2**, 5.6)
1836 (6.0)	**1833 (6.5)**, 1830 (5.6)
1831 (5.5)	**1827 (6.5)**
1826 (5.8)	1823 (5.7)

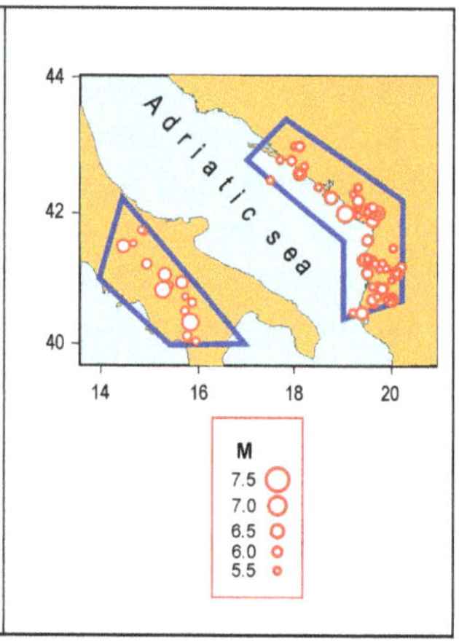

Fig. 2 Geometry of the zones implied in the presumed interrelation between Southern Dinaric and Southern Apennine seismic sources and list of the major seismic events occurred after 1810 (the shocks with M ≥ 6.0 are in *bold*). Data sources as in Appendix

the Southern Apennines is relatively low. A more detailed description of the seismic correlation cited above and a discussion about its possible uncertainties are reported in previous papers (Viti et al. 2003; Mantovani et al. 2010, 2012).

3 Interaction Between Calabrian and Hellenic Seismic Sources

Another significant correlation has been recognized between the major earthquakes of Calabria and those of the Hellenides sector lying between the Ionian islands and Albania (Fig. 3). The hypothesis that a strong seismic activation of the Hellenic thrust zone may increase the probability of major earthquakes in Calabria is consistent with the structural/tectonic setting sketched in Fig. 3. Indeed, such scheme suggests that a significant seismic slip at the Hellenic thrust zone is expected to produce a reduction of the upward vertical flexure of the Adriatic lithosphere, which may help the overthrusting of the Calabrian wedge, that is the process which is

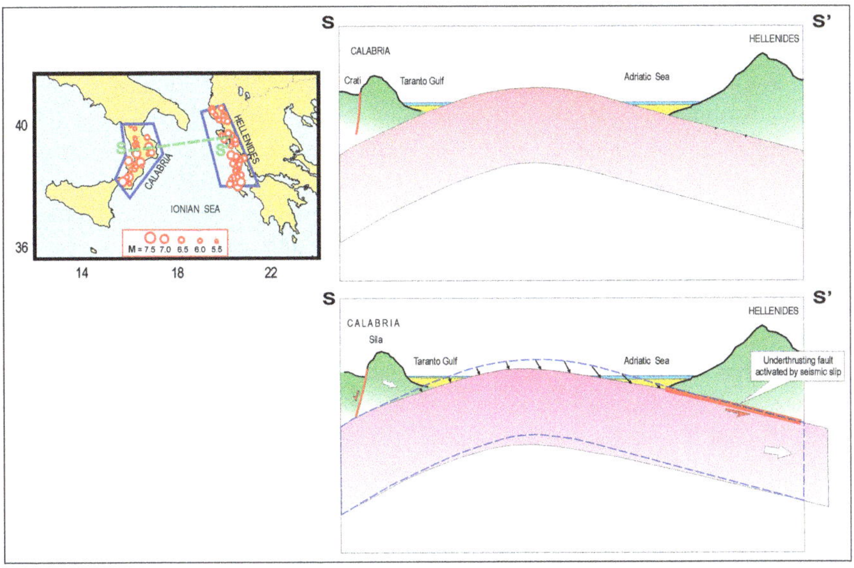

Fig. 3 The map shows the geometry of the two presumably interrelated Calabrian and Hellenic seismic zones, the trace of the section and the epicenters of the earthquakes occurred in the two zones since 1600 A.D. (*red circles*). The events considered are listed in Table 1. The *top section* shows a schematic reconstruction of the structural setting in the southern Adriatic area from Calabria to the Northern Hellenides. The *bottom section* illustrates a tentative reconstruction (*vertically exaggerated*) of the reduction of vertical flexure of the Adriatic plate (*dashed line*), which may occur in response to a strong decoupling earthquake at the Hellenic thrust zone. This effect may favour the outward escape of the Calabrian wedge towards the Ionian domain, presumably accompanied by seismic activation of its main fault systems

accommodated by the seismic activation of normal and strike-slip faults in Calabria (Mantovani et al. 2008, 2009, 2012).

The above interpretation and its implications on the interaction of the Calabrian and Hellenic seismic sources is consistent with the quantification of the effects of post-seismic relaxation induced by strong earthquakes in the Hellenides (Mantovani et al. 2008, 2012), which indicates that such phenomenon may have influenced the Calabrian shocks that occurred some years after the Hellenic triggering events.

The possibility that the above phenomenon has a systematic character is supported by the comparison of the seismic histories of the two zones involved (Table 1), which points out that all Calabrian shocks with M ≥ 6.0 have been preceded, within 10 years, by at least one event with M ≥ 6.5 in the Hellenides. If lower magnitudes (M ≥ 5.5) are considered, the correspondence remains fairly significant, since only 2 (out of 26) Calabrian events have not been preceded by equivalent shocks in the Hellenides. The above evidence supports the hypothesis that a major Calabrian earthquake can hardly occur without being preceded by significant seismic activity in the Hellenic zone (Mantovani et al. 2012).

Concerning the opposite aspect of the presumed interrelation, one can note that only 12, out of 20, Hellenic seismic crises were followed by a Calabrian earthquake with M ≥ 6.0. This indicates that the role of the Hellenic events as precursors of Calabrian shocks may be affected by uncertainty. In particular, it is worth noting that since 1948 no Hellenic events with M ≥ 6.5 have been followed by an event in Calabria. This drastic increase of false alarms coincides with the latest period (since 1947) during which no earthquakes with M ≥ 5.5 have occurred in Calabria (Table 1). Such long quiescence (66 years) is rather anomalous with respect to the previous behavior, in particular with the fact that from 1626 to 1947 the average inter-event time between M ≥ 5.5 shocks was of 12 years and in any case it was never longer than 41 years.

In order to find a possible explanation of such long quiescence and of the fact that since the middle of the 20th century the correspondence between Hellenic and Calabrian earthquakes has undergone a considerable worsening, we advance the hypothesis that such anomalous behavior is an effect of a very rare major tectonic event that has drastically changed the strain and stress fields in the above zones. It concerns the large westward displacement that the Anatolian-Aegean-Balkan system has undergone in response to the series of very strong earthquakes that since 1939 activated the entire North Anatolian fault system (NAF) (e.g., Barka 1996). While activations of the easternmost (Erzincan zone) and westernmost (Marmara zone) sectors of the NAF have occurred other times in the past centuries (e.g., Ambraseys and Jackson 1998), the rare event was the fact that the post-1939 crisis involved an activation of the NAF central sector, that had been almost silent for several centuries. This event favored the migration of the whole Anatolian wedge, which noticeably strengthened the E-W compressional regime in the Aegean region (squeezed between Anatolia and Adriatic-Africa blocks). Considering the minimum work principle, it is reasonable to suppose that the fast shortening required by such dynamics was mainly accommodated by the outward extrusion of the Aegean zones (Peloponnesus and central Aegean) which face the Ionian oceanic domain. The extrusion of the northern Hellenides (facing the Adriatic continental domain) would

Table 1 List of major Calabrian and Hellenic events, with M ≥ 5.5 and M ≥ 6.0 respectively, occurred since 1600 A.D. in the zones depicted in Fig. 3

Calabria (M ≥ 5.5)	Hellenides (M ≥ 6.0)
	2003(6.2)
	1983(6.7, 6.0)
	1953(6.2, 7.0, 6.6)
	1948(6.5, 6.5)
1947(5.7)	
1928(5.8)	1920(6.0), 1915(6.1, 6.3, 6.0)
1913(5.7), 1908(7.1), 1907 (5.9), 1905(7.0)	1912(6.1), 1897(6.6), 1895(6.2, 6.2,6.5, 6.2)
1894(6.1)	1893(6.6)
1887(5.5)	1885(6.0)
	1872(6.0)
1870(6.1)	1869(6.7, 6.0), 1867(7.2), 1866(6.4, 6.3, 6.6), 1865 (6.3) 1862(6.2. 6.4). 1860(6.4)
	1859(6.0), 1858(6.4, 6.2, 6.0)
1854(6.2)	1851(6.8)
1836(6.2)	1833(6.5)
1835(5.8), 1832(6.6)	1825(6.7), 1823(6.3)
	1820(6.6), 1815(6.3)
1791(6.0)	1786(6.5), 1783(6.6, 6.5)
1783(7.0, 6.6, 7.0)	1773(6.5)
	1772(6.1), 1769(6.8)
1767(6.0)	1767(6.7), 1766(6.6), 1759(6.3)
	1745(6.0)
1744(5.7), 1743(5.7)	1743(6.9), 1741(6.3), 1736(6.0)
	1732(6.6), 1723(6.3), 1722(6.3), 1714(6.3), 1709(6.2)
1708(5.5)	1704(6.4), 1701(6.6)
1693(5.7)	
	1674(6.3), 1666(6.2)
1659(6.6)	1658(6.7), 1650(6.2)
	1638(6.3)
1638(7.0, 6.9)	1636(72), 1630(6.5)
1626(6.0)	1625(6.5)
	1613(6.3), 1612(6.3)
	1601(6.3)

Data sources as in Appendix

instead have involved much higher resistance. This hypothesis may explain why since about 1945 most seismic activity in such area has affected the Aegean structures lying south of the Cephalonia fault system and of the North Aegean trough, while a much lower seismic activity has occurred in the Northern Hellenides (Fig. 4).

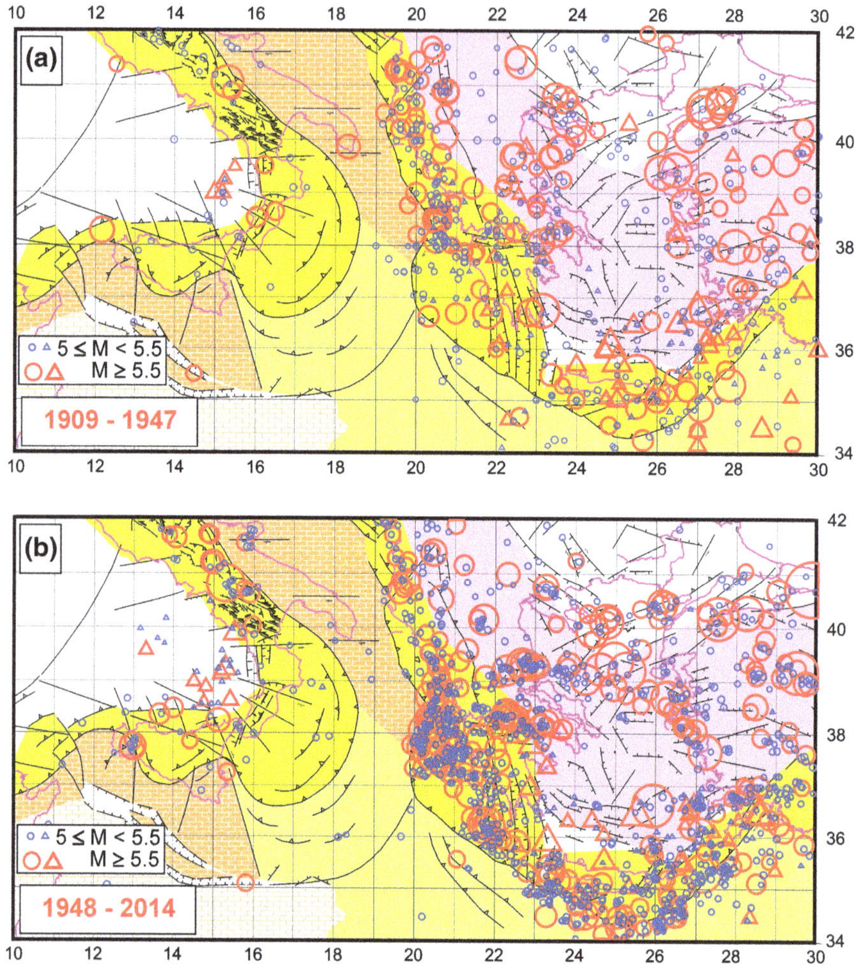

Fig. 4 Distribution of major earthquakes (M > 5) in the Anatolian-Aegean-Balkan system in two time intervals (**a** 1909–1947 and **b** 1948–2013) which respectively preceded and followed the arrival in the Aegean area of the effects of the large westward jump of Anatolia (see text for comments). *Circles* and *triangles* indicate earthquake epicentres with depth ≤60 and >60 km respectively. Seismic data have been taken from: Ergin et al. (1967), Rothé (1971), Ben-Menahem (1979), Makropoulos and Burton (1981), Iannaccone et al. (1985), Comninakis and Papazachos (1986), Ambraseys and Finkel (1987), Anderson and Jackson (1987), Eva et al. (1988), Jackson and McKenzie (1988), Godey et al. (2006), ISC Catalogue: http://www.isc.ac.uk/iscbulletin/. Other references as in Appendix

Since the activation of the Northern Hellenides thrust zone is supposed to be a necessary triggering of Calabrian seismicity (Fig. 3 and Table 1), the above evidence (Fig. 4) can explain why no major earthquakes have occurred in Calabria since 1947 (Table 1). The same interpretation helps understanding why in the period 1850–1908 (Table 1), that was characterized by very high seismic activity in the Northern Hellenides, very strong earthquakes have instead occurred in Calabria.

The evidence and arguments described above suggest that at present the probability of strong earthquakes in Calabria is relatively low and it is expected to remain so until seismic activity in the Northern Hellenides again undergo a significant increase.

This case may offer an interesting example of the different predictions of seismic hazard in Calabria that can be derived by a probabilistic evaluation, based on the local seismic history, and by the above deterministic approach based on the knowledge of the seismotectonic context. Assuming that seismicity is a Poissonian process (e.g., Stucchi et al. 2011), the probabilistic approach would predict a relatively high probability of earthquakes at present, due to the fact that the time elapsed since the last strong shock (about 70 years) is considerably longer than the average return period in that zone (about 12 years). The deterministic approach, instead, predicts a relatively low probability, considering that the present strain rate field in the zone involved (induced by a rare and presumably long living tectonic event) may prevent the development of the process (the outward escape of the Calabrian wedge) that is expected to favour the activation of main fault systems in Calabria.

4 Migration of Seismicity Along the periAdriatic Zones

The present knowledge about the geodynamics and tectonic setting in the central Mediterranean area (e.g., Mantovani et al. 2006, 2007a, b, 2009; Viti et al. 2006, 2009, 2011) suggests that the Adriatic plate (Adria hereafter) is stressed by the convergence of the confining plates (Africa, Eurasia and Anatolian-Aegean system) and tends to move roughly northward (Fig. 5).

This plate motion is accommodated by tectonic activity at the eastern (Hellenides, Dinarides), northern (eastern Southern Alps) and western (Apennines) boundaries of that plate, involving fairly different strain styles. Underthrusting of Adriatic lithosphere mainly occurs beneath the Hellenides (from the Ionian islands to Albania) and southern Dinarides (e.g., Louvari et al. 2001; Aliaj 2006; Benetatos and Kiratzi 2006; Kokkalas et al. 2006). Seismotectonic activity is highest in the Hellenic sector since such zone marks the collision zone between converging blocks (Adria and the Anatolian Aegean system), while in the Southern Dinarides tectonic activity is only due to the motion of Adria with respect to the almost fixed Carpatho-Pannonian system. The activation of the Cephalonia fault (e.g., Louvari et al. 1999) allows the relative motion between two Hellenic sectors, the Peloponnesus wedge, facing the Ionian oceanic lithosphere, and the Epirus, facing the Adriatic continental domain (Mantovani et al. 2006).

In the northern Dinarides the relative motion of Adria with respect to the adjacent structures is mainly accommodated by dextral transpression at the fault system recognized in Istria and Slovenia (e.g., Markusic and Herak 1999;

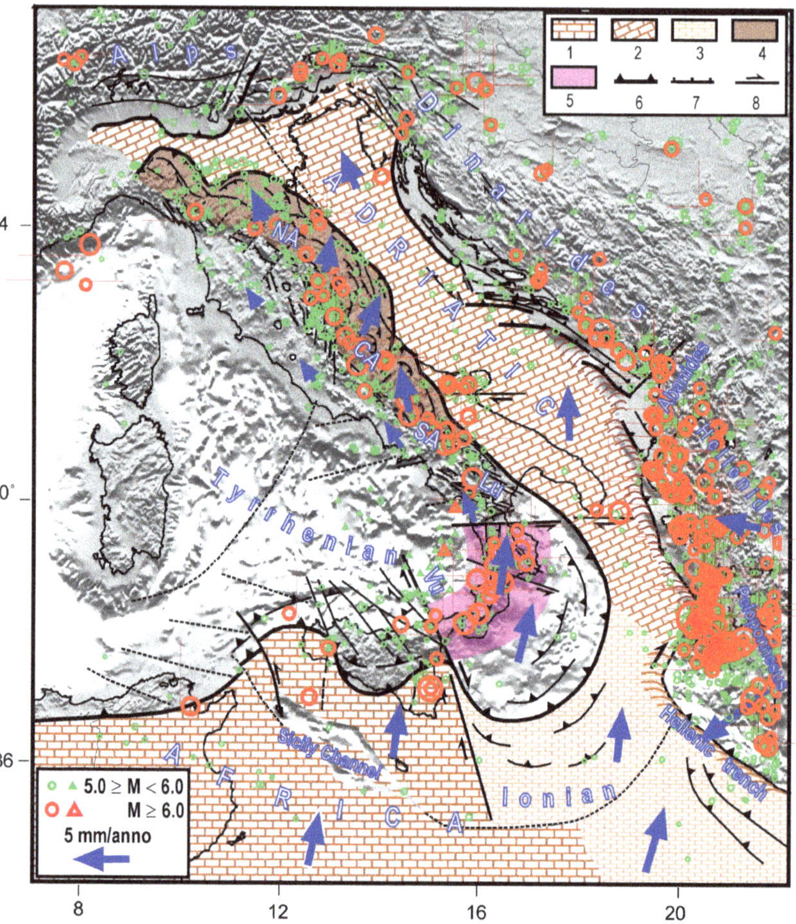

Fig. 5 Post-early Pleistocene kinematic and tectonic patterns in the central Mediterranean region (from Cenni et al. 2012; Mantovani et al. 2012). *1*, *2* African and Adriatic continental domains *3* Ionian oceanic domain *4* Outer sector of the Apennine belt carried by the Adriatic plate *5* Calabrian Arc *6–8* Major compressional, extensional and transcurrent tectonic features. *Blue arrows* show a tentative reconstruction of the Quaternary kinematic pattern with respect to Eurasia (from Mantovani et al. 2007b). *Circles* indicate earthquake epicentres in the period 1600–2013, taken from: Shebalin et al. (1974), Makropoulos and Burton (1981), Papazachos and Comninakis (1982), Comninakis and Papazachos (1986), Anderson and Jackson (1987), Jackson and McKenzie (1988), Papazachos and Papazachos (1989), Albini (2004), Guidoboni and Comastri (2005), Godey et al. (2006), Rovida et al. (2011), Makropoulos et al. (2012), Global CMT Catalog (Ekström et al. 2012); CATGR1900 at www.geophysics.geol.uoa.gr. *CA* Central Apennines; *Lu* Lucanian Apennines, *NA* Northern Apennines; *SA* Southern Apennines; *Vu* Vulcano-Syracuse fault system

Kuk et al. 2000; Poljak et al. 2000; Burrato et al. 2008). In the eastern Southern Alps, the Adriatic lithosphere underthrusts the Alpine edifice (e.g., Bressan et al. 2003; Galadini et al. 2005).

On the western side of Adria, mainly corresponding to the Apennine belt, the tectonic context is more complex (Fig. 5), mainly due to the fact that the outer sector of that chain is forced by belt-parallel compression (induced by the Adriatic plate) to separate from the inner Tyrrhenian side of the chain and to extrude laterally, at the expense of the adjacent Adriatic domain (Viti et al. 2006, 2011; Mantovani et al. 2009; Cenni et al. 2012). Such more mobile, deforming and uplifting part of the belt is constituted by the Molise-Sannio wedge (in the southern Apennines), the eastern side of the Lazio-Abruzzi carbonate platform (ELA), in the central Apennines, and the Romagna-Marche-Umbria (RMU) and Toscana-Emilia (TE) wedges, in the northern Apennines. The separation of those escaping wedges from the inner almost fixed belt is accommodated by extensional and sinistral transtensional deformation, mainly concentrated in the axial part of the chain, where a series of basins has developed in the Quaternary (e.g., Piccardi et al. 2006 and references therein). Compressional deformation develops at the outer front of the extruding wedges, where they overthrust the Adriatic domain (e.g., Scisciani and Calamita 2009). In the central Apennines, the decoupling between the outer ELA block and the western side of that platform is accommodated by two major SE-NW sinistral transtensional fault systems (L'Aquila and Fucino, e.g., Piccardi et al. 2006; Elter et al. 2012).

The outward extrusion (at the expense of the Ionian domain) and uplift of the Calabrian wedge is driven by belt-parallel compression (Mantovani et al. 2009). This interpretation is fairly consistent with the structural tectonic features evidenced by seismic surveys (Finetti 2005; Del Ben et al. 2008).

The relative motion between the outward extruding Calabrian wedge (at the expense of the Ionian domain, and the Molise-Sannio wedge (moving roughly NE ward, in connection with Adria) is accommodated by the system of NW-SE sinistral strike-slip faults recognized in the Lucanian Apennines (e.g., Catalano et al. 2004; Ferranti et al. 2009).

The motion of Adria is very slow during quiescent periods, while it locally accelerates during co-seismic and post-seismic phases, in response to major decoupling earthquakes at the eastern, western and northern periAdriatic boundaries.

Considering the tectonic context described above and the fact that the seismic activation of a periAdriatic sector may influence the probability of strong shocks in nearby sectors (Viti et al. 2003, 2012, 2013; Mantovani et al. 2008, 2010, 2012), one could expect to observe regularities in the time-space distribution of seismicity along the periAdriatic zones. This hypothesis is corroborated by the time patterns of seismicity at the main periAdriatic sectors for the period following 1400 (Fig. 6).

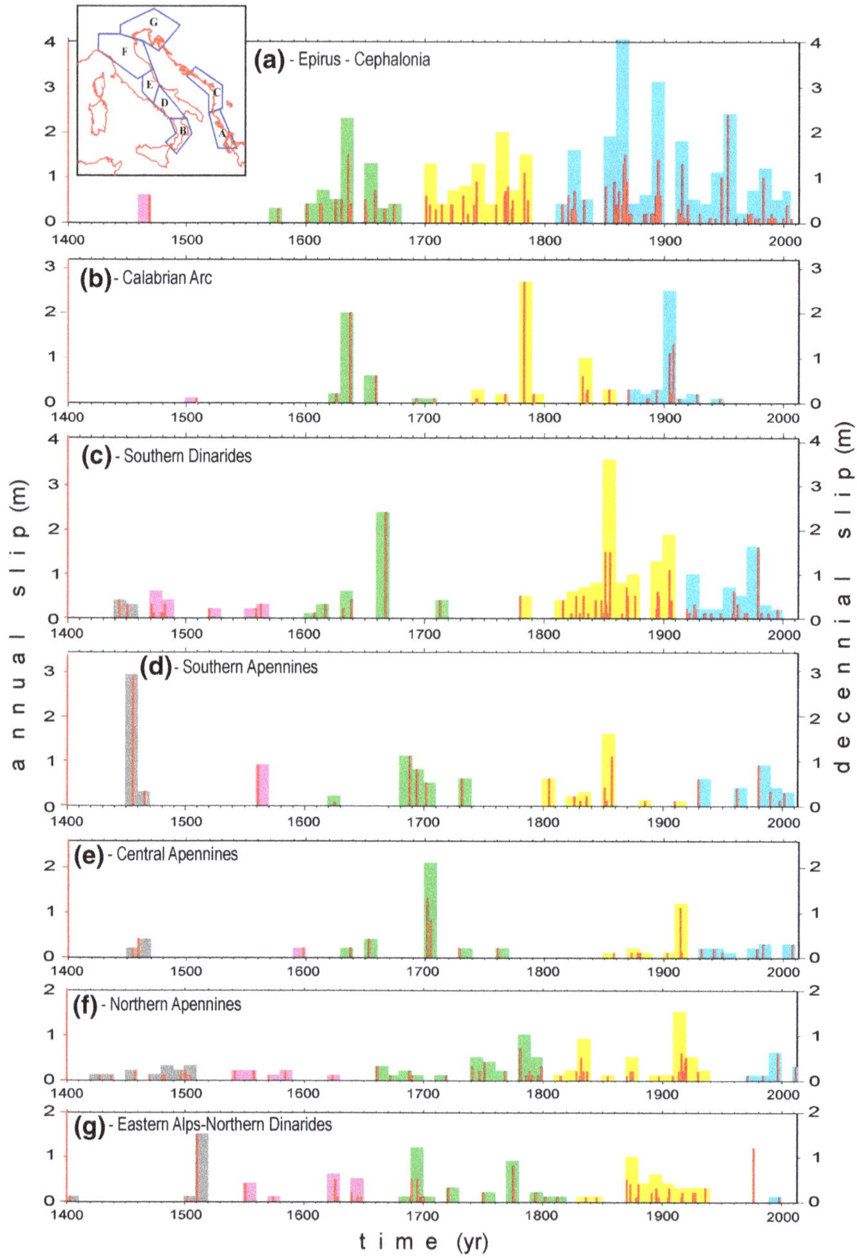

Fig. 6 Time patterns of major seismicity (M ≥ 5.5) in the main periAdriatic seismic zones since 1400 A.D. The geometries of the zones considered are shown in the inset. *Red bars* indicate the total seismic slip (meters) occurred during the related year, computed by the relation of Wells and Coppersmith (1994) between average slip and earthquake magnitude. The height of *vertical boxes* indicates the sum (meters) of seismic slips over the related decades. *Colours* help to recognize the presumed migrations of seismic phases from the southern to the northern periAdriatic zones (see text for comments). Seismicity data are listed in the Appendix along with references

In order to provide an information more representative of the effects that major decoupling earthquakes produce on local plate kinematics, the diagrams in Fig. 6 show the time pattern of the annual sum of seismic slips, computed by the Wells and Coppersmith (1994) relation:

$$\log_{10} u = -4.8 + 0.69\, M \tag{1}$$

where u is the average seismic slip (in meters) on the fault and M is the earthquake magnitude. The same diagram also aims at giving insights into the time concentration of seismic slips, by reporting the total seismic slip over intervals of ten years. In particular, this parameter may be useful to recognize how rapidly the surrounding structures may have been stressed by the effects of the triggering earthquakes (post-seismic relaxation), as discussed in Sect. 2.

The time patterns shown in Fig. 6 point out that in the periAdriatic zones seismicity is mostly discontinuous over time, with periods of high activity separated by almost quiescent phases. Furthermore, it may be recognized that seismic phases tend to progressively migrate over time from the southern to northern zones, through the eastern (Hellenides and Dinarides) and western (Apennines) boundaries of Adria, up to reach the northern compressional front (Eastern Alps) of that plate. In Fig. 6, the presumed migrating sequences are tentatively marked by different colours (grey, green, yellow and blue). One might suppose that the periAdriatic decoupling earthquakes involved in each sequence may have allowed the whole plate to make a further step (roughly 1–2 m) towards Europe.

The first presumed sequence (grey in Fig. 6) can only be recognized for the central and northern periAdriatic zones, where a significant increase of seismic activity took place, from about the middle of 15th century in Albania and Southern Dinarides, to the beginning of the 16th century in the northern Adriatic front. Since very low seismicity is documented in the Hellenides and Calabria before 1600 AD, it is not possible to recognize when this sequence may have started in such zones. Anyway, the comparison of the seismicity pattern that occurred since 1444 (Fig. 7a) and the one that took place in the previous period (Fig. 7b) points out the considerable increase of activity that the central and northern periAdriatic zones underwent after the occurrence of major seismic crises in the Albania-Southern Dinarides (1444–1451) and Southern Apennines (1456).

The quantification, by numerical experiments, of the effects of post-seismic relaxation induced by the strong 1456 earthquakes in the southern and central Apennines (Mantovani et al. 2012; Viti et al. 2013) indicates that such phenomenon may have influenced the occurrence of the major shock that took place in the northern Apennines roughly 2 years later (Upper Tiber Valley, 1458 M = 5.8). The seismic sequence cited above was followed in most periAdriatic zones by a long period of moderate activity (white bands in Fig. 6), when only few strong shocks occurred in the Albania-Southern Dinarides and Northern Apennines.

The first presumably complete sequence (green in Fig. 6) started with a considerable increase of seismic activity at the Hellenides during the first decades of the

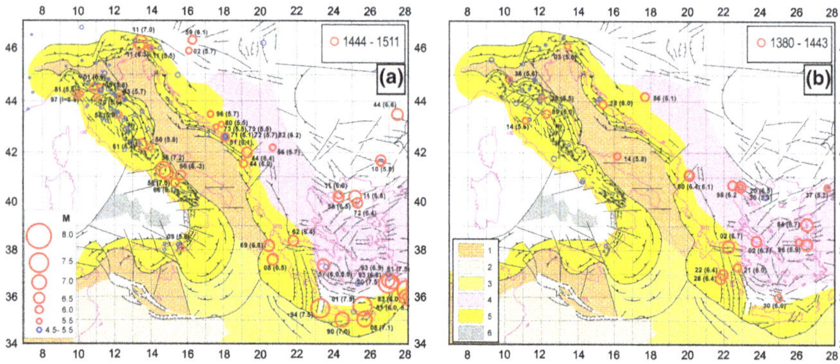

Fig. 7 Distribution of major shocks in the central Mediterranean area during the time interval 1444–1511 (**a**), presumably related to the first seismic sequence (*grey* in Fig. 6). In order to give information on how seismic activity during this phase was considerably stronger than the one occurred in the previous period, the seismicity pattern in the time interval 1380–1443 (**b**) is also reported. Symbols and data sources as in Fig. 4

17th century. This crisis was followed by a significant increase of seismic activity in all other periAdriatic zones, up to reach the northern front of Adria in the first half of the 18th century. In particular, the Calabrian Arc was hit by one of the major seismic crises (1638, M = 6.9, 7.0) of its known seismic history. It is worth noting that in the northern Apennines seismic activity that occurred during this phase was characterized by a fairly clear northward migration, as evidenced in the three pictures of Fig. 8.

At the northern Adriatic front, major seismic activity lasted until about the end of the 18th century, after which it underwent a drastic reduction for a relatively long period, until 1870.

The beginning of a new seismic sequence (yellow in Fig. 6) was marked by a drastic increase of seismic activity in the Hellenides in the last decades of the 18th century. In this case too, the Balkan crisis was accompanied by strong earthquakes in Calabria (1783, M = 7.0, 7.0, 6.6). Soon after the first crisis, a new seismic period occurred in the Hellenides, from 1815 to 1826, that was again followed by other major shocks in Calabria (1854, M = 6.2 and 1870 M = 6.1). That sequence then continued with several major events in the Albania-Southern Dinarides and Southern Apennines.

In the central Apennines, seismic activity was moderate for a relatively long period, until the occurrence of the strongest shock ever recorded in this zone (Fucino 1915, M = 7.0). This major decoupling event was then followed by a series of strong shocks in the northern Apennines in the following 5 years (1916–1920). As argued in some papers (Mantovani et al. 2010, 2012; Viti et al. 2012, 2013), the

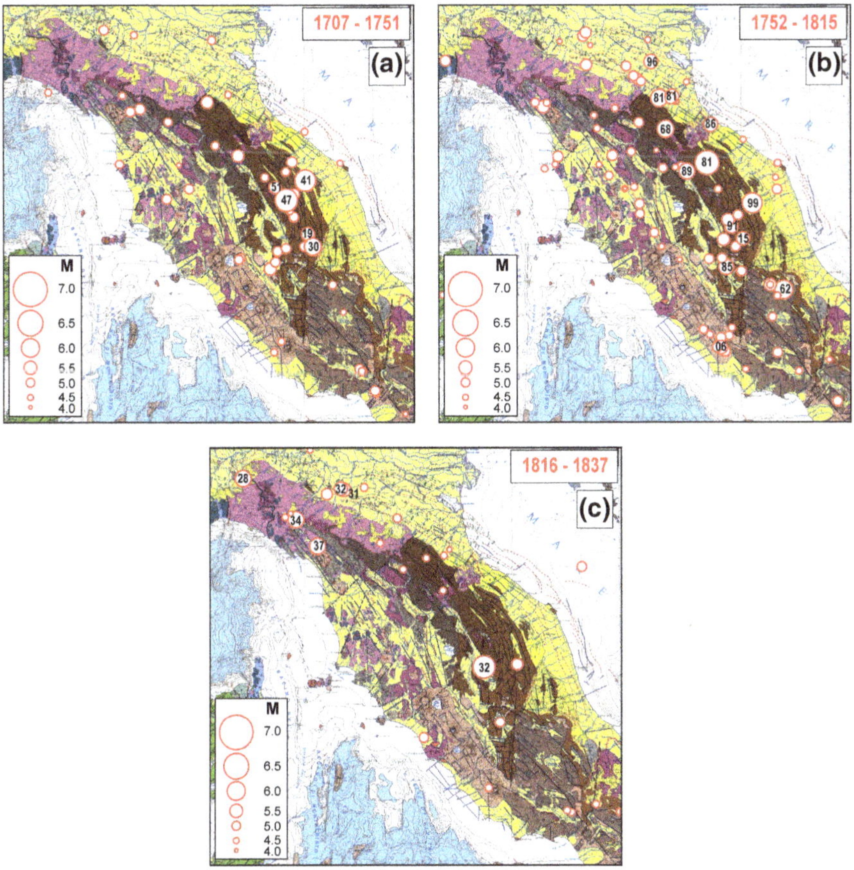

Fig. 8 Distribution of major earthquakes in the northern Apennines during the green sequence in Fig. 6. The three pictures point out the progressive northward migration of seismicity, during three consecutive periods, indicated in the frames. For the events with M ≥ 5.5 the year of occurrence is given. The underlying tectonic map is taken from Funiciello et al. (1981)

space-time distribution of major shocks during the above seismic sequence (1915–1920) is consistent with the tectonic implications of the proposed tectonic context in the Apennine belt (Viti et al. 2012, 2013). Furthermore, numerical modelling of the effects of the post-seismic relaxation induced by the 1915 Fucino and subsequent (1916–1920) strong earthquakes (Viti et al. 2012, 2013) shows that each event of such series just occurred when the respective source zone was reached by the highest values of the strain and strain rate perturbation induced by the previous shocks. Finally, the strain regime of the predicted post-seismic perturbations generally agrees with the style of seismic faulting recognized in the Apennine zones activated during the 1916–1920 sequence.

The last seismic sequence (blue in Fig. 6) started around 1850, with a period of very high activity (from about 1850 to 1872) in the Northern Hellenides, which was

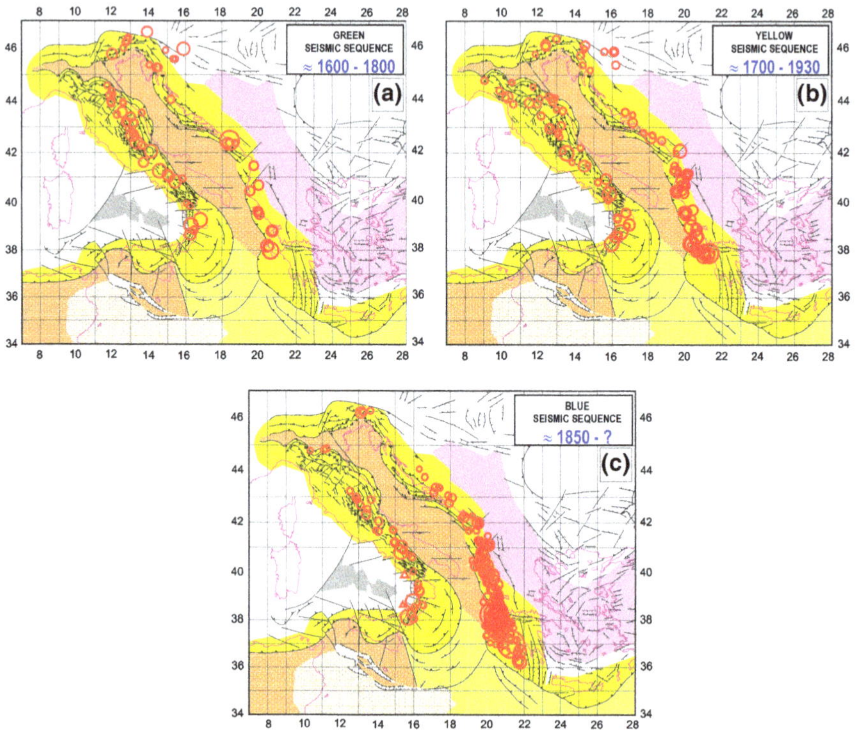

Fig. 9 Spatial distribution of seismicity during the last 3 sequences shown in Fig. 6 (*yellow*, *green* and *blue*). Data sources as in Appendix

soon followed by another crisis roughly lasting from 1885 to 1903. As in previous cases, these crises were accompanied by major earthquakes in Calabria (1870 M = 6.1, 1905 M = 6.9, 1908 M = 7.2). Subsequently, seismic phases underwent a progressive northward migration through the periAdriatic zones, up to reach the central sectors of the Dinarides and Apennines, while scarce seismicity has so far affected the northern zones. No major shocks (M > 5.5) have so far occurred in the northern Dinarides and only one major seismic crisis has affected the Eastern Alps (1976). Scarce seismic activity has also interested the northern Apennines, with only 3 moderate shocks (1971 M = 5.6 and 2012 M = 5.8, 5.9).

Further insights into the fact that the ongoing sequence has not yet undergone a full development may be derived from the space distribution of major earthquakes in the last three sequences (Fig. 9). The first two sequences involved a fairly homogeneous covering of most seismic periAdriatic zones, whereas the last sequence presents a very scarce covering of the northern zones.

Thus, if one could always rely on the fact that seismic activity in the periAdriatic zones is affected by a systematic tendency to progressively migrate from south to north, the evidence shown in Figs. 6 and 9 would indicate that the probability of hosting the next major shock is higher in the northern zones (northern Apennines, northern Dinarides and eastern Southern Alps) than in the southern zones (Calabria and Southern Apennines). As concerns the central Apennines, the recognition of the present probability of major shocks is not easy, since during the last sequence this zone has been affected by a number of earthquakes, but not as strong as the ones occurred in the previous sequences. So one cannot exclude that a further seismic activation of that sector will occur in the next future. Thus, an intermediate probability, higher than the one in the southern zones and lower than the one in the Northern Apennines, could tentatively be assumed for this zone. Furthermore, considering that in the proposed seismotectonic scheme the activation of the Northern Apennines is expected to precede major earthquakes in the northern Adriatic front, in line with what has happened in the known history, one might tentatively assume that at present the probability of hosting the next shock is higher in the northern Apennines than in the northern front of Adria.

A synthesis of the relative probabilities discussed above is shown in Fig. 10. It is opportune to clarify that the prediction here proposed only aims at providing a possible scale of priorities for eventual initiatives for risk mitigation and that no information is provided about the time of occurrence of the next event.

Notwithstanding that the relationships between strong earthquakes and the pattern of minor seismicity are not yet clear (e.g., Marsan 2005 and references therein), it may be useful to know the present level of such activity in the periAdriatic zones (Fig. 10). One major evidence in this pattern is that in the Northern Apennines, i.e. the zone here indicated as the most prone to the next strong earthquake, the number of minor events is much higher than in the other zones of the Italian peninsula. This proportion does not change significantly if larger thresholds of magnitude are considered (Fig. 10).

One could try some considerations about the possible implications of the above evidence. A high level of the so-called instrumental seismicity in a zone evidently indicates that such zone is actually undergoing a significant level of stress and strain, that is presumably accommodated by minor sliding at the faults which are most favorably oriented. The fact that such phenomenon is more evident in the Northern Apennines could support the hypothesis that at present such zone is the one undergoing the most intense stress. However, one must be aware that such evidence does not necessarily imply that the zone involved will be affected by a strong shock in the next future. For instance, it might occur that after a long series of minor sliding the fault involved reaches a configuration which inhibits any further sliding (e.g., Wesnousky 2006). This could considerably slowdown seismic activity even for a long time. On the other hand, one cannot obviously exclude that the fracture reaches a zone of favored sliding, causing a major shock (e.g., Marsan 2005).

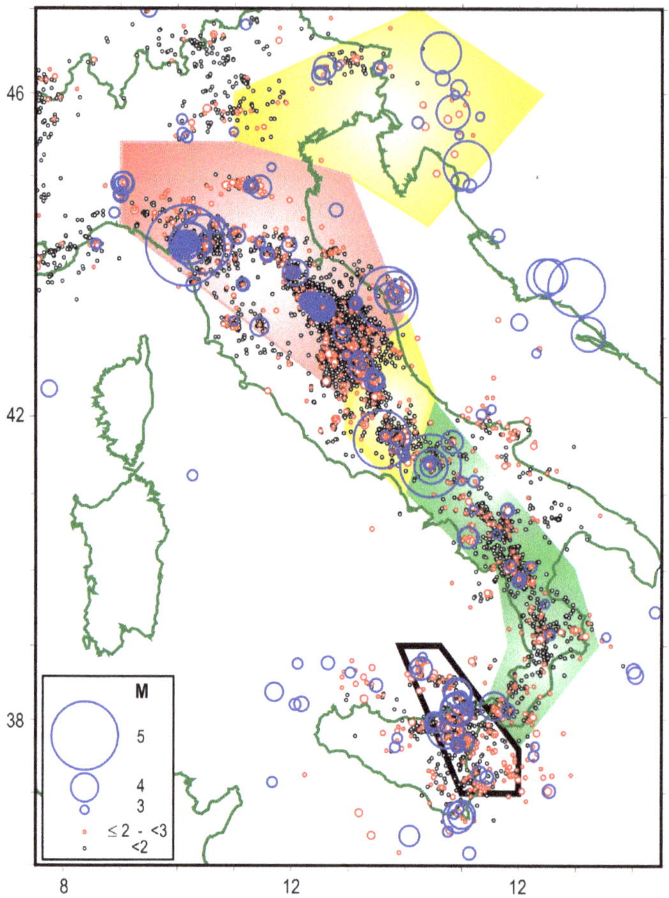

Seismicity between 1-1-2013 and 28-2-2014 with h ≤ 30 Km			
	whole	M ≥ 2	M ≥ 3
Eastern Alps and Northern Dinarides	152	60	16
Northern Apennines	18608	1284	102
Central Apennines	1435	186	10
Southern Apennines	1097	197	20
Calabrian Arc	1345	164	7
Eastern Sicily	1335	296	21

Fig. 10 Relative probability of major earthquakes in the Italian zones considered in this study. *Red* indicates the zone with the highest expected probability (Northern Apennines), *orange* identifies the intermediate probability (Central Apennines and Eastern Alps) and *green* indicates the lowest probability (Southern Apennines and Calabria). No prediction is provided for Sicily. The *table* reports the number of minor shocks occurred in 2013 in the above zones. Data from ISIDe Working Group (INGV 2010)

5 Conclusions and Discussion

It is suggested that the next major earthquake (M ≥ 5.5) in the Italian peninsula will most probably be located in the northern part of the territory, with particular regard to the region here identified as Northern Apennines. This prediction is obtained by a deterministic approach, based on the expected short-term development (tens of years) of the proposed seismotectonic context. The model adopted provides that Adria, stressed by the convergence of the confining blocks, tends to move roughly northward. The development of this plate kinematics is rather complex in space and time, since displacement mainly accelerates locally, in response to major decoupling earthquakes in periAdriatic zones. Each shock triggers a perturbation of the strain field that, propagating through the plate, may increase the probability of seismic activation in the next sectors of the Adria boundary.

This interpretation may plausibly account for the time patterns of seismicity that has developed in the periAdriatic zones since 1400 A.D. In particular, it accounts for the fact that seismic energy release in each zone is mostly discontinuous, with main seismic phases separated by almost quiescent periods, and for the general tendency of main seismic phases to progressively migrate from south to north, through the eastern and western Adria boundaries, up to reach the northern front of the plate (eastern Southern Alps).

At least two presumably complete migration sequences, roughly 200 years long, may tentatively be recognized in the period considered. A further, still ongoing, sequence presumably started around the second half of the 19th century and has so far involved major seismic activations of the Hellenides, Calabria, Southern Dinarides, Southern Apennines and Central Apennines, while only few earthquakes have so far affected the northern zones (Northern Apennines, Northern Dinarides and eastern Southern Alps). This evidence might imply that the next step of the proposed seismotectonic pattern will most probably involve the seismic activation of fault systems located in the northern periAdriatic zones. The development of the previous sequences would suggest a higher probability for the Northern Apennines with respect to the Eastern Alps.

This prediction is consistent with the implications of two significant interrelations recognized between Italian and Dinaric-Hellenic zones, which suggest that at present the probability of strong earthquakes is relatively low in the Southern Apennines and Calabria. This prediction relies on the hypothesis that the seismic activation of such Italian zones is conditioned by the occurrence of major seismic crises in the Southern Dinarides and Hellenides respectively, where no significant precursory events have occurred in the last tens of years (Fig. 6 and Table 1).

An intermediate probability, comparable to that of the eastern Southern Alps, is proposed for the Central Apennines, where the release of seismic energy during the ongoing sequence has not reached the levels reached in previous sequences (Fig. 6).

On the basis of the evidence and arguments discussed in this work and in previous publications (Mantovani et al. 2010, 2012; Viti et al. 2012, 2013), we have tentatively defined the scale of priorities given in Fig. 10.

We are obviously aware that the reliability of the proposed prediction cannot easily be demonstrated, mainly due to the shortness and incompleteness of the seismic history now available, which only allows the recognition of few migration patterns in the periAdriatic zones. However, one should take into account that the approach here adopted is not only based on a mere empirical analysis of the seismic history, but relies on a deterministic scheme which is based on the expected short-term behaviour of a tectonic model inferred by a long and accurate analysis of a large amount of Earth Science data (e.g., Mantovani et al. 2006, 2009; Viti et al. 2006, 2011, 2012). Furthermore, it must be pointed out that geodetic observations (GPS) in the Italian region indicate that the present kinematic pattern of the Apennine belt is fairly similar to the one that may be deduced by the post-early Pleistocene deformation pattern (Cenni et al. 2012).

This study does not provide any prediction for some Italian seismic zones, such as Sicily, Apulia and Northwestern Italy, since for those zones we do not have sufficient information to afford hypotheses about the future short-term development of seismic activity (e.g., Mantovani et al. 2009). The tentative prediction synthetized in Fig. 10 is based on two main aspects, one is deterministic, relying on the knowledge of the ongoing tectonic setting and its possible connection with the spatio-temporal distribution of major earthquakes, and the other is mainly empirical, concerning the significant seismicity regularity patterns deduced by the analysis of the seismic histories of the zones involved. For the 3 zones mentioned above the second aspect is not available.

Concerning the deterministic aspect, some considerations about the present seismic hazard in Sicily can be tentatively made. As discussed earlier, the westward jump of the Anatolian-Aegean system (after 1939) has most probably strengthened E-W compression in the zone comprising the northern Ionian zone, the Calabrian Arc and the Hyblean block (including Sicily). This effect is expected to inhibit the outward extrusion of the Calabrian wedge, due to the presumed thickening and upward flexure of the Ionian domain. This hypothesis might explain the scarce seismic activity in Calabria since the middle of the last century (Fig. 4). On the other hand, the above compressional context could have favoured the northward escape of the Hyblean wedge, which would imply an increased probability of seismic activation of the main lateral guide of that extrusion, that is the Vulcano-Siracusa fault system, crossing the northeastern part of Sicily (Fig. 5).

However, since the jumps of the Anatolian-Aegean system are very rare tectonic events, with presumed recurrence times of several centuries (e.g., Ambraseys and Jackson 1998), it is not possible to recognize in the known seismic history eventual significant regularity patterns concerning the interaction of seismic sources in such kind of situations. Thus, not having any empirical confirmation of the expected tectonic processes, we cannot afford any reliable prediction about seismic activity of Sicily in the next future. In this regard, it could be worth noting that in the last decades of the previous century, i.e. the period that followed the westward jump of the Anatolian Aegean system, seismic activity in eastern Sicily has undergone a significant increase with respect to the previous period (Fig. 4). Furthermore, one could also note that the in the last year instrumental seismicity in eastern Sicily has been higher than in other zones (Fig. 10), except the northern Apennines.

A last consideration may be devoted to the fact that in January 2014 some earthquakes with an intermediate magnitude (5.5 < M < 6.3) occurred at the Cefalonia fault system, the zone of possible precursors of Calabrian earthquakes. This level of magnitude is low with respect to the events that have triggered strong earthquakes in Calabria (Table 1), but these signals may suggest the opportunity of improving the seismological and geodetic monitoring of the zones involved.

6 Possible Social Impact of the Proposed Prediction

On the basis of the past seismic history, one can reasonably suppose that in the next decades Italy will be hit by one or two major earthquakes. This implies that the problem of risk mitigation will primarily concern a limited part of the territory. This consideration, however, cannot be of much help if no information is available about which zones may be involved. Thus, any effort to get significant insights about the most probable spatial distribution of major earthquakes in the next future should be encouraged.

This work describes an attempt in this direction, carried out by exploiting the present knowledge on the ongoing seismotectonic setting in the central Mediterranean region and its possible connection with the time-space distribution of past major earthquakes. The results of this investigation suggest that the probability of strong shocks (M > 5.5) is presently highest in the Northern part of Italy, with particular regard to the Northern Apennines.

Notwithstanding the possible uncertainties, we think that the above prediction could be useful for practical purposes, such as the planning of the initiatives for seismic risk mitigation in Italy. In case of a successful prediction, the concentration of resources in the zones proposed would allow a not negligible improvement of safety. In case of failed prediction, the resources eventually employed in that area would not be wasted, since they would have allowed an improvement of safety in a zone which can plausibly be considered as most prone to next shocks. On the other hand, the adoption of a *blind* strategy (no priority zone) would imply that each of the numerous seismic areas of Italy would only benefit of a very limited portion of the available resources. Moreover, given the plausibility of the evidence and arguments presented in this work, we think that the probability of a successful result of the proposed strategy is higher than the ones of its failure.

The information provided in this work may also have implications for scientific activity. For instance the zones identified could become sites of specific monitoring (geodetic and seismological), aimed, for instance, at identifying eventual perturbations of the velocity and strain fields, possibly connected with impending shocks.

Appendix

List of earthquakes used for the diagrams of Fig. 6. M is the magnitude, Cat is the reference to seismic catalogues listed as follows: (1) Albini (2004); (2) Ambraseys (1990); (3) Global CMT Catalog (Ekström et al. 2012); (4) Working Group CPTI (2004); (5) Rovida et al. (2011); (6) Guidoboni and Comastri (2005); (7) ISIDe Working Group (INGV) (2010); (8) Karnik (1971); (9) Mariotti and Guidoboni (2006); (10) Seismological Catalogues of Greece; (11) Makropoulos et al. (2012); (12) Margottini et al. (1993); (13) Comninakis and Papazachos (1986); (14) Papazachos and Papazachos (1989); (15) Ribaric (1982); (16) Shebalin et al. (1974); (17) Stucchi et al. (2012); (18) Shebalin et al. (1998); (19) Toth et al. (1988).

Hellenides		
Date (y-m-d)	M	Cat
1278-2-25	6.6	17
1469	6.6	17
1577	6.2	17
1601-4-26	6.3	17
1612-5-26	6.3	17
1613-10-12	6.3	17
1625-6-28	6.5	17
1630-7-2	6.5	17
1636-9-20	7.2	10
1638-7-16	6.3	17
1650	6.5	14
1651-2-26	5.9	17
1658-8-24	6.7	17
1666-11	6.2	18
1674-1-16	6.3	17
1701-4-5	6.6	17
1704-11-22	6.4	17
1709	6.2	17
1714-9-8	6.3	17
1722-6-5	6.3	17
1723-2-22	6.3	17
1732-11	6.6	17
1736	6	10
1741-6-23	6.3	17
1743-2-20	6.9	10
1759-6-13	6.3	17
1766-7-24	6.6	17

(continued)

(continued)

Hellenides		
Date (y-m-d)	M	Cat
1767-7-22	6.7	17
1769-10-12	6.8	14
1772-5-12	6.1	18
1773-5-23	6.5	10
1783-3-23	6.6	17
1783-6-7	6.5	10
1786-2-5	6.5	17
1815	6.3	17
1820-2-21	6.6	17
1823-6-19	6.1	17
1825-1-19	6.7	17
1826-1-26	5.8	17
1833-1-19	6.5	14
1851-10-12	6.8	17
1858-4-5	6	10
1858-9-20	6.2	17
1858-10-10	6.4	17
1859-9-12	6	18
1860-4-10	6.4	18
1862-3-14	6.4	17
1862-10-4	6.2	17
1865-10-10	6.3	17
1866-1-2	6.6	17
1866-3-2	6.3	18
1866-12-4	6.4	17
1867-2-4	7.2	17
1869-8-14	6	18
1869-12-28	6.7	17
1871-4-9	5.8	17
1872-2-11	6	18
1883-6-27	5.5	10
1883-8	5.5	10
1885-12-14	6	10
1889-4-1	5.9	17
1890-5-21	5.9	17
1891-6-27	5.8	17
1893-6-14	6.6	17
1895-5-13	6.2	17
1895-5-14	6.5	17

(continued)

(continued)

Hellenides		
Date (y-m-d)	M	Cat
1895-5-15	6.2	17
1895-6-16	6.2	18
1896-2-10	5.5	18
1896-3-18	5.8	18
1897-1-17	6.6	17
1912-1-24	6.1	11
1914-11-27	5.9	11
1915-1-27	6.1	11
1915-8-7	6.3	11
1915-8-10	5.6	11
1915-8-10	6	11
1915-8-11	5.7	11
1915-8-19	5.9	11
1917-5-23	5.6	11
1920-10-21	5.6	11
1920-11-26	6	18
1920-11-29	5.5	18
1921-9-13	5.5	11
1930-11-21	5.8	18
1938-3-13	5.7	11
1939-9-20	5.5	11
1943-2-14	5.6	11
1948-4-22	6.5	11
1948-6-30	6.5	11
1953-8-9	5.9	11
1953-8-11	6.6	11
1953-8-12	7	11
1953-8-12	5.7	11
1953-8-12	5.9	11
1953-10-21	6.2	11
1960-11-5	5.7	11
1967-2-9	5.5	11
1970-7-2	5.8	11
1972-9-17	5.8	11
1973-11-4	5.8	11
1976-1-18	5.6	11
1979-11-6	5.6	11
1983-1-17	6.7	11
1983-1-19	5.5	11

(continued)

(continued)

Hellenides		
Date (y-m-d)	M	Cat
1983-3-23	6	11
1987-2-27	5.6	3
1988-5-18	5.5	3
1990-6-16	5.8	3
1992-1-23	5.6	3
1993-6-13	5.7	3
1994-2-25	5.5	3
2000-5-26	5.6	3
2003-8-14	6.2	3
2003-8-14	5.5	3
2007-3-25	5.7	3

Calabrian arc		
Date (y–m–d)	M	Cat
1509-2-25	5.6	5
1626-4-4	6	5
1638-3-27	7	5
1638-6-8	6.9	5
1659-11-5	6.6	5
1693-1-8	5.7	5
1708-1-26	5.5	5
1743-12-7	5.7	5
1744-3-21	5.7	5
1767-7-14	6	5
1783-2-5	7	5
1783-2-7	6.6	5
1783-3-28	7	5
1791-10-13	6	5
1832-3-8	6.6	5
1835-10-12	5.8	5
1836-4-25	6.2	5
1854-2-12	6.2	5
1870-10-4	6.1	5
1886-3-6	5.6	5
1887-12-3	5.5	5
1894-11-16	6.1	5
1905-9-8	7	5
1907-10-23	5.9	5
1908-12-28	7.1	5

(continued)

(continued)

Calabrian arc		
Date (y–m–d)	M	Cat
1913-6-28	5.7	5
1928-3-7	5.8	5
1947-5-11	5.7	5

South dinarides		
Date (y–m–d)	M	Cat
1237-3	6.2	18
1270-3	6.5	6
1273-3	6.5	10
1359	6	6
1380	6.1	17
1444	6.4	18
1451	6.1	18
1471	6.1	18
1472	5.7	18
1473-1-20	5.5	18
1479-10-20	5.5	18
1480-10-18	5.5	18
1482-2-15	6.2	18
1520-5-17	6	17
1559-6-24	6	1
1563-6-13	6.1	18
1608-7-25	5.6	17
1617	6.2	18
1632	6	17
1639-7-28	6.4	18
1667-4-6	7.5	18
1713-1-0	6.3	17
1780-9-21	6.5	18
1816	6.3	18
1823-8-7	5.7	18
1827-4-17	6.5	18
1830	5.6	18
1833-1-19	6.5	14
1837-10-4	5.5	18
1843-9-5	6.2	17
1843-9-26	5.6	18
1848	6.4	17
1850-4-13	5.6	17

(continued)

(continued)

South dinarides		
Date (y–m–d)	M	Cat
1851-1-20	6	18
1851-10-17	6.2	18
1851-10-17	6.6	17
1851-10-20	6.3	18
1851-12-29	5.5	18
1852-8-26	6.2	17
1853-12-11	5.7	18
1855-7-3	6.6	10
1855-7-5	6.8	18
1855-7-16	5.5	18
1855-8-14	5.5	18
1865-10-10	5.5	18
1869-1-10	5.6	18
1869-3-18	6	18
1869-4-14	5.5	17
1869-9-1	6.2	18
1870-9-28	6.5	17
1876-6-4	6.3	18
1876-6-5	5.6	18
1894-4-6	5.9	17
1895-5-14	5.5	18
1895-6-21	5.5	18
1895-8-6	6.2	17
1895-10-8	5.5	18
1896-2-10	5.9	17
1896-2-10	6.2	17
1905-6-1	6.6	18
1905-6-1	5.5	18
1905-6-3	5.5	18
1905-8-4	6	18
1905-8-6	5.5	18
1906-3-1	6.4	18
1907-8-1	5.7	18
1907-8-16	6.2	18
1920-11-29	5.6	13
1920-12-18	5.6	18
1921-3-30	5.6	11
1922-4-11	5.6	18
1926-12-17	5.8	18

(continued)

(continued)

South dinarides		
Date (y–m–d)	M	Cat
1926-12-17	5.8	18
1927-2-14	6	18
1934-2-4	5.6	18
1935-3-31	5.6	11
1940-2-23	5.5	18
1948-8-27	5.5	18
1958-4-3	5.6	18
1959-8-17	5.9	18
1959-9-1	6.2	18
1959-10-7	5.6	18
1962-3-18	6.2	18
1968-11-3	5.5	18
1969-4-3	5.5	11
1970-8-19	5.5	18
1979-4-15	7	2
1979-4-15	5.7	11
1979-5-24	6.4	18
1982-11-16	5.7	18
1988-1-9	5.7	18
1996-9-5	6	3

South apennines		
Date (y–m–d)	M	Cat
1273-12-18	5.8	5
1293-9-4	5.8	5
1361-7-17	6	5
1456-12-5	7.2	5
1456-12-5	7	6
1456-12-5	6.3	6
1466-1-15	6.1	5
1561-7-31	5.6	5
1561-8-19	6.8	5
1625-9-0	5.8	5
1688-6-5	7	5
1694-9-8	6.8	5
1702-3-14	6.5	5
1732-11-29	6.6	5
1805-7-26	6.6	5
1826-2-1	5.8	5

(continued)

(continued)

South apennines		
Date (y–m–d)	M	Cat
1831-1-2	5.5	5
1836-11-20	6	5
1851-8-14	6.4	5
1853-4-9	5.6	5
1857-12-16	7	5
1885-12-26	5.5	5
1910-6-7	5.7	5
1930-7-23	6.6	5
1962-8-21	5.7	5
1962-8-21	6.1	5
1980-11-23	6.9	5
1990-5-5	5.8	5
1990-5-5	5.5	5
1998-9-9	5.6	5
2002-10-31	5.7	5
2002-11-1	5.7	5

Central apennines		
Date (y-m-d)	M	Cat
1120-3-25	5.8	5
1170-5-9	5.6	5
1209	6	6
1315-12-3	5.6	5
1348-9-13	5.6	5
1349-9-9	5.9	5
1349-9-9	6	6
1349-9-9	6.6	5
1456-12-5	5.8	6
1461-11-27	6.4	5
1599-11-6	6	5
1639-10-7	5.9	5
1654-7-24	6.3	5
1703-1-14	6.7	5
1703-1-16	5.9	17
1703-2-2	6.7	5
1706-11-3	6.8	5
1730-5-12	5.9	5
1762-10-6	6	5
1859-8-22	5.5	5

(continued)

(continued)

Central apennines		
Date (y-m-d)	M	Cat
1874-12-6	5.5	5
1879-2-23	5.6	5
1881-9-10	5.6	5
1904-2-24	5.6	5
1915-1-13	7	5
1916-11-16	5.5	5
1933-9-26	6	5
1943-10-3	5.8	5
1950-9-5	5.7	5
1979-9-19	5.9	5
1984-5-7	5.9	5
1984-5-11	5.5	5
2009-4-6	6.2	5

North apennines		
Date (y-m-d)	M	Cat
1269-9	5.6	5
1277	5.6	5
1279-4-30	5.6	5
1279-4-30	6.3	5
1293-3	5.6	5
1298-12-1	6.2	5
1328-12-1	6.4	5
1352-12-25	6.4	5
1353-1-1	6	6
1389-10-18	6	5
1428-7-3	5.5	5
1438-6-11	5.6	5
1458-4-26	5.8	5
1470-4-11	5.6	5
1481-5-7	5.6	5
1483-8-11	5.7	5
1497-3-3	5.9	6
1501-6-5	6	5
1505-1-3	5.6	5
1542-6-13	5.9	5
1558-4-13	5.8	5
1570-11-17	5.5	5
1584-9-10	5.8	5

(continued)

(continued)

North apennines		
Date (y-m-d)	M	Cat
1624-3-19	5.5	5
1661-3-22	6.1	5
1672-4-14	5.6	5
1688-4-11	5.8	5
1690-12-23	5.6	5
1719-6-27	5.5	5
1741-4-24	6.2	5
1747-4-17	5.9	5
1751-7-27	6.3	5
1768-10-19	5.9	5
1781-4-4	5.9	5
1781-6-3	6.4	5
1781-7-17	5.6	5
1786-12-25	5.6	5
1789-9-30	5.8	5
1791-10-11	5.5	5
1796-10-22	5.6	5
1799-7-28	6.1	5
1815-9-3	5.5	5
1828-10-9	5.8	5
1831-9-11	5.5	5
1832-1-13	6.3	5
1832-3-13	5.5	5
1834-2-14	5.8	5
1837-4-11	5.8	5
1854-2-12	5.6	5
1870-10-30	5.6	5
1873-3-12	6	5
1875-3-17	5.9	5
1897-9-21	5.5	5
1909-1-13	5.5	5
1914-10-27	5.8	5
1916-5-17	6	5
1916-8-16	6.1	5
1916-8-16	5.5	5
1917-4-26	5.9	5
1918-11-10	5.9	5
1919-6-29	6.3	5
1920-9-7	6.5	5

(continued)

(continued)

North apennines		
Date (y-m-d)	M	Cat
1930-10-30	5.8	5
1971-7-15	5.6	5
1984-4-29	5.7	5
1997-9-26	5.7	5
1997-9-26	6	5
1997-10-6	5.5	5
1997-10-14	5.7	5
2012-5-20	5.9	7
2012-5-29	5.8	7

Eastern Alps–Northern dinarides		
Date (y-m-d)	M	Cat
1323	6	5
1348-1-25	7	5
1403-9-6	5.6	5
1502-3-26	5.7	18
1511-3-26	7	5
1511-6-25	5.6	17
1511-8-8	6.3	18
1551-3-26	6.3	18
1574-8-14	5.6	5
1626-7-3	6.5	18
1628-6-17	5.6	4
1640	6	4
1645	5.6	4
1648	5.7	18
1689-3-10	5.6	4
1690-12-4	6.5	5
1695-2-25	6.5	5
1697-3-15	5.6	4
1699-2-11	5.6	4
1700-7-28	5.6	5
1721-1-12	6.1	5
1750-12-17	5.9	18
1775-10-13	6.8	19
1776-7-10	5.8	5
1794-6-7	6	5
1802-1-4	5.6	18
1812-10-25	5.7	5

(continued)

(continued)

Eastern Alps–Northern dinarides		
Date (y-m-d)	M	Cat
1836-6-12	5.5	5
1845-12-12	5.7	18
1870-3-1	6.4	18
1870-3-1	5.6	17
1873-6-29	6.3	5
1878-9-23	5.6	17
1880-11-9	6.3	18
1891-6-7	5.9	5
1895-4-14	6.2	5
1897-5-15	5.6	5
1905-12-17	5.6	18
1906-1-2	6.1	18
1916-3-12	5.6	5
1917-1-29	5.8	4
1926-1-1	5.9	5
1928-3-27	5.8	5
1936-10-18	6.1	5
1976-5-6	6.5	5
1976-9-11	5.6	5
1976-9-15	5.9	5
1976-9-15	6	18
1976-9-15	6	5
1998-4-12	5.7	3

References

Albini, P. (2004). A survey of past earthquakes in the Eastern Adriatic (14th to early 19th century). *Annals of Geophysics, 47*, 675–703.

Aliaj Sh. (2006). The Albanian orogen: Convergence zone between Eurasia and the Adria microplate. In: N. Pinter, G. Grenerczy, J. Weber, S. Stein & D. Medak (Eds.), *The Adria Microplate: GPS geodesy, tectonics and hazard, NATO science series IV-earth and environmental sciences* (Vol. 61, pp. 133–149). Springer, Netherlands.

Ambraseys, N., & Finkel, C. F. (1987). Seismicity of Turkey and neighbouring regions, 1899–1915. *Annales Geophysicae, 5B*, 501–726.

Ambraseys, N. N. (1990). Uniform magnitude re-valuation of European earthquakes associated with strong-motionrecords. *Earthquake Engineering & Structural Dynamics, 19*, 1–20. doi:10.1002/eqe.4290190103.

Ambraseys, N. N., & Jackson, J. A. (1998). Faulting associated with historical and recent earthquakes in the eastern Mediterranean region. *Geophysical Journal International, 133*, 390–406.

Anderson, D. L. (1975). Accelerated plate tectonics. *Science, 167*, 1077–1079.

Anderson, H., & Jackson, J. (1987). The deep seismicity of the Tyrrhenian Sea. *Geophysics Journal of Royal Astronomical Society, 91*, 613–637.

Ascione, A., Caiazzo, C., & Cinque, A. (2007). Recent faulting in Southern Apennines (Italy): Geomorphic evidence, spatial distribution and implications for rates of activity. *Bollettino Della Società Geologica Italiana. (Italian Journal of Geosciences), 126*, 293–305.

Barka, A. A. (1996). Slip distribution along the North Anatolian Fault associated with the large earthquakes of the period 1939 to 1967. *Bulletin of the Seismological Society of America, 86*, 1238–1254.

Ben-Menahem, A. (1979). Earthquake catalogue for the Middle East (92 B.C.–1980 A.D.). *Bollettino Geofisica Teoreticheskaya Applied, 84*, 245–310.

Benetatos, C., & Kiratzi, A. (2006). Finite-fault slip models for the 15 April 1979 (Mw 7.1) Montenegro earthquake and its strongest aftershock of 24 May 1979 (Mw 6.2). *Tectonophysics, 421*, 129–143.

Bressan, G., Bragato, P., & Venturini, C. (2003). Stress and strain tensors based on focal mechanisms in the seismotectonic framework of the Eastern Southern Alps. *Bulletin of the Seismological Society of America, 93*, 1280–1297.

Brodsky, E. E. (2009). The 2004–2008 worldwide superswarm. In *Eos transaction AGU fall meet* (Suppl. 90, S53B).

Burrato, P., Poli, M. E., Cannoli, P., Zanferrari, A., Basili, R., & Galadini, F. (2008). Sources of Mw5 + earthquakes in northeastern Italy and western Slovenia: an updated view based on geological and seismological evidence. *Tectonophysics, 453*, 157–176. doi:10.1016/j.tecto.2007.07.009.

Catalano, S., Monaco, C., & Tortorici, L. (2004). Neogene-quaternary tectonic evolution of the Southern Apennines. *Tectonics 23:TC2003*. doi:10.1029/2003TC001512.

Cenni, N., Mantovani, E., Baldi, P., & Viti, M. (2012). Present kinematics of Central and Northern Italy from continuous GPS measurements. *Journal of Geodynamics, 58*, 62–72. doi:10.1016/j.jog.2012.02.004.

Comninakis, P. E., & Papazachos, B. C. (1986). Catalogue of earthquakes in Greece and surrounding area for the period 1901–1985. *Geophysics Lab* (Pub. 1). Greece: University of Thessaloniki.

Crowley, H., Colombi, M., Borzi, B., Faravelli, M., Onida, M., Lopez, M., et al. (2009). A comparison of seismic risk maps for Italy. *Bulletin Earthquake Engineering, 7*, 149–180. doi:10.1007/s10518-008-9100-7.

De Alteriis, B. (1995). Different foreland basins in Italy: examples from the central and southern Adriatic Sea. *Tectonophysics, 252*, 349–373.

Del Ben, A., Barnaba, C., & Toboga, A. (2008). Strike-slip systems as the main tectonic features in the Plio-Quaternary kinematics of the Calabrian Arc. *Marine Geophysical Researches, 29*, 1–12.

Di Pasquale, G., Orsini, G., & Romeo, R. W. (2005). New developments in seismic risk assessment in Italy. *Bulletin Earthquake Engineering, 3*, 101–128. doi:10.1007/s10518-005-0202-1.

Durand, V., Bouchon, M., Karabulut, H., Marsan, D., Schmittbuhl, J., Bouin, M.-P., et al. (2010). Seismic interaction and delayed triggering along the north anatolian fault. *Geophysical Reseach Letters, 37*, L18310. doi:10.1029/2010GL044688.

Ekström, G., M. Nettles, and A. M. Dziewonski. (2012). The global CMT project 2004-2010: Centroid-moment tensors for 13,017 earthquakes. *Phys. Earth Planet. Inter.*, 200–201, 1–9. doi:10.1016/j.pepi.2012.04.002

Elter, F. M., Elter, P., Eva, C., Eva, E., Kraus, R. K., Padovano, M., & Solarino, S. (2012). An alternative model for the recent evolution of Northern-Central apennines (Italy). *Journal of Geodynamics, 54*, 55–63.

Ergin, K., Guclu, U., Uz, Z. (1967). *A catalogue of earthquakes for Turkey and surrounding area (11 A.D. to 1964 A.D.)*. Instambul, Turkey.

Eva, C., Riuscetti, M., & Slejko, D. (1988). Seismicity of the Black sea region. *Bollettino Geofisica Teoreticheskaya Applied, 117*(118), 53–66.

Ergintav, S., McClusky, S., Hearn, E., Reilinger, R., Cakmak, R., Herring, T., et al. (2009). Seven years of postseismic deformation following the 1999, M = 7.4 and M = 7.2, Izmit-Duzce, Turkey earthquake sequence. *Journal Geophysical Research, 114*, B07403. doi:10.1029/2008JB006021.

Ferranti, L., Santoro, E., Gazzella, M. E., Monaco, C., & Morelli, D. (2009). Active transpression in the northern Calabria Apennines, southern Italy. *Tectonophysics, 476*, 226–251. doi:10.106/j.tecto.2008.11.010.

Finetti, I. R. (2005). The calabrian arc and subducting ionian slab from new CROP seismic data. In I. R. Finetti (Ed.), *Deep seismic exploration of the Central Mediterranean and Italy* (pp. 393–412). CROP PROJECT: Elsevier.

Finetti, I. R., & Del Ben, A. (2005). Ionian tethys lithosphere roll-back sinking and back-arc tyrrhenian opening from new CROP seismic data. In I. R. Finetti (Ed.), *Deep seismic exploration of the Central Mediterranean and Italy* (pp. 483–504). CROP PROJECT: Elsevier.

Freed, A. M. (2005). Earthquake triggering by static, dynamic, and postseismic stress transfer. *Annual Review of Earth and Planetary Sciences, 33*, 335–367.

Freed, A. M., Ali, S. T., & Burgmann, R. (2007). Evolution of stress in southern California for the past 200 years from coseismic, postseismic and interseismic stress changes. *Geophysical Journal International, 169*, 1164–1179.

Funiciello, R., Parotto, M., Praturlon, A. (1981). *Carta Tettonica d'Italia/Tectonic Map of Italy, scala 1:1500000. CNR-PFG* (Pubbl. n. 269). Grafica Editoriale Cartografica, Roma, Italy.

Galadini, F., Poli, M. E., & Zanferrari, A. (2005). Seismogenic sources potentially responsible for earthquakes with M ≥ 6 in the Eastern Southern Alps (Thiene-Udine sector, NE Italy). *Geophysical Journal International, 161*, 739–762.

Godey, S., Bossu, R., Guilbert, J., & Mazet-Roux, G. (2006). The Euro-Mediterranean bulletin: A comprehensive seismological bulletin at regional scale. *Seismological Research Letters, 77*, 460–474.

Guidoboni, E., & Comastri, A. (2005). *Catalogue of earthquakes and tsunamis in the Mediterranean area from the 11th to the 15th century*. Roma, Italy: Istituto Nazionale di Geofisica e Vulcanologia.

Guidoboni, E., Ferrari, G., Mariotti, D., Comastri, A., Tarabusi, G., & Valensise, G. (2007). *CFTI4Med, catalogue of strong earthquakes in Italy (461 B.C.-1997) and mediterranean area (760 B.C.-1500)*. Istituto Nazionale di Geofisica e Vulcanologia-Storia Geofisica Ambiente. http://storing.ingv.it/cfti4med/.

Iannaccone, G., Scarcella, G., & Scarpa, R. (1985). Subduction zone geometry and stress patterns in the Tyrrhenian Sea. *Pure and Applied Geophysics, 123*, 819–836.

ISC—International seismological centre. http://www.isc.ac.uk/iscbulletin/.

ISIDe Working Group (INGV). (2010). Italian seismological instrumental and parametric database. http://iside.rm.ingv.it.

Jackson, J., & McKenzie, D. (1988). The relationship between plate motions and seismic moment tensors and the rates of active deformation in the Mediterranean and Middle East. *Geophysical Journal, 93*, 45–73.

Karnik, V. (1971). Seismicity of the European area, Part II. Dordrecht, Germany.

Kokkalas, S., Xypolias, P., Koukouvelas, I., & Doutsos, T. (2006). Postcollisional contractional and extensional deformation in the Aegean region. In Y. Dilek, & S. Pavlides (Eds.), *Postcollisional tectonics and magmatism in the Mediterranean region and Asia, Geological Society of America Special Paper* (Vol. 409, pp. 97–123). doi: 10.1130/2006.2409(06).

Kuk, V., Prelogovic, E., & Dragicevic, I. (2000). Seismotectonically active zones in the Dinarides. *Geol. Croatica, 53*, 295–303.

Lay, T., Kanamori, H., Ammon, C. J., Hutko, A. R., Furlong, K., & Rivera, L. (2009). The 2006–2007 Kuril Islands great earthquake sequence. *Journal Geophysical Research, 114*, B11308. doi:10.1029/2008JB006280.

Louvari, E., Kiratzi, A. A., & Papazachos, B. C. (1999). The Cephalonia Transform Fault and its extension to western Lefkada Island (Greece). *Tectonophysics, 308*, 223–236.

Louvari, E., Kiratzi, A. A., Papazachos, B. C., & Katzidimitriou, P. (2001). Fault-plane solutions determined by waveform modelling confirm tectonic collision in the eastern Adriatic. *Pure and Applied Geophysics, 158*, 1613–1637.

Luo, G., & Liu, M. (2010). Stress evolution and fault interactions before and after the 2008 Great Wenchuan earthquake. *Tectonophysics, 491*, 127–140. doi:10.1016/j.tecto.2009.12.019.

Makropoulos, K. C., & Burton, P. W. (1981). A catalogue of seismicity in Greece and adjacent areas. *Geophysical Journal Royal Astronomical Society, 65*, 741–762.

Makropoulos, K., Kaviris, G., & Kouskouna, V. (2012). An updated and extended earthquake catalogue for Greece and adjacent areas since 1900. *Natural Hazards and Earth Systems Sciences, 12*, 1425–1430. doi:10.5194/nhess-12-1425-2012.

Mantovani, E., Viti, M., Babbucci, D., Tamburelli, C., Albarello, D. (2006). Geodynamic connection between the indentation of Arabia and the Neogene tectonics of the central-eastern Mediterranean region. In Y. Dilek, & S. Pavlides (Eds.), *Post-collisional tectonics and magmatism in the mediterranean region and Asia, Geological Society of America Special Paper* (Vol. 490, pp. 15–49).

Mantovani, E., Viti, M., Babbucci, D., & Albarello, D. (2007a). Nubia-Eurasia kinematics: an alternative interpretation from Mediterranean and North Atlantic evidence. *Annals of Geophysics, 50*, 341–366.

Mantovani, E., Viti, M., Babbucci, D., & Tamburelli, C. (2007b). Major evidence on the driving mechanism of the Tyrrhenian-Apennines arc-trench-back-arc system from CROP seismic data. *Bollettino Della Società Geologica Italiana, 126*, 459–471.

Mantovani, E., Viti, M., Babbucci, D., & Vannucchi, A. (2008). Long-term prediction of major earthquakes in the Calabrian Arc. *Environmental Semeiotics, 1*, 190–207. doi:10.3383/es.1.2.3.

Mantovani, E., Babbucci, D., Tamburelli, C., & Viti, M. (2009). A review on the driving mechanism of the Tyrrhenian-Apennines system: Implications for the present seismotectonic setting in the Central-Northern Apennines. *Tectonophysics, 476*, 22–40.

Mantovani, E., Viti, M., Babbucci, D., Albarello, D., Cenni, N., & Vannucchi, A. (2010). Long-term earthquake triggering in the southern and Northern Apennines. *Journal of Seismology, 14*, 53–65.

Mantovani, E., Viti, M., Babbucci, D., Cenni, N., Tamburelli, C., & Vannucchi, A. (2012). Middle term prediction of earthquakes in Italy: Some remarks on empirical and deterministic approaches. *Bollettino Geofisica Teoreticheskaya Applied, 53*, 89–111.

Margottini, C., Ambraseys, N. N., & Screpanti, A. (1993). *La magnitudo dei terremoti italiani del XX secolo*. ENEA, Rome: Italy.

Mariotti, D., & Guidoboni, E. (2006). Seven missing damaging earthquakes in Upper Valtiberina (Central Italy) in 16th–18th century: Research strategies and historical sources. *Annals of Geophysics, 49*, 1139–1155.

Markusic, S., & Herak, M. (1999). Seismic zoning of Croatia. *Natural Hazards, 18*, 269–285.

Marsan, D. (2005). The role of small earthquakes in redistributing crustal elastic stress. *Geophysical Journal International, 163*, 141–151. doi:10.1111/j.1365-246X.2005.02700.x.

Mikumo, T., Yagi, Y., Singh, S. K., & Santoyo, M. A. (2002). Coseismic and post-seismic stress changes in a subducting plate: Possible stress interactions between large interplate thrust and intraplate normal-faulting earthquakes. *Journal Geophysical Research, 107*, 2023. doi:10.1029/2001JB000446.

Moretti, I., & Royden, L. (1988). Deflection, gravity anomalies and tectonics of doubly subducted continental lithosphere: Adriatic and Ionian Seas. *Tectonics, 7*, 875–893.

Ozawa, S., Nishimura, T., Suito, H., Kobayashi, T., Tobita, M., & Imakiire, T. (2011). Coseismic and post-seismic slip of the 2011 magnitude-9 Tohoku-oki earthquake. *Nature, 475*, 373–376. doi:10.1038/nature10227.

Papazachos, B. C., Comninakis, P. E. (1982). *A catalogue of historical earthquakes in Greece in and surrounding area. Geophysics Lab* (Publ. 5). Greece: University of Thessaloniki.

Papazachos, B. C., & Papazachos, C. B. (1989). *The earthquakes of Greece*. University of Thessaloniki, Greece: Geophysics Lab Publication.

Piccardi, L., Tondi, G., & Cello, G. (2006) Geo-structural evidence for active oblique estension in South-Central Italy. In N. Pinter, G. Grenerczy, J. Weber, S. Stein, & D. Medak (Eds.), *The Adria microplate: GPS geodesy, tectonics and Hazard, NATO Science Series IV-Earth and Environmental Sciences* (Vol. 61, pp 95–108). Berlin: Springer

Poljak, M., Zivcic, M., & Zupancic, P. (2000). The seismotectonic characteristics of Slovenia. *Pure and Applied Geophysics, 157*, 37–55.

Pollitz, F. F., Burgmann, R., & Romanowicz, B. (1998). Viscosity of oceanic asthenosphere inferred from remote triggering of earthquakes. *Science, 280*, 1245–1249.

Pollitz, F., Bakun, W. H., & Nyst, M. (2004). A physical model for strain accumulation in the San Francisco Bay region: Stress evolution since 1838. *Journal Geophysical Research, 109*, B11408. doi:10.1029/2004JB003003.

Pollitz, F. F., Burgmann, R., & Banerjee, P. (2006). Postseismic relaxation following the great 2004 Sumatra-Andaman earthquake on a compressible self-gravitating Earth. *Geophysical Journal International, 167*, 397–420. doi:10.1111/j.1365-246X.2006.03018.x.

Pollitz, F. F., Stein, R. S., Sevilgen, V., & Burgmann, R. (2012). The 11 April 2012 east Indian Ocean earthquake triggered large aftershocks worldwide. *Nature, 490*, 250–253. doi:10.1038/nature11504.

Ribaric, C. (1982). *Seismicity of Slovenia: Catalogue of earthquakes (792 A.D.–1981)*. Ljubljana, Slovenia.

Rothé, J. P. (1971). Seismicité de l'Atlantique oriental, de la Méditerranée occidentale et de ses bordures. *Revue de Géographie Physique et de Géologie Dynamique, 13*, 419–428.

Rovida, A., Camassi, R., Gasperini, P., & Stucchi M (Eds.). (2011). *CPTI11, the 2011 version of the Parametric Catalogue of Italian Earthquakes*. Milano, Bologna. http://emidius.mi.ingv.it/CPTI.

Rydelek, P. A., & Sacks, I. S. (1990). Asthenospheric viscosity and stress diffusion: a mechanism to explain correlated earthquakes and surface deformation in NE Japan. *Geophysical Journal International, 100*, 39–58.

Ryder, I., Parsons, B., Wright, T. J., Funning, G. J. (2007). Post-seismic motion following the 1997 Manyi (Tibet) earthquake: InSAR observations and modelling. *Geophysical Journal International*. http://dx.doi.org/10.1111/j.1365–246X.2006.03312.x.

Scisciani, V., & Calamita, F. (2009). Active intraplate deformation within Adria: Examples from the Adriatic region. *Tectonophysics, 476*, 57–72. doi:10.1016/j.tecto.2008.10.030.

Seismological Catalogues of Greece. http://www.geophysics.geol.uoa.gr/.

Shebalin, N.B., Karnik, B., Hadzievki, D. (Eds.). (1974). *UNDP/ENESCO Survey of the Balkan region. Catalogue of earthquakes Part I 1901–1970*. Paris, France: Unesco.

Shebalin, N. V., Leydecker, G., Mokrushina, N. G., Tatevossian, R. E., Erteleva, O. O., & Vassiliev, V. Yu. (1998). *Earthquake catalogue for Central and Southeastern Europe 342 BC-1990 AD. Final report to contract n° ETNU-CT93–0087*. Brussels, Belgium.

Stucchi, M., Meletti, C., Montaldo, V., Crowley, H., Calvi, G. M., & Boschi, E. (2011). Seismic hazard assessment (2003–2009) for the Italian building code. *Bulletin of the Seismological Society of America, 101*, 1885–1911.

Stucchi, et al. (2012). The SHARE European Earthquake Catalogue (SHEEC) 1000–1899. *Journal of Seismology,*. doi:10.1007/s10950-012-9335-2.

Toda, S., Stein, R. S., & Sagiya, T. (2002). Evidence from the AD 2000 Izu islands earthquake swarm that stressing rate governs seismicity. *Nature, 419*, 58–61.

Toth, L., Zsiros, T., & Monus, P. (1988). Earthquake history of the Pannonian basin and adjacent territories from 456 A.D. *European Earthquake Engineering, 3*, 44–50.

Viti, M., D'Onza, F., Mantovani, E., Albarello, D., & Cenni, N. (2003). Post-seismic relaxation and earthquake triggering in the southern Adriatic region. *Geophysical Journal International, 153*, 645–657.

Viti, M., Mantovani, E., Babbucci, D., & Tamburelli, C. (2006). Quaternary geodynamics and deformation pattern in the Southern Apennines: implications for seismic activity. *Bollettino Della Società Geologica Italiana, 125*, 273–291.

Viti, M., Mantovani, E., Babbucci, D., & Tamburelli, C. (2009). Generation of trench arc-back arc systems in the Western Mediterranean region driven by plate convergence. *Bollettino Della Società Geologica Italiana, 128*, 89–106.

Viti, M., Mantovani, E., Babbucci, D., & Tamburelli, C. (2011). Plate kinematics and geodynamics in the Central Mediterranean. *Journal of Geodynamics, 51*, 190–204. doi:10.1016/j.jog.2010.02.006.

Viti, M., Mantovani, E., Cenni, N., & Vannucchi, A. (2012). Post-seismic relaxation: An example of earthquake triggering in the Apennine belt (1915–1920). *Journal of Geodynamics, 61*, 57–67. doi:10.1016/j.jog.2012.07.002.

Viti, M., Mantovani, E., Cenni, N., & Vannucchi, A. (2013). Interaction of seismic sources in the Apennine belt. *Journal of Physics and Chemistry of the Earth, 63*, 25–35. doi:10.1016/j.pce.2013.03.005.x.

Wells, D. L., & Coppersmith, K. J. (1994). New empirical relationships among magnitude, rupture length, rupture width, rupture area and surface displacement. *Bulletin of the Seismological Society of America, 84*, 974–1002.

Wesnousky, S. (2006). Predicting the endpoints of earthquake ruptures. *Nature, 444*, 358–360. doi:10.1038/nature05275.

Working Group CPTI. (2004). *Catalogo Parametrico dei Terremoti Italiani, versione 2004 (CPTI04)*. Bologna: INGV. doi: 10.6092/INGV.IT-CPTI04.

Forecasting Moderate Earthquakes in Northern Algeria and Morocco

José A. Peláez, Mohamed Hamdache, Carlos Sanz de Galdeano, Rashad Sawires and Mª Teresa García Hernández

1 Introduction

The studied region, in the northern border of the African Plate, has suffered moderate to strong earthquakes in the last decades. Among them, the September 9, 1954, and the October 10, 1980, El Asnam (formerly known as Orléansville), Algeria, earthquakes, with magnitudes M_S 6.7 and 7.3, respectively, the February 29, 1960, Agadir, Morocco, M $\sim$ 6.0 earthquake, or the most recent May 21, 2003, Zemmouri-Algiers, Algeria, M_W 6.8 earthquake. The 1954, 1980 and 2003 Algerian earthquakes caused a large loss of lives (1200, 5000–20,000, according to different estimates, and 2300 people killed, respectively), as well as the 1960 Moroccan earthquake ($\sim$12,000 people killed).

In all cases, there were a myriad of injured people, left homeless and heavily damaged and destroyed homes due to structural inadequacies of the buildings. Moreover, critical facilities as hospitals or schools were damaged or destroyed in all quoted earthquakes. Only during the 2003 Algerian earthquake, 130 schools suffered extensive to complete damage in the Algiers region (Bendimerad 2004).

Apart from these large earthquakes, a large amount of small to moderate earthquakes has been also recorded in this area. A review can be read in the works by Peláez et al. (2007) and Hamdache et al. (2010).

J.A. Peláez (✉) · R. Sawires · Mª T. García Hernández
Department of Physics, University of Jaén, Jaén, Spain
e-mail: japelaez@ujaen.es

M. Hamdache
Département Études et Surveillance Sismique, CRAAG, Algiers, Algeria

C. Sanz de Galdeano
IACT, CSIC-University of Granada, Granada, Spain

R. Sawires
Department of Geology, University of Assiut, Assiut, Egypt

S. D'Amico (ed.), *Earthquakes and Their Impact on Society*,
Springer Natural Hazards, DOI 10.1007/978-3-319-21753-6_3

Forecasting earthquakes, in this or other regions, is a crucial task mainly for two reasons. One of them merely scientific: forecasting it is one of the main points of the scientific knowledge. The second one practical in itself: it is an important component for the seismic risk mitigation (Marzocchi and Zechar 2011).

Several authors consider that earthquakes are not only natural events but social events too. The impact of earthquakes is not only a serious brake on economic and social development from the resulting damages, but they can generate serious social impact on the population by generating important traumas, limiting or even completely changing the fundamental process of operation of the social function. Although poorly studied, the societal response to forecasting studies differs from place to place. The question for developing this type of studies arises from the hope that people may someday be able to be warned, and then do things to protect themselves and their properties before earthquakes strike. It is well known that nowadays every warning system is still under debate and, according to the societal response study developed by Mileti and Darlington (1995), people are more likely to believe scientific earthquake predictions when they overlap or converge with nonscientific forecasts. Belief in a prediction will vary along different ethnic and social classes and age lines. Credibility is shaped by the general sense of trust that people have in government when prediction information is released by government authorities. In fatalistic cultures, prediction and forecasting must be viewed as credible but preparedness actions do not necessarily follow. Also, as pointed by Stallings (1982), the nature of prediction itself may have some influence on its credibility. Predictions may be stated either in terms of likelihood of an earthquake in a certain place during a certain period of time, or of the absence of earthquakes above a certain magnitude in a region between two epochs. There is some indication that prediction of the former type (the presence of some events) is inherently more credible than those of the latter type (the absence of some events). These considerations are only some of the reasons that advocate the earthquake forecasting and seismic hazard studies.

The main goal of the earthquake forecasting process is to reduce the loss of lives and property damages. It starts identifying the areas prone to be the site of moderate to strong earthquakes in the future. From this point, it is necessary to select the areas with higher seismic risk level and to establish an effective land and physical planning. In these areas, the physical planners must incorporate all the available required actions in order to create a safe environment. How much reduce the risk and at what additional cost are two key questions for the politicians, and must be answered based on the different options given by the professional planners. In any case, some actions are simple and easy, for example, act in a planned way versus act randomly. This also requires professional experience and political determination.

After select the areas with higher seismic risk level, the main actions must be focused into decrease the risk level and reduce the impact of the disaster. The list of the most important things to be protected could be the following: (a) the human lives, (b) the economic activities, including facilities and equipment, (c) the operational capacity of the region, including transport and communication networks, energy, water, etc., (d) health services and facilities, and (e) housing services, among other.

Other question is to define areas with different risk level in order to select future development areas. The lower risk level areas must be where must be placed the most important components of the city, that is, residential and industrial zones. In higher risk level areas must be limited the growth of existing settlements, as well as avoiding build, for example, new warehouses or manufacturing plants. In several countries it is now impossible to completely leave earthquake prone areas. Then, it is of paramount importance to select the less dangerous places among available and to develop measures for disaster prevention. Although it is not possible to avoid all the earthquake consequences, planning actions can help to reduce them to some acceptable level.

One of the most important planning actions is to update the seismic actions in the current seismic provisions. Earthquakes don't kill people, buildings do, is the most repeated saying in Earthquake Engineering. In order to save lives and reduce losses during an earthquake, it must be crucial the adoption and enforcement of up-to-date building codes (FEMA). Taking into account that it is not possible to reduce the seismic hazard of a region, it is necessary to reduce the seismic risk. First at all, building codes must be enforced, especially in areas with moderate to high seismic hazard and in areas where there is a significant probability that a moderate to high earthquake happens. Moreover, local governments must fall on the seismic retrofitting of old buildings and vulnerable structures, the biggest contributors to seismic hazard in mostly countries, making them more resistant to earthquakes.

Thereby, forecasting earthquakes studies, like presented in this work, are key studies in order to delineate the areas in which it is more valuable to provide a certain level of protection for buildings and other facilities. This and other previous studies on seismic hazard carried out in this region (Peláez et al. 2003, 2005b, 2006; Hamdache et al. 2012), demonstrate the need for to make some changes in order to improve the seismic input in current Moroccan and Algerian building codes. Conducted studies on earthquake engineering research, reported in the scientific literature and defended in international meetings, must be serious reasons to update seismic provisions.

The most important problem in the forecasting is to obtain the best model. One of the used approaches is to consider that moderate earthquakes follow the same spatial distribution that small to moderate earthquakes. This is the way followed in this work, a model based in the spatially smoothed seismicity method.

The approach proposed by Frankel (1995) in order to compute seismic hazard values has been widely used in the last years, being that used nowadays, for example, to compute the USGS National Seismic Hazard Maps. This method models the seismicity that cannot be assigned to specific geological structures, termed as distributed or background seismicity, depending on authors. In this type of studies, the region is divided into square cells, and the number of earthquakes above a certain reference or threshold magnitude is counted. This count, that is, the total number of events observed above this reference magnitude is the maximum likelihood estimate of the *a*-parameter in the Gutenberg-Richter relationship

(Weichert 1980). Then, it is smoothed spatially, thus, including the uncertainty in the earthquake location in the final seismic hazard results. To perform the smooth, it is usual to use a Gaussian filter, because it has the characteristic that preserves the total number of earthquakes. Finally, the seismic hazard computation is based in the well-known total probability theorem, expressed in terms of rate of exceedance of a certain level of ground motion.

Using this approach, our research group has carried out seismic hazard assessments, among other places, in Portugal and Spain (Peláez and LópezCasado 2002) and Algeria (Peláez et al. 2003). Afterwards, these assessments have been updated and extended not only to peak ground horizontal acceleration (PGA) values but to spectral acceleration (SA) values, uniform hazard spectra (UHS) and even Arias intensity (AI) values (Peláez et al. 2005a, b, 2006; Hamdache et al. 2012).

Among other sources of uncertainty in any probabilistic seismic hazard study, we must consider the characterization of seismic sources (both faults and source zones), and specially, the ground-motion attenuation models (SSHAC 1997). This is the reason because in this work we present a preliminary time-independent forecasting model, the first component in any seismic hazard assessment, overlooking the second one, the question of the estimation of the ground-motion level. Thereby, the final results obtained in the assessment do not include the typical epistemic and aleatory (also named modelling and parametric) uncertainties in ground-motion predictions.

To develop this model, we have used a procedure similar to the one used in the spatially smoothed seismicity approach to compute seismic hazard values: each square cell in the study area is assumed to be a source. Our contribution is in the same line, for example, of the forecast conducted for the Italian territory by Akinci (2010). This simple study/analysis does not considers explicitly neither the geology of the region (including geophysical, paleoseismic or geodetic studies) nor physical models concerning processes associated with earthquake triggering.

Initially, we will test in our region if minor earthquake locations could be used to forecast future places where will happen moderate to strong earthquakes, which is the main assumption in this type of studies (Frankel 1995). The authors of this work do not agree with a uniform background zone used as an applicable model in seismic hazard or forecasting studies (Peláez and LópezCasado 2002; Peláez et al. 2003), that is, to assume that certain type and size of events can occur anywhere with equal probability. In any case, this assumption will be also checked.

2 Geological Setting of the Area

The Maghrebian region (Fig. 1) occupies the NW part of the African (Nubia) Plate in what is referred to its continental crust. The oceanic crust of this plate continues till the area of the Azores Islands. North the Maghreb is the Eurasian Plate and to the south is the Saharan Shield. Presently, the convergence between the Nubia Plate

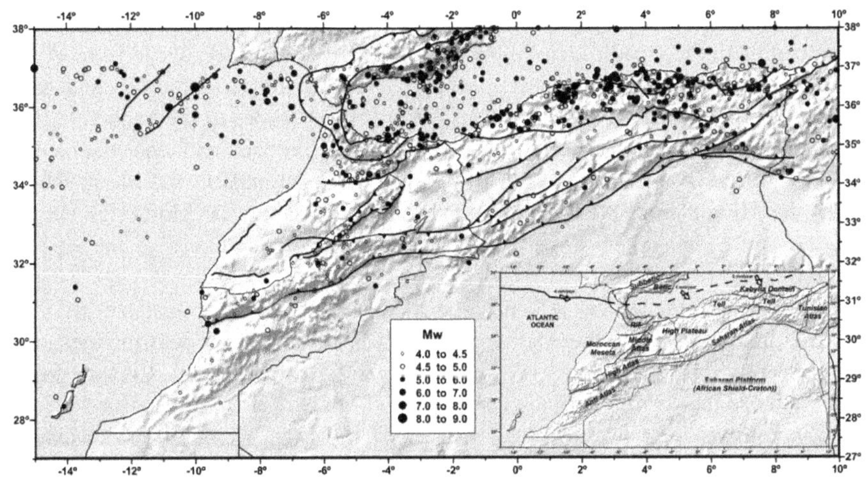

Fig. 1 Tectonic sketch showing the main tectonics domains and the seismicity included in the used catalog

and Iberia has an approximate NNW-SSE direction, with relative velocity values of the order of 3 to 5 cm/year, according the places (DeMets et al. 1990).

Between the Maghreb and the Eurasian Plate, from the Gibraltar Arc to the south of Italy, an intermediate complex domain is intercalated. This domain is formed by some oceanic basins, as is the Algero-Provençal Basin and the Thyrrehnian Basin, and by an old region, presently disintegrated and forming the Internal zones of the Betic-Rifean Chain, of the Kabylias (in Algeria), the Peloritani Mountains (Sicily) and the Calabrian area in Italy (AlKaPeCa domain; Bouillin et al. 1986). This area underwent from the early Miocene the northward subduction of Africa in its NW border. Owing to this process were formed the two main cited oceanic basins, causing simultaneously the disintegration of the AlKaPeCa domain (Durand Delga and Fontboté 1980).

The Saharan Shield, deformed by the Pan-African Orogeny (Precambrian to early Cambrian), is in contact with the Atlasic Mountains (Fig. 1). To the west and NW of the Atlas is situated the Moroccan Meseta (in which the Paleozoic series are very well exposed). To the NE and east of the Atlas are the High Plateaux in Algeria (also partially formed Palaeozoic rock, moreover the Mesozoic and Tertiary cover). Both areas, the Moroccan Meseta and the High Plateaux, contact respectively to the north with the Rif and Tell mountains. The Saharan Shield forms part of the Precambrian areas of Africa, clearly cratonized and generally not affected by later important deformations. It is a stable area. In Morocco, the northern part of the Saharan Shield is called Antiatlas, and corresponds to a Precambrian and, mainly, Paleozoic area, making a tectonic transition between the shield and the Atlas.

The Atlasic Mountains correspond to an intracontinental chain, partly formed by Palaeozoic rocks, moreover other ones of Jurassic and Cretaceous age. It is divided in High Atlas and Middle Atlas, and farther continued in Algeria in the so called

Saharan Atlas. The High Atlas corresponds to the southern part of this chain. This chain begins in the Agadir area, in the coast of Morocco (practically in front of the Canary Islands). From this place, the High Atlas continues to the NE and east, where finally is divided in two branches. One of them passes to the Saharan Atlas, although with lesser heights, cross Algeria in an E-W direction and reach the central part of Tunisia. Another branch of the High Atlas continues to the north, where forms the Middle Atlas with a NE-SW direction, separating the Moroccan Meseta and the High Plateaux in Algeria, and finally reaching the coast to the east of Melilla.

On the whole, the present Atlasic areas have been tectonically unstable from the Triassic times, and along the Alpine orogeny suffered important deformations, and more recently, also important volcanism, reaching the Quaternary. Considering a wider geotectonic scenery, these volcanic rocks form part of the volcanism crossing the Alboran Sea (the extreme western part of the Mediterranean) (Jacobshagen 1992) from the SE of Spain and continued more to the north, while to the SW reach the Canary Islands. Nevertheless, on the whole, the Atlas can be considered as an aulacogen bordering the northwestern part of the Saharan Shield. Along this chain, important and deep faults exist, even cutting the crust, as indicate the long cited line of volcanism.

The Rif and Tell chains are composed by a great number of nappes, advancing to the south in the Tell and to the south and southwest in the north of Morocco. They locally thrust part of the northern areas of the Moroccan Meseta and the High Plateaux, and even in some places part of the Atlasic Mountains. They are formed by sedimentary External zones (slightly affected by metamorphism in some places) and by Internal zones. Great part of the Internal zones (at its time divided in several tectonic complexes) are affected by alpine metamorphism, moreover the previous existence of Hercinian and even older deformations. In any case, their present structure has being formed during the Alpine Orogeny. In the Rif, they appear mainly in the area of Tetuan, north and south of this town, in Morocco, and in the both Kabylias (Grande Kabylia, to the west, and Petite Kabylia, to the east), in Algeria.

These Alpine chains (Rif and Tell) have being structured from the Cretaceous to the Oligocene-early Miocene. After this structuring, and in some places during it, numerous Neogene, mainly Miocene, basins were formed and disposed above any previous domain, covering many previous structures.

In the Rif and Tell chains, from the middle Miocene, and particularly from the late Miocene to the present, the previous deformation of AlCaPeKa practically ended and predominate a near N-S to NNW-SSE compression between the Nubia plate and Eurasia. This compression, accompanied in many areas by a near perpendicular tension (Galindo Zaldívar et al. 1993), facilitated the formation of new systems of faults. The most important is a set of NE-SW faults crossing the Alboran Sea. These faults have a predominant sinistral strike-slip displacement, while another set, with NW-SE direction, has a dextral character. Additionally, reverse faults, many of which have N70°E to E-W direction, were formed.

In many cases, these strike-slip faults moved mainly as normal faults, releasing by means of this way the regional tension. Even the near E-W faults in some places can move as normal faults, owing to local stresses in areas suffering process of uplift or subsidence. The stress in some of these cases even produces a radial extension (Sanz de Galdeano et al. 2010).

Considered on the whole, the Maghrebian area is a complex region owing to the deformation produced by the interaction of two plates, an intermediate domain pinched in between, and the propagation of the deformations to the south, forming the Atlas Mountains, limiting the cratonic northwestern African areas. To these deformations must be added the great line of fractures that from the north of Europe, passing by the Rhin Basin and the Iberian eastern border, cross the Alboran Sea, reach the Atlas and continues in the Canary Islands (Sanz de Galdeano 1990).

3 Seismic Catalog and Spatial Distribution of Earthquakes

To know if small earthquakes delimit the areas where large earthquakes will happens, as was quoted previously, is the basis of the spatially smoothed approach both in seismic hazard and in this type of forecasting studies. Here we check both spatial distributions ('small' and 'large' earthquakes) in order to confirm this hypothesis in our region. Kafka and Ebel (2011) consider that this is the 'least-astonishing null hypothesis', being a standard of comparison for other more complex spatial forecast methods (*v.g.*, Zechar and Jordan 2010; Falcone et al. 2010). Kafka and Ebel (2011) call this method cellular seismology.

The reliability of this analysis is related to the reliability of the used catalog, *i.e.*, it is related to its completeness and homogeneity. To develop our study, we have used two unified catalogs in terms of moment magnitude, including only main (Poissonian) events, compiled specifically for seismic hazard and forecasting studies in the region, one of them covering northern Morocco (Peláez et al. 2007) and the second one northern Algeria (Hamdache et al. 2010). The catalog for northern Morocco includes earthquakes in the area between 27° to 37°N and 15°W to 1°E. Initially, it spans the years 1045 to 2005. The catalog for northern Algeria covers the area between 32° to 38°N and 3°W to 10°E, initially spanning the years 856 to 2008. These catalogs have been updated to 2011 and aggregated. Duplicated earthquakes in the overlapped areas as well as non-crustal events (events with depth below 30 km) have been erased (Fig. 1).

Overall, the final catalog can be considered complete above magnitude M_W 5.0 since 1900, with a mean rate of 2.15 events/year, and above magnitude M_W 6.0 since 1885, with a mean rate of 0.21 events/year. Earthquakes above M_W 4.0 are completes only since 2003, with a mean rate of 29.82 events/year; in the preceding period 1925–2003, the mean rate was only 7.71 events/year (Fig. 2). We must take into account that Moroccan and Algerian seismological networks have not covered

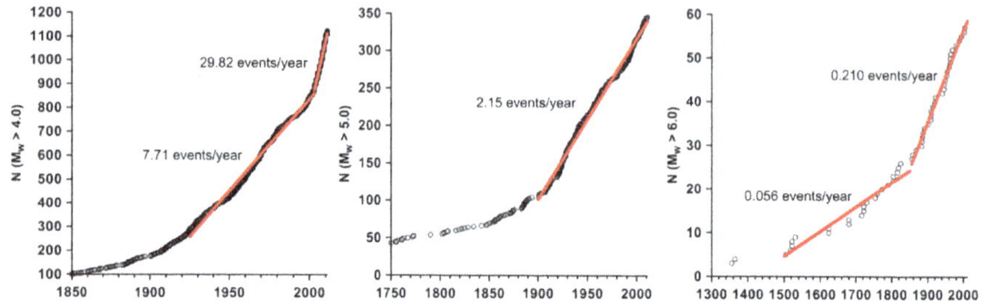

Fig. 2 Accumulative number of earthquakes in the catalog above magnitudes M_W 4.0, 5.0 and 6.0 versus time. Completeness periods are displayed

Fig. 3 Percentage of hits (see text)

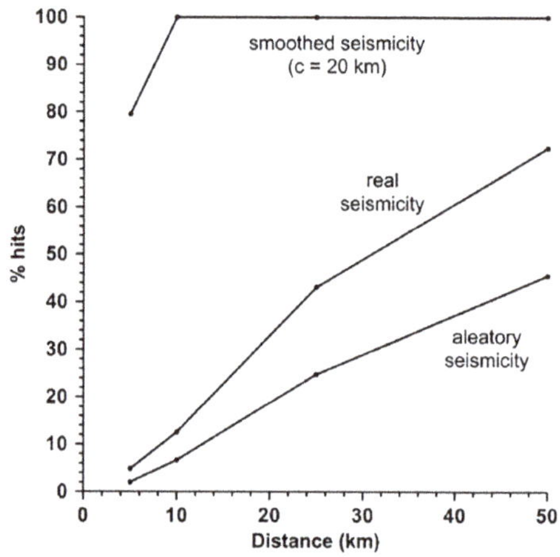

efficiently this region, and mostly earthquakes included in both catalogs are those located by the Spanish National Geographic Institute network, distant of the southern and eastern parts of the study region (Peláez et al. 2007; Hamdache et al. 2010).

Considering only earthquakes since 1900 with magnitude above M_W 5.0, we have obtained an overall *b*-value equal to 0.88 ($\sigma = 0.02$). Although there are some spatial variations of this parameter in the study region, for this study we will consider that it is constant.

Using this catalog, we have checked what is the percentage of real events with magnitude equal to or above M_W 5.0 happened within a certain distance, for different distances, of at least one previous event with magnitude in the range M_W 4.0–4.9 (Fig. 3). We call it percentage of hits, in the line of the works by Kafka and Walcott (1998), Kafka and Levin (2000) and Kafka (2002). Moreover, we have also computed the percentage of hits for an aleatory distribution of simulated events with magnitude equal to or above M_W 5.0 (Fig. 3). In order to compute both percentages, we have used earthquakes with magnitude in the range M_W 4.0–4.9 since 1925 in a region 0.5° smaller in extent than extension of the used catalog, to avoid boundary effects. Must be taken into account that seismicity in the range M_W 4.0–4.9 from 1925 to 2003 is not complete, then, the computed percentage of hits in both cases (real and simulated events with magnitude equal to and above M_W 5.0) must be lesser than the true.

Obviously, the computed percentage of hits is dependent on the specified distance to be considered related small and large events (Fig. 3). The main conclusion is that, independently of the considered distance (Fig. 3 shows results for distances less than or equal to 50 km), the percentage of hits is significantly greater in the real case that when considering aleatory simulated events, double in the case of specified distances less than 25 km. For instance, there is a 72.5 % probability that a real earthquake with magnitude greater than or equal to M_W 5.0 could happen at less than 50 km of at least a previous earthquake in the range M_W 4.0–4.9.

In an aleatory (uniform) distribution of simulated earthquakes with magnitude equal to or above M_W 5.0, the probability is only 45.6 %. For this last computation, all earthquakes in the range M_W 4.0–4.9 since 1925, and aleatory locations for earthquakes with magnitude equal to or above M_W 5.0 have been considered. Considering in a more accurate simulation not all earthquakes in the range M_W 4.0–4.9 but only previous earthquakes to the aleatory events, the percentage of hits must be significantly lower. This implies to consider time-aleatory and space-aleatory events, not only space-aleatory events as we have done.

From our catalog, we can proclaim that in the study region it can be observed a relatively good agreement between locations for these two data sets. In fact, considering the previously discussed percentage, nearly 3 out of every 4 earthquakes with magnitude above M_W 5.0 happened since 1925 were located less than 50 km of at least a previous earthquake with magnitude in the range M_W 4.0–4.9. This is approximately the same result that has been obtained in other regions using different data sets and magnitude intervals (i.e., Helmstetter et al. 2006, 2007; Werner et al. 2011). This result supports the use of the proposed forecasting approach.

Then, we perform a spatial smooth of the seismicity pertaining to the first dataset, that is, earthquakes in the range M_W 4.0–4.9 from 1925 to 2011. Considering the catalog uncompleteness for the epoch 1925–2003 for this range of earthquakes (Fig. 3), we have completed the number of earthquakes in each cell proportionately to the counted number of events in the range M_W 4.0–4.9 in the time period 1925–2003, in order to extent the current mean rate of events until 1925. This process is not necessary to compute the percentage of hits, but it is essential in order to assess forecasts. We have used square cells with dimensions 10 km × 10 km for which we count directly the number of earthquakes recorded in each of them, and then, we smooth it using a Gaussian filter with a correlation distance c of 20 km. This implies that we spread each earthquake between its own cell and the neighbouring cells: the Gaussian filter cause that, if we associate to the own cell a weight equal to 1.0, the cell distant the correlation distance c contributes with a weight equal to 0.37, and the cell distant $2c$ contributes only with 0.02. Thereby, using this filter, we can take into account uncertainties in the earthquake locations. As it was stated in the introduction section, the main characteristic of this filter is that it preserves the total number of earthquakes.

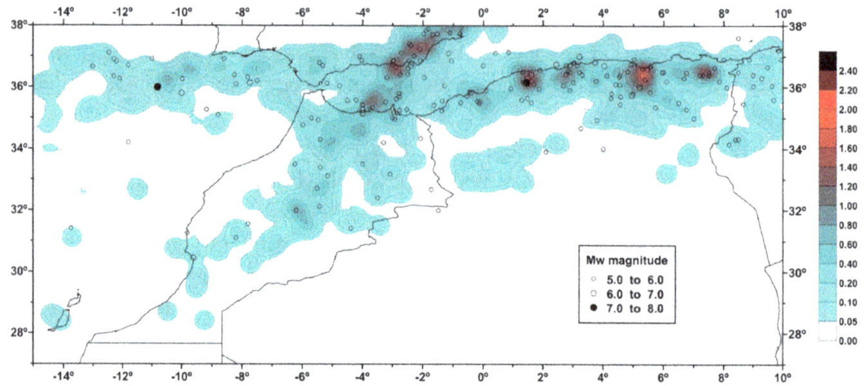

Fig. 4 Smoothed number of events in the range M_W 4.0–4.9 since 1925. Single events with magnitude above M_W 5.0 in the used catalog are also displayed

After counting earthquakes, completing and filtering this data set, we have computed again the percentage of hits. It can be also observed in Fig. 3. In order to compare both distributions, earthquakes in the range M_W 4.0–4.9 and earthquakes above M_W 5.0, the resulting smoothed number of earthquakes as well as earthquakes above magnitude M_W 5.0 are displayed in Fig. 4. In this case, were considered and smoothed all earthquakes in the range M_W 4.0–4.9 since 1925.

4 Forecasting Locations of Moderate Earthquakes

The final stage in our assessment is to forecast where will happen moderate and moderate to strong ($M_W \geq 5.0$ and $M_W \geq 6.0$) events from the spatially smoothed seismicity. We will consider the smoothed number of events with magnitude above M_W 4.0 in order to compute the yearly number of events in each cell above a certain magnitude value, by using the well-known Gutenberg-Richter recurrence relationship. The a-value is obtained directly from smoothed number of events (using a correlation distance equal to 20 km), and the used b-value will be the computed regional value. Once yearly number of events above a certain magnitude is obtained for each cell (n), assuming a Poissonian process, the probability of exceedance (P) for a certain exposure time (T) for the selected magnitude can be obtained from the relationship

$$P = 1 - e^{-nT}$$

Taking into account cell dimensions, probability per square kilometer is obtained multiplying by 0.01.

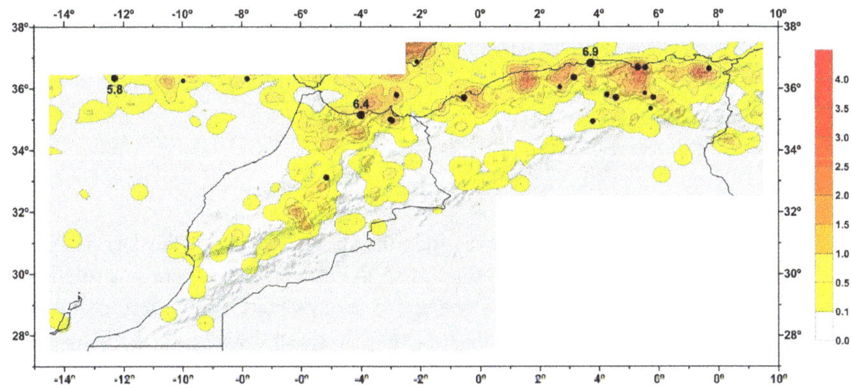

Fig. 5 Probability of exceedance per cell for earthquakes with magnitude above M_W 5.0 in the next 10 years. Single events with magnitude above M_W 5.0 located in the last 10 years (2003–2013) are also displayed

Probabilities for an exposure time of 10 years for earthquakes with magnitude $M_W \geq 5.0$ can be observed in Fig. 5, and for earthquakes with magnitude $M_W \geq 6.0$ in Fig. 6.

In Fig. 5 can be observed maximum values in Northern Algeria in the Tell, and in Northern Morocco mainly in the eastern part of the Rif. Specifically, the following significant maximum values are obtained in different areas (from east to west): 4.1 % in Guelma, 3.7 % in the N of Setif, 2.6 % in Blida, 3.8 % in El Asnam, and 2.2 % in the NW of Oran regions, all of them in Algeria, and 2.9 % in the NE of Al Hoceima, and 2.5 % in the S of the Middle Atlas regions, both of them in Morocco. Other little relative maximum values can be observed throughout the Northern Algeria and Moroccan regions.

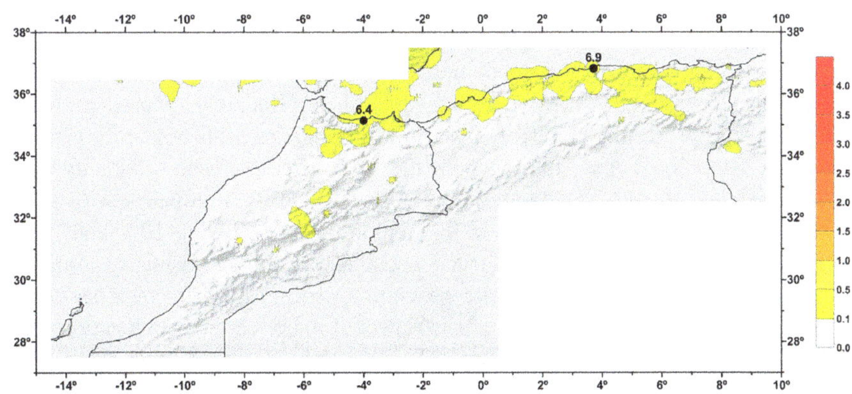

Fig. 6 Probability of exceedance per cell for earthquakes with magnitude above M_W 6.0 in the next 10 years. Single events with magnitude above M_W 6.0 located in the last 10 years (2003–2013) are also displayed

Table 1 Computed probabilities per cell in the next 10 years, for the locations of the 2003 Algiers, M_W 6.8, and 2004 N Tamassint, M_W 6.4, earthquakes

	2003 Algiers (%)	2004 N Tamassint (%)
P ($M_W \geq 5.0$)	2.0	1.3
P ($M_W \geq 6.0$)	0.3	0.2

To check these results, events with magnitude $M_W \geq 5.0$ in the last 10 years (from 2003 to 2013) have been also displayed. As can be seen, there is a relatively good agreement between locations of these events and forecasted maximum probabilities. The three events with biggest magnitude in the region (the May 2003 Algiers, Algeria, M_W 6.8, the February 2004 N Tamassint, Morocco, M_W 6.4, and the July 2007 Azores-Cape Sant Vincent, M_W 5.8 earthquakes) have been enhanced.

The 2004 N Tamassint, Morocco, earthquake is located in the easternmost part of the Rif, near the area quoted above where maximum probabilities have been obtained in this region. The 2003 Algiers, Algeria, earthquake (Hamdache et al. 2004) is located in the Tell, between two of the areas with relative maximum probabilities, the N of Setif and Blida regions. As can be seen in Fig. 6, these two earthquakes are also included in the areas with biggest probability that earthquakes with magnitude $M_W \geq 6.0$ could happen. In this figure, mostly the Tell and the eastern part of the Rif regions are demarcated clearly as regions where probabilities per cell are in the range 0.1–0.5 %. Individual probabilities for these two earthquakes are displayed in Table 1.

5 Conclusions

In Figs. 5 and 6 are displayed potential areas from a probabilistic point of view to host future earthquakes with magnitudes $M_W \geq 5.0$ and $M_W \geq 6.0$, respectively, in our region of study. The maximum computed values per cell in the region are 4.1 % in Guelma, Algeria, and 2.9 % in the NE of Al Hoceima, Morocco, when computing the probability of exceedance for earthquakes with $M_W \geq 5.0$ in the following 10 years. Taking into account the known recent mean seismicity rate, 21–22 earthquakes with magnitude above M_W 5.0 and 2 earthquakes with magnitude above M_W 6.0 will happens in this region in the following 10 years.

These results have been obtained using a spatially smoothed seismicity approach and a Poissonian process (earthquake generation process have no memory) considering earthquakes above M_W 4.0 after to check if minor earthquake locations can be used to forecast future places where will happen moderate to large events. An updated earthquake catalog was specifically used for this computation. Although it has a good behaviour concerning its completeness and homogeneity, certainly it is the main source of uncertainty in the reliability of the final results.

Certain future refinements (for example, a regionalization of the *b*-value) will be considered in the future in order to improve this model.

Considering the areas that will host moderate and moderate to high earthquakes from the computed probabilities, jointly with the obtained results in previous seismic hazard assessments in the region conducted by the authors of this work, they show that it is necessary to improve the seismic action in the current Moroccan and Algerian building codes. It is unacceptable to continue using as seismic input in the Algerian code (RPA-99, and the different updated versions) a political regionalization, called seismic zoning, based in the seismicity, neither based in the seismic hazard nor in the probability that a certain earthquake can happen. To divide the country in four "seismic zones", assigning them a constant design acceleration value, is a poor approach that cannot still be used. The next version (RPA-2015) appears to contain the same weakness and shortcomings, despite the efforts to change such rules. The same happened in the previous Moroccan code (RPS2000). In the current code (RPS2011), it has been established a new seismic zoning, in this case based in previous seismic hazard results, but also assigning constant design acceleration values for each one of the five delimited zones.

As it was stated before, the recent studies carried out in this region on earthquake engineering research, must be already used to update seismic provisions.

Acknowledgement This research was supported by the Spanish Seismic Hazard and Active Tectonics research group, the Algerian C.R.A.A.G., and the grant CGL2011-30153-C02-02 of the Spanish Ministerio de Ciencia e Innovación.

References

Akinci, A. (2010). HAZGRIDX: Earthquake forecasting model for ML ≥ 5.0 earthquakes in Italy based on spatially smoothed seismicity. *Annals of Geophysics, 53*, 51–61.

Bendimerad, F. (2004). Earthquake vulnerability of school buildings in Algeria. In *OECD Program on Educational Building and Geohazards International. Ad Hoc Experts' Group Meeting on Earthquake Safety in Schools*, OECD, Paris, pp. 35–44.

Bouillin, J., Durand Delga, M., & Olivier, P. (1986). Betic-Rif and Tyrrhenian distinctive features, genesis and development stages. In F. C. Wezel (Ed.), *The origin of arcs* (pp. 281–304). Amsterdam: Elsevier.

DeMets, C., Gordon, R. G., Argus, D. F., & Stein, S. (1990). Current plate motions. *Geophysical Journal International, 101*, 425–478.

Durand Delga, M., & Fontboté, J. M. (1980). Le cadre structural de la Méditerranée occidentale. In: *Géologie des chaînes alpines issues de la Téthys* (Aubouin, J., Debelmas, J., et Latreille, M., *dir.*), *Mémoire B.R.G.M.* Vol. 115, pp. 67–85.

Falcone, G., Console, R., & Murru, M. (2010). Short-term and long-term earthquake occurrence models for Italy: ETES, ERS and LTST. *Annals of Geophysics, 53*, 41–50.

Frankel, A. (1995). Mapping seismic hazard in the Central and Eastern United States. *Seismological Research Letters, 66*(4), 8–21.

Galindo Zaldívar, J., González Lodeiro, F., & Jabaloy, A. (1993). Stress and palaeostress in the Betic-Rif cordilleras (Miocene to the present). *Tectonophysics, 227*, 105–126.

Hamdache, M., Peláez, J. A., Talbi, A., & López Casado, C. (2010). A unified catalog of main earthquakes for Northern Algeria from A.D. 856 to 2008. *Seismological Research Letters, 81*, 732–739.

Hamdache, M., Peláez, J. A., Talbi, A., Mobarki, M., & López Casado, C. (2012). Ground motion hazard values for Northern Algeria. *Pure and Applied Geophysics, 169*, 711–723.

Hamdache, M., Peláez, J. A., & YellesChauche, A. K. (2004). The Algiers, Algeria earthquake (M_W 6.8) of 21 May 2003: Preliminary report. *Seismological Research Letters, 75*, 360–367.

Helmstetter, A., Kagan, Y. Y., & Jackson, D. D. (2006). Comparison of short-term and time-independent forecast models for Southern California. *Bulletin of the Seismological Society of America, 96*, 90–106.

Helmstetter, A., Kagan, Y. Y., & Jackson, D. D. (2007). High-resolution time-independent grid-based forecast for M ≥ 5 earthquakes in California. *Seismological Research Letters, 78*, 78–86.

Jacobshagen, V. (1992). Major fracture zones of Morocco: The South Atlas and the Transalboran fault systems. *Geologische Rundschau, 81*, 185–197.

Kafka, A. L. (2002). Statistical analysis of the hypotheis that seismicity delineates areas where future large earthquakes are likely to occur in the Central and Eastern United States. *Seismological Research Letters, 73*, 992–1003.

Kafka, A. L., & Ebel, J. E. (2011). Proximity to past earthquakes as a least-astonishing hypothesis for forecasting locations of future earthquakes. *Bulletin of the Seismological Society of America, 101*, 1618–1629.

Kafka, A. L., & Levin, S. Z. (2000). Does the spatial distribution of smaller earthquakes delineate areas where larger earthquakes are likely to occur? *Bulletin of the Seismological Society of America, 90*, 724–738.

Kafka, A. L., & Walcott, J. R. (1998). How well does the spatial distribution of smaller earthquakes forecast the locations of larger earthquakes in the Northeastern United States? *Seismological Research Letters, 69*, 428–440.

Marzocchi, W., & Zechar, J. D. (2011). Earthquake forecasting and earthquake prediction: different approaches for obtaining the best model. *Seismological Research Letters, 82*, 442–448.

Mileti, D. S., & Darlington, J. D. (1995). Societal response to revised earthquake probabilities in the San Francisco Bay area. *International Journal of Mass Emergencies and Disasters, 13*, 119–145.

Peláez, J. A., Chourak, M., Tadili, B. A., Brahim, L. A., Hamdache, M., LópezCasado, C., & Martínez Solares, J. M. (2007). A catalog of main Moroccan earthquakes from 1045 to 2005. *Seismological Research Letters, 78*, 614–621.

Peláez, J. A., Delgado, J., & López Casado, C. (2005a). A preliminary probabilistic seismic hazard assessment in terms of Arias intensity in Southern Spain. *Engineering Geology, 77*, 139–151.

Peláez, J. A., Hamdache, M., & López Casado, C. (2003). Seismic hazard in Northern Algeria using spatially smoothed seismicity. *Results for peak Ground Acceleration. Tectonophysics, 372*, 105–119.

Peláez, J. A., Hamdache, M., & López Casado, C. (2005b). Updating seismic hazard values of Northern Algeria with the 21 May 2003 M 6.8 Algiers earthquake included. *Pure and Applied Geophysics, 162*, 2163–2177.

Peláez, J. A., Hamdache, M., & López Casado, C. (2006). Seismic hazard in terms of spectral acceleration and uniform hazard spectra in Northern Algeria. *Pure and Applied Geophysics, 163*, 119–135.

Peláez, J. A., & López Casado, C. (2002). Seismic hazard estimate at the Iberian Peninsula. *Pure and Applied Geophysics, 159*, 2699–2713.

Sanz de Galdeano, C. (1990). La prolongación hacia el sur de las fosas y desgarres del Norte y Centro de Europa: Una propuesta de Interpretación. *Rev Soc Geol España, 3*, 231–241.

Sanz de Galdeano, C., Shanov, S., Galindo Zaldívar, J., Radulov, A., & Nikolov, G. (2010). A new tectonic discontinuity in the Betic Cordillera deduced from active tectonics and seismicity in the Tabernas Basin. *Journal of Geodynamics, 50*, 57–66.

SSHAC. (1997). *Senior seismic hazard analysis committee—Recommendations for probabilistic seismic hazard analysis: Guidance on uncertainty and use of experts*. Lawrence Livermore National Laboratory Report (NUERG/CR-6372, UCRL-ID-122160), Livermore.

Stallings, R. A. (1982). Social aspects related to the dissemination and credibility of earthquake predictions in cross-cultural perspective. In *Earthquake prediction: Proceedings of the International Symposium on Earthquake Prediction*, Terra Scientific Pub. Co., pp. 59–67.

Weichert, D. H. (1980). Estimation of the earthquake recurrence parameters for unequal observation periods for different magnitudes. *Bulletin of the Seismological Society of America, 70*, 1337–1346.

Werner, M. J., Helmstetter, A., Jackson, D. D., & Kagan, Y. Y. (2011). High-resolution long-term and short term earthquake forecast for California. *Bulletin of the Seismological Society of America, 101*, 1630–1648.

Zechar, J. D., & Jordan, T. H. (2010). Simple smoothed seismicity earthquake forecast for Italy. *Annals of Geophysics, 53*, 99–105.

An Earthquake Catalogue (2200 B.C. to 2013) for Seismotectonic and Seismic Hazard Assessment Studies in Egypt

Rashad Sawires, José A. Peláez, Raafat E. Fat-Helbary and Hamza A. Ibrahim

1 Introduction

Of all natural hazards, earthquakes are those which historically have caused the most extensive impact and disruption in terms of damage to infrastructure, human-casualties and economic losses. They are the expression of a continuing evolution of the Earth Planet and a reshaping of the Earth's surface. They are the most deadly of all natural disasters affecting the human environment. Every year more than one million earthquakes shake different regions of the world, some so feeling and gentle that only the most sensitive instruments can detect the motion, and others so violent that whole communities are shattered and large sections of terrain are shifted in this process that can start landslides, block rivers, cause floods, and set massive sea waves surging across the oceans.

The amount of damage and number of fatalities at a certain location caused by an earthquake depends on various factors: the magnitude and characteristics of the earthquake focus, distance from the epicenter, soil characteristics, density of buildings and population, and structural design of buildings and infrastructures, among others. These facts are playing an important role in decreasing or increasing the number of victims in recent earthquakes, especially in the developing countries. The increasing population in the earthquake-prone cities, poor construction quality and lack of building code enforcement are major reasons why the vulnerability due to earthquake is also increasing.

R. Sawires · H.A. Ibrahim
Geology Department, Faculty of Science, Assiut University, Assiut 71516, Egypt

R. Sawires · J.A. Peláez (✉)
Department of Physics, University of Jaén, Jaén, Spain
e-mail: japelaez@ujaen.es

R.E. Fat-Helbary
Aswan Regional Earthquake Research Center, 152, Aswan, Egypt

S. D'Amico (ed.), *Earthquakes and Their Impact on Society*,
Springer Natural Hazards, DOI 10.1007/978-3-319-21753-6_4

Today scientists, technicians and engineers know a lot of details and information about earthquakes, where they are most likely to occur, how deep they originate, and how they affect land. Researchers are applying this knowledge to future programs for predicting when and where the next earthquake might occur and for constructing buildings and installations that might be better able to withstand earthquake violence. Man may never be able to control or even predict earthquakes with satisfactory accuracy, but he can learn to live with them in relative safety. Occurrences of earthquakes in different parts of the world and the resulting losses, especially human lives, have highlighted the structural inadequacy of many buildings to support seismic loads.

Short and mid-term earthquake forecasting may one day be able to reduce significantly casualties associated with catastrophic earthquakes. However, a long-term preventive policy is the only possible way for the reduction of life losses and socioeconomic impact associated with them. Such preventive policy should be based on: (i) the assessment of seismic hazard and risk, (ii) the implementation of safe building construction codes, (iii) the increased public awareness on natural disasters, and (iv) a strategy of land-use planning taking into consideration the seismic hazard and other natural disasters (Riad et al. 2000).

Seismic hazard estimation is an essential component for earthquake-resistant design, and for the preparation of seismic zoning maps which provide the necessary input information for the design of ordinary structures. The preparation of basic data for the seismic hazard assessment (SHA) starts immediately after evaluating the seismic record (a complete, reliable, and processed earthquake catalogue).

Earthquake catalogues are the starting point for any SHA. Catalogues consist of estimates of past earthquake locations, described by three spatial and one temporal coordinates, and the magnitudes of events that have occurred in or near the region of interest. The quality, consistency and homogeneity of these data are directly reflected in the accuracy of the results of a SHA. Earthquake catalogues, along with a good understanding of the geology and seismotectonic environment are the fundamental bases for constructing the seismic source model, which is the first element needed to carry out the assessment.

A reliable knowledge of the earthquakes that occurred at a site of interest is the key stone for seismic hazard estimates and seismotectonic studies. The seismic history available is the result of a collection of quite heterogeneous pieces of information spanning from recent instrumental records to macroseismic observations obtained from old documentary sources. In general, this collection is characterized by different levels of reliability due to heterogeneity in sampling procedures, data processing, and availability of relevant information. As concerns the instrumental part of the catalogue, changes in the monitoring networks (both in terms of geometry and density of stations), and the procedures for earthquake parameterization, could result in seismic compilations characterized by different levels of completeness and reliability.

A complete and consistent catalogue of earthquakes can provide good data for studying the distribution of earthquakes in a region as a function of space, time, and magnitude. However, most catalogues do not report the magnitude of earthquakes consistently over time, in addition to varying uncertainties in hypocenter locations. This may pose as an obstacle for delineating seismicity patterns or for assessing seismic hazards. Because earthquake magnitude has become an indispensable source parameter of earthquakes since its inception, it is important to convert the original magnitudes based on various scales in different time periods to a common magnitude scale throughout the whole period.

A unified catalogue is crucial for statistical analysis of earthquakes, even more for seismic hazard studies. A reliable earthquake catalogue is containing accurate information (e.g., epicentre locations, focal depths, and earthquake magnitudes) which is of utmost importance in order to determine various parameters needed for the SHAs.

Although Egypt is affected by moderate seismic activity compared to other countries, it is exposed to high seismic risk. This is due to many factors: (1) the population in Egypt, as well as all important and archaeological sites, are concentrated within a narrow belt along the Nile Valley and Delta, (2) most of earthquakes occurred near overpopulated cities and villages, (3) the methods of construction vary between old (as those still being used in the villages) and new buildings with poor construction practice, and finally, (4) the soil characteristics in different localities in Egypt and their influence on seismic amplification. A damaging earthquake is a real, as well as a current, threat to the safety, social integrity, and economic wellbeing of the population in the region. Thus, SHA studies in Egypt are greatly needed to identify areas with different degrees of vulnerability. They will serve for further risk studies, construction codes, and also for land-use planning.

The published seismic databases covering Egypt until now are inadequate to study the long-term seismicity and recurrence interval for large earthquakes in the region. Thus, the primary goal of this work was to catalogue all known events from every available published or unpublished source for the area between 21° to 38°N and 22° to 38°E. A unified earthquake catalogue was obtained (includes all the historical, instrumental and focal mechanism solutions data), using for this purpose several empirical relationships among reported magnitudes, macroseismic intensity, and moment magnitude (M_W). Finally, all dependent events were removed as well as earthquakes with magnitudes smaller than M_W 3.0. The final catalogue covers the period from 2200 B.C. to 2013 and includes 16,642 mainshocks. Its development and main characteristics are discussed below.

2 Tectonic Setting of Egypt

Egypt is situated in the northeastern corner of the African Plate, along the southeastern edge of the Eastern Mediterranean region. It is interacting with the Arabian and Eurasian Plates through divergent and convergent plate boundaries, respectively (Fig. 1). Egypt is bounded by three active tectonic plate margins: the African-Eurasian plate margin, the Gulf of Suez-Red Sea plate margin, and the Gulf of Aqaba-Dead Sea Transform Fault (DST). The seismic activity of Egypt is due to the interaction and the relative motion between the plates of Eurasia, Africa and Arabia. Within the last decade, some areas in Egypt have been struck by significant earthquakes causing considerable damage. Such events were interpreted as the result of this interaction.

The primary features of active plate boundaries in the vicinity of Egypt have been discussed in details by many authors (e.g., McKenzie 1970, 1972; Neev 1975; Ben-Menahem et al. 1976; Garfunkel and Bartov 1977; Ben-Avraham et al. 1987; Woodward-Clyde Consultants 1985; Mesherf 1990; Kebeasy 1990). Also, the relationship between those plate boundaries and shallow seismicity was studied by different authors (e.g., Sofratome Group 1984; Abou Elenean 1997; Abou Elenean and Hussein 2007). A summary of the most important tectonic features in the vicinity of Egypt (Fig. 1) is given in the following paragraphs.

(a) *Africa-Eurasia Plate Margin*: The African and Eurasian Plates are converging across a wide zone in the Mediterranean Sea. The effects of the plate interaction are mainly north and remote from the Egyptian coastal margin. The zone is characterized by folding within the Mediterranean Sea floor and subduction of the Northeastern African Plate beneath Cyprus and Crete (Maamoun et al. 1980). Some evidences of secondary deformation appear to be occurring along the Northern Egyptian coast as represented by moderate earthquake activity. The earthquake activities constitute a belt parallel to the Hellenic and Cyprian Arcs (Abou Elenean 2007). Some of the largest events located to the south of Crete and Cyprus Islands were felt and caused few damage on the northern part of Egypt (e.g., August 8, 1303, intensity VIII offshore Mediterranean earthquake, February 13, 1756, intensity VI and June 26, 1926, M_S 7.4 Hellenic Arc earthquakes, October 9, 1996, M_W 6.8 Cyprus earthquake, and October 12, 2013, M_S 6.4 Crete earthquake).

(b) *Gulf of Suez-Red Sea Plate Margin*: The Arabian Plate is continuing to rotate away from the African Plate along the Red Sea spreading center. The earthquake activity along that boundary is related to the Red Sea rifting, plutonic activity and the intersection points of the NW (Gulf of Suez-Red Sea Faults) with the NE DST Faults (Abou Elenean 2007). Active sea-floor spreading has been identified as far as about 20° to 22°N latitudes, from the continuous presence of basaltic crust in the axial rift of the Red Sea and the geophysical signatures of newly emplaced oceanic crust (Cochran 1983). The extension of this deformation zone toward the north (Suez-Cairo shear zone) is considered as the most active part of Northern Egypt. Some evidences of an extension of the Suez-Cairo-Alexandria shear zone to the north, towards the Mediterranean Sea, were described (Kebeasy et al. 1981; Ben-Avraham et al. 1987). The largest earthquakes along this zone caused some damage in Northern Egypt (e.g., July 11, 1879, intensity VI and March 6, 1900, M_S 6.2 Gulf of Suez earthquakes, and March 31, 1969, M_W 6.8 Shedwan Island earthquake).

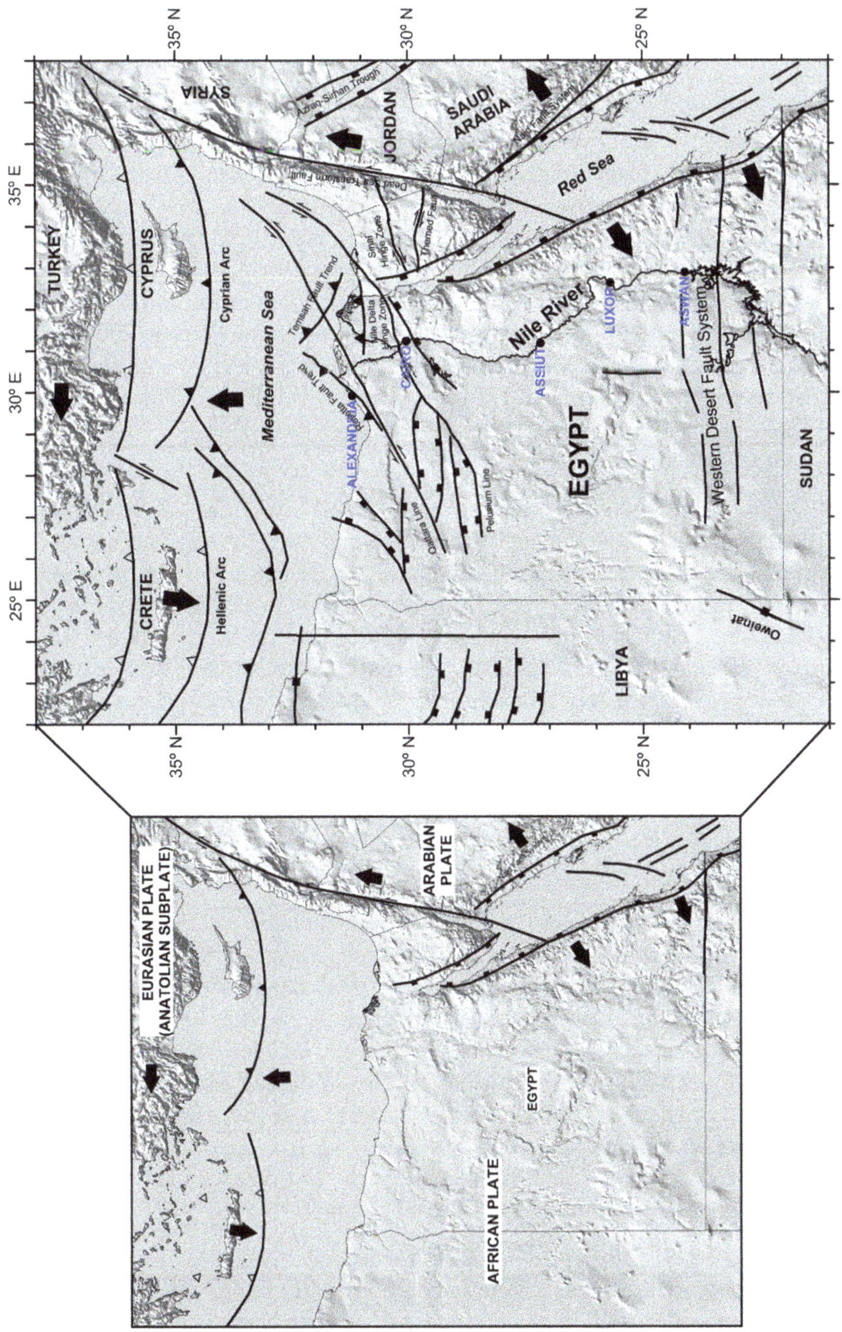

Fig. 1 Tectonic sketch for Egypt and its vicinity (modified after Ziegler 1988; Meulenkamp et al. 1988; Dewey et al. 1989; Guiraud and Bosworth 1999)

(c) *Gulf of Aqaba-Dead Sea Transform Fault*: It is a major left-lateral strike-slip fault that accommodates the relative motion between Africa and Arabia (Salamon et al. 2003). It connects a region of extension in the Northern Red Sea with the Taurus collision zone to the north. This fault zone consists of *en echelon* faults with extensional jogs, with the largest such step over being the Dead Sea pull-apart basin. The main faults of this zone are trending N-S to NNE-SSW. They are found on the Sinai and Arabian deformed coastal areas, as well as within the Gulf of Aqaba (Ben-Avraham 1985). The seismic activity of this shear boundary is relatively high and appeared to be clustered in some places where there is intersection of two or more faults (NNE-SSW and WNW-ESE) or attributed to upwelling of magma (Abou Elenean 2007). Some larger events were reported along this fault trend and caused damage in Northern Egypt (e.g., March 18, 1068, intensity VIII Elat earthquake, May 20, 1202, intensity VIII Lebanon earthquake, and November 22, 1995, M_W 7.2 Gulf of Aqaba earthquake).

Based on the geophysical studies in the territory of Egypt, Youssef (1968) classified the main structural elements of Egypt into the following fault categories: (a) Gulf of Suez-Red Sea, (b) Gulf of Aqaba, (c) E-W, (d) N-S, and (e) N45°W trends. However, Meshref (1990), from the magnetic tectonic trend analysis, showed the tectonic trends which influenced Egypt throughout its geologic history as: (a) NW (Gulf of Suez-Rea Sea), (b) NNE (Aqaba), (c) E-W (Tethyan or Mediterranean Sea), (d) N-S (Nubian or East African), (e)WNW (Drag), (f) ENE (Syrian Arc), and (g) NE (Aualitic or Tibesti) trends.

3 Seismicity of Egypt

The seismicity of Egypt was also studied by many authors (e.g., Sieberg 1932a, b; Ismail 1960; Gergawi and El-Khashab 1968; Maamoun et al. 1984; Kebeasy 1990; Ambraseys et al.1994; Abou Elenean 1997, 2007; Badawy 1999, 2005). Although Egypt is an area of relatively low to moderate seismicity, it has experienced some damaging local shocks throughout its history, as well as the effects of larger earthquakes in the Hellenic Arc and Eastern Mediterranean area. It has also been affected by earthquakes in Southern Palestine and the Northern Red Sea (Ambraseys et al. 1994).

Some of the most significant earthquakes in the last decades that took place in Egypt are discussed in some detail in the following:

- *September 12, 1955 offshore Alexandria earthquake (M_S 6.8)*: This earthquake was felt in the entire Eastern Mediterranean Basin. In Egypt, it was felt strongly and led to the loss of 22 lives and damage in the Nile Delta, between Alexandria and Cairo. Destruction of more than 300 buildings of old brick construction was reported in Rosetta, Idku, Damanhour, Mohmoudya and Abu-Hommos. A maximum intensity (I_{max}) of VII was assigned to a limited area in the Bihira province, where five people died and 41 were injured. In addition, intensity between V and VII was reported in 15 other localities (Kebeasy 1990).

- *March 31, 1969 Shedwan earthquake (M_S 6.9)*: This event occurred to the northwest of Shedwan Island, in the surroundings of the Gulf of Suez. In this island, fissures and cracks in the soil were found. Ten kilometers west of the fractured area in the sea, a submerged coral reef was raised above sea level after the event. This earthquake was preceded by 35 large foreshocks during the last 15 days of March 1969. This event was also felt in Saudi Arabia (Kebeasy 1990; El-Sayed et al. 1994).
- *December 9, 1978 Gilf El-Kebeir earthquake (m_b 5.3)*: It is the largest instrumental earthquake in the southwestern region of Egypt. Its epicenter was located in the Gilf El-Kebeir area. The southern part of Egypt was an unpopulated desert and the intensity distribution of this earthquake was not estimated. The nearest station to the epicenter was Helwan, 850 km away. Neither foreshocks nor aftershocks were recorded for this earthquake (Riad and Hosney 1992).
- *November 14, 1981 Aswan (Kalabsha) earthquake (m_b 5.3, M_S 5.6)*: This earthquake occurred in the Nubian Desert of Aswan. It is of great significance because of its possible association with Nasser's Lake. Its effects was strongly felt to the north of its epicenter (Kalabsha, about 60 km southeast of Aswan) up to Assiut (440 km to the north) and to the south up to Khartoum (870 km to the south). The epicentral intensity was VII–VIII. Eleven buildings were damaged and surface faulting was reported in the epicentral area. Several cracks on the western bank of the Nasser's Lake, and several rock-falls and minor cracks on the eastern bank were reported. The largest crack was about 1 m in width and 20 km in length. This earthquake was preceded by three foreshocks and followed by a large number of aftershocks. This event caused a great alarm due to its proximity to the Aswan High Dam (Kebeasy 1990; El-Sayed et al. 1994).
- *July 2, 1984 Abu-Dabbab earthquake (m_b 5.1)*: It was felt strongly in Aswan, Qena and Quseir. Large numbers of foreshocks and a large sequence of aftershocks were recorded (Kebeasy 1990).
- *October 12, 1992 Cairo (Dahshour) earthquake (m_b 5.8, M_S 5.9)*: Its epicenter was about 40 km south of Cairo, in Dahshour, causing a disproportional amount of damage (estimated at more than 500 million of Egyptian Pounds) and the loss of many lives. The shock was strongly felt and caused occasional damage and life loss in the Nile Delta, around Zagazig. Damage extended throughout Fayoum and as far as south of Beni-Suef and Minia. The area mostly affected was Cairo, in particular the old suburbs of Cairo, Bulaq and the region to the south, along the western bank of the Nile to Gerza (Jirza) and El-Rouda. In all, 350 buildings collapsed completely and 9000 were irreparably damaged, 545 people died and 6512 people were injured. Many of the causalities in Cairo were victims of panic-stricken stampedes of pupils rushing out of schools. About 350 schools and 216 mosques were ruined, and about 50,000 people were left homeless. It was considered as one of the largest earthquakes in Egypt during the last century (El-Sayed et al. 1994; Abou Elenean et al. 2000).

- *November 22, 1995 Gulf of Aqaba earthquake* (M_W7.2): It was a strong earthquake located in the southern part of the Gulf of Aqaba, 350 km southeast of Cairo. Most of the reported damage was concentrated in Sinai Peninsula, where a certain number of hotels were damaged leading to loss of three lives and the injury of ten people (Al-Ahram newspaper, 23/11/1995) in Nuweiba, on the Gulf of Aqaba. Damage was also reported in the platforms of the port facilities in Nuweiba (Fat-Helbary 1999; Riad and Yousef 1999).
- *May 28, 1998 Ras El-Hikma earthquake (m_b 5.5)*: It was located offshore, on the northwestern part of Egypt, and it was widely felt in Northern Egypt. The I_{max} of VII was reported at Ras El-Hikma village (~300 km west of Alexandria) and intensity of V–VI at Alexandria. Ground fissures were observed along the beach, and some cracks also in the concrete buildings (Hassoup and Tealeb 2000).

In Egypt, mostly population settlements are concentrated along the Nile Valley and Nile Delta, so, the seismic risk is generally related to the occurrence of moderate-size earthquakes at near distances (e.g., 1992, M_S 5.9 Cairo earthquake), rather than bigger earthquakes that are known to occur at far distances along the Northern Red Sea, Gulf of Suez, and Gulf of Aqaba (e.g., 1969, M_S 6.9 Shedwan, and 1995, M_W 7.2 Gulf of Aqaba earthquakes), as well as the Mediterranean offshore (e.g., 1955, M_S 6.8 Alexandria earthquake) (Abou Elenean et al. 2010).

4 Seismic Record in Egypt

The assessment of seismic hazard and mitigation of seismic risk in Egypt are very important tasks due to the fast growth, and the development of giant and strategic projects (e.g., New Suez Canal). The preparation of basic data for the SHA starts immediately after evaluating the seismic record (complete and unified earthquake catalogue). In evaluating earthquake hazards for a given region, it is necessary to know its earthquake history. For Egypt, the seismic record is mainly divided into two main periods: (a) the historical or the pre-instrumental part (prior to 1900), and (b) the instrumental period (from 1900 until the moment). The last period can be subdivided at the same time into: (i) the early instrumental era (1900–1997), until the establishment of the Egyptian National Seismic Network (ENSN), and (ii) the modern instrumental era (post 1997 until the moment).

4.1 The Pre-instrumental Era (Pre-1900)

Instrumental record of earthquakes has been started by the beginning of the twentieth century. The earthquake information up to 1900 is compiled mainly from historical documents and books. Therefore, this kind of records is known as historical earthquake information. The location of those earthquakes provides a reasonable indicator for future events, and in order to forecast earthquake activity, it is necessary to determine the ancient history of faults. The completeness and accuracy of available information on earthquakes have evolved with time. Since the large-magnitude earthquakes with large recurrence periods are rare events, it is important to extend the seismic catalogues as far as possible back in time.

The chief sources of historical earthquake data in Egypt are inscriptions, papyri and archeological evidences provided by temples and monuments in the pre-Islamic Period (before 622 A.D.). Arabic chronicles and literature are also an important and rich source of macro-seismic data in the Islamic Period (from 622 A.D.). Other sources also include diplomatic correspondence, especially during the Ottoman Rule over Egypt, starting from 1517. Hence, it is obvious that, Egypt is rich in historical sources to furnish the mass of macro-seismic data, thanks to its prosperous culture, strategic location at the intersection of trade routes, in addition to its political importance since ancient times. Due to these reasons, Egypt is one of the few regions of the world which has long been known to have seismological records dating back to 2200 B.C.

The historical seismicity of Egypt was studied by many authors. Among the most earliest and most important of these studies are those given by Lyons (1907), Sieberg (1932a, b), Karnik (1968), Maamoun (1979), Poirier and Taher (1980), Maamoun et al. (1984), Kebeasy (1990), Ambraseys et al. (1994), El-Sayed and Wahlström (1996), Badawy (1999), Fat-Helbary (1999), Riad and Yousef (1999), Riad et al. (2004), and Badawy et al. (2010). According to them, many earthquakes were reported to have occurred in and around Egypt causing great damage in different localities.

The following is a brief discussion for some important historical earthquake databases mentioned above, showing their limitations:

- As-Souty, an Egyptian polymath, collected a catalogue of about 130 earthquakes in the Islamic World (from Spain to Transoxania) in his work "Kashf El-Salsala fi Wasf El-Zalzala", considered the first published chronology of historical earthquakes in the Middle East region. As-Souty frequently names his information sources, and cites them accurately. This work contains an earthquake list for the period between the years 712 A.D. to 1499 A.D.; which translated into English by Spenger in 1843 from Arabic manuscripts of the National Library at Paris. Thanks to the devotion of two of As-Souty's disciples, an invaluable continuation extends the list of events affecting Egypt down to 1588 (Badawy 1999).
- The first catalogue devoted to the earthquakes in Egypt is Lyons (1907) "preliminary list", which included 27 events between 27 B.C. and 1906, some of them are duplicated, generally giving its sources of information. Lyons catalogue has a large gap between 1303 and 1698, into which Sieberg (1932a, b) was able to insert only two events, both erroneous.
- Willis (1928) presented a list of earthquakes, which contains 130 shocks compiled on the authority of As-Souty. Willis's list was compiled in part by Sieberg (1932a, b) and others, but they did not bother to convert into A.D. the

Muslim Calendar which were given by As-Souty. Ambraseys (1962) correlated these calendar dates against another manuscript of the same work at British Museum, and suggested that this list is dated about six centuries too early.

- Rothé (1969) adopted Sieberg's work that is quoted again by Maamoun (1979) and Maamoun and Ibrahim (1978).
- Taher (1979) presents a full corpus of text and a summary from Arabic sources. Taher's work is the starting point for the retrieval and reassessment of historical information. Summary results of Taher's research were published by Poirier and Taher (1980), whose catalogue unfortunately passes on the inaccuracies in the original work.
- Ambraseys et al. (1994) made an attempt to include an accurate compilation of macroseismic information for a region defined at its greatest extent from 0° to 34°N and 10° to 60°E. Concerning Egypt, many of the events included in the previous catalogues have been excluded by Ambraseys et al. (1994) and have been proven to be false.

According to different authors (e.g., Sieberg 1932a, b; Karnik 1968; Maamoun 1979; Poirier and Taher 1980; Savage 1984; Maamoun et al. 1984; Ambraseys et al. 1994; Badawy 1999), about 83 historical earthquakes were reported to have occurred in the Egyptian territory, and have caused some damage in different localities. The uneven distribution of population in Egypt creates inaccuracy in the proper identification of the origin and effects of the Egyptian historical earthquakes. This yields some spurious events which have not seismic origin.

Badawy (1999) describes the time distribution of those earthquakes: (a) seven earthquakes have been reported in the period Before Christ (B.C.), (b) a number of three events were reported up to the end of the ninth century, (c) eight earthquakes have been reported in the tenth century, (d) a dramatic decline in earthquake number has been notified in the eleventh and twelfth centuries (Fatimid Period), (e) in the fifteenth and sixteenth centuries (Mamluk Period), the number of earthquakes re-increased and reaching ten events, (f) when Egypt was a province of the Ottoman Empire, in the seventeenth and eighteenth centuries, another dramatic decline was present, and finally, (g) the maximum number of the reported earthquakes was seventeen in the nineteenth century.

Furthermore, Badawy (1999) describes the spatial distribution of those earthquakes. Earthquake epicenters are located almost exclusively in Cairo, the Nile Valley and the Nile Delta. He referred that most of the earthquakes affected these areas originated from epicenters at the subduction zone in the north and rifting zone in the east, but the distribution of population in a narrow band along the Nile Valley and Delta creates challenging problems toward locating and assessing the origin and true effects of the historical earthquakes in Egypt.

In the present study, all the references mentioned previously are chronologically searched and all the available historical events were inserted and compiled into the new catalogue. This is to go back as far as possible into historical times, critically review and summarize the pre-instrumental seismic data in Egypt and its surroundings, and create a new unified version of the historical earthquake catalogue for Egypt.

4.2 The Instrumental Period (Post-1900)

The early instrumental period of 1900–1997 is still poorly understood, even for basic parameters such as earthquake locations. In some cases, this is the result of inherent limitations in the distribution, response characteristics, and timing of the instruments. Therefore, locations for most of the pre-1964 earthquakes are poorly determined.

Recording of instrumental earthquakes in Egypt started as early as 1899 with the establishment of Helwan Observatory (HLW: 29.85°N, 31.33°E, 115 m elevation on limestone bedrock). This site was selected for both geophysical and astronomical investigation. An E-W component of a Milne-Shaw seismograph was the only instrument used initially. Another N-S component of a Milne-Shaw and a vertical component of a Galitzin-Willip seismograph were added in 1922 and 1938, respectively, in the same location. In 1951, the first episode of modernization, started by adding another set of short-period Sprengnether seismographs in Helwan. Before 1960, the seismic monitoring was carried out by using an individual seismograph station consisting of a three-component sensor. The observations of each individual seismograph station were collected in a data analysis center for location and source parameters determination of every seismic event. The time lag between the recording and processing was so long (Badawy 2005).

In May 1962, the Helwan Seismological Station was chosen by the U.S. Coast and Geodetic Survey to be part of the World-Wide Standardized Seismograph Network (WWSSN). All systems were then replaced by the standardized set of Benioff short-period and Sprengnether long-period seismographs. This station is still on operation nowadays. In December 1972, a Japanese seismograph system with visual recording and frequency analyzer was installed also in Helwan. In late 1975 the first seismograph was installed in Southern Egypt that was able to record small local earthquakes. This seismograph was one of three stations installed at the Egyptian territory at Aswan (ASW), Abu-Simbel (ABS), and Matruh (MAT). These stations were equipped with Russian standard short-period SKM-3 seismometers and GK-VII M galvanometers (Abou Elenean et al. 2000).

Following the November 14, 1981, M_S 5.3 -m_b 5.1 Aswan earthquake, portable micro-earthquake recorders were installed around the northern part of Aswan reservoir area by the Egyptian Geological Survey (EGS) from December 1981 to June 1982 to study the possible induced seismicity in Nasser's Lake. In late June 1982, and after the occurrence of the Nasser's Lake earthquake on 14 November 1981, the portable seismic field stations were replaced by a telemetered network including eight stations. Those seismic stations were erected by the Helwan Observatory and Lamont-Doherty Geological Observatory (USA) around the northern part of Aswan Reservoir. Furthermore, it was enlarged to eleven stations in 1984 and to thirteen stations in 1985. Complete playback and analysis systems were installed at Helwan for analysis of the digital data. An analog strong-motion accelerograph network also was installed at different levels of the High Dam and Old Aswan Dam (Abou Elenean et al. 2000).

On July 2, 1984 a m_b 5.1 earthquake happened in Abu Dabbab area, along the Red Sea. It was the maximum recorded magnitude of the earthquake swarm observed since 1970. During the period from June 19, 1984 to January 4, 1985, the National Research Institute of Astronomy and Geophysics (NRIAG) installed a temporary three-station network (MEQ-800) including Abu Dabbab station, around Wadi Abu Dabbab, along the Red Sea coast, to study earthquake-swarms activity. Four short-period (MEQ equipped with SS-1 ranger) single vertical-component seismograph stations were added to the national network during 1986–1990 at Kottamia (KOT), Hurghada (HUR), Tell-El-Amarna (TAS), and Marsa Alam (MRS) (Badawy 2005).

In August 1991, a very broad-band station (KEG) was erected at Kottamia, as a part of MEDNET Project. In cooperation with the International Institute of Seismology and Earthquake Engineering (IISEE), from Japan, and the NRIAG, a local network including ten telemetered short period (L4C, Markproduct) seismic stations was installed in August 1994, around the southern part of the Gulf of Suez (Hussein et al. 2008).

After the occurrence of the October 12, 1992, M_W 5.8 earthquake in Dahshour area, 35 km to the southwest of Cairo, which caused 561 deaths, 9832 injured and left a damage of more than 35 million US$ (Abou Elenean et al. 2000), the Egyptian government financed the NRIAG to construct the ENSN, which covers the whole Egyptian territory to detect and record mostly of local and regional earthquakes, as well as teleseismic events. The institute upgraded the data communication system from telephone lines to satellite to increase the efficiency of the ENSN.

By the end of 2002, the installation of the ENSN was completed covering whole Egypt, and five sub-centers have been constructed and equipped. Moreover, a new Earthquake Disaster Reduction Data Center (EDRDC) was established and supported by Geographic Information System (GIS) technology. This network had to be a technologically-sophisticated system to meet the needs of public safety and emergency management, providing improved data for better quantification of hazard and risk associated with both natural and artificial seismic sources and related engineering applications, as well as basic research. The ENSN is a digital network with duplex communication that stands essentially on three major elements: a high-resolution digitizer (HRD series) providing a resolution of 24 bits (132-dB dynamic range) and Global Positioning System (GPS) timing, NAQS32-P acquisition and monitoring software, and monitoring of the required technical parameters of the remote and repeater stations (Abou Elenean et al. 2000).

The ENSN consists of 60 remote stations transmitting data to the main center at Helwan and to five subcenters at Burg El-Arab, El-Kharga, Marsa Alam, Hurghada and Aswan. All instruments are short-period velocity sensors (43 stations have one component and 13 stations have three-component sensors) with 24-bit digital recording systems and a sampling rate of 100 Hz. It also has four broadband seismograph stations equipped with STS-1 seismometers. Most of the data are received at the main center from remote stations and subcenters via satellite communication. Data from stations close to processing centers are sent through telemetry or telephone lines. The distribution of the seismic stations and strong-motion units was chosen to cover known seismicity sources, as well as cover some regions with known historical earthquakes but little instrumental seismicity (e.g., Siwa Oasis). With this dense network of instruments it is possible to record most ongoing earthquake activity in Egypt (Badawy 2005).

5 Earthquake Catalogue Compilation

In the current study, the collection, analysis and completeness of the historical and instrumental earthquake data, compiled from different local and international sources, is performed. The initial data suffered from incompleteness, duplication, and large epicentral and hypocentral location errors. Large efforts and time were required to evaluate different datasets, eliminating duplicate records after reconciling differences, and sorting out aftershocks.

As mentioned above, in developing this catalogue, the authors investigated and employed published and unpublished sources, covering different time periods with different magnitude scales, and several papers and reports on historical seismicity. The following are the different used sources arranged according to their preference:

- The regional catalogue given by Ambraseys et al. (1994), which cover the period from 1900 to 1992. Thier investigation is concerned with a large and irregular area defined at its greatest extent by the coordinates 0° to 34°N latitude and 10° to 60°E longitude. They attempt to provide a uniform account of the seismicity of the region mentioned above, based on the retrieval and assessment of original sources of macroseismic information.
- EHB (Engdahl et al. 1998) catalogue, which is a revised version of the International Seismological Centre (ISC) bulletin, containing data from 1960 to 2008 (http://www.isc.ac.uk/ehbbulletin/). The Engdahl et al. (1998) algorithm has been used to improve routine hypocenter determinations made by the International Seismological Summary (ISS), the ISC and the Preliminary Determination of Epicenters (PDE), before a new and location algorithm (Bondár and Storchak 2011) was introduced. The EHB algorithm does not recalculate magnitudes. Mostly body-wave (m_b) and surface-wave (M_S) values are taken from the ISC bulletin, and M_W values from the global Global Centroid Moment Tensor (CMT) catalogue.
- The ISC online bulletin (http://www.isc.ac.uk/), which includes earthquake data in digital format from 1964 until 2014. It is updated periodically. It has been the basic international seismic bulletin, completed and corrected with the other sources. It includes checked and unchecked data from national and local agencies from around the world. These data, which include hypocenters, phase arrival-times, focal mechanism solutions, etc., are automatically grouped into events, which form the basis of the ISC Bulletin.
- The National Earthquake Information Center (NEIC) global earthquake bulletin, also called the PDE bulletin. The PDE is an online bulletin covering the period

from 1900 to 2007 (http://earthquake.usgs.gov/earthquakes/). The word "Preliminary" was used for the final bulletin because the bulletin of the ISC is considered to be the final global archive of parametric earthquake data (phase pick times and amplitudes).

- Global CMT catalogue provided by Harvard University, which has been the main source for the focal mechanism solutions data (http://www.globalcmt.org/). Its main database runs from January, 1976, until the present moment (Dziewonski et al. 1981; Ekström et al. 2012). Furthermore, it includes the M_W computed according to Kanamori (1977).
- Another important source of the focal mechanism solutions data is the European-Mediterranean Regional Centroid Moment Tensor (RCMT) Catalogue (http://www.bo.ingv.it/RCMT/) for the European and Mediterranean area. The main product of RCMT is a routinely updated catalogue of seismic moment tensors (Pondrelli et al. 2002, 2004, 2006, 2007, 2011).
- All the available published and unpublished texts both local and international (Poirier and Taher 1980; Maamoun et al. 1984; Riad and Meyers 1985; Kebeasy 1990; Ambraseys et al. 1994; El-Sayed and Wahlström 1996; Badawy and Horváth 1999; Ambraseys 2001; Riad et al. 2004) were used to cover the pre-instrumental and the early-instrumental periods, in addition to provide macroseismic intensity for the historical events.
- The annual bulletin of the ENSN for events occurred after 1982 till the end of 2010, and the annual bulletin of the Aswan Regional Earthquake Research Centre (from 1982 until the end of 2012).

When merging different catalogues it is necessary to avoid the duplication of events eventually reported in more than one of the source catalogues. This can be achieved by carefully checking the possible double events (i.e., records which could be associated to the same earthquake) in the obtained catalogue (Primakov and Rotwain 2003). Accordingly, the merging procedure has been performed as follows. The possible common events, with origin time differences less than one minute and location differences less than one degree for latitude and longitude, have been first identified. All the records satisfying such conditions have been examined manually, to analyse specific cases. If the same event was listed with different coordinates and origin time, the parameters estimated from local records (i.e., national and regional catalogues) have been used. Otherwise, the parameters from the global catalogues have been considered. The depth wasn't taken into consideration, due to the large errors affecting this quantity (Hussein et al. 2008).

Moreover, the straightforward merging of the data sources mentioned above would yield a heterogeneous earthquake catalogue, with different magnitude types, not always comparable. For Egypt and its surroundings, the most frequently reported size estimates are the magnitude types given by the local catalogues, that is, local (M_L) and duration (m_D) magnitudes, which come from the annual bulletins of the ENSN and Aswan Regional Earthquake Research Centre. In addition to these magnitudes, m_b and M_S, as reported in the United States Geological Survey (USGS) and ISC global catalogues, are also listed for moderate to large events. Furthermore, M_W values for the biggest events, from CMT and EHB bulletins, were obtained.

In order to unify magnitudes, the different magnitude scales were selected according to the following preference: M_W, M_S, m_b and local magnitudes (M_L and m_D), respectively. The initial compilation spanning a spatial region from 21° to 38° N and from 22° to 38°E, and includes all the events having an assigned magnitude of 3.0 and above for international sources, and any magnitude for local sources, on any magnitude scale. The initially compiled catalogue comprised more than 64,000 earthquakes (historical and instrumental events), covering the time period from 2200 B.C. to the beginning of 2014.

6 Catalogue Analysis

In order to produce an earthquake catalogue with a unified magnitude scale, it is necessary to follow and perform the following two important steps, upon the de-clustered earthquake data (Peresan and Rotwain 1998): (a) study the relationships between the different kinds of magnitude reported in the catalogue, in order to have a formal rule for the choice of any relationship that can be applied for magnitude conversion, resulting into an acceptable unity of the catalogue, and (b) study the completeness of the catalogue to find the magnitude thresholds above which the different data sets, as well as of the resulting catalogue over the investigated time period 2200 B.C.–2013, are complete. The completeness analysis can be checked using the following: (i) drawing plots of the cumulative number of earthquakes against time, and (ii) drawing plots of the frequency of earthquakes versus magnitude, in accordance with the magnitude-frequency relationship by Gutenberg and Richter (1942):

$$log_{10}N(m) = a - bm \tag{1}$$

6.1 Converting Intensities and Reported Magnitudes to Moment Magnitude

One of the main goals of this work was to obtain a unified earthquake catalogue. The M_W was used as the unifying magnitude scale, because it is the most commonly used in recent seismic hazard studies. Also, the reason is that this type of magnitude do not saturates when increase earthquake size or seismic moment. Several empirical regression relationships between the different reported magnitudes, I_{max} and M_W have been employed. These relationships are those considered after establishing specific relationships from our database or to use the most reliable ones after studying the available magnitude relationships in the scientific literature. In the final catalogue, in addition to the unified M_W, the initially reported magnitude has been included. This will allow users to use other type of magnitude to unify the catalogue or to use other relationships to calculate unified magnitude if they wish.

The equivalent moment magnitude ($M_W{}^*$), i.e., the final unified M_W, was computed for each set of reported magnitude data. For earthquakes with reported M_W, this value was the equivalent magnitude. Events which have reported M_S or m_b magnitudes, a conversion of the reported magnitude to M_W is performed using empirical relationships (Eqs. 2 and 3, respectively) developed directly from the current catalogue. The first relationship is a second degree polynomial fit between reported M_S and M_W magnitudes, using to it 355 events ($4.0 \leq M_S \leq 7.3$). The second one is a linear fit between m_b and M_W, using 816 events ($3.5 \leq m_b \leq 6.7$). Both of them show a general agreement with widely used relationships as developed by Johnston (1996a) (Figs. 2 and 3).

$$M_W^* = 3.97(\pm 0.61) - 0.13(\pm 0.24)M_S + 0.080(\pm 0.023)M_S^2 \quad (2)$$

$$M_W^* = -1.314(\pm 0.097) + 1.262(\pm 0.020)m_b \quad (3)$$

For events in the catalogue with reported m_D or M_L magnitudes (Fig. 4), the relationships of Hussein et al. (2008) for the local magnitude scales were used here to convert m_D and M_L values to M_W (Eqs. 4 and 5, respectively). The authors prefer to use the Hussein et al. (2008) relationships rather than developing their own

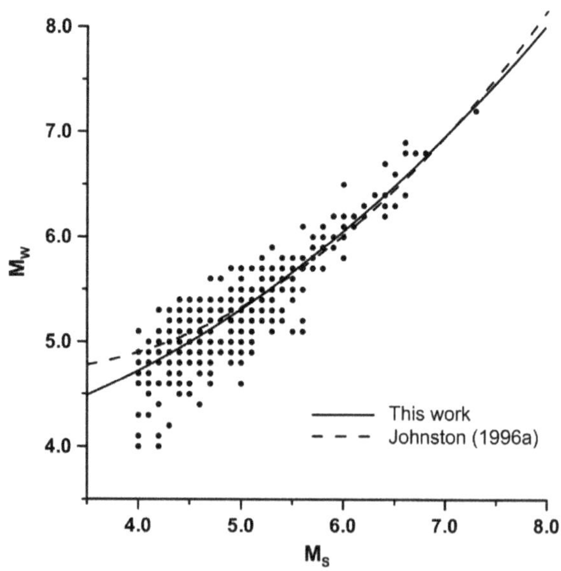

Fig. 2 Relationship between M_S and M_W magnitudes

Fig. 3 Relationship between m_b and M_W magnitudes

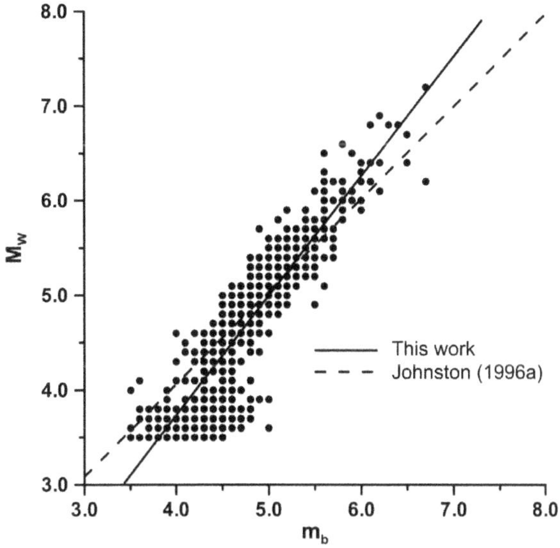

Fig. 4 Relationship between m_D and M_L, and M_W magnitudes

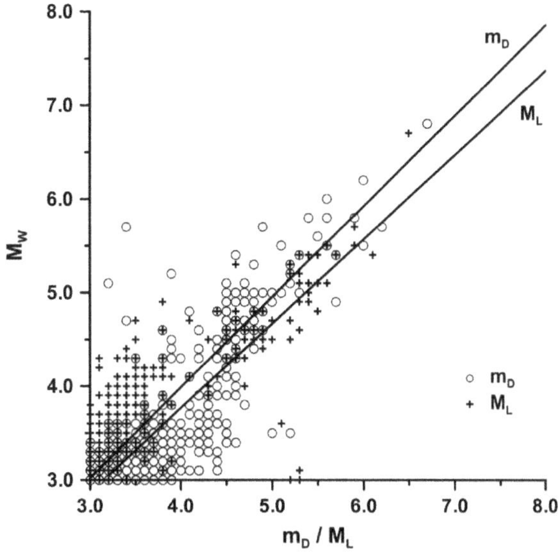

relations, because the current data have some gaps in the magnitudes distribution that make an obstacle to construct a good fit.

$$M_W^* = \frac{2}{3}[1.35(\pm 0.11)M_L + 16.3(\pm 0.53)] - 10.7 \tag{4}$$

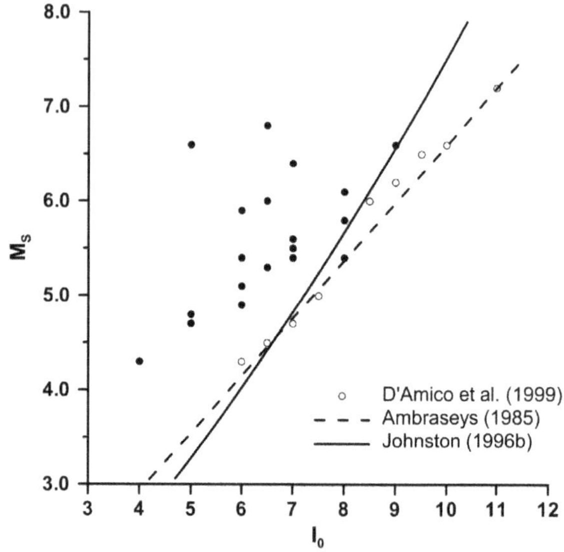

Fig. 5 Relationship between I_{max} and M_S

$$M_W^* = \frac{2}{3}[1.45(\pm 0.07)M_D + 16.3(\pm 0.30)] - 10.7 \qquad (5)$$

Finally, when I_{max} was the reported earthquake size, $M_W{}^*$ was computed from the empirical polynomial second-degree relationship between I_{max} and seismic moment developed by Johnston (1996b) (Eq. 6 and Fig. 5).

$$M_W^* = \frac{2}{3}\left(19.36 + 0.48 \cdot I_{\max} + 0.0244 \cdot I_{\max}^2\right) - 10.7 \qquad \sigma = 0.77 \qquad (6)$$

In the compiled database, there are few earthquakes with assigned I_{max} and reported magnitude. Mostly large earthquakes having these two values have a M_S macroseismic magnitude computed by Ambraseys et al. (1994), that do not allows to establish a robust regression analysis (Fig. 5). Furthermore, the relationships developed by Ambraseys (1985) and D'Amico et al. (1999) appear to be unsuitable for the current study because of its behaviour at high intensity values.

In the final catalogue, a code is employed to inform users about the method used to obtain the $M_W{}^*$ for each event.

6.2 Catalogue Declustering

Declustering process attempts to identify time and space-dependent earthquakes (aftershocks, foreshocks and swarm-type activity). After compiling the unified catalogue, dependent (non-Poissonian) earthquakes were removed. This is a necessary step in any time-independent SHA. For most hazard-related studies, it is assumed that the seismicity behaves in a time-independent fashion (e.g., Reiter 1990; Giardini et al. 1999; Frankel 1995).

Two main declustering algorithms are normally in use: (a) the approach proposed by Gardner and Knopoff (1974), which identify dependant events when they are included in the same space and time window, and (b) the approach provided by Reasenberg (1985), who defines the interaction windows in space-time in a somewhat more sophisticated way attempting to introduce physical properties behind triggering. The spatial and temporal extent of a cluster is not fixed, as it is in the windowing method, but it depends on the development of an individual sequence.

In the current work, the authors follow the same procedure as the Moroccan and Algerian main earthquake catalogues prepared by Peláez et al. (2007) and Hamadche et al. (2010), respectively. All dependent earthquakes were identified using the classic routine, and essentially using the same parameters, proposed by Gardner and Knopoff (1974). Given an earthquake with a certain M_W, a scan within a characteristic distance L(M_W) and time T(M_W) was performed for the entire catalogue. The largest earthquake in this search is considered to be the mainshock. In the current study, window sizes of 900 days and 100 km were used for a given M_W 8.0 event, and 10 days and 20 km for a M_W 3.0 event. For in-between magnitudes, proportional values for L and T were used. After this process, the catalogue was cut off below magnitude M_W 3.0. These magnitudes are not significant for SHA studies and its period of completeness is very low. For example, Table 1 shows earthquakes above M_W 6.0 in the Egyptian territory and surroundings, that is, the most energetic ones in the final catalogue. Furthermore, maps showing the distribution of earthquakes of the declustered catalogue, for events $M_W \geq 3.0$, are depicted in Figs. 6 and 7.

6.3 Catalogue Completeness

Modelling the seismicity in each seismic zone needs knowledge on the magnitude of completeness below which only a fraction of all events that have been taken place are documented (e.g., Kijko and Graham 1999; Rydelek and Sacks 2003; Wiemer and Wyss 2000, 2003). The authors estimated the degree of completeness firstly for the whole set of date in the entire catalogue,and then for the Egyptian territory and surroundings (area between latitudes 22° to 33.5°N and longitudes 22° to 36°E), without data from the Mediterranean Sea region. The procedure used to identify the completeness levels of the catalogue is the usual one, to plot the cumulative number of events above a certain magnitude versus time. This permits to identify the epochs in which the rates of events are constant.

Table 1 Catalogued earthquakes with magnitude equal to or above M_W 6.0

Date mm/dd/yyyy	Time hh:mm:ss	Longitude (°)	Latitude (°)	Depth (km)	Reported magnitude	I_{max}	Location	Final M_W	Reference
590 B.C.	–	35.200	33.300	–	–	IX	Sidon District, Syria	6.4[b]	Maamoun et al. (1984)
12 B.C.	–	35.000	32.000	–	–	IX	North Jerusalem, Palestine	6.4[b]	Maamoun et al. (1984)
–/–/0019	–	35.500	33.000	–	–	X	Southern Lebanon	7.0[b]	Riad et al. (2004)
–/–/0030	–	35.200	31.800	–	–	VIII–X	Jerusalem, Palestine	6.1[b]	Maamoun et al. (1984)
–/–/0332	–	34.000	33.500	–	–	IX	Coastal Lebanon	6.4[b]	Riad et al. (2004)
–/–/0419	–	35.500	33.000	–	–	IX	Southern Lebanon	6.4[b]	Riad et al. (2004)
–/–/0746	–	35.600	32.000	–	–	XI	Balqa, Jordan	7.7[b]	Riad et al. (2004)
–/–/0854	–	35.320	32.480	–	–	IX	Jenin, Palestine	6.4[b]	Badawy and Horváth (1999)
–/–/0857	–	31.000	28.000	–	M_S 6.1[a]	VIII	SE Minia, Egypt	6.1[c]	Ambraseys et al. (1994)
10/04/0935	–	31.200	30.500	–	M_S 6.1[a]	VIII	Banha, Nile Delta, Egypt	6.1[c]	Ambraseys et al. (1994)
09/15/0951	18:00:00	29.550	31.130	–	M_S 6.1[a]	VIII	Near Alexandria, Egypt	6.1[c]	Ambraseys et al. (1994)
01/04/1034	–	35.320	32.480	–	–	X–XI	Jenin, Palestine	7.4[b]	Badawy and Horváth (1999)

(continued)

Table 1 (continued)

Date mm/dd/yyyy	Time hh:mm:ss	Longitude (°)	Latitude (°)	Depth (km)	Reported magnitude	I_{max}	Location	Final M_W	Reference
08/31/1111	–	31.000	31.000	–	M_S 6.1[a]	VIII	Gharbia, Nile Delta, Egypt	6.1[c]	Badawy and Horváth (1999)
05/02/1212	–	34.570	29.330	–	–	VIII–IX	Eastern Sinai. Egypt	6.1[b]	Badawy and Horváth (1999)
–/–/1262	–	31.150	30.030	–	–	IX–X	Northern Cairo, Egypt	6.7[b]	Badawy and Horváth (1999)
02/20/1264	–	31.000	29.000	–	M_S 6.1[a]	VIII	Beni Suef, Egypt	6.1[c]	Ambraseys et al. (1994)
–/–/1269	–	35.000	32.000	–	–	IX	NW Jerusalem, Palestine	6.4[b]	Riad et al. (2004)
–/–/1287	–	35.270	32.570	–	–	VIII–IX	Northern Jerusalem, Palestine	6.1[b]	Badawy and Horváth (1999)
07/30/1303	–	31.150	30.030	–	–	IX	Cairo, Egypt	6.4[b]	Badawy and Horváth (1999)
05/–/1341	–	29.550	31.130	–	–	VIII–IX	Near Alexandria, Egypt	6.1[b]	Badawy and Horváth (1999)
01/14/1546	16:00:00	35.100	32.000	–	–	X	Northern Jerusalem, Palestine	7.0[b]	Ambraseys et al. (1994)
04/09/1588	–	31.550	30.030	–	–	IX	Near Cairo, Egypt	6.4[b]	Badawy and Horváth (1999)
10/–/1754	–	32.250	29.600	–	M_S 6.6[a]	V	SW Suez, Egypt	6.6[c]	Ambraseys et al. (1994)

(continued)

Table 1 (continued)

Date mm/dd/yyyy	Time hh:mm:ss	Longitude (°)	Latitude (°)	Depth (km)	Reported magnitude	I_{max}	Location	Final M_W	Reference
07/11/1879	18:00:00	33.000	29.000	–	M_S 5.9[a]	VI	Gulf of Suez, Egypt	6.0[c]	Ambraseys et al. (1994)
03/06/1900	17:58:00	33.000	29.000	–	M_S 6.2[a]	–	Gulf of Suez, Egypt	6.2[c]	Ambraseys et al. (1994)
07/11/1927	13:04:00	35.300	32.200	15	M_S 6.1, m_b 5.4	–	Northern Jerusalem, Palestine	6.1[c]	Ambraseys (2001)
09/12/1955	06:09:24	29.610	32.200	20	M_S 6.4, m_b 6.5	VII	Mediterranean Sea, NW Alexandria	6.4[c]	Ambraseys (2001)
03/31/1969	07:15:51	33.938	27.513	6	M_W 6.8, M_S 6.6	IX	Shedwan Island, Red Sea	6.8	ISC/EHB
08/03/1993	12:43:08	34.548	28.708	18	M_W 6.1, M_S 5.8	–	Eastern Sinai, Egypt	6.1	ISC/EHB
11/22/1995	04:15:15	34.809	28.769	19	M_W 7.2, M_S 7.3	–	Gulf of Aqaba, Egypt	7.2	ISC/EHB

[a]Ambraseys et al. (1994), using macroseismic data
[b]From I_{max}, using the Johnston (1996b) relationship between I_{max} and M_W
[c]From M_S, using the relationship established in this work

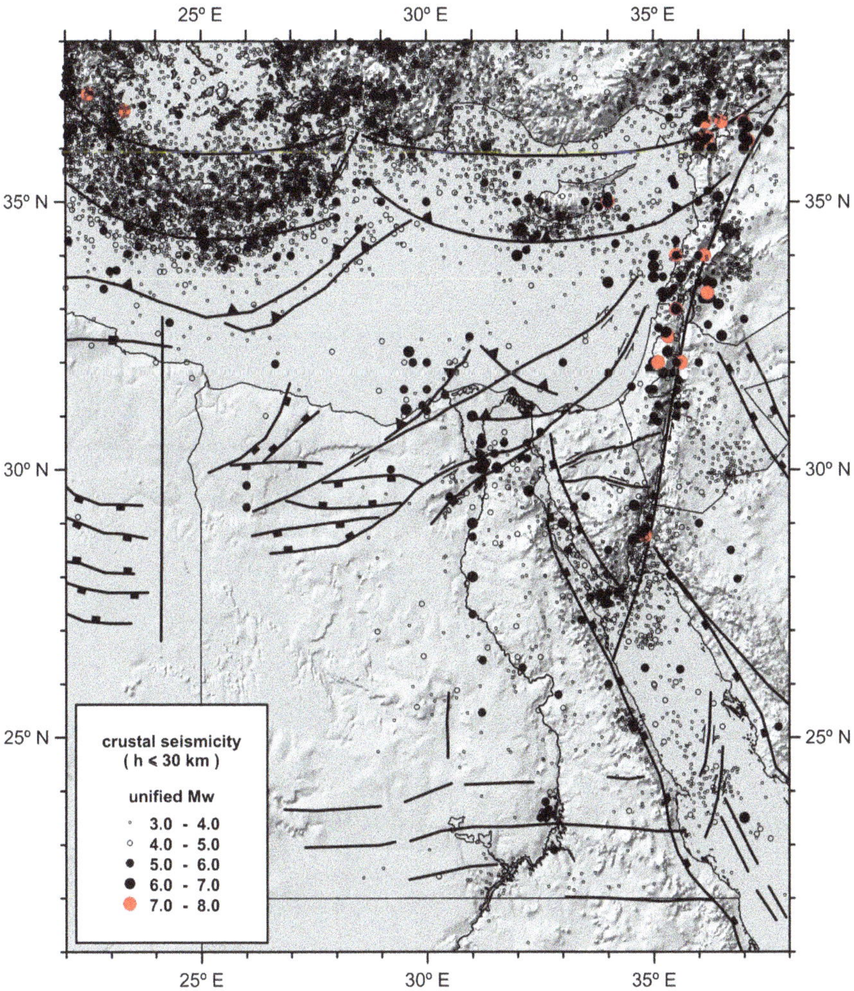

Fig. 6 Catalogued crustal ($h \leq 30$ km) earthquakes

The results of the completeness analysis, for both the whole catalogue and the Egyptian territory and its surroundings are shown on Table 2, and Figs. 8 and 9. Those figures depict the cumulative number of earthquakes above magnitudes M_W 3.0 to M_W 7.0 at intervals of 0.5. For example, we obtained that the earthquake catalogue of the Egyptian territory and surroundings (Fig. 9) is complete for earthquakes above M_W 3.0, 4.0, and 5.0 since 1993, 1983 and 1950, approximately, with seismicity rates of 67.0, 5.74, and 0.646 events/year, respectively. However, the whole catalogue appears to be complete, for the previous magnitude values, since 2003, 1993 and 1980, approximately, with rates of 1071, 69.0, and 9.76 events/year, respectively.

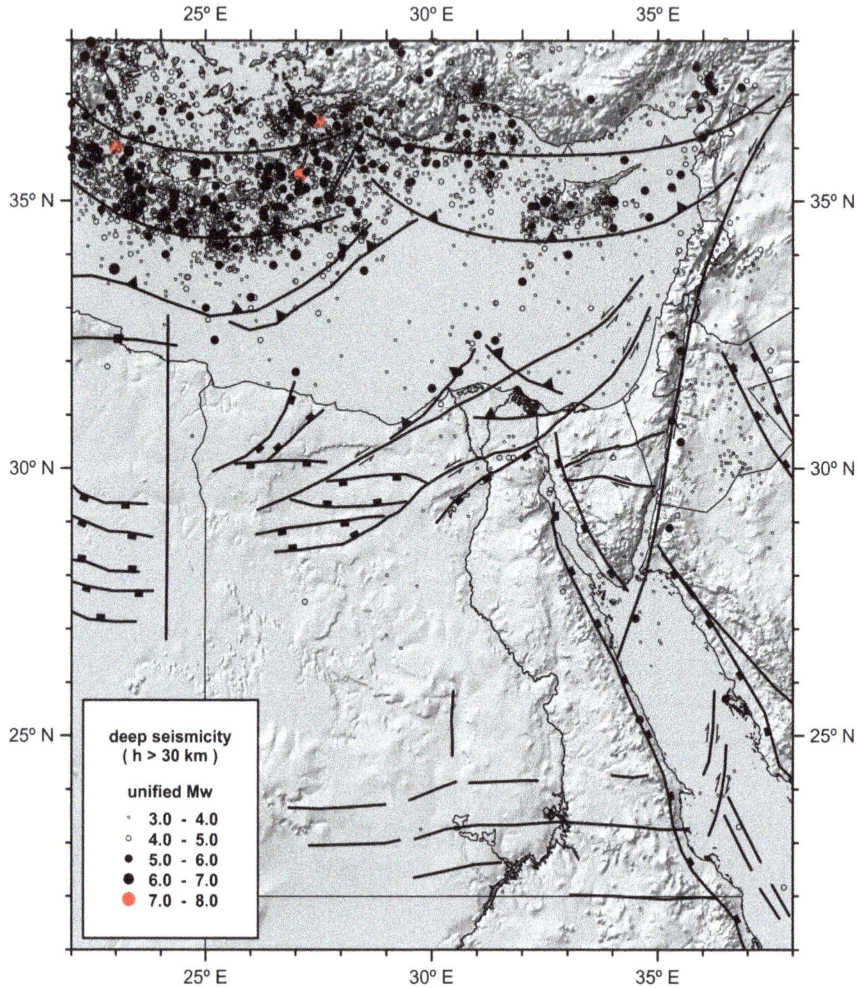

Fig. 7 Catalogued deep (h > 30 km) earthquakes

Some of the observed completeness periods are clearly related to the establishment and improving of the local and international networks: (a) 1900 is the appropriate date in which is established the Helwan Observatory, (b) 1960 coincides approximately with the deployment of the WWSSN, (c) 1983 is the year related to the installation of the Aswan Seismological Network after November 14, 1981, M_S 5.3 Kalabsha earthquake, (d) 1993 coincides with the installation of a large number of seismic stations in Egypt after October 12, 1992, M_W 5.8 Cairo earthquake, and finally, (e) 2003 reflects the final improvement and development of the ENSN.

Table 2 Completeness period and seismicity rate for different magnitude values

M_W	The whole catalogue		The Egyptian territory	
	Year	Rate (events/year)	Year	Rate (events/year)
≥3.0	2003	1071	1993	67.0
≥3.5	1993	206	1983	16.9
≥4.0	1993	69.0	1983	5.74
≥4.5	1980	30.4	1977	1.86
≥5.0	1980	9.76	1950	0.646
≥5.5	1960	2.46	1950	0.187
≥6.0	1920	0.591	700	0.0166
≥6.5	1900	0.186	700	0.00446
≥7.0	200 B.C.	0.00911	0	0.00207

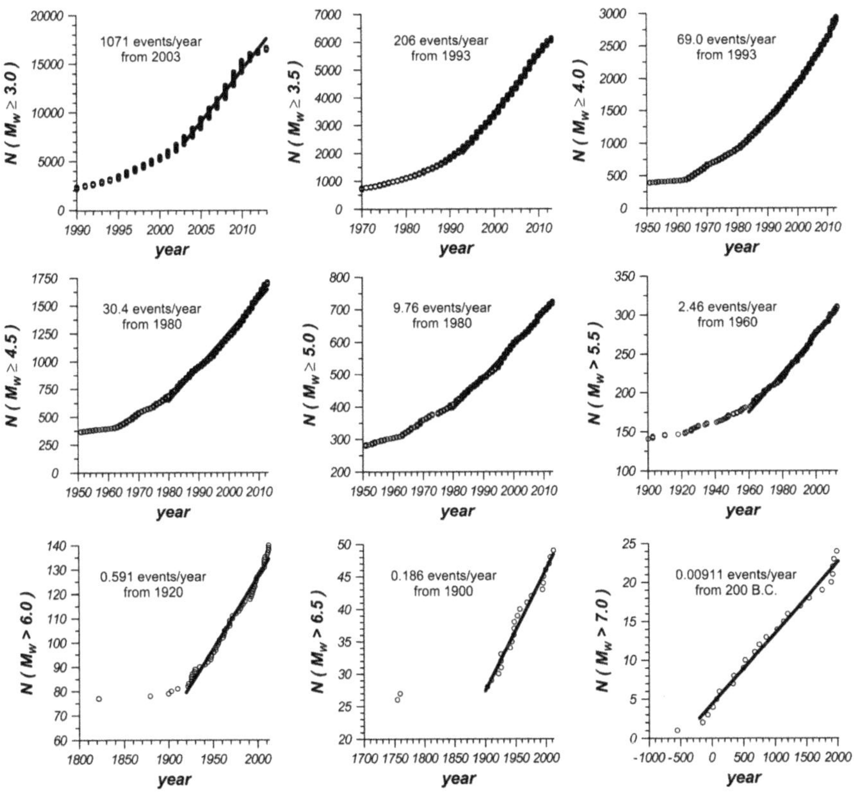

Fig. 8 Cumulative number of earthquakes above different magnitude values for the whole catalogue

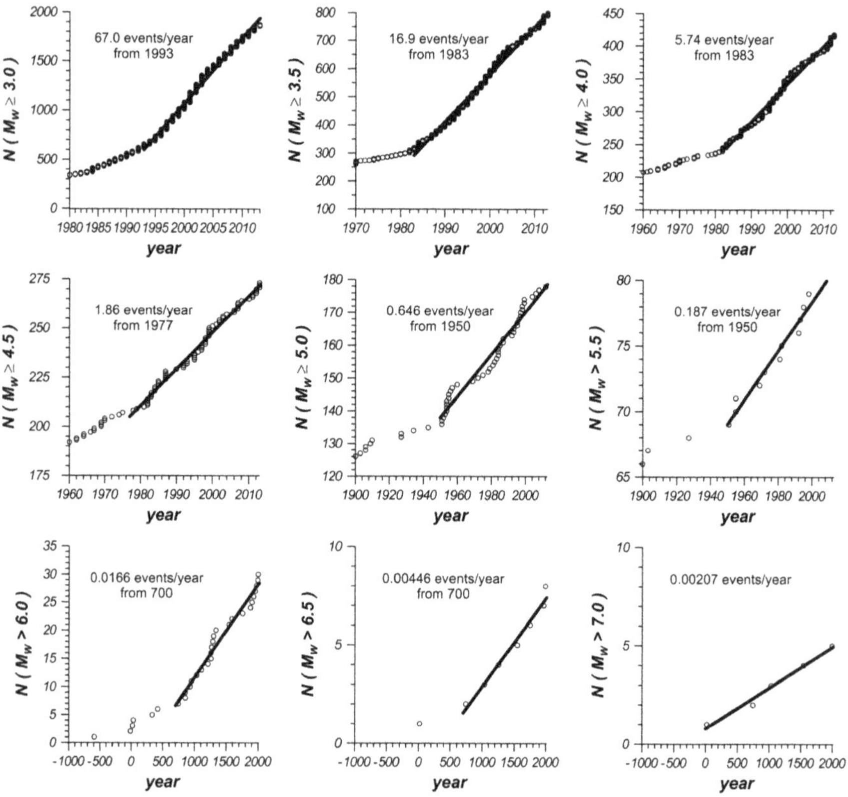

Fig. 9 Cumulative number of earthquakes above different magnitude values for the Egyptian territory and its surroundings

7 Magnitude-Frequency Relationship

As it was mentioned before, Gutenberg and Richter (1942) noted that the magnitude-frequency relationship obeys a power law given by Eq. 1, where N is the number of earthquakes of magnitude M or larger, and a and b are parameters. The *a-value* depends on the period of observation, the size of the investigated area, and the level of seismicity. The *b-value* is generally assumed to be related to the degree of the fracturing and the heterogeneity of the materials and stress regime, among other factors, depicting the relationship among large and small earthquakes. A *b-value* equals to 1.0 means that the number of earthquakes in the area decreases by tenfold when magnitude increases in a unit.

Figure 10 displays the recurrence (magnitude-frequency) relationship for earthquakes in both the whole catalogue and the Egyptian territory, in the time period likely complete for magnitudes above M_W 3.5, from 1993 and 1983, respectively. There is a good fit in both plots, with a typical *b-value* equal to 0.94

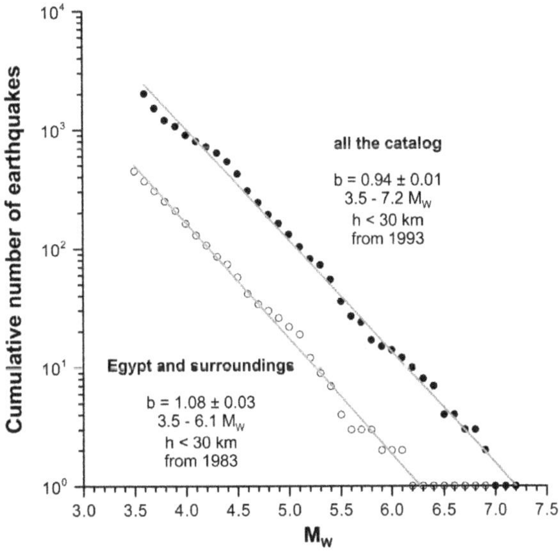

Fig. 10 Cumulative number of earthquakes versus magnitude for the entire catalogue for Egypt and its surroundings

and 1.08, respectively. These values emphasize the fact that, for those time intervals, the catalogue is likely complete and Poissonian, for the quoted magnitudes. In the recurrence relationship for Egypt and its surroundings (Fig. 10), the November 22, 1995, M_W 7.2 Gulf of Aqaba earthquake do not follow the magnitude-frequency relationship. It must be investigated if it could be considered a characteristic earthquake (Youngs and Coppersmith 1985).

8 Focal Mechanism Solutions Database (1940–2013)

Earthquake source mechanisms or fault-plane solutions are of prime importance in monitoring local, regional and global seismicity and seismotectonics. They have proven to be of great importance in defining the nature of earthquake faulting and its causative stresses in many regions of the world. They reflect the stress pattern acting in the area under study and may help to map its tectonic structure, which causes the earthquakes. Instrumental recordings provide an ever-expanding source of data for understanding earthquakes and their locations, source properties, and radiated seismic waves. Focal mechanism solutions help us to identify the fault plane (with the aid of geological information and/or aftershocks distribution), the type of faulting, the direction of slip and the compatibility of our solution with the general sense of ground motion in the area. Furthermore, the use of earthquake focal mechanisms or seismic stress tensors for analysis of seismotectonic deformation is a fundamental goal of the modern geodynamics. As such, it has been widely used to evaluate, in a more or less independent manner, the nature of recent crustal deformations on scales ranging from global to regional and local scale.

The insufficient coverage of seismic stations until the 1980's limited the number of fault-plane solutions that have been computed in the Egyptian territory (Badawy 2005). In the current study, different local and international sources were examined, and focal mechanism solutions data were compiled into a single database. Those sources are the following: Constantinescu et al. (1966), Huang and Solomon (1987), Hussein (1989), Eck and Hofstetter (1990), Riad and Hosney (1992), Abou

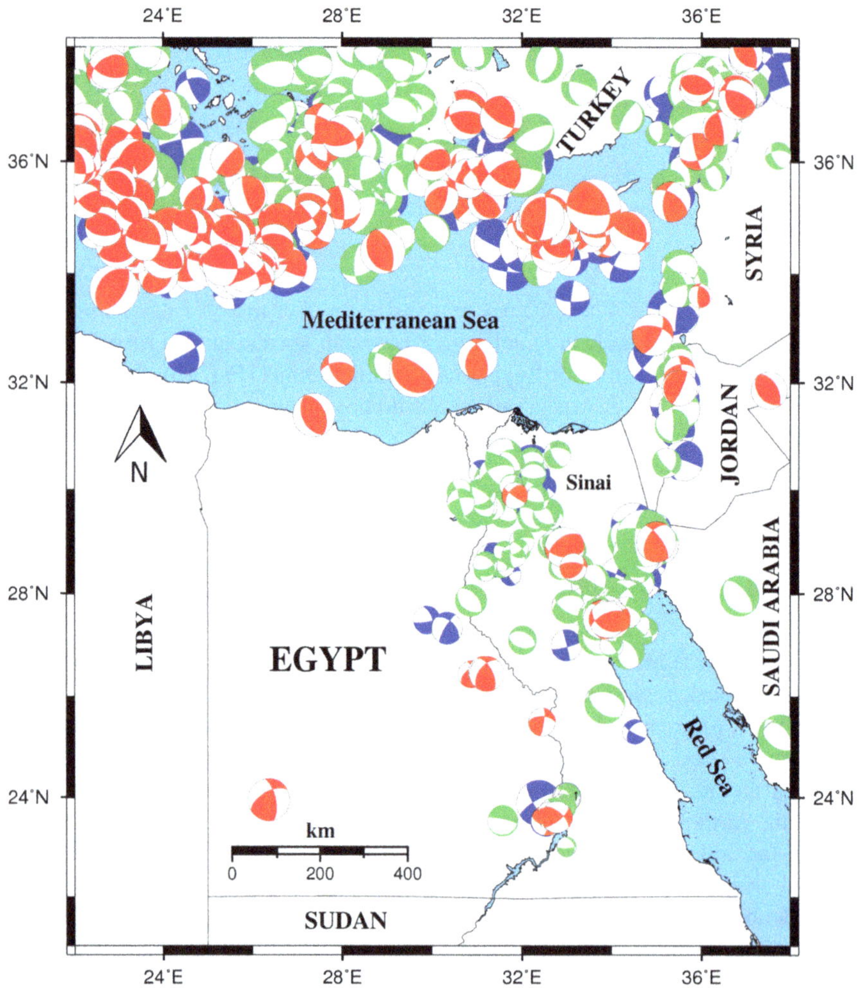

Fig. 11 Distribution of the catalogued focal mechanism solutions in and around Egypt for earthquake events included in the current database. *Sphere sizes* are in proportion to the M_W. *Different colours* refer to different faulting mechanisms (*Blue* strike slip. *Green* normal. *Red* reverse)

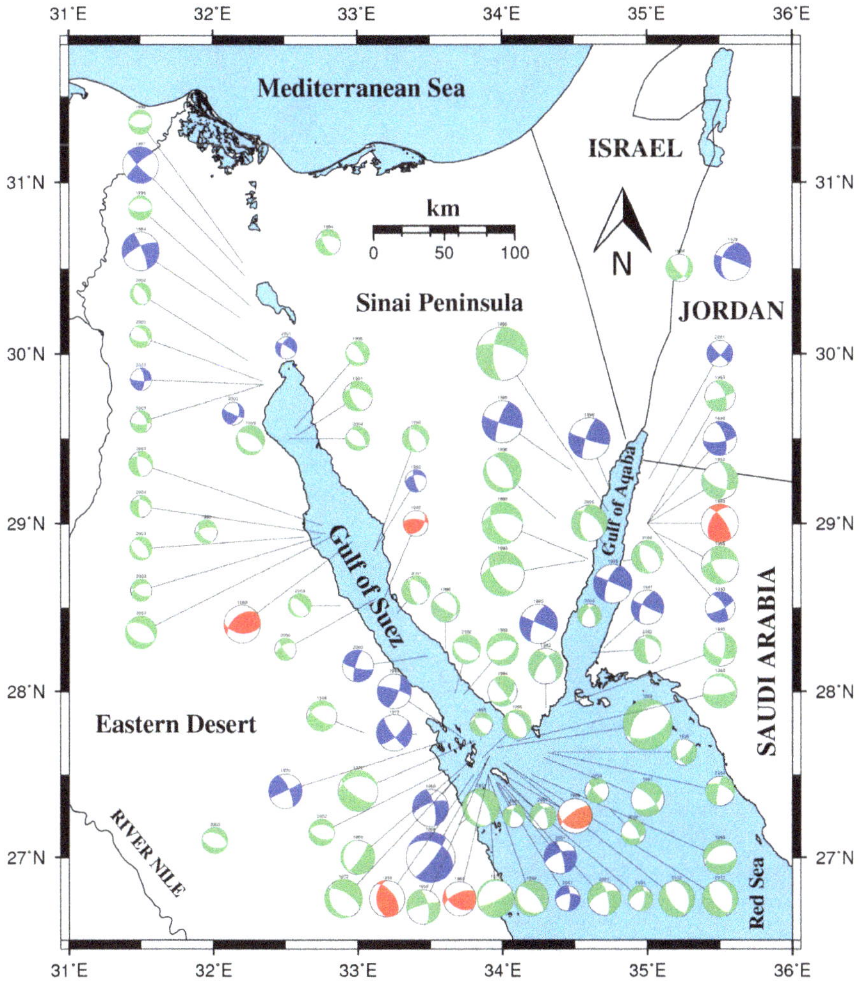

Fig. 12 Distribution of the catalogued focal mechanism solutions in Sinai Peninsula. *Sphere sizes* are in proportion to the M_W. *Different colours* refer to different faulting mechanisms

Elenean (1993, 1997, 2007), Abdel Fattah et al. (1997), El-Sayed et al. (2001), Badawy (2001, 2005), Badawy and Abdel Fattah (2001, 2002), Salamon et al. (2003), Hofstetter (2003), Hofstetter et al. (2003), Fat-Helbary and Mohamed (2003), Abou Elenean et al. (2004), El-Sayed et al. (2004), Korrat et al. (2005), Badawy et al. (2006, 2008), Hussein et al. (2006, 2013), Marzouk (2007), Abdel-Rahman et al. (2009), Abou Elenean and Hussein (2007, 2008), ENSN (1998–2010), Morsy et al. (2011, 2012), Abu El-Nader (2013).

In addition, the solutions of the Global CMT Harvard catalogue, the International Seismological Centre (2011), the NEIC, the RCMT Catalogue in the Mediterranean region, as well as ZUR-RMT catalogue of the Institute of

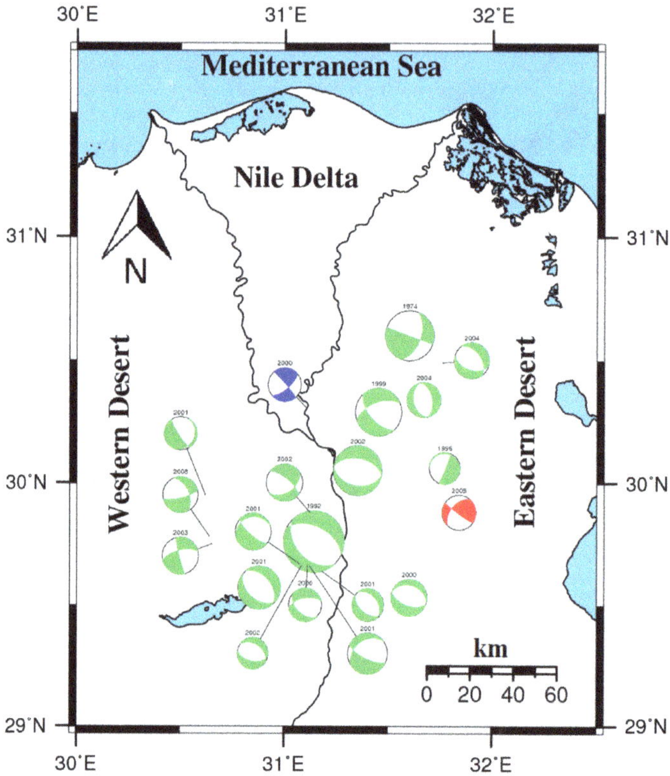

Fig. 13 Distribution of the catalogued focal mechanism solutions around Cairo area. *Sphere sizes* are in proportion to the M_W. *Different colours* refer to different faulting mechanisms

Technology (ETH) of Zurich, were also included in the final catalogue. Hence, a total number of 688 focal mechanism solutions were collected covering different active seismogenic zones (Figs. 11, 12, 13 and 14) in Egypt and surroundings, including Eastern Mediterranean, spanning the spatial area from 21° to 38°N, and from 22° to 38°E. Most of them have a magnitude greater than or equal to M_W 3.0, occurring in the time period 1940–2013. Table 3 shows the focal mechanism parameters for some selected biggest events which taken place in and around Egypt, and their related M_W values.

9 Discussion

The seismicity distribution of the compiled catalogue is depicted in Figs. 6 and 7. Most of the crustal seismicity (Fig. 6) is concentrated and released within specific seismogenic belts that are mainly related to the different plate boundaries

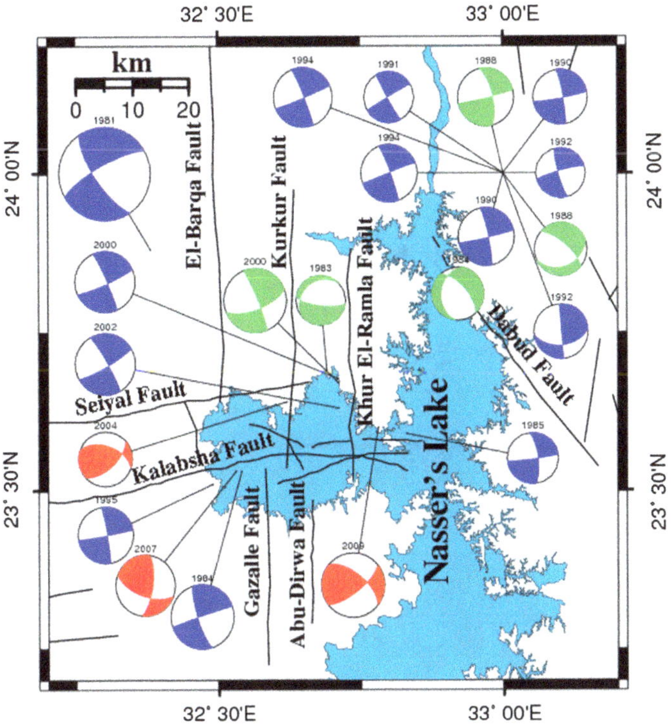

Fig. 14 Distribution of the catalogued focal mechanism solutions in Aswan region. *Sphere sizes* are in proportion to the M_W. *Different colours* refer to different faulting mechanisms

surrounding Egypt. These seismic belts are: the DST, the Cyprian and the Hellenic Arcs, and the Gulf of Suez-Red Sea Rift. Thus, Egypt is characterized by a unique tectonic situation including the convergence of the African and Eurasian Plates across a wide zone of deformation in the Eastern Mediterranean to the north, strike-slip movement along the DST to the east and the Gulf of Suez-Red Sea Rift. In addition, there is a moderate seismic activity inside the Egyptian territory and away from the surrounding plate boundaries. This seismicity is mainly due to different fault systems: (a) around Nasser's Lake, (b) Red Sea Coast, especially in Abu Dabbab area, (c) Cairo-Suez region, and (d) Mediterranean Sea Coastal zone.

However, there is a significant difference between the distribution of the deep (h > 30 km) seismicity (Fig. 7) and the crustal one (Fig. 6). Deeper earthquake events are concentrated along the Hellenic and the Cyprian Arcs, in the Mediterranean region. This is due to the convergence and subduction which take place between the African and the European Plates. In addition, a few deep earthquakes are also observed along the DST, and in the triple junction point, at the northern part of the Red Sea, which represents the intersection between the Gulf of Suez and Gulf of Aqaba. Furthermore, there is a few activity along the Pelusium

Table 3 Focal mechanism solution parameters for the most energetic earthquakes ($M_W \geq 5.5$) that have taken place in Egypt and its surroundings

Date mm/dd/yyyy	Time hh:mm:ss	Longitude (°)	Latitude (°)	Depth (km)	M_W	Strike (°)	Dip (°)	Rake (°)	Reference
01/30/1951	23:07:24	33.400	32.400	10.0	5.7	295	65	−116	Constantinescu et al. (1966)
09/12/1955	06:09:24	29.600	32.200	33.0	6.4	125	25	090	Salamon et al. (2003)
03/16/1956	19:32:00	35.300	33.300	33.0	5.9	280	90	−15	Salamon et al. (2003)
03/31/1969	07:15:51	33.960	27.660	06.2	6.8	220	45	−110	Salamon et al. (2003)
06/28/1972	09:49:35	33.800	27.700	06.1	5.5	260	30	−130	Huang and Solomon (1987)
11/14/1981	09:05:35	32.380	23.880	15.0	5.8	146	72	−015	CMT catalogue
09/17/1982	22:22:30	22.900	33.700	23.4	6.3	219	34	093	CMT catalogue
10/12/1992	13:09:55	31.140	29.760	22.0	5.8	136	42	−075	CMT catalogue
08/03/1993	12:43:06	34.570	28.780	15.0	6.1	145	57	−043	AbdelFattah et al. (1997)
08/03/1993	16:33:24	34.590	28.790	15.0	5.7	149	55	−047	AbdelFattah et al. (1997)
11/22/1995	04:15:26	34.730	29.070	18.4	7.2	181	61	−027	AbdelFattah et al. (1997)
11/22/1995	18:07:15	34.740	29.200	10.0	5.7	199	77	007	CMT catalogue
11/23/1995	18:07:26	34.480	29.310	15.0	5.7	199	77	007	CMT catalogue
05/28/1998	18:07:26	27.360	31.390	39.0	5.5	154	44	089	CMT catalogue
05/19/2009	17:35:02	37.760	25.200	12.0	5.7	143	42	−088	CMT catalogue

and Qattara lines, which extend in the direction NE-SW to the northern part of Egypt.

On the other hand, regarding the focal mechanism solution plots (Figs. 11, 12, 13 and 14), it can be concluded that:

- The focal mechanism solutions of the events occurred in the southern side of the Hellenic Arc (Fig. 11) show a behaviour either pure reverse faults or reverse faults with strike-slip components. However, the dominant mechanisms of the northern side of the arc are normal focal mechanisms which are related to the extensional stress field, due to the back arc activity. Some events, which occur either in the southern side or in the northern side, show normal faulting or normal faulting with strike-slip component behaviour.
- The Cyprian Arc (Fig. 11) is generally characterized by thrusting and shear mechanisms. Reverse faults or reverse faults with small strike-slip components are also obtained for some events in Southern Cyprus. Strike-slip mechanisms with small reverse or normal components are located to the west of the Cyprian Arc.
- Earthquakes along the DST zone (Figs. 11 and 12) have left-lateral to normal mechanisms, in general agreement with the tectonic model given by Mart and Hall (1984).
- Earthquake focal mechanisms in Northern Egypt (Fig. 13) indicate normal-faulting mechanisms with strike-slip component, suggesting a probable extension of the stress field of the Red Sea and Suez Rifts beneath the Nile Delta (Meshref 1990; Badawy 1998, 2001; Badawy and Horváth 1999; El-Sayed et al. 2001).
- In Southern Egypt (Fig. 14), focal mechanisms show relatively pure right-lateral with a few normal-faulting mechanisms, perhaps suggesting a relatively homogeneous stress field. This is in a good agreement with recent GPS analyses (Badawy et al. 2003; Mahmoud 2003).

10 Conclusions

In the current study, a Poissonian catalogue of 16,642 main shocks, with a magnitude above or equal to moment magnitude 3.0 was obtained after compiling all the available national and international sources. The catalogue span the years from 2200 B.C. to the end of 2013, within a region bounded by 21° to 38°N and 22° to 38°E. This study represents an extension and upgrading of different databases on Egyptian seismicity. Tabulated data contain origin time, coordinates, depth, reported magnitudes and/or maximum intensity, and unified moment magnitude. The reported size is also included in the database in reference to those scientists who might prefer to use relationships other than those employed in the current work.

From the compilation of the entire catalogue, the following conclusions can be drawn:

(a) The occurrence of both aftershocks and swarm-type activities represents a large number of events in the initial compilation of the current catalogue.
(b) It is clearly appeared that after the deployment of both the World-Wide Standardized Seismograph Network, Aswan Seismological Network, and the establishment of the Egyptian National Seismic Network, the number of recorded earthquakes increased abruptly, and the magnitude threshold was reduced, which it is shown clearly in the catalogue completeness analysis in the different epochs.
(c) A general concentration of the historical earthquake activity is quite clear around the Nile Valley and Nile Delta. This is due to the settlement patterns, as well as a potential amplification of sediments.
(d) Both of historical and instrumental earthquakes show a clear concentration in Northern Egypt, being distributed in relatively similar ways, showing that these areas have witnessed activity for many centuries.
(e) Egypt is suffering both interplate and intraplate earthquakes. Intraplate earthquakes are less frequent, but still represent an important component of seismic risk in Egypt. Shallow seismicity is concentrated mainly in the surrounding plate boundaries and on some active seismic zones, like Aswan, Abu Dabbab, and Cairo-Suez regions, while the deeper activity is concentrated mainly along the Cyprian and Hellenic Arcs due to the subduction process between Africa and Europe.
(f) Different fault-plane solutions are distributed in different locations in and around Egypt, and all have a general agreement with the geology and tectonics of the studied regions, and also with previous studies.

In conclusion, the authors are confident that the resulting databases cover some gaps and lack of homogeneity observed in previous catalogues for the region. Compiled catalogues are available to download from http://www.ujaen.es/investiga/rnm024/Egypt-catalog.pdf.

Acknowledgments This research work is supported by the Egyptian Ministry of the Higher Education (Cultural Affairs and Missions Sector, Cairo) and the Spanish Seismic Hazard and Active Tectonics research group.

References

AbdelFattah, A. K., Hussein, H. M., Ibrahim, E. M., & Abu El-Atta, A. S. (1997). Fault plane solutions of the 1993 and 1995 Gulf of Aqaba earthquakes and their tectonic implications. *Annali di Geofisica, 40*, 1555–1564.

Abdel-Rahman, K., Al-Amri, A. M. S., & Abdel-Moneim, E. (2009). Seismicity of Sinai Peninsula, Egypt. *Arabian Journal of Geosciences, 2*, 103–118.

Abou Elenean, K. (1993). *Seismotectonics of the Mediterranean region north of Egypt and Libya*. M.Sc. Thesis, Faculty of Science, Mansoura University, Egypt.
Abou Elenean, K. (1997). *Seismotectonics of Egypt in relation to the Mediterranean and Red Sea tectonics*. Ph.D. Thesis, Faculty of Science, Ain Shams University, Egypt.
Abou Elenean, K. (2007). Focal mechanism of small and moderate size earthquakes recorded by the Egyptian National Seismic Network (ENSN), Egypt. *NRIAG Journal of Geophysics, 6*, 119–153.
Abou Elenean, K., Arvidsson, R., & Kulhanek, O. (2004). Focal mechanism of smaller earthquakes close to VBB Kottamia station, Egypt. *Annals of the Geological Survey of Egypt, XXVII*, 357–368.
Abou Elenean, K. M., & Hussein, H. M. (2007). Source mechanism and source parameters of May 28, 1998 earthquake, Egypt. *Journal of Seismology, 11*, 259–274.
Abou Elenean, K. M., & Hussein, H. M. (2008). The October 11, 1999 and November 08, 2006 Beni Suef Earthquakes, Egypt. *Pure and Applied Geophysics, 165*, 1391–1410.
Abou Elenean, K. M., Hussein, H. M., El-Ata, A., & Ibrahim, E. M. (2000). Seismological aspects of the Cairo earthquake, 12th October, 1992. *Annali di Geofisica, 43*, 485–504.
Abou Elenean, K. M., Mohamed, M. E., & Hussein, H. M. (2010). Source parameters and ground motion of the Suez-Cairo shear zone earthquakes, Eastern Desert, Egypt. *Natural Hazards, 52*, 431–451.
Abu El-Nader, I. (2013). Source parameters and moment magnitude of 30 January 2012 earthquake, Northern Red Sea. *Seismological Research Letters, 84*, 805–809.
Ambraseys, N. N. (1962). A note on the chronology of Willis's list of earthquakes in Palestine and Syria. *Bulletin of the Seismological Society of America, 52*, 77–80.
Ambraseys, N. N. (1985). Intensity-attenuation and magnitude-intensity relationships for Northwest European earthquakes. *Earthquake Engineering and Structural Dynamics, 13*, 733–778.
Ambraseys, N. N. (2001). Far-field effects of Eastern Mediterranean earthquakes in Lower Egypt. *Journal of Seismology, 5*, 263–268.
Ambraseys, N. N., Melville, C. P., & Adams, R. D. (1994). *The seismicity of Egypt, Arabia and Red Sea*. Cambridge: Cambridge University Press.
Badawy, A. (1998). Earthquake hazard analysis in northern Egypt. *Acta Geodaetica et Geophysica Hungarica, 33*, 341–357.
Badawy, A. (1999). Historical seismicity of Egypt. *Acta Geodaetica et Geophysica Hungarica, 34*, 119–135.
Badawy, A. (2001). Status of the crustal stress as inferred from earthquake focal mechanisms and borehole breakouts in Egypt. *Tectonophysics, 343*, 49–61.
Badawy, A. (2005). Seismicity of Egypt. *Seismological Research Letters, 76*, 149–160.
Badawy, A., & Abdel-Fattah, A. K. (2001). Source parameters and fault plane determinations of the 28th December, 1999 Northeastern Cairo earthquake. *Tectonophysics, 343*, 63–77.
Badawy, A., & Abdel-Fattah, A. K. (2002). Analysis of the southeast Beni-Suef northern Egypt earthquake sequence. *Journal of Geodynamics, 33*, 219–234.
Badawy, A., Abdel-Monem, S. M., Sakr, K., & Ali, Sh M. (2006). Seismicity and kinematic evolution of middle Egypt. *Journal of Geodynamics, 42*, 28–37.
Badawy, A., Al-Gabry, M., & Girgis, M. (2010). Historical seismicity of Egypt, a study for previous catalogues producing revised weighted catalogue. In *The Second Arab Conference for Astronomy and Geophysics, Cairo, Egypt*.
Badawy, A., El-Hady, Sh, & Abdel Fattah, A. K. (2008). Microearthquakes and neotectonics of Abu-Dabbab, Eastern Desert of Egypt. *Seismological Research Letters, 79*, 55–67.
Badawy, A., & Horváth, F. (1999). Seismicity of the Sinai subplate region: Kinematic implications. *Journal of Geodynamics, 27*, 451–468.
Badawy, A., Issawy, S., Hassan, G., & Tealeb, A. (2003). Kinematics engine of the ongoing deformation field around Cairo, Egypt. *Acta Geophysica Polonica, 51*, 447–458.
Ben-Avraham, Z. (1985). Structural framework of the Gulf of Elat (Aqaba). *Journal of Geophysical Research, 90*, 703–726.

Ben-Avraham, Z., Nur, A., & Cello, G. (1987). Active transcurrent fault system along the north African passive margin. *Tectonophysics, 141*, 260–294.

Ben-Menahem, A., Nur, A., & Vered, M. (1976). Tectonics, seismicity and structure of the Afro-Eurasian junction-the breaking of an incoherent plate. *Physics of the Earth and Planetary Interiors, 12*, 1–50.

Bondár, I., & Storchak, D. (2011). Improved location procedures at the International Seismological Centre. *Geophysical Journal International, 186*, 1220–1244.

CMT, Global Centroid Moment Tensor Catalogue. http://www.globalcmt.org/.

Cochran, J. (1983). A model for the development of the Red Sea. *American Association of Petroleum Geologists Bulletin, 67*, 41–69.

Constantinescu, L., Ruprechtova, L., & Enescu, D. (1966). Mediterranean-Alpine earthquake mechanisms and their seismotectonic implications. *Geophysical Journal of the Royal Astronomical Society, 10*, 347–368.

D'Amico, V., Albarello, D., & Mantovani, E. (1999). A distribution-free analysis of magnitude-intensity relationships: an application to the Mediterranean Region. *Physics and Chemistry of the Earth (A), 24*, 517–521.

Dewey, J. F., Helman, M. L., Turco, E., Hutton, D. H. W., & Knott, S. D. (1989). Kinematics of the western Mediterranean. In M. P. Coward, D. Dietrich, & R. G. Park (Eds.), *Alpine tectonics* (pp. 265–283). London: Geological Society. (Special Publication).

Dziewonski, A. M., Chou, T. A., & Woodhouse, J. H. (1981). Determination of earthquakes source parameters from waveform data for studies of global and regional seismicity. *Journal of Geophysical Research, 86*, 2825–2852.

Eck, T. V., & Hofstetter, A. (1990). Fault geometry and spatial clustering of microearthquakes along the Dead Sea-Jordan rift fault zone. *Tectonphysics, 180*, 15–27.

Egyptian National Seismic Network (ENSN) Bulletins. (1998–2010). *Earthquakes in and around Egypt*. National Research Institute of Astronomy and Geophysics (NRIAG), Cairo, Egypt.

Ekström, G., Nettles, M., & Dziewonski, A. M. (2012). The global CMT project 2004-2010: Centroid-moment tensors for 13,017 earthquakes. *Physics of the Earth and Planatary Interiors, 200–201*, 1–9.

El-Sayed, A., Korrat, I., & Hussein, H. M. (2004). Seismicity and seismic hazard in Alexandria (Egypt) and its surroundings. *Pure and Applied Geophysics, 161*, 1003–1019.

El-Sayed, A., Vaccari, V., & Panza, G. F. (2001). Deterministic seismic hazard in Egypt. *Geophysical Journal International, 144*, 555–567.

El-Sayed, A., & Wahlström, R. (1996). Distribution of the energy release, b-values and seismic hazard in Egypt. *Natural Hazards, 13*, 133–150.

El-Sayed, A., Wahlström, R., & Kulhánek, O. (1994). Seismic hazard of Egypt. *Natural Hazards, 10*, 247–259.

Engdahl, E. R., Van der Hilstand, R., & Buland, R. (1998). Global teleseismic earthquake relocation with improved travel times and procedures for depth determination. *Bulletin of the Seismological Society of America, 88*, 722–743.

Fat-Helbary, R. E. (1999). *Investigation and assessment of seismic hazard in Egypt*. Final Report Submitted to Fundacion MAPFRE.

Fat-Helbary, R. E., & Mohamed, H. H. (2003). Seismicity and seismotectonics of the West Kom Ombo area, Aswan, Egypt. *Journal of Applied Geophysics, 2*, 253–260.

Frankel, A. (1995). Mapping seismic hazard in the central and eastern United States. *Seismological Research Letters, 66*, 8–21.

Gardner, J. K., & Knopoff, L. (1974). Is the sequence of earthquakes in Southern California, with aftershocks removed, Poissonian? *Bulletin of the Seismological Society of America, 64*, 1363–1367.

Garfunkel, Z., & Bartov, Y. (1977). The tectonics of the Suez Rift, Israel. *Geological Survey Bulletin, 71*, 44.

Gergawi, A., & Khashab, A. (1968). Seismicity of U.A.R. *Helwan Observatory Bulletin, 76*, 27.

Giardini, D., Grünthal, G., Shedlock, K. M., & Zhang, P. Z. (1999). The GSHAP global seismic hazard map. *Annali di Geofisica, 42*, 1225–1230.

Guiraud, R. A., & Bosworth, W. (1999). Phanerozoic geodynamic evolution of Northeastern Africa and the North-western Arabian platform. *Tectonophysics, 315*, 73–108.

Gutenberg, B., & Richter, C. F. (1942). Earthquake magnitude, intensity, energy and acceleration. *Bulletin of the Seismological Society of America, 32*, 163–191.

Hamdache, M., Peláez, J. A., Talbi, A., & Casado, C. L. (2010). A unified catalogue of main earthquakes for northern Algeria from A.D. 856 to 2008. *Seismological Research Letters, 81*, 732–739.

Hassoup, A., & Tealab, A. (2000). Attenuation of intensity in the northern part of Egypt associated with the May 28, 1998 Mediterranean earthquake. *Acta Geophysica Polonica, 48*, 79–92.

Hofstetter, A. (2003). Seismic observations of the 22/11/1995 Gulf of Aqaba earthquake sequence. *Tectonophysics, 369*, 21–36.

Hofstetter, A., Thio, H. K., & Shamir, G. (2003). Source mechanism of the 22/11/1995 Gulf of Aqaba earthquake and its aftershock sequence. *Journal of Seismology, 7*, 99–114.

Huang, P., & Soloman, S. (1987). Centroid depth and mechanisms of mid-ocean ridge. *Journal of Geological Research, 92*, 1361–1383.

Hussein, H. M. (1989). *Earthquake activities in Egypt and adjacent region and its relation to geotectonic features in A.R.E.* M.Sc. Thesis, Faculty of Science, Geology Department, Mansoura University, Egypt.

Hussein, H. M., Abou Elenean, K. M., Marzouk, I. A., Korrat, I. M., Abu El-Nader, I. F., Ghazala, H., & El Gabry, M. N. (2013). Present-day tectonic stress regime in Egypt and surrounding area based on inversion of earthquake focal mechanisms. *Journal of African Earth Sciences, 81*, 1–15.

Hussein, H. M., Abou Elenean, K. M., Marzouk, I. A., Peresan, A., Korrat, I. M., Abu El-Nader, E., et al. (2008). Integration and magnitude homogenization of the Egyptian earthquake catalogue. *Journal of Natural Hazards, 47*, 525–546.

Hussein, H. M., Marzoul, I., Moustafa, A. R., & Hurukawa, N. (2006). Preliminary seismicity and focal mechanisms in the southern Gulf of Suez: August 1994 through December 1997. *Journal of African Earth Sciences, 45*, 48–60.

International Seismological Centre. (2011). *On-line Bulletin*, http://www.isc.ac.uk. International Seismological Centre, Thatcham, United Kingdom.

Ismail, A. (1960). Near and local earthquakes at Helwan from 1903–1950. *Helwan Observatory Bulletin, 49*, 33.

Johnston, A. C. (1996a). Seismic moment assessment of earthquakes in stable continental regions-I. Instrumental seismicity. *Geophysical Journal International, 124*, 381–414.

Johnston, A. C. (1996b). Seismic moment assessment of earthquakes in stable continental regions-II. Historical seismicity. *Geophysical Journal International, 125*, 639–678.

Kanamori, H. (1977). The energy release in great earthquakes. *Journal of Geophysical Research, 82*, 2981–2987.

Karnik, V. (1968). *Seismicity of the European area*. Academia, Polishing House of the Czechoslovak Academy of Science, Praha.

Kebeasy, R. M. (1990). Seismicity. In R. Said (Ed.), *The geology of Egypt* (pp. 51–59). Rotterdam, Netherlands: A.A. Balkerma.

Kebeasy, R. M., Maamoun, M., Albert, R. N. H., & Megahed, M. (1981). Earthquake activity and earthquake risk around Alexandria, Egypt. *Bulletin of International Institute of Seismology and Earthquake Engineering, 19*, 93–113.

Kijko, A., & Graham, G. (1999). Parametric-historic procedure for probabilistic seismic hazard analysis-part II: assessment of seismic hazard at specified site. *Pure and Applied Geophysics, 154*, 1–22.

Korrat, I. M., El Agami, N. L., Hussein, H. M., & El Gabry, M. N. (2005). Seismotectonics of the passive continental margin of Egypt. *Journal of African Earth Sciences, 41*, 145–150.

Lyons, H. G. (1907). Earthquakes in Egypt. *Survey notes, Cairo, 1*, 277–286.

Maamoun, M. (1979). Macroseismic observation of principal earthquakes in Egypt. *Bulletin of Helwan Institute of Astronomy and Geophysics, 183*.

Maamoun, M., Allam, A., & Megahed, A. (1984). Seismicity of Egypt. *Bulletin of Helwan Institute of Astronomy and Geophysics*, 109–160.

Maamoun, M., Allam, A., Megahed, A., & El-Ata, A. (1980). Neotectonic and seismic regionalization of Egypt. *Bulletin of International Instrumental Seismology and Earthquake Engineering (Cairo), 18*, 27–39.

Maamoun, M., & Ibrahim, E. M. (1978). Tectonic activity in Egypt as indicated by earthquake. *Helwan Observatory Bulletin, 170.*

Mahmoud, S. M. (2003). Seismicity and GPS-derived crustal deformation in Egypt. *Journal of Geodynamics, 35*, 333–352.

Mart, Y., & Hall, J. K. (1984). Structural trends in the northern Red Sea. *Journal of Geophysical Research, 89*, 11352–11364.

Marzouk, I. A. (2007). Collecting and evaluating the focal mechanism catalogue of Egypt. *Journal of Applied Geophysics, 6*, 383–398.

McKenzie, D. (1970). Plate tectonics of the Mediterranean region. *Nature, 326*, 239–243.

McKenzie, D. (1972). Active tectonics in the Mediterranean region. *Geophysical Journal of the Royal Astronomical Society, 30*, 109–185.

Meshref, W. (1990). Tectonic framework. In R. Said (Ed.), *The Geology of Egypt* (pp. 113–155). Rotterdam, Netherlands: A. A. Balkema.

Meulenkamp, J. E., Wortel, M. J. R., Van Wamel, W. A., Spakman, W., & Hoogerduyn-Starting, E. (1988). On the Hellenic subduction zone and the geodynamic evolution of Crete since the late Middle Miocene. *Tectonophysics, 146*, 203–216.

Morsy, M., El-Hady, Sh, & Abd El-Meneam, E. (2012). Source parameters of some recent earthquakes in the Gulf of Aqaba, Egypt. *Arabian Journal of Geosciences, 5*, 943–952.

Morsy, M., Hussein, H. M., Abou Elenean, K. M., & El-Hady, Sh. (2011). Stress field in the central and northern parts of the Gulf of Suez area, Egypt from earthquake fault plane solutions. *Journal of African Earth Sciences, 60*, 293–302.

Neev, D. (1975). Tectonic evolution of the Middle East and the Levantine basin (Easternmost Mediterranean). *Geology, 3*, 683–686.

PDE, Preliminary Determination of Epicentre. USGS National Earthquake Information Center (NEIC), http://earthquake.usgs.gov/earthquakes/.

Peláez, J. A., Chourak, M., Tadili, B. A., Aït Brahim, L., Hamdache, M., López Casado, C., & Martínez Solares, J. M. (2007). A catalogue of main Moroccan earthuakes from 1045 to 2005. *Seismological Research Letters, 78*, 614–621.

Peresan, A., & Rotwain, I. M. (1998). *Analysis and definition of magnitude selection criteria for NEIC (PDE) data, oriented to the compilation of a homogeneous updated catalogue for CN monitoring in Italy*. ICTP, Miramare, Trieste, Italy: The Abdus Salam International Centre for Theoretical Physics.

Poirier, J., & Taher, M. (1980). Historical seismicity in the Near and Middle East, north Africa, and Spain from Arabic documents (VIIth–XVIIth century). *Bulletin of Seismological Society of America, 70*, 2185–2201.

Pondrelli, S., Morelli, A., & Ekström, G. (2004). European-Mediterranean regional centroid moment tensor catalog: Solutions for years 2001 and 2002. *Physics of the Earth and Planetary Interiors, 145*, 1–4, 127–147.

Pondrelli, S., Morelli, A., Ekström, G., Mazza, S., Boschi, E., & Dziewonski, A. M. (2002). European-Mediterranean regional centroid-moment tensors: 1997–2000. *Physics of the Earth and Planetary Interiors, 130*, 71–101.

Pondrelli, S., Salimbeni, S., Ekström, G., Morelli, A., Gasperini, P., & Vannucci, G. (2006). The Italian CMT dataset from 1977 to the present. *Physics of the Earth and Planetary Interiors, 159*(3–4), 286–303.

Pondrelli, S., Salimbeni, S., Morelli, A., Ekström, G., & Boschi, E. (2007). European-Mediterranean regional centroid moment tensor catalog: Solutions for years 2003 and 2004. *Physics of the Earth and Planetary Interiors, 164*, 1–2, 90–112.

Pondrelli, S., Salimbeni, S., Morelli, A., Ekström, G., Postpischl, L., Vannucci, G., & Boschi, E. (2011). European-Mediterranean regional centroid moment tensor catalog: Solutions for 2005–2008. *Physics of the Earth and Planetary Interiors, 185*, 74–81.

Primakov, I., & Rotwain, I. (2003). The package for analysis of earthquake catalogues (EDCAT, CATAL and AFT). In *Seventh Workshop on Non-linear Dynamics and Earthquakes Prediction, ICTP, Trieste, Italy*.

RCMT, European-Mediterranean RCMT Catalogue. http://www.bo.ingv.it/RCMT/.

Reasenberg, P. A. (1985). Second-order moment of Central California seismicity. *Journal of Geophysical Research, 90*, 5479–5495.

Reiter, L. (1990). *Earthquake hazard analysis*. Columbia: Columbia University Press.

Riad, S., Ghalib, M., El-Difrawy, M. A., & Gamal, M. (2000). Probabilistic seismic hazard assessment in Egypt. *Annals of the Geological Survey of Egypt, XXIII*, 851–881.

Riad, S., & Hosney, H. (1992). Fault plane solution for the Gilf Kebir earthquake and the tectonics of the southern part of the Western Desert of Egypt. *Annals of the Geological Survey of Egypt, 18*, 239–248.

Riad, S., & Meyers, H. (1985). *Earthquake catalogue for the Middle East countries (1900–1983)*. In National Geophysical Data Centre, World Data Centre A for Solid Earth Geophysics. Rep. SE-40. National Oceanic and Atmospheric Administartion (NOAA), US Department of Commerce, Boulder, Colorado, USA.

Riad, S., Taeleb, A. A., El Hadidy, S., Basta, N. Z., Abou Elela, A. M., Mohamed, A. A., & Khalil, H. A. (2004). *Ancient earthquakes from some Arabic sources and catalogue of Middle East historical earthquakes* (pp. 71–91). EGSMA, NARSS, UNDP, UNESCO.

Riad, S., & Yousef, M. (1999). *Earthquake hazards assessment in the southern part of the Western Desert of Egypt*. Final Report Submitted to the National Authority for Remote Sensing and Space Sciences.

Rothé, J. P. (1969). *The seismicity of the Earth 1953–1965*. Paris: UNESCO.

Rydelek, P. A., & Sacks, I. S. (2003). Comment on "minimum magnitude of completeness in earthquake catalogs: Examples from Alaska, the Western United States, and Japan" by Stefan Wiemer and Max Wyss. *Bulletin of Seismological Society of America, 93*, 1862–1867.

Salamon, A., Hofstetter, A., Garfunkel, Z., & Ron, H. (2003). Seismotectonics of the Sinai subplate-the Eastern Mediterranean region. *Geophysical Journal International, 155*, 149–173.

Savage, W. (1984). *Evaluation of regional seismicity*. Woodward and Clyde Consultants, (Unpublished). Internal Report to Aswan High Dam Authority.

Sieberg, A. (1932a). Handbuch der Geophysik. *Erdbeben-geographie* (Band IV, pp. 527–1005). Berlin: Borntraeger.

Sieberg, A. (1932b). Erdbeben und Bruchschollenbau in Östlichen Mittelmeergebiet. *Denkschriften der Medizinisch-Naturwissenschaftlichen Gesellschaft zu Jena, 18*(2).

Sofratome Group. (1984). *El-Daba nuclear power plant*. Unpublished Report, NPPA Ministry of Electricity, Egypt.

Taher, M. A. (1979). *Corpus des textes arabes relatifs aux tremblements de terre et autres catastrophes naturelles, de la conquête arabe au XII H/XVIII JC*. LLD Thesis, University Paris I.

Wiemer, S., & Wyss, M. (2000). Minimum magnitude of complete reporting in earthquake catalogues: examples from Alaska, the Western United States, and Japan. *Bulletin of Seismological Society of America, 90*, 859–869.

Wiemer, S., & Wyss, M. (2003). Reply to "Comment on 'minimum magnitude of completeness in earthquake catalogues: Examples from Alaska, the western United States, and Japan' by Stefan Wiemer and Max Wyss" by Paul A. Rydelek and I.S. Sacks. *Bulletin of Seismological Society of America, 93*, 1868–1871.

Willis, B. (1928). Earthquakes in the Holy Land. *Bulletin of Seismological Society of America, 18*, 73–103.

Woodward-Clyde Consultants. (1985). *Earthquake activity and stability evaluation for the Aswan High Dam*. Unpublished Report, High and Aswan Dam Authority, Ministry of Irrigation, Egypt.

Youngs, R. R., & Coppersmith, K. J. (1985). Implications of fault slip rates and earthquake recurrence models to probabilistic seismic hazard estimates. *Bulletin of the Seismological Society of America, 75*, 939–964.

Youssef, M. I. (1968). Structural pattern of Egypt and its interpretation. *The American Association of Petrolum Geologists Bulletin, 53*, 601–614.

Ziegler, P. A. (1988). Evolution of the Arctic-North Atlantic and Western Tethys. *American Association of Petroleum Geologists, 43*, 1–206.

Probabilistic Seismic Hazard Assessment for Romania

Radu Vacareanu, Alexandru Aldea, Dan Lungu, Florin Pavel, Cristian Neagu, Cristian Arion, Sorin Demetriu and Mihail Iancovici

1 Introduction

The probabilistic seismic hazard analysis (PSHA) for a specific site is performed by considering the contribution of all possible earthquakes having any possible magnitudes (ranging from a lower bound minimum magnitude to an upper bound maximum magnitude, if any) and occurring, within the considered seismic source (s), at any possible sources-to-site distances, along with the associated uncertainties. The PSHA integrates both the aleatory uncertainties (associated to the seismicity and to the ground motions parameters) and the epistemic uncertainties (through the use of the logic tree approach).

The major part of the seismic hazard in Southern and Eastern Romania is dominated by the Vrancea subcrustal seismic source (Lungu et al. 2000). In addition to this intermediate-depth seismic source, the seismicity of Romania is attributed to thirteen other crustal seismic sources (Vacareanu et al. 2013a). More details on the seismic sources considered in this study are presented in Sect. 4 of this chapter.

This chapter presents a seismic hazard model for Romania and the results obtained within the framework of the BIGSEES national research project (http://infp.infp.ro/bigsees/default.htm) financed by the Romanian Ministry of Education and Scientific Research in the period 2012–2016. One of the most important objective of the BIGSEES Project is to provide a refined and updated seismic hazard map for a further revision of the seismic design code in Romania. The content of this chapter represents an updated, revised and collectively integrated view of the papers

R. Vacareanu (✉) · A. Aldea · D. Lungu · F. Pavel · C. Neagu · C. Arion · S. Demetriu · M. Iancovici
Seismic Risk Assessment Research Center, Technical University of Civil Engineering, 124 Lacul Tei Blvd., Bucharest 020396, Romania
e-mail: radu.vacareanu@utcb.ro

C. Arion
e-mail: arion@utcb.ro

S. D'Amico (ed.), *Earthquakes and Their Impact on Society*,
Springer Natural Hazards, DOI 10.1007/978-3-319-21753-6_5

of Aldea et al. (2014), Pavel et al. (2014a) and Vacareanu et al. (2014a) that were published in Romania in the *Proceedings of the 5th National Conference of Earthquake Engineering and the 1st National Conference on Earthquake Engineering and Seismology* (Material used with kind permission by CONSPRESS.

The first part of this chapter deals with the selection and the grading of the ground motion prediction equations applicable to strong ground motions generated by both crustal and subcrustal sources that contribute to the seismic hazard of Romania. The influence of the input data on the results of the probabilistic seismic hazard analysis is evaluated in the second part of this chapter. The seismicity parameters for all the concerned seismic sources are obtained and discussed. The main assumptions, the input data and the structure of the logic tree used in the seismic hazard analysis are presented in the third part along with the main results expressed in terms of mean hazard curves and seismic hazard maps.

2 Selection of GMPEs

The first part of this chapter focuses on an evaluation of the four ground motion prediction models proposed within the SHARE project for the Vrancea subcrustal seismic source (Delavaud et al. 2012). These four ground motion prediction equations (GMPEs) are: Youngs et al. (1997)—YEA97, Zhao et al. (2006)—ZEA06, Atkinson and Boore (2003)—AB03 and Lin and Lee (2008)—LL08. A previous testing of the four models was performed in (Vacareanu et al. 2013b) using a strong ground motion database of 109 recordings. The grading of the candidate GMPEs was performed using the goodness-of-fit parameters given by Scherbaum et al. (2004, 2009) and Delavaud et al. (2012). In addition, the distribution of the inter-event and intra-event residuals was also checked using the procedure given in Scassera et al. (2009). The results showed that the Youngs et al. (1997) and Zhao et al. (2006) GMPEs produced a good fit with the available strong ground motion database.

A summary of the main characteristics of the four above-mentioned ground motion prediction models is given in Table 1. Some of the data shown in Table 1 were taken also from Douglas (2011). In addition to these four ground motion prediction models, a GMPE (Vacareanu et al. 2014b)—VEA14—derived using strong ground motion recorded during intermediate-depth Vrancea and alogene earthquakes is tested, as well. This ground motion model has distinct coefficients for the region in-front of the Carpathian Mountains (termed fore-arc region, situated to the East and to the South with respect to the Carpathians) and for the region behind the Carpathian Mountains (termed back-arc region, situated to the West and to the North of the Carpathians).

An additional testing of two ground motion prediction models for the Vrancea subcrustal seismic source—YEA97 and ZEA06 was also performed in Pavel et al. (2014b) using an increased strong ground motion database of 233 recordings.

Table 1 Summary of the characteristics of the considered GMPES for Vrancea subcrustal seismic source

GMPE	Database	No. of horizontal records	No. of earthquakes	Magnitude range	Source-to-site range (km)	Depth range (km)	No. of site classes
YEA97	Global	476	164	5–8.2	8.5–550.9	10–229	2
AB03	Global	>1200	43	5.5–8.3	11–550	<100	4
ZEA06	Japan + overseas	4518 + 208	249 + 20	5–8.3	0–300	<162	5
LL08	NE Taiwan + foreign	4244 + 139	44 + 10	4.3 (6)–7.3 (8.1)	15–630	4 (15)–146 (161)	2
VEA14	Vrancea + foreign	344 + 360	9 + 29	5.1–8.0	2–399	60–173	3

Subsequently, an investigation on the performance of several GMPEs developed for crustal seismic sources is also performed in this section. In this respect, three GMPEs are tested—Cauzzi and Faccioli (2008)—CF08, Akkar and Bommer (2010)—AB10 and Idriss (2008)—I08. The characteristics of the considered GMPEs for crustal sources are given in Table 2. These three ground motion prediction models were selected because the parameters of their functional form are readily available. More recent state of the art ground motion models were not selected for testing because parameters like depth-to-top of rupture (Z_{Tor}), down-dip rupture width (W), average shear wave velocity over the top 30 m of subsurface ($v_{s,30}$) or the depth to v_s = 1.0 or 2.5 km/s ($Z_{1.0}$ and $Z_{2.5}$) could not be computed or the data needed for their estimation (Kaklamanos et al. 2011) are not available at this moment.

All the candidate GMPEs are tested using the goodness-of-fit parameters proposed by Scherbaum et al. (2004, 2009) and Delavaud et al. (2012). In addition, the ranking procedure proposed by Kale and Akkar (2013) is also used for the evaluation of the candidate ground motion prediction models.

2.1 Strong Ground Motion Database

The strong ground motion database used for testing and grading of GMPEs consists of 431 recordings from 10 intermediate-depth Vrancea earthquakes and of 125 recordings in Romania from 25 crustal earthquakes. Only subcrustal seismic events with $M_W > 5.0$ were selected in the database since smaller magnitude events have relatively minor structural effects. The distribution of the epicentral distance vs. magnitude of the recorded strong ground motions is represented in Fig. 1 for both subcrustal and crustal earthquakes. All the analysed strong ground motions were collected for the BIGSEES national research project and were recorded by three seismic networks: INCERC (Building Research Institute), INFP (National Institute for Earth Physics) and CNRRS (National Centre for Seismic Risk Reduction). The soil conditions for the recording seismic stations were taken from borehole data assembled within the BIGSEES national research project and, for the sites with no borehole information, were inferred from the topographic slope method proposed by Wald and Allen (2007). The seismic stations were divided into three categories—A, B or C—according to the corresponding soil conditions defined as in EN 1998-1 (CEN 2004). The database consists of both analogue and digital recordings (the later corresponding to earthquakes after 1999). The raw recordings were available only for the digital data. The processing of all the raw analogue strong ground motion recordings was performed originally with an Ormsby band pass filter having the low-cut frequency of 0.15–0.25 Hz and the high-cut frequency of 25–28 Hz. The digital recordings were processed according to the procedures given in the literature (Akkar and Bommer 2006; Boore and Bommer 2006) and using a band-pass Butterworth filter of 4th order with cut-off frequencies of 0.05 Hz and 50 Hz.

Table 2 Summary of the characteristics of the considered GMPEs for crustal seismic sources

GMPE	Database	No. of horizontal records	No. of earthquakes	Magnitude range	Source-to-site range (km)	Depth range (km)	No. of site classes
CF08	Global	1164	60	5.0–7.2	6–150	2–22	4
AB10	Europe + Middle East	532	131	5.0–7.6	0–100	0–28	3
I08	Global	942	72	4.5–7.7	0–200	2–31	2

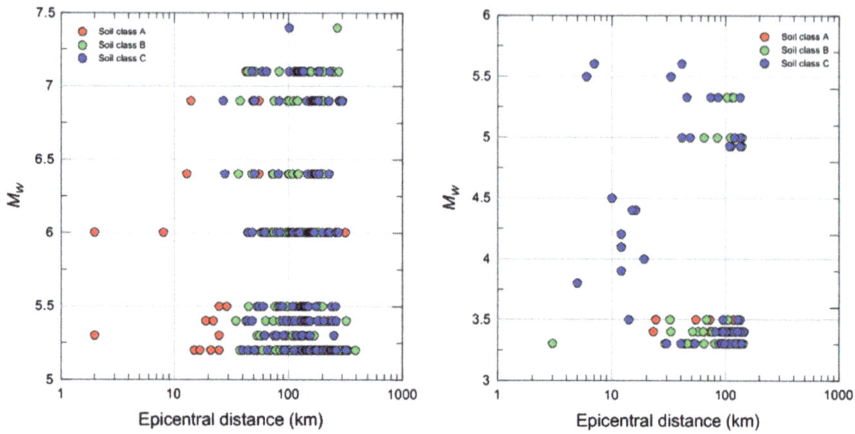

Fig. 1 Distribution of the epicentral distances of the recording seismic stations with respect to the earthquake magnitude for subcrustal earthquakes (*left*) and crustal earthquakes (*right*). The recording seismic stations are divided according to their corresponding soil conditions

2.2 Evaluation of the GMPEs

The study of Vacareanu et al. (2013b) grades the four ground motion prediction models recommended within the SHARE research project (Delavaud et al. 2012) using a strong ground motion database of 109 recordings from Vrancea subcrustal earthquakes and using the goodness-of-fit parameters given in Scherbaum et al. (2004, 2009) and Delavaud et al. (2012). The results show that the Youngs et al. (1997) GMPE fits the best with the available strong ground motion database. In addition, in the study of Pavel et al. (2014b), two ground motion models—Youngs et al. (1997) and Zhao et al. (2006)—are again tested using an increased strong ground motion database of 233 recordings from Vrancea intermediate-depth earthquakes. The results show again that the YEA97 GMPE is graded better than the ZEA06 GMPE.

In this study, the four GMPEs recommended in SHARE project are re-evaluated and, in addition, the ground motion prediction model of Vacareanu et al. (2014b)—VEA14—is also tested. The grading is performed using some of the goodness-of-fit parameters given in (Scherbaum et al. 2004, 2009; Delavaud et al. 2012); moreover, the testing procedure of Kale and Akkar (2013) is employed, as well. The methodology of Kale and Akkar (2013) is called Euclidean distance-based ranking (*EDR*) and it relies on the Euclidean distance concept (*DE*). This methodology considers separately the ground motion variability (standard deviation) and the bias between the median predictions of the model and the observed data. The testing is performed separately for the fore-arc region and for the back-arc region for the strong ground motions recorded during Vrancea subcrustal earthquakes. The same testing procedures are applied subsequently for the ground motion prediction models developed for the crustal seismic sources.

The average results of the grading process using all the GMPEs' spectral periods for some of the goodness-of-fit measures are shown in Table 3, separately for the

Table 3 Values of grading parameters for GMPEs applied for the Vrancea subcrustal seismic source—1

Grading parameter	Ground motion prediction model									
	Fore-arc					Back-arc				
	VEA14	YEA97	ZEA06	AB03	LL08	VEA14	YEA97	ZEA06	AB03	LL08
MEANNR	0.317	−0.075	0.224	1.500	0.455	0.265	−1.256	−1.109	0.005	−1.758
MEDNR	0.316	−0.028	0.281	1.521	0.510	0.284	−1.300	−1.214	0.094	−1.663
STDNR	0.995	0.919	1.223	1.524	1.278	1.216	1.215	1.525	2.010	2.022
MEDLH	0.508	0.559	0.382	0.107	0.335	0.437	0.180	0.163	0.152	0.077
LLH	1.506	1.273	1.987	4.629	2.260	1.949	3.351	3.748	4.136	6.617
Model ranking based on LLH	**2**	**1**	**3**	**5**	**4**	**1**	**2**	**3**	**4**	**5**

Bold represents the rank of the corresponding GMPE

Table 4 Values of grading parameters for GMPEs applied for the crustal seismic sources—1

Grading parameter	Ground motion prediction model		
	Crustal sources		
	CF08	I08	AB10
MEANNR	0.304	0.078	1.697
MEDNR	0.494	−0.088	2.214
STDNR	1.530	1.920	2.379
MEDLH	0.283	0.202	0.010
LLH	2.741	2.748	7.821
Model ranking based on LLH	**1**	**2**	**3**

Bold represents the rank of the corresponding GMPE

fore-arc region and back-arc region; the average results for crustal sources are reported in Table 4. The parameters are the following: median of the likelihood *LH-MEDLH*, the mean *MEANNR*, median *MEDNR* and standard deviation *STDNR* of the normalized residuals and the average sample log-likelihood *LLH*. *LLH* can be defined as a measure of the distance between the tested model and the actual data.

Tables 5 and 6 show the results of the grading procedure using the methodology of Kale and Akkar (2013). $\sqrt{MDE}$ parameter takes into account the variability, while $\sqrt{\kappa}$ accounts for the bias between the predicted and the observed values. The overall goodness-of-fit parameter is termed Euclidean distance-based ranking (*EDR*).

In order to investigate in-detail how well the selected ground motion prediction models fit with the available dataset of observed strong ground motions, the variations of the goodness-of-fit parameters—$\sqrt{MDE}$, $\sqrt{\kappa}$ and *EDR* with spectral period are plotted in Figs. 2, 3 and 4.

One can notice the large variation of the goodness-of-fit parameters with the spectral period of the AB03 model for the fore-arc region, as well as the large values of *EDR* computed for all GMPEs applied for the back-arc region, except AB03 and VEA14 models. In the case of the crustal GMPEs, it appears that the lowest variability is encountered for the CF08 and I08 ground motion models.

Subsequently, the analysis of the residuals (inter-event and intra-event) for the spectral period $T = 0.0$ s is performed with respect to the earthquake magnitude and epicentral distance (Scassera et al. 2009). The distributions of the peak ground acceleration inter-event residuals with the earthquake moment magnitude (magnitude scaling) and of the intra-event residuals with the epicentral distance (distance scaling) are checked for both fore-arc and back-arc regions (for subcrustal earthquakes) and crustal earthquakes, as well, in Figs. 5 and 6.

The bias in the magnitude scaling is checked through the slopes of the linear trendlines fitted against the inter-event residuals. One can notice from Fig. 5 that the VEA14 ground motion prediction model has the lowest slope of the fitted trendline for both fore-arc and back-arc regions. In the case of the GMPEs for crustal seismic sources, it appears that the CF08 and I08 models have the lowest slopes of the fitted trendlines.

The variation of the intra-event residuals with the epicentral distance of the recording seismic station (distance scaling) is checked in Fig. 6 for the same

Table 5 Values of grading parameters for GMPEs applied for the Vrancea subcrustal seismic source—2

Grading parameter	Ground motion prediction model									
	Fore-arc					Back-arc				
	VEA14	YEA97	ZEA06	AB03	LL08	VEA14	YEA97	ZEA06	AB03	LL08
$\sqrt{MDE^2}$	0.93	1.04	1.01	1.39	1.02	1.04	1.70	1.40	1.31	1.73
$\sqrt{\kappa}$	1.09	1.21	1.26	1.28	1.37	1.08	1.70	1.47	1.17	1.89
EDR	1.01	1.27	1.27	1.84	1.41	1.13	2.90	2.10	1.55	3.28
Model ranking based on *EDR*	**1**	**2**	**2**	**5**	**4**	**1**	**4**	**3**	**2**	**5**

Bold represents the rank of the corresponding GMPE

Table 6 Values of grading parameters for GMPEs applied for the crustal seismic sources—2

Grading parameter	Ground motion prediction model		
	Crustal sources		
	CF08	I08	AB10
$\sqrt{MDE^2}$	1.38	1.82	2.21
$\sqrt{\kappa}$	1.02	1.03	1.12
EDR	1.47	1.88	2.47
Model ranking based on *EDR*	**1**	**2**	**3**

Bold represents the rank of the corresponding GMPE

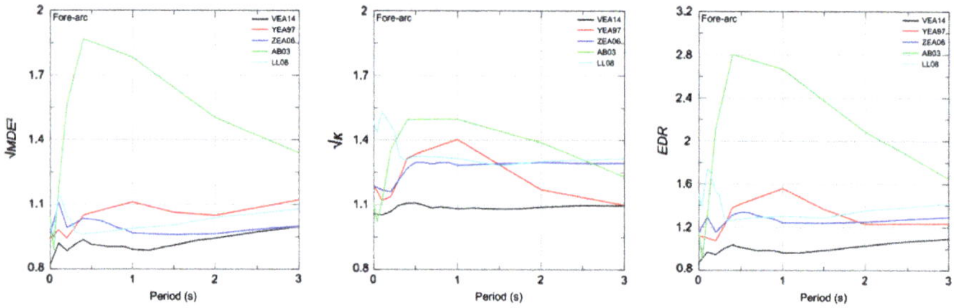

Fig. 2 Variation of $\sqrt{MDE^2}$, $\sqrt{\kappa}$ and *EDR* parameters with spectral period for the ground motion prediction models used in the fore-arc region

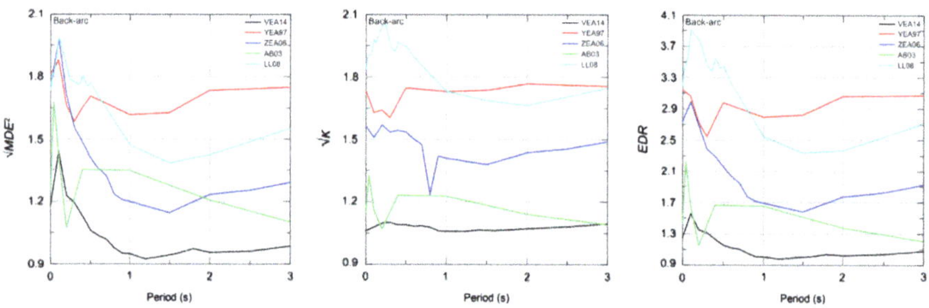

Fig. 3 Variation of $\sqrt{MDE^2}$, $\sqrt{\kappa}$ and *EDR* parameters with spectral period for the ground motion prediction models used in the back-arc region

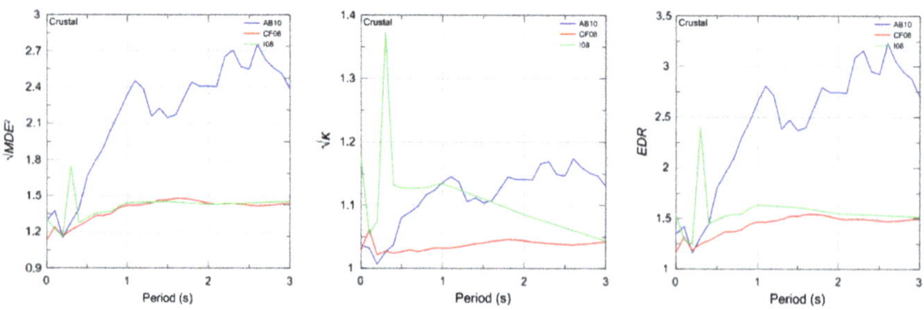

Fig. 4 Variation of $\sqrt{MDE^2}$, $\sqrt{\kappa}$ and *EDR* parameters with spectral period for the ground motion prediction models used for the crustal seismic sources

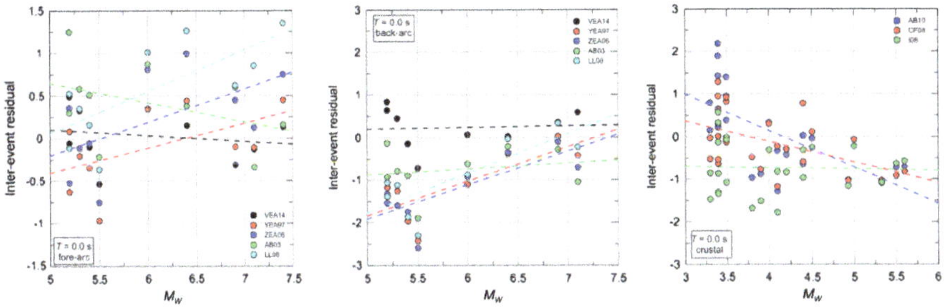

Fig. 5 Variation of inter-event residuals of peak ground acceleration with the earthquake magnitude—magnitude scaling (the inter-event trendlines are shown with *dotted lines*)

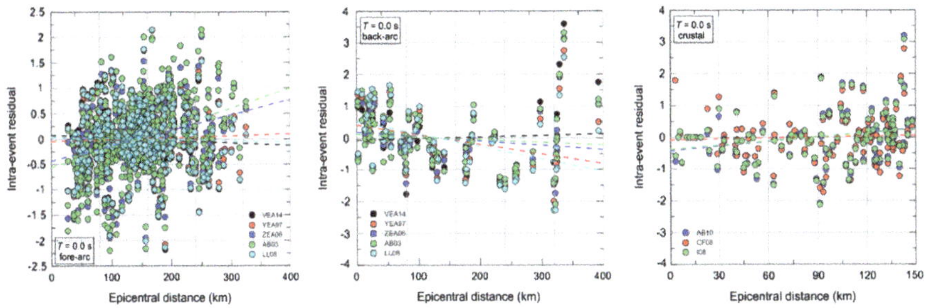

Fig. 6 Variation of intra-event residuals of peak ground acceleration with the epicentral distance of the recording seismic station—distance scaling (the intra-event trendlines are shown with *dotted lines*)

spectral period of $T = 0.0$ s. The differences in the slopes of the fitted trendlines are much lower than in the case of the inter-event residuals. Neither of the analysed ground motion prediction models either for the fore-arc, back-arc or crustal sources does not appear to grossly under-predict or over-predict the observed ground motion. The largest slopes are encountered for the AB03 GMPE in the fore-arc region, for LL08 GMPE in the back-arc region and AB10 GMPE for the crustal seismic sources. As such, it appears that the magnitude scaling determines mainly whether a particular GMPE can be reliably applied or not for Vrancea subcrustal seismic source and for the crustal seismic sources, as well.

Several weighting factors have been proposed for the GMPEs selected in the SHARE project for the Vrancea subcrustal seismic source, in order to be used in the probabilistic seismic hazard assessment (Delavaud et al. 2012; Vacareanu et al. 2013b). The weighting factors shown in the two references are reported in Table 7.

In this study, a new weighting scheme involving the tested ground motion prediction equations is proposed. The weighting factors given in Table 8 are obtained based on the values of the goodness-of-fit parameters *LLH* and *EDR* shown previously and on the evaluation of the distribution of inter-event and intra-event residuals for each spectral period.

Table 7 Weighting factors for PSHA from previous studies

Ground motion prediction model	Weighting schemes		
	(Delavaud et al. 2012)		(Vacareanu et al. 2013b)
	WS 1	WS 2	
ZEA06	0.40	0.25	0.30
AB03	0.20	0.25	0.15
YEA97	0.20	0.25	0.40
LL08	0.20	0.25	0.15

Table 8 Proposed weighting factors for PSHA

Fore-arc		Back-arc		Crustal	
GMPE	Weighting factors	GMPE	Weighting factors	GMPE	Weighting factors
VEA14	0.40	VEA14	0.60	AB10	0.15
YEA97	0.25	AB03	0.20	CF08	0.45
ZEA06	0.25	YEA97	0.10	I08	0.40
LL08	0.10	ZEA06	0.10		

3 Sensitivity Analysis

This section presents an overview of the sensitivity of the PSHA results for 4 cities in Romania (Bucharest, Focsani, Iasi and Craiova), located in the fore-arc of the Carpathian Mountains, in the area of influence of Vrancea subcrustal seismic source. The sensitivity analysis is performed for different assumptions and values of the input data, aiming at the proper selection of the final choices for the PSHA presented in Sect. 4.

The sensitivity analysis considers PSHA computations performed solely for Vrancea subcrustal (intermediate depth) seismic source. The earthquake catalogue considered for the Vrancea source is ROMPLUS (http://www.infp.ro/catalog-seismic) revised for SHARE Project (Stucchi et al. 2013). The *a* and *b* seismicity parameters of Vrancea subcrustal source are obtained through maximum likelihood method given in (McGuire 2004).

The ground motion prediction equation (GMPE) used in the sensitivity analysis is the one developed for Vrancea subcrustal source by Vacareanu et al. (2014b). In the present sensitivity analysis the GMPE is applied for the geometric mean of the two orthogonal horizontal components of peak ground acceleration (*PGA*). The selected GMPE depends on the following parameters: moment magnitude (M_W), source to site (hypocentral) distance, focal depth and soil class. The selected GMPE takes into consideration the location of the analyzed site with respect to the Carpathian Mountains (back-arc and fore-arc sites) and considers the uncertainties in predicting the ground motion parameters through the inter-event standard deviation (representing the earthquake to earthquake variability of ground motions) and the intra-event standard deviation (representing the variability of ground motions within the earthquakes).

The sensitivity analysis focuses on the effect of different choices of PSHA input data: earthquake catalogue duration, minimum and maximum magnitude, focal depth and the number (ε) of logarithmic standard deviations (σ) by which the logarithmic ground motion deviates from the mean of the predicted logarithmic value.

The probability of exceedance of a certain level of *PGA* is obtained by integrating the probabilities of all possible magnitudes, source-to-site distances and focal depth and associated exceedance probabilities of *PGA*, through the total probability formula (McGuire 1999). It must be mentioned that the peak ground acceleration and/or spectral acceleration values given in this chapter represent the geometric mean of the two orthogonal horizontal components of motion, as provided by the GMPEs used in the analysis.

3.1 Influence of Earthquakes Catalogue Duration

Even the earthquake catalogue for Vrancea subcrustal source has data starting with year 984, the early centuries have a lack of reported seismic events due to the scarcity of written information; consequently, for satisfying the catalogue completeness criteria, only the more recent period can be used. Two possible choices were considered: (a) period 1802–2014 and (b) period 1901–2014. The Maximum Curvature technique from Wiemer and Wiss (2000) is applied for assessing the magnitude of completeness for each earthquake catalogue duration. The maximum earthquake magnitude, $M_{\max}$ (i.e., the upper limit of magnitude that cannot be exceeded) was evaluated based on seismic data, using the procedure presented in Kijko (2004). The obtained results are as follows: (a) 1802–2014—$M_{W,min} = 5.7$; $M_{W,max} = 8.2$, and (b) 1901–2014—$M_{W,min} = 4.8$; $M_{W,max} = 8.1$. The seismic hazard curves (probabilities of exceedance in 50 years as function of *PGA*) are presented in Fig. 7 for Bucharest (average epicentral distance to Vrancea subcrustal source, $\Delta = 135$ km), Focsani ($\Delta = 45$ km), Iasi ($\Delta = 190$ km) and Craiova ($\Delta = 260$ km).

In civil engineering, the domain of interest of probability of exceedance ranges down to the value of 2 % in 50 years (corresponding to a 2475 yr mean return period). The standard recommended probability of exceedance in 50 yr is 10 % (corresponding to a mean return period of 475 yr). In all the comparisons shown in Fig. 7, the results indicate that in this range of interest the seismic hazard curves are practically identical, regardless of the catalogue duration. The differences start to be significant at very low probabilities of exceedance, lower than 1 % in 50 years (corresponding to 4950 yr mean return period).

Finally, the weights for the catalogue duration are decided based on expert judgment, considering a larger weight for the catalogue with the seismic events starting from 1901 up to 2014, due to the more reliable information on position of focus and magnitude of earthquakes.

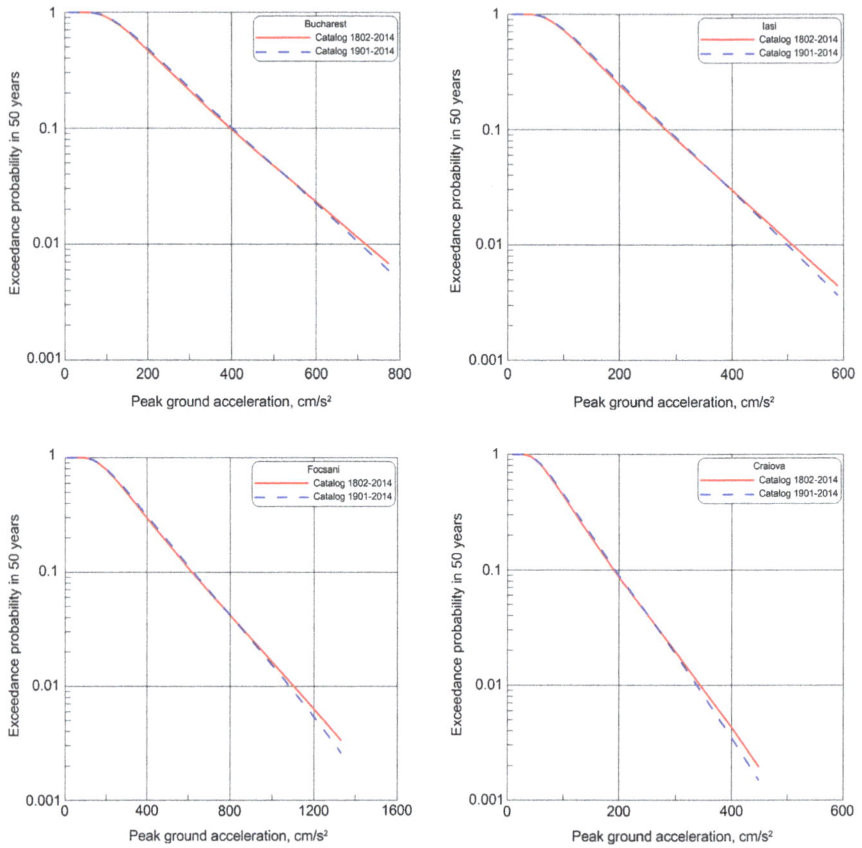

Fig. 7 Influence on the PSHA results (*PGA*) of the earthquake catalogue duration for four selected sites

3.2 *Influence of the Maximum Magnitude*

Since the period 1802–2014 includes the strongest known earthquake in Romania's documented history (October 26th, 1802 eq., $M_W \approx 7.9$), the sensitivity analysis for the maximum magnitude was performed using this catalogue period. The minimum magnitude is the completeness one: $M_{W,min} = 5.7$.

The results obtained using the analitically evaluated $M_{W,max} = 8.2$ are compared in Fig. 8 with the less conservative values of the maximum magnitude: $M_{W,max} = 8$ and $M_{W,max} = 8.1$.

As expected, higher the maximum magnitude, higher the hazard values. One can remark that for exceedance probabilities in 50 yr down to 10 %, the hazard curves have quite close values; after that, the difference start to be somehow significant (approaching 10 %).

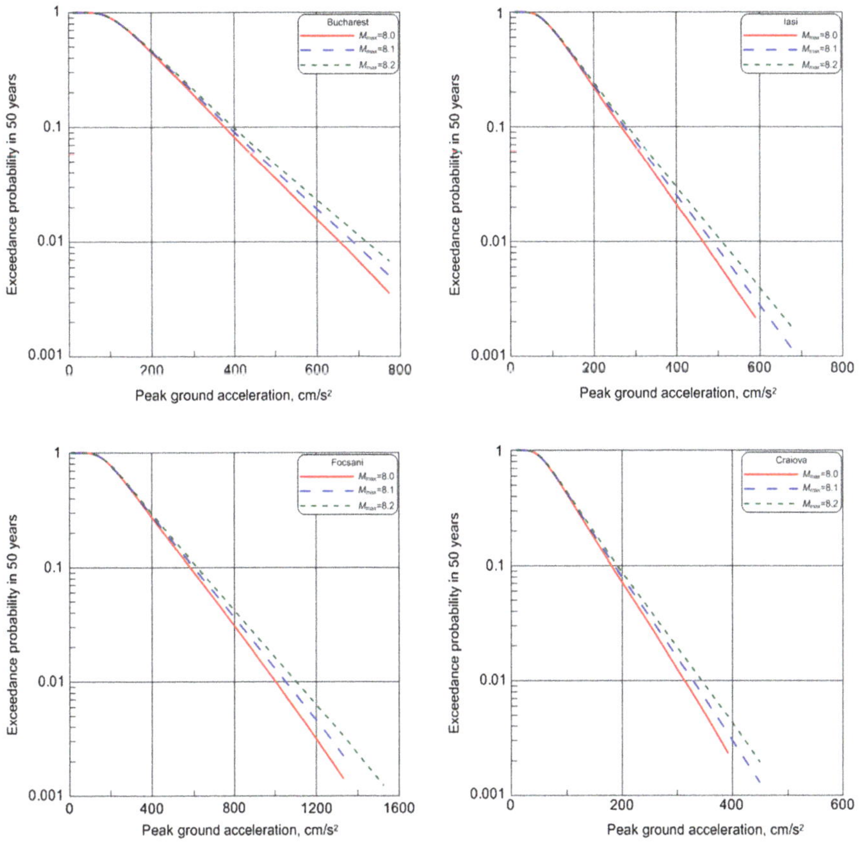

Fig. 8 Influence on the PSHA results (*PGA*) of the maximum magnitude for four selected sites: Bucharest (*top left*), Iasi (*top right*), Focsani (*bottom left*) and Craiova (*bottom right*)

This result is incorporated in the PSHA logic tree, presented in Sect. 4, by considering a branch corresponding to $M_{W,max}$ = 8.1, supplementary to the one for the computed $M_{W,max}$ = 8.2. Since the later value is considered as the best estimate, it received a higher weighting factor, while the branch with a smaller maximum magnitude received a lower weighting factor. In a similar way, for the branch corresponding to the catalogue period 1901–2014, two branches for the maximum magnitude were considered: $M_{W,max}$ = 8.1 (higher weighting factor) and $M_{W,max}$ = 8.0 (lower weighting factor).

3.3 Influence of the Minimum Magnitude

A study of the influence of the minimum magnitude was performed using the 1802–2014 catalogue. The computations are compared for 3 values of minimum magnitude: the magnitude of completeness $M_{W,min} = 5.7$, a magnitude slightly higher $M_{W,min} = 5.8$ and a magnitude slightly lower $M_{W,min} = 5.6$. The maximum magnitude is the same in all three cases $M_{W,max} = 8.2$. The results are comparatively presented in Fig. 9.

The results from the cases with $M_{W,min}$ = 5.7 and $M_{W,min}$ = 5.8 are quite close (slightly smaller in the second case), while the seismic hazard is significantly higher when using $M_{W,min}$ = 5.6.

The hypothesis $M_{W,min}$ = 5.8 satisfies the catalogue completeness criteria, but it reduces the number of seismic events in the catalogue. Because of this and of the limited effect on the seismic hazard results, this hypothesis was disregarded for further PSHA computations. The hypothesis $M_{W,min}$ = 5.6 does not satisfy the

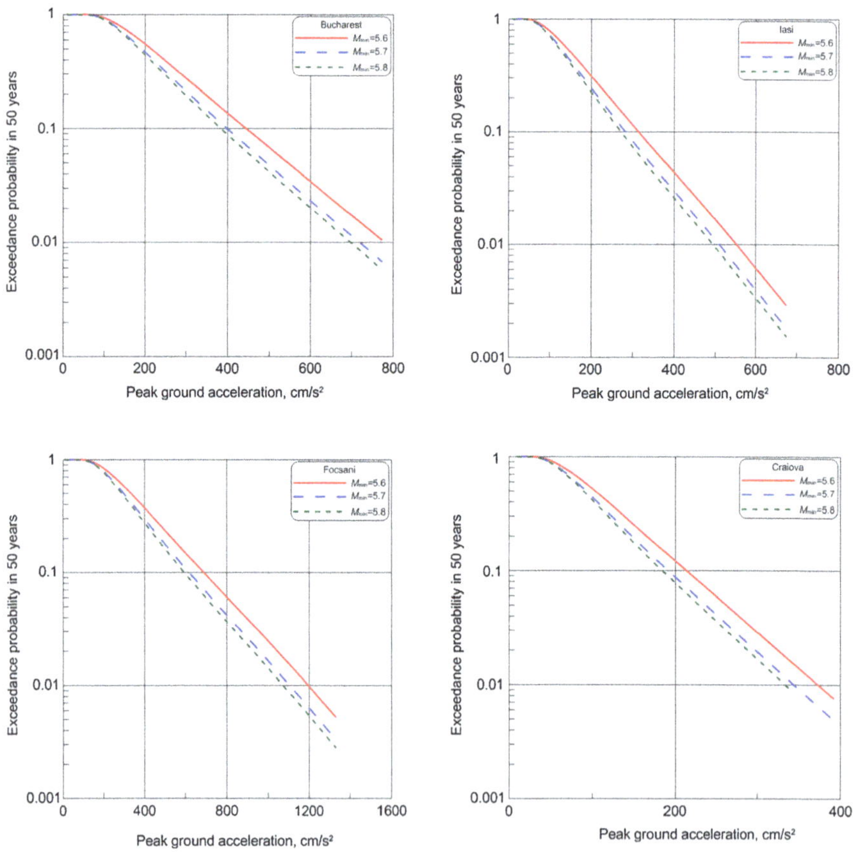

Fig. 9 Influence on the PSHA results (*PGA*) of the minimum magnitude for four selected sites: Bucharest (*top left*), Iasi (*top right*), Focsani (*bottom left*) and Craiova (*bottom right*)

catalogue completeness criteria, the empirical distribution largely departing from the Gutenber-Richter magnitude distribution law; thus, it was also disregarded for further PSHA computations.

In the PSHA computations presented in Sect. 4, the minimum magnitude was not considered as a logic tree parameter/branch. For the catalogues considered in the PSHA, the minimum magnitude is the completeness magnitude.

3.4 *Influence of the Focal Depth Range*

Earthquakes from Vrancea subcrustal source have focal depths mainly in between 60 and 180 km. The two hazard curves presented in Fig. 10 for the four selected cities are computed in two hypothesis:

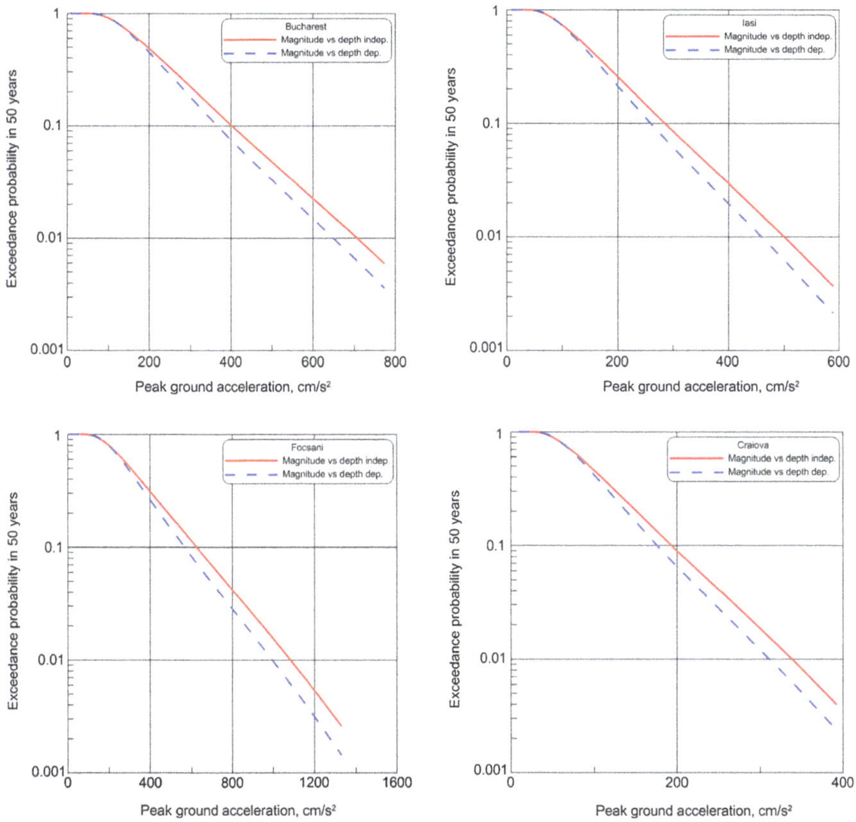

Fig. 10 Influence on the PSHA results (*PGA*) of the focal depth for four selected sites: Bucharest (*top left*), Iasi (*top right*), Focsani (*bottom left*) and Craiova (*bottom right*)

1. "magnitude versus depth independent" (the earthquake catalogue is considering all events with focal depths in the range 60 ÷ 180 km; the completeness magnitude, the maximum magnitude and the frequency-magnitude distribution analysis are based on the full data set), and
2. "magnitude versus depth dependent" (the earthquake catalogue is splitted in four bins corresponding to four depth ranges: 60 ÷ 90 km, 90 ÷ 120 km, 120 ÷ 150 km and 150 ÷ 180 km; the completeness magnitude, the maximum magnitude and the frequency-magnitude distribution analysis are based on data in each bin).

When one splits the earthquake catalogue in bins based on the focal depth, the seismic hazard decreases almost for all the range of exceedance probabilities. Considering such an approach might be an attractive option in some instances, since the effects at ground surface are lower. But, in the present case, the bins resulting from splitting the earthquake catalogue have low numbers of events in each category; consequently, the uncertainties related to the completeness magnitude and to the maximum magnitude are increased; thus it is unreliable to propose weighting factors for each logic tree branch corresponding to each depth bin. Due to these reasons and to the fact that the focal depth influence is directly incorporated in the GMPE (through the terms involving the hypocentral distance and the focal depth itself), the hypothesis of dividing the seismic events catalogue in bins based on the focal depth was disregarded in the PSHA study presented in Sect. 4.

3.5 Influence of ε

An important parameter that influences the PSHA results is the number (ε) of logarithmic standard deviations (σ) by which the logarithmic ground motion deviates from the mean of the logarithmic value of the observed predicted value (i.e. the normalized residual). In Fig. 11 are presented the seismic hazard curves for the four selected cities for three values: $\varepsilon = 1$, $\varepsilon = 2$ and $\varepsilon = 3$. One can notice the remarkable numerical influence on the results of the number of logarithmic standard deviations considered in the analysis, for the whole range of exceedance probabilities. These results are in agreement with those from the paper of Strasser et al. (2008).

3.6 PSHA Results Versus Observed Values

This sub-section is about the possibility (if any) to compare the PSHA results with observed data over a limited period of time.

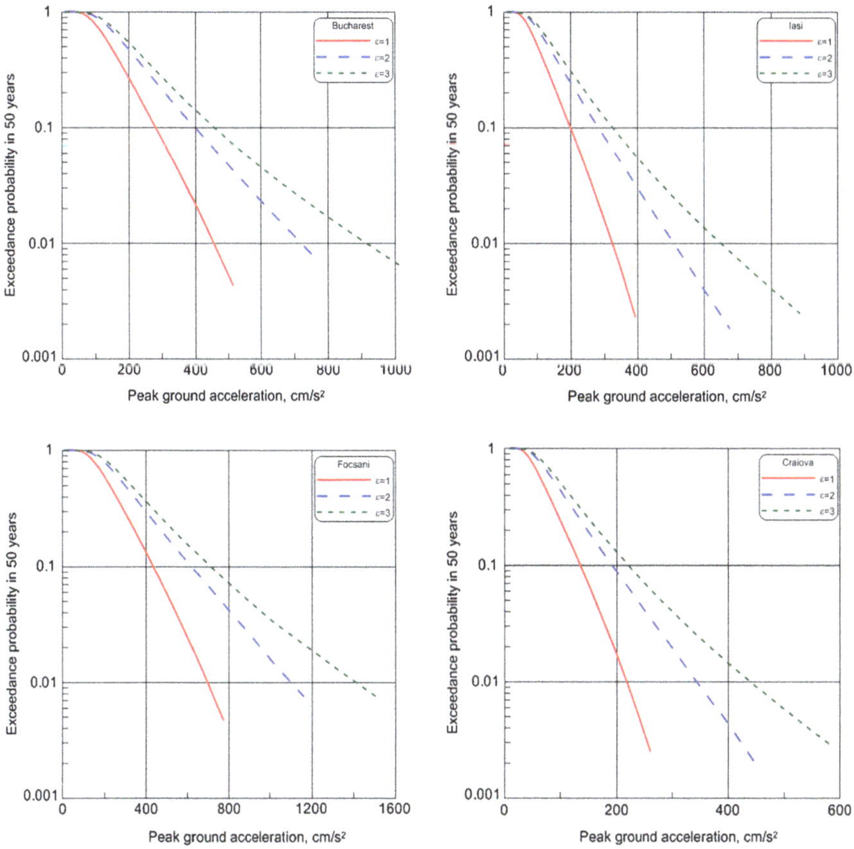

Fig. 11 Influence on the PSHA results (*PGA*) of the considered number of normalized residuals ε for four selected sites: Bucharest (*top left*), Iasi (*top right*), Focsani (*bottom left*) and Craiova (*bottom right*)

Iervolino (2013) reminds in his paper that the hazard level for a given site located in the epicentral region may be exceeded during individual seismic events. However, one has to consider the fact that PSHA accounts for all possible epicenters within a certain seismic zone. So, in the case of comparisons of the seismic hazard analysis results with observed values in individual seismic events, the proper comparison is between the observed ground motions and the predicted ground motion levels from an earthquake originating in the given epicenter (obtained through a deterministic approach using ground motion prediction models and by taking into consideration the variability of the obtained results).

Beauval et al. (2008) obtained the minimum time windows required for a reliable estimation (with a specified degree of uncertainty) of the occurrence rate of peak ground accelerations for a given site, values that are valid irrespective of the level of seismicity of the area.

These values range from a two year time window for peak ground accelerations with mean return period of one month up to 250,000 years for peak ground accelerations with a mean return period of 10,000 years. In the case of a mean return period of 475 years, the minimum observation time is 12,000 years (for an uncertainty level of 20 %). Consequently, the comparison of predicted and observed values for a given site can be performed only for peak ground accelerations with short mean return periods, since the first strong ground motion recordings were obtained in the later first half of the 20th century.

If one considers the peak ground acceleration corresponding to a mean return period of 100 years, the minimum time windows for comparison with recorded data are (Beauval et al. 2008):

- 1000 years for a 30 % uncertainty level,
- 2500 years for a 20 % uncertainty level and
- 10,000 years for a 10 % uncertainty level.

The analysis performed by Beauval et al. (2008) for El Centro site in California, operating since 1934, and for Lefkada site in Greece, operating since 1973, show acceptable uncertainty levels (up to 30 %) for peak ground accelerations up to 0.1 g. As such, comparisons between observed and predicted values (rates) can be performed up to these levels of peak ground acceleration. The same type of analysis was performed for a seismic station in Mexico-City by Ordaz and Reyes (1999) and the results yield an even lower level of peak ground acceleration—0.05 g.

A similar analysis is performed in this study for two seismic stations in Romania—Vrancioaia (epicentral region) and Carcaliu (south-east from Vrancea subcrustal seismic source), Fig. 12 (left). The results show for a 40 % uncertainty level a *PGA* value around 0.05 g.

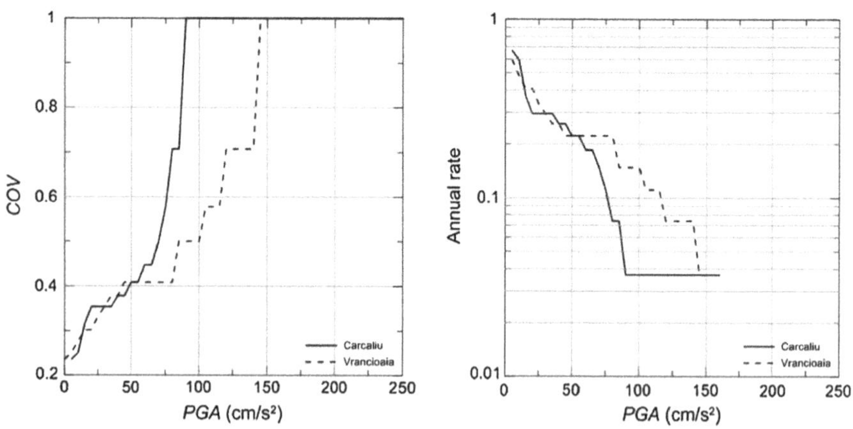

Fig. 12 Analysis of recorded ground motions for two seismic stations in Romania—Carcaliu and Vrancioaia: uncertainty level for *PGA*—*left*, and annual rate of occurrence for *PGA*—*right*

One can also notice in Fig. 12 (right) that the corresponding levels of *PGA* for an annual rate of occurrence of 0.1 (i.e., a mean return period of 10 years) are 0.07 g for Carcaliu seismic station and 0.12 g for Vrancioaia seismic station.

The results in Fig. 12 are in agreement with the studies of Ordaz and Reyes (1999), Beauval et al. (2008) and Iervolino (2013). Even if ground motion recordings at a given site are available for a very long time span, a significant uncertainty level still remains. The study of Beauval et al. (2008) "demonstrates that the comparisons between observations and predictions can provide only limited constraints on probabilistic seismic hazard estimates".

4 Seismic Hazard Analysis

The seismic hazard analysis can be approached in a deterministic or a probabilistic manner, the latter being fully established and described in some seminal textbooks and papers such as: Cornell (1968), McGuire (1976, 1999, 2004), Reiter (1990), Kramer (1996). The PSHA for a site is performed by considering all the ground motions occurring from earthquakes having any possible magnitudes (ranging from a lower bound magnitude to an upper bound magnitude, if any) and/or sources-to-site distances within the seismic source(s) influencing the site, along with their associated uncertainties.

The probabilistic seismic hazard analysis PSHA has the advantage of fully integrating all the aleatory uncertainties arising from seismicity and ground motions parameters expected in a future earthquake on a particular site. The epistemic uncertainties are included in the analysis, as well, through the use of the logic tree (Bommer et al. 2005; Bommer and Scherbaum 2008).

According to our limited knowledge, several studies on deterministic and probabilistic seismic hazard assessment for Romania, in terms of horizontal peak ground acceleration and/or macro-seismic intensities, were published by Lungu et al. (1999, 2006), Musson (2000), Mäntyniemi et al. (2003, 2004), Radulian et al. (2000b), Ardeleanu et al. (2005), Leydecker et al. (2008), Mârza et al. (1991), Sokolov et al. (2009).

The first type of uncertainties we considered in the *PSHA* are the aleatory ones. To this aim, we revisited the seismicity analysis of the sources that influence the Romanian territory. The magnitude of completeness and the maximum magnitude are obtained for each seismic source based on the *ROMPLUS* catalogue (http://www.infp.ro/catalog-seismic) revised in the framework of the *SHARE* Project (Stucchi et al. 2013). Then, the influence of the ground motion prediction equations *GMPE*s on the seismic hazard assessment is accounted using the best graded ground motion models for Vrancea intermediate-depth seismic source and crustal sources selected in Sect. 2.

The epistemic uncertainty is accounted for in this study using a logic tree (Bommer et al. 2005; Bommer and Scherbaum 2008). The sensitivity of the outcome of the *PSHA* due to various models and parameters' values (i.e. earthquake catalogue, completeness and maximum magnitudes etc.) is presented in Sect. 3. Finally, the results of the *PSHA* for Romania are presented in terms of seismic hazard curves and maps. In the following, the amplitude of the ground motion parameter obtained in PSHA represents the geometric mean of the two orthogonal horizontal components of the motion.

4.1 Seismic Sources Influencing Romanian Territory

The seismic sources contributing to the earthquake hazard of Romania are defined in the studies of the National Institute for Earth Physics, *INFP*. In Fig. 13, the seismic sources influencing the Romanian territory are presented: 13 sources of crustal depths and one of intermediate-depth seismicity in Vrancea region. The contours of the seismic source's areas, established in BIGSEES project by INFP, are refined, keeping the same stress field characteristics, starting from the study of Radulian et al. (2000a), such as to take into account the distribution of recent seismicity and the revision of historical earthquakes, recently carried out within SHARE Project.

The Serbia source is defined taking into account the known fault distribution and the epicenters of events as reported in SHARE catalogue (Stucchi et al. 2013). The sources in North Eastern Bulgaria are defined after Simeonova et al. (2006).

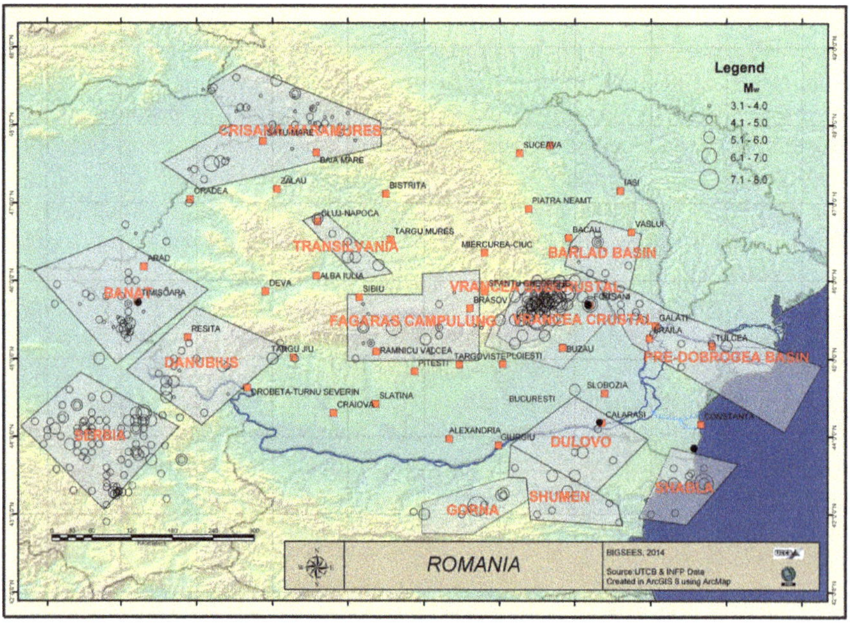

Fig. 13 Sources contributing to the seismic hazard of Romania

Out of the 14 seismic sources presented in Fig. 13, Vrancea subcrustal seismic source is the most active and powerful and is influencing more than two thirds of the Romanian territory as well as parts of Republic of Moldova and Bulgaria. The Vrancea subcrustal seismic source, located at the bend of the Carpathian Mountains in the Eastern part of Romania, is a region of concentrated intermediate-depth seismicity, far from any active plate boundaries (Ismail-Zadeh et al. 2012). This seismic source is limited to a volume having a horizontal area of about 30 × 70 km^2 (Ismail-Zadeh et al. 2012) and spanning in depth between 60 and 170 km (Marmureanu et al. 2010). Beyond the depth of 170 km the seismicity decreases very sharply. The Vrancea seismic zone is bounded to the east by an area of shallow seismicity which extends in the Focsani basin (Ismail-Zadeh et al. 2012). The distribution of the epicentres is elongated in the NE-SW direction, bordered to the NE by the Trotus fault and to the SW by the Intramoesic fault (Ismail-Zadeh et al. 2012). A more thorough description of the seismological features of the Vrancea intermediate-depth seismic source is beyond the scope of this chapter and can be found elsewhere (Ismail-Zadeh et al. 2012; Radulian et al. 2000a).

Beauval et al. (2006) proposed a measure called fractal dimension (*D*-value) for quantifying the degree of clustering of the earthquakes within a certain seismic zone. The values of *D* range from 0, corresponding to a point source, to 2, corresponding to a uniform distribution of earthquakes in space. In the case of a line source, the value of *D* is around 1. The computations performed for the catalogues of the Vrancea crustal and subcrustal seismic zones reveal *D*-values around 2, confirming the assumption of uniform distribution of seismicity in space adopted for PSHA in this study. *D*-values around 1.8 are obtained for Banat and Fagaras crustal seismic zones, while in the case of the catalogue for the Pre-Dobrogea basin a value around 1.3 is obtained, showing a more clustered distribution of earthquakes. However, the influence of the Vrancea subcrustal seismic source is by far dominant for the sites situated in the Pre-Dobrogea basin, thus rendering as negligible the influence of the uniformly distributed seismicity assumed for the Pre-Dobrogea Basin seismic source.

The earthquake catalogues used in the seismicity analysis are recently revised by the National Institute of Earth Physics of Romania, INFP for the SHARE Project (Stucchi et al. 2013).

Assessing the magnitude of completeness, M_C of earthquake catalogues is an essential and compulsory step in the seismicity analysis. The completeness magnitude, M_C is theoretically defined as the lowest magnitude at which 100 % of the earthquakes in a space-time volume are detected (Rydelek and Sacks 1989). M_C is often estimated by fitting a Gutenberg-Richter model to the observed frequency-magnitude distribution. The magnitude at which the lower end of the frequency-magnitude distribution departs from the Gutenberg-Richter law is taken as an estimate of M_C (Zuniga and Wyss 1995). The Maximum Curvature technique (Wiemer and Wyss 2000) is applied in this study for assessing the magnitude of completeness for each earthquake catalog used in PSHA.

The maximum earthquake magnitude, M_{max}, is defined as the upper limit of magnitude for a given seismic source so that, by definition, no earthquakes are

Table 9 Earthquake catalogues, number of seismic events, magnitude of completeness, maximum magnitude and seismicity parameters used in PSHA

Catalogue of seismic source	Year of completeness	# of earthquakes	Magnitude of completeness, M_C	Maximum magnitude, M_{max}	*a*	*b*
Banat	1843	57	3.8	6.4	2.19	0.69
Barlad Basin	1894	40	3.2	5.8	2.96	1.07
Crisana	1823	57	3.5	6.6	1.78	0.65
Danubius	1879	54	3.2	6.0	0.73	0.34
Fagaras	1826	31	3.5	6.8	1.01	0.51
Pre-Dobrogea Basin	1900	54	3.1	5.7	2.51	0.91
Serbia	1901	122	4.2	6.1	5.37	1.26
Transilvania	1523	11	4.5	6.2	0.81	0.52
Vrancea crustal	1893	40	3.8	6.3	3.04	0.89
Vrancea intermediate-depth	1802	97	5.7	8.2	4.40	0.83
Dulovo	1892	21	3.2	7.1	0.59	0.43
Shabla	1900	17	4.5	7.8	2.88	0.82
Gorna	1900	46	4.1	7.4	2.56	0.70
Shumen	1850	19	4.5	6.7	2.33	0.70

possible with a magnitude exceeding M_{max}. A procedure for the evaluation of M_{max}, which is free from subjective assumptions and depends only on seismic data, is given in Kijko (2004) and is applied in this study.

The magnitude and year of completeness, the maximum magnitude and the number of seismic events considered in the analysis are given in Table 9.

The *a* and *b* seismicity parameters, obtained for all the seismic sources, through the maximum likelihood method (McGuire 2004), are also reported in Table 9.

4.2 GMPEs Used in the PSHA

The ground motions prediction equations (GMPEs) initially considered for the probabilistic seismic hazard analysis for Vrancea intermediate-depth seismic source are the ones given in (Delavaud et al. 2012), where four GMPEs selected within the SHARE regional project of Global Earthquake Model (GEM) are recommended for Vrancea subcrustal seismic source, namely: Youngs et al. (1997)—YEA97, Zhao et al. (2006)—ZEA06, Atkinson and Boore (2003)—AB03 and Lin and Lee (2008)—LL08. Besides these GMPEs, the ground motion prediction model proposed in Vacareanu et al. (2014b), named VEA14, for intermediate-depth earthquakes and average soil conditions, is used for the analysis of the seismic hazard of Romania. The GMPEs are applied with different weights in the fore-arc region of Romania

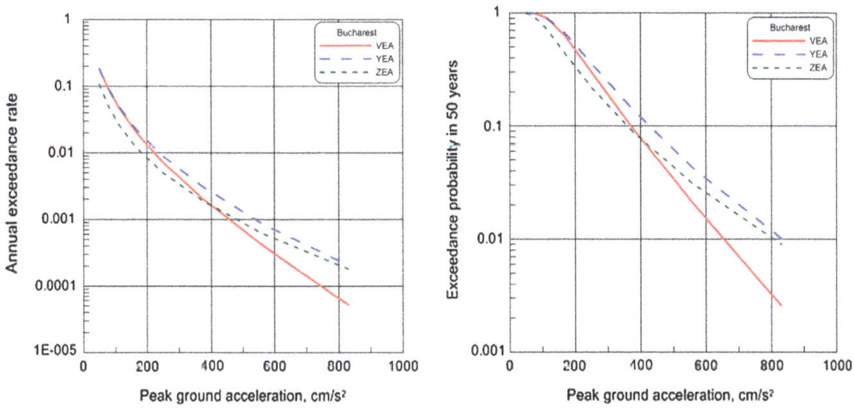

Fig. 14 The influence of GMPE on PSHA results for Bucharest. The results are given as mean annual rate of exceedance (*left*) and exceedance probability in 50 years (*right*). The GMPEs considered are YEA97, ZEA06 and VEA14

(to the east and to the south with respect to the Carpathians Mountains) and in the back-arc region (to the north and to the west with respect to the Carpathians Mountains). The first generation of GMPE (Lungu et al. 1994) developed purposely for Vrancea intermediate depth seismic source was not considered in this study since it is a ground motion model only for peak ground acceleration and the present analysis is considering the spectral accelerations as well.

The influence of the GMPEs used in the analysis of seismic hazard from Vrancea intermediate-depth seismic source is shown in Fig. 14 for the city of Bucharest (situated in the fore-arc region). The ground motion parameter used for comparison in Fig. 14 is the geometric mean of the two orthogonal horizontal components of the peak ground acceleration. The GMPEs used in this comparison are: Youngs et al. (1997)—YEA97, Zhao et al. (2006)—ZEA06 and Vacareanu et al. (2014b)—VEA14. One can notice from Fig. 14 that ZEA06 and VEA14 produce the same *PGA* value for 10 % exceedance probability in 50 years, while YEA97 and VEA14 lead to almost equal *PGA* values for 1 % exceedance probability in 50 years.

For the crustal seismic sources, the GMPEs of Cauzzi and Faccioli (2008)—CF08, Akkar and Bommer (2010)—AB10 and Idriss (2008)—I08 are considered.

4.3 Logic Tree

The logic trees are used in PSHA to incorporate the epistemic uncertainties in the modeling of seismic sources and ground motion prediction (Bommer et al. 2005; Bommer and Scherbaum 2008). The branches of the logic trees represent alternative models or values of the parameters considered in the analysis. The epistemic uncertainties are expressed as branches' weights representing either the goodness-of-fit of,

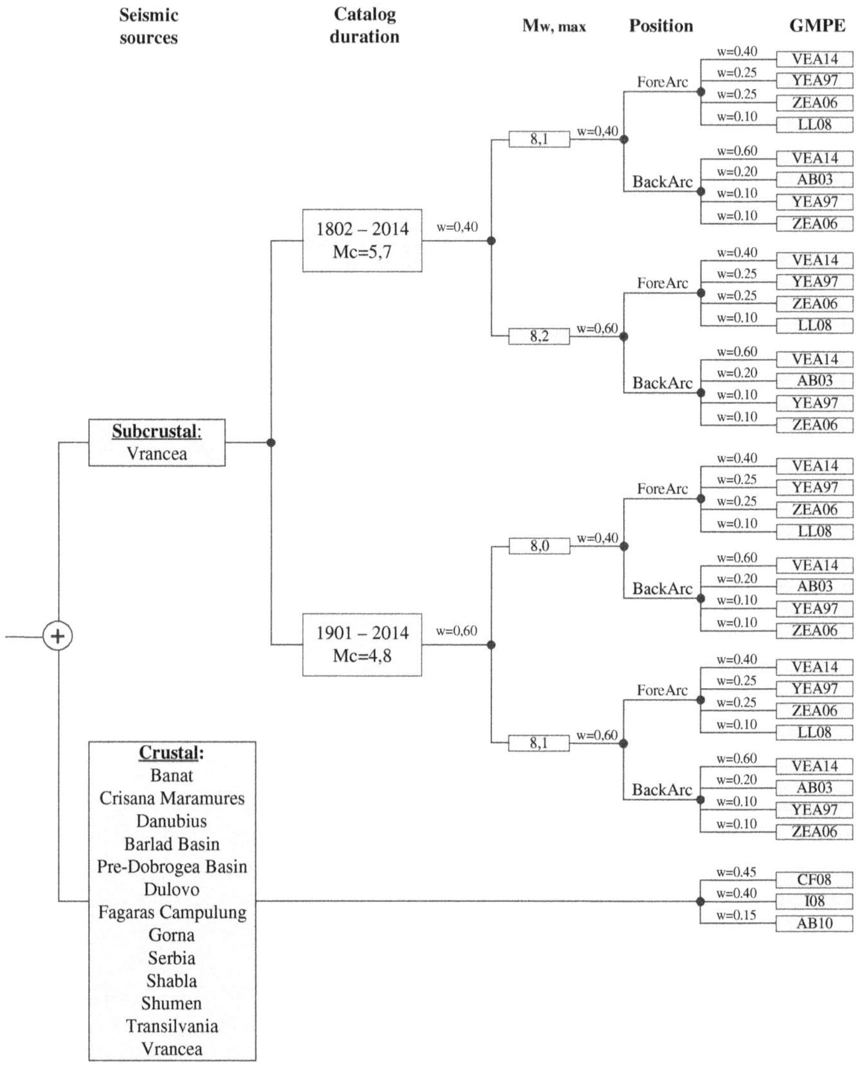

Fig. 15 Aggregate logic tree used in the *PSHA* for Romania

or the degree of belief in the models and/or parameters assigned to the corresponding branches. The logic tree approach was introduced in the PSHA for the first time by Kulkarni et al. (1984).

The logic tree used in this study is presented in Fig. 15 and aggregates altogether the crustal and subcrustal seismic sources, as well as the situation when a given site is situated in the fore-arc or back-arc region with respect to Vrancea subcrustal seismic source. The weighted branches of the logic tree refers to the catalogue duration and the maximum magnitude for Vrancea intermediate-depth seismic source, and to the GMPEs used in PSHA for all the seismic sources. The weights for the catalogue duration are based on expert judgment, considering a larger weight for the catalogue with the seismic events from 1901 up to 2014 due to the more reliable information on position of focus and magnitude of earthquakes. The weights for the GMPEs are based on goodness-of-fit indicators and are detailed in Sect. 2. The GMPEs used in the analysis and their weights given in Sect. 3 are shown in Fig. 15.

Also, one can notice in Fig. 15 that there are branches without weights assigned, as it is the case for the seismic sources and position of the analyzed site. These branches are just differentiating, in the overall PSHA, between the crustal and intermediate-depth seismic sources and between the sites that are considered to be in the fore-arc region or in the back-arc region with respect to the Carpathians Mountains for Vrancea subcrustal source.

4.4 PSHA Results and Seismic Hazard Maps

The PSHA is performed for the following ground motion parameters: geometric mean of the two orthogonal horizontal components of either peak ground acceleration or spectral accelerations at vibration periods $T = 0.2$ s; 0.4 s; 1.0 s; 2.0 s and 3.0 s. The selected vibration periods are common for all the GMPEs used in the analysis. The computer program CRISIS2008 (Ordaz et al. 2013) was used for performing PSHA. The following results for the seismic hazard of Romania are based on the largest database of earthquakes and strong ground motions used so far towards this aim.

The seismic mean hazard curves for two selected cities, Bucharest (average epicentral distance with respect to Vrancea subcrustal source of 130 km towards S-SW) and Galati (average epicentral distance with respect to Vrancea subcrustal source of 115 km towards E-SE) are presented in Fig. 16.

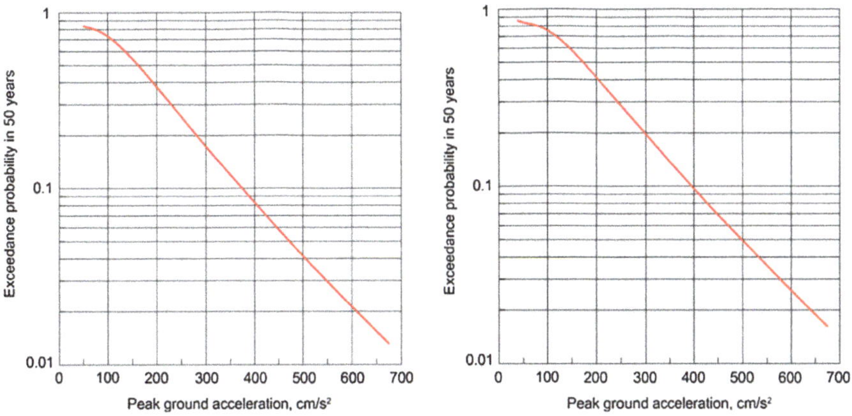

Fig. 16 Seismic hazard curves for two selected cities: Bucharest (*left*) and Galati (*right*)

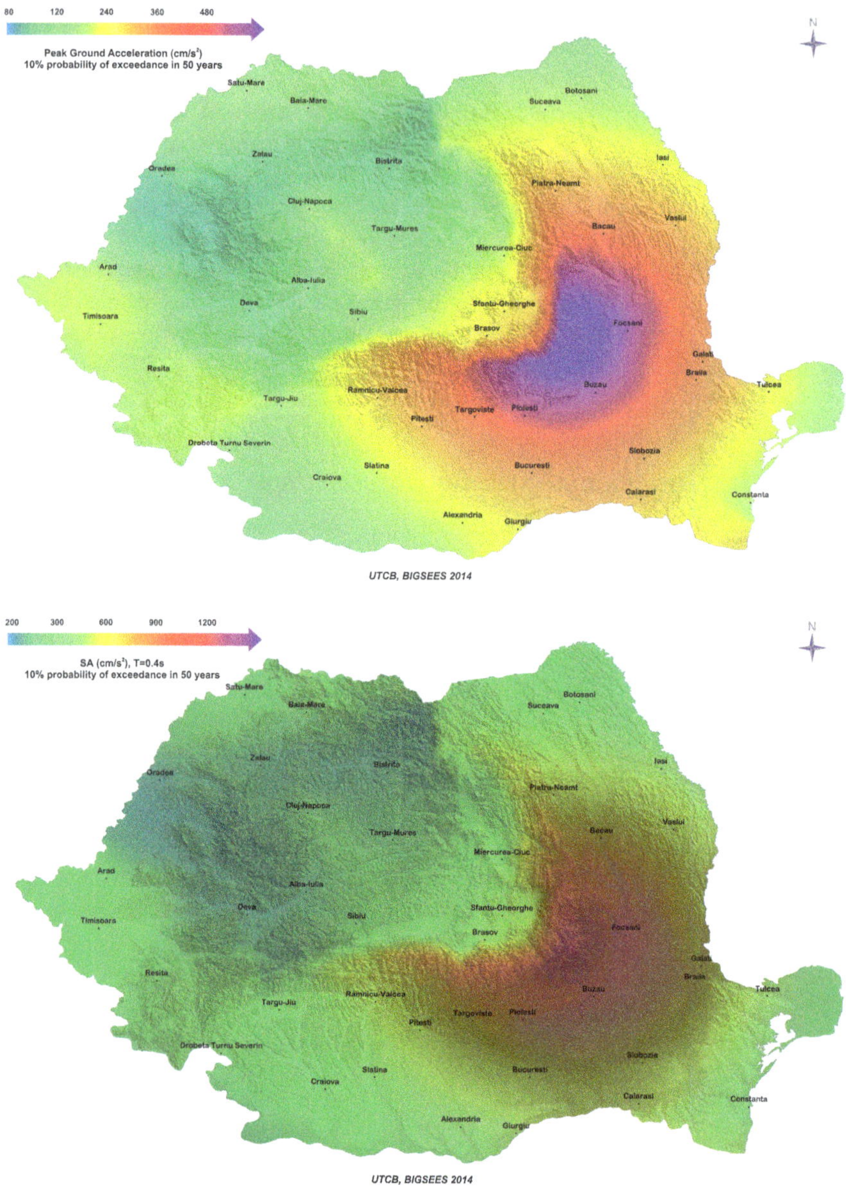

Fig. 17 Seismic hazard maps of Romania. The results are given as peak ground acceleration (*top*) and spectral acceleration for $T = 0.4$ s (*bottom*) with 10 % exceedance probability in 50 years (the acceleration values are expressed as the geometric mean of the two orthogonal horizontal components of the motion). The contribution to seismic hazard of all seismic sources that influence the Romanian territory is considered

Some results of the PSHA for Romania are presented in a concise form in Fig. 17. The seismic hazard maps for the geometric mean of the two orthogonal horizontalcomponents of the peak ground acceleration and of the spectral acceleration for $T = 0.4$ s with 10 % exceedance probability in 50 years are given in Fig. 17, considering the aggregate contribution of all the seismic sources included in the analysis. The strong influence of Vrancea intermediate-depth seismic source for the S-E of Romania is obvious. The contribution to the seismic hazard of the sources situated in N-E of Bulgaria and of the Fagaras-Campulung and Transilvania seismic sources is noticeable from Fig. 17.

5 Conclusions

This study focuses on the probabilistic seismic hazard assessment for Romania. Four GMPEs recommended within the SHARE project (Delavaud et al. 2012), as well as a GMPE developed purposely for Vrancea subcrustal seismic source (Vacareanu et al. 2014b), are tested using some procedures given in the literature (Delavaud et al. 2012; Scherbaum et al. 2004, 2009; Kale and Akkar 2013). The testing is performed separately for the region in front of the Carpathian Mountains (fore-arc region) and for the region behind the Carpathian Mountains (back-arc region). The strong ground motion database consists of over 400 strong ground motion recordings from ten intermediate-depth Vrancea seismic events. The strong ground motions recorded in October 6, 2013 Vrancea earthquake with $M_W = 5.2$ are also included in the database. In addition, three ground motion prediction models developed for crustal seismic sources are evaluated using the same testing procedures against a strong ground motion database of 125 recordings in Romania from 25 earthquakes. The results show that for Vrancea subcrustal seismic source the VEA14, YEA97 and ZEA06 ground motion prediction models are graded the best in the case of the fore-arc region, while in the back-arc region the VEA14 and AB03 GMPEs provide the most reliable estimates of the observed ground motions. In the case of the crustal seismic sources, the CF08 and I08 ground motion models fit the best with the available strong ground motion database. A new weighting scheme to be used for the probabilistic seismic hazard assessment (*PSHA*) of Romania is given, considering only the best four ground motion prediction models in both the fore-arc and the back-arc regions for subcrustal earthquakes and the three GMPEs selected for crustal erathquakes. The weighting schemes are based on the computed values of several goodness-of-fit measures (*LLH* and *EDR*), as well as on the distribution of the inter-event and intra-event residuals and on expert judgement.

Next, a sensitivity analysis performed for four cities in Romania (in the fore-arc region of the Carpathian Mountains) allowed the proper selection of the input data for the PSHA presented in Sect. 4. For the exceedance probabilities of interest in civil engineering, the earthquake catalogue duration proved to have a negligible influence, the minimum and maximum magnitudes a relatively small influence, and the number of normalized residuals (ε) a significant influence.

The results presented in this chapter are obtained within the framework of BIGSEES national research project (http://infp.infp.ro/bigsees/default.htm) and represents a preliminary outcome, as of 2014. The final results are expected by the end of 2016 and will represent the basis for the revision of the seismic action calibration in the earthquake resistant design code of Romania.

Acknowledgments Funding for this research was provided within BIGEES Project by the Romanian Ministry of Education and Scientific Research under the Grant Number 72/2012. This support is gratefully acknowledged. The authors would like to extend their gratitude to Professor Mario Ordaz for providing the CRISIS2008 computer code.

References

Akkar, S., & Bommer, J. J. (2006). Influence of long-period filter cut-off on elastic spectral displacements. *Earthquake Engineering and Structural Dynamics, 35*, 1145–1165.

Akkar, S., & Bommer, J. J. (2010). Empirical equations for the prediction of PGA, PGV and spectral accelerations in Europe, the Mediterranean and the Middle East. *Seismological Research Letters, 81*(2), 195–206.

Aldea, A., Vacareanu, R., Lungu, D., Pavel, F., & Demetriu, S. (2014). Probabilistic seismic hazard for Romania. Part II: sensitivity analysis. In: Vacareanu R, Ionescu C, editors, *Proceedings of the 5th National Conference on Earthquake Engineering & 1st National Conference on Earthquake Engineering and Seismology* (pp. 221–228), Bucharest, Romania.

Ardeleanu, L., Leydecker, G., Bonjer, K.-P., Busche, H., Kaiser, D., & Schmitt, T. (2005). Probabilistic seismic hazard map for Romania as a basis for a new building code. *Natural Hazards and Earth Systems Sciences, 5*, 679–684.

Atkinson, G., & Boore, D. (2003). Empirical ground-motion relations for subduction-zone earthquakes and their application to Cascadia and other regions. *Bulletin of the Seismological Society of America, 93*(4), 1703–1729.

Beauval, C., Bard, P. Y., Hainzl, S., & Guéguen, P. (2008). Can strong-motion observations be used to constrain probabilistic seismic-hazard estimates? *Bulletin of the Seismological Society of America, 98*(2), 509–520.

Beauval, C., Hainzl, S., & Scherbaum, F. (2006). The impact of the spatial uniform distribution of seismicity on probabilistic seismic-hazard estimation. *Bulletin of the Seismological Society of America, 96*(6), 2465–2471.

Bommer, J. J., & Scherbaum, F. (2008). The use and misuse of logic-trees in probabilistic Seismic hazard analysis. *Earthquake Spectra, 24*, 997–1009.

Bommer, J. J., Scherbaum, F., Bungum, H., Cotton, F., Sabetta, F., Abrahamson, N. A., et al. (2005). On the use of logic trees for ground-motion prediction equations in seismic-hazard analysis. *Bulletin of the Seismological Society of America, 95*, 377–389.

Boore, D., & Bommer, J. J. (2006). Processing of strong motion accelerograms: Needs, options and consequence. *Soil Dynamics and Earthquake Engineering, 25*, 93–115.

Cauzzi, C., & Faccioli, E. (2008). Broadband (0.05 to 20 s) prediction of displacement response spectra based on worldwide digital records. *Journal of Seismology, 12*(4), 453–475.

CEN. (2004). *Eurocode 8: Design of structures for earthquake resistance, Part 1: General rules, seismic actions and rules for buildings*. EN 1998-1:2004. Brussels, Belgium.

Cornell, C. A. (1968). Engineering seismic risk analysis. *Bulletin of the Seismological Society of America, 58*, 1583–1606.

Delavaud, E., Cotton, F., Akkar, S., Scherbaum, F., Danciu, L., Beauval, C., et al. (2012). Toward a ground-motion logic tree for probabilistic seismic hazard assessment in Europe. *Journal of Seismology, 16*(3), 451–473.

Douglas, J. (2011). *Ground-motion prediction equations 1964–2010*. PEER Report 2011/102, College of Engineering, Berkeley.

Idriss, I. M. (2008a). An NGA empirical model for estimating the horizontal spectral values generated by shallow crustal earthquakes. *Earthquake Spectra, 24*(1), 217–242.

Idriss, I. M. (2008b). An NGA empirical model for estimating the horizontal spectral values generated by shallow crustal earthquakes. *Earthquake Spectra, 24*(1), 217–242.

Iervolino, I. (2013). Probabilities and fallacies: Why hazard maps cannot be validated by individual earthquakes. *Earthquake Spectra, 29*(3), 1125–1136.

Ismail-Zadeh, A., Matenco, L., Radulian, M., Cloetingh, S., & Panza, G. (2012). Geodynamics and intermediate-depth seismicity in Vrancea (the south-eastern Carpathians): Current state-of-the art. *Tectonophysics, 530–531*, 50–79.

Kaklamanos, J., Baise, L. G., & Boore, D. M. (2011). Estimating unknown input parameters when implementing the NGA ground-motion prediction equations in engineering practice. *Earthquake Spectra, 27*(4), 1219–1235.

Kale, Ö., & Akkar, S. (2013). A new procedure for selecting and ranking ground-motion prediction equations (GMPEs): The Euclidean distance-based ranking (EDR) method. *Bulletin of the Seismological Society of America, 103*(2A), 1069–1084.

Kijko, A. (2004). Estimation of the maximum earthquake magnitude, m_{max}. *Pure Applied Geophysics, 161*, 1655–1681.

Kramer, S. (1996). *Geotechnical earthquake engineering*. Upper Saddle River: Prentice Hall.

Kulkarni, R. B., Youngs, R. R., & Coppersmith, K. J. (1984). Assessment of confidence intervals for results of seismic hazard analysis. In *Proceedings, Eighth World Conference on Earthquake Engineering* (Vol. 1, pp. 263–270), International Association for Earthquake Engineering, Tokyo.

Leydecker, G., Busche, H., Bonjer, K.-P., Schmitt, T., Kaiser, D., Simeonova, S., et al. (2008). Probabilistic seismic hazard in terms of intensities for Bulgaria and Romania—Updated hazard maps. *Natural Hazards and Earth Systems Sciences, 8*, 1–9.

Lin, P. S., & Lee, C. T. (2008). Ground-motion attenuation relationships for subduction-zone earthquakes in Northeastern Taiwan. *Bulletin of the Seismological Society of America, 98*(1), 220–240.

Lungu, D., Cornea, T., & Nedelcu, C. (1999). Hazard assessment and site dependent response for Vrancea earthquakes. In F. Wenzel, et al. (Eds.), *Vrancea earthquakes: Tectonics, hazard and risk mitigation* (pp. 251–267). Dordrecht: Kluwer.

Lungu, D., Demetriu, S., Radu, C., & Coman, O. (1994). Uniform hazard response spectra for Vrancea earthquakes in Romania. *Proceedings of Tenth European Conference on Earthquake Engineering* (Vol. 1, pp. 365–370).

Lungu, D., Vacareanu, R., Aldea, A., & Arion, C. (2000). *Advanced structural analysis*. Bucharest: Conspress.

Lungu D., Demetriu, S., Aldea, A., & Arion, C. (2006). Probabilistic seismic hazard assessment for Vrancea earthquakes and seismic action in the new seismic code of Romania. *Proceedings of First European Conference on Earthquake Engineering and Seismology*, (September 3–8, 2006, Geneva, CD_ROM, paper 1003).

Mäntyniemi, P., Marza, V. I., Kijko, A., & Retief, P. (2003). A new probabilistic seismic hazard analysis for the Vrancea (Romania) Seismogenic Zone. *Natural Hazards, 29*, 371–385.

Marmureanu, G., Cioflan, C. O., & Marmureanu, A. (2010). *Research regarding the local seismic hazard (microzonation) of the Bucharest metropolitan area (in Romanian)*. Iasi: Tehnopress.

Marmureanu, G., Popescu, E., Popa, M., Moldovan, A. I., Placinta, A. O., & Ralulian, M. (2004). Seismic zoning characterization for the seismic hazard assessment in South-Eastern Romania territory. *Acta Geod Geoph Hung, 39*, 259–274.

Mârza, V. I., Kijko, A., & Mäntyniemi, P. (1991). Estimate of earthquake hazard in the Vrancea (Romania) region. *Pure and Applied Geophysics, 136*, 143–154.

McGuire, R. (1976) FORTRAN computer program for seismic risk analysis. *U.S. Geological Survey Open-File Report* 76–67.

McGuire, R. (1999). Probabilistic seismic hazard analysis and design earthquakes: Closing the loop. *Bulletin of the Seismological Society of America, 85*(5), 1275–1284.

McGuire, R. (2004). Seismic hazard and risk analysis. *Earthquake Engineering Research Institute MNO-10*.

Musson, R. M. W. (2000). Generalized seismic hazard maps for the pannonian basin using probabilistic methods. *Pure and Applied Geophysics, 157*, 147–169.

Ordaz, M., & Reyes, C. (1999). Earthquake hazard in Mexico-City; observations versus computations. *Bulletin of the Seismological Society of America, 89*(5), 1379–1383.

Ordaz, M., Martinelli, F., D'Amico, V., & Meletti, C. (2013). CRISIS2008: A Flexible Tool to Perform Probabilistic Seismic Hazard Assessment. *Seismological Research Letters, 84*(3), 495–504.

Pavel, F., Vacareanu, R., Neagu, C., & Arion, C. (2014a). Probabilistic seismic hazard for Romania. Part I: selection of GMPEs. In: Vacareanu R, Ionescu C, editors, *Proceedings of the 5th National Conference on Earthquake Engineering & 1st National Conference on Earthquake Engineering and Seismology*, (pp. 213–220) Bucharest, Romania.

Pavel, F., Vacareanu, R., Arion, C., & Neagu, C. (2014b). On the variability of strong ground motions recorded from Vrancea earthquakes. *Earthquakes and Structures, 6*(1), 1–18.

Radulian, M., Mandrescu, N., Popescu, E., Utale, A., & Panza, G. (2000a). Characterization of Romanian seismic zones. *Pure and Applied Geophysics, 157*, 57–77.

Radulian, M., Vaccari, F., Mandrescu, N., Panza, G. F., & Moldoveanu, C. L. (2000b). Seismic hazard of Romania: Deterministic approach. *Pure and Applied Geophysics, 157*, 221–247.

Reiter, L. (1990). *Earthquake hazard analysis: Issues and insights*. New York: Columbia University Press.

Rydelek, P. A., & Sacks, I. S. (1989). Testing the completeness of earthquake catalogs and the hypothesis of self-similarity. *Nature, 337*, 251–253.

Scassera, G., Stewart, J., Bazzurro, P., Lanzo, G., & Mollaioli, F. (2009). A comparison of NGA ground-motion prediction equations to Italian data. *Bulletin of the Seismological Society of America, 99*(5), 2961–2978.

Scherbaum, F., Cotton, F., & Smit, P. (2004). On the use of response spectral-reference data for the selection and ranking of ground-motion models for seismic-hazard analysis in regions of moderate seismicity: The case of rock motion. *Bulletin of the Seismological Society of America, 94*(6), 2164–2185.

Scherbaum, F., Delavaud, E., & Riggelsen, E. (2009). Model selection in seismic hazard analysis: An information theoretic perspective. *Bulletin of the Seismological Society of America, 99*(6), 3234–3247.

Simeonova, S. D., Solakov, D. E., Leydecker, G., Bushe, H., Schmitt, T., & Kaiser, D. (2006). Probabilistic seismic hazard map for Bulgaria as a basis for a new building code. *Natural Hazards and Earth Systems Sciences, 6*, 881–887.

Sokolov, V Yu., Wenzel, F., & Mohindra, R. (2009). Probabilistic seismic hazard assessment for Romania and sensitivity analysis: A case of joint consideration of intermediate-depth (Vrancea) and shallow (crustal) seismicity. *Soil Dynamics and Earthquake Engineering, 29*, 364–381.

Strasser, F., Bommer, J. J., & Abrahamson, N. A. (2008). Truncation of the distribution of ground-motion residuals. *Journal of Seismology, 12*, 79–105.

Stucchi, M., Rovida, A., Gomez Capera, A. A., Alexandre, P., Camelbeeck, T., Demircioglu, M. B., et al. (2013). The SHARE European Earthquake Catalogue (SHEEC) 1000–1899. *Journal of Seismology, 17*, 523–544.

Vacareanu, R., Lungu, D., Marmureanu, G., Cioflan, C., Aldea, A., & Arion, C., et al. (2013a). Statistics of seismicity for Vrancea subcrustal source. In *Proceedings of the International Conference on Earthquake Engineering* SE-50 EEE, 29–31 May 2013 (Paper no. 138), Skopje.

Vacareanu, R., Pavel, F., & Aldea, A. (2013b). On the selection of GMPEs for Vrancea subcrustal seismic source. *Bulletin of Earthquake Engineering, 11*(6), 1867–1884.

Vacareanu, R., Lungu, D., Aldea, A., Demetriu, S., Pavel, F., Arion, C., Iancovici, M., & Neagu, C. (2014a). Probabilistic seismic hazard for Romania. Part III: seismic hazard maps. In: Vacareanu R, Ionescu C, editors, *Proceedings of the 5th National Conference on Earthquake*

Engineering & 1st National Conference on Earthquake Engineering and Seismology, (pp. 229–236) Bucharest, Romania.

Vacareanu, R., Radulian, M., Iancovici, M., Pavel, F., & Neagu, C. (2014b). Fore-arc and back-arc ground motion prediction model for Vrancea intermediate depth seismic source. *Journal of Earthquake Engineering*, DOI: 10.1080/13632469.2014.990653.

Wald, D. J., & Allen, T. I., (2007). Topographic slope as a proxy for seismic site conditions and amplification, *Bulletin of the Seismological Society of America* , *97*(5): 1379–1395.

Wiemer, S., & Wyss, M. (2000). Minimum magnitude of complete reporting in earthquake catalogs: Examples from Alaska, the western United States, and Japan. *Bulletin of the Seismological Society of America, 90*, 859–869.

Youngs, R. R., Chiou, S. J., Silva, W. J., & Humphrey, J. R. (1997). Strong ground motion attenuation relationships for subduction zone earthquakes. *Seismological Research Letters, 68*(1), 58–73.

Zhao, J., Zhang, J., Asano, A., Ohno, Y., Oouchi, T., Takahashi, T., et al. (2006). Attenuation relations of strong ground motion in Japan using site classification based on predominant period. *Bulletin of the Seismological Society of America, 96*(3), 898–913.

Zuniga, F. R., & Wyss, M. (1995). Inadvertent changes in magnitude reported in earthquake catalogs: Their evaluation through b-value estimates. *Bulletin of the Seismological Society of America, 85*, 1858–1866.

Practicality of Monitoring Crustal Deformation Processes in Subduction Zones by Seafloor and Inland Networks of Seismological Observations

Keisuke Ariyoshi and Yoshiyuki Kaneda

1 Introduction: A Review of Previous Studies on the Locked Region off Tokai

1.1 The Seismic Gap off Tokai, Central Japan

The seismic gap off Tokai (Ishibashi 1981) is well known, and the last Tokai earthquake (M8.4) occurred there in 1854. This Tokai seismic gap has had a large social impact on seismologists worldwide and also the general public in Japan, and many researchers have investigated the risk occurrence of another earthquake from seismological and geological aspects.

Figure 1b reviews previous studies of the seismic gap. Ishibashi (1981) proposed fault models for the 1944 Tonankai earthquake (F1, F2) and the next Tokai earthquake (F3) on the basis of seismological tectonics. Yoshioka et al. (1993), Sagiya (1999) estimated the back slip (Savage 1983) region of the Tokai earthquake using benchmarks and triangulation and GPS, respectively. Matsumura (1997) estimated the locked region on the basis of fault mechanisms for small earthquakes, as shown in Fig. 1d. Taking these results into account, the Central Disaster Management Council (2001) proposed the source region of the next Tokai earthquake. The Geographical Survey Institute has continued dense observations of inland leveling changes in the Tokai district since 1979.

Some seismologists have considered these continuous leveling observations useful data to evaluate long-term changes in crustal deformation resulting from the

K. Ariyoshi (✉) · Y. Kaneda
Earthquake and Tsunami Research Project for Disaster Prevention,
Japan Agency for Marine-Earth Science and Technology, Yokohama 236-0001, Japan
e-mail: ariyoshi@jamstec.go.jp

Y. Kaneda
Disaster Mitigation Research Center, Nagoya University, Nagoya 464-8601, Japan

S. D'Amico (ed.), *Earthquakes and Their Impact on Society*,
Springer Natural Hazards, DOI 10.1007/978-3-319-21753-6_6

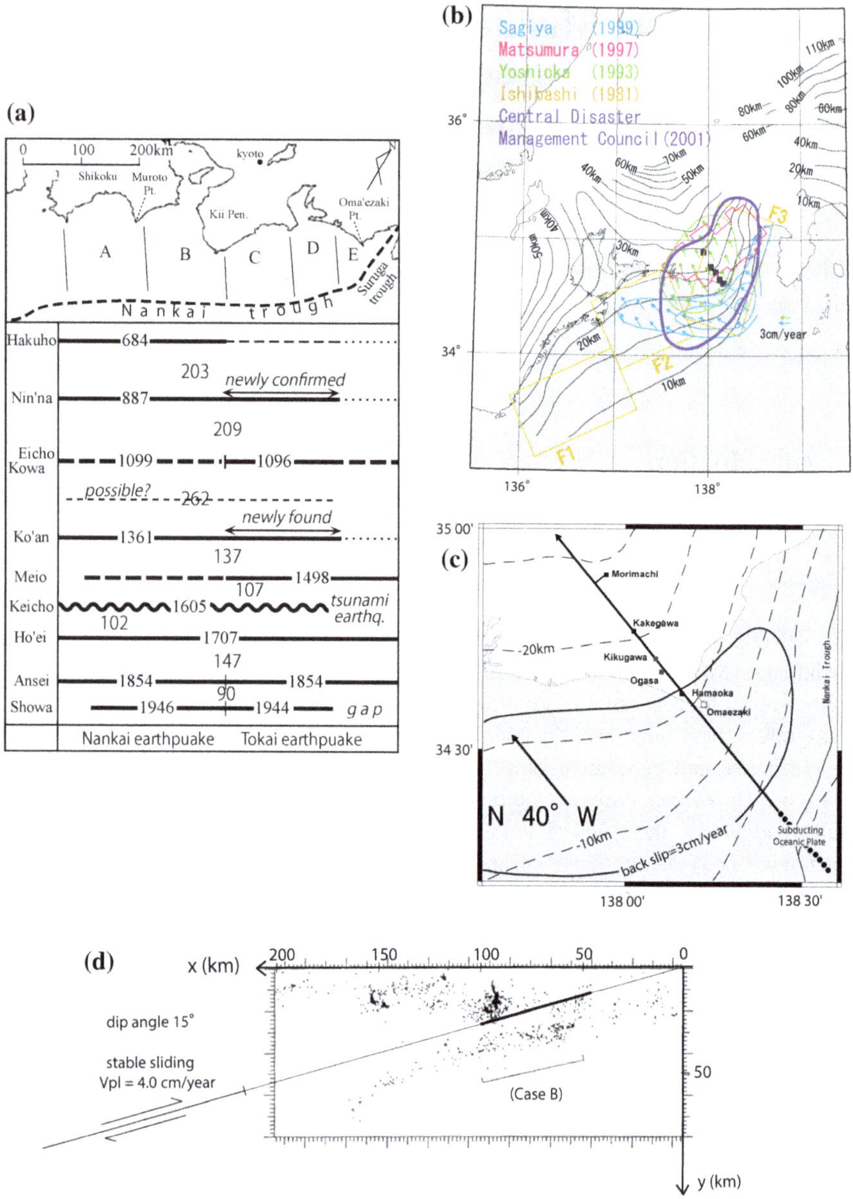
(a)
0 100 200km
kyoto
Shikoku
Muroto Pt.
Kii Pen.
Oma'ezaki Pt.
A
B
C
D
E
Suruga trough
Nankai trough
Hakuho 684
203
newly confirmed
Nin'na 887
209
Eicho Kowa 1099 1096
possible? 262
newly found
Ko'an 1361
137
Meio 1498
107
Keicho 1605 tsunami earthq.
102
Ho'ei 1707
147
Ansei 1854 1854
90
Showa 1946 1944 gap
Nankai earthpuake
Tokai earthpuake
(b)
Sagiya (1999)
Matsumura (1997)
Yoshioka (1993)
Ishibashi (1981)
Central Disaster
Management Council (2001)
F1
F2
F3
3cm/year
36°
34°
136°
138°
(c)
35 00'
34 30'
138 00'
138 30'
Morimachi
Kakegawa
Kikugawa
Ogasa
Hamaoka
Omaezaki
-20km
-10km
N 40° W
back slip=3cm/year
Subducting Oceanic Plate
Nankai Trough
(d)
x (km)
200
150
100
50
0
y (km)
dip angle 15°
stable sliding
Vpl = 4.0 cm/year
(Case B)

◀ **Fig. 1** **a** Historical megathrust earthquake activity along the Nankai–Suruga trough (slightly modified from Hori (2006)). Roman and italic numerals indicate earthquake occurrence years and time intervals between two successive series, respectively. *Thick solid*, thick broken, and thin *broken lines* show certain, probable, and possible rupture zones, respectively. *Thin dotted lines* mean unknown. **b** Overview of the Tokai district with previous studies superimposed. Broken *contour lines* indicate the depth of the plate interface fixed by Noguchi (1996). *Rectangles* (F1–F3) represent the fault model of the Tonankai earthquake (F1, F2) and the next Tokai earthquake (F3) proposed by Ishibashi (1981). The enclosed regions indicated by *blue*, *magenta*, *green*, and *purple lines* represent the locked region, as estimated by Sagiya (1999), Matsumura (1997), Yoshioka et al. (1993), and the Central Disaster Management Council (2001). **c** Close-up of the target area showing the horizontal dip direction (X-axis) and benchmark points. **d** The 2D model for plate subduction in the Tokai district used in the numerical simulation. The plate boundary is shown by a solid line, the thick line represents the zone of $(a - b) < 0$. Hypocenters and the locked region on the plate interface determined by Matsumura (1997) are shown by dots and by the segment shown as a line along the dip, respectively. Stable sliding at 4.0 cm/year is assumed for plate boundary deeper than 60 km. Figure 1b, c, and d are modified from Ariyoshi et al. (2003)

Tokai earthquake because GPS and InSAR observations have only recently started, whereas the recurrence interval of the Tokai earthquake is expected to be more than 100 years. In this section, we review an example of trial estimation to forecast crustal deformation on the basis of the leveling data set and discuss problems from previous studies that need to be solved.

1.2 2D Preliminary Model of the Tokai Earthquake

To understand the present coupling state of the plate interface and forecast crustal deformation, we formulate a 2D model (Fig. 1d) of recurrent megathrust earthquakes in the Tokai area along dip direction, as in Fig. 1c, based on the back slip direction. Using results from Sagiya (1999), Matsumura (1997), as shown in Fig. 1b, we erected two models (Cases A and B), as shown in Fig. 2, for which the frictional parameters of the rate- and state-dependent aging law (Dieterich 1979; Ruina 1983) are listed in Table 1. We adopt a quasi-dynamic equilibrium (Rice 1993) between frictional resistance and shear stress driven by dip slip on each divided subfault. As all cases represent characteristic megathrust earthquakes, we calculate the recurrence interval (T_r) and coupling ratio (*CR*) (slip faster than 1 cm/s over total slip), as listed in Table 1.

1.3 Trial Fitting of Simulated Leveling Changes to Observed Data at Benchmarks

In this study, we use two data sets of leveling changes at Hamaoka, Ogasa, Kikugawa, Kakegawa, and Morimachi (Fig. 1c) from 1994 to (i) 2001 and (ii) 2003

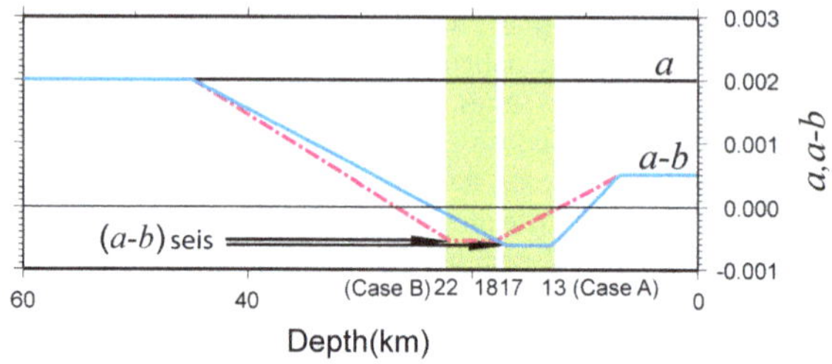

Fig. 2 Depth variations in friction parameters after Ariyoshi et al. (2001). The value of $(a-b)_{\text{seis}}$ is defined as the most negative and different among models. The depth range with this value is fixed from 13 to 17 km and from 18 to 22 km for Cases A and B, respectively. Cases A-2 and B-2 are shown as examples

Table 1 Assumed values of frictional parameters and results of numerical simulations (T_r and CR)

Case	$(a\text{–}b)_{\text{seis}}$ [10^{-4}]	d_c [cm]	T_r [year]	CR [%]
A-2	6.1	1.0	115.3	70.2
A-3	6.02	3.0	115.7	40.3
B-2	5.38	3.0	116.7	50.4

in which Kakegawa is treated as a reference point. By converting the horizontal distance between the five benchmarks along the X-axis, we also calculate the time history of the simulated leveling changes at the five benchmarks using a grid search of two unknown parameters (D_x, T_{2001} and T_{2003}), as shown in Fig. 3.

Before fitting to the observed data, we extract trend components using a stationary autoregressive component model, as shown in Fig. 4. To estimate the values of D_x, T_{2001}, and T_{2003} with the residual prediction, we adopt the least squares method between the trend component and observed data at Hamaoka and Morimachi relative to Kakegawa.

The results are summarized in Table 2, and Fig. 5 shows the fitting data at Hamaoka as an example. As the Tonankai earthquake has not yet occurred, our trial estimation of both T_{2001} and T_{2003} in Table 2 fails; this indicates several problems as follows:

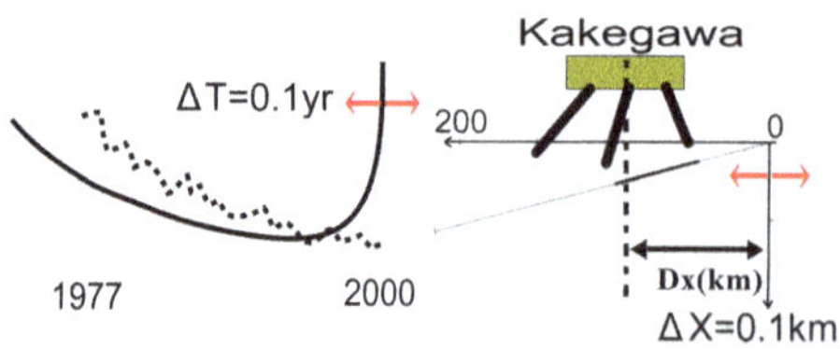

Fig. 3 Schematic illustration of the fitting method used. There are two unknown parameters: (I) distance between the oceanic trench and benchmark point of Kakegawa (D_x) and (II) occurrence time of the next megathrust earthquake when using the observed data until 2001 and 2003 (T_{2001} and T_{2003})

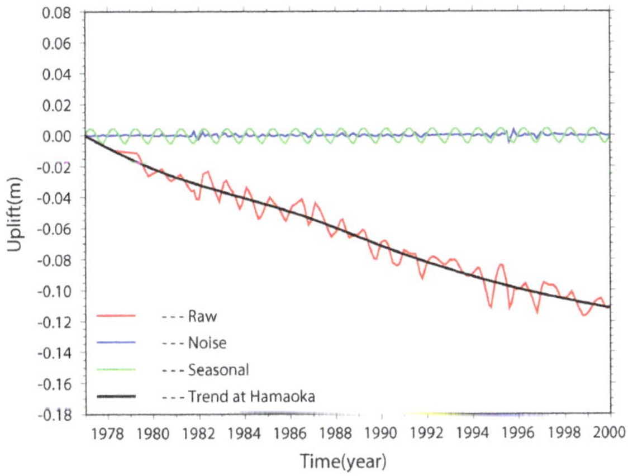

Fig. 4 An example of filtering of leveling data between Hamaoka and Kakegawa (reference) classified from raw data (*red*) into trend (*black*), seasonal (*green*) and noise (*blue*) components

Table 2 Results of estimated unknown parameters with residual minimum square (R.m.s.)

Case	D_x [km][a]	$T_{2001} \pm 2\sigma$ [A.D.]	$T_{2003} \pm 2\sigma$ [A.D.]	R.m.s. [cm][a]
A-2	71.4	2001.1–2002.4	2003.4–2003.8	1.0
A-3	65.0	2010.9–2011.4	2011.5–2011.9	0.2
B-2	80.5	2006.5–2008.5	2007.7–2009.3	1.0

[a] D_x and R.m.s. values are almost the same between T_{2001} and T_{2003}

(i) The values of T_{2001} and T_{2003} tend to be shorter in cases with shorter critical distance because pre-seismic slip becomes smaller for a shorter critical distance (Kato and Hirasawa 1999).

(ii) When the rate of change of uplift is quite small, the difference between T_{2001} and T_{2003} should be 2003–2001 = 2. This means that the reliability of estimation should be based on the convergence of T_{xxxx} (*xxxx*: the latest year of available data) rather than the R.m.s.

(iii) This long-term uplift is driven by slow slip events in the western Tokai region centered on Lake Hamana, adjacent to the anticipated Tokai earthquake source area from 2000 to 2005 (Ozawa et al. 2002), which is not considered in our 2D model.

Therefore, it is necessary to formulate a 3D model of the subduction plate boundary and fit various types of observational data in addition to leveling changes. We develop this model in the following sections.

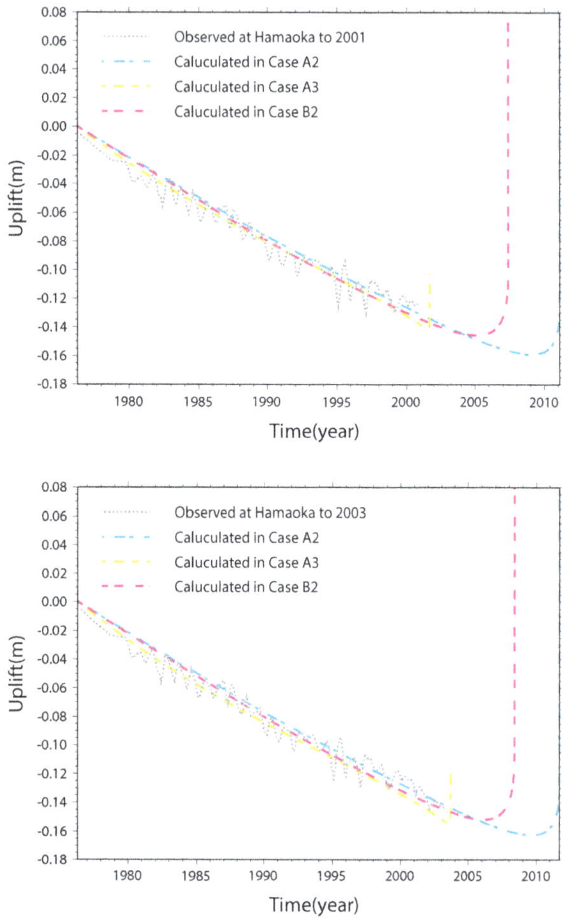

Fig. 5 Comparison of observed (*gray lines*) and calculated (*colored lines*) changes in uplift at Hamaoka from 1994 to 2001 (*top*) and 2003 (*bottom*)

2 Trial Estimation of the Stress Field Using Inland Observation Networks

To assess the validity of the fitting analysis, we also calculate the stress changes of small intraplate earthquakes during the observational period by applying the T_{2001} (or T_{2003}) value to the numerical simulation. Figure 6 shows a comparison between the P-axes of microearthquakes observed between 1980 and 1992 and simulated principal stress changes during that time using T_{2001}. From Fig. 6, we find that most of the small intraplate earthquakes reflect the estimated stress field change in the interseismic stage of the Tokai earthquake, which means that the assumed location of the locked region of the Tokai earthquake and shape of the subduction plate boundary in our 2D model are largely consistent with the observational results and analysis of small intraplate earthquakes is useful to monitor the stress field change.

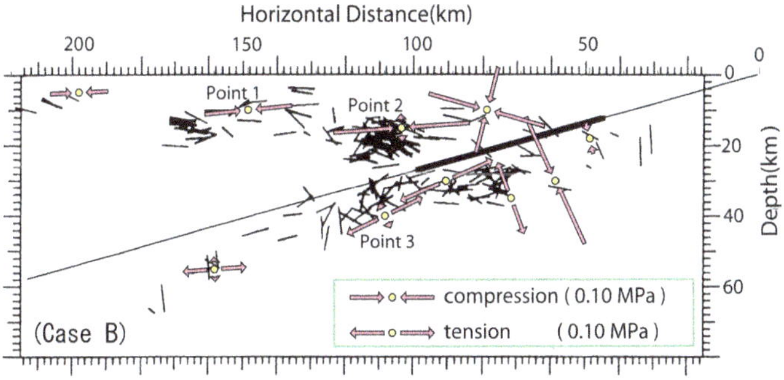

Fig. 6 Comparison of the P-axis directions of small intraplate earthquakes observed from 1980 to 1992 (*solid lines*, estimated by Matsumura (1997)) and principal vectors of stress change (*arrows*) calculated in Case B with T_{2001} = 2007.4, as listed in Table 2, after Ariyoshi et al. (2001)

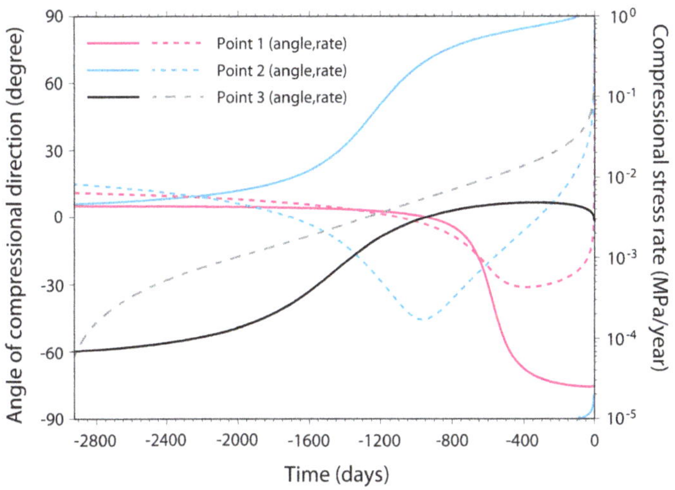

Fig. 7 Precursory changes in stress rate at Points 1–3 of Fig. 6. Compressional directions are shown by *solid lines* with the *left* vertical scale. The rates are shown by *dashed lines* with the *right* vertical scale. This figure is modified after Ariyoshi et al. (2001)

This result is largely similar for all cases in Table 2. As this result is also similar when changing the value of T_{2001} (or T_{2003}) from 2003 to 2060, the stress field in Fig. 6 is sufficiently robust because of the interseismic stage for the Tokai earthquake.

On the basis of this consistency at Points 1–3 in Fig. 6, where small intraplate earthquakes are locally active, Ariyoshi et al. (2001) also calculated the stress change in the pre-seismic stage to discuss detectability in advance of the occurrence. Figure 7 shows the stress rate with rotation along the dip direction as the principal direction at the three points in the pre-seismic stage. At Points 2 and 3, approximately five years before the occurrence, Fig. 7 shows that the principal stress at the down-dip edge of the locked region is nearly tensional and compressional along the dip direction in the upper and lower parts of the intraplate, respectively. As its change rate increases exponentially several years before the occurrence, large foreshocks may occur, as observed for the 1978 earthquake off Miyagi for which several M6-class foreshocks occurred about one year to six months before the earthquake (Takagi 1980). At point 1, Fig. 7 shows that the rate of stress change has decreased several years before the occurrence, which suggests that shallow earthquakes in the continental plate become inactive or quiescent. From these results, monitoring of intraplate earthquakes is important to estimate the state of plate coupling around megathrust earthquakes.

3 Modeling of the Earthquake Cycle for the 1944 Tonankai Earthquake

Following the 2004 Sumatra-Andaman Earthquake (e.g., Lay et al. 2005), the Japanese government has established the Dense Oceanfloor Network system for Earthquakes and Tsunamis (DONET) along the Nankai Trough (Fig. 8). In the Tonankai district, *M*8-class (*M* 8.1–8.5) megathrust earthquakes will probably occur in the near future (e.g., The Headquarters for Earthquake Research Promotion 2013); DONET-I has now operated since August 2011.

In this section, we develop the modeling of a subduction plate boundary around the Tonankai earthquake source region by modeling shallower slow earthquakes, as well as deeper ones around a megathrust earthquake in order to understand the long-term cycle of the shallower slow earthquake in terms of detectability by DONET-I.

From the possible location of the Nankai Trough estimated by Coffin et al. (1998), Bird (2003), we use two ways of setting a model region, as shown in Fig. 8. For simplicity, we assume that the direction of the relative plate motion is in a purely dip direction at a rate of 4 cm/year over the whole model region. Although the observed direction of relative plate motion differs horizontally by as much as 20–30° from the dip direction (Sella 2002) and estimated back slip rate in the corresponding region appears to be in the range 3–5 cm/year (Ito et al. 1999; Miyazaki and Heki 2001), we ignore these differences in the present simple model. Using the result of Nakanishi et al. (2008) for the dip angle of the plate boundary, we consider a 3D model of the Tonankai district in a uniform elastic half-space, as shown in Fig. 9a, b. Because of the short length along the trench direction, as an approximation we assume that the shape of the subduction plate boundary is bent only along the dip direction. We also assume as an approximation a periodic boundary condition along the strike direction because the source region of the Tonankai earthquake is sandwiched between the Tokai and Nankai earthquakes and

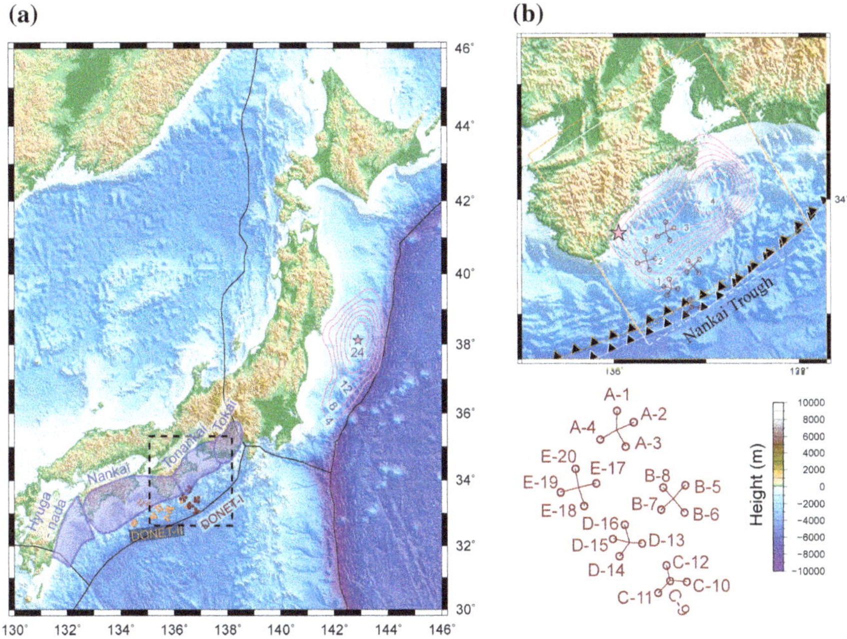

Fig. 8 **a** Map of Japan showing DONET-I (*red*) and DONET-II (*orange*) observation points. *Trench lines* are based on Bird (2003). The *pink star* and contours show the epicenter and slip isograms (4 m increment) of the 2011 Tohoku earthquake (Ozawa et al. 2011). The four *blue* regions along the Nankai Trough represent the estimated source regions of (from east to west) the Tokai, Tonankai, Nankai, and Hyuganada earthquakes. **b** Map of the area around the Tonankai district showing the target area of this study (*rectangle* with *dashed line* in **a**). Trench marks with *orange* and *white colors* are based on Coffin et al. (1998), Bird (2003), respectively. *Rectangles* with *dashed lines* along the trench represent the target area of this study. *Pink stars* and contours show the epicenter and slip isograms (0.5 m increment) of the 1944 Tonankai earthquake estimated by Kikuchi et al. (2003). *Red circles* with lines represent the observation points and science nodes of DONET-I. This figure is slightly modified from Ariyoshi et al. (2014)

because VLF event migration along the strike direction straddles the Tokai and Tonankai areas (Ito et al. 2007; Obara and Sekine 2009; Obara 2010).

As shown in Fig. 9c, the length along the strike direction and width along the dip direction are 200 and 215 km, respectively. These are discretized into 2048 (strike) and 452 (dip) cells in which the cell size along the strike component is uniform (200 km/2048 cell) and the size along the dip component changes based on the frictional stability (L_b) (Rubin and Ampuero 2005). The slip in the target area of the plate boundary is assumed to obey the quasi-static equilibrium relationship between shear and frictional stresses,

$$\mu_i \sigma_i = \sum_{j=1}^{N} K_{ij}(u_j(t) - V_{\mathrm{pl}} t) - \frac{G}{2\beta} \frac{\mathrm{d}u_i}{\mathrm{d}t}, \tag{1}$$

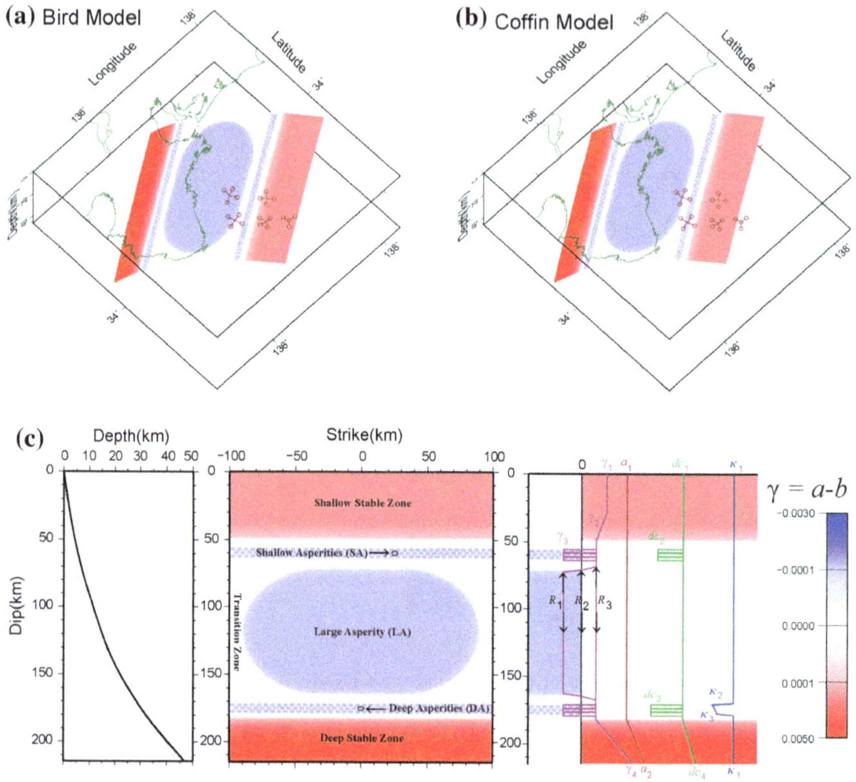

Fig. 9 (*top*) (**a**) (**b**) 3D views of a subduction plate boundary model with frictional parameter $\gamma = a - b$ for trench locations based on **a** Bird (2003) and **b** Coffin et al. (1998). **c** Depth and frictional parameters as functions of distance along the dip direction from the surface, where $(a_1, a_2) = (2, 5)$ [$\times 10^{-3}$], $(\gamma 1, \gamma 2, \gamma 3, \gamma 4) = (0.7, 0.01, -0.3, 4.9)$ [$\times 10^{-3}$], $(d_{c1}, d_{c2}, d_{c3}, d_{c4}) = (10, 0.3, 0.4, 400)$ [mm], and $(\kappa_1, \kappa_2, \kappa_3) = (1.0, 0.07, 0.13)$. The values for frictional parameters are based on rock laboratory results. This figure is based on Ariyoshi et al. (2014)

where i and j denote location indices of a receiver and source cell, respectively. The left hand side of the equation describes frictional stress, where μ and σ are the friction coefficient and effective normal stress, respectively. The right hand side describes the shear stress in the i-th cell due to dislocations, where K_{ij} is the Green's function for the shear stress (Okada 1992) in the i-th cell, N is the total number of cells, V_{pl} is the relative speed of the continental and oceanic plates, t denotes time, G denotes rigidity, and β is the shear wave speed. K_{ij} is calculated from the quasi-static solution for uniform pure dip-slip u relative to the average slip $V_{\text{pl}}t$ (Savage 1983) over a rectangular dislocation in the j-th cell. The part of the plate boundary deeper than 47 km is assumed to slip at the constant rate of V_{pl}, which approximately ignores shear stress on the right hand side of the equation

(Savage 1983). Parts of the first term on the right hand side are written as convolutions under a periodic condition on the planar surface divided into an even size along the strike direction, which allows us to take advantage of a fast Fourier transform convolution on the strike component, thus saving the calculation cost (Rice 1993; Ariyoshi et al. 2007, 2009, 2012).

In Eq. (1), the effective normal stress σ is given by,

$$\sigma_i(z) = \kappa(z)(\rho_{\mathrm{rock}} - \rho_{\mathrm{w}})gz, \tag{2}$$

where ρ_{rock} and ρ_{w} are densities of rock and water, respectively, g is the acceleration due to gravity, and z is the depth. The function $\kappa(z)$ is a superhydrostatic pore pressure factor, as given in Fig. 9c.

The frictional coefficient μ is assumed to obey the RSF law given by,

$$\mu = \mu_0 + a\log(V/V_0) + b\log(V_0\theta/d_{\mathrm{c}}), \tag{3}$$

$$\mathrm{d}\theta/\mathrm{d}t = 1 - V\theta/d_{\mathrm{c}}, \tag{4}$$

where a and b are friction coefficient parameters, d_c is the characteristic slip distance associated with b, θ is a state variable for the plate interface, V is the slip velocity, and μ_O is a reference friction coefficient defined at a constant reference slip velocity of V_0.

The constant parameters used in this study are V_{pl} = 4.0 × 10^{-2} m/yr (or 1.3 × 10^{-9} m/s), G = 30 GPa, β = 3.75 km/s, ρ_{rock} = 2.75 × 10^3 kg/m^3, ρ_{w} = 1.0 × 10^3 kg/m^3, g = 9.8 m/s^2, V_0 = 1 μm/s, μ_0 = 0.6, and Poisson's ratio ε = 0.25.

Our simulation results show that megathrust earthquakes occur periodically and that the recurrence interval (T_r) and magnitude (M_w) are nearly constant (T_r = 113 years, M_w = 7.9). One of the resulting spatial distributions of co-seismic slip (>3 cm/sec) accompanied by a megathrust earthquake is shown in Fig. 10. The asymmetric slip distribution is caused by the asymmetric distribution of SA and DA, as shown in Fig. 9. The fluctuation of co-seismic slip behavior in a recurrent megathrust earthquake is so small that the locations of the slip peak and rupture initiation point can almost be considered characteristic. The small amount of slip on the edge of the southwest side is due to the periodic boundary condition along the strike component, which is negligible in calculations of the seismic moment magnitude.

Compared with the co-seismic slip distribution of the 1944 Tonankai earthquake shown in Fig. 8, the simulated co-seismic slip distribution in Fig. 10 has a similar peak value (~3.5 m) approximately (34°N, 137°E). In addition, the 1944 Tonankai earthquake has a similar magnitude (M_w 7.9; Kikuchi et al. (2003)) and its recurrence interval is approximately 100–150 years (Ishibashi 1981). From these results, we consider this simulated megathrust earthquake to be the likely scenario for a future Tonankai earthquake to investigate the crustal deformation caused by slow earthquakes for the megathrust earthquake cycle.

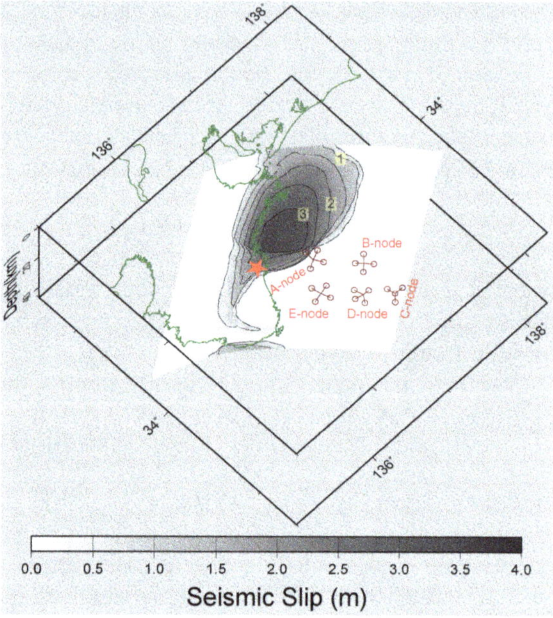

Fig. 10 (*left*) Slip isograms (0.5 m increments) of the simulated Tonankai earthquake. The *red star* represents the epicenter of the 1944 Tonankai earthquake. This figure is based on Ariyoshi et al. (2014)

4 Detectability of Crustal Deformation Driven by Slow Earthquakes

Around the source region of the simulated Tonankai earthquake, if shallower slow earthquakes occur in the belt of small asperities along a depth of approximately 5 km, DONET-I is expected to respond significantly to these slow earthquakes because of the short distance between the source of VLF events and observation points, as shown in Fig. 9. We analytically estimate the leveling change at DONET-I observation points from

$$H_k = \sum_{j=1}^{N} L_{kj} u_j(t), \tag{5}$$

where H is the amount of leveling change and L is the Green's function for the leveling change due to dip slip in the j-th source cell (Okada 1992) on the k-th DONET-I observation point depending on the Coffin or Bird Model, as shown in Fig. 9a, b. Note that the product of $L_{kj}u_j(t)$ is significant only in the source regions of shallower slow earthquakes in the pre-seismic stage of LA.

Figure 11 shows the leveling change at DONET-I observation points for the Bird and Coffin models (Fig. 9a, b) from ten years before to the onset of the simulated Tonankai earthquake. In Fig. 11, long-term uplift is derived from the pre-seismic slip of the Tonankai earthquake, while short-term uplift/depression is from

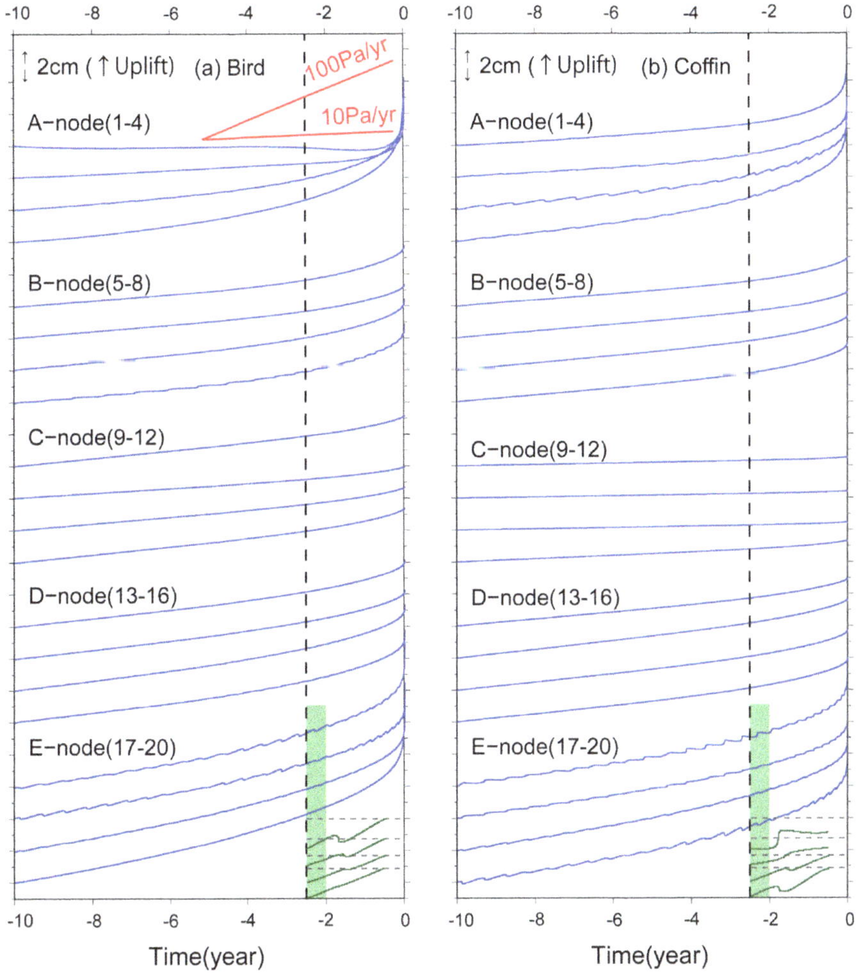

Fig. 11 Leveling change at DONET-I from ten years before to the onset of the simulated Tonankai earthquake for the (**a**) Bird (2003) and (**b**) Coffin et al. (1998) models. Green time windows for the E-node are shown in Fig. 8. *Dark green curves* show the time history in the green time windows magnified by four. The two red lines represent rates of 10 and 100 Pa/year. This figure is based on Ariyoshi et al. (2014)

shallower slow earthquakes. In this section, we investigate the characteristics of each type of crustal deformation.

Figure 11 illustrates the differences between science nodes (A–E in Fig. 8b): long-term (longer than several years) leveling changes and temporal acceleration of the uplift at nodes closer to the LA (A- and E-nodes) are greater than at nodes closer to the trench (C-node), mainly because of the shorter distance from the LA as shown in Fig. 10.

The node closer to the LA shows greater differences between leveling observation points in the same node (E-18 and E-20 in Fig. 8b) than for points closer to the trench (C-9 and C-12), which is more significant after the time indicated by broken lines in Fig. 11. This is because the difference in distance from the LA to leveling observation points in the same node closer to the trench (<C-9 to LA> and <C-12 to LA>) is relatively smaller than in a node closer to the LA (<E-18 to LA> and <E-20 to LA>).

These characteristics are common to both Coffin and Bird models. As the Bird model has a shorter distance from the LA to observational points compared with the Coffin model, crustal deformation in the Bird model tends to be greater than that in the Coffin model. This result suggests that precise determination of the subduction plate structure plays an important role in quantitative evaluation of the crustal deformation at ocean floor observation points.

For the E-node, Fig. 11 shows short-term (shorter than several days) leveling changes due to slow earthquake swarms for both Coffin and Bird models. However, its characteristics are different between the two models. In the B-node, significant short-term leveling changes appear only in the Bird model (B-8), and leveling changes are only seen in the A-node (A-2, A-3, and A-4) for the Coffin model. These results mean that the crustal deformation is localized because of the short distance between the sources of the slow earthquake swarms and DONET-I receivers. A denser network is required around the epicenters of shallower slow earthquakes for detection by stacking data on the short-term leveling changes.

At points E-17 and E-18, the short-term leveling change is uplift for the Coffin model and depression for the Bird model. These differences can be generally explained on the basis of a dislocation theory for dip slip on a buried fault (Segall 2010), where the vertical displacement on the free surface is uplift with a relatively steep slope around the up-dip edge of a fault and depression with a relatively slight slope around the down-dip edge of a fault. For the Coffin model, as shown in Fig. 9b, E-17 and E-18 are located on the up-dip part of SA and E-19 and E-20 (which show no significant change because of being too far from the SA belt) are on the down-dip part of the SA belt. Conversely, all the observation points for the E-node are on the down-dip part for the Bird model, as shown in Fig. 9a, which is consistent with the theoretical analysis. Therefore, precise determination of slow earthquake hypocenters and trench locations is very important for advance estimation of short-term leveling changes at DONET using numerical simulations.

For nodes responding significantly to slow earthquake swarms in Fig. 11, the rate of leveling change tends to be higher toward the origin of the simulated Tonankai earthquake. In the pre-seismic stage of a megathrust earthquake, the moment release rate of slow earthquakes, which is expected to be proportional to the average slip velocity in SA or DA under the condition of a fixed fault area, becomes higher as a result of pre-seismic slip around the LA (Ariyoshi et al. 2012), which explains the higher rate of leveling change.

To detect leveling changes using hydraulic pressure gauges, it is necessary to remove noise and drift components from raw data. From previous long-term seafloor measurements using a Paroscientific pressure gauge (Polster et al. 2009), which is used for DONET, the estimated noise level is approximately 10–20 Pa and the drift rate is approximately 5–10 kPa/year with a time constant of approximately 50–200 days. Figure 11 illustrates the long-term leveling change at DONET-I: the rate of the long-term leveling change is shown to be approximately 10 (C-node) to 100 (A, E-nodes) Pa/year, which is lower than the estimated drift component. Figure 11 also shows that the time constant of long-term leveling change is significantly longer than one year. These results mean that it is possible to estimate the drift component by subtracting the crustal deformation calculated in Fig. 11 if the simulation results can quantitatively reproduce the pre-seismic slip, while it may be difficult to extract long-term crustal deformation from hydraulic pressure gauge data at a single DONET-I point if the simulation results only explain crustal deformation qualitatively. As the trend of long-term change in all nodes is uplift, stacking data in the same node and/or combination of several nodes (which should remove sensor drift and noise components individually for each hydraulic pressure gauge) may be helpful for detection of long-term crustal deformation.

Figure 12 illustrates the short-term leveling changes caused by slow earthquake swarms and shows a close-up of the leveling change for the E-node from 2.5 to 2.0 years before the onset of the simulated Tonankai earthquake. Focusing on E-17, the simulation result shows a maximum rate of leveling change of about 1.5×10^{-5} m/s ($4.4 \times 10^{+3}$ kPa/year) with a duration time of about 0.74 s, which is a

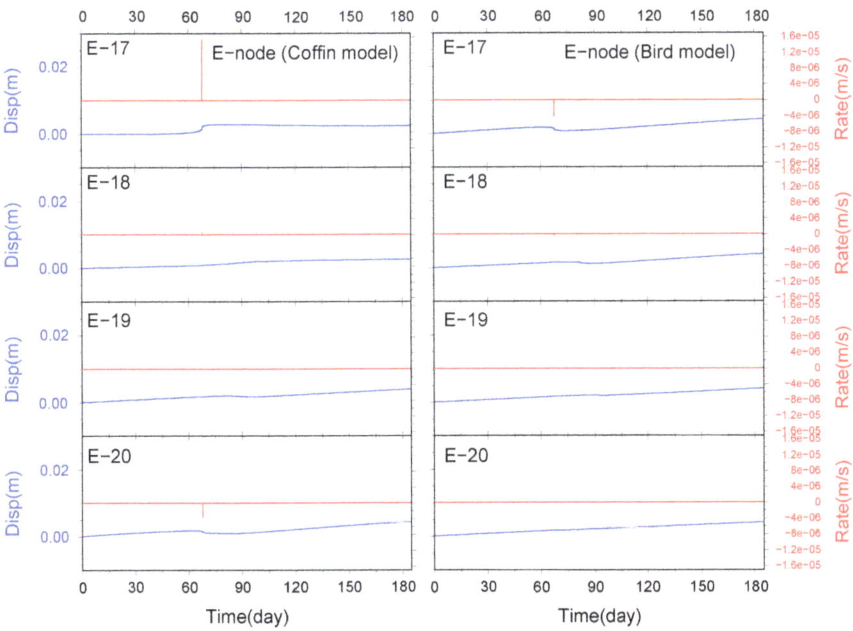

Fig. 12 Leveling changes (*blue lines*) and their rates (*red lines*) at the E-node of DONET-I for about half a year as a close-up of the green colored time windows in Fig. 11. The origin time is set as the start time of the green windows. This figure is based on Ariyoshi et al. (2014)

sufficiently higher rate and shorter duration than the estimated drift component (Polster et al. 2009), as described above. However, the amount of short-term leveling change at E-17 is expected to be about 2 mm (20 Pa), which suggests that the short-term leveling change could be obscured by the noise component. This small leveling change driven by shallower VLF swarms is caused by a low dip angle (about 5° on SA, as shown in Fig. 9c) in our model.

For real-time monitoring of crustal deformation near the trench before the occurrence of the next Tonankai earthquake in the near future, it would be desirable to deal with DONET data rather than conduct detailed analysis of long-term measurements.

Figure 13 shows a test sample of the daily processed data at DONET-I for one month. The data processing interval will be much shorter than 1 day when the processed data are used for real-time monitoring. In this test, we calculate a moving average of 100 Hz data detided by using the Tide4n code (Tamura et al. 1991) at intervals of 6 h in a time window of 12 h (hereafter referred to as filtered data). Next, we obtain stacked data by averaging the filtered data at four points in the same node. The stacked data is used to reduce electromagnetic noise and amplify the systematic response to changes in the sea temperature and ocean flow, in addition to long-term leveling changes resulting from large pre-seismic slip around the LA. The stack data is more robust for use as a reference than single filtered data at a nearby observation point because of the short range between four observation points in the same node. Then, we construct a differential component (hereafter referred to as differential data) by removing the stacked data from the filtered data. The differential data are used to extract local crustal deformation, such as slow earthquake swarms, as shown for the E-node in Fig. 11, and relative long-term leveling change, which is significantly observed for the A-node of the Bird model in Fig. 11. As both the stack and differential data are updated individually from new sampling, it is possible to conduct these processes in real time.

Focusing on the differential data in Fig. 13, we find that the fluctuation is about several tens of Pa (several millimeters of leveling change), which is comparable with the leveling amount of short-term change shown in E-17 of the Coffin model in Fig. 8 (about 2 mm: 20 Pa). From Fig. 12, we observe that the short-term leveling change is incoherent in the same node. If a hydraulic pressure gauge responds to shallower VLF swarms at only one observation point in a node, as seen for E-17 of the Coffin model in Fig. 12, the differential data would be expected to contain three-quarters of its value (15 Pa in the case of 20 Pa at E-17 in Fig. 12 when the change at E-20 is negligible). This estimate suggests that short-term local leveling change driven by shallower VLF swarms might be concealed by fluctuations when using only hydraulic pressure gauges.

As shown in Fig. 13, we now have a continued daily solution of hydraulic pressure gauge data as a simple test run using a moving average from 100 Hz sampling data in a 6 h time-interval. As some VLF events may have a duration time shorter than 1 s, as shown at E-17 in Fig. 12, in the future it will be necessary to enhance the sampling of the differential data from daily to higher than 1 Hz to detect the temporal and local leveling changes caused by shallower VLF swarms.

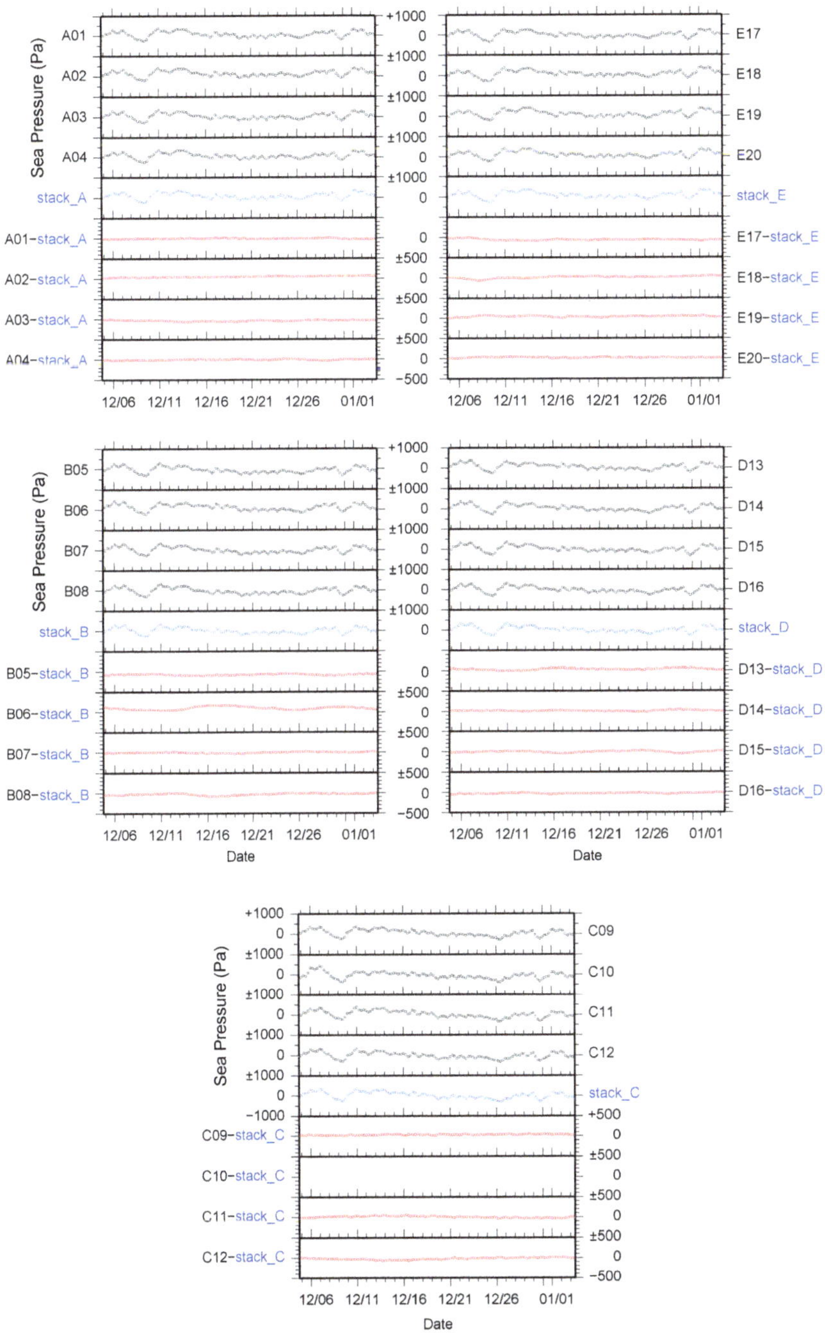
Sea Pressure (Pa)
A01
A02
A03
A04
stack_A
A01−stack_A
A02−stack_A
A03−stack_A
A04−stack_A
+1000
0
±1000
±500
−500
12/06 12/11 12/16 12/21 12/26 01/01
E17
E18
E19
E20
stack_E
E17−stack_E
E18−stack_E
E19−stack_E
E20−stack_E
B05
B06
B07
B08
stack_B
B05−stack_B
B06−stack_B
B07−stack_B
B08−stack_B
Date
D13
D14
D15
D16
stack_D
D13−stack_D
D14−stack_D
D15−stack_D
D16−stack_D
C09
C10
C11
C12
stack_C
C09−stack_C
C10−stack_C
C11−stack_C
C12−stack_C
−1000
+500

◀ **Fig. 13** Daily analyzed data from hydraulic pressure gauges observed at DONET-I for one month. The time series plotted as black (*upper four columns*), *blue* (*middle column*), and *red* (*lower four columns*) circles represent filtered data (moving average of 100 Hz data at intervals of 6 h in a time window of 12 h), stacked data (average of the filtered data at the same node), and differential data (removal of the filtered data from the stack data), respectively. This figure is based on Ariyoshi et al. (2014)

Water temperature gauges are also installed in the DONET hydraulic pressure gauges, which assists in reducing the noise component in the stacked and differential data by temperature correction (Inazu and Hino 2011).

DONET has broadband seismometers buried in the ocean floor at nearly the same locations as the hydraulic pressure gauges (Nakano et al. 2013a, b), which enables estimation of fault parameters of shallow VLF events. As the dip angle of the basal plane of the overriding wedge (décollement) is steeper than that of the oceanic plate near the trench (Ikari and Saffer 2012), shallow VLF events on a décollement (Sugioka et al. 2012) or megasplay faults (Ito and Obara 2006) are expected to cause leveling changes higher than our simulation results (Fig. 12) for a dip angle of a VLF event significantly steeper than five degrees. When we also use the origin time and hypocenter location of a VLF event automatically detected by seismometers, the accuracy of detection of VLF events based on leveling changes of DONET is expected to become higher.

5 Discussion and Conclusions: Social Impact of Seismological Studies on the Public and Toward Disaster Mitigation of Megathrust Earthquakes

On the basis of our simulation results, we discuss the social impact of seismological studies on the public. From Table 2 and Fig. 5, our trial forecast of the next Tokai earthquake has an estimation range of about 1.5–2 years. On the other hand, economic losses are estimated to be about 7 billion dollars per day, if economic activities were to be temporarily halted in the Tokai district. If economic activities were to cease for two years, the loss would increase to at most about 5 trillion dollars, which is too great for Japanese society in reality.

Conversely, let us discuss the necessity of precision in the forecast. The Cabinet Office reported that the economic loss resulting from megathrust earthquakes along the Nankai Trough, including the Tokai and Tonankai earthquakes, is estimated to be between 0.37 and 2 trillion dollars. If we adopt the strategy of disaster mitigation by simply adopting the lower economic cost, the estimation range of forecasting earthquake should be shorter than 1.8–9.6 months at least. This is an example of why forecasting of megathrust earthquakes has been difficult in practice so far. To enhance the forecasting precision, it is important to develop numerical modeling of earthquake cycles and observation detectability, including seismicity and crustal deformation, by comparing both types of data quantitatively, as shown in Fig. 3.

Our conclusions can be summarized as follows.

(i) Multidirectional analyses including GPS, inland benchmark, triangulation, and seismometers are necessary to enhance the validity of prediction analysis.
(ii) Geological surveys of the 3D subduction plate structure and location of the oceanic trench are important to estimate crustal deformation due to plate motion from numerical simulations for data assimilation.
(iii) Analysis of the fault mechanism for intraplate earthquakes, as well as interplate ones is useful to monitor the stress field caused by a megathrust earthquake in both the pre-seismic and interseismic stage, which was also pointed out by Hasegawa et al. (2011) after the 2011 Tohoku earthquake.
(iv) Ocean floor observations, such as DONET and inline cable off Tohoku, are useful for monitoring crustal deformation near the trench on the basis of slow earthquake activity and leveling changes at hydraulic pressure gauges. As DONET-II and the inline cable off Tohoku are scheduled to be installed in a few years (Fig. 14), detectability of pre-seismic change near the trench is expected to be enhanced.

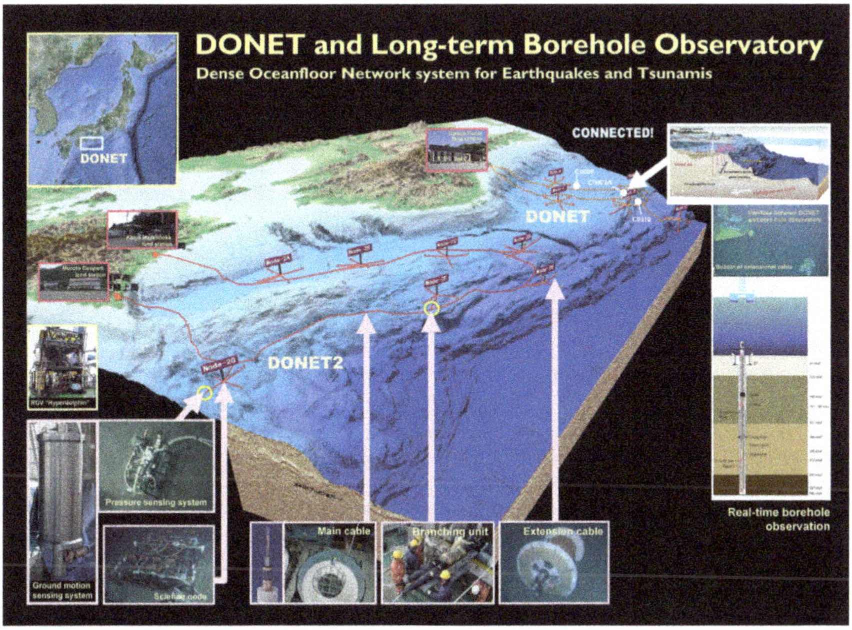

Fig. 14 Overview of dense oceanfloor network system for earthquake and Tsunamis (DONET) in the Tonankai and Nankai regions (DONET-II in Fig. 8)

Acknowledgements Some figures were drawn using GTOPO30 (available from the U.S. Geological Survey) and GMT software (Wessel and Smith 1998). Hydraulic pressure data shown in Fig. 13 and their map in Fig. 14 were obtained by the DONET program of the Ministry of Education, Culture, Sports, Science and Technology (MEXT). Publication of this chapter was partly supported by MEXT for Young Scientists (B), 23710212, 2013, Grant-in-Aid for Scientific Research (B), 15H04228, 2015, and by the Geodynamics program. The authors would like to thank Enago (www.enago.jp) for the English language review.

References

Ariyoshi, K., Kato, N., & Hasegawa, A. (2001). Numerical simulation study on recent changes in crustal deformation and seismicity in the Tokai area, central Japan. *Journal of Geography, 110*, 557–565 (in Japanese with English abstract and captions).

Ariyoshi, K., Kato, N., & Hasegawa, A. (2003). Numerical simulation study on recent changes in crustal deformation and seismicity in the Tokai Area, Central Japan II, *International Union of Geodesy and Geophysics, XXIII General Assembly (Sapporo, Japan), SS02*.

Ariyoshi, K., Matsuzawa, T., Hino, R., & Hasegawa, A. (2007). Triggered non-similar slip events on repeating earthquake asperities: Results from 3D numerical simulations based on a friction law. *Geophysical Research Letters, 34*. doi:10.1029/2006GL028323.

Ariyoshi, K., Hori, T., Ampuero, J. P., Kaneda, Y., Matsuzawa, T., Hino, R., & Hasegawa, A. (2009). Influence of interaction between small asperities on various types of slow earthquakes in a 3-D simulation for a subduction plate boundary. *Gondwana Research, 16*(3–4), 534–544. doi:10.1016/j.gr.2009.03.006.

Ariyoshi, K., Matsuzawa, T., Ampuero, J. P., Nakata, R., Hori, T., Kaneda, Y., & Hasegawa, A. (2012). Migration process of very low-frequency events based on a chain-reaction model and its application to the detection of preseismic slip for megathrust earthquakes. *Earth Planets Space, 64*(8), 693–702. doi:10.5047/eps.2010.09.003.

Ariyoshi, K., Nakata, R., Matsuzawa, T., Hino, R., Hori, T., Hasegawa, A., & Kaneda, Y. (2014). The detectability of shallow slow earthquakes by the Dense Oceanfloor network system for earthquakes and Tsunamis (DONET) in Tonankai district, Japan. *Marine Geophysical Research, 35*(3), 295–310. doi:10.1007/s11001-013-9192-6.

Bird, P. (2003). An updated digital model of plate boundaries. *Geochemistry Geophysics Geosystems, 4*(3), 1027. doi:10.1029/2001GC000252.

Central Disaster Management Council. (2001). *Report of the specialized investigation committee about Tokai earthquake (in Japanese)*. http://www.bousai.go.jp/kaigirep/chuobou/20011218/pdf/siryou2-1.pdf, (2014-01-25).

Coffin, M. F., Gahagan, L. M., & Lawver, L. A. (1998). Present-day plate boundary digital data compilation. *University of Texas Institute for Geophysics Technical Report, 174*, 5.

Dieterich, J. H. (1979). Modeling of rock friction: 1. *Experimental results and constitutive equations, Journal of Geophysical Research, 84*, 2161–2168.

Hasegawa, A., Yoshida, K., & Okada, T. (2011). Nearly complete stress drop in the 2011 Mw9.0 off the pacific coast of Tohoku Earthquake. *Earth Planets Space, 63*(7), 703–707. doi:10.5047/eps.2011.06.007.

Hori, T. (2006). Mechanisms of separation of rupture area and variation in time interval and size of great earthquakes along the Nankai Trough, southwest Japan. *Journal of Earth Simulator, 5*, 8–19.

Ikari, M. J., & Saffer, D. M. (2012). Permeability contrasts between sheared and normally consolidated sediments in the Nankai accretionary prism. *Marine Geology, 295–298*, 1–13. doi:10.1016/j.margeo.2011.11.006.

Inazu, D., & Hino, R. (2011). Temperature correction and usefulness of ocean bottom pressure data from cabled seafloor observatories around Japan for analyses of tsunamis, ocean tides, and low-frequency geophysical phenomena. *Earth, Planets and Space, 63*, 1133–1149. doi:10.5047/eps.2011.07.014.

Ishibashi, K. (1981). Specification of a soon-to-occur seismic faulting in the Tokai district, central Japan, based upon seismotectonics. In D. W. Simpson & P. G. Richards (Eds.), *Earthquake prediction: An international review* (pp. 297–332). Washington, D. C: American Geophysical Union.

Ito, T., Yoshioka, S., & Miyazaki, S. (1999). Interplate coupling in southwest Japan deduced from inversion analysis of GPS data. *Physics of the Earth and Planetary Interiors, 115*, 17–34.

Ito, Y., & Obara, K. (2006). Dynamic deformation of the accretionary prism excites very low frequency earthquakes. *Geophysical Research Letters, 33*, L02311. doi:10.1029/2005GL025270.

Ito, Y., Obara, K., Shiomi, K., Sekinie, S., & Hirose, H. (2007). Slow earthquakes coincident with episodic tremors and slow slip events. *Science, 315*, 503–506. doi:10.1126/science.1134454.

Kato, N., & Hirasawa, T. (1999). A model for possible crustal deformation prior to a coming large interplate earthquake in the Tokai distinct. *Central Japan, Bulletin of the Seismological Society of America, 89*, 1401–1417.

Kikuchi, M., Nakamura, M., & Yoshikawa, K. (2003). Source rupture processes of the 1944 Tonankai earthquake and the 1945 Mikawa earthquake derived from low-gain seismograms. *Earth Planets Space, 55*, 159–172.

Lay, T., Kanamori, H., Ammon, C. J., Nettles, M., Ward, S. N., Aster, R. C., et al. (2005). The great Sumatra-Andaman earthquake of 26 December 2004. *Science, 308*, 1127–1133. doi:10.1126/science.1112250.

Matsumura, N. (1997). Focal zone of a future Tokai earthquake inferred from the seismicity pattern around the plate interface. *Tectonophysics, 273*, 271–291.

Miyazaki, S., & Heki, K. (2001). Crustal velocity field of southwest Japan: Subduction and arc-arc collision. *Journal of Geophysical Research, 106*, 4305–4326. doi:10.1029/2000JB900312.

Nakanishi, A., Kodaira, S., Miura, S., Ito, A., Sato, T., Park, J. O., et al. (2008). Detailed structural image around splay-fault branching in the Nankai subduction seismogenic zone: Results from a high-density ocean bottom seismic survey. *Journal of Geophysical Research, 113*, B03105.

Nakano, M., Nakamura, T., Kamiya, S., Ohori, M., & Kaneda, Y. (2013a). Intensive seismic activity around the Nankai trough revealed by DONET ocean-floor seismic observations. *Earth, Planets and Space, 65*(1), 5–15. doi:10.5047/eps.2012.05.013.

Nakano, M., Nakamura, T., Kamiya, S., & Kaneda, Y. (2013b). Seismic activity beneath the Nankai trough revealed by DONET ocean-bottom observations. *Marine Geophysical Research, 35*(3), 271–284. doi:10.1007/s11001-013-9195-3.

Noguchi, S. (1996). Geometry of the Philippine Sea Slab and the convergent tectonics in the Tokai District, Japan. *Journal of the Seismoogical Society of Japan, 49*, 295–325 (in Japanese with English abstract and figure captions).

Obara, K. (2010). Phenomenology of deep slow earthquake family in southwest Japan: Spatiotemporal characteristics and segmentation. *Journal of Geophysical Research, 115*, B00A25. doi:10.1029/2008JB006048.

Obara, K., & Sekine, S. (2009). Characteristic activity and migration of episodic tremor and slow-slip events in central Japan. *Earth Planets Space, 61*, 853–862.

Okada, Y. (1992). Internal deformation due to shear and tensile faults in a halfspace. *Bulletin of the Seismogical Society of America, 82*, 1018–1040.

Ozawa, S., Murakami, M., Kaidzu, M., Tada, T., Sagiya, T., Hatanaka, Y., et al. (2002). Detection and monitoring of ongoing aseismic slip in the Tokai region, central Japan. *Science, 298*, 1009–1012.

Ozawa, S., Nishimura, T., Suito, H., Kobayashi, T., Tobita, M., & Imakiire, T. (2011). Coseismic and postseismic slip of the 2011 magnitude-9 Tohoku-Oki earthquake. *Nature, 475*, 373–376. doi:10.1038/nature10227.

Polster, A., Fabian, M., & Villinger, H. (2009). Effective resolution and drift of Paroscientific pressure sensors derived from long-term seafloor measurements. *Geochemistry, Geophysics, Geosystems, 10*, Q08008. doi:10.1029/2009GC002532.

Rice, J. R. (1993). Spatio-temporal complexity of slip on a fault. *Journal of Geophyical Research, 98*, 9885–9907.

Rubin, A. M., & Ampuero, J. P. (2005). Earthquake nucleation on (aging) rate and state faults. *Journal of Geophysical Research, 110*, B11312. doi:10.1029/2005JB003686.

Ruina, A. (1983). Slip instability and state variable friction laws. *Journal of Geophysical Research, 88*, 10359–10370.

Sagiya, T. (1999). Interplate coupling in the Tokai district, central Japan, deduced from continuous GPS data. *Geophysical Research Letters, 26*, 2315–2318.

Savage, J. C. (1983). A dislocation model of strain accumulation and release at a subduction zone. *Journal of Geophysical Research, 88*, 4984–4996.

Segall, P. (2010). *Earthquake and volcano deformation*. Oxford: Princeton University Press.

Sella, G. F., Dixon, T. H., & Mao, A. (2002) REVEL: A model for recent plate velocities from space geodesy. *Journal of Geophysical Research, 107(B4 ETG11)*, 1–30. doi:10.1029/2000JB000033.

Sugioka, H., Okamoto, T., Nakamura, T., Ishihara, Y., Ito, A., Obana, K., et al. (2012). Tsunamigenic potential of the shallow subduction plate boundary inferred from slow seismic slip. *Nature Geoscience, 5*, 414–418. doi:10.1038/NGEO1466.

Takagi, A. (1980). Concluding remarks and precursory seismic activity of the 1978 Miyagi-Oki Earthquake. In *Proceedings of Earthquake Prediction Research Symposium, Seismological Society of Japan and Subcommittee of Earthquake Prediction, National Committee of Geophysics, Science Council of Japan*, 231–241 (in Japanese with English abstract and figure captions).

Tamura, Y., Sato, T., Ooe, M., & Ishiguro, M. (1991). A procedure for tidal analysis with a Bayesian information criterion. *Geophysical Journal International, 104*, 507–516.

The Headquarters for Earthquake Research Promotion. (2013). *Evaluations of occurrence potentials or subduction-zone earthquakes to date (written in Japanese)*. http://www.jishin.go.jp/main/p_hyoka02_kaiko.htm.

Wessel, P., & Smith, W. H. F. (1998). New, improved version of the generic mapping tools released. *EOS Transactions, AGU, 79*, 579.

Yoshioka, S., Yabuki, T., Sagiya, T., Tada, T., & Matsu'ura, M. (1993). Interplate coupling and relative plate motion in the Tokai district, central Japan, deduced from geodetic data inversion using ABIC. *Geophysical Journal International, 113*, 607–621.

Neo-deterministic Definition of Seismic and Tsunami Hazard Scenarios for the Territory of Gujarat (India)

A. Magrin, I.A. Parvez, F. Vaccari, A. Peresan, B.K. Rastogi, S. Cozzini, D. Bisignano, F. Romanelli, Ashish, P. Choudhury, K.S. Roy, R.R. Mir and G.F. Panza

1 Introduction

Seismic risk mitigation is a worldwide concern and the development of effective mitigation strategies requires sound seismic hazard assessment. The performances of the classical probabilistic approach to seismic hazard assessment (PSHA), currently in use in several countries worldwide, turned out fatally inadequate when considering the earthquakes occurred worldwide during the last decade. When dealing with the protection of critical structures (e.g. nuclear power plants) and

A. Magrin · F. Vaccari · A. Peresan (✉) · D. Bisignano · F. Romanelli · G.F. Panza
Department of Mathematics and Geosciences, University of Trieste, Trieste, Italy
e-mail: aperesan@units.it

A. Magrin · F. Vaccari · A. Peresan · D. Bisignano · F. Romanelli · G.F. Panza
The Abdus Salam International Centre for Theoretical Physics, ICTP SAND Group, Trieste, Italy

I.A. Parvez · Ashish · R.R. Mir
CSIR Fourth Paradigm Institute, C-MMACS, NAL Belur Campus, Bangalore, India

A. Peresan
Istituto Nazionale di Oceanografia e di Geofisica Sperimentale, CRS, Udine, Italy

B.K. Rastogi · P. Choudhury · K.S. Roy
Institute of Seismological Research, Raisan, Gandhinagar, Gujarat, India

S. Cozzini
Consiglio Nazionale delle Ricerche, Istituto Officina dei Materiali, CNR/IOM uos Democritos, Trieste, Italy

G.F. Panza
Institute of Geophysics, China Earthquake Administration, Beijing Peoples' Republic of China

G.F. Panza
International Seismic Safety Organisation, ISSO, Arsita, Italy

S. D'Amico (ed.), *Earthquakes and Their Impact on Society*, Springer Natural Hazards, DOI 10.1007/978-3-319-21753-6_7 cultural heritage, where it is necessary to consider extremely long time intervals, the standard PSHA estimates are by far unsuitable, due to their basic heuristic limitations. Moreover, it is nowadays widely recognized by the engineering community that probabilistic Peak Ground Acceleration estimates alone are not sufficient for the adequate design of special buildings and infrastructures, since displacements play a critical role and the dynamical analysis of the structure response requires complete time series of ground motion. In view of the mentioned limits of PSHA estimates, it appears preferable to resort to a scenario-based approach to seismic hazard assessment that may turn out to be necessary to complement and validate the results that will be eventually produced by large scale projects like GEM (http://www.globalquakemodel.org/).

Current computational resources and physical knowledge of the seismic waves generation and propagation processes, along with the improving quantity and quality of geophysical data (spanning from seismological to satellite observations), allow nowadays for viable numerical and analytical alternatives to the use of probabilistic approaches. A set of scenarios of expected ground shaking due to a wide set of potential earthquakes can be defined by means of full waveforms modelling, based on the possibility to efficiently compute synthetic seismograms in complex laterally heterogeneous anelastic media. In this way a set of scenarios of ground motion can be defined, either at national and local scale, the latter considering the 2D and 3D heterogeneities of the medium travelled by the seismic waves.

The considered scenario-based approach to seismic hazard assessment, namely the Neo-Deterministic Seismic Hazard Assessment approach (NDSHA) (e.g. Panza et al. 2012, 2013), builds on rigorous theoretical basis and exploits the currently available computational resources that permit to compute realistic synthetic seismograms. The NDSHA approach intends to provide a fully formalized operational tool for effective seismic hazard assessment, readily applicable to compute complete time series of expected ground motion (i.e. the synthetic seismograms) for seismic engineering analyses and other mitigation actions.

The NDSHA methodology has been successfully applied to strategic buildings, lifelines and cultural heritage sites, and for the purpose of seismic microzoning in several urban areas worldwide (e.g. Panza et al. 1999, 2002; Mourabit et al. 2014). Recently the NDSHA method has been enabled on different computational platforms, ranging from GRID computing infrastructures to HPC dedicated cluster up to Cloud computing. Such e-infrastructures provide an innovative and unique approach to address this problem (Magrin et al. 2012). They demonstrated to be an efficient way to share and access resources of different types, which can effectively enhance the capability to define realistic scenarios of seismic ground motion. Intensive usage of these infrastructure enable scientists to compute a wide set of synthetic seismograms, dealing efficiently with variety and complexity of the potential earthquake sources, and the implementation of parametric studies to characterize the related uncertainties.

The need for a reliable assessment of the earthquake hazards is especially relevant in Gujarat, which is the industrial hub of India (hosting one of the world's largest refineries, chemical industries and large maritime facilities) and is now developing a number of special economic and investment area. Gujarat region witnessed many destructive earthquakes in the past. The severe seismic hazard to the region has been confirmed by the 2001 Bhuj earthquake, which caused more than 20,000 victims and extensive damage (Shanker 2001). We describe here the preliminary results obtained for the definition of NDSHA scenarios for the Gujarat region. The applied methodology, integrated with advanced computational infrastructures, naturally supplies realistic time series of ground motion, which represent also reliable estimates of ground displacement readily applicable to seismic design and isolation techniques, useful to preserve historical monuments and relevant man-made structures. The study aims to provide the community (e.g. authorities and engineers) with advanced information for seismic and tsunami risk mitigation, which is particularly relevant to Gujarat in view of the rapid development and urbanization of the region.

2 Regional Scale Neo-deterministic Seismic Hazard Assessment in Gujarat

2.1 *Method*

The procedure for the neo-deterministic seismic zoning (Panza et al. 2001, 2012) is based on the calculation of synthetic seismograms (ground motion scenarios). Starting from the available information on Earth structure, seismic sources, and the level of seismicity of the investigated area, it is possible to compute complete synthetic seismograms and the related estimates on peak ground acceleration (PGA), velocity (PGV) and displacement (PGD) or any other parameter relevant to seismic engineering (such as design ground acceleration, DGA) which can be extracted from the computed theoretical signals. NDSHA defines the hazard from the envelope of the values of ground motion parameters determined considering a wide set of scenario earthquakes; accordingly, the simplest product of this method is a map where the maximum of a given seismic parameter is associated to each site.

In the NDSHA approach, the definition of the space distribution of seismicity uses only the largest events reported in the earthquake catalogue at different sites. At regional scale, on account of the quality of the available data we discretize the study area with a $0.2° \times 0.2°$ regular grid. Earthquake epicenters reported in the catalogue are grouped into $0.2° \times 0.2°$ cells, assigning to each cell the maximum magnitude recorded within it. A smoothing procedure is then applied to account for spatial uncertainty and for source dimensions (Panza et al. 2001). Only cells located within the seismogenic zones are retained. This procedure for the definition of earthquake locations and magnitudes for NDSHA makes the method pretty robust against uncertainties in the earthquake catalogue, which is not required to be complete for magnitudes lower than 5. A double-couple point source is placed at the center of each cell, with a focal mechanism consistent with the properties of the corresponding seismogenic zone and a depth, which is a function of magnitude (10 km for $M < 7$, 15 km for $7 \leq M < 8$, 20 km for $M \geq 8$).

To define the physical properties of the source-site paths, the territory is divided into a set of structural models composed of flat, parallel anelastic layers that represent the average lithosphere properties at regional scale. Synthetic seismograms are then computed by the modal summation technique for sites placed at the nodes of the grid that covers the territory, considering the average structural model associated to the regional polygon that includes the site. Seismograms are computed for an upper frequency content of 1 Hz, which is consistent with the level of detail of the regional structural models. In Parvez et al. (2003), to reduce the number of computed seismograms, the source-site distance was kept below an upper threshold, which is taken to be a function of the magnitude associated with the source. Specifically, the maximum source-receiver distance was set equal to 300 km for $M \geq 8$. The optimization of modal summation programs has made possible to extend the distance threshold up to 150, 200, 400 and 800 km respectively for $M < 6$, $6 \leq M < 7$, $7 \leq M < 8$ and $M \geq 8$.

In order to take account of source dimensions, in the standard procedure (e.g. Parvez et al. 2003) seismograms are scaled using the spectral scaling laws proposed by Gusev (1983), as reported in Aki (1987). Here we adopted a more realistic source model, in which phase spectrum accounts for the duration and other features of the rupture process (Parvez et al. 2011). In this representation, a combination of extended (ES) and point sources exploiting the algorithm developed by Gusev and Pavlov (2009) and Gusev (2011), is used. The time functions of the subsources that constitute the ES are summed in order to obtain the equivalent single source representative of the entire space and time structure of the ES and the related Green's function. A neutral directivity has been chosen in the computation of the synthetic seismograms.

From the set of complete synthetic seismograms, various maps of seismic hazard describing the maximum ground shaking at the bedrock can be produced. The parameters representative of earthquake ground motion are maximum displacement, velocity, and acceleration. The acceleration parameter in the NDSHA is given by the design ground acceleration (DGA). This quantity is obtained by scaling the chosen normalized design response spectrum (normalized elastic acceleration spectra of the ground motion for 5 % critical damping) with the response spectrum computed at frequencies below 1 Hz (for further details see Panza et al. 1996). Besides DGA maps (which allow extrapolating the estimated acceleration from 1 Hz to higher frequencies), acceleration maps can also be defined from synthetic seismograms computed up to 10 Hz, as shown by Panza et al. (2012); however more detailed information about structural models should be used for a realistic modelling of the ground shaking in this case. For the seismic assessment and design of special buildings and infrastructures, a detailed local scale modelling can be carried out by NDSHA accounting for the lateral heterogeneities of the medium traveled by seismic waves and for the properties of seismic sources. The local scale scenarios provide complete ground motion description over a wide range of frequencies, from the low frequencies relevant to tall buildings and seismically isolated structures, and up to high frequencies, eventually exceeding 10 Hz.

2.2 Input Data

The computation of realistic synthetic seismograms in the process of NDSHA is naturally dependent on the knowledge of the source and propagation effects. Therefore, the input parameters describing the structural models and seismic sources must be properly defined and assigned to the studied area, exploiting all significant available information.

Earthquake catalogues are the most essential and important parameter for any kind of seismic zoning or hazard studies. In the NDSHA approach the definition of the space distribution of seismicity accounts essentially for the damaging events ($M \geq 5$) reported in the earthquake catalogues. Therefore catalogues completeness at moderate-to-low magnitudes is not necessary, contrary to PSHA. In the present study, the earthquake data set spanning the time interval from 25 A.D. to 2011 has been used for Gujarat from the catalogue compiled for the whole India. Starting from the databases made available by the international agencies like NOAA, ISC, NEIC, CNSS, CMT and national agencies like ISR, IMD and NDMA and several published research papers (e.g. Rao and Rao 1984; Oldham 1883; Tandon and Chaudhury 1968; Iyengar et al. 1999; Kumar et al. 2001), a comprehensive catalogue in terms of M_W has been compiled (Fig. 1a).

The seismogenic zones defined by Parvez et al. (2003) for Gujarat region have been modified, accounting for recent publications and new data. Fault plane solutions in the present study are mainly taken from Harvard CMT Catalogues (Dziewoński et al. 1981; Ekström et al. 2012). However, published focal mechanism solutions by Fitch (1970), Molnar et al. (1973) and Chandra (1977, 1978) have been used for the large earthquakes that occurred before 1977 and for the focal mechanism solutions of Peninsular India events. A representative fault plane solution is defined for each seismogenic zone by the average mechanism obtained from the available moment tensor solutions (Fig. 1b).

In this study, the velocity structure used by Parvez et al. (2003) has been substituted by a relatively higher resolution cellular model, with structures defined for cells of $1° \times 1°$, based on updated bibliographic research. In order to propose a suitable structural model, all available geophysical and geological information for the investigated territory have been considered after an extensive bibliographic research (Acton et al. 2010; Mandal 2006; Tewari et al. 2009). Most of the cells have square shape except at the edges, which were terminated along political borders of the country (Fig. 1c). Those cells where updated velocity models are not available have been completed using the structural model given by Parvez et al. (2003) checking for the similarity of the geological conditions. Figure 1c shows the shear wave velocity (Vs) for the topmost layer of each cell. We have also updated the attenuation model with respect to the previous study based on the attenuation

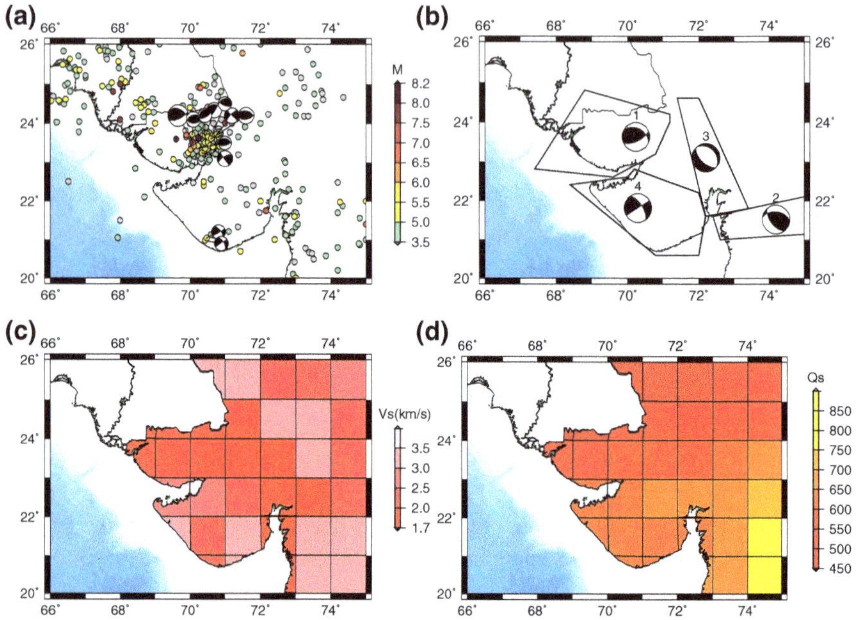

Fig. 1 Input parameters describing the seismic sources and structural models: **a** seismicity map of Gujarat region (25 A.D.–2011), with magnitudes expressed in terms of M_W; **b** seismogenic zones and representative fault plane solutions considered in this study; **c** shear wave velocity (Vs) in the topmost layer for each cell of the structural models; **d** quality factor (Qs) in the uppermost layer for each cell

tomography across Eurasia for the lithosphere (Mitchell et al. 2008) and further below from the study of Gung and Romanowicz (2004). Figure 1d illustrates the Qs in the uppermost crustal layer considered for each cell. The high crustal Qs values characterising Indian subcontinent, particularly in shield regions, are discussed in Mitchell et al. (2008)

2.3 *Results*

The spatial distribution of design ground acceleration (DGA), peak ground velocity (PGV) and peak ground displacement (PGD) extracted from the complete synthetic seismograms are shown in Fig. 2a–c. The maximum values of ground shaking are in north-western part of Gujarat where the biggest historical earthquake occurred. The values reported in the maps are mostly due to sources in seismogenic zone number 1. The main difference with maps shown in Parvez et al. (2003) is the increase of expected ground motion in the south-eastern part of Gujarat, mainly in Bhavnagar and surroundings.

The DGA values were converted to intensities using the relation between acceleration and EMS intensity of Lliboutry (2000), as shown in Fig. 2d. We

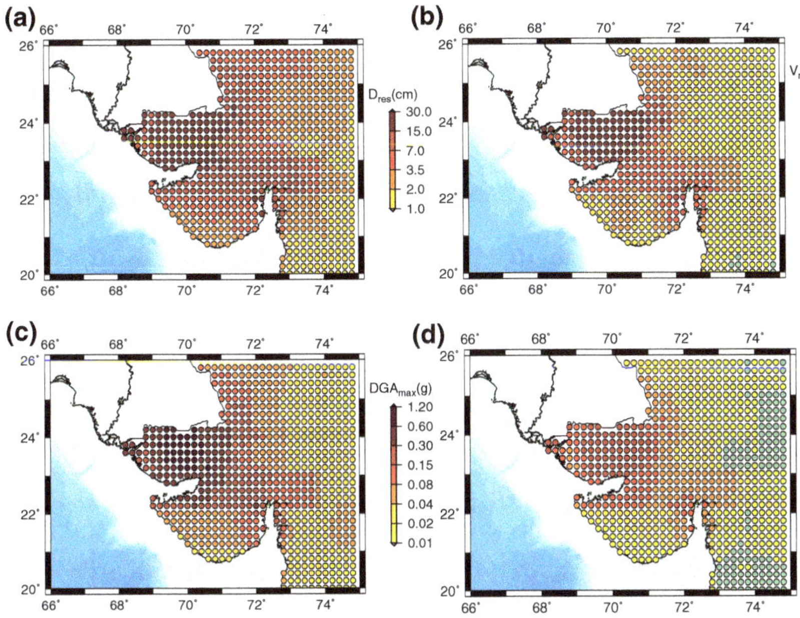

Fig. 2 Spatial distributions of the estimated peak ground displacement (PGD) in cm (**a**), peak ground velocity (PGV) in cm/s (**b**), design ground acceleration (DGA) in g (**c**) and of EMS intensities computed from DGA shown in (**c**)

compared these preliminary results with the maximum observed intensities reported in EMS scale by Martin and Szeliga (2010). For each cell of the same regular 0.2° × 0.2° mesh on the territory of Gujarat, used for the modelling, we find the maximal observed intensity in the square centered at this point (Fig. 3a). We can see that the modelled intensities are almost never exceeded by maximum observed intensities. In a further step we applied a smoothing of three cells to observed

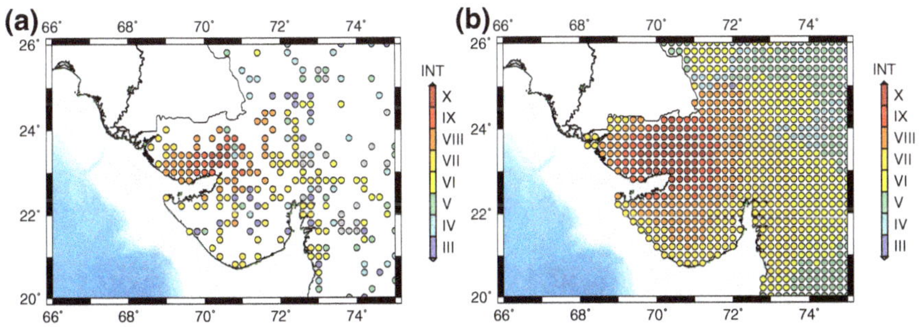

Fig. 3 Spatial distribution of maximum observed intensities in EMS scale taken from Martin and Szeliga (2010), discretized into cells of 0.2° × 0.2°: **a** and smoothed distribution of maximum observed intensities in EMS scale taken from Martin and Szeliga (2010) **b**

intensities, i.e. the same smoothing applied to magnitudes for the synthetic seismic sources; the results of this procedure (Fig. 3b) and the map of DGA converted to intensity (Fig. 2d) are very similar.

2.4 Comparing Neo-deterministic and Probabilistic Ground Shaking Estimates

A preliminary comparative analysis is performed amongst the neo-deterministic map proposed in this study and the probabilistic seismic hazard maps defined by ISR (ISR 2012) for Gujarat territory. The comparison is carried out considering the PGA given by the probabilistic method and the DGA given by the neo-deterministic one, following Zuccolo et al. (2011). Specifically, the PSHA hazard maps defined for 10 and 2 % probability of being exceeded in 50 years (i.e. return period of 475 years and 2475 years, respectively) are considered. The PGA values provided by the probabilistic maps for the Gujarat territory (ISR 2012) are sampled at each node of the 0.2° × 0.2° grid (Fig. 4) where DGA values were computed (Fig. 2c).

The NDSHA maps are generally intended to provide an upper bound for expected ground motion, compatible with seismic history and seismotectonic of the region. Therefore a better agreement is naturally expected with PSHA map defined with a lower probability of exceedance, and thus corresponding to a rather long return period (i.e. 2475 years).

The differences between NDSHA and PSHA hazard maps, calculated on the common grid points, are shown in Fig. 5a, b. The upward triangles indicate that the difference between NDSHA and PSHA is positive, while the downward triangles indicate that the same difference is negative. In general, NDSHA is giving larger values than PSHA in the areas where the strongest events have been observed; this fact supports the idea that probabilistic estimates tend to underestimate the hazard

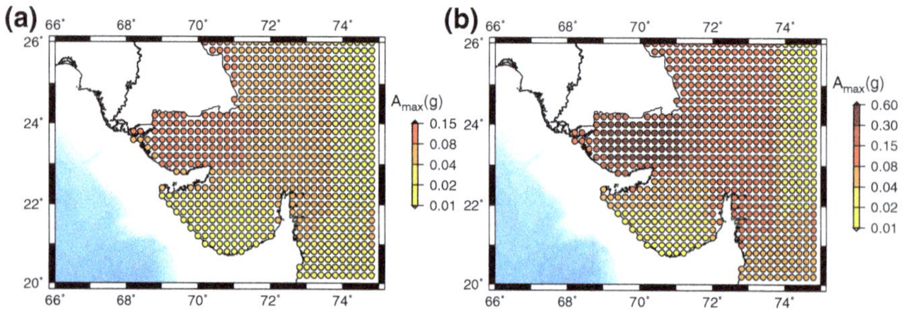

Fig. 4 Probabilistic seismic hazard map of Gujarat (ISR 2012) expressed as PGA with: **a** 10 % and **b** 2 % probability of being exceeded in 50 years. PGA values (in g) are plotted considering the same grid nodes and *color palette* as in Fig. 2c

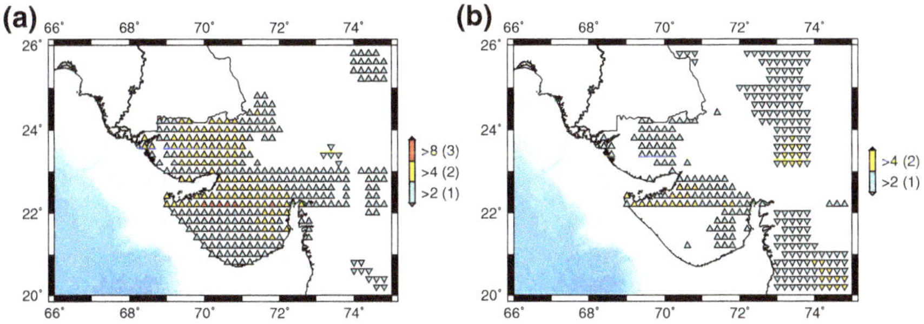

Fig. 5 Ratio between the NDSHA and PSHA maps obtained for: **a** 10 % probability of being exceeded in 50 years; **b** 2 % probability of being exceeded in 50 years. The *upward triangles* indicate a ratio greater than one, while the *downward triangles* indicate a ratio lower than one. The *color palette* refers to the ratio between NDSHA and PSHA estimates (intensity difference is given in *bracket*)

where the largest earthquakes, which are characterized by a longer return period, may occur.

The scatter plots showing the distribution of the PSHA values of Fig. 4 with respect to the NDSHA values of Fig. 2c, as computed at each grid node, are shown in Fig. 6a, b. The black thick line corresponds to identical values estimated by PSHA and NDSHA. The map and the scatter plot obtained for a return period of 2475 years (Figs. 5b and 6b) highlight the fact that the PSHA provides estimates larger than those given by the NDSHA for small values of ground shaking, while the PSHA estimates are comparatively low where the largest events occurred in the

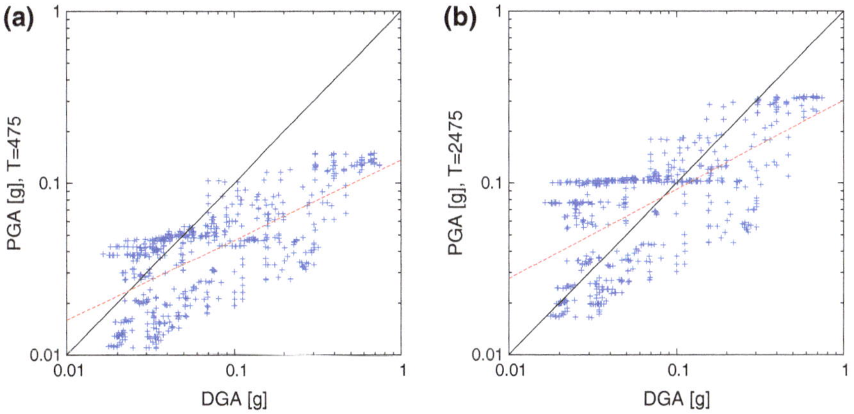

Fig. 6 Scatter plots comparing the PGA from PSHA analysis and DGA values from NDSHA for the common sites. PGA values correspond to estimates obtained for: **a** 10 % probability of being exceeded in 50 years; **b** 2 % probability of being exceeded in 50 years. The *solid black line* corresponds to the values for which PGA and DGA estimations coincide. The linear regression line (*dashed red line*) is shown as well

past and where the most severe ground shaking is expected. Therefore the standard NDSHA, which does not depend upon the sporadicity of the earthquakes, gives conservative estimates in high-seismicity areas. Such estimates are closer to PSHA estimates for long return periods ($T \geq 2475$ years), with differences mostly within one degree of intensity (i.e. a factor of about two in the ground shaking). The differences between the maps obtained from the two different approaches are due to many elements, among which we can mention the different criteria followed in the compilation of the input earthquake catalogue, the simplified ground motion prediction equations used in PSHA and the seismicity parametrisation adopted in the two approaches. The NDSHA approach, in fact, accounts mainly for the largest past earthquakes, whereas PSHA requires considering also moderate size earthquakes for which the catalogue completeness is not always guaranteed.

3 Tsunami Hazard Assessment for Coastal Region of Gujarat

3.1 Methodology

Modelling hazard scenarios has the main purpose to assess the maximum threat expected from a studied phenomenon in a certain area and to give specific directives to the local authorities in order to prevent and mitigate serious consequences on the population, the infrastructures and the environment. To build scenario-based tsunami hazard maps for a specific coastal area one has first to characterize the seismic sources (or other tsunamigenic events, not considered in this study, like e.g. landslides at sea) and select the earthquake scenarios that can drive the hazard. By means of the modelling we then calculate the maximum amplitude of the vertical displacement of the water particles at the sea surface since it is the most relevant aspect of the tsunami wave and also is the only characteristic routinely recorded in the chronicles and therefore in catalogues. The horizontal displacement field is calculated too, and, on average, it exceeds the vertical one by an order of magnitude approximately (this accounts for the great inundating power (run-up) of tsunami waves with respect to wind driven ones). The extremely efficient analytical modelling techniques (computation times are of the order of seconds and are bound to decrease with the natural rate of improvement of computers) considered here for simulations can be utilized also for a tsunami warning system. A modelling can be performed as soon as real time incoming open-sea level data become available, in order to validate, or close, an impending alarm.

Earlier studies about tsunami hazard in the region (e.g. Singh et al. 2008; Okal and Synolakis 2008; Jaiswal et al. 2009; Patel et al. 2013) have been focused on detailed numerical computations of inundation estimates, however they mostly simulate and analyze single scenario events, such as the historical one (27th November 1945) or a maximum credible one. The method applied in this study is based on the efficient analytical computation of a large number of synthetic mareograms (Panza et al. 2000) from different sources and evaluates the tsunami hazard for Gujarat in terms of offshore peak wave amplitude. This method permits to account for a large number of possible tsunamigenic sources and to carry out a number of parametric studies, which allow identifying the sources that can significantly affect the studied zone, as well as the relevant magnitudes. The approach, where the ocean and the solid Earth are fully coupled, is the extension, performed by Panza et al. (2000), to the case of tsunami propagation, of the well-known modal theory and therefore we simply refer to it as "modal method".

From the mathematical point of view, in the modal approach the equations of motion are solved for a multi-layered model structure, so the set of equations is converted into a matrix problem where to look for eigenvalues and eigenfunctions. In general, the modal theory gives a solution corresponding to the exact boundary conditions, and so it can be easily extended to models with slightly varying thickness of the water layer. Therefore, the modal method allows us to calculate synthetic signals for both laterally homogeneous (1D) and laterally heterogeneous (2D) structures. For the 2D case, the structural model is parameterized by a number of 1D structures put in series along the profile from the source to the receiving site (Panza et al. 2000). The parameterization of the bathymetry could be important for the longer source-site paths, since it can significantly influence travel times. However, in this study, in order to efficiently compute a large number of synthetic signals, we adopted a simplified description of 2D media; specifically, for every source-receiver path, we used just two different structures: one at the source and one at the site. Our aim is to provide quick indications about the level of the wave we can expect from different possible earthquake scenarios, rather than to compute detailed inundation (run-up) maps. At this first order level, the fast calculation of a very large number of scenarios, is more relevant than a very high level of precision. Due to the use of simplified models, the mareograms are computed only along straight segments from the source to the sites of interest, neglecting all three-dimensional effects, such as refraction and diffraction. When analyzing the results it is necessary to consider that the applied approach is linear, whereas topographic and bathymetric effects not necessarily are; therefore the results have to be considered as open sea approximation (i.e. for a fluid layer with thickness ranging from deep ocean to about 50 m), which can be efficiently used as initial conditions by numerical approaches for the computation of detailed inundation scenarios. Moreover, in proximity of the coast a number of local effects can generally occur, due to the thinning of the liquid layer, strongly influencing both travel time and maximum amplitude. The ensemble of these phenomena is often called shoaling and it is the main responsible for the final tsunami run-up. The major contribution is the amplification of the wave approaching the coast due to the progressive thinning of the water layer. The principle of conservation of energy requires that the wave energy, when the tsunami reaches shallow waters, is redistributed into a smaller volume, this results in a growth of the maximum amplitude. The linear theory gives for the shoaling amplification factor a simple expression, known as Green's law. Typically the shoaling factor ranges from 1 (no growth) up to several units (amplification) depending on the considered domain (e.g. Ward 2011). Shoaling amplification acts approximately until the wave amplitude is less than half the sea depth (Ward and Day 2008). In such a case, nonlinear phenomena cause the waves to break and eventually turn them backward. Ward and Day (2008) suggest that, due to complications of wave refraction and interference, run-up is best considered as a random process that can be characterized by its statistical properties. Models and observations hint that run-up statistics follow a single skewed distribution spreading between 1/2 and 2 times its mean value.

An example of this methodology, with a validation with Tohoku tsunami of 11th march 2011 and the computation of tsunami hazard for Vietnam coasts, can be found in Bisignano et al. (2011).

3.2 Selected Sources and Results

To calculate tsunami hazard scenarios we first investigate the available historical data from the earthquake catalogues. Since for some areas the available earthquake catalogues could be incomplete or contain errors, the choice of earthquake scenarios has to include the geological and seismotectonic information as well: this leads to a definition of a set of selected tsunamigenic zones with a distribution of focal mechanisms. Once the earthquake scenarios and the structural models are chosen, the modal method is used to compute the synthetic tsunami grams for the selected sites, or for a gridded zone, to compute hazard maps. These maps report the maximum wave amplitude, computed in the linear regime. The efficiency of the modal method permits also to map the tsunami propagation in extended basins for many earthquake scenarios and therefore permits to easily obtain a wide database of pre-computed signals, which can be used with an offshore warning system (e.g. Dart System) to be compared with real-time data.

The possibility to compute efficiently different tsunami scenarios is very important in zones like the western coast of India and, in particular, for the Gujarat coast, considering the rapid development of its activities; in fact, in this zone there are about 20 active ports, 3 refineries, and a number of jetties, several oil storage installations and chemical industries (Gujarat Maritime Board website) boosting the economy of the region. For this zone there are few instrumental records and related studies for tsunami impacts, and, to our best knowledge, the only well documented and studied event is that of 27th November 1945, when an earthquake located in the Makran region, with an estimated magnitude equal to 8.1, generated a tsunami that hit the Gulf of Kachchh, with waves of 11–11.5 m and Mumbai, with waves with maximum run-up of 0.7 m (Singh et al. 2008; Heidarzadeh et al. 2008; Jaiswal et al. 2009; Patel et al. 2013). The historical chronicles record other tsunamis that did hit Gujarat region and, in particular, some authors (e.g. Singh et al. 2008; Rastogi and Jaiswal 2006) indicate evidences of a seismic gap in the western part of Makran region, suggesting a high tsunamigenic potential for this zone.

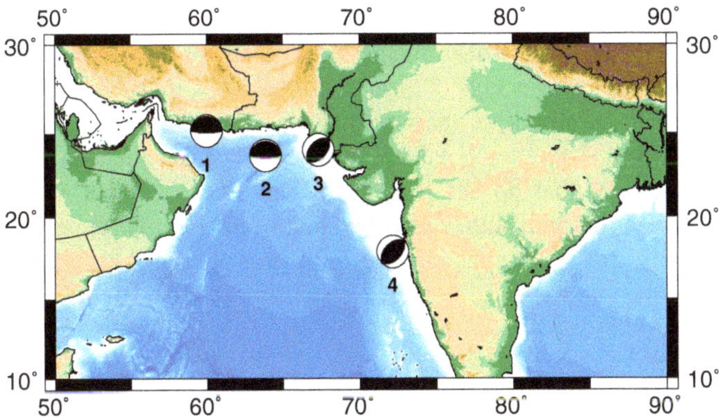

Fig. 7 Location and focal mechanism of selected sources. Source 1 is located in correspondence of the supposed location of the earthquake of 1008 that generated a tsunami in the northern Indian Ocean. The location of Source 2 is related to the relatively well studied event of the 27th November 1945. Sources 3 and 4 have been chosen considering the possible location of two older events, 326 B.C. and 1524, respectively

Considering historical events and literature (in particular Heidarzadeh et al. 2008; Jaiswal et al. 2009) we have selected the 4 potential tsunamigenic sources that are shown in Fig. 7. Some historical reports indicate other tsunami events (e.g. 1668, 1819, 1845); however, following again Heidarzadeh et al. (2008), we have considered them not directly originated by earthquakes but by other tsunamigenic events and, thus, they are not considered in this work, even if they can be relevant for the aim of local tsunami hazard assessment.

Source 1 is located in correspondence of the supposed location of the earthquake of 1008 that generated a tsunami in the northern Indian Ocean. The choice of Source 2 location is related to the more studied event of the 27th November 1945. In simulating this event we have used magnitude and focal mechanism indicated by Dr. A.P. Singh (personal communication), i.e.: $M_W = 8.0$, Strike = 246°, Dip = 7° and Rake = 89°. As we will show later our results for off shore tsunami height are compatible with other simulations and historical reports (e.g. Heidarzadeh et al. 2008; Jaiswal et al. 2009). Sources 3 and 4 have been chosen considering two older events, 326 B.C. and 1524 respectively. In spite of the obvious difficulties to study such old events both of them seem to be quite well confirmed as tsunami earthquake events. In particular, for the 326 B.C. event Heidarzadeh et al. (2008) fix a magnitude range between 7 and 8. Our simulation confirms that, due to their proximity to the studied zone, these two sources can increase significantly the tsunami hazard for the Gujarat coast and that many efforts should be done to understand their actual seismic and tsunamigenic potential. As a starting point for our simulations we have chosen three values of magnitude, 7, 7.5, 8, with focal depths equal to 15, 20 and 25 km, respectively. The strike angles of the chosen focal mechanisms follow the known fault orientation for the region. For dip and rake we have chosen for all the 4 sources an inverse dip-slip mechanism, following both available information on the

1945 event and a conservative approach that suggested us to use, in absence of other information, the most tsunamigenic mechanism for the computations.

The results are shown in Figs. 8, 9 and 10: the maps indicate the values of the maximum vertical amplitude of the tsunami wave (off shore) estimated for every point in the grid, and the color palette emphasizes peak values exceeding about 40 cm; the scenarios that do not give significant values are not presented here. Actually, a coastal amplification of one or more unit is to be expected for these values, especially for what concerns the Gulf of Kachchh where the near-shore bathymetry is steep up to −5 m, so the waves can be strongly amplified, and where

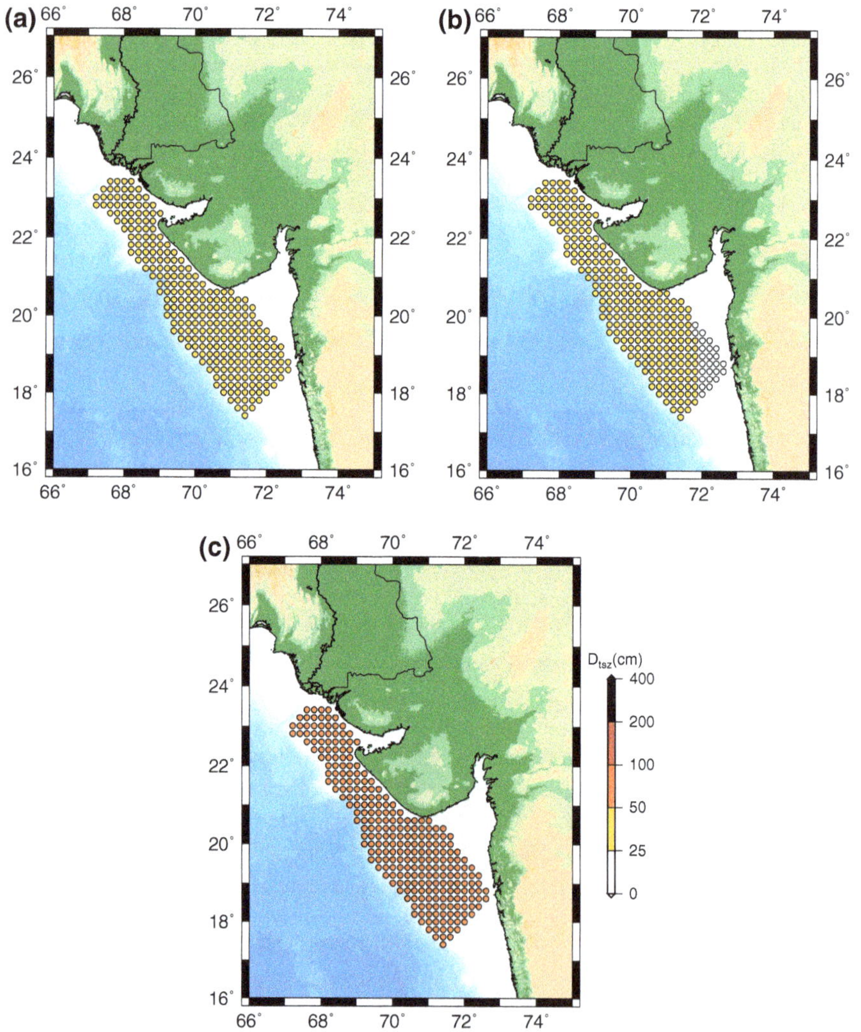

Fig. 8 Hazard scenario due to Source 1 with M = 8.0 (**a**) and hazard scenarios due to Source 2 with M = 7.5 (**b**) and M = 8.0 (**c**)

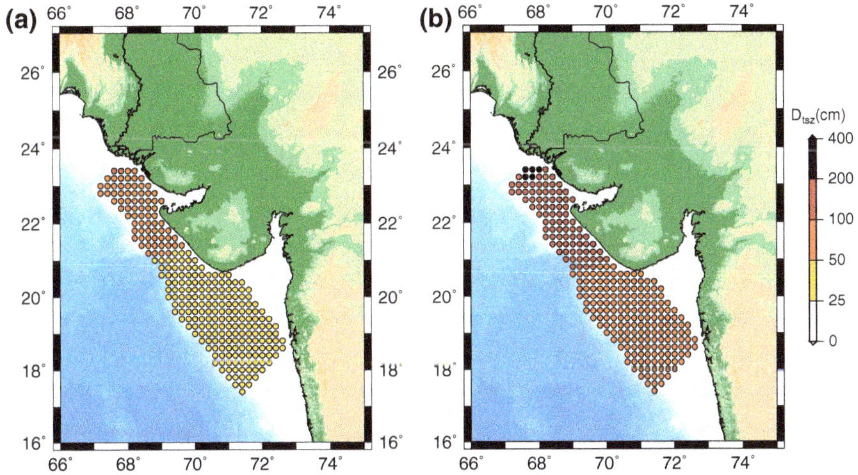

Fig. 9 Hazard scenarios due to Source 3 with M = 7.5 (**a**) and M = 8.0 (**b**)

the near-shore topography is very low (1–5 m above msl) and covered with vast intertidal areas, making those areas particularly prone to inundation (as confirmed by the fact that here normal spring tides occasionally flood the region).

The presented scenarios indicate that the two sources that can mostly affect Gujarat coasts are Sources 3 and 4; Source 3, in particular, may cause open sea vertical amplitudes larger than 50 cm even for an earthquake with magnitude 7 (Fig. 9); these values can lead, due to bathymetric and topographic effects, to run-up and inundation values that must be considered potentially dangerous for a populated and industrialized zone like Gujarat. For Source 2, when considering an earthquake with magnitude 8.1, our computations produce results that are in quite good agreement with historical records and other studies of the 27th November 1945 giving offshore peak values of about 1 m (e.g. Heidarzadeh et al. 2008; Jaiswal et al. 2009; Patel et al. 2013).

Detailed geological and tectonic studies should be performed on the considered tsunamigenic sources, especially when ancient and not instrumentally recorded tsunami are reported by chronicles, to better constrain their tsunamigenic potential. Accordingly, parametric studies are envisaged on the historical and other potential sources (considering different magnitude, focal depth and rupture mechanism), including the use of extended source models for earthquakes close to the study zone (Bisignano et al. 2011), in order to have a more complete picture of the hazard. Our results can also be used as input for the calculation of run-up and inundation values integrating the detailed information for the local bathymetry and topography (e.g. Singh et al. 2008; Patel et al. 2013) along the Gujarat coasts. Thus, the analytical and numerical approaches can be usefully integrated to get a comprehensive description of the tsunami hazard and to identify the appropriate measures (e.g. warning systems, tsunami barriers, evacuation plans) to mitigate the related risk.

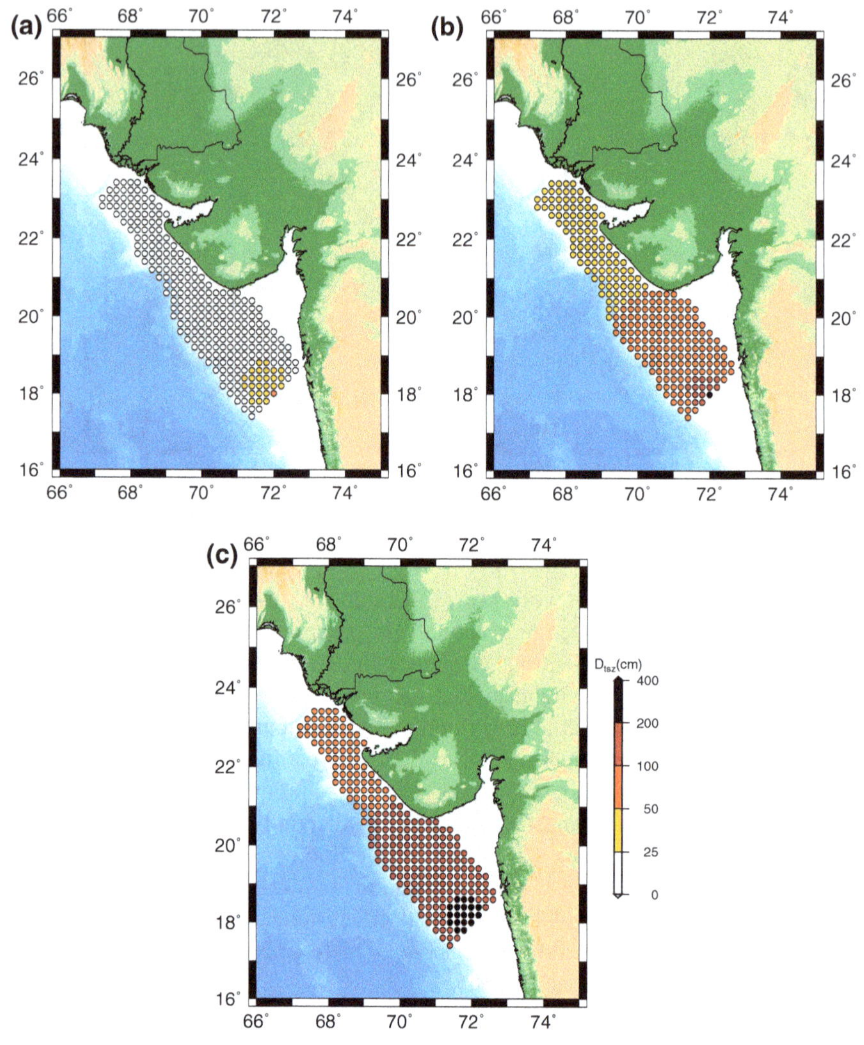

Fig. 10 Hazard scenarios due to Source 4 with M = 7.0 (**a**), M = 7.5 (**b**) and M = 8.0 (**c**)

4 Discussion and Conclusion

The seismic hazard maps available before 2001 Gujarat earthquake proved dramatically inadequate in predicting the losses of life and property, resulting into several costly disasters. During the last decade a huge amount of high quality geophysical and geotechnical data for the Gujarat territory have been collected at ISR and have been already used for seismic hazard analysis by the standard methodologies for probabilistic seismic hazard assessment (PSHA), widely applied in many countries.

However, since the performances of PSHA in anticipating ground shaking turned out very unsatisfactory for most of the large earthquakes worldwide (Wyss et al. 2012), it seems essential for a reliable hazard estimate to consider the NDSHA approach, to cross-check and validate the obtained maps.

The NDSHA methodology has been successfully applied to strategic buildings, lifelines and cultural heritage sites, and for the purpose of seismic microzoning in several urban areas worldwide, including India. The first national scale NDSHA map developed for the entire territory of India (Parvez et al. 2003) adequately described the ground shaking recorded for the 2001 Bhuj earthquake, though such earthquake was not yet included in the earthquake catalogue when the map was originally computed. Recent strong earthquakes, occurred in other areas where neo-deterministic maps were available, successfully confirmed the NDSHA predicted ground shaking (e.g. Boumerdes (Algeria) earthquake, 2003), while the probabilistic estimates turned out underestimated. The PSHA, by forecasting the expected value of shaking over a specified time interval, underestimates the prediction of the actual shaking if earthquakes with longer recurrence times occur (Zuccolo et al. 2011) and often it does not even describe shaking from past, yet observed, large earthquakes. In other words PSHA estimates are dramatically controlled by how sporadic are large events.

When dealing with the protection of critical structures (e.g. nuclear power plants), high rise buildings and cultural heritage, where it is necessary to consider extremely long time intervals, the NDSHA approach appears preferable, since it provides a more complete (i.e. realistic synthetic seismograms) and reliable description of the ground shaking that may affect a given region. Moreover, the NDSHA maps, which do not rely on the estimation of seismicity rates, allows the use of independent indicators of the seismogenic potential of a given area like active faults data and the seismogenic nodes (e.g. Panza et al. 2013), since the available information from past events may well not be representative of future earthquakes.

In this study the NDSHA has been applied to the Gujarat territory, considering a refined structural model and making use of an updated and comprehensive earthquake catalogue, which combines information from different sources. The comparison of the seismic hazard maps produced for Gujarat by the probabilistic (PSHA) and neo-deterministic (NDSHA) approaches shows that NDSHA provides values larger than those given by the PSHA in high-seismicity areas and in areas identified as prone to large earthquakes. Comparatively smaller values are obtained in low-seismicity areas. This is a natural consequence of the smoothing property of PSHA. The NDSHA values exceed PSHA estimates even for rather long return periods ($T \geq 2475$ years); differences however do not exceed one degree of intensity.

In addition a similar modelling approach (Panza et al. 2001) has been applied for tsunami hazard assessment, based on the computation of a large number of synthetic mareograms from different sources. The computed scenarios allowed us to single out the tsunamigenic sources, along with the relevant magnitudes, which may significantly affect the Gujarat coasts. The estimated wave amplitudes can be used as the boundary conditions for the numerical methods, allowing for a faster modelling of the run-up and inundation maps that strongly depend on non-linear effects and detailed bathymetry and topography data.

The results from this study provide the community (e.g. authorities and engineers) with advanced information for seismic and tsunami risk mitigation, which is particularly relevant to Gujarat in view of the rapid development and urbanization of the region. The applied methodology naturally supplies realistic time series of ground motion, which represent reliable estimates of ground displacement readily applicable to seismic design and isolation techniques, useful to preserve historical monuments and relevant man-made structures.

Acknowledgements The results shown in this work have been obtained within the project "Definition of seismic hazard scenarios and microzoning by means of Indo-European e-infrastructures" funded by Regione autonoma Friuli Venezia Giulia in the framework of the interventions aimed at promoting, at regional and local level, the cooperation activities for development and international partnership (Progetti Quadro, L.R. 19/2000). For the tsunami computations, the work has been supported by the RITMARE Flagship Project funded by Italian Ministry of University and Research.

References

Acton, C., Priestley, K., Mitra, S., & Gaur, V. (2010). Crustal structure of the Darjeeling-Sikkim Himalaya and southern Tibet. *Geophysical Journal International, 184*, 829–852.

Aki, K. (1987). Strong motion seismology. In M. Erdik & M. Toksöz (Eds.), Strong ground motion seismology, no. 204 in NATO ASI Series, Series C: Mathematical and Physical Sciences. Berlin: Springer.

Bisignano, D., Romanelli, F., Peresan, A. (2011). Modeling scenarios of earthquake-generated tsunami for Vietnam coast. In *Proceeding for The International Symposium of Grids and Clouds and the Open Grid Forum*, Academia Sinica, Taipei, Taiwan, 19–25 March 2011.

Chandra, U. (1977). Earthquakes of peninsular India—A seismotectonic studio. *Bulletin of Seismological Society of America, 57*(5), 1387–1413.

Chandra, U. (1978). Seismicity, earthquake mechanisms and tectonics along the Himalayan mountain range and vicinity. *Physics of the Earth and Planetary Interiors, 16*(2), 109–131.

Dziewoński, A., Chou, T. A., & Woodhouse, J. (1981). Determination of earthquake source parameters from waveform data for studies of global and regional seismicity. *Journal of Geophysical Research, 86*, 2825–2852.

Ekström, G., Nettles, M., & Dziewoński, A. (2012). The global CMT project 2004–2010: Centroid-moment tensors for 13,017 earthquakes. *Physics of the Earth and Planetary Interiors, 200*, 1–9.

Fitch, T. (1970). Earthquake mechanisms in the Himalayan, Burmese, and Andaman regions and continental tectonics in central Asia. *Journal of Geophysical Research, 75*(14), 2699–2709.

Gung, Y., & Romanowicz, B. (2004). Q tomography of the upper mantle using three component long period waveforms. *Geophysical Journal International, 157*, 813–830.

Gusev, A. (1983). Descriptive statistical model of earthquake source radiation and its application to an estimation of short-period strong motion. *Geophysical Journal of the Royal Astronomical Society, 74*(3), 787–808.

Gusev, A. (2011). Broadband kinematic stochastic simulation of an earthquake source: a refined procedure for application in seismic hazard studies. *Pure and Applied Geophysics, 168*(1), 155–200.

Gusev, A., & Pavlov, V. M. (2009). Broadband simulation of earthquake ground motion by a spectrum-matching, multiple-pulse technique. *Earthquake spectra, 25*(2), 257–276.

Heidarzadeh, M., Pirooz, M. D., Zaker, N. H., Yalciner, A. C., Mokhtari, M., & Esmaeily, A. (2008). Historical tsunami in the Makran subduction zone off the southern coasts of Iran and Pakistan and results of numerical modeling. *Ocean Engineering, 35*(8–9), 774–786.

ISR. (2012). *ISR annual report*. ISR: Tech. rep. http//:www.isr.gujarat.gov.in

Iyengar, R., Sharma, D., & Siddiqui, J. (1999). Earthquake history of India in medieval times. *Indian Journal of history of science, 34*(3), 181–237.

Jaiswal, R., Singh, A., & Rastogi, B. (2009). Simulation of the Arabian Sea tsunami propagation generated due to 1945 Makran earthquake and its effect on western parts of Gujarat (India). *Natural Hazards, 48*(2), 245–258.

Kumar, S., Wesnousky, S. G., Rockwell, T. K., Ragona, D., Thakur, V. C., & Seitz, G. G. (2001). Earthquake recurrence and rupture dynamics of Himalayan Frontal Thrust, India. *Science, 294*, 2328–2331.

Lliboutry, L. (2000). Quantitative geophysics and geology. Berlin: Springer.

Magrin, A., Peresan, A., Vaccari, F., Cozzini, F., Rastogi, B., Parvez, I., Panza, G. F. (2012). Definition of seismic and tsunami hazard scenarios by exploiting EU-India Grid e-infrastructures. In *Proceedings of the International Symposium on Grids and Clouds (ISGC 2012)*, February 26–March 2, 2012.

Mandal, P. (2006). Sedimentary and crustal structure beneath Kachchh and Saurashtra regions, Gujarat, India. *Physics of the Earth and Planetary Interiors, 115*, 286–299.

Martin, S., & Szeliga, W. (2010). A catalog of felt intensity data for 570 earthquakes in India from 1636 to 2009. *Bulletin of the Seismological Society of America, 100*(2), 562–569.

Mitchell, B. J., Cong, L., Ekström, G. (2008). A continent-wide map of 1-Hz Lg coda Q variation across Eurasia and its relation to lithospheric evolution. *Journal of Geophysical Research, 113* (B04303).

Molnar, P., Fitch, T. J., & Wu, F. T. (1973). Fault plane solutions of shallow earthquakes and contemporary tectonics in Asia. *Earth and Planetary Science Letters, 19*, 101–112.

Mourabit, T., Elenean, K. A., Ayadi, A., Benouar, D., Suleman, A. B., Bezzeghoud, M., et al. (2014) Neo-deterministic seismic hazard assessment in North Africa. *Journal of seismology, 18* (2), 301–318.

Okal, E., & Synolakis, C. (2008). Far-field tsunami hazard from mega-thrust earthquakes in the Indian Ocean. *Geophysical Journal International, 172*, 995–1015.

Oldham, T. (1883). A catalogue of Indian earthquakes from the earliest times to the end of 1869 A. D. *Memoirs of Geological Survey of India, 29*, 163–215.

Panza, G. F., Alvarez, L., Aoudia, A., Ayadi, A., Benhallou, H., Benouar, D., et al. (2002). Realistic modeling of seismic input for megacities and large urban areas (the unesco/iugs/igcp project 414). *Episodes, 25*, 160–184.

Panza, G. F., Mura, C., Peresan, A. (2013). Seismic Hazard and Strong Ground Motion: an Operational Neo-deterministic Approach from National to Local Scale. In UNESCO-EOLSS Joint Commitee (Eds.), *Encyclopedia of life support systems (EOLSS)—Geophysics and geochemistry*. Oxford: Eolss Publishers.

Panza, G. F., Mura, C., Peresan, A., Romanelli, F., & Vaccari, F. (2012). Seismic hazard scenarios as preventive tools for a disaster resilient society. In R. Dmowska (Ed.), *Advances in geophysics* (Vol. 53, pp. 93–165). San Diego: Academic Press.

Panza, G. F., Romanelli, F., & Vaccari, F. (2001). Seismic wave propagation in laterally heterogeneous anelastic media: theory and applications to seismic zonation. In R. Dmowska & B. Saltzman (Eds.), *Advances in geophysics* (Vol. 43, pp. 1–95). San Diego: Academic Press.

Panza, G. F., Romanelli, F., & Yanovskaya, T. B. (2000). Synthetic tsunami mareograms for realistic oceanic models. *Geophysical Journal International, 141*, 498–508.

Panza, G. F., Vaccari, F., Costa, G., Suhadolc, P., & Fah, D. (1996). Seismic input modelling for zoning and microzoning. *Earthquake Spectra, 12*, 529–566.

Panza, G. F., Vaccari, F., & Romanelli, F. (1999). IGCP project 414: Realistic modeling of seismic input for megacities and large urban areas. *Episodes, 22*, 26–32.

Parvez, I. A., Romanelli, F., & Panza, G. F. (2011). Long period ground motion at bedrock level in Delhi city from Himalayan earthquake scenarios. *Pure and Applied Geophysics, 168*, 409–477.

Parvez, I. A., Vaccari, F., & Panza, G. F. (2003). A deterministic seismic hazard map of India and adjacent areas. *Geophysical Journal International, 155*, 489–508.

Patel, V., Dholakia, M., & Singh, A. (2013). Tsunami risk 3D visualizations of Okha Coast, Gujarat (India). *International Journal of Engineering Science and Innovative Technology (IJESIT), 2*, 130–138.

Rao, B. R., & Rao, P. S. (1984). Historical seismicity of peninsular India. *Bulletin of the Seismological Society of America, 74*(6), 2519–2533.

Rastogi, B., & Jaiswal, R. (2006). A catalog of tsunamis in the Indian Ocean. *Science of Tsunami Hazards, 25*(3), 128–143.

Shanker, R. (2001). Seismotectonics of Kutch Rift basin and its bearing on the Himalayan seismicity. *ISET Journal of Earthquake Technology, 38*(2–4), 59–65.

Singh, A., Bhonde, U., Rastogi, B., & Jaiswal, R. (2008). Possible inundation map of coastal areas of Gujarat with a tsunamigenic earthquake. *Indian Minerals, 61*(3–4), 59–64.

Tandon, A., Chaudhury, H. (1968). Koyna earthquake of December, 1967. Technical report 59, India Meteorological Department.

Tewari, H., Rao, G. S. P., & Prasad, B. R. (2009). Uplifted crust in parts of western India. *Journal of the Geological Society of India, 73*(4), 479–488.

Ward, S. N. (2011). Tsunami. In *Encyclopedia of Solid Earth Geophysics* (pp. 1473–1493). Berlin: Springer.

Ward, S., & Day, S. (2008). Tsunami balls: A granular approach to tsunami runup and inundation. *Communications in Computational Physics, 3*(1), 222–249.

Wyss, M., Nekrasova, A., & Kossobokov, V. (2012). Errors in expected human losses due to incorrect seismic hazard estimates. *Natural Hazards, 62*, 927–935.

Zuccolo, E., Vaccari, F., Peresan, A., & Panza, G. F. (2011). Neo-deterministic and probabilistic seismic hazard assessments: A comparison over the Italian territory. *Pure and Applied Geophysics, 168*(1), 69–83.

Geophysical Characterization of Liquefied Terrains Using the Electrical Resistivity and Induced Polarization Methods: The Case of the Emilia Earthquake 2012

Nasser Abu Zeid

1 Introduction

Massive surface fracturing, sand upwelling, sand volcanoes, limited blister formation and lateral spreading liquefaction features took place immediately after the main shock of May 20th, 2012 earthquake (ML = 5.9, depth: 6.3 km, max/min hypocentral distance to the test sites: 15–20 km), that struck the Emilia-Romagna Region, Northern Italy (Fig. 1). These phenomena, induced by the liquefaction of a shallow, water-saturated sand and silty-sand layer, have resulted in the damage of several buildings as well as of roads and sidewalks. In particular, the liquefied layer (s) were clustered, mainly, along a narrow zone located between Sant'Agostino and Vigarano Mainarda towns (situated in the southwestern portion of Ferrara Province, North Italy), in correspondence to the paleo-ridge of the old Reno River (Papathanassiou et al. 2012).

Liquefaction in the last century had occurred in many parts of the globe following moderate to strong earthquakes. However, this phenomenon gained the attention of scientist and engineers only after the dramatic events following the 1964 earthquakes that hit Alaska and Japan (Seed and Lee 1966; Seed and Idriss 1967). These events provoked Seed (1979) to comment on these effects as follows: "*...these events, more than anything else, probably did more to stimulate geotechnical engineering studies of earthquake-induced liquefaction than any other single factor*".

Liquefaction of saturated sand and silty-sand sediments represents one of the most important co-seismic site effects in geotechnical studies that must be investigated for the correct mitigation of their effects (i.e. microzonation). In the recent years, geotechnical engineers, in collaboration with other scientific disciplines, started to particularly focus their attention on the liquefaction potential of sands

N. Abu Zeid (✉)
Department of Physics and Earth Sciences, University of Ferrara,
Via Saragat 1 Blocco B, 44122 Ferrara, Italy
e-mail: nsa@unife.it

S. D'Amico (ed.), *Earthquakes and Their Impact on Society*,
Springer Natural Hazards, DOI 10.1007/978-3-319-21753-6_8

with considerable fines content (of 25–45 %) (Seed et al. 2003; Boulanger and Idriss 2006; Bray and Sancio 2006). Liquefaction is a well-known process that refers to the loss of shear strength in saturated cohesionless sediments due to cyclic stresses induced by moderate to strong earthquakes. The accumulation of the induced shearing stress, under certain soil and shaking conditions, may trigger one or more liquefaction modes due to a build-up of excess pore water pressure that exceeds the total stress (or overburden pressure), hence, reducing the effective stress to zero.

The phenomenon occurs in the ground below the water table, generally, at depths less than 20 m. Liquefaction rarely occurs at greater depths due to the increasing overburden pressure (and hence total stress), making it difficult to reduce effective stress to zero as a result (Krinitzsky et al. 1993). However, exception do exist for sites that contain exceptionally loose soils highly susceptible to liquefaction at greater depths. In these cases special effective stress analysis techniques are required to evaluate the potential for deeper liquefaction. In our study area this is not the case. Densification is another mechanism by which a dry loose sand sediment (i.e. above water table) may undergo volume changes when cyclic shaking changes its volume (i.e. contracts), while on the contrary, a densely packed sand increases its volume (Ishihara 1974) as its dilate on shearing.

Undesired liquefaction effects are represented by ground failure that represents the main cause of damage to the built environment. These effects have been observed since historical times and provided clues about the past occurrence of moderate to strong earthquakes (Galli 2000). The author analyzed moderate to strong historical earthquakes occurring in Italy in the period 1117 AD–1990 (Ms range: 4.2–7.5 and estimated magnitude “Me” of 4.8–7.5) that was recently reviewed after the 2012 Emilia earthquake (Galli et al. 2012). As an example the reader’s attention is drawn to the well-documented 1570 earthquake that struck Ferrara, North Italy, some 450 years ago (Me: 5.5, depth: 7 km) (Guidoboni 1997; Guidoboni et al. 2007). Other, well-documented sites include, among many others, the 1811/1812 (M: 7.7/7.5) New Madrid strong earthquake sequence (Ernstson and Neumair 2011) and the 1906 San Francesco earthquakes (Mw: 7.7/7.8).

The liquefaction phenomenon has also been studied by many authors in order to understand its causes and to provide tools for its prediction and mitigation. Liquefaction related literature agrees on the principal triggering factors, related mainly to soil conditions and dynamic characteristics of the seismic shaking. The former include grain-size distribution, mineralogy, uniformity coefficient, fabric and depositional environment (e.g., Mitchell and Soga 2005). The authors suggest that the geotechnical properties of fine-grained sediments, typically the plasticity index, can be used as a discriminating factor for judging the susceptibility of a stratum to liquefaction.

Field subsurface characterization methods require the execution of at least one borehole and/or a number of CPT/SPT/CPT in order to be able to determine a safety factor against liquefaction potential (e.g. Seed and Idriss 1982; Robertson et al. 1985, 1992; Youd et al. 2001). These techniques are well-established for site investigation normally disciplined in standard building codes (see also Eurocode 8).

Although these methods are necessary, the gathered information has a local significance that limits its spatial and depth extrapolation. The availability of punctual and detailed information about sites that underwent liquefaction has led to the development of a well-documented database, originally collected by Andrus et al. (1999), where 225 case histories are reported. Of these, 96 belonging to liquefied sites and 129 to non-liquefied ones. The analyzed sites were struck by 26 earthquakes (ML: 5.3–8.3) that occurred in the period between 1906 and 1995. However, this study focused, among other parameters, on the analysis of the shear wave velocities (Vs) to predict the liquefaction potential of soils in other areas of similar lithology and hazard. The Vs is used to evaluate the shear resistance as the determination of the small-strain rigidity modulus is straight forward (Stokoe and Nazarian 1985; Tokimatsu and Uchida 1990; Yanguo and Yumin 2007; Yanguo et al. 2009). Accordingly, the availability of indirect and non-invasive geophysical methods that can help in sounding large subsurface volumes are highly appreciated. Such methods can help in siting sites for direct collection of subsurface information. Perhaps, it worth's mentioning that shear wave velocities are influenced by initial soil conditions, namely density and rigidity of the solid skeleton. These provide the resistance to cyclic shearing induced by wave propagation. This explains why shear wave velocity based seismic methods have become widely used and available for site characterization, especially in highly urbanized areas. These are used for routine site characterization or during the post-seismic reconstruction phase.

Other geophysical methods, sensitive to texture (porosity), fabric and saturation are represented by geoelectromagnetic methods as these methods can easily capture volume contrasts in the subsurface physical properties related to subsurface lithological heterogeneities. These comprise Electrical Resistivity Tomography (ERT), Induced Polarization Tomography (IPT) and Ground Penetrating Radar (GPR). As the name suggests, the former methods are used to map spatial and vertical variation of the following physical properties: resistivity (ρ), polarzability (IP) and dielectric permittivity (ε). In the following, I shall discuss the first two methods as the last one has been widely treated in the paleo-seismological/paleo-liquefaction related literatures (e.g. Wolf et al. 1998; Tuttle et al. 1999; Wolf and Tuttle 2006; Gross et al. 2002; Improta et al. 2010; Ercoli et al. 2013; Nobes et al. 2013). Today, both ERT and GPR methods can be considered consolidated tools for subsurface imaging in palaeo-seismological investigations (McCalpin 2009), however, their employment in the subsurface investigation of potentially liquefiable sites has received less attention (e.g. al-Shukri et al. 2006; Liu and Li 2001; Maurya et al. 2006). *An interesting work was presented by* Nakazawa et al. (2012), where both shear wave velocity and Electrical Capacitive Tomography techniques (Timofeev 1994; Kuras et al. 2006) were employed to investigate possible post-seismic liquefaction effects beneath a runway in the Matsuyama Airport and the taxiway in Tokyo International Airport. However, no IP measurements where conducted as the capacitive-based resistivity equipment can't register this effect.

To my knowledge, the use of the IP method in liquefaction related studies is poorly documented in the literature, if not completely absent. The discussion shall focus on the use of high resolution ERT/IPT geophysical techniques as a tool in aiding geologists to better reconstruct the conceptual subsurface geological model and to pin point zones of diverse lithology that are prone to liquefaction. This shall enhance, from one side our ability to forecast probable future liquefaction-induced ground failure sites (i.e. zoning) and from the other to help engineers in considering the reconstruction of existing non-critical structures in a way to withstand modest deformations that may result in anticipated limited damage. Nevertheless, the ERT/IPT images can also be useful for subsurface characterization of areas where urban and/or industrial expansions are to be planned.

In this way, I think that the use of non-invasive geophysical methods can provide a step forward towards filling the existing gap between available information about the subsurface retrieved at a single point and usually employed for liquefaction potential prediction (i.e. CPT, CPTu, SCPTu, borehole sampling) and the subsurface spatial continuity and distribution of lithological strata potentially prone to liquefaction. To this end, results obtained from three case studies carried out at sites that underwent liquefaction following the Emilia earthquake (May 20th, 2012) shall be presented and discussed. The sites include three typologies of buildings: agricultural warehouse, industrial and residential building located in the southwestern part of Ferrara's province, North Italy (Fig. 1).

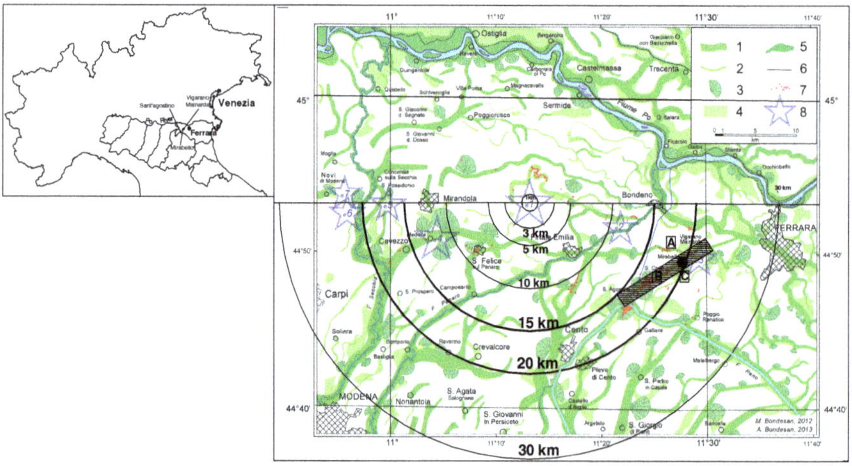

Fig. 1 Simplified geomorphological map of the area showing, in *green color*, the traces of the paleo-river ridges known to exist. *Red dots* indicate the position of observed liquefaction points, *shaded rectangle* location of the test site: A, B and C, *semi-circles* equal distance from the epicenter of the main shock of May 20, 2012 (ML:5.9), *1* Paleo-river ridge, *2* Paleo-river, *3* Sandy fluvial forms (paleo-channels, fans), *4* other areas with prevalent sand deposits, *5* active flood plains, *6* river banks, *7* observed liquefaction point and *8* epicenter of the main events (ML > 5.0) occurred in Emilia during the period: 20 May–3 June, 2012 (e1: main event of ML: 5.9 occurred on 20th of May 2012 and e4: the second event occurred on 29th of May 2012 ML: 5.8)

2 Electrical Resistivity and Induced Polarization Tomography (ERT/IPT)

The electrical resistivity method is one of the most widely used geophysical techniques for the investigation of the near subsurface especially in alluvial plain territories such as the Po River Plain in northern Italy. The method was developed at the beginning of the 19th century by Schlumberger. Since then, it has been used as a non-invasive exploration tool in many disciplines. These include: geology, environment, engineering, archaeology, cultural heritage and lately also in forensic investigations (e.g. Telford et al. 1990; Reynolds 2011). Schlumberger, also, was the first to notice the associated capacitive effect during field electrical resistivity measurements that he called as "provoked polarization". In short, he noticed that most often the measured potential difference did not vanish after switching off the energizing current source, instead, it showed a gradual decrease with time. This behavior later on led to the development of the Induced Polarization (IP) method. In short, the method makes use of the capacitive action of the saturated subsurface to locate zones where polarizable minerals (e.g. Pyrite, Chalcopyrite, Copper, Graphite) and disseminated clays. In addition, the method was successfully employed in tracing leachates from waste management facilities (e.g. Abu Zeid et al. 2004) and in the localization of recently disrupted subsurface materials in otherwise homogeneous undisturbed lithology. This latter fact is very important as it indirectly invokes modifications in the sediments' structure: porosity and tortuosity which indirectly control permeability. The influence of a wide range of alluvial sediments (clay to sand lithologies) on the chargeability was thoroughly studied by Iliceto et al. (1982). The authors showed, through laboratory tests on loose samples, that the IP technique can be used to differentiate between sediments of different texture.

From the technical point of view, the IP method is based on measuring the level of polarizability of the subsurface materials (Sumner 1976; Schoen 1996). Polarizability is defined as the ratio of the charging voltage to the measured voltage after current switch off. This definition implicitly takes elapsed time in consideration. Consequently, the IP is an electrical phenomenon which is independent of the medium's electrical properties. It can be measured either in time or in frequency domain. In the former case, the polarizability, defined as chargeability (M), represents the time integration of the decaying voltage at discrete time-windows, following current switch-off (Fig. 2). In the latter, polarizability is defined as a "Phase Shift" measured (in mrad) at a pre-defined frequency or as a "Percent Frequency Effect—PFE" (in %) where resistivity is measured at two different current injection frequencies (normally 1 and 10 Hz). The "Phase" is simply obtained by taking the arctangent of the ratio of the imaginary to real components of the complex resistivity (Sumner 1976; Reynolds 2011). Modern resistivity meters can easily handle the measurement of both properties simultaneously with a modest increment in data acquisition and processing phases.

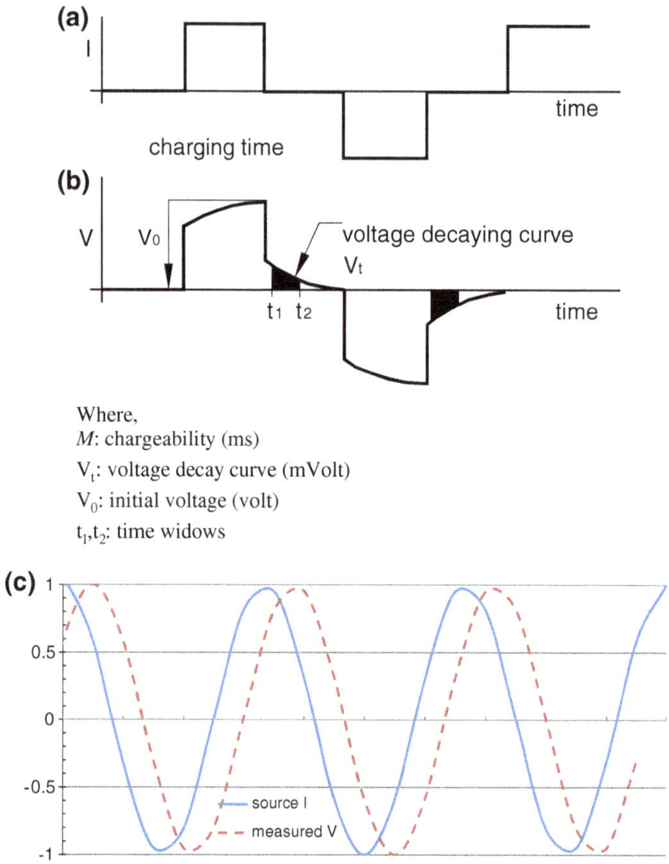

Fig. 2 Induced Polarization measurements: **a** in the time domain: chargeability (M), **b** and **c** in frequency domain: PFE and phase

The origin of the IP phenomenon can be explained by three mechanisms of which the first two are widely accepted: (1) electrode polarization, linked to the presence of mineral grains in water filled pores; (2) membrane polarization, linked to the presence of clay distributed within a coarse grained sediment saturated with fresh water and (3) polarization due to constrictivity of fresh water saturated pores, i.e. to the variation of their diameters (Schoen 1996; Zadorozhnaya 2008). The last mechanism seems promising although more research has still to be done, especially at field scale.

Both ERT and IPT data can be acquired using a set of 32 or more electrodes disposed along a profile or distributed aerially in case of 2D and 3D surveys respectively. Measurements are accomplished by sending a DC or very low frequency AC currents into the ground through a couple of stainless steel electrodes (AB) and measuring the developed potential between a second couple of electrodes

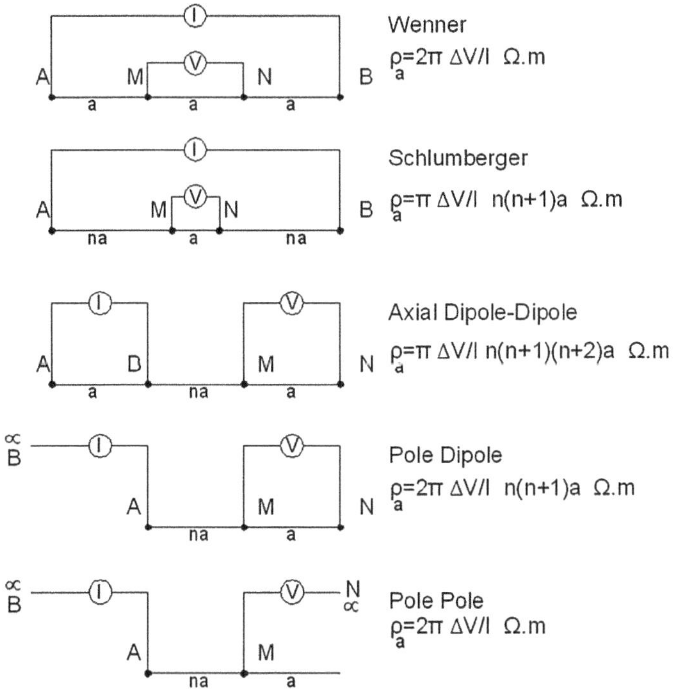

Fig. 3 Electrode arrays. The one used in this study was the combined Wenner-Schlumberger. *AB* current dipole, *MN* potential dipole, *a* electrode spacing, *K* geometric factor, ρ_a apparent resistivity

(MN). To this end, different electrode arrays were developed with the aim to favor the possibility to capture lateral and/or vertical resistivity/IP variations (Fig. 3). The possibilities offered by modern georesistivity meters allow for the acquisition of a large amount of data that can be contoured to produce the so-called apparent resistivity/IP pseudo-sections (Fig. 4). Such representation is only informative and provides only qualitative information about the subsurface, hence data must by processed numerically (i.e., inversion) in order to get the best estimate of the subsurface distribution of the model parameters (i.e. resistivity and chargeability). To this end, numerous well-developed 2D/3D inversion codes are available that can handle surface topography (e.g. Loke and Barker 1996) and complex data acquisition geometries (e.g. Eartlab, http://www.ertlab.com/ last accessed on November 2014). The quality of the obtained models can be judged by observing the RMS error that describes the difference between field and theoretically calculated data based on model parameters (i.e. real resistivity and chargeability physical properties). Lateral resolution of the tomographic images corresponds roughly to half the electrode spacing used in the near sub-surface layers.

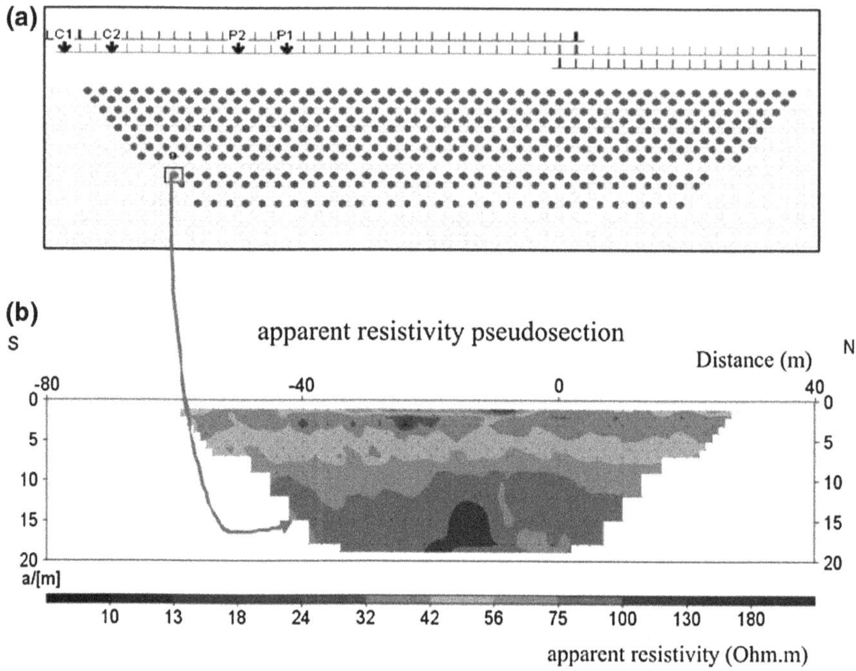

Fig. 4 2D pseudosection representation of the measurement points of the parameter of interest **a** measured points **b** contouring section of resistivity and/or IP: chargeability, PPF or phase

3 Geological and Paleo-Geographical Outline

The main earthquake of 20 May 2012 initiated a seismic sequence that comprised 7 earthquakes of ML > 5.0 occurred in an area of 50 km length and 10 km width. The main shock (ML > 5.9) involved the frontal sector of the Northern Apennines and in particular the buried front of the Romagna and Ferrara northward-verging active thrust belt. It is represented by a fold-and-thrust belt underlying the Po Plain, buried under a thick wedge-like Pliocene-Quaternary depositional succession. The geometry of this hidden chain is relatively well defined by numerous seismic reflection profiles and deep wells. The most important tectonic structures are represented by the Ferrara Arc and the minor Adriatic and Romagna Arcs, these blind thrusts are likely to be still active (Toscani et al. 2009).

Superficial sediments of the test sites are composed mainly of alluvial materials that have been deposited in different environments. These include channel and proximal and distal levee, inter-fluvial, meander and swamp. These sediments also form the main hydrogeological units overlying the Middle to Upper Pliocene bedrock, which according to available subsurface information derived from past seismic reflection sections and deep wells (Pieri and Groppi 1981) was placed at 700–800 m below ground level. As a consequence, the outcropping deposits are everywhere Holocene in age, substantially loose or poorly compacted in the first 15–20 m and in terms of grain size, could vary from clay to coarse sand with local presence of organic materials, pebbles and gravels (Fig. 5).

The Reno paleo-river channel was formed after important crevasse episodes during late Medieval times. From that time to the late 18th century, the Reno water was able neither to reach directly the Adriatic Sea nor to directly inflow the Po River, running about 10 km to the north of the study area, therefore generating wide inland marshes and lakes. The depositional evolution of the area was punctuated by a large number of unsuccessful artificial embankments and land reclamation efforts. But it was only towards the end of the 18th century that the Reno River was successfully diverted towards the Adriatic Sea through a former southern distributary channel of the Po River just to the south of the investigated site and any water flow northwards of Sant'Agostino was impeded (Cremonini 1988; Bondesan 1989).

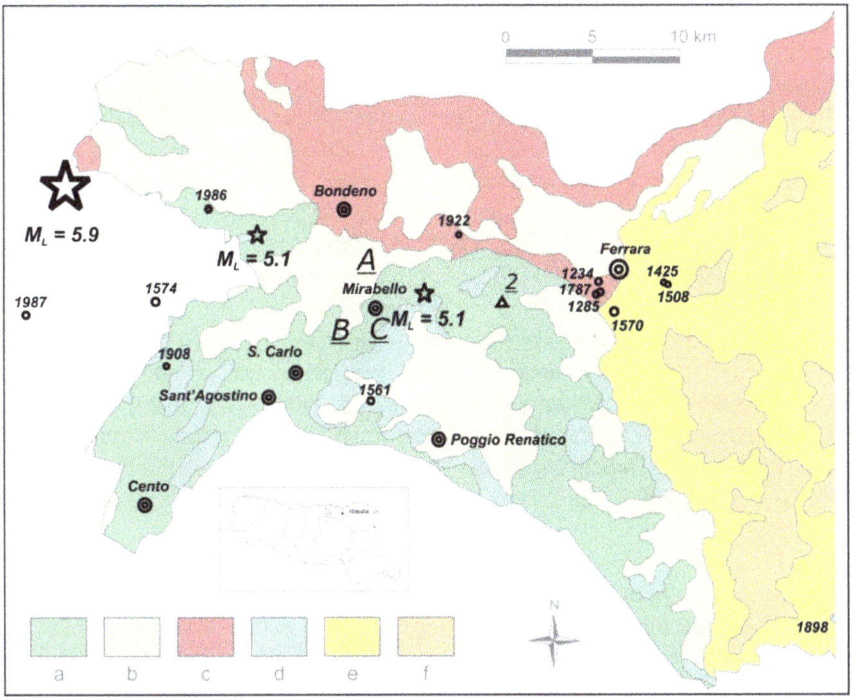

Fig. 5 Simplified geological map of the western portion of the Ferrara of the Province of Ferrara. The map shows the location of the main earthquakes of the 20th of May 2012 (*stars*) and the location of the Mirabello Town test site discussed in this paper (1) as well as the location of a paleo-liquefaction site to the SW of Ferrara (Torre Fossa Village). The main lithologic units of Holocene age are: *a* medium to fine sand (channel and proximal levee deposits); *b* silty clay, clay and clayey silt (interfluvial and swamp deposits); *c* sandy silt, fine sand and silty clay (distal levee deposits); *d* medium to coarse grained sand (alluvial plain and meander deposits); *e* medium to fin grained sand (distribution channel and levee deposits); *f* silt, clayey silt (swamp deposits). Modified after Data Base of the Emilia-Romagna Region (URL: geo.regione.emilia-romagna.it/geocatalogo/)

Consequently, the former channel dried up leaving the ancient depositional morphology well-preserved. Locally, it was wiped-out to accommodate industrial and residential constructions. This act, however, contributed in incrementing the damage severity. The former levee crests are still some 5 m higher than the surrounding plain which contributed in inducing the formation of surface ruptures and lateral spreading.

4 Data Collection and Processing

The geoelectrical survey was carried out at three sites that exhibit one form or another of liquefaction induced features. The first site (A: agricultural warehouse), located some 700 m NE of the paleo-river ridge, was investigated by two orthogonal ERT profiles of 93 m length each. Here, the site suffered liquefaction with no evident surface ruptures, however, one warehouse was heavily damaged leading to its complete demolition. The second (B: industrial zone), located on the paleo-river ridge course of the Reno River, was investigated by one ERT/IPT profile of 126 m length. Here, surface ruptures were observed that lead to the damage of a portion of the industrial building that was demolished later on. The third site (C: residential building located in the eastern plain) was investigated by three ERT/IPT profiles of 62 m length. Here, the site is believed to have undergone both densification and liquefaction. These sites belongs to Vigarano Mainarda and Mirabello towns, located to the SW of Ferrara, North Italy (A, B, C Figs. 1 and 5). Field observations carried out after the earthquake showed that more than 50 % of liquefaction features occurred in these municipalities.

Geoelectrial resistivity and IP data were collected using The ABEM SAS1000/ES464 geo-resistivity-meter (Sweden, http://www.abem.se/products). This equipment acquires, simultaneously the resistance data and the IP information in the "time-domain" mode, i.e. by measuring the apparent chargeability (Fig. 2). In all sites, the Wenner-Schlumberger electrode array was used for data acquisition. This array represents an acceptable compromise between lateral and vertical resolution. Electrode spacing ranged between 0.75 and 1.5 m. This allows to maintain an horizontal resolution of less than 0.75 m. Resistance data was acquired with two cycles of energization, while the IP were acquired using one time window of 100 ms length starting after a 10 ms delay, following current switch-off. It is to be taken into consideration that all test sites are characterized by shallow water depth (less than 2 m). This fact is important as the absence of saturated sediments shall neither result in measurable IP nor liquefaction, although densification of loose sediments due to shaking may still occurs (Toshiyasu et al. 2008).

The apparent resistivity and chargeability data were inverted using the RES2DINV (2012) commercial software that implements an inversion algorithm based on the smoothness-constrained least-squares method and uses the quasi-Newton approximation for optimization (Loke and Barker 1996a). The inversion strategy is essentially of the Occam's type (deGroot-Hedlin and Constable

1990; Sasaki 1992). The inverted resistivity and chargeability 2D images showed RMS relative errors, describing the discrepancy between field and predicted apparent resistivity data, generally less than 2 and 1 % (absolute RMS) for resistivity and chargeability inversion models respectively. This low fitting error was achieved thanks to the good quality of field data.

5 Results

The inversion 2D images of the three test sites are reported in Figs. 6, 7 and 8, In Fig. 6a–c, two resistivity models carried out at site A are shown. In this site, the agricultural warehouse (rectangle in Fig. 6a) was heavily damaged following the earthquake which led to its demolition. However, the nearby buildings, although being constructed in the same period, suffered less damage. The question to answer was if the damage was caused by the high vulnerability of the "A1" building with respect to the others, or whether it was related to a different site response.

The inversion 2D resistivity model of the longitudinal profile (Fig. 6b) shows the presence of a low resistivity layer (clay) of lenticular geometry sandwiched between two high resistivity layers (S/SS). The perpendicular profile (Fig. 6c) pointed out the absence of the clay layer between 15 and 39 m. In fact, the resistivity values of this particular zone, located beneath the damaged building, indicate the presence of silt/fine sand sediments that most probably underwent liquefaction. The nearby buildings did not suffer a similar level of damage probably due to two factors: the buildings are less vulnerable or in the shallow subsurface the clay layer, as confirmed by the resistivity model of ERT-2, played an important role in containing the level of deformation. It is known among geotechnical engineers, that the thickness of the crust above potentially liquefiable sand layer may reduce liquefaction effects (Sonmez 2005).

The 2D resistivity and chargeability inversion models in the second test site (B) are illustrated in Fig. 7a–c. The resistivity model (Fig. 7b) identified two resistivity levels denoted (a1, a2). The first (a1) is characterized by resistivity values greater than 25 Ω m, while the second (a2) shows low resistivity values intercalated by local lateral heterogeneities (A). The low resistivity values (a2) can be associated with a succession of clay and silt sediments, while the lateral heterogeneities (A?) with resistivity values between 20 and 40 Ω m, refer to the presence of silt and fine sand. This interpretation is in accordance with the subsurface lithology obtained from nearby boreholes where a sandy silt layer has been encountered at 10.5–11 m b.g.l. (borehole level at −1.8 m with respect to the ground surface of the ERT/IPT profile). It is believed that this layer has undergone liquefaction although very modest quantities of sand reached the surface.

Concerning the first level, the resistivity model suggests the presence of lateral resistivity heterogeneities at the following distances: 12 m, 35–65 m and around 96 m. These are believed to be associated with subsurface fractures whose traces were visible on the surface at the moment of data collection. One of the main

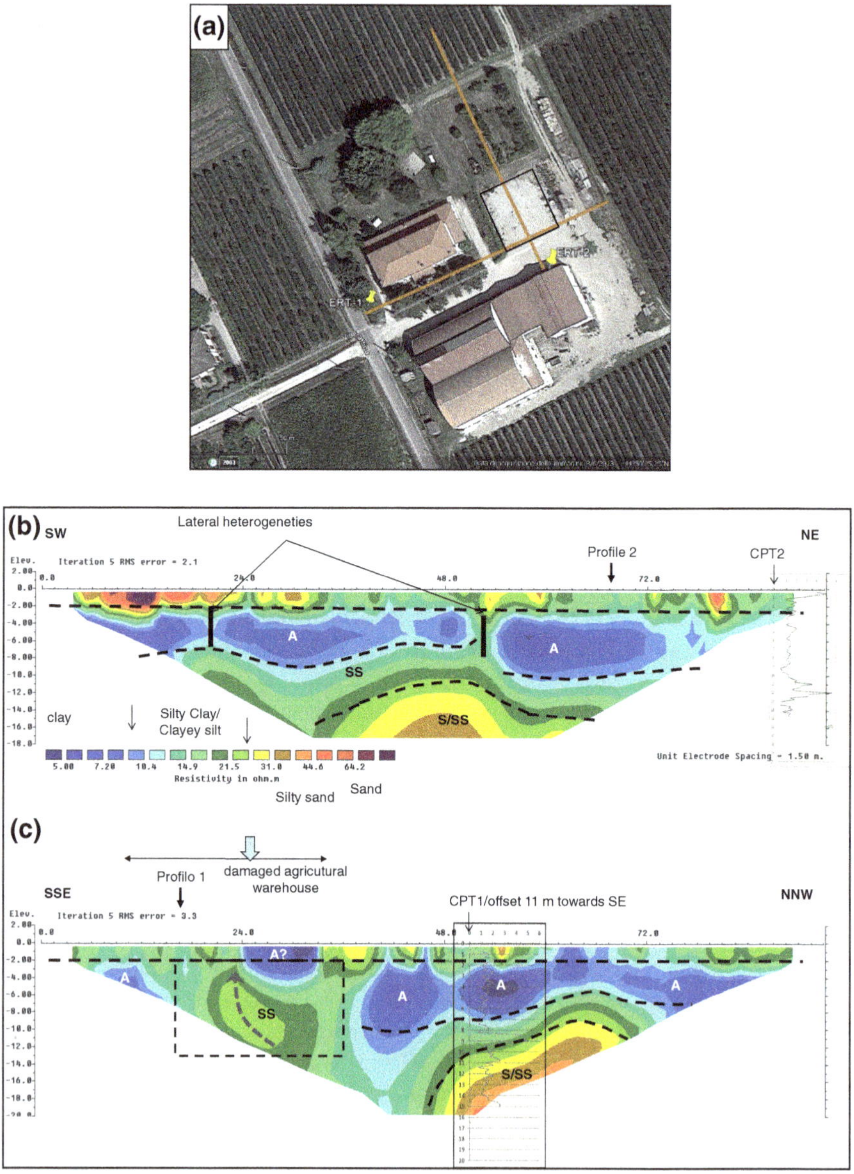

Fig. 6 **a** Detailed location map of the two ERT profiles carried out at test site (*A*). **b** 2D resistivity images of profile ERT-1 (**b**) and ERT-2 (**c**). *Rectangle* outlines the warehouse building limits that was heavily damaged after the earthquake. Two CPT tests carried out before the geoelectrical survey are projected on the ERT images

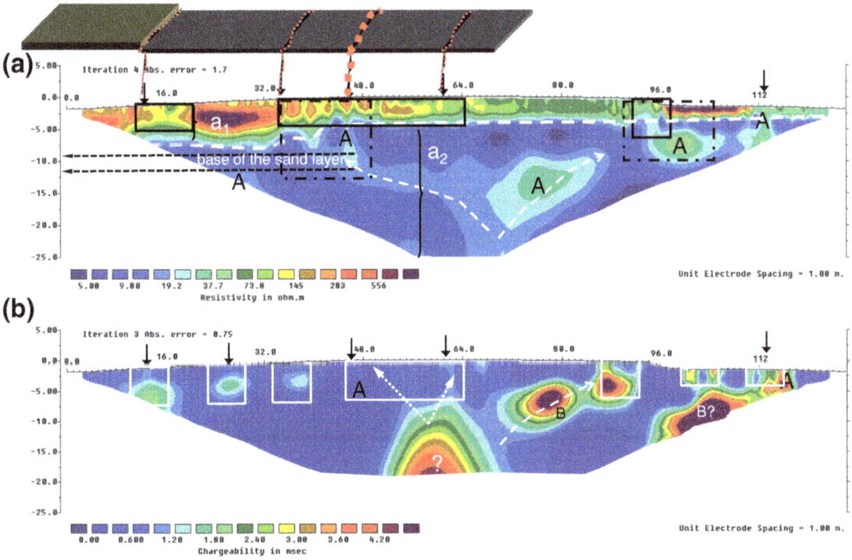

Fig. 7 **a** Resistivity and Induced Polarisation inversion results of the ERT/IPT profile carried out at the second test site (*B* in Fig. 1). 2D resistivity image (**a**) and chargeability (**b**). *Rectangle outlines* the locations where liquefaction features are most visible (*A*). Indications about liquefaction occurrence can be clearly seen on the IPT section (*B*)

Fig. 8 Paleo-liquefaction feature encountered in a trench excavated on the same branch of the paleo-river-ridge about 3.6 km to the south of test site *B*. The feature has been linked to the historical earthquake that hit Ferrara area on 17/11/1570 (Me: 5.5). More information about the trench details can be found in Abu Zeid et al. (2012) and Papathanassiou et al. (2012). Photograph courtesy of Riccardo Caputo (Papathanassiou et al. 2012)

fractures (located at 35 m in Fig. 7a) has caused major damage to the industrial building nearby, located some 25 m to the north. Moreover, the resistivity anomaly, located at 48 m along the profile was associated with the expulsion of a modest amount of fine sand that reached the surface. It is interesting to note that the liquefied sand has moved horizontally towards NW and SE as suggested by the resistivity and chargeability models (explained below). Towards these directions, a high density of surface fissures were observed (rectangles in Fig. 7b, c). It worth's mentioning that the local ground topography slopes slightly towards NW where a thick layer of saturated sand sediment is present between 35 and 95 m. Moving towards SE, the resistivity model evidences the presence of at least two meters thickness of low resistivity sediments (clay to clayey silt) beneath the surficial sediments. It can be postulated that their presence has contributed in contrasting pore pressure increment (i.e. being impermeable these sediments formed a barrier which contributed in the dissipation of excess pressure horizontally towards NW and SE).

The corresponding chargeability model (rectangles in Fig. 7c) evidences the presence of anomalies, which are caused by variations in sediment texture (i.e. presence of silt and fine sand). The location of these anomalies is very near to the fractures (dykes) indicating the probable pathway followed by the sand/water mixture following the liquefaction of the sand layer. The most significant feature is the one located between chainages 45 and 64 m, where the trace of the nearly vertical fractures can be clearly followed. Similar features indicating possible upward movements of sediments can be seen at 7 m depth between chainages 75 and 90 m.

This fracture system has developed within the right levee of the Reno paleo-river whose course in the past was artificially modified by the construction of numerous hydraulic works to mitigate flood hazard and to maintain navigation active. These hydraulic works are believed to have resulted in the deposition of intercalated sand, sandy slits and clays resulting in the formation of local aquifers with limited extension. These sand and silty sand layers have undergone liquefaction. According to a recent paleoseismological study, Papathanassiou et al. (2012) have found evidences of paleo-liquefaction features observed along the walls of a trench excavated purposely 300 m to the south of the test site (Fig. 8). The features are believed to have been produced following the historical earthquake that struck Ferrara on 17/11/1570 AC.

The resistivity and chargeability inversion models of one profile running parallel to the residential building in the third test site (C) are shown in Fig. 9a, b. This residential building is situated outside the ancient course of the Reno river towards east. The resistivity model (Fig. 9a) shows laterally articulated resistivity values in the first two meters, especially beneath the right end of the profile (A? in Fig. 9b). This horizon is followed by a laterally heterogeneous one showing alternating resistive and conductive structures (A? in Fig. 9b). The corresponding chargeability model (Fig. 9c) confirms the presence of IP anomalies at the same locations that are indicated by the letter B? These anomalies are geometrically elongated in the vertical/sub-vertical direction suggesting the probable occurrence of densification of

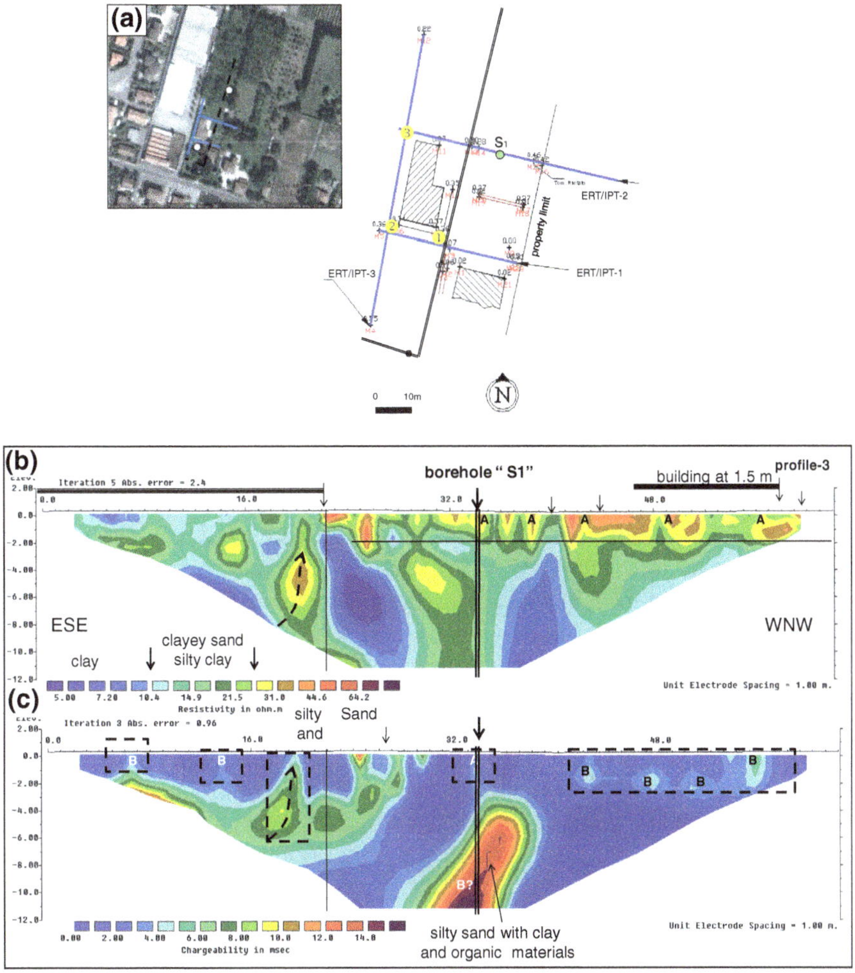

Fig. 9 **a** Detailed location map of the two ERT profiles carried out at the third (*C*) test site. 2D resistivity images of profile ERT-2 (**b**) and IPT-2 (**c**). *Rectangle outlines* the locations where liquefaction features are most visible (*A*). Indications about liquefaction occurrence can be clearly seen on the IPT section (*B*). *S*1 borehole, *dashed arrow* probable sand ejection pathway

the dumped surface sand sediments used to raise the ground level and most probably were not optimally compacted. Other IP anomalies (dashed rectangle with arrow), delineate ruptures due to liquefaction that occurred at shallow depths (around 5 m) beneath the first 30 m of the profile. A similar effect is observed also at 33 m with a probable link to the high IP anomaly (B? Fig. 9c). This high chargeability anomaly is likely to be caused by the presence of fine sand mixed with dispersed organic clay. In fact, the borehole drilled in the same site indicated the presence of silty clay sediments with sand mixed with organic materials between 2 and 5 m depth.

6 Summary and Conclusions

Liquefaction phenomenon results in high water pressure being exerted on the soil particles due to earthquake shaking. Under certain condition this may result in permanent deformations leading to the formation of fractures, sand boils, lateral spreading on gently sloping landforms and landslides in highly sloping terrains. Some factors are known to affect liquefaction of saturated sands and silty sands such as materials type (i.e. compositional characteristics: particle size, shape, gradation, and relative density), in addition to, duration and amplitude of the dynamic excitation of one or several successive earthquakes. The geo-lithological factors are known also to affect other physical properties such as shear wave velocity, electrical resistivity and polarizability which constitute the backbone of seismic and electric geophysical methods. In this work, I highlighted the importance of subsurface electrical imaging in liquefaction related studies. Special emphasis has been given to usefulness of the IPT technique in localizing lithologic heterogeneities in alluvial plain lands where their presence is the rule and not the exception.

The employment of the IPT completes and adds more independent information about the subsurface lithology which aid in the reconstruction of realistic geo-hydrogeological models. In addition, the gained information help in capturing post-seismic modifications. In particular, the IPT succeeded in mapping ruptures depth extension as well as the probable pathways followed during excess pore water pressure dissipation. It was interesting to note, in the second example (site B), how the liquefied sediments had moved horizontally due to the presence of a thick impermeable crust. This crust has facilitated the lateral dissipation of excess pore water pressure. Of course, this crust was removed to allow for the construction of the industrial building that had suffered a lot of damage leading to its demolition. The observed resistivity and IP anomalies are related to porosity and permeability modifications following liquefaction. These modifications can be easily explained by invoking the Archie (1942) empirical equation which explains that effective porosity, related to permeability, is inversely proportional to the formation factor (F defined as the ratio of formation resistivity and water resistivity) of clean sand (fine percent <10 %). The intrusion of liquefied sand mixed with water and fine materials has contributed in the increment of the IP response.

The ERT/IPT may also be used to map and control consolidation effects normally used to mitigate the liquefaction risk. This last issue, however, requires further tests and analyses in order to evaluate possibilities and limitations. I think that the IPT technique may be used as a permanent monitoring tool to complement standard tools used for pressure monitoring in the future. The advantage of using these techniques resides in the fact that they can cover large volumes of subsurface materials. Hence reducing costs the society has to pay for post-seismic event reconstruction phase. I do believe that the best way to mitigate earthquake risk is to increase citizens perception to this type of risk that is related not only to the construction and its continuous maintenance but also to the spatial distribution of subsurface sediments. Such integration contributes in the process of smart underground monitoring of sites highly vulnerable to liquefaction.

Last, the possibility to use the IPT to investigate the subsurface conditions of existing building is feasible and can be easily realized by installing the electrodes around the building itself. In this case, full 3D data acquisition protocols can be set up and acquired data can be processed using available 3D codes for data inversion.

References

Abu Zeid, N., Bianchini, G., Santarato, G., & Carmela, V. (2004). Geochemical characterisation and geophysical mapping of Landfill leachates: The Marozzo canal case study (NE Italy). *Environmental Geology, 45*, 439–447.

Abu Zeid N., Bignardi S., Caputo R., Santarato G., Stefani M. (2012). Electrical resistivity tomography investigations on liquefaction and fracturing phenomena at San Carlo. In Anzidei M., Maramai A. and Montone P. (Eds), Annals Of Geophysics, *The Emilia (northern Italy) seismic sequence of May-June, 2012: preliminary data and results*, (Vol. 55(4), pp. 713–716), Annals Of Geophysics.

Al-Shukri, H., Hanan, M., & Tuttle, M. (2006). Three-dimensional imaging of earthquake-induced features with ground penetrating radar. *Near Marianna, Arkansas, Seismo, Res, Lett, 77*, 505–513.

Andrus, R. D., Stokoe, K. H., & Chung, R. M. (1999). *Draft guidelines for evaluating liquefaction resistance using shear wave velocity measurements and simplified procedures*. NISTIR 6277, National Institute of Standards and Technology, Gaithersburg, Md.

Archie G. E. (1942). The electrical resistivity log as an aid in determining some reservoir characteristics. *Trans A.I.M.E. 146*:54–62.

Barker, R. D. (1990). Investigation of groundwater salinity by resistivity methods. In S. H. Ward (Ed.), *Geotechnical and environmental geophysics*, *2*, 201–212. Investigations in Geophysics Tulsa (Oklahoma): SEG.

Bondesan M. (1989). Geomorphological hazards in the Po delta and adjacent areas. *Suppl. Geografia Fisica e Dinamica Quaternaria*, 2:25–33.

Boulanger, R. W., & Idriss, I. M. (2006). Liquefaction susceptibility criteria for silts and clays. *Journal of Geotechnical and Geoenvironmental Engineering, ASCE, 132*(11), 1413–1426.

Bray, J. D. & Sancio, R. B. (2006). Assessment of the liquefaction susceptibility of finegrained soils. *Journal of Geotechnical and Geoenvironmental Engineering, ASCE, 132*(9), 1165–1177.

Burrato, P., Ciucci, F., & Valensise, G. (2003). An inventory of river anomalies in the Po Plain, Northern Italy: Evidence for active blind thrust faulting. *Annales Geophysicae, 46*(5), 865–882.

Caputo, R., Piscitelli, S., Oliveto, A., Rizzo, E., & Lapenna, V. (2003). The use of electrical resistivity tomography in active tectonic. Examples from the Tyrnavos Basin, Greece. *Journal of Geodynamics, 36*, 19–35.

Carminati, E., Martinelli, G., & Severi, P. (2003). Influence of glacial cycles and tectonics on natural subsidence in the Po Plain (Northern Italy): Insights from 14C ages. *Geochemistry, Geophysics, Geosystems, 4*(10), 1–14.

Constable, S. C., Parker, R. L., & Constable, C. G. (1987). Occam's inversion. *A practical algorithm for generating smooth models from electromagnetic sounding data, Geophysics, 52*, 289–300.

Cremonini S. (1988). Specificità dell'Alto Ferrarese nella problematica evolutiva dell'antica idrografia padana inferiore. In: Grafis (Ed.), Bondeno ed il suo territorio dalle origini al Rinascimento, Bologna, pp. 17–24.

deGroot-Hedlin, C., & Constable, S. (1990). Occam's inversion to generate smooth, two dimensional models form magnetotelluric data. *Geophysics, 55*, 1613–1624.

ErtLab. (2014). Multi-phase technolgies and geoastier web pages (http://www.mpt3d.com/ and http://www.geostudiastier.it).

Erchul, R. A., & Gularte, R. C. (1982). Electrical resistivity used to measure liquefaction of sand. *Journal of the Geotechnical Engineering Division, 108*(5), 778–782.

Ercoli, M., Pauselli, C., Frigeri, A., Forte, E., & Federico, C. (2013). Geophysical paleoseismology" through high resolution GPR data: A case of shallow faulting imaging in Central Italy. *Journal of Applied Geophysics, 90*, 27–40.

Galli, P. (2000). New empirical relationships between magnitude and distance for liquefaction. *Tectonophysics, 324*(3), 169–187.

Galli P., Castenetto S. & Peronace E. (2012). The MCS macroseismic survey of the Emilia 2012 earthquakes. *Annals of Geophysics, 55*, 663–672.

Gross, R., Green, A. G., Horstmeyer, H., Holliger, K., & Baldwin, J. (2003). 3-D georadar images of an active fault: Efficient data acquisition, processing and interpretation strategies. *Subsurface Sensing Technologies and Applications, 4*, 19–40.

Guidoboni E. (1997). An early project for anti-seismic house in Italy. Pirro Ligorio's manuscript treatise of 1570–74, in "European Earthquake Engineering", vol. 4:1–18.

Guidoboni E., Ferrari G., Mariotti D., Comastri A., Tarabusi G., & Valensise G. (2007). Catalouge of strong earthquakes in Italy from 461 BC to 1997 and in the Mediterranean area, from 760 BC to 1500. An advanced laboratory of historical seismolgy. http://strong.ingv.it/cfti4med/.

Iliceto, V., Santarato, G., & Veronese, S. (1982). An approach to the identification of fine sediments by induced polarization laboratory measurements. *Geophysical Prospecting, 30*(3), 331–347.

Improta, L., Ferranti, L., De Martini, P. M., Piscitelli, S., Bruno, P. P., Burrato, P., et al. (2010). Detecting young, slow-slipping active faults by geologic and multidisciplinary high-resolution geophysical investigations: A case study from the Apennine seismic belt, Italy. *Journal of Geophysical Research, 115*, B11307.

Ishihara, K. (1974). Liquefaction of subsurface soils during earthquakes, Technocrat, *7*(3), 1–31.

Ishihara, K., Tatsuoka, F., & Yasuda, S. (1975). Undrained deformation and liquefaction of sand under cyclic stresses. *Soils and Foundation, 15*(1), 29–44.

Jinguujia, M., Toprakb, S., & Kunimatsua, S. (2007). *Soil Dynamics and Earthquake Engineering, 27*, 191–199.

Krinitzsky E. L., Gould J. P. & Edinger P. H. (1993). Fundamentals of earthquake resistant construction. John Wiley & Sons., New York.

Kuras, O., Beamish, D., Meldrum, P. I., & Ogilvy, R. D. (2006). Fundamentals of the capacitive resistivity technique: G152. *Geophysics, 71*, G135.

Kuras O., Beamish D., Meldrum P. I. & Ogilvy R. D. (2006). Fundamnetals of the capacitive resistivity technique. *Geophysics, 71*(3), 135–152.

Liu, L., & Li, Y. (2001). Identification of liquefaction and deformation features using ground penetrating radar in the New Madrid seismic zone, USA. *Journal of Applied Geophysics, 47*, 199–215.

Loke, M. H., & Barker, R. D. (1996). Rapid least-squares inversion of apparent resistivity pseudosections by a quasi-Newton method. *Geophysical Prospecting, 44*, 131–152.

Maurya, D. M., Goyal, B., Patidar, A. K., Mulchandani, N., Thakkar, M. G., & Chamyal, L. S. (2006). Ground penetrating radar imaging of two large sand blow craters related to the 2001 Bhuj earthquake, Kachchh, Western India. *Journal of Applied Geophysics, 60*, 142–152.

McCalpin, J. P. (2009). *Paleoseismology*. Burlington, MA (USA): Academic Press.

Mitchell, J. K., & Soga, K. (2005). *Fundamentals of soil behaviour* (3rd ed.). New York: Wiley.

Nakazawa H., Sugano T., & Kohama E. (2012). Case studies on evaluation of liquefaction resistance in terms of combination of surface wave exploration and electrical prospecting. In: *Proceedings of the WCEE 2012*, Lisbon.

Nobes, D. C., Bastin, S., Charlton, G., Cook, R., Gallagher, M., Graham, H., et al. (2013). Geophysical imaging of subsurface earthquake-induced liquefaction features at Christchurch Boys High School, Christchurch, New Zealand. *Journal of Environmental and Engineering Geophysics, 18*, 255–267.

Papathanassiou G., Caputo R. & Rapti-Caputo D. (2012). Liquefaction-induced ground effects triggered by the 20th May, 2012 Emilia-Romagna (Northern Italy) earthquake. *Annals of Geophysics, 55*(4), 735–742.

Pieri M., & Groppi G. (1981). Subsurface geological structure of the PoPlain, Italy. Consiglio Nazionale delle Ricerche, Progetto finalizzato Geodinamica, sotto progetto Modello Strutturale (Vol. 414, pp. 13). Roma (in Italian).

Reynolds, J. M. (2011). *An Introduction to Applied and Environmental Geophysics* (2nd ed.). Wiley Blackwell.

Robertson, P. K., Campanella, R. G., Gillespie, D., & Rice, A. (1985). Seismic CPT to measure in-Situ Shear Wave Velocity. *Proceedings of Measurement and Use of Shear Wave Velocity for Evaluating Dynamic Soil Properties. Proceedings of a session held in conjunction with the ASCE Convention.*, ASCE, Denver, CO, USA, pp. 34–48.

Robertson P. K., Woeller D. J. & Finn W. D. L. (1992). Seismic cone penetration test for evaluating liquefaction potential under cyclic loading. *Canadian Geotechnical Journal, 29*, 686–695.

Sasaki Y. (1992). Resolution of resistivity tomography inferred from numerical simulation. *Geophysical Prospecting, 40*, 453–464.

Schoen J. H. (1996). Physical properties of rocks. Handbook of geophysical exploration (Vol. 18, pp. 583). Pergamon Press, Inc.

Seed, H. B. (1979). Soil liquefaction and cyclic mobility evaluation for level ground during earthquakes. *Journal of the Geotechnical Engineering Division, ASCE, 105*(GT2), 201–255.

Seed, H. B., & Idriss, I. M. (1967). Analysis of soil liquefaction: Niigata Earthquake. *Journal of the Soil Mechanics and Foundation*, No. SM6, 108–134.

Seed, H. B., & Idriss, I. M. (1982). *Ground motions an soil liquefaction during earthquakes.* Oakland, CA: Earthquake Engineering Research Institute Monograph.

Seed, H. B., & Lee, K. L. (1966). Liquefaction of saturated sands during cyclic loading. *Journal of Soil mechanics and Soil Division, ASCE, 92*(SM6), 105–134.

Seed, H. B., Idriss, I. M., & Arango, I. (1983). Evaluation of liquefaction potential using field performance data. *Journal of Geotechnical Engineering, 109*, 458–482.

Seed, R .B., Cetin, O. K., Moss, R. E. S., Kammerer, A. M., Wu, J., Pestana, J. M., Riemer, M. F., Sancio, R. B., Bray, J. D., Kayen, R.E., & Faris, A. (2003). Recent advances in soil liquefaction engineering: a unified and consistent framework. *26th Annual ASCE Los Angeles Geotechnical Spring Seminar*: Long Beach, California, April 30, (pp. 71).

Sheriff R. E. (2002). Encyclopedic dictionary of applied geophysics (Vol. 13, pp. 429). SEG (Geophysical References No. 13).

Sonmez, H., & Gokceoglu, C. (2005). A liquefaction severity index suggested for engineering practice. *Environmental Geology, 48*, 81–91.

Stokoe, K. H., & Nazarian, S. (1985). Use of Rayleigh waves in liquefaction studies. In R. D. Woods (Ed.), *Measurement and use of shear wave velocity for evaluating dynamic soil properties* (pp. 1–17). ASCE: New York.

Sumner, J. S. (1976). Principles of induced polarization for geophysical exploration: Elsevier, p. 277.

Sumner, J. S. (1976). *Principles of induced polarization for geophysical exploration. Devlopments in economic geology*, *5*, 277. Amsterdam: Elsevier.

Telford W. M., Geldart L. P. & Sheriff R. E. (1990). Applied Geophysics. Cambridge University Press, Cambridge (U.S.A.), pp. 770.

Timofeev V. M. (1994). A new ground resistivity method for engineering and geophysics (pp. 701–715). In *Proceedings of Symposium on the Applications of Geophysics to Engineering and Environmental Problems, 6th, Boston*, 27–31 March 1994. Denver, CO: Environmental and Engineering Geophysical Society.

Tokimatsu, K., & Uchida, A. (1990). Correlation between liquefaction resistance and shear wave velocity. *Soils and Foundation, 30*, 33–42.

Toscani, G., Burrato, P., Di Bucci, D., Seno, S., & Valensise, G. (2009). Plio-quaternary tectonic evolution of the Northern Apennines thrust fronts (Bologna-Ferrara section, Italy): Seismotectonic implications. *Boll. Soc. Geol. It., 128*(2), 605–613.

Toshiyasu, U., Motoki, K., Ryosuke, U., & Noriaki, S. (2008). Liquefaction of unsaturated sand considering the pore air pressure and volume compressibility of the soil particle skeleton. *Soils and Foundation, 48*(1), 87–99.

Tuttle, M. P., Collier, J., Wolf, L. W., & Lafferty, R. H. (1999). New evidence for a large earthquake in the New Madrid seismic zone between A.D. 1400 and 1670. *Geology, 27*(9), 771–774.

Wolf, L. W., Collier, J., Tuttle, M., & Bodin, P. (1998). Geophysical reconnaissance of earthquake-induced liquefaction features in the New Madrid seismic zone. *Journal of Applied Geophysics, 39*, 121–129.

Wolf, L. W., Tuttle, M. P., & Park, S. (2006). Geophysical surveys of earthquake-induced liquefaction deposits in the New Madrid seismic zone. *Geophysics, 71*(6), B223–B270.

Yanguo Z., & Yunmin C. (2007). Laboratory investigation on assessing liquefaction resistance of sandy soils by shear wave velocity. *Journal of Geotechnical and Geoenvironmental Engineering, ASCE, 133*(8), 959–972.

Yanguo Z., Yunmin C., & Daosheng L. (2009). Shear wave velocity-based liquefaction evaluation in the great Wenchuan earthquake: a preliminary case study. *Earthquake Engineering and Engineering Vibration, 8*, 231–239.

Youd, T. L., Idriss, I. M., Andrus, R. D., Arango, I., Castro, G., Christian, J. T., et al. (2001). Liquefaction resistance of soils: Summary report from the 1996 NCEER and 1998 NCEER/NSF workshops on evaluation of liquefaction resistance of soils. *Journal of Geotechnical and Geoenviromental Engineering, 127*, 817–833.

Zadorozhnaya V. Y. (2008). Resistivity measured by direct and alternating current: why are they different? *Advances in Geosciences, 19*, 45–59.

Working Strategies for Addressing Microzoning Studies in Urban Areas: Lessons from the 2009 L'Aquila Earthquake

Giovanna Vessia, Mario Luigi Rainone and Patrizio Signanini

1 Introduction

The Italian territory is affected by frequent seismic events producing high intensity in near-field areas. Although the maxima moment magnitudes are about 7, several disasters were associated with historical strong earthquakes. Hence, the seismic effects in Italy "can be considered as historical variables of the degree of intensity: historical building techniques, economic levels and population size" (Guidoboni and Ferrari 2000). Many are the historical documents collected by local authorities and the Church since the Middle Age (among others Baratta 1901; Guidoboni et al. 2007) that systematically witness the consequences of the past earthquakes in terms of fatalities, disruptions in urban centers and sometimes the economic losses. Recently, Guidoboni et al. (2007) collected all this information in a Catalogue of strong earthquakes in Italy from 461B.C. to 2000. This catalogue shows that catastrophic seismic events occur every 5½ year, on average, from the Middle Age up to now: 4509 sites were hurt by seismic events causing Intensity degrees higher than VII (MCS scale) (Guidoboni 2013), whereas over the whole Italian territory 11,440 sites suffered heavy damages and 4800 victims. Some crossed analyses on economic and social effects of Italian earthquakes suggested that after strong seismic events, the affected human communities suffered economic crises (Guidoboni et al. 2011). For instance, the seismic event that occurred at Ferrara city in 1570 (Guidoboni and Comastri 2005) destroyed many palaces and towers boosting the first "Seismic Building code" written by the famous architect Pirro

G. Vessia (✉) · M.L. Rainone · P. Signanini
Department of Engineering and Geology, University "G. D'Annunzio" of Chieti-Pescara, Chieti, Italy
e-mail: g.vessia@unich.it

G. Vessia
Institute of Research for Hydrogeological Protection, National Research Council (CNR-IRPI), Bari, Italy

S. D'Amico (ed.), *Earthquakes and Their Impact on Society*,
Springer Natural Hazards, DOI 10.1007/978-3-319-21753-6_9

Ligorio (ed. Guidoboni 2006). In addition, five great events known as 1783 Calabrian earthquakes and 1908 Messina earthquake caused 80,000 fatalities on both sides of Reggio Calabria and Messina plus a tsunami wave. This latter event definitively changed the face of Messina city and produced a long lasting economic crisis (Guidoboni 2013).

In recent years, the Department of Civil Protection estimated the economic losses due to the strong earthquakes that occurred in Italy from 1968 to 2003 (Fig. 1). It is worth noticing that these losses rapidly increase over time maybe because of the rapid urbanization of several portions of the territory where the methods used to estimate the seismic hazard and risk were not effective at all. The urgent need to improve the methods used for determining seismic hazard and risk assessments is supported by a recent event: the April 6, 2009 L'Aquila earthquake (Equivalent Moment Magnitude Mw = 6.3). It is the third largest event recorded by the Italian accelerometer seismic network after the Friuli earthquake in 1976 (Mw = 6.4) and the Irpinia earthquake in 1980 (Mw = 6.9). It can be considered one of the most mournful seismic event occurred in recent time in Italy: 308 fatalities, 60.000 people displaced (data source http://www.protezionecivile.it) and estimated damages for 1894 M€ (data source http://www.ngdc.noaa.gov). Nonetheless, the signals recorded by the fixed and temporary seismic stations represent a valuable source of information for earthquake scientists to test the knowledge about the near field (the directivity, high frequency content and high vertical components of the propagated seismic waves) and local amplification effects studied so far.

Forty years ago, the 1976 Friuli earthquake boosted quantitative microzoning studies associated with the macroseismic maps of damages where differential local site effects in near-field conditions were highlighted. Subsequently, after the 1980 Irpinia earthquake, the project "Geodinamica" was financed: it was the first research project, in Italy, that tried to introduce numerical one- and two-dimensional (1D and 2D) finite element simulations to assess site effects (Postpischl et al. 1985). Then, after the Umbria-Marche earthquake in 1997 (Mw = 5.9), some evidences of local seismic effects (e.g. Nocera Umbra case study), due to both seismic impedance contrasts between neighbor geological formations and their surface and subsurface geometric shapes, suggested to deal with microzoning studies in order to reduce the costs of strong Italian earthquakes in terms of structural damages and economic

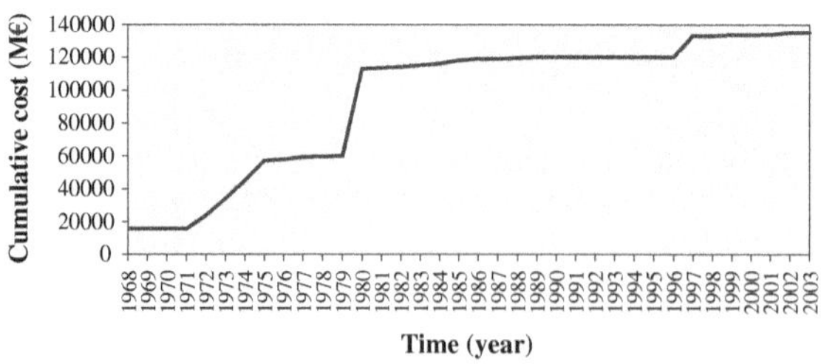

Fig. 1 Cumulative losses due to the strong earthquakes in Italy from 1968 to 2003 (in M€) (after Guidoboni et al. 2013 modified)

losses. Later on, the National Working Group against Earthquakes (GNDT), within the framework of the project "ProgettoEsecutivo 1998", undertook novel studies on probabilistic national seismic hazard (Galadini et al. 2000) and on microzonation (Marcellini 1999). The 2002 Molise earthquake ($M_w = 5.7$) highlighted the need of an up-to-date technical code for designing structures and infrastructures against earthquakes as well as an up-to-date national reference map of seismic hazard. Hence, the first edition of the New Italian Building Codes was published in 2003 by the issue of the Decree of the President of the Council of Ministers (DPCM) 3274. It implemented by law the New Italian Seismic Hazard Map calculated for a return period of 475 years. In 2008, by means of the issue of the Ministerial Decree (DM 14 gennaio 2008), the new technical provisions became law after several revisions were implemented from their first issue on 2003. At the same time, the Italian Department of Civil Protection (DPC) published in 2008 the "national guidelines for executing microzoning studies" (Gruppo di Lavoro 2008) related to liquefaction, instability and amplification phenomena. The working scales vary between 1:5000 and 1:10,000 depending on the size of the municipal territory. Such activities can be performed at three detail levels: level 1, by mapping homogeneous geological units with respect to seismic behavior; level 2, by mapping numerical indexes for homogeneous susceptible areas; level 3, by mapping homogeneous seismic responses carried out by means of 1D, 2D and even three-dimensional (3D) numerical analyses. The L'Aquila earthquake of 2009 occurred when the discussion on the best techniques for both measuring the seismic properties at the site and representing seismic hazard scenarios at selected return periods was at its apex. Soon after the terrible unpredicted main shock on April 6th, 2009, all the financial efforts of the Italian government were devoted on one hand to face the emergency, and, on the other hand, to check the effectiveness of the technical guidelines already published. Many strong and weak motion records were available for the main shock and its strong and weak after shocks. These records are now published on two public databases: ITACA (for strong motion events) at http://itaca.mi.ingv.it/ItacaNet/itaca10_links.htm and ISIDE (for weak motion and instrumental events) http://iside.rm.ingv.it/iside/standard/index.jsp managed by the Italian National Institute of Geophysics and Volcanology (INGV). In addition, all the microzonation studies undertaken in the most damaged areas of the Aterno Valley have been collected in a book published in 2010 (Working Group MS-AQ 2010). It aimed at providing microzonation maps in terms of amplification factors for reconstruction purposes. Twelve macro areas within the Aterno Valley that experienced higher than the VI degree of seismic intensity measured on the Mercalli-Cancani-Sieberg (MCS) scale (Fig. 2) were chosen for the microzonation. The present study focuses on microzoning studies addressing the Aterno Valley after the 2009 L'Aquila earthquake. The authors discuss the role played by geo-lithological conditions in the amplification effects occurred at different sections of the Aterno Valley and the most efficient techniques to investigate them. Moreover, the authors provide suggestions to quantify surface geology contribution to the overall amplification effects in near-field conditions by means of 1D and 2D numerical simulations. Furthermore, different aspects of the Italian strategy to tackle microzoning studies are discussed by

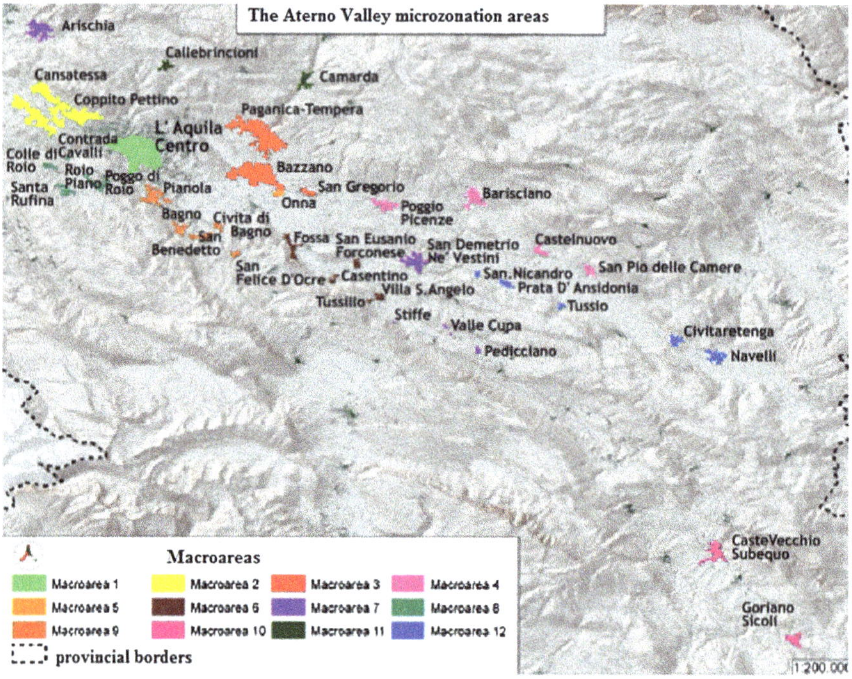

Fig. 2 Twelve macroareas map where microzoning activities have been performed

means of seismic monitoring activities and noise measurements that are carried on in five macroareas of the Aterno Valley, that are the Macroarea 1 (L'Aquila city center), Macroarea 5 (Onna city), Macroarea 6 (Villa Sant'Angelo-Tussillo), Macroarea 7 (Arischia) and Macroarea 8 (Poggio di Roio) (Fig. 2).

Accordingly, after the Introductory Sects. 1, 2 briefly describes the features of the recent and historical seismicity of L'Aquila and Onna; Sect. 3 outlines the surface geology of the Aterno Valley. In particular, litho-technical descriptions of the first 30 m under the five seismic stations that recorded the main shock on April 6th are reported. Section 4 presents the results of the seismic monitoring performed between April and July 2009 in 5 sites of the Aterno Valley (Onna, Arischia, Villa Sant'Angelo, Tussillo, Poggio di Roio and L'Aquila city). These measures are discussed with respect to the microzoning maps drawn at the corresponding macroareas in terms of amplification factors. Section 5 collects the authors' experiences on numerical simulations of local amplification phenomena—in free-field conditions occurred in two parts of the Aterno Valley: 1D analyses performed at L'Aquila city center and 2D analyses along the Monticchio-Onna-San Gregorio alignment. Finally, Sect. 6 presents noise measures recorded at Villa Sant'Angelo and Tussillo sites and four seismic stations (near L'Aquila city). The latter are used to derive the natural frequencies at the sites through the Nakamura H/V ratios calculated for both weak motions after the 2009 L'Aquila earthquake and the noise measures recorded some days after the main shock.

2 Main Features of the Aterno Valley Seismicity

The Aterno Valley is an intermountain quaternary basin generated by the normal fault tectonic regime within the central sector of the Apennine chain. It is located between the Gran Sasso Mountain and the Monti Ocre-Velino-Sirente structural units. The Aterno Valley is elongated NW-SE and it can be subdivided into three areas: northern, southern and central area where L'Aquila city is located. Before the 2009 L'Aquila earthquake, no active faults had been identified within the Aterno Valley according to the Database of Individual Seismogenic Sources DISS3 (DISS Working group 2009), that is a geo-referenced repository of seismogenic source models for Italy. Nonetheless, Meletti and Valensise (2004) included the Paganica Fault (responsible for the 2009 L'Aquila earthquake) within the inventory of the active faults belonging to the central sector of the Apennine chain. On the 6th of April, 2009, at 01:32 UTC, a MW = 6.3 earthquake struck the Aterno Valley, especially L'Aquila city center and its surroundings causing 306 fatalities and the displacement of more than 60,000 people, due to the heavy damages to civil structures and buildings. Moreover, a large part of the historical buildings and churches were severely affected: the old town of L'Aquila was strongly damaged and Onna city (Fig. 3), South of L'Aquila, was fully destroyed. The MCS intensity reached IX-X degree at the Onna site and L'Aquila city center. The main shock followed a seismic sequence that had started in October 2008 and was followed by thousands of aftershocks. The aftershock distribution (Chiarabba et al. 2009), as well as DinSAR analyses (Atzori et al. 2009) in Fig. 4a, showed the reactivation of Paganica fault, 15 km long, NW-SE striking (about 150°) and SW dipping structure (about 50°). The fault plane was 28 and 17.5 km wide and the hypocenter was located at 9.5 km (Ciriella et al. 2009). Furthermore, the joint inversion of SAR interferometric and GPS data performed by Atzori et al. (2009), Fig. 4b, showed an asymmetric deformation field with maximum displacements no greater than 10–12 cm, compared to a maximum slip of about 1 m recognized SE of L'Aquila town. Accordingly, Milana et al. (2011) recognized a decreasing trend in the amplification factors from South to North. Such behavior suggested a relevant role of the fault directivity effects (Atzori et al. 2009).

Fig. 3 Onna city after the April 6, 2009 seismic event struck (**a**, **b**); L'Aquila city (**c**)

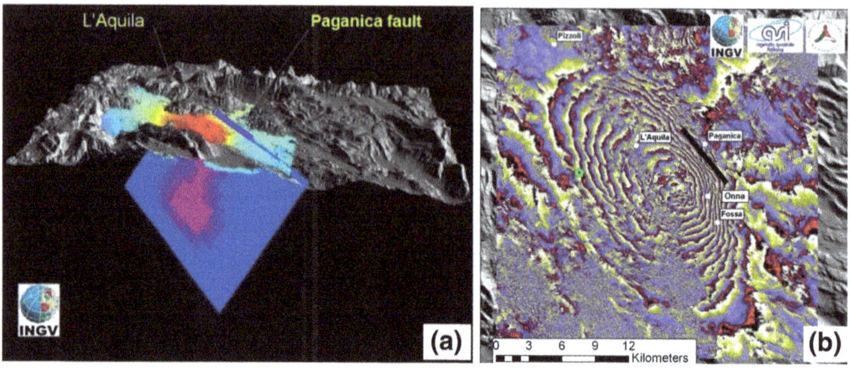

Fig. 4 **a** Fault plane solution by InSAR data inversion, corresponding to the theory of a seismogenic and capable Paganica fault (Salvi et al. 2009); **b** COSMO-SkyMed ascending interferogram (post-event image April 13), normal baseline 430 m, look angle 36° from an ascending orbit. *Green symbol* indicates the Mw = 6.3 main shock location (INGV 2009)

Looking at Fig. 4b, each concentric fringe quantifies 1.5 cm of co-seismic vertical deformation; in addition, Onna is placed where the settlement lines are denser while L'Aquila city shows lower vertical displacement. As can be seen, Onna city is less than 5 km far from the Paganica fault, whereas L'Aquila city is 5.7 km far from it: the two cities fall within the Paganica fault hanging wall. Unfortunately, main shock registrations are only available at L'Aquila city center and in the northern part of the Aterno Valley. Nonetheless, Onna seems to have experienced higher regime of vertical deformation than L'Aquila city. Does it depend on its vicinity to the Paganica fault or could it be due to the deformability of the sediments where Onna is settled? In the following, some clues will be collected to show the prominent role of sediments on the damages occurred at Onna and L'Aquila. Finally, it is worth noticing that the epicentral projection of the hypocenter area (see Fig. 4b, the green point) doesn't correspond to the most solicited area. Thus, hereafter, the epicentral distance of seismic stations will not be taken into consideration. Instead, the intensity values drawn from a Macroseismic map (Fig. 5) will be taken as the reference for the following microzoning studies.

Considering the historical seismicity of Onna (Table 1) and L'Aquila (Table 2) there are only four seismic events that heavily struck both cities: two out of four had Intensities as high as 9 MCS. Baratta (1901) collected documentary information on the most ancient seismic events at Onda (the ancient name of Onna): in 1461 and 1703 he reported two strong earthquakes that caused several deaths and a complete destruction of buildings at Onda. Baratta plotted the boundaries of the major damaged areas due to some historical seismic events that affected the southern part of the Aterno River Valley. Figure 6 shows the elliptical boundary of the 1703 earthquake centered at L'Aquila city. As reported by the author, this event was characterized by several shocks with major damage registered South-eastward of L'Aquila city. Thus, it seems to be an historically based idea that, as observed after 2009 L'Aquila earthquake, the damage features can be associated with amplified

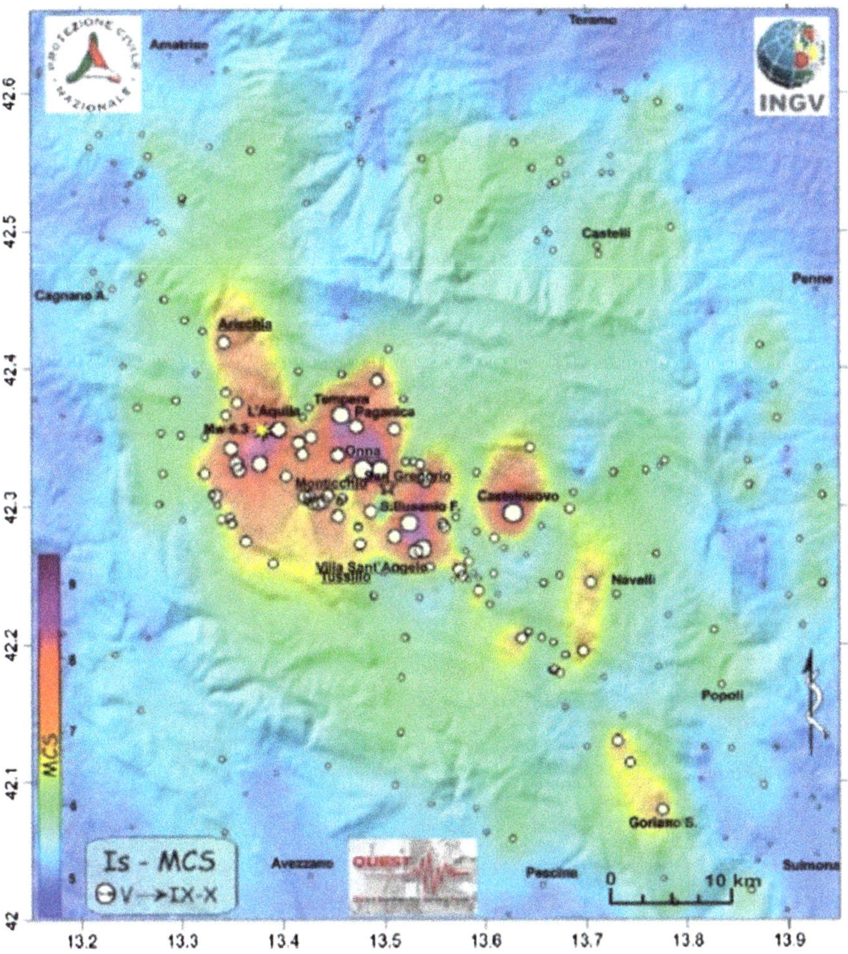

Fig. 5 Macroseismic map of the main shock on April 6, 2009 (Galli and Camassi 2009 modified)

Table 1 Historical seismicity at Onna

Intensity at the Onna site (MCS)	Year/month/day/hour	Epicentral area	Intensity at the epicentral site (MCS)	Magnitude moment (Mw)
7–8	1915/01/13/06:52	Avezzano	11	6.99
7–8	1958/06/24/06:07	Aquilano	7–8	5.17
8	1703/02/02/11:05	Aquilano	10	6.65
10	1461/11/26/21:30	Aquilano	10	6.46

Source http://emidius.mi.ingv.it/DBMI11

Table 2 Historical seismicity at L'Aquila

Intensity at the L'Aquila city (MCS)	Year/month/day/hour	Epicentral area	Intensity at the epicentral site (MCS)	Magnitude moment (Mw)
9	1349/09/09	Aquilano	9	5.88 ± 0.31
9	1461/11/27/21:05	Aquilano	10	6.41 ± 0.34
9	1703/02/02/11:05	Aquilano	10	6.72 ± 0.17
8	1315/12/03	Castelli dell'Aquilano	8	5.57 ± 0.34
7–8	1791/01	L'Aquila	7–8	5.35 ± 0.34
7–8	1915/01/13/06:52	Avezzano	11	7.00 ± 0.09
7	1703/01/14/18:00	Appennino umbro-reatino	11	6.74 ± 0.11
7	1762/10/06/12:10	Aquilano	9	5.99 ± 0.34
7	1786/07/31	L'aquila	6	4.94 ± 0.36
7	1958/06/24/06:07	L'Aquila	7–8	5.21 ± 0.11
6–7	1750/02/01	L'aquila	6–7	4.93 ± 0.34
6–7	1805/07/26/21:00	Molise	10	6.62 ± 0.11
6–7	1916/04/22/04:33	Aquilano	6–7	5.10 ± 0.25

Source http://emidius.mi.ingv.it/DBMI11

Fig. 6 The most affected areas surrounding L'Aquila city after historical seismic events by Baratta (1901)

seismic signals on soft and thick surficial sediments and where morphology and subsurface complex geometries are detected. Nonetheless, Table 2 shows 10 out of 13 seismic events that damaged L'Aquila with an Intensity degree higher than 7 and up to 9. Among others, the ones that damaged both L'Aquila and Onna cities caused higher degree of damages at Onna than at L'Aquilacity.

3 Main Features of the Surface Geology of the Aterno Valley

The Aterno Valley is a river tectonic depression hosting continental deposits since the Early Pleistocene. These deposits are constituted by mainly lacustrine, alluvial and colluvial sediments as well as landslide accumulations (Bosi and Bertini 1970). The Quaternary tectonic basin is elongated in the NW-SE direction and is bounded by predominately NW-SE-striking and SW-dipping active normal faults (Bagnaia et al. 1992) that generated elongated ridges in pre-Quaternary carbonate rocks and emerging from the continental deposits. As shown in Fig. 7, continental sediments were deposited inside the Valley, mainly belonging to lacustrine, fluvial and slope environments. The geological sequences observed within the Aterno Valley (Working Group MS-AQ 2010) consist of a sequence of units that can be detected only by drilling. Recently, a borehole at the L'Aquila downtown hill near the AQK seismic station has been performed up to 300 m. It showed a 250 m thick homogeneous sequence formed by clayey silts and sands laid upon carbonate bedrock (Amoroso et al. 2010). On the contrary, the Middle Pleistocene variably cemented calcareous breccias and dense calcareous gravels (the so-called L'Aquila Breccias) outcrop in several areas of the Aterno Valley and form the L'Aquila downtown hill. They are superimposed to the clayey-sandy-igniferous upper unit and the clayey-sandy unit. The L'Aquila Breccias are composed by clasts, whose size may reach even several cubic meters that came northward from the Gran Sasso chain. The Quaternary sequence continues with terraced fluvial deposits from the Aterno paleo-river (Vetoio Stream unit) that are laterally in contact with alluvial-debris fan

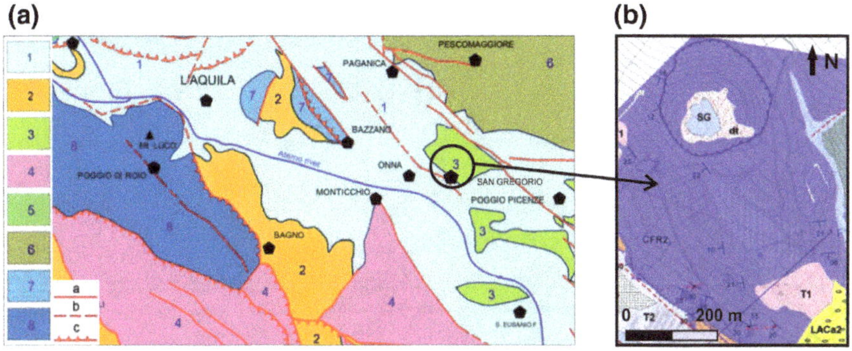

Fig. 7 **a** Essential geological outlines of the Aterno Valley. *1* Continental deposits. *2* terrigenous units (upper Miocene). *3* Mesozoic platform and mioceniclimestones (area of Aterno River). *4* Cretaceous platform and mioceniclimestones (Ocre Mts.) and calcareous-marly-detrital lithologies. *5* Mt. Ruzza area. *6* Filetto-Pescomaggiore area. *7* Mt. Pettino area. *8* HypSlash>Roio-Tornimparte area. **a** Normal fault. **b** Presumed normal fault. **c** Overthrust or transgressive fault. **b** Geologicalmap 1:5000 scale at San Gregorio site. SG: San Gregorio alluvial fan. T2: sandy silts and gravel. T1: gravel and conglomerate. LACa2: gravel. CFR2: calcarenite (After Working Group MS-AQ 2010 modified)

deposits (Mt. Pettino unit) corresponding to the sediment surface of Mt. Pettino. Finally, the youngest sediments exposed along the whole Aterno valley are dated to the Holocene-upper Pliocene age. Such deposits (unit 1, Fig. 7) are alluvial deposits consisting of alternations of more or less coarse gravels, sands and silty clays of fluvial and alluvial-fan environments organized in lenticular bodies associated with the catchment basin feeding the Aterno River. The lithological and geophysical characters of these recent continental deposits have been investigated in the framework of recent projects funded by the Italian Department of Civil Protection, over the last 5 years, as well as additional geotechnical and geophysical investigations for the microzonation studies performed in the most affected areas from the 2009 earthquake (Working Group MS-AQ 2010; Monaco et al. 2009 and Lanzoet al. 2011).

Figure 8 shows the location of five seismic stations installed in the Aterno Valley by the INGV before the 6th of April that recorded the main shock. Figures 9, and 10 display the shear wave velocity profiles derived from cross-hole and down-hole tests performed at those 5 seismic stations, that are: AQA, AQV and AQG in the upper Aterno valley and AQK at L'Aquila city center (Fig. 8). The results of lithological and seismic tests are available on the website http://itaca.mi.ingv.it/ItacaNet/itaca10_links.htm. AQA (Fig. 9) and AQV (Fig. 10) stations are founded on Holocene deposits of alluvial origin, whose variable thickness is due to the deepening of the basin shaped carbonate bedrock. AQK station (Fig. 9a, b) is placed on the Aielli-Pescina Supersynthem, in particular the landslide carbonate breccias exposed at L'Aquila city center. Here, the breccias reach a depth of 40 m overlying stiff silts that show constant shear wave velocity value equal to 650 m/s up to the

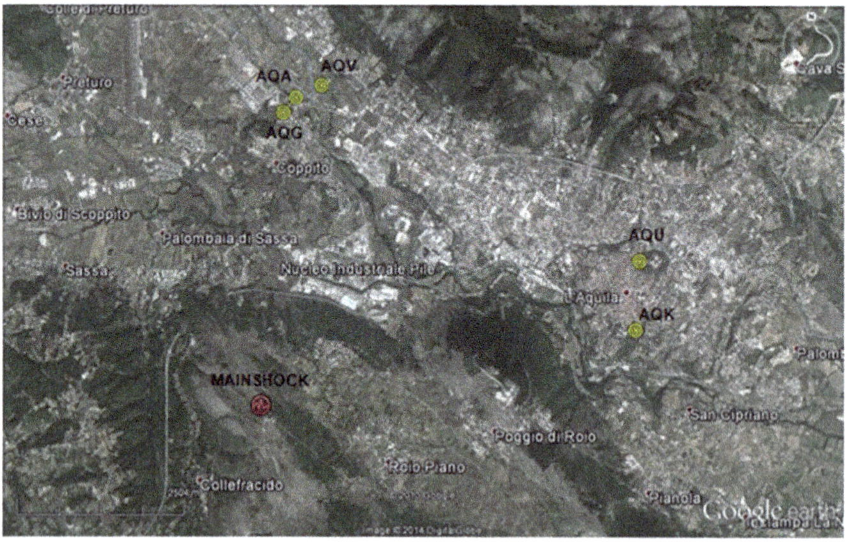

Fig. 8 Location of the seismic stations that recorded the main shock; (in *red*) locations of the epicenters of the main shock and some strong aftershock of 2009 L'Aquila earthquake

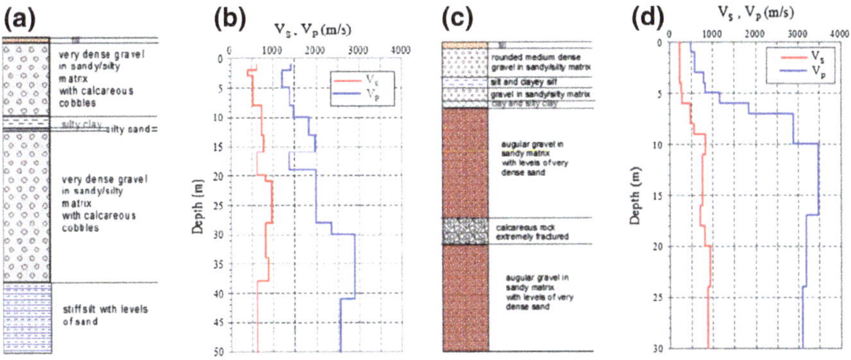

Fig. 9 AQK seismic station characterization: **a** lithology; **b** seismic wave velocity profile. AQA seismic station characterization: **c** lithology; **d** seismic wave velocity profile

end of the sounding. This drilling doesn't intercept the bedrock that is the Mesozoic carbonate bedrock. Nonetheless, as mentioned before, a further deeper drilling has been recently performed near the AQK station (Amoroso et al. 2010): it found this formation at a depth of 250 m.

At AQA (Fig. 9c, d) the breccias show typical rocky shear wave velocity value since a 9 m depth, that is 800 m/s. At AQV station, at the central sector of the valley (Fig. 10c, d), the carbonate bedrock is intercepted at a depth of 50 m.

Finally, the AQG station (Fig. 10a, b) is placed on the edge of a valley to which AQV and AQA belong: AQV is placed at the center, AQA in the middle between AQV and AQG. The lithological characteristics of the sequence under AQG show fractured carbonate rock where shear wave velocity is 700 m/s as a mean value over the first 25 m thickness.

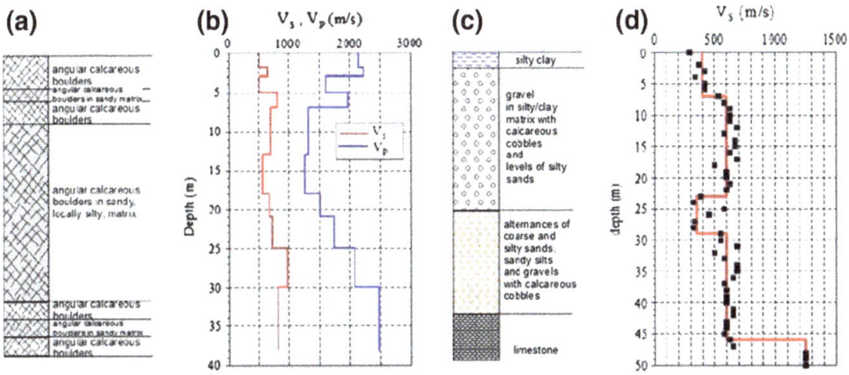

Fig. 10 AQG seismic station characterization: **a** lithology; **b** seismic wave velocity profile. AQV seismic station characterization: **c** lithology; **d** seismic wave velocity profile

Such lithological description by shear wave velocity profiles is useful, although preliminary, in local site response description. As the next section shows, the characteristics of the seismic records of the main shock at those stations cannot be fully understood if only "near field and source" effects are considered. On the contrary, when the local seismic behavior due to the complex surface geology is taken into account, the observed amplifications can be explained and even predicted by means of 1D numerical simulations, at a preliminary stage.

4 Seismic Monitoring Throughout the Southern Portion of the Aterno Valley

4.1 Seismic Monitoring at Urban Sites

Seismic monitoring has been carried out soon after the main shock of April 6th. Some weak aftershocks were registered at Onna, Monticchio and San Gregorio, as illustrated in Figs. 12, 13, 14 and at Villa Sant'Angelo and Tussillo (Figs. 15, 16, 17 and 18) An Array of four SSU Micro-Seismograph (GeoSonics Inc.) seismometers and two DAQLink III (Seismic Source Co.) seismometers with arrays of HS-1–LT 3C (Geo Space Corporation) geophones were used. The monitoring activity was undertaken by programming the velocimeters with a trigger threshold of 0.25 mm/s, a pre-trigger of 1 s, a sampling step of 0.003 s and a recording length of 15 s. The lower bound frequency that can be recorded by means of these velocimeters is 5 Hz, although 2 Hz is the lowest frequency that can be detected by these geophones according to their technical features (Fig. 11, Table 3). For the seismometers, the acquisition parameters were set as follows: sampling window of 0.004 s, recording duration of 80 s, the trigger threshold of 0.02 in./s and pre-trigger time duration of 10 s.

Few events recorded during the seismic monitoring were classified by the INGV office (with Magnitude and hypocenter depth), such as those in Figs. 12 and 13; however, all the records have been compared with the microzoning maps drawn by the Working Group MS-AQ (2010) that are hereafter (Figs. 16, 17, 18 and 19).

Figures 12, 13 and 14 relate to the three recorded events: on 21/04/2009 at 23:35 UTC with Ml = 2.6 and hypocentre depth of 9.1 km (Fig. 12); on 21/04/2009 at 05:03UTC with Ml = 2.5 and hypocentre depth of 10.5 km (Fig. 13); on 22/04/2009 at 04:46 UTC with Ml = 2.6 and hypocentre depth of 10.1 km (Fig. 14). Although the geophones were installed only 3 km away from each other, the peaks of the pseudo-velocity spectra (PPV) measured during these three seismic events show significant differences in the peak magnitudes and the amplified frequencies. In Fig. 12, sensor 4823, installed over the silty-clayey and sandy-gravelly alluvial deposits at Onna, measured the highest peaks in the area: from PPV = 1.27 to 6.35 mm/s. At the same time, sensor 4824 installed at mid-slope on Monticchio calcareous rocks measured the lowest PPV values, varying from 0.46 to 0.56 mm/s.

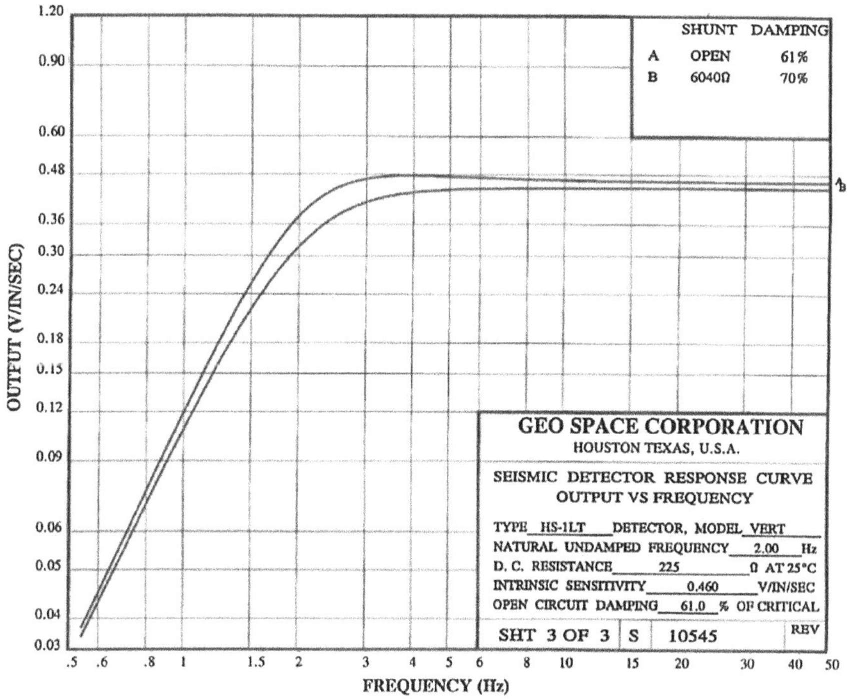

Fig. 11 Technical features of HS-1–LT 3C (Geo Space Corporation) geophones

Table 3 Technical features of SSU micro-seismograph (Geosonics Inc.)

Resolution	0.25 mm/s
Frequency range	2–250 Hz
Record length	15 s (1 s step)
Memory	Up to 20 events or (in continuous) 11 days
Accuracy	5 %

Furthermore, sensor 4826 installed at the top of the slope on calcareous rocks at San Gregorio measured a high PPV equal to 1.27 mm/s. An additional sensor (4825)was then installed over incoherent deposits at the entrance of a narrow valley, approximately 1 km far from Monticchio, South-westwards measured a PPV of 1.78 mm/s. Relating to the seismic event on 21/04/2009 at 23:35 UTC, the signal recorded at Onna shows a peak at a frequency of 7 Hz, whereas at Monticchio and San Gregorio the peak arises at 9-15 Hz. Similar responses are shown in Figs. 13 and 14 for the other two weak seismic events. Amplifications at low frequencies (lower than 2 Hz) at these sites could not be measured due to the instrumental limitations; however, the recorded velocity spectra are discussed here to quantify the local seismic effects. Thus, with respect to the earthquake event on 21 April 2009 at 23:35 UTC, the Onna PPV is 5 and 11 times higher than the San Gregorio

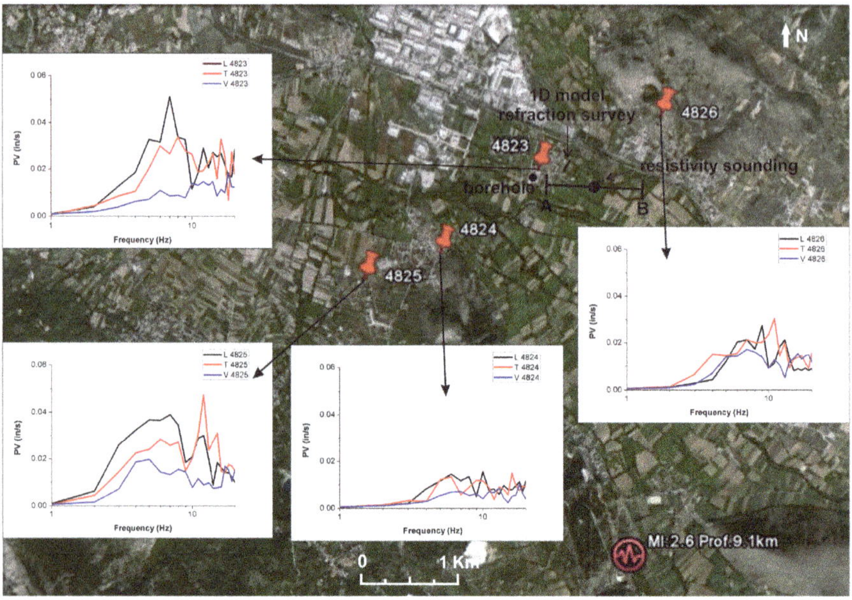

Fig. 12 Earthquake event on 21 April 2009 at 23:35 UTC: Onna (4823) and Monticchio (4824 and 4825) monitoring site locations with velocity spectra (PV); geophysical surveys and borehole locations at the Onna site

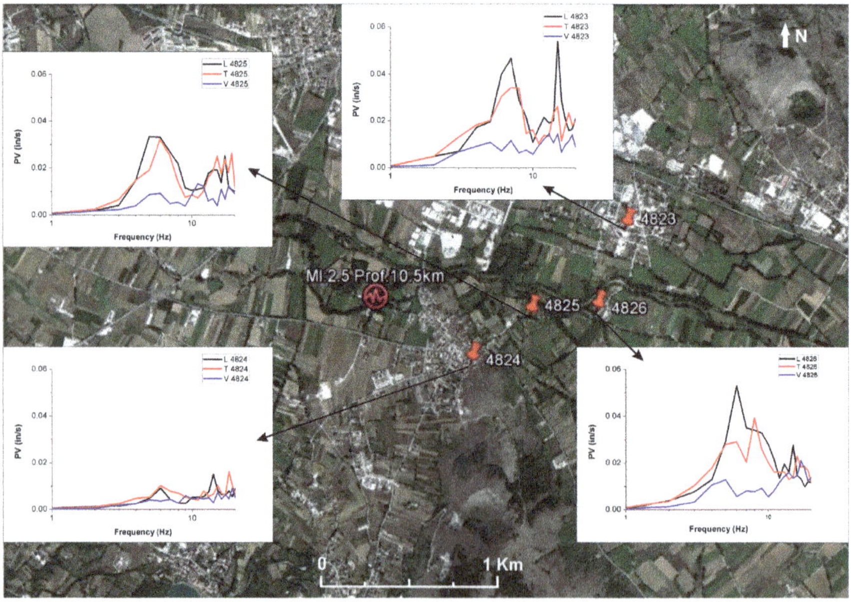

Fig. 13 Earthquake event on 21 April 2009 at 05:03 UTC: Onna (4823), Monticchio (4824) and sites in between (4825 and 4826) monitoring sites locations with velocity spectra (PV); geophysical surveys and borehole locations at the Onna site

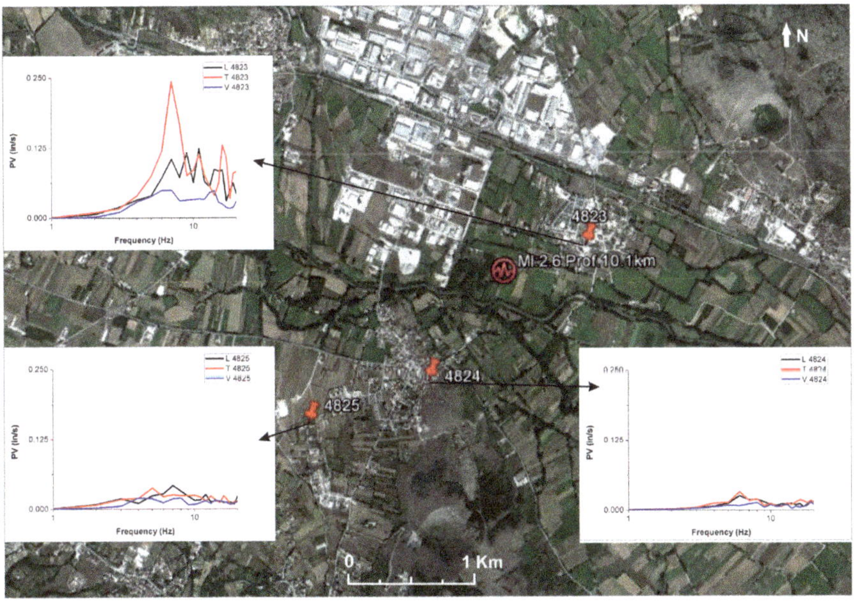

Fig. 14 Earthquake event on 22 April 2009 at 04:46 UTC: Onna (4823) and Monticchio (4824 and 4825) monitoring site locations with velocity spectra (PV); geophysical surveys and borehole locations at the Onna site

and Monticchio ones. These differences in the measured spectra can be associated with different litho types, seismic impedance contrasts and possibly shear wave focalizations due to basin shape (Vessia et al. 2011) at Onna site and topographic relief at San Gregorio site. Moreover, the recorded PPV values reflect the differences in suffered damages. As a matter of fact, the difference in seismic responses among the three municipalities (Onna, Monticchio and San Gregorio) grows up to 3.5° on MCS intensity scale during the April 6th main shock event: Onna shows an I_{MCS} = 9.5, Monticchio an I_{MCS} = 6 and San Gregorio an I_{MCS} = 9 (Galli and Camassi 2009) (Fig. 5).

The microzoning map at Onna (Fig. 23b) has only been drawn. As a matter of fact, Monticchio was not involved in microzoning activities because of the low damages and seismic intensity. San Gregorio was investigated up to the first level of microzoning; however, 2D numerical analyses carried out in Sect. 5 (Fig. 31b) showed amplifications in the period range 0.35–0.45 s (2–3 Hz) (Fig. 36a).

The seismic monitoring has been developed also at Villa Sant'Angeloand Tussillo city centers by the same geophones used at Monticchio-Onna–San Gregorio site. These sensors where located in the historical center at Villa Sant'Angelo and on the outcropping rock at Tussillo. In these sites, the main shock of April 6, 2009 caused differentiated damages although the two centers are 800 m from each other: Villa Sant'Angelo suffered I_{MCS} = 9, whereas Tussillo suffered an I_{MCS} = 8 (Galli and Camassi 2009). The sensors were installed during the period

June 18–July 19, and on July 7, 2009 at 10:15 local time, a small event was recorded by sensors 4823 and 4824 (Fig. 15).

As can be seen in Fig. 15, comparing the three components recorded at Villa Sant'Angelo and Tussillo, it can be noted that the first site shows higher signal ordinates than Tussillo. The differences in the responses can be appreciated in Fig. 16, where the velocity response spectra are reported on the amplification map, drawn by the Working Group MS-AQ (2010). In this map, Fa is the Amplification factor calculated according to Eq. (1). Tussillo village is divided into two parts: one stable, that is Fa = 1 and the other one that is amplified as Fa = 2.7. The event was recorded by the writers in the stable part. Looking at Fig. 16, the Villa Sant'Angelo site shows a higher amplification than the stable portion of Tussillo, although this latter amplifies the signal. The Fa at this point is 1. In particular, the peak of the velocity response spectrum is 0.23 in/s at Villa Sant'Angelo that is located over the sandy-gravelly and silty-sandy alluvial deposits whereas the peak at Tussillo is equal to 0.08 in./s on the outcropping rock. At both sites the peaks are about 15 Hz (0.07 s) although at Villa Sant'Angelo there is another peak in the velocity response spectra at about 10 Hz (0.1 s).

After this event, a stronger event on July 24, 2009 at 19:36 local time was recorded by means of two seismometers DAQLink III. They were located at the same points at Villa Sant'Angelo (historical center, in the municipal garden), whereas at Tussillo it was moved in the urbanized area (the one with a Fa = 2.7 in

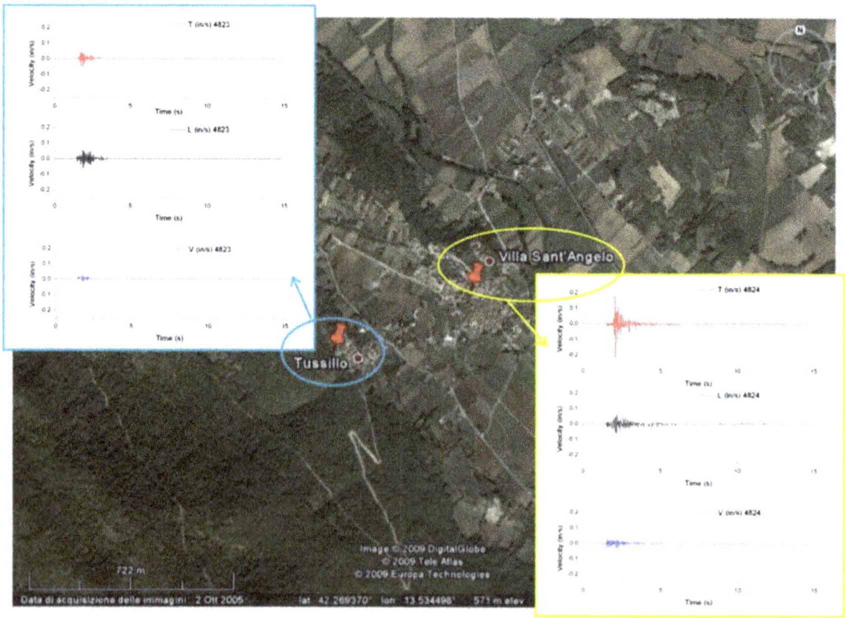

Fig. 15 Seismic monitoring at Villa Sant'Angelo—Tussillo: the time history records relate to the small event occurred on July 7, 2009 at 10:15 local time

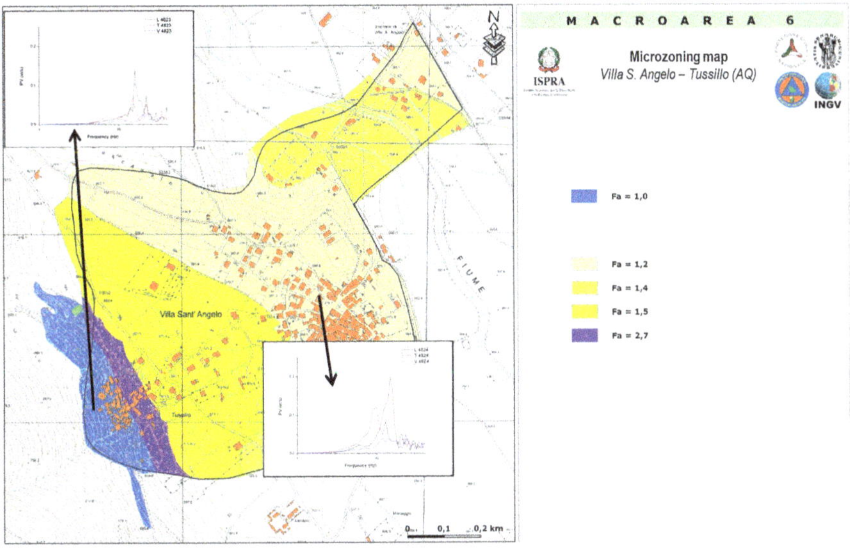

Fig. 16 Velocity response spectra from July 7, 2009 at 10:15 local time records at both Villa Sant'Angelo and Tussillo compared with microzonation map drawn from Working Group (2010)

Fig. 17). Figure 17 shows the Fourier spectra of the three components at the two sites. In this second event, again, it is evident that points placed in amplified areas ranked Fa = 2.7 or 1.2 show amplifications in the same ranges of frequencies and for the same amplitude. Maybe the use of the amplification factor cannot predict the differences either in amplified ranges of frequency nor in amplification amplitudes.

Similar evidences in recorded signals can be noted at the sites: Arischia (Macroarea 7, Fig. 18a, b)and Poggio di Roio (Macroarea 8, Fig. 19a, b). In the case of Arischia, three sensors DAQlink III have been set at three points in the urban area. As can be seen, from Fig. 18b, the three points are located on the three areas associated with three different amplification factors: the yellow one is Fa = 1.4 and accordingly shows a lower amplification with respect to the other ones; the violet that is Fa = 2.7 and the orange one that is Fa = 1.5. In spite of this classification, the velocimeters showed similar high amplifications, in terms of Fourier velocity spectrum over a range of periods 0.01–0.14 s (7–100 Hz). This is actually in contrast with the classification based on Fa.

Finally, two velocimeters were located at Poggio di Roio city center less than 200 m from one another (Fig. 19a). All the urban area is classified as stable with Fa ≤ 1 (Fig. 19b). The records of the three components showed a different behavior of the two points: the first one does not amplify, the other one amplifies in the range of periods 0.05–0.1 s (10–20 Hz). This is, again, in contrast with the classification based on Fa calculated values.

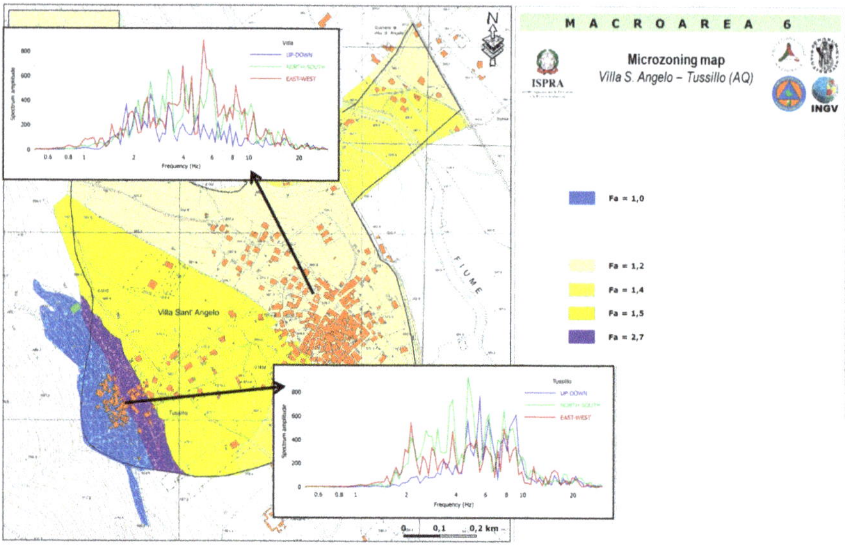

Fig. 17 Velocity Fourier Spectrum of the event occurred on July 24, 2009 at 19:36 local time records at both Villa Sant'Angelo and Tussillo compared with microzonation map drawn from Working Group (2010)

4.2 *Concluding Remarks on Seismic Monitoring*

The seismic monitoring can be useful in microzoning studies for back-analysing the microzoning maps drawn through numerical methods. Amplification indexes, used for microzoning purposes, should be validated through the seismic monitoring with respect to (1) the amplified ranges of periods and (2) the amplification amplitudes compared site to site. From the monitoring activity performed at some sites within the Aterno Valley, the experimental evidences did not fit often with the seismic zonation expressed in terms of amplification factors Fa. Many reasons might be suggested: among others, (1) the inadequacy of the calculated amplification factor to interpret the filtering role of the surface geology on seismic records; (2) the need of validating 1D and 2D numerical simulations by means of weak and strong motion records at a site before using them to microzonation activities. Finally, the writing authors propose to use the *amplification functions* (both measured and calculated) to represent the "typical response" of a site, whether it is amplified or not.

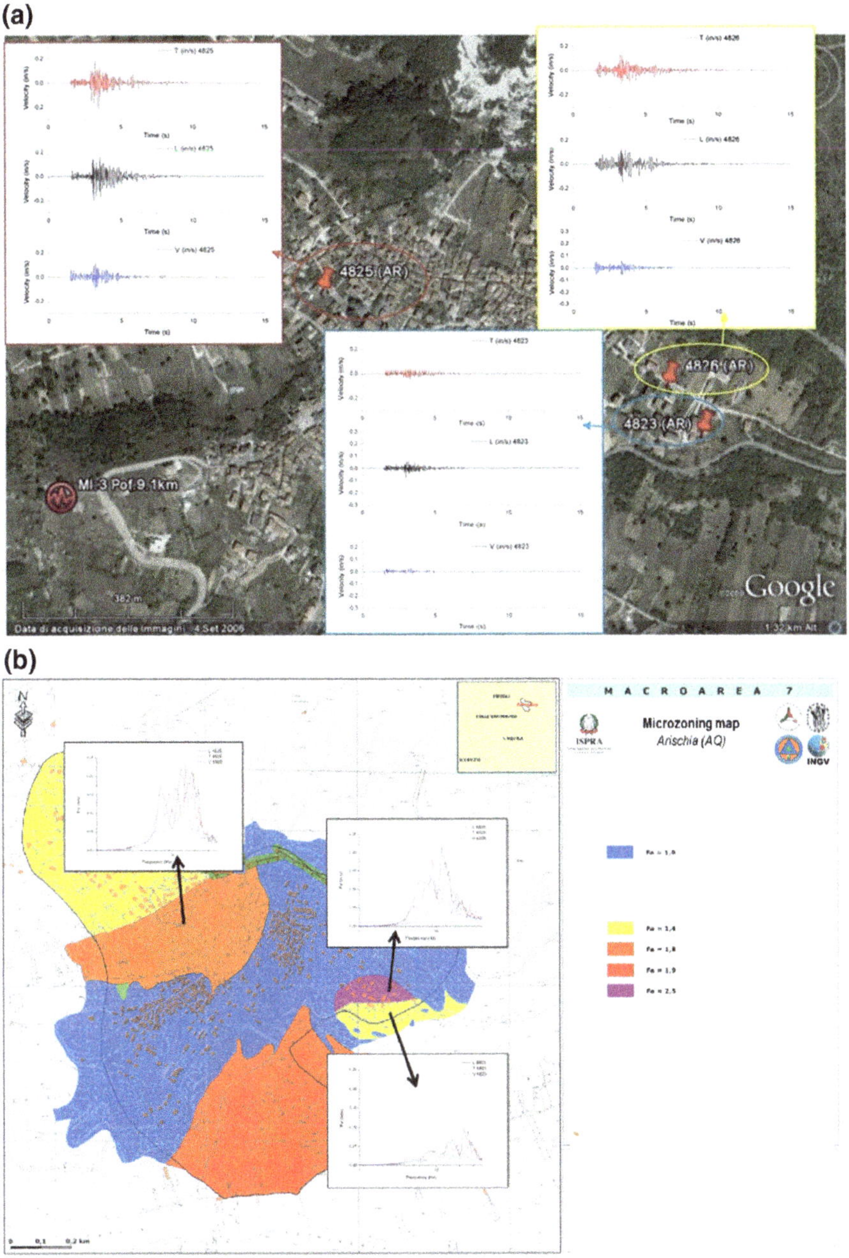

Fig. 18 Seismic monitoring at Arischia city center. Three velocimeters recorded the weak motion event on April 25, 2009 at 13:13 local time (Ml 3; Prof. 9.1 km)

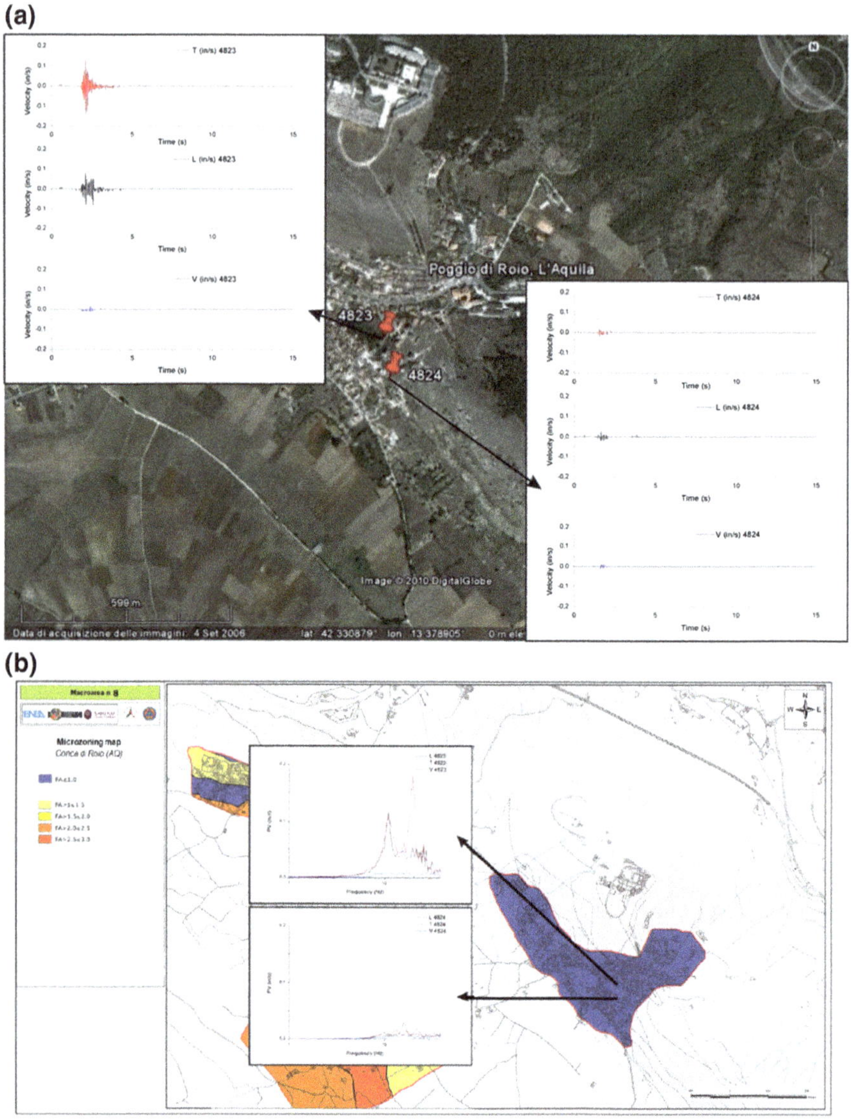

Fig. 19 Seismic monitoring at Poggio di Roio city: two seismographs recorded the seismic event on April 18, 2009 at 20:40 local time. The two sensors were located at less than 200 m of distance

5 Numerical Simulations of Seismic Amplification Within the Aterno Valley

5.1 Some Evidences from the Main Shock Records

Five seismic stations recorded the main shock event on April 6, 2009 at 1:32 UTC within the Aterno Valley: they are shown in Fig. 8. Three out of five (AQA, AQG and AQV) stations are placed in the northern part of the Aterno Valley, along a basin shaped seismic bedrock, whereas the others (AQU and AQK) are placed at L'Aquila city center. The two groups of seismic stations show similar distances from the epicenter, as shown in Fig. 8 (varying between 4.4 and 6.0 km), although the first group (AQA, AQV and AQG) is placed more than 10 km far from Paganica fault. Four out of five stations have been seismically characterized by means of wave velocity profiles (Figs. 9 and 10): only AQU station, placed in the castle of L'Aquila city, is not characterized. Nonetheless, in the present study the acceleration spectrum registered at AQU has been plotted with the others in Fig. 20: the color identifies the station, the line type the component: solid line for UP component, dotted line for NS and broken line for WE components.

Despite the distance from the fault, Fig. 20a, b show that:

1. with respect to the vertical component, the maximum peak in the acceleration spectra is shown at AQV, then at AQA, AQK and AQU. These three sites show Holocene sediments with decreasing thickness. This is true also for the horizontal components, as can be easily noticed from Table 4;
2. the vertical peaks are all aligned on 19–20 Hz except for AQG (Table 5). If "near field" or "directivity" effects are relevant, the responses at AQG, AQV and AQA will be quite the same. On the contrary, the lowest peak and the lowest frequency is shown at the seismic station where fractured rock outcrops;
3. the horizontal peaks seem to occur at lower frequencies than the vertical ones: it could be due to the seismic properties of the lithological sequences of soil layers and geometric effects. Accordingly, from Table 5, along the valley that is oriented SW-NE, the AQG, AQA and AQV stations show their peaks at different frequencies. Moreover, the peaks are comparable between AQA and AQV, whereas at AQG, they are far lower. Such behavior recalls the basin effect (among others Vessia et al. 2011; Vessia and Russo 2013; Rainone et al. 2013).
4. from the inspection of Fig. 20a, b it can be noticed that the vertical components amplify shorter ranges of periods 0.04–0.2 s (2–25 Hz) with respect to the horizontal ones 0.08–0.6 s (1.7–12.5 Hz). On average, the peaks of the acceleration response spectra in vertical and horizontal components are similar.

As already mentioned, all the stations but AQU, have been characterized by means of a sounding and a Down Hole test published on the website: http://itaca.mi.ingv.it/ItacaNet/itaca10_links.htm. Here, the main features of this characterization have been summarized in Table 6. The seismic bedrock has not been detected under all the stations, although at different depths as well as for different shear wave

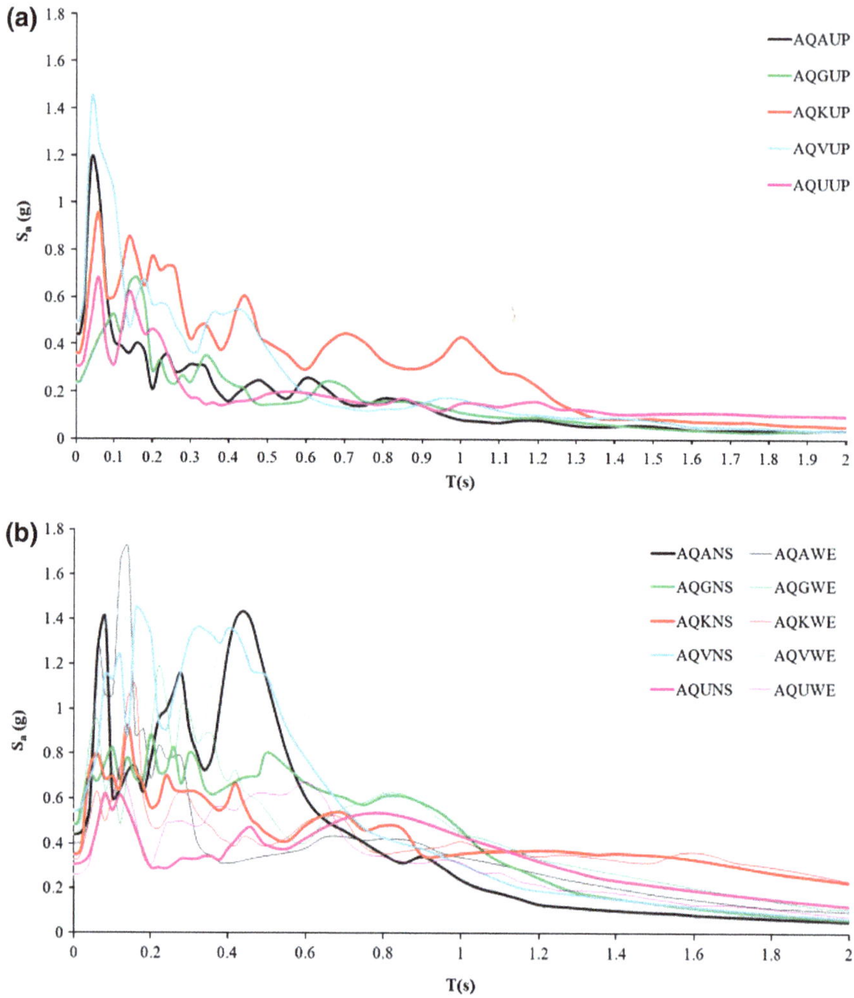

Fig. 20 Acceleration spectra of the three components of the main shock registrations at the five seismic stations considered: **a** Vertical component (UP); **b** Horizontal components (NS-WE)

Table 4 Peak of the acceleration spectra (g) for each main shock component at seismic stations within the Aterno Valley

Station code	WE	NS	UP
AQK	1.1	0.94	0.99
AQA	1.74	1.65	1.55
AQV	1.79	1.52	2.36
AQG	1.20	0.89	0.79
AQU	0.85	0.70	1.29

Table 5 Amplified frequencies (Hz) in the acceleration spectra related to the three components of the main shock at five seismic stations

Station code	WE	NS	UP
AQK	6.2	25	20.8
AQA	7.2	17.8	18.9
AQV	9.0	5.4	20.8
AQG	4.5	5.0	6.6
AQU	7.8	7.6	19.6

Table 6 Seismic properties of the soil under the seismic stations that recorded the main shock

Seismic station	V_{S30} (m/s)	Bedrock depth (m)	Mean Vs over the bedrock depth (m/s)	Shear velocity contrast CV
AQK	717	13	622	1.25
AQA	552	9	342	2.36
AQV	474	47	523	2.42
AQG	685	25	651	1.5
AQU	–	–		–

velocity seismic profiles, V_S mean values have been calculated. Moreover, the shear velocity contrasts have been calculated (that is the ratio between the shear wave velocities of the bedrock and the overlying sediments) and reported in Table 6.

Considering the weighted mean value for shear wave velocity over the bedrock depth (fourth column in Table 6), it can be seen that AQV and AQA show the first and the second highest velocity contrasts C_v respectively. Such parameter plays a relevant role in surficial amplifications.

Comparing Tables 4 and 6 it can be noted that the highest horizontal peaks are registered at those stations that are far away from the Paganica Fault but are characterized by higher velocity contrast with respect to AQK and AQU, placed at L'Aquila city center (nearer to the Paganica fault). AQK shows a first seismic bedrock at 20 m, made up of L'Aquila "megabrecce" formation, that is very dense gravel in sandy matrix with calcareous cobbles: the mean value of shear wave velocity V_S within the first 20 m depth is 900 m/s. The megabrecce overlies the stiff silts that probably reach a 250 m depth and shows V_S = 622 m/s as a mean value over the bedrock (Table 6).

Comparing the peaks at AQK and AQG in a W-E direction, it can be noted that the second station is more amplified, although the shear wave mean value over the bedrock is similar (Table 6). On the contrary, the velocity contrast at AQG is higher than at AQK. The seismic bedrock at AQG is 25 m deep, whereas at AQK there are two seismic bedrocks: one at 20 m and the other at about 250 m. From the response characters highlighted, the role of surface geology seems to be significant although difficult to identify and quantify. For instance, the highly amplified frequencies can be related to the high frequency content of the seismic signals in "near-field" conditions, but further studies are needed for shedding lights on the influence of the local seismic response of heterogeneous surface geological conditions at high frequency contents. However, the amplified high frequencies seem to be also affected by seismic properties of soils and structures or buildings, that can be estimated by measuring the resonance frequencies of both soil and structures (Rainone et al. 2010; Vessia et al. 2012). Finally, further evidences from the Lanzo et al. (2010) study highlighted the relevance of local amplifications in the 2009 L'Aquila earthquake. Accordingly, Fig. 21 shows 5 % damped pseudo-acceleration response spectra for four stations on the hanging wall of the fault. Such spectra are related to the motions rotated into fault normal and fault parallel directions, based on the 147° strike reported in the Earthquake setting and source characteristic section. According to Lanzo et al. (2010), the plot shows:

1. no evidences of significant polarization of ground motion in the fault normal direction, that is related to the rupture directivity;
2. a significant energy content at high frequencies, 10–20 Hz. Focusing on the period range 0.3–1.0 s, the spectral ordinates at AQA and AQV stations, overlying soft sediments, are larger and relatively more energetic than AQG and AQK stations, founded on firmer soils.

This study strengthens the role played by local and surface geology on amplification phenomena in near field conditions. Some 1D simulations have been performed at Onna site and L'Aquila city center, at the seismic station called MI03 and AQK. Unfortunately, MI03 didn't register the main shock. The present numerical simulations have two main objectives: (1) to stress the influence of the litho-technical description of surficial geological sequences for predicting amplification phenomena and (2) to suggest an operating approach for characterizing the amplification at the site in microzoning studies.

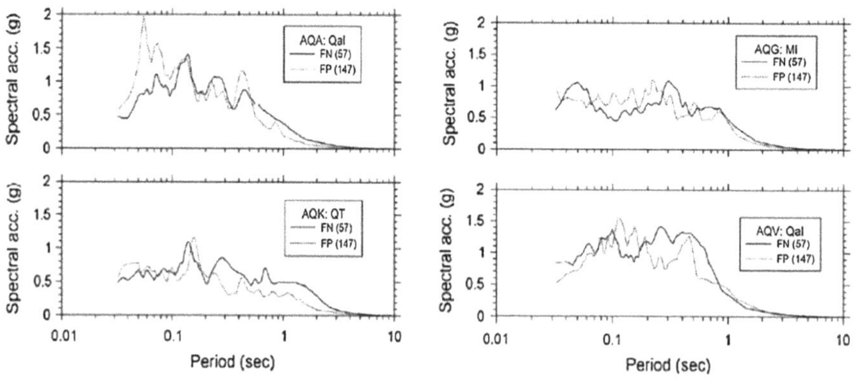

Fig. 21 Acceleration spectra from main shock records on the hanging wall of the Paganica fault (after Lanzo et al. 2010)

5.2 Predicting Local Seismic Response at Onna (Macroarea 5) and L'Aquila City Centers (Macroarea 1) by Means of Numerical Simulations

5.2.1 One-Dimensional Analyses

One-dimensional (1D) numerical simulations have been performed to overpass some inconsistencies between the strong-motion records and the proposed strategies for local seismic response prediction at L'Aquila city center, identified as Macroarea 1 (Fig. 2).

The first seismological studies pointed out: (1) the impulsive character of the recorded signals, (2) their high frequency content, (3) the south-eastern direction of the peak ground accelerations and increasing velocities according to the directivity effects of the rupture propagation along the Paganica fault, (4) the recorded vertical components higher than the horizontal ones according to near-field conditions. Despite the common agreement on the abovementioned phenomena, some relevant exceptions have been observed at some seismic stations: AQV registered anomalous highly amplified horizontal components with respect to AQK or AQU. Moreover, some anomalous heterogeneous damages occurred at neighboring sites, suggested to investigate the role of local geo-lithological conditions on the heterogeneous damaging. For instance, the main shock provoked severe damages and many fatalities throughout Onna city (MCS = IX − X), whereas at the Monticchio site, that is less than 2 km far from Onna, a few damages caused a MCS VI seismic intensity. Figure 12 shows the great differences in amplifications registered for a minor aftershock (Mw 2.6 and hypocenter depth at 9.1 km) that occurred on April 21, 2009 at 23:35 UTC, at the two sites. Thus, for those areas where the MCS intensity was higher than VI, microzonation maps have been drawn (Working group MS-AQ 2010).

At level three, microzoning maps are derived from the calculated amplification factor Fa. The Fa is defined according to the following procedure (Gruppo di Lavoro 2008).

1. To identify the period corresponding to the maximum ordinate P_M of the Input (TA_i) and the Output (TA_o) acceleration spectra.
2. Centered on P_M, the following mean values of input ($SA_{m,i}$) and output ($SA_{m,o}$) acceleration spectra are calculated (Gruppo di Lavoro 2008):

$$Fa = \frac{SA_{output}}{SA_{input}} = \frac{\frac{1}{T_a}\int_{0.5P_M}^{1.5P_M} SA_{output}(T)dT}{\frac{1}{T_a}\int_{0.5P_M}^{1.5P_M} SA_{input}(T)dT} \tag{5.1}$$

where SA(T) is the acceleration spectrum: the output is related to the results of 1D and 2D numerical simulations, and the input is the strong motion signal used as input motion in the same simulations. Ta is the amplified period within the acceleration spectra. Although quantitative micro-zoning procedures have been carefully defined and explained, many critical points arise from the microzoning maps reported in Fig. 22. Figure 22a shows the undefined Fa value within the blue area where the AQK station is placed. At the same time, Onna city is classified as Fa = 1.8 with a complete destruction of small buildings whereas some portions of L'Aquila city are ranked Fa = 1.9 or 2.0 with no correspondence with damages.

Thus, the authors try to suggest a simple method for relating the litho-technical properties of a sequence of surface sediments to the amplified seismic response. To this end, 1D simulations have been carried out on a few possible sedimentary sequences at AQK and MI03. The code used is EERA (Bardet et al. 2000) that simulates the non-linear seismic response of soils by means of the equivalent linear constitutive law. This seismic behavior model needs shear modulus reduction curve $G(\gamma)/G_{max}$ and damping increasing curve $D(\gamma)$ measured by laboratory tests for each litho type, e.g. resonant column, cyclic triaxial equipment. For the present simulations, the curves measured by Working Group MS-AQ (2010) for investigated soils at macroareas 1 and 5 have been used. Moreover, the input motions considered are the ones illustrated in Fig. 23, provided by Pace et al. (2011) for microzonation studies carried out in the mentioned macroareas:

1. the uniform hazard spectrum of the Italian Building code (NTC08). It is a lower bound spectrum of 2009 L'Aquila earthquake;
2. a probabilistic uniform hazard spectrum with a return period of 475 years.; it has been obtained by two source models named LADE1 and SP96 GMPE (Pace et al. 2011). This spectrum gives an upper bound spectrum to L'Aquila earthquake, reaching a maximum spectral value of about 1 g at 0.25 s;
3. a deterministic spectrum obtained from SP96 GMPE for a magnitude-distance pair (Mw 6.7, Repi = 10 km). The deterministic spectrum gives intermediate spectral values ranging from 0.35 g at 0 s, to 0.8 g at 0.25 s and 0.4 g at 1 s.

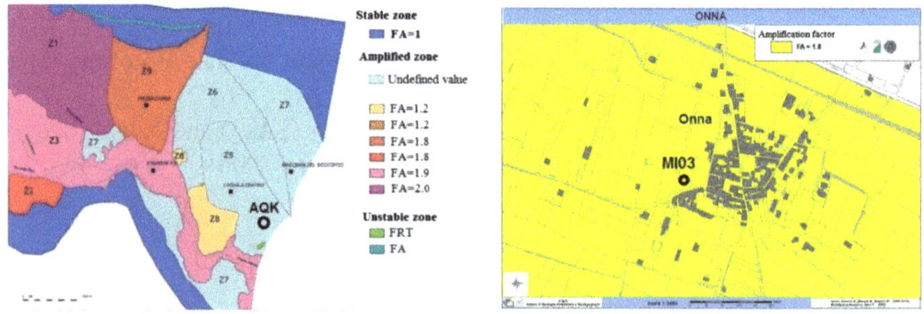

Fig. 22 **a** Macroarea 1: seismic microzonation of level three at L'Aquila center (after Working Group MS-AQ 2010 modified); **b** Macroarea 5: seismic micro-zonation of level three at Onna city

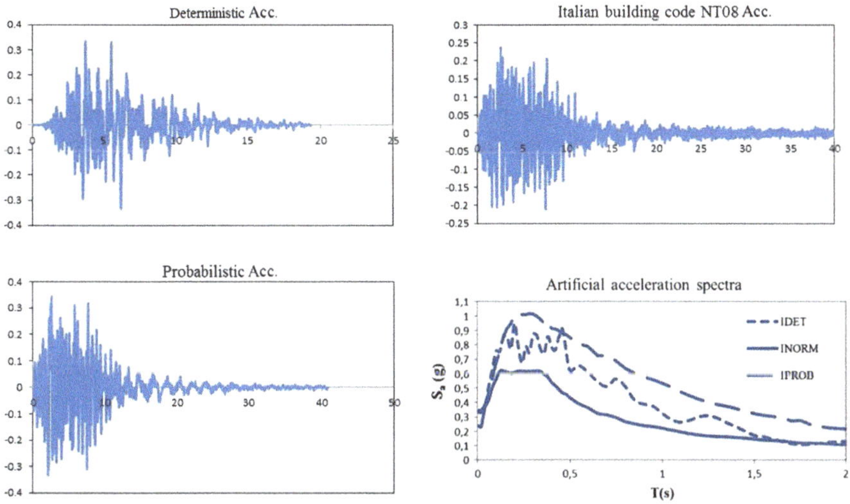

Fig. 23 Artificial input motions simulating 6th April 2009 main shock used for numerical analyses within Macroareas 1 and 5

The 1D numerical study presented here considered two models for the sediment sequence under the AQK station (Fig. 24a) and two other models under the MI03 station (Fig. 24b). The two models at AQK concern different points of view about

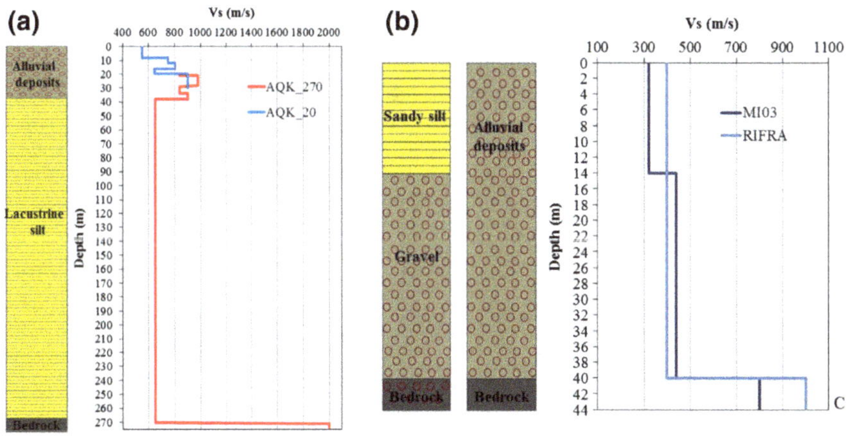

Fig. 24 **a** SH wave velocity profiles under AQK station for 1D numerical analyses. On the *left*: lithological model; on the right, two velocity profiles: Model 1(in *blue*), bedrock at 20 m depth; Model 2 (in *red*), bedrock at 270 m depth. **b** SH wave velocity profiles under MI03 station for 1D numerical analyses. On the *left*, two lithological models; on the *right* the corresponding velocity profiles: Model 1(in *black*): sandy silt up to 14 m and gravels up to the bedrock at 40 m depth. Model 2(in *blue*): lithological succession of sand, silt and gravels called "alluvial deposits" up to the bedrock at a 40 m depth

the role of the velocity contrasts within the Holocene sequences, here named "alluvial deposits". As mentioned before, the heterogeneity of the most recent and surficial sediments result in wave velocity inversions within the first 20 m depth. Such fast strata, according to the present authors, can be considered as preferential paths for seismic waves, playing the role of "hanging bedrocks". This idea is an alternative to the common standpoint that poses the seismic bedrock approximately at about 250 m depth that is the geological bedrock. Here, at AQK, such a bedrock has been found at 270 m (Amoroso et al. 2010). Similarly, under the MIO3 station at Onna city center (Fig. 24b), the seismic bedrock has been reached at 40 m without any velocity inversions. The sedimentary sequence published by Working Group MS-AQ (2010) shows a downwards-graded increase in shear wave velocity values. The authors performed a refraction test, in 2009, from which the sequences of sands and gravels show a mean value of 400 m/s up to the bedrock depth where a 1000 m/s has been found: the bedrock shear wave velocity value is higher than the one suggested by the Working group MS-AQ (2010). Thus, the second model suggested by the writing authors put in evidence the presence of a velocity contrast Cv = 2.5 instead of 1.8. Such condition, from the author standpoint, can influence the local response of the sediments by increasing surficial amplification. The results of the abovementioned 1D numerical analyses concerning AQK are presented in Fig. 25a, b and Fig. 26a, b and concerning MI03 in Fig. 27. Figure 25a shows, in blue, the output acceleration spectra calculated for the 20 m-bedrock model, whereas in black, the input acceleration spectra deconvolved at the depth of the seismic bedrock. In the same plot the acceleration spectra of the main shock horizontal components recorded at the AQK station have been reported. Model 1 amplifies the same periods keeping the shape of the input acceleration spectra; thus, such model does not change the input frequency content. Looking at Fig. 25b, from the shape of the amplification function Af can be seen that the model amplifies the period range 0.05–0.2 s (5–20 Hz) that is the amplified range of periods in the NS and WE components of the main shock at AQK station.

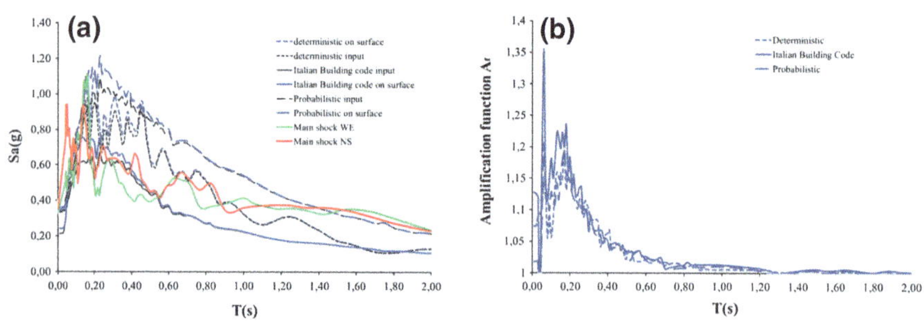

Fig. 25 1D numerical analyses results at AQK, for Model 1: **a** Acceleration spectra; **b** Amplification function Af

Hence, the Model-1 response is influenced by the input motion, but it amplifies those high frequencies as shown in the strong motion registrations. The amplification extent varies from 1.2 to 1.35; thus, according to Model 1, the AQK site is strongly influenced by the input frequency content that is moderately amplified. Figure 26a shows the output acceleration spectra of Model 2, where the seismic bedrock is set at a 270 m depth. In this case, the frequency content of the output spectra is changed and the amplified periods are the longer ones. As can be better understood by Fig. 26b, the recorded strong motion NS and WE components show amplified periods lower than the ones amplified by Model 2. As a matter of fact, the periods amplified are higher than 0.20 s. This is true for the three inputs. Finally, although the site response is conditioned by the input motion, the two analyzed models show how relevant can be the contribution of the litho-technical model in modifying the amplified ranges of periods/frequencies. In the case of AQK, the model of the "hanging seismic bedrock" fits better than the model with the "geological bedrock" to the measured records at the AQK site (Fig. 26).

For the case of the MI03 station at Onna city, Fig. 27a shows the site responses of two models: one suggested in this paper (RIFRA, in blue) and the other accepted by Working Group MS-AQ (2010) (MI03, in red). The three input motions have been deconvolved at 40 m: both models show higher amplifications with respect to the AQK station but Model RIFRA amplifies a broader range of periods, from 0.2 to 0.7 s. It is true for the three inputs and the amount of this amplification can be read in Fig. 27b: up to 1.5 at 0.2 s (5 Hz) and up to 2.4 at 0.5–0.7 s (2–1.43 Hz). The other model (MI03) produces lower amplifications (up to 2) in a narrower period range (0.2–0.55 s). Unfortunately, at MI03 no main shock records are available. Nonetheless, Di Giulio et al. (2011) analyzed three strong aftershocks registered at MI03 that occurred on 9th April with Ml 5.4, 5.1 and 5. They show a large spectral content in horizontal components for T < 0.3 s, and a spectral peak at 0.4 s with relative amplification level of 5, on average. Moreover, the shapes of spectral ratios between MI03 registrations and a reference bedrock site indicate amplifications in a large frequency band, up to 10 Hz (0.1 s). At MI03, the resonance frequency derived from the spectral ratio between the recorded horizontal and vertical components, for the case of the Ml 5.4 aftershock, shows a spectral peak at 1.8 Hz (0.6 s). Thus, in the case of the MI03 seismic station, the Model RIFRA proposed by the writing authors better explains the main characters of the strong motion registrations. Accordingly, the role of the velocity contrasts seems to be relevant at both sites. Finally, in order to suggest a practical method for measuring the possible amplifications at the site, the amplification function Af can be employed joined to the litho-technical models that correctly interpret, according to 1D simulations, the role of the velocity contrasts and the velocity inversions in the sediment sequences. Moreover, the AQK model here proposed can be used at the L'Aquila site where deep geological bedrock is associated with surficial velocity inversion. Finally, an accurate seismic characterization of the surface sediment behavior is needed for reproducing the relevant features of the local seismic response. As can be seen, at MI03, two-dimensional numerical simulations are needed.

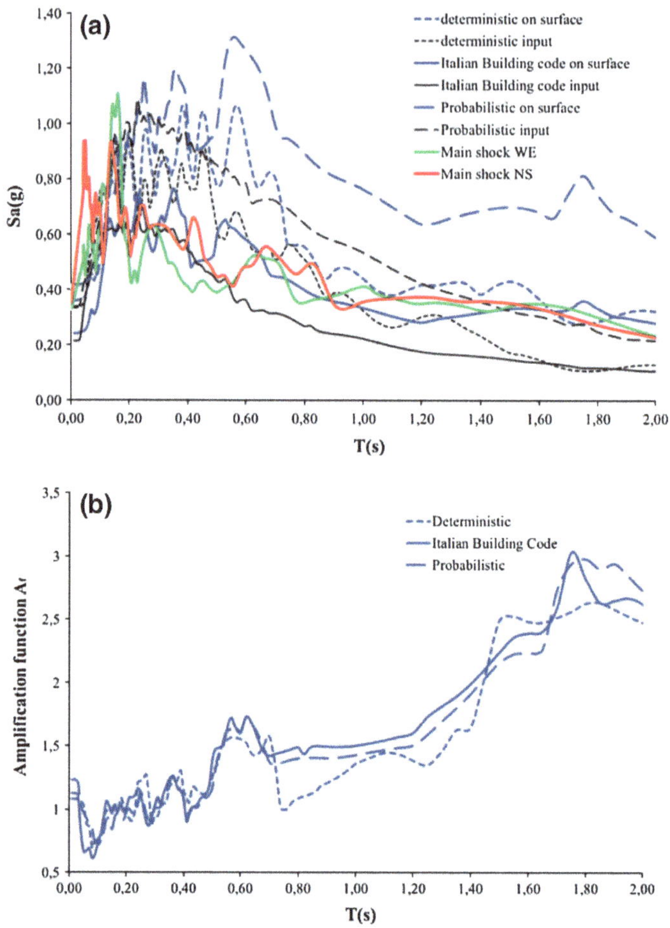

Fig. 26 1D numerical analyses results at AQK, for Model 2: **a** Acceleration spectra; **b** Amplification function Af

5.2.2 Two-Dimensional Analyses at Macroarea 5—Onna-Monticchio-SanGregorio Alignment

A geological cross section of the Monticchio-Onna-San Gregorio alignment is shown in Fig. 28, suggested by Working group MS-AQ (2010). In the central portion of this cross section, where the mean elevation is 580 m, the following lithological sequence can be sketched from the surface to the basement: a very thin layer of Holocene deposit (unit 3), calcareous alluvial and fluvial deposits of sand and gravel with interbedded clay and silt, overlying Pleistocene silty-clay sediments (unit 2); approximately a 50 m depth of lower Pleistocene conglomerate substrate (unit 1). The deep basement is assumed to be the Mesozoic limestone/Flysch that

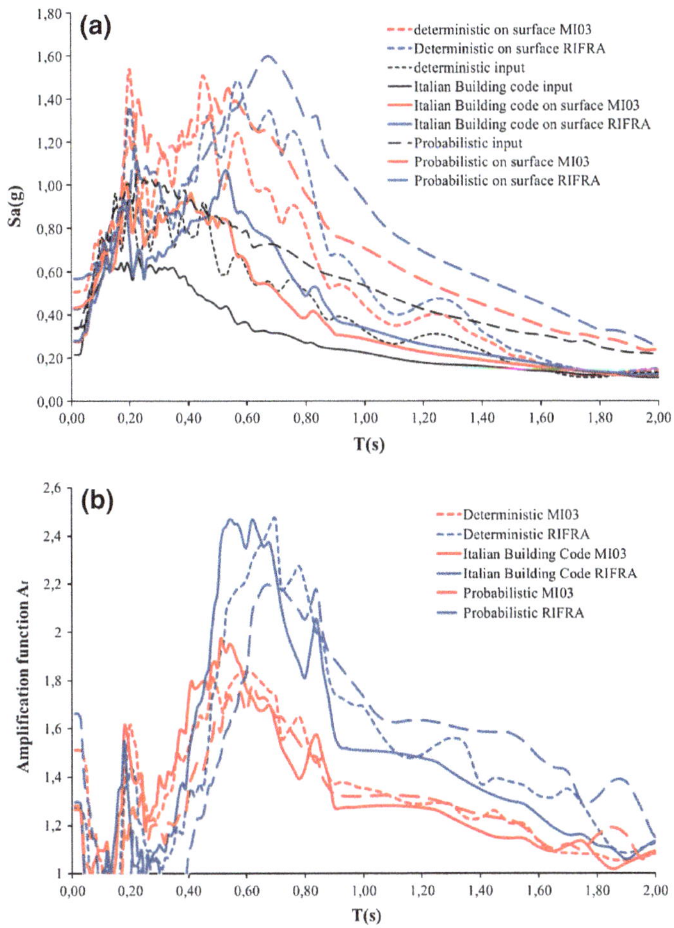

Fig. 27 1D numerical analyses results at MI03, for Model 1 (in *red*) and Model 2 (in *blue*): **a** Acceleration spectra; **b** Amplification function Af

outcrops in the uppermost Monticchio village (Di Giulio et al. 2011). This village is built on a gentle slope at the toe of the northern part of the Cavalletto Mountain at an approximately 600 m elevation.

Monaco et al. (2009) recently drilled a boring on the western plain of Monticchio for geotechnical characterization (Fig. 29c, borehole in red nearby Monticchio). Calcareous gravel and cobbles were found down to the bottom of the borehole, at 40 m (Fig. 29a). The eldest continental deposits (early Pleistocene) consist of very stiff or cemented lacustrine carbonate silt with inter-bedded sand that crop out at San Gregorio (APAT 2005). The geological relieves and two- electrical resistivity tomography (ERT) carried out at San Gregorio, show the well-stratified

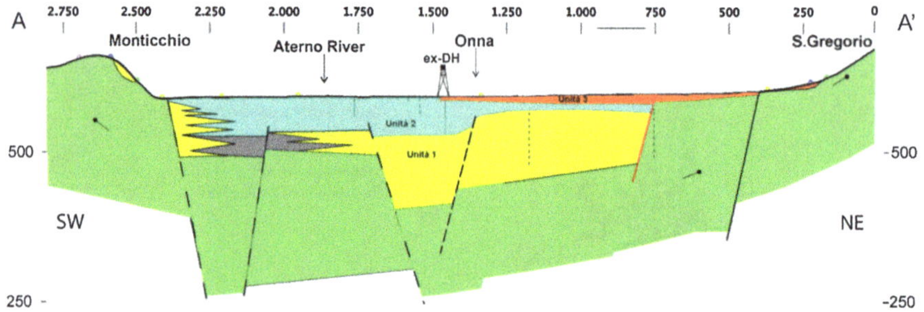

Fig. 28 Geological setting along a section crossing the middle Aterno Valley (After Working group MS-AQ 2010 modified). *Unit 1* conglomerates and gravel. *Unit 2* alternation of gravel, sand and silt. *Unit 3* eluvium-colluvium made up of soft silt and clay with scattered gravel grains. *Green unit* limestone or flysch bedrock. The cross section vertical scale is magnified by 2

bio-clastic calcarenite bedrock outcropping on the western flank of the San Gregorio hillside (named "calcareniti a foraminiferi", CFR2 in Fig. 7b).

This unit is locally intensely fissured. Geological and geophysical investigations were performed at Onna (Macroarea 5) for microzoning studies. Figure 29b shows some of the numerous field tests performed by Working group MS-AQ: one seismic refraction line (L7-SRZ1), two down-holes, soundings at 30 m depth, sounding at 60 m depth, micro-tremor measures (Nakamura technique), three geo-electric lines (ERT), among others. Accordingly, at Onna the quaternary continental deposits are made up of eluvium-colluvium of a few- meter thickness overlaying fluvial and alluvial units, more than 60 m deep. The geological bedrock was not intercepted by the soundings although a seismic bedrock was found by means of the refraction line, characterized by a shear wave velocity equal to 1000 m/s (Rainone et al. 2010). It is reported in Fig. 29c (refraction survey in black).

The refraction seismic surveys in P and SH waves, were processed by tomographic techniques (Fig. 30a). Furthermore, a resistivity depth sounding was carried out to identify the depth of the geological bedrock. The results of the two geophysical surveys are reported in Fig. 30a, b. They show that alluvial deposits can be detected up to a 100 m depth, under the first 2 m of cover. The deposits are characterized by two electrical horizons that are 63 Ω m resistivity as deep as 45 m and 48 Ω m up to 106 m. Furthermore, a high resistivity unit of 240 Ω m has been detected downwards. It is assumed as the geological bedrock. Nonetheless, the refraction line shows the seismic bedrock at about 40 m with a shear wave velocity equal to 1000 m/s (Fig. 30a). Moreover, the writing authors performed a borehole up to a 60 m depth near the MIO3 station (Fig. 29d). The stratigraphy shows, according to the velocity profile "RIFRA" (Fig. 30c) a new lithological horizon at about 40 m characterized by small carbonate clasts called "gravel" and sand in Fig. 29d. This 5 m stratum overlays stiff silty sand and sandy silt that show higher V_S values higher than 800 m/s up to 50 m. These V_S varies between 750 and

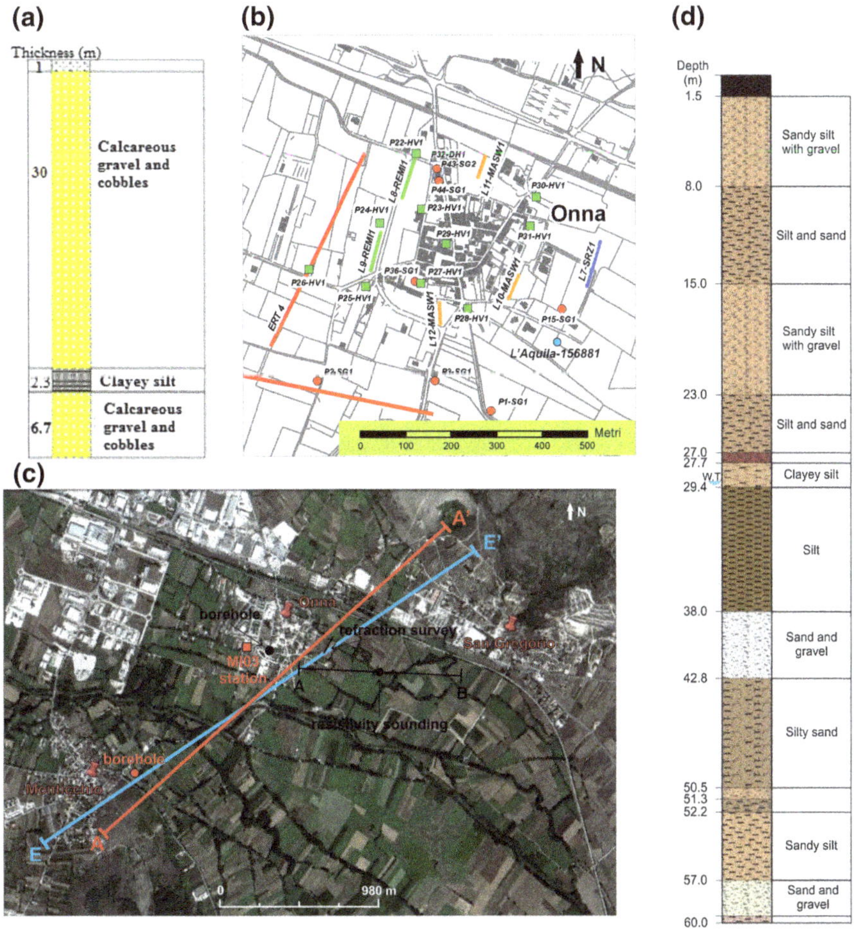

Fig. 29 **a** Stratigraphic profile from a 40 m sounding at Monticchio site (After Monaco et al. 2009 modified). **b** Map of investigation at Onna (After Working group MS-AQ 2010 modified): boreholes (*red circles*); microtremors measurements (*green squares*); well drilled by ISPRA (*blue square*); electrical resistivity tomography (ERT) surveys (*red lines*); are refraction microtremor surveys (ReMi), (*green lines*); seismic refraction survey (*blue line*); multichannel analysis of surface waves (MASW) surveys (*yellow lines*). **c** Traces of two cross-sections within the Aterno River Valley at Onna site: section EE′ is used for the present 2D numerical analyses; section AA′ refers to the geological section. **d** Stratigraphic succession observed by the writing authors along the 60 m borehole performed nearby MIO3 station at Onna city (borehole in *black* in Fig. 29c)

1000 m/s along the range depths 40–50 m. Accordingly, the borehole shows the stiffer lithologies up to 57 m and up to 60 m the carbonate clasts and sand are found again. Hence, 40 m can be considered a “seismic bedrock” according to the seismic soil classification of the building code.

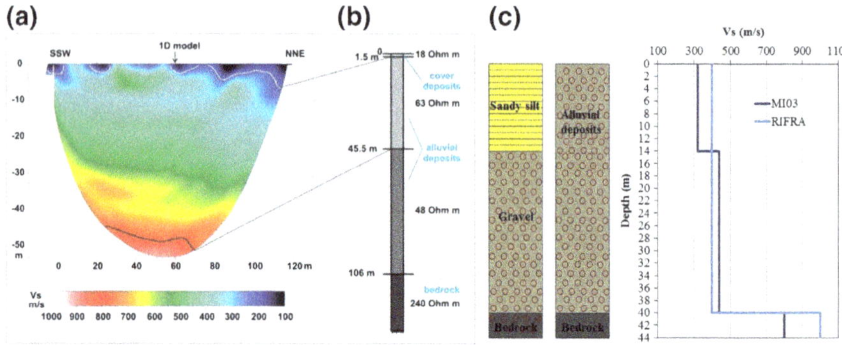

Fig. 30 Shear waves tomographic section carried out at the Onna site (**a**); resistivity depth-sounding carried out at the Onna site (**b**); **c** Litho-seismic model at Onna seismic station named MI03: in *red*, the shear wave velocity profile measured by Working group MS-AQ (2010); in *blue*, the velocity profile measured by the present authors

Figure 30c shows a comparison between two seismic vertical profiles: on the left, the one measured by means of ESAC-FK method (Working group MS-AQ, 2010) under the MI03 seismic station (Fig. 30b, MI03 station in red) and, on the right, the seismic profile from seismic refraction lines performed at Onna by the writing authors (Fig. 30c). The two sites are 1 km away from each other, thus slight differences in lithological sequences can be appreciated. Nonetheless, with respect to seismic characterization of alluvial deposits, named gravels in the MI03 stratigraphy, agreement on the mean shear wave velocity of the first 40 m has been drawn. Accordingly, the new seismic vertical profile, named RIFRA, has been used for 2D numerical simulations: a mean shear wave velocity of 400 m/s up to a 40 m depth and a seismic bedrock, with a mean value of 1000 m/s, made of gravels and conglomerates in silty lacustrine matrix (varying point to point according to the cross- section model in Fig. 26). It is worth noticing that within the Onna sector of the Aterno River Valley, the geological bedrock deepens according to a host-graben shape; thus, the deposit thickness varies from point to point. Accordingly, the model maximum depth varies from 35 to 70 m from the center of the valley towards San Gregorio. Such seismic model of the near-surface sediments fits well with the results from Di Giulio et al. (2011) and Bergamaschi et al. (2011) studies. Finally, hydrogeological investigations performed by ISPRA (http://sgi1.isprambiente.it/GeoMapViewer/index.html) have been considered. The static water level is −75 m as measured at L'Aquila-156881 well (Fig. 29b). Another well drilled between Bazzano and Onna, shows the static water level at −87 m (L'Aquila-156727). Phreatic groundwater was measured at −11 and −17 m in July 2005 in the Pliocene-Quaternary deposits. At Onna, the writing authors performed a measure of groundwater level at a piezometer located in the borehole drilled near the MIO3 station (Fig. 29d): the water level was detected at a 29 m depth. Nonetheless, due to the variable characters of the continental deposit permeability, a detailed hydrogeological study is needed to assess point by point the groundwater level in the Aterno Valley.

Hereafter, 2D numerical simulations of the Monticchio-Onna-San Gregorio cross section were performed according to the model shown in Fig. 31. The aims of this study are twofold:

1. to build adequate numerical models for simulating local seismic effects. To this end, comparisons among 1D and 2D numerical results and actual records at MI03 seismic stations (at Onna) have been addressed;
2. to try to estimate the contribution to the total amplification due to the seismic bedrock variable geometry and to the seismic impedance contrasts of soil deposits at Onna. The focus on the Onna site is strictly related to three strong aftershocks recorded at that site and available on the Itaca website. The 2D numerical model used hereafter is derived from the geological section published by Working group MS-AQ (2010), shown in Fig. 31a and modified based on geophysical investigations performed by Rainone et al. (2010) and described in this section. At the bottom of the proposed numerical model, unit 1 or gravels from Table 7 are considered as the seismic bedrock because the depth of the geological bedrock is not known. The Monticchio-Onna-SanGregorio cross-section under study is rotated with respect to the reference section. In doing this, the authors aligned the studied section with the geophone locations 4824, 4823 and 4826 (Fig. 14). It is traced in Fig. 29c.

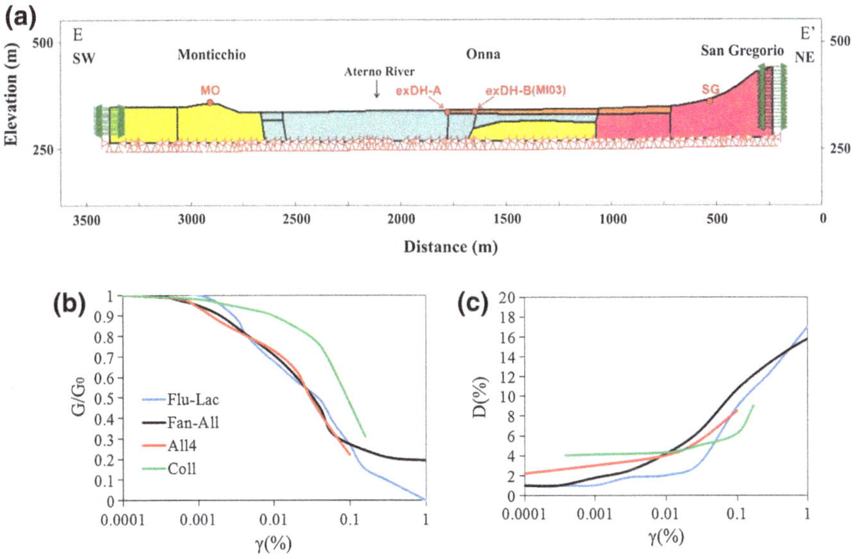

Fig. 31 **a** 2D seismic model of the Onna sector of the Aterno Valley; **b** Normalized shear modulus reduction and **c** damping ratio curves with shear strain used for linear-equivalent constitutive law within the numerical analyses

Table 7 Values of physical and mechanical properties of soil and rock units within the simulated section

Lithologic unit	Description	γ (kN/m^3)	V_S (m/s)	G (MPa)	Shear modulus reduction and damping curve in Fig. 34
1 (orange)	Eluvium-colluvium deposits	18	350	200	Coll
2 (light blue)	Alternations of gravel, sand and clay	19	450	390	All4
3 (yellow)	Gravels	21	1000	2100	Fan-All
4 (fuxia)	Weathered carbonate bedrock	21	800	1300	Rock (by Idriss[a])

[a]After Shake91 (Schnabel et al. 1991)

Dynamic numerical analyses were performed based on the soil properties at Onna (Macroarea 5) and San Gregorio (Macroarea 3) measured through several research works published during the last three years (Monaco et al. 2009; Di Giulio et al. 2011; Bergamaschi et al. 2011) and from the authors field experience at the Onna site (geophysical investigation in Fig. 29b). The equivalent linear model was used to simulate the response of all the formations; the elastic dynamic properties are reported in Table 7. Figure 31c, d shows the variation of the shear modulus reduction curves and the damping ratio curves with the shear strain measured by Compagnoni et al. (2011). For the weathered carbonate bedrock outcropping at San Gregorio, curves from literature are used: such curves are retrieved from Shake91 code (Schnabel et al. 1991). Although Working group MS-AQ (2010) reported the presence of terraced deposits of Valle Majelama Synthem superimposed on the calcareous bedrock, the authors on-site observations at the geophone locations identified only a fissured calcarenite unit. Thus a mean value of 800 m/s for shear wave velocity is used within numerical analyses (Table 7 and Fig. 31a). Moreover, according to the authors' knowledge on the hydrogeological characteristics of the Onna sector of the Aterno River Valley, the groundwater level within the numerical analysis has been considered at a 40 m depth, although locally the presence of hanging superficial groundwater can be found due to the depositional system.

The investigated section (Fig. 31a) covers 3250 m in length and 80 m in elevation; the numerical method used is the finite element implemented in QUAKE/W (Krahn 2004). Rectangular shaped elements have been used in the finite element domain discretisation. The optimum maximum size of the finite elements has been derived from the following rule, after Kuhlemeyer and Lysmer (1973):

$$l_{\max} = \frac{V_S}{7 \cdot f_{\max}} \tag{5.2}$$

where V_S is the shear wave velocity of the element material and f_{max} is the maximum frequency value of the input signal to be propagated toward the surface, which is assumed to be 20 Hz. The boundary conditions applied to lateral cut-off boundaries are nodal zero vertical displacements and nodal horizontal dampers with a viscosity coefficient, D_{node}, the values of which have been derived from the relationship:

$$D_{node} = \rho V_S \cdot {}^{L}\!/_{2} \cdot 1 \tag{5.3}$$

where ρ is the density and V_S is the shear wave velocity of the material, and L/2 is half the distance between the nodes times a unit distance into the section. Finally, the boundary conditions at the bottom of the model are nodal zero vertical displacements and the horizontal input time history accelerations.

The input motions considered and applied to the bottom of the model (Fig. 31a) are the time histories plotted in Fig. 24. In Fig. 32 the acceleration response spectra of these input motions are plotted according to the following naming:

1. the uniform hazard spectrum of the Italian Building code (NTC08) is named NORM;
2. the probabilistic uniform hazard spectrum with a return period of 475 years is named PROB;
3. the deterministic spectrum obtained from SP96 GMPE for a magnitude-distance pair (M_w = 6.7, R_{epi} = 10 km) is named DET.

It is worth noticing that, at the Onna seismic station (MI03) the main shock records are not available although three strong aftershocks have been recorded. In Fig. 32 acceleration spectra of the artificial input motions and the aforementioned aftershocks have been plotted together. It can be seen that the maxima spectral ordinates of these records vary from 0.1 to 0.42 g, whereas the ones of the artificial input motions range between 0.6 and 1 g. From this evidence and in order to

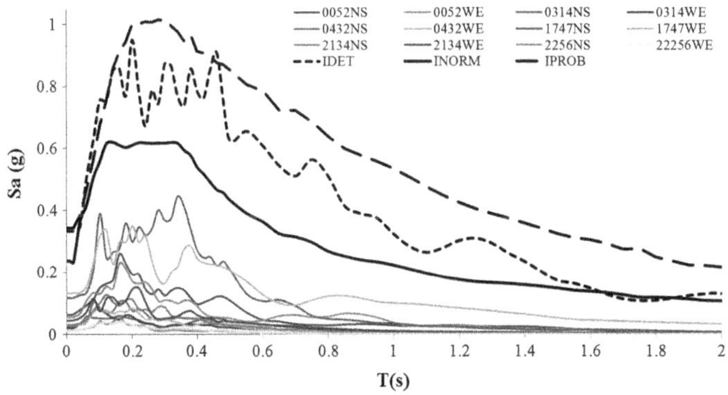

Fig. 32 Acceleration spectra of artificial input motions simulating 6th April 2009 main shock and the strongest aftershocks recorded at Onna seismic station (MI03)

compare the calculated 1D and 2D acceleration spectra with the available records at the Onna site, the preceding three artificial inputs have been scaled at three peak acceleration values PGA: 0.05, 0.15 and 0.25 g. Then, these time histories have been deconvolved by means of 1D EERA code (Bardet et al. 2000) at the bottom of the section model, that is 70 m depth, through the gravels (reported in Table 7). The spectra of nine deconvolved input signals are reported in Fig. 33. These deconvolved acceleration spectra are all characterized by a valley at about 0.3 s, that means the fundamental resonance period of this 70 m soil column corresponds to that period, as can be easily confirmed by means of the formula:

$$T_0 = \frac{4 \cdot H}{V_S} \tag{5.4}$$

Moreover, two ranges of periods are amplified: the first one is narrow and corresponds to 0.15 s (6.7 Hz), the second one is larger varying from 0.5 to 0.75 s (1.4–2 Hz). These modifications to the artificial spectra are related to the deconvolution through the gravels, assumed as the seismic bedrock. For the purpose of this numerical study carried out along the Monticchio-Onna-San Gregorio alignment, this seismic bedrock model seems to be the most representative, provided that (1) the input motions are artificially calculated and take into account the 2009 L'Aquila earthquake main shock not recorded at the Onna seismic station and (2) accurate geotechnical and geological characterization of deep deposits is not available.

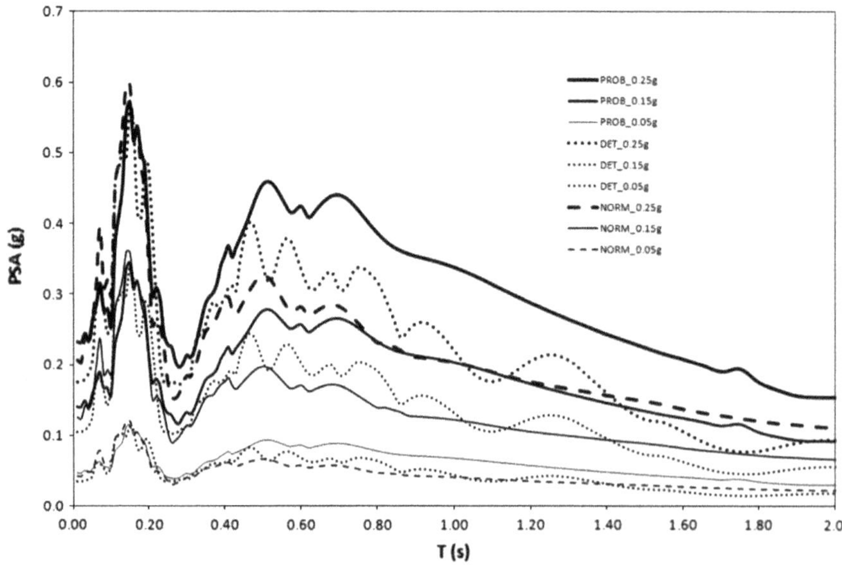

Fig. 33 Artificial input signals called NORM, DET and PROB scaled at 0.05, 0.15 and 0.25 g and deconvolved to the bedrock depth 70 m for performing the 2D analyses within Onna sector of the Aterno Valley (Fig. 32)

Moreover, 1D and 2D numerical analysis results in terms of acceleration response spectra have been compared with three strong aftershocks recorded at the Onna seismic station (MI03): (1) on 07 April 2009 at 17:47 UTC Mw = 5.6 with epicentral distance R_{epi} = 5.9 km; (2) on 07 April 2009 at 21:34 UTC Mw = 4.6 with epicentral distance R_{epi} = 10 km; and (3) on 09 April 2009 at 00:52 UTC Mw = 5.4 with epicentral distance R_{epi} = 20.5 km. This seismic station corresponds to the point ExDH-B on the 2D numerical model (Fig. 31a). Another point is ExDH-A that can be representative of the Onna site, is also considered within the discussion of calculated results: at this point, the alluvial deposit deepens to 70 m whereas the surficial eluvium-colluvium deposits have constant thickness. The two points are 125 m away from each other and their differences in spectral accelerations are discussed. Figure 34a, b shows acceleration spectra calculated by means of 1D and 2D simulations, plotted together with the actual acceleration spectra recorded during the aforementioned three strong aftershocks at the Onna site (MI03). 1D Onna model for MI03 has been shown in Fig. 31c (on the left), whereas 2D point is ExDH-B. Both 1D and 2D simulations in Fig. 34a, b are related to the three artificial input motions scaled at 0.05 g. In Fig. 34a East-West record components are plotted whereas in Fig. 34b the North-South ones are shown. The studied cross section is oriented South-West and North-East, thus the two components of the records can be relevant for the comparison. As Fig. 34a shows, the 1747 record seems to be well predicted in the period range 0.05–0.3 s by the 2D simulations. Then, the valley at 0.3 s and the second amplified period, at 0.4 s are also caught. On the contrary, the "numerical" amplification from 0.4 to 0.75 s is not actually present in the aftershocks. Such difference can be due to the artificial input motions that are not built based on the Onna site response but according to the criteria summarized in Sect. 5. However, these two amplified ranges of periods are not representative for the three aftershocks under study but could be useful to predict the site response due to seismic events characterized by both larger frequency content and greater energy content. Accordingly, the authors suggest to discuss, in Fig. 34a, b, the portion of the acceleration spectra from 1D and 2D analyses that are representative for the considered records. The aim of the comparison is to point out the need of 2D simulations at Onna. In Fig. 34a the two weaker aftershocks show the same trend of the strongest one: (1) the highest acceleration peak at periods lower than 0.2 s; (2) a valley at 0.3 s and (3) smaller peaks up to 0.4 s. As can be seen, the longer the periods corresponding to the acceleration peaks the smaller the seismic event magnitude and the smaller the acceleration peaks. Such a trend is not caught by the 1D results: 1D acceleration spectra underestimate the strongest aftershocks whereas overestimate the weakest (2134 record).

Figure 34b shows that the North-South record components have different spectral acceleration trends with respect to the East-West. However, 2D results are more suitable than 1D to predict the amplifications at periods lower than 0.2 s based on 0052 and 2134 records. The North-South components seem to be influenced by a different geological setting with respect to the cross section modeled that, on the contrary, seems to be representative for the East-West components. Due to several

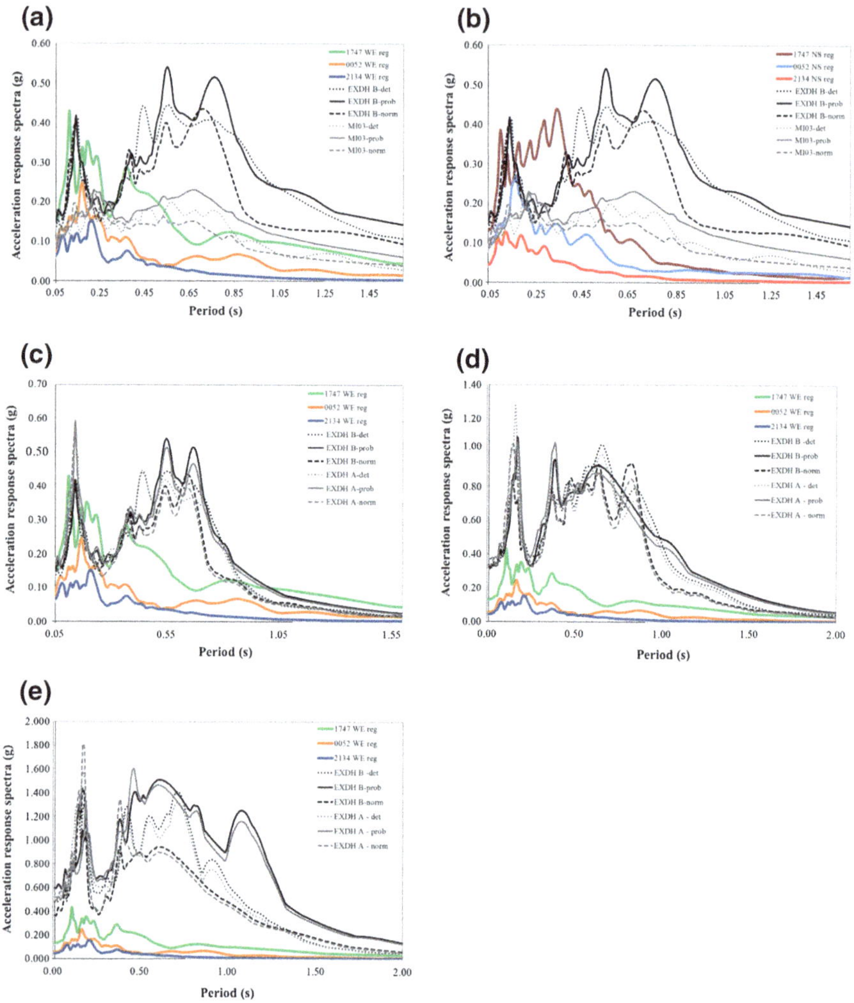

Fig. 34 Aftershock records at MI03 (Onna seismic station), 1D (named MI03, *grey*) and 2D (named EXDH B, *black*) acceleration spectra calculated for DET, PROB and NORM artificial inputs scaled at 0.05 g, at MI03 seismic station and at the node EXDH B respectively: **a** WE record components; **b** NS record components. WE components of the aftershocks recorded at MI03 (Onna seismic station) and 2D acceleration spectra calculated at the nodes ExDH-A and ExD-B for DET, PROB and NORM artificial inputs scaled at: **c** 0.05 g, **d** 0.15 g and **e** 0.25 g

approximations in 2D numerical modeling (e.g. the 2D numerical reduction from 3D actual phenomena and the use of artificial input motions), the correspondence between the calculated and recorded acceleration spectra seem to be good for the East-West components. The 2D model used in this study enables to quantify the maximum acceleration peak and the corresponding periods both for periods lower than 0.2 s and periods up to 0.4 s. Moreover, the comparison between calculated and recorded spectra at the Onna site confirms two assumptions proposed by the present study: the role of "seismic bedrock" played by the gravels at a 40 m depth and the need of a 2D model for a correct prediction of amplification where a basin shaped valley is detected. In the case study, the need of 2D simulations is shown at the Onna site because it is placed near the valley border. Figure 34c, e shows results from 2D simulations only. The acceleration spectra calculated at two points ExDH-A and ExDH-B on the model surface are plotted together and compared with the W-E components of the strong aftershocks recorded at the MI03 seismic station. The two points are both representative for the variable stratigraphy of the Onna city site. Thus, in order to simulate the seismic response throughout the Onna municipal territory both points are taken into account. Three artificial inputs have been considered, as in the preceding section. For calculating the acceleration spectra three PGAs have been considered. Figure 35c, e show smaller differences in the peak acceleration amplitudes at the two points for longer periods and higher differences for periods lower than 0.3 s. ExDH-A and ExDH-B are characterized by different bedrock depth. In particular, for periods lower than 0.2 s, the highest acceleration peaks are attained at ExDH-A for DET and NORM input motions. The shape of the acceleration spectra at the two points, especially at ExDH-B (with the same stratigraphy under MI03 where the aftershocks have been recorded) are in good agreement with the WE components of the strong aftershocks when the input motions are scaled at PGA = 0.05 g. It well predicts the 17:47 aftershock relative to a magnitude M_W = 5.6 up to 0.4 s. However, the main shock magnitude was M_W = 6.3 and possibly induced higher amplifications at longer periods at the Onna city site. In order to estimate the local amplification magnitude due to both the subsurface geometry and the deposit impedance contrasts the amplification functions have been calculated at ExDH-A and ExDH-B for the input motions scaled at

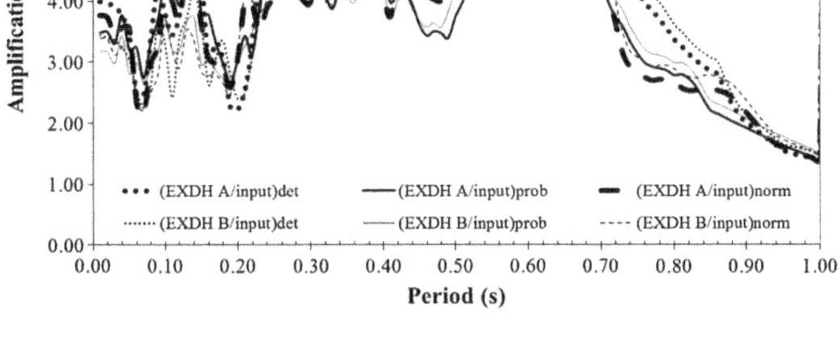

Fig. 35 Amplification function calculated for scaled input signals at 0.05 g at Onna site (EXDH A and B)

PGA = 0.05 g. All of the amplified period ranges shown in Fig. 34c, e have been considered in order to provide suggestions on amplification functions at the Onna site taking into account a larger number of seismic records. Looking at Fig. 34d, e where input motions are scaled at PGA = 0.15 and 0.25 g respectively, a shift in amplified periods can be observed with respect to Fig. 34a: at higher PGA values, the peak at 0.12 s (8.3 Hz) is shown for PGA = 0.15 g and one between 0.1 and 0.2 s (5–10 Hz) for PGA = 0.25 g. Correspondingly, amplified higher periods vary between 0.4 and 0.8 s (1.25–2.5 s) for PGA = 0.15 g and between 0.4 and 1.2 s (0.8–2.5 Hz) for PGA = 0.25 g. The amplification functions plotted in Fig. 35 are calculated as the ratios between the numerical responses at ExDH-A and ExDH-B and input signals deconvolved to the bedrock. In the plot, the surficial amplification due to the continental deposits increases for longer periods (lower frequencies). The amplification functions vary between 2 and 5 in the period range 0.05–0.20 s (5–20 Hz), between 3.5 and 6 in the period range 0.23–0.55 s (1.8–4.3 Hz) and between 4.5 and 7.5 in the period range 0.55–0.75 s (1.3–4.3 Hz). Based on the calculated acceleration spectra, the mean amplification function at the Onna site for the period range 0.05–0.75 s is 5. This factor confirms the value suggested by Di Giulio et al. (2011) based on the analyses of spectral ratio of the strong aftershock registrations at Onna site. Whether periods lower than 0.2 s are only considered, a mean value of 3.5 can be considered for the amplification function. Moreover, at Onna site, the local amplification calculated by means of amplification function is 5, much higher than the amplification factor derived from the microzonation study at the Onna site, that is 1.8 (Fig. 36a) although it is calculated by means of a different expression (Working group MS-AQ 2010).

2D numerical results calculated at Monticchio and San Gregorio are shown in Fig. 36a. At these sites no strong motion records from the Itaca website are available, thus comparisons with velocity spectra from Sect. 4 (Figs. 12, 13 and 14) have been accomplished in terms of amplified frequencies at the three sites: Monticchio, San Gregorio and Onna. In Fig. 36a, the acceleration spectra calculated for the three artificial inputs scaled at 0.05 g at the preceding sites are plotted together. As can be seen, at the Monticchio site two amplified periods are calculated: 0.11 s (9 Hz) and 0.3 s (3.3 Hz) with the lowest acceleration ordinates with respect to San Gregorio and Onna. At San Gregorio, two amplified period ranges can be detected: 0.12–0.18 s (5.6–8.3 Hz) and 0.37–0.5 s (3–2.7 Hz). Finally, at Onna two amplified period ranges are identified: 0.14 s (7.1 Hz) and 0.55–0.67 s (1.5–1.8 Hz).

With respect to the general trend of these spectra, at Monticchio where stiffer units outcrop, the extension of the amplified period ranges is limited as well as the acceleration ordinates, whereas at the San Gregorio site the response is intermediate between Monticchio and Onna: the amplified period ranges are larger than Monticchio but narrower than Onna; on the contrary, the acceleration ordinates are higher than Monticchio but lower than Onna. All of the three sites show amplification within the period range 0.1–0.2 s. This is in agreement with the results from monitoring activity reported in Figs. 12, 13 and 14. Although the seismic events recorded can be considered weak motions, amplified frequencies higher than 5 can

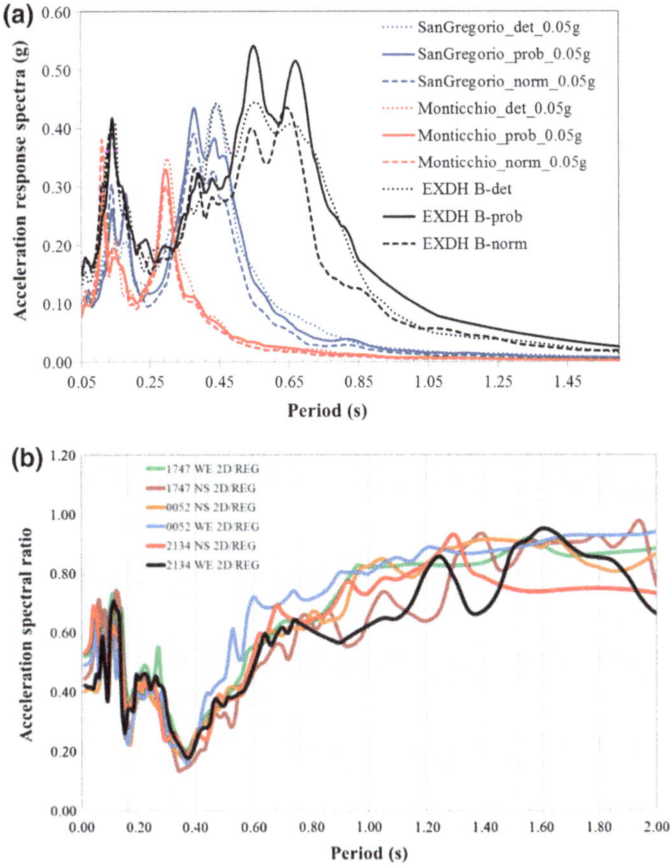

Fig. 36 **a** Acceleration spectra calculated by means of 2D numerical analyses at Monticchio, San Gregorio and Onna for the three artificial input motions scaled at PGA = 0.05 g. **b** Spectral ratio between the aftershock records at MI03 and the corresponding deconvolved signals at 40 m depth, according to the 1D profile in Fig. 32

be detected: at Monticchio they are 8–10 Hz (0.1–0.13 s) and 2.8–3.3 Hz (0.3–0.35 s); at San Gregorio they are 8–10 Hz and 1.8–2.4 Hz; at Onna they are 7 Hz (0.14 s) and 1.3–2 Hz. These calculated responses at the three sites can help explaining the different damage levels suffered from structures. Considering the simplified expression for calculating the predominant period of a multistory building (NTC 2008):

$$T_1 = C_1 \cdot H^{3/4} \quad (5.5)$$

where H is the height of the building from the foundation level and C_1 is the coefficient depending on the type of the structure: for steel structures $C_1 = 0.085$, for reinforced cemented structures $C_1 = 0.075$, whereas for other structures $C_1 = 0.050$.

Taking into account that at Monticchio village the buildings are prevalently masonry type with $6 \leq H \leq 10$ m, it can be calculated that the predominant period ranges between 0.19 and 0.28 s. The amplified periods (or high frequencies) at Monticchio do not correspond to the estimated fundamental periods of the buildings. This circumstance, in addition to the lower amplification suffered by Monticchio village, can help understand the lower intensity felt here. On the contrary, at San Gregorio and at Onna, the highest peaks correspond to higher periods that likely provoked double resonance effects.

Finally, in order to address the second objective of this numerical study, that is to calculate the contribution of 2D geometry versus 1D impedance contrast on the amplification function, the strong aftershock records at Onna are deconvolved from the surface through 40 m alluvial deposits at a mean shear wave velocity of 400 m/s down to the bedrock. It is made up of gravel and conglomerate with shear wave velocity at 1000 m/s. Such a litho-seismic profile under MI03 seismic station has already been used for assessing the amplified frequency of the station (Rainone et al. 2013): here, this profile gives a possible "transfer function" for separating the geometrical amplification effects (2D effect) from those due to the soil layer impedance contrasts (1D effect). 1D analysis has been performed by EERA (Bardet et al. 2000) according to the litho-seismic profile drawn from the seismic refraction line executed by the writing authors Fig. 30c (RIFRA). The 1D deconvolution involved all of the horizontal components of the strong aftershock records at MI03 (Fig. 36b). Results are plotted in terms of acceleration spectral ratios, that are the ratio between the deconvolved signal at the bedrock and the ground records: this ratio can be regarded as the ratio between the 2D amplification effect over the total amplification on the surface.

The acceleration spectral ratios plotted in Fig. 36b show that, approximately in the period range 0.05–0.15 s (6.7–20 Hz) the geometry of the valley contributes as much as 50 % on average. As the periods increase there is a reduction in this contribution corresponding to a minor amplification effects (see Figs. 34c, d) up to 0.4 s. In the range 0.4–2 s the 2D contribution to the amplification increases: from 0.6 to 2 s it increases up to 90 %. This means that, for the soils considered, the impedance contrast plays a relevant role up to 0.15 s because it is responsible for the 50 % of the recorded amplification; for higher periods, its contribution reduces to minimum values at 2 s. These outcomes are consistent with the quite simple mean seismic profile measured over 40 m at the Onna site within the alluvial deposits and it can explain the two amplified period ranges observed: at low periods (high frequency). The sediment/bedrock impedance and the geometric amplification contribute at 50 %; on the contrary, the geometric amplification is prevalently responsible for high peaks at longer periods (lower frequency).

5.3 *Concluding Remarks from 1D and 2D Numerical Simulations*

The present section discusses some critical points raised from the great work undertaken from the Working Group MS-AQ (2010) aiming at microzoning macroareas 1 and 5 in the Aterno Valley. Results from 1D simulations performed by the writing authors at two seismic stations at L'Aquila city center and at Onna city, highlight the relevance of the local seismic effects confirmed by strong motion records of the main shock on April 6th and some strong aftershocks occurred soon after. Nonetheless, at L'Aquila city 1D simulations are able to catch the main features of the local response only if "hanging seismic bedrock" is considered. Considering the Amplification function calculated as the ratio between the output and the deconvolved input of the 1D numerical analyses, it can be noted that a 35 % amplification occurred at 0.05 s (20 Hz) and up to 25 % over the narrow period range 0.15–0.20 s (5–6.7 Hz). These values are confirmed by the records and can be used where microzoning maps do not define the amplification (Fig. 23). At Onna city, 1D simulations cannot explain the records. Thus, 2D numerical analyses have been performed along the Monticchio-Onna-San Gregorio alignment. Results in terms of acceleration response spectra have been compared with three strong aftershocks recorded at the MI03 seismic station (Onna). These numerical analyses addressed two main objectives: (1) to calculate the mean amplification at the Onna site through the amplification functions of the acceleration spectra; and (2) to assess the contribution to the local amplification of the subsurface geometry with respect to the seismic impedance contrast. The 2D numerical results, at Onna, correctly model the strongest aftershock just in the case of the input motion scaled at PGA = 0.05 g. There, a mean amplification function of 5 has been calculated over the period range 0.1–0.75 s. This study suggests to use the *amplification function of the acceleration response spectra* (that is the ratio between the output and the input spectra) instead of the amplification factor for microzoning purpose: this function can be both used as a whole, whenever amplified period ranges are searched for or it could be considered by portions, whether the mean amplification values are considered in limited period ranges. Finally, the contribution of 2D geometry has been estimated at Onna site by means of deconvolving the strong aftershocks through the 1D stratigraphy at MI03: it weighs 50 % over the total amplification for periods lower than 0.4 s, whereas for higher periods it increases up to 90 %. These results contribute to shed light on the key role of subsurface geometry in local amplification effects in "near field" conditions, although limited to a few records and to the Onna site. These numerical analyses suggest that numerical models don't need to be complicated but effective on average due to the high heterogeneity and complexity of the actual subsurface settings in alluvial filled valleys.

6 Microzoning by Means of Ambient Noise Measurements

6.1 Theoretical Aspects and Limitations of Nakamura Method

Nakamura (1989) technique uses ambient noise measures to derive seismic properties of the site. The method was first introduced by Nogoshi and Igarashi (1971) based on the initial studies of Kanai and Tanaka (1961) and it is known as the "H/V method". Among the empirical methods the H/V spectral ratios on ambient vibrations (microtremors) is probably one of the most common approach and it is actually the recommended technique to derive the natural frequency of the site in microzoning studies in Italy (ICMS 2008, 2010). The H/V method assumes that microtremors are made up of Rayleigh waves that propagate in the soft half infinitive stratum of soil that is responsible of the local amplification (Fig. 37). The advantage of this method consists on separating the contribution of the source to path from the site in terms of frequency content of the signals through the ratio among recorded signal components avoiding the troublesome selection of a reference site.

Nakamura's method is applied to four spectra in the frequency domain, that are two pairs of Horizontal (H_S, H_B) and Vertical (V_S, V_B) components related to the soft layer overlaying the rigid one according to the model in Fig. 37. In order to give some outlines of the Nakamura method, the hypotheses underneath the procedure are listed:

- Microtremors are generated by local sources;
- Surficial sources of microtremors do not affect the deep sources;
- The vertical component of the motion is not affected by the local amplification of the surficial soft layer.

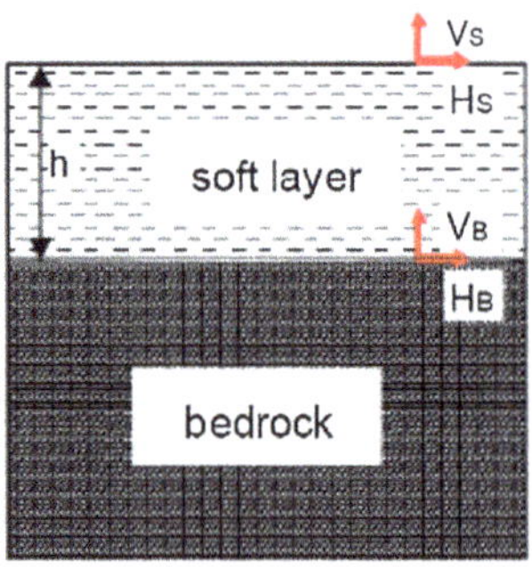

Fig. 37 Reference model for H/V method: V_S and H_S are the vertical and horizontal components of the motion on the surficial layer; V_B and H_B are the vertical and horizontal components of the motion on the bedrock

Under these hypotheses, the ratio between the vertical components of the motion is affected by local source effects, whereas the ratio between the horizontal components of the motion is affected by both local source and amplification. Thus, multiplying the two ratios the transfer function S(f) can be calculated:

$$\frac{H_S(f)}{H_B(f)} \cdot \frac{V_S(f)}{V_B(f)} = S(f) \tag{6.1}$$

Within the hypothesis that at the bottom of the soft layer, the following expression is satisfied:

$$\frac{V_B(f)}{H_B(f)} = 1 \tag{6.2}$$

Then, the expression of the transfer function in Eq. (7) gives:

$$\frac{H_S(f)}{V_S(f)} = S(f) \tag{6.3}$$

The simplified approach that Nakamura's method relies on shows advantages and disadvantages when actually applied to the complex environments that urbanized centers represent. Thus, some guidelines have been issued from the SESAME project (Site Effects Assessment using Ambient Excitations, European Commission, no. EVG1-CT-2000-00026, Coordinator: Pierre-Yves Bard, May 1st, 2001–December 31th, 2004, Keywords: Earthquake—Site effects—Noise—Urban Area). These guidelines provide some recommendations to be taken into account when the H/V technique is used to deal with microzoning studies. The recommendations given apply basically to the case where the method is used alone on noise measures for assessing the natural frequency of sites of interest and are therefore based on a rather strict set of criteria. They are made of two parts: (1) experimental conditions and (2) criteria for reliability of results. In particular, some advices are provided when H/V technique is used within the urban centers:

(a) Users are advised that recording near structures such as buildings, trees, etc., may influence the results: there is clear evidence that movements of structures due to the wind may introduce strong low-frequency perturbations in the ground. *Unfortunately, it is not possible to quantify the minimum distance from the structure where the influence is negligible, as this distance depends on too many external factors* (structure type, wind strength, soil type, etc.).
(b) Avoid measuring above underground structures such as car parks, pipes, sewer lids, etc.; these structures may significantly alter the amplitude of the vertical motion.
(c) Wind probably has the most frequent influence and we suggest avoiding measurements during windy days. Even a slight wind (approx. >5 m/s) may strongly influence the H/V results by introducing large perturbations at low

frequencies (below 1 Hz) that are not related to site effects. A consequence is that wind only perturbs low frequency sites.

(d) Measurements during heavy rain should be avoided, while slight rain has no noticeable influence on H/V results.

(e) Extreme temperatures should be treated with care, following the manufacturer's recommendations for the sensor and recorder; tests should be made by comparing night/day or sun/shadow measurements.

(f) Low pressure meteorological events generally raise the low frequency content and may alter the H/V curve. If the measurements cannot be delayed until quieter weather conditions, the occurrence of such events should be noted on the measurement field sheet.

6.2 *H/V Ratios Measured at Macroarea 6 in the Aterno Valley*

The writing authors observed all the preceding advices and guidelines in the study presented hereafter. With the purpose of validating the Nakamura method at Villa Sant'Angelo and Tussillo site (Macroarea6), microtremor acquisitions have been performed by means of two devices:

- The Tromino, that is composed of three high-definition electrodynamic velocimeters set orthogonal to one another and put in a box. These velocimeters are characterized by a frequency range of acquisition varying between 0.1 and 256 Hz. The data are recorded in a memory card of 256 Mb with no cables inside the box;
- TheDAQLink III, that is a seismometer with a 24 bit acquisition system (Seismic source Co.) and a three components geophone. This type of seismometer records continuously with a sampling window ranging from 0.0208 to 16 ms and a sampling frequency varying between 48,000 and 64.5 sample/s; the Bandwidth varies in the range 0–15 kHz and the analogic filter has a flat response up to 8000 Hz.

On January 20, 2010, 8 noise measurements through the Tromino and 4 through DAQLink have been performed nearby the points investigated at Villa Sant'Angelo and Tussillo by the National Department of Civil Protection (DPC) and Prof. Mucciarelli through the Nakamura method (Fig. 38a).

The following acquisition parameters have been used:

- Time windows longer than 30 min with sampling frequency higher than 125 Hz and sampling time lower than 8 ms;
- The transient portions of the record have been erased according to the SESAME guidelines;

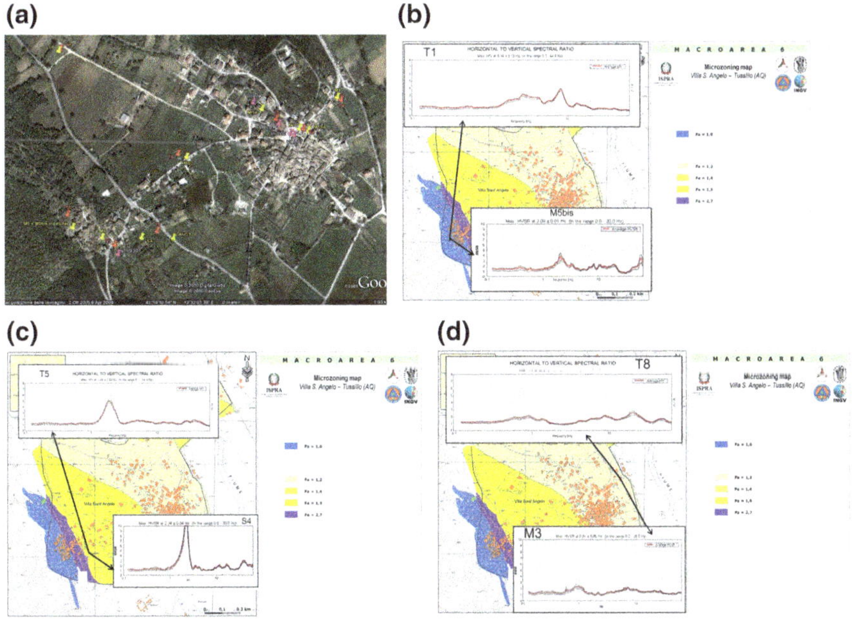

Fig. 38 **a** Noise measurements locations performed at Villa Sant'Angelo-Tussillo: (in *red*) by the authors of this study; (in *yellow*) DPC; (in *pink*) by Prof. Mucciarelli. **b** H/V measurements at two neighbor points at Tussillo center: T1 (this study) and M5bis (DPC). **c** H/V measurements performed by the Tromino at: T5 (this study) and at S4 (DPC) under similar ground conditions. **d** Noise measurements recorded at Tussillo center: at T8 (this study); at M3 (DPC)

- Fast Fourier Transform has been used to calculate the H/V with the tapering step;
- The spectral smoothing has been performed by means of the Konno-Ohmachi smoothing window: where b parameter is equal to 40 (for the DaqLink records) and Triangular smoothing window (for the Tromino records)

The H/V ratios have been calculated for each sub-windows of 20 s, then the mean and the standard deviation of all ratios have been calculated and plotted (Fig. 38b, c).

Figure 38b shows the H/V measured at two neighbor points at Tussillo center: T1 (this study) and M5bis (DPC). These acquisitions have been done by the Tromino. For the present study T1 point is set on the ground (to avoid inferences with any stiff material) whereas the DPC point is set on a concrete slab. As can be noted, the two plots are different: T1 evidences two peaks at 2–3 Hz and at 8 Hz; on the contrary the M5bis shows four peaks at 2, 10–20 Hz, 40 Hz and 55 Hz. In the presence of these peaks, the operator selects the most representative one. This selection is highly subjective. With respect to the plots, from the T1 8 Hz can be taken whereas from M5bis 2 Hz should be reasonable. On the contrary, Fig. 38c shows the H/V ratios measured at two points on the same ground type at Villa Sant'Angelo center: T5 (this study) and S4 (DPC) (see Fig. 38a).

In this case, the two plots show an evident peak at 2 Hz although the peak ordinate is double at T5 with respect to S4. It could be due to the presence of Love waves that do not have vertical components contributing to the amplification of the horizontal components. Figure 38d reports the H/V ratios measured at Villa Sant'Angelo at: T8 (this study) and M3 (DPC). Both plots are almost flat and no peak frequency can be detected. Figure 39 shows the ratios measured by the Tromino at: T2 (this study), M4 (DPC) and TUSS01 (Prof. Mucciarelli). As can be seen, the three curves show similar trends but only at M4 a clear peak in the ratio can be detected at 2 Hz. The others show two smaller peaks at two frequency intervals, that are 2–3 Hz and 10–15 Hz. No clear single peak can be detected but at the M4 point (Fig. 39).

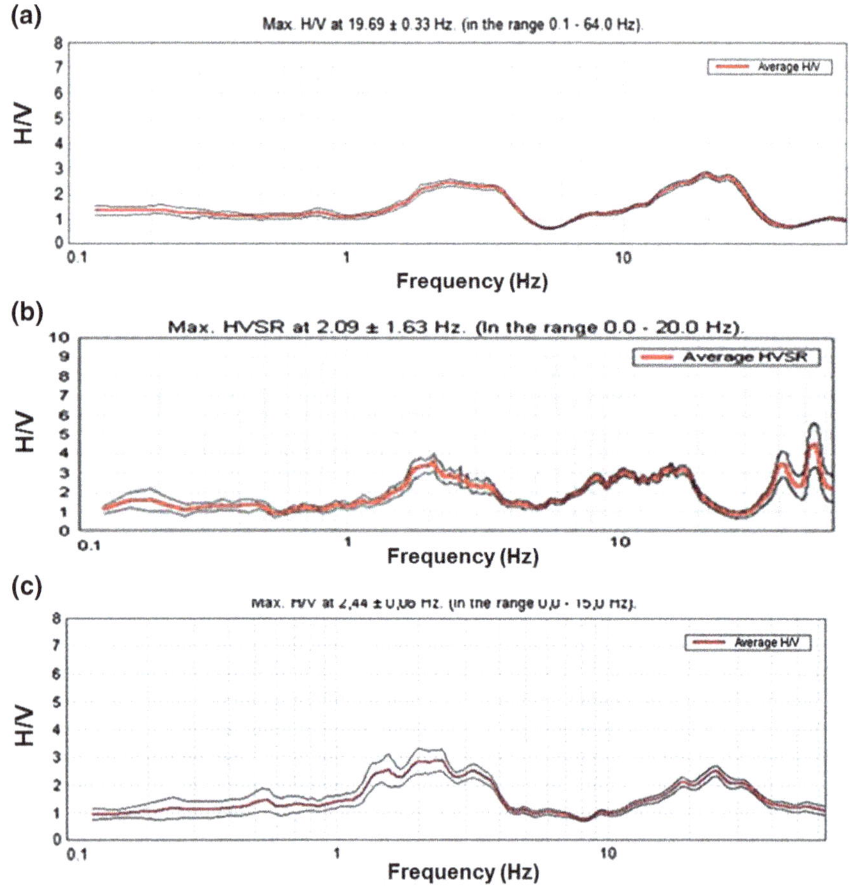

Fig. 39 H/V ratios at neighbor points at Tussillo center: **a** T2 (this study); **b** M4 (DPC); **c** TUSS01 (Prof. Mucciarelli)

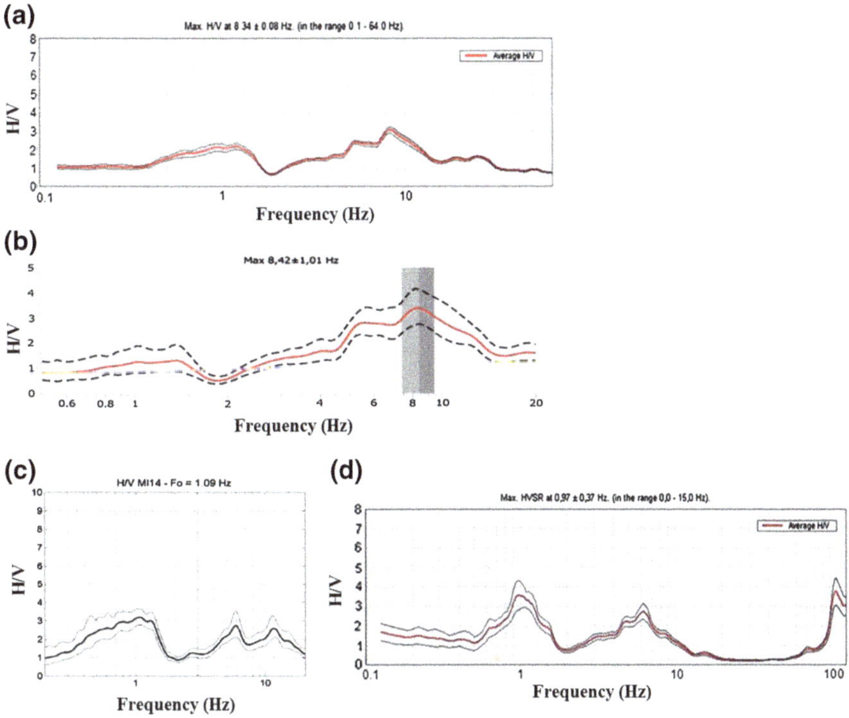

Fig. 40 H/V ratios at neighbor points at Villa Sant'Angelo historical center (at Municipal garden): **a** at T3 (this study) recorded by the Tromino; **b** at T3 (this study) recorded by the DAQLink; **c** at MI14 (DPC) recorded by the Tromino; **d** at VILL03 (Prof. Mucciarelli) recorded by the Tromino

Figure 40a, b show the ratios measured at T3 (this study): by the Tromino (Fig. 40a) and by the DAQLink (Fig. 40b). Thus, at the same point, H/V functions show quite similar trends and peaks: the highest is recorded at 8–9 Hz and the lowest at about 1 Hz. Comparing Fig. 40a, c, d shows that there are some differences, but peaks at 1 Hz and 6 Hz are measured at VILL03, whereas at MI14 three peaks are evident at 1, 6 and 10 Hz. Again, in these cases, different operators can derive different peak frequencies because the experimental measurements are ambiguous. Figure 41a–d compare H/V ratios measured at Villa Sant'Angelo center. In this case, the comparison between the Tromino and the DAQLink devices show similar trends and peaks. Figure 41a shows a peak at 1 Hz and 8–9 Hz, whereas the DAQLink (Fig. 41b) at 1 and 7 Hz. The measurements in Fig. 41c) shows only one peak at 1 Hz. Finally, Fig. 41d shows three peaks at 0.9–1.1 Hz, at 5–6 Hz and 8–9 Hz. No unique values from the comparisons can be drawn, indeed.

Therefore, from the comparisons of the measurements from this study at Villa Sant'Angelo and Tussillo it can be noted that the Nakamura ratio is sensitive to the site conditions and can give different predictions of the natural frequency according to the operator because often the functions show no clear peaks. Furthermore,

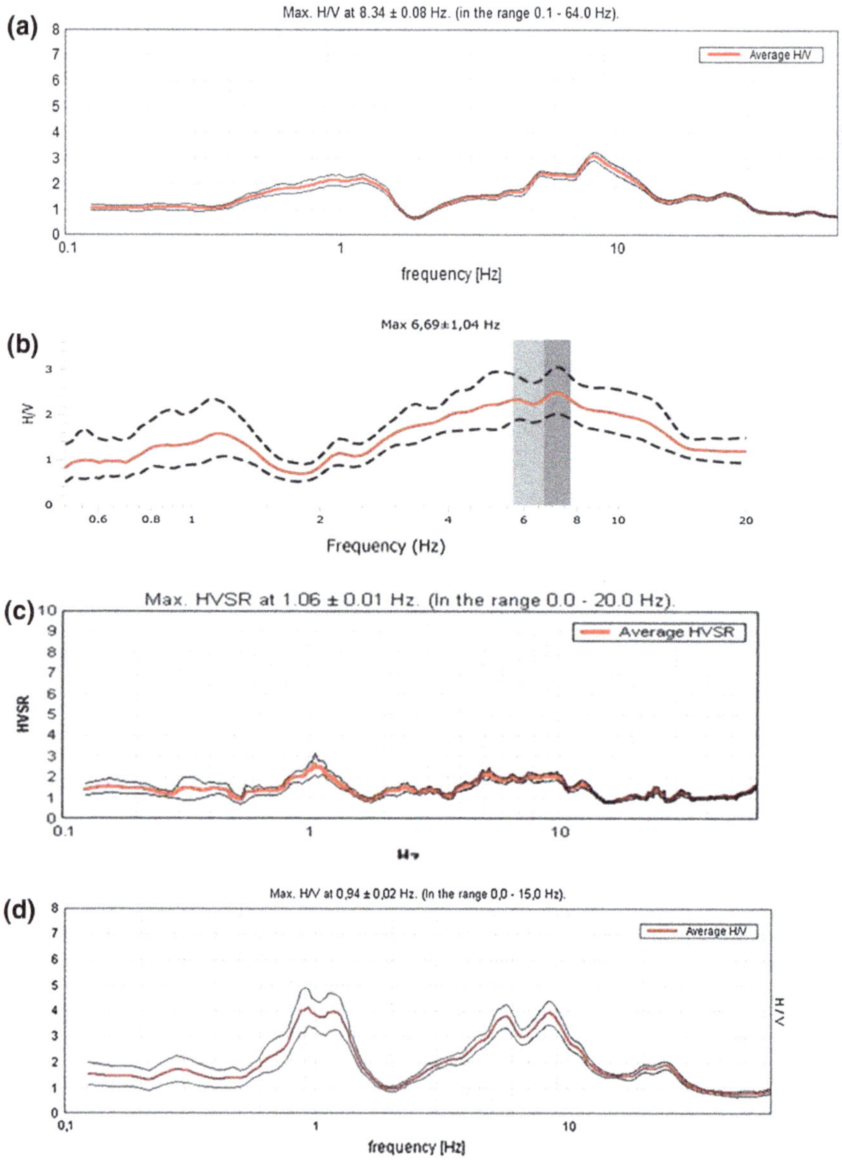

Fig. 41 Location of noise measurements at Villa Sant'Angelo center: **a** at T4 (this study) by the Tromino device; **b** at T4 by the DAQLink device; **c** at M1 (DPC) by the Tromino device; **d** at VILL02 (Prof. Mucciarelli) by the Tromino device

measures from the two devices, Tromino and DAQLinkIII show similar ratios although variable from point to point. The noise measurements are rarely confirmed by actual records. The authors used the records from the seismic monitoring illustrated in Sect. 5, to verify the prediction performed by the H/V ratios on the natural frequency at the site. Figure 42a, b shows the Nakamura method applied to the final portion of the records after the seismic event recorded on July 7, 2009 at 10:15 local time. As can be seen, the natural frequency derived from the peaks of the H/V ratio do not match with the peaks from the weak seismic motion records. This result comes from the comparison of weak motion and noise measurements. Through the same procedure, the authors calculated the H/V ratios by means of the final portions of the records of the main shock on April 6, 2009 at four stations: AQA, AQK, AQM and AQV (Fig. 43). These portions are related to a few minutes after the strong motion time window. They are used to back analyze the H/V ratios from noise measurements at the seismic station points, according to D'Intinosante et al. (2007) and De Ferrari et al. (2008).

As can be pointed out, none of the peaks measured by means of the Nakamura method correspond to the peak of the strong motion records but the one calculated at AQK station. This result cannot be predicted in advance. Moreover, the peaks from H/V correspond to higher natural frequencies in these four cases, than the ones from the weak motion records at AQK. Finally, from the studies presented it can be drawn that Nakamura method: (1) often provides more than one peak corresponding to different natural frequencies; (2) the peaks are heavily affected especially in urban areas that can not be easily disregarded by filtering the measurements; (3) the peaks in H/V ratios are not commonly related to both weak and strong motion amplified frequencies. Thus, the use of noise measurements for microzoning activities should be discouraged especially when it is used alone and it is not validated by other geophysical testing.

6.3 *Concluding Remarks on Noise Measurements for Predicting the Natural Frequency of the Site*

In this last section noise measurements have been performed in order to verify: (1) their repeatability by using different acquisition devices; and (2) their ability to measure the natural frequency of the site. As shown, acquisitions repeated at the same point by the same operator through the Tromino and the Seismometer DAQLink III show similar H/V ratios. On the contrary, H/V measures acquired by means of the Tromino by different operators at different times although at neighbor points can show peaks at different frequencies or can show the same amplified frequencies but with a very different amplitude. When the noise measures are, finally, compared with the weak motion tails of actual seismic events they always show different amplified ranges of frequencies. According to the operator or the weather conditions, noise measures can provide relevant differences in amplified frequencies and cannot give reliable information on the amplification extent. Other techniques, like seismic monitoring, or the shear wave velocity measures should be preferred to estimate the natural frequency of the site. Finally, it is worth to keep in

(a)

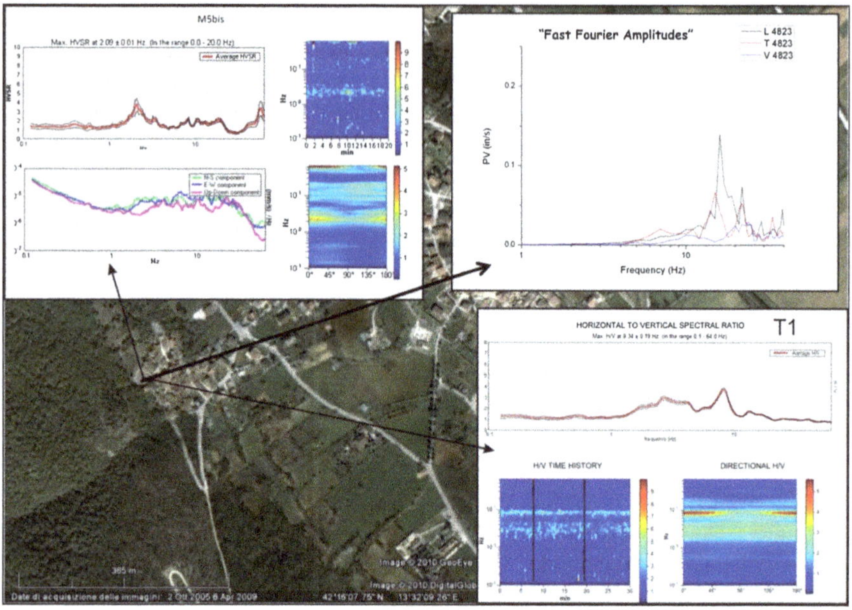

(b)

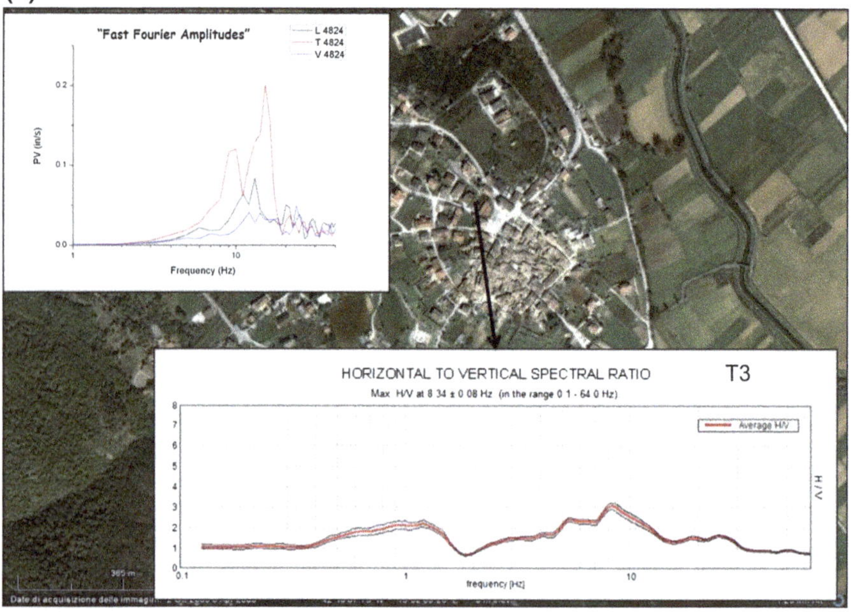

Fig. 42 Location of noise measurements from the seismic event occurred on July 7, 2009 at 10:15 local time: the noise measures are related to the final portion of the signal after the event

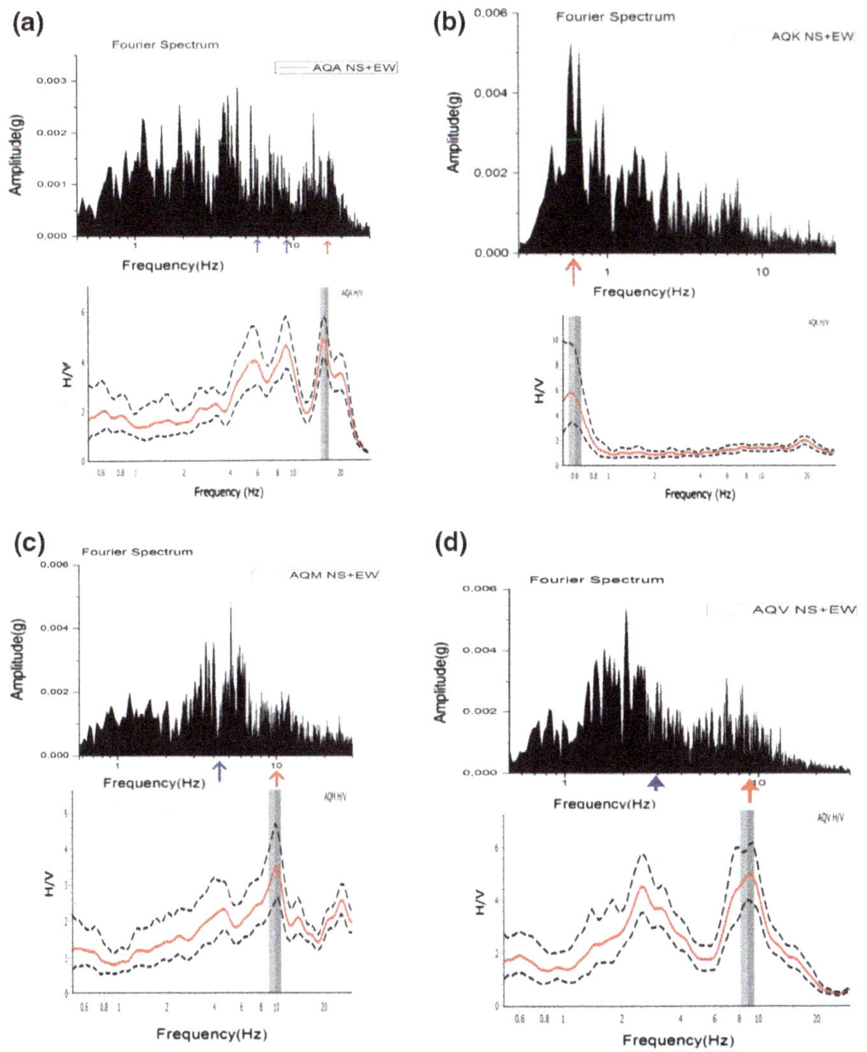

Fig. 43 Pairs of Fourier spectrum and H/V ratios calculated by Nakamura's method, related to the main shock records on April 6, 2009 at four seismic stations: **a** AQA, **b** AQK, **c** AQM, **d** AQV. The Fourier spectra is calculated from the strong motion records and the H/V ratios are calculated from the tails of the records (weak motions)

mind that soil behaviors are strain dependent. This means that their natural frequency at small strain level (microtremors) differs to a certain extent from the one at medium strain level (weak motions) and at high strain level (strong motion). Then, depending on the purpose of the natural frequency measure, different strain levels will be investigated to make the correct measure.

References

Amoroso, S., Del Monaco, F., Di Eusebio, F., Monaco, P., Taddei, B., Tallini, M., et al. (2010). *Campagna di indagini geologiche, geotecniche e geofisiche per lo studio della risposta sismica locale della città dell'Aquila: la stratigrafia dei sondaggi (giugno–agosto 2010).* Report CERFIS no. 1 (in Italian).

APAT. (2005). *Carta Geologica d'Italia, F.o 359 L'Aquila, scale 1:50,000.* APAT (Servizio Geologico d'Italia) and Regione Abruzzo. S.EL.CA, Firenze.

Atzori, S., Hunstad, I., Chini, M., Salvi, S., Tolomei, C., Bignami, C., et al. (2009). Finite fault inversion of DIn SAR coseismic displacement of the 2009 L'Aquila earthquake (central Italy). *Geophysical Research Letters, 36*, L15305. doi:10.1029/2009GL039293

Bagnaia, R., D'Epifanio, A., & SylosLabini, S. (1992). Aquila and subaequan basins: An example of Quaternary evolution in central Apennines, Italy. *Quaternaria Nova, II*, 187–209.

Baratta, M. (1901). *I terremoti d'Italia. Saggio di storia, geografia e bibliografia sismica italiana con 136 sismocartogrammi.* Arnaldo Forni editore.

Bardet, J. P., Ichii, K., & Lin, C. H. (2000). *EERA: A computer program for equivalent-linear earthquake site response analyses of layered soil deposits.* University of Southern California, Department of Civil Engineering.

Bergamaschi, F., Cultrera, G., Luzi, L., Azzara, R. M., Ameri, G., Augliera, P., et al. (2011). Evaluation of site effects in the Aterno river valley (Central Italy) from aftershocks of the 2009 L'Aquila earthquake. *Bulletin of Earthquake Engineering, 9*, 697–715.

Bosi, C., & Bertini, T. (1970). *Geologia della Media Valle dell'Aterno,* Memorie della Società Geologica Italiana (Vol. IX, pp. 719–777) (in Italian).

Chiarabba, C., Amato, A., Anselmi, M., Baccheschi, P., Bianchi, I., Cattaneo, M., et al. (2009). The 2009 L'Aquila (central Italy) Mw 6.3 earthquake: main shock and aftershocks. *Geophysical Research Letters, 36*, L18308. doi:10.1029/2009GL039627

Ciriella, A., Piatanesi, A., Cocco, M., Tinti, E., Scognamiglio, L., Michelini, A., et al. (2009). Rupture history of the 2009 L'Aquila (Italy) earthquake from non-linear joint inversion of strong motion and GPS data. *Geophysical Research Letters, 36*, L19304. doi:10.1029/2009GL039795

Compagnoni, M., Pergalani, F., & Boncio, P. (2011). Microzonation study in the Paganica-San Gregorio area affected by the April 6, 2009 L'Aquila earthquake (central Italy) and implications for the reconstruction. *Bulletin of Earthquake Engineering, 9*, 181–198.

De Ferrari, R, Ferretti, G., Spallarossa, D. (2008). Utilizzo di misure di rumore ambientale per la definizione degli effetti di sito: limiti di applicabilità della metodologia Nakamura. *Geologia Tecnica& Ambientale, 2*, 5–18, ISSN 1722-0025.

Di Giulio, G., Marzorati, S., Bergamaschi, F., Bordoni, P., Cara, F., D'Alema, E., et al. (2011). Local variability of the ground shaking during the 2009 L'Aquila earthquake (April 6, 2009 Mw 6.3): The case study of Onna and Monticchio villages. *Bulletin of Earthquake Engineering, 9*, 783–807.

DISS Working Group. (2009). *Database of individual seismogenic sources (DISS), Version 3.1.0: A compilation of potential sources for earthquakes larger than M 5.5. in Italy and surrounding areas.* Technical report, Istituto Nazionale di Geofisica e Vulcanologia (INGV).

D'Intinosante, V., Ferrini, M., Eva, C., & Ferretti, G. (2007). Valutazione degli effetti di sito mediante l'utilizzo dirumore ambientale in alcuni siti ad elevata sismicità della Toscana Settentrionale (Garfagnana e Lunigiana), XIICongresso Nazionale "L'ingegneria Sismica in Italia", Pisa 10–14 giugno 2007.

Galadini, F., Meletti, C., & Rebez, A. (Eds.). (2000). *Ricerche del GNDT nel campo della pericolosità sismica (1996–1999).* Roma: CNR-GNDT, Gruppo Nazionale per la Difesa dai Terremoti.

Galli, P., & Camassi, R. (2009). *Rapporto sugli effetti del terremoto aquilano del 6 aprile 2009.* Dipartimento della Protezione Civile Istituto Nazionale di Geofisica e Vulcanologia QUEST Team. http://emidius.mi.ingv.it/DBMI08/aquilano/query_eq/quest.pdf

Gruppo di Lavoro, M. S. (2008). *Indirizzi e criteri per la Microzonazione sismica*. Roma: Conferenza delle Regioni e delle Province autonome—Dipartimento della protezione civile.
Guidoboni, E., & Ferrari, G. (2000). Historical variables of seismic effects: Economic levels, demographic scales and building techniques. *Annali di Geofisica, 43*(4), 687–705.
Guidoboni, E. (2006). Libro di diversi terremoti, Edizione Nazionale delle Opere di Pirro Ligorio. Edizione critica, Introduzione e Apparato storico a cura di E. Guidoboni, Roma, Editore De Luca (p. 260).
Guidoboni E., & Comastri A. (2005). *Catalogue of earthquakes and tsunamis in the Mediterranean areas from the 11th century to the 15th century* (1037 pp). Bologna: INGV-SGA.
Guidoboni, E., Ferrari, G., Mariotti D., Comastri, A., Valensise G., & Tarabusi, G. (2007). *CFTI4Med: Catalogue of strong earthquakes in Italy from 461 BC to 2000 and in the Mediterranean area, from 760 BC to 1500, an advanced laboratory of historical seismology*. http://storing.ingv.it/cfti4med/
Guidoboni, E., Valensise, G. (2011). Il peso economico e sociale dei disastri sismici in Italia negli ultimi 150 anni (1861–2001) (www.buponline.com), INGV, Centro Euro-Mediterraneo di documentazione, Una storia Fuori dai manuali, BononiaUniversity Press.
Guidoboni, E. (2013). *Un'altra storia: terremoti e ricostruzioni in Italia*. Centro EEDIS—Centro Euro-Mediterraneo di documentazione, 6 Dicembre, Rovereto.
INGV—Istituto Nazionale di Geofisica e Vulcanologia. (2009). *Measurement and modeling of co-seismic deformation during the L'Aquila earthquake. Preliminary results*. Synthetic report of the activities of ASI-SIGRIS personnel and Earthquake Remote Sensing Group of the Remote Sensing Laboratory, National Earthquake Center, INGV, Rome.
Kanai K. & Tanaka T. (1961). *On Microtremor VIII*, Bull.Earthq. Res. Inst., Tokyo University, vol.39, pp.97–114.
Krahn J. (2004). *Dynamic modeling with quake/W*. Geo-Slope International Ltd.
Kuhlemeyer, L., & Lysmer, J. (1973). Finite element method accuracy for wave propagation problems. *Journal of Soil Mechanics and Foundations Division, 99*, 421–427.
Lanzo, G., Di Capua, G., Kayen, R. E., Kieffer, D. S., Button, E., Biscontin, G., et al. (2011). *The Aterno valley strong-motion array: seismic characterization and determination of subsoil model. Bulletin of Earthquake Engineering, 9*, 1855–1875.
Lanzo, G., Di Capua, G., Kayen, R. E., Kieffer, D. S., Button, E., Biscontin, G., et al. (2010). Seismological and geotechnical aspects of the Mw = 6.3 l'Aquila earthquake in central Italy on 6 April 2009. *International Journal of Geoengineering Case Histories, 1*(4), 206–339.
Marcellini A. (coord). (1999). *Rapporto Progetto UMSEG. GNDT*. http://it.scribd.com/doc/183624479/criteri-per-microzonazione-sismicaMS-VOLUME-2-pdf
Meletti C. and Valensise G. - Gruppo di Lavoro MPS (2004). *Redazione della mappa di pericolosità sismica prevista dall'Ordinanza PCM3274*. Zonazione sismogenetica ZS9–App. 2 al rapporto conclusivo.
Milana, G., Azzara, R., Bertrand, E., Bordoni, P., Cara, F., Cogliano, R., et al. (2011). The contribution of seismic data in microzonation studies for downtown L'Aquila. *Bulletin of Earthquake Engineering, 9*(3), 741–759.
Monaco, P., Totani, G., Barla, G., Cavallaro, A., Costanzo, A., D'Onofrio, A., et al. (2009). Geotechnical aspects of the L'Aquila earthquake. *Proceedings of 17th International Conference on Soil Mechanics and Geotechnical Engineering, Earthquake Geotechnical Engineering Satellite Conference*, 2–3 October 2009, Alexandria, Egypt.
Nakamura, Y. (1989). A method for dynamic characteristics estimation of subsurface using microtremor on the ground surface. *Railway Technical Research Institute, Quarterly Reports, 30*(1), 25–33.
Nogoshi M. & Igarashi T. (1971). On the amplitude characteristics of microtremor (part 2) (in Japanese with english abstract). *Journal of seismological Society of Japan, 24*, 26–40.
NTC08. (2008). *Norme Tecniche per le Costruzioni (DM 14 gennaio 2008)*. Gazzetta Ufficiale n. 29 del 4 febbraio 2008—Suppl. Ordinario no. 30.

Pace, B., Albarello, D., Boncio, P., Dolce, M., Galli, P., Messina, P., et al. (2011). Predicted ground motion after the L'Aquila 2009 earthquake (Italy, Mw 6.3): Input spectra for seismic microzoning. *Bulletin of Earthquake Engineering*, doi:10.1007/s10518-010-9238-y

Postpischl, D., Faccioli, E., & Barberi, F. (1985). *Progetto Finalizzato 'Geodinamica': Monografie Finali.* Quaderni de "La ricerca scientifica", Consiglio Nazionale delle Ricerche, Editore Consiglio Nazionale delle Ricerche.

Rainone, M. L., Torrese, P., Pizzica, F., Greco, P., & Signanini, P. (2010). *Measurement of seismic local effects and 1D numerical modeling at selected sites affected by the 2009 seismic sequence of L'Aquila".* Keystone, Colorado: SAGEEP.

Rainone, M. L., Signanini, P., Vessia, G., Greco, P., Di Benedetto, S. (2013). L'Aquila seismic event on 6th April 2009: site effects and critical points in microzonation activity within the Aterno Valley municipalities. *Proceedings of 7th International Conference on Case Histories in Geotechnical Engineering*, April 29–May 4, Chicago.

Salvi, S., & Working Group. (2009). *Risultati Preliminari SAR.* INGV report (in Italian).

Schnabel, P. B., Lysmer, J., & Seed, H. B. (1991). *SHAKE—A computer program for earthquake response analysis of horizontally layered sites.* Berkeley: University of California.

Vessia, G., Russo, S., & Lo Presti, D. (2011). A new proposal for the evaluation of the amplification coefficient due to valley effects in the simplified local seismic response analyses. *Italian Geotechnical Journal, 4*, 51–77.

Vessia, G., & Russo, S. (2013). Relevant features of the valley seismic response: The case study of Tuscan Northern Apennine sector. *Bulletin of Earthquake Engineering, 11*(5), 1633–1660.

Vessia, G., Rainone, M. L., Signanini, P., Greco, P., Di Benedetto, S., & Torrese, P. (2012). Integrated Geophysical and Geotechnical seismic techniques for dynamic site characterization at urbanized centers. *8th Alexander von Humboldt International Conference, Natural Disasters, Global Change, and the Preservation of World Heritage Sites*, Cusco, Peru, 12–16 November.

Working Group MS–AQ (2010). *Microzonazione sismica per la ricostruzione dell'area aquilana.* Regione Abruzzo—Dipartimento della Protezione Civile, L'Aquila 3 vol. and Cd-rom (in Italian).

Earthquake-Induced Reactivation of Landslides: Recent Advances and Future Perspectives

Salvatore Martino

1 Introduction

Earthquake-induced landslides are generally responsible for severe damages and losses as proved by records of last seven years which demonstrate that more than 50 % of the total losses due to landslides in the World are due to co-seismic slope failures (Petley 2012). Moreover, as reported by Bird and Bommer (2004), the greatest damage caused by earthquakes are often related to landslide events. Several historical earthquake-induced landslides demonstrated the severity of such events as they often involved areas which have been intensely damaged by the seismic shaking. This was, among others, the case of Las Colinas landslide, triggered by the January 13th 2001 M_w 7.6 El Salvador earthquake, which caused about 585 losses (Evans and Bent 2004). Earthquake-induced landslides can also trigger co-related phenomena, by a sort of "domino-effect", among which river damming and tsunamis. An extraordinary example of such an effect is reported in some historical chronics that testify the catastrophic scenario due to earthquake-induced Scilla rock avalanche triggered by the 6th February 1783 earthquake in Southern Italy (Bozzano et al. 2011b) which killed almost 1500 persons as it produced a 16 m height tsunami wave along the coastline where people found refuge after the mainshock occurred one day before (Mazzanti and Bozzano 2011).

The last published version of the database of earthquake-induced ground effects in Italy CEDIT by Martino et al. (2014; http://www.ceri.uniroma1.it/cn/gis.jsp), whose main peculiarity is to be constructed on the basis of several historical documents over a period of approximately one millennium from 1000 AD to present, demonstrates that landslides represent the most documented type of ground failures.

S. Martino (✉)
Department of Earth Sciences and Research Center for Geological Risks (CERI), "Sapienza" University of Rome, Rome, Italy
e-mail: salvatore.martino@uniroma1.it

S. D'Amico (ed.), *Earthquakes and Their Impact on Society*,
Springer Natural Hazards, DOI 10.1007/978-3-319-21753-6_10

These effects correspond to more than 40 % of the inventoried ones, which include ground-cracks, liquefactions and superficial faulting.

Earthquake-induced landslides should be distinguished in "first time slope failures" and "reactivated landslides" as very different approaches are requested for providing failure scenarios as well as for identifying the main predisposing conditions.

Earthquake-induced first time slope failures mainly involve jointed rock masses since they are often represented by disrupted landslides (Keefer 1984), i.e. rock-falls, topples or block-sliding. Such effects are expected to occur up to some tens of kilometers also in case of $4.0 \leq M_w \leq 5.0$ earthquakes (Keefer 1984; Rodriguez et al. 1999); several studies demonstrated the significant role of local seismic effects (i.e. due to topography, slope orientation, near field conditions) in triggering disrupted landslides (Sepulveda et al. 2005; Alfaro et al. 2012 and references therein). Nevertheless, a part for back-analysis approaches, several critical features still remain in forecasting first time rock failures as: (i) diffused geomechanical characterization should be available or suitable criteria for extrapolating them, (ii) very detailed geological features should be collected on generally impervious areas, (iii) high resolution digital elevation models (DEM) should be provided for steep slopes or cliffs.

Earthquake reactivated landslides more commonly involve coherent soils or debris; generally, it is possible to inventory these landslides as they are active or quiescent phenomena, which can be recognized by typical landforms or historical chronics that document their past activations. Although already existing landslide masses are a priori recognizable, a great effort is requested to evaluate how their stability conditions change due to earthquake occurrence as well as to quantify their co-seismic or post-seismic mobility in terms of expected displacements. This difficulty depends on the complex interactions between the seismic waves and the landslide mass that is conditioned by several features among which the slope geometry, the landslide mass properties, the physical characteristics of the seismic waves (Lenti and Martino 2012, 2013).

With respect to the curves of magnitude versus maximum epicentral distance that are available in literature, the most of the outliers (i.e. far field landslide events) inventoried in the "instrumental age" (i.e. with available accelerometric records) are represented by coherent landslides which involve stiff clays or soils (Delgado et al. 2011). The majority of the documented outlier cases regards the reactivation of large to very large landslides, whose volume is in the order of ten millions of cubic meters. Such a high landslide intensity, combined with the unexpected far-field occurrence, is particularly relevant for risk management in high seismicity areas.

This paper is focused on earthquake-induced landslide reactivation and report a state-of-art and some recent advances in slope stability analysis as well as in co-seismic displacement evaluation for pre-existing landslide masses. As discussed in the following, most of the criticism in analyzing slope stability under seismic action regards the dynamic inputs considered for computing the safety factor

(SF) and the uncertainness in assuming mechanical properties necessary for evaluating the available soil strength. On the other hand, the main criticism in computing co-seismic landslide displacements depends on the rheological assumptions (i.e. rigid vs. deformable soil masses) as well as on the complexity of the physical interactions between seismic waves and landslide masses (i.e. 1D or 2D seismic amplification, incidence angle of seismic waves related to the slope geometry, seismic wave polarization within landslide mass).

2 Slope Stability Under Seismic Action

Slope stability analyses under seismic action are traditionally performed by pseudostatic approaches, assuming that a constant equivalent seismic action, expressed by the pseudostatic seismic coefficient (k), is applied to the landslide mass in addition to the gravity force. The safety factor (SF) is computed by the ratio between the available strength along the sliding surface and the acting forces. The most general formulation for the slope stability analysis assumes that both the horizontal (k_x) and vertical (k_y) components of the equivalent seismic action can be considered; nevertheless, the force equilibrium analysis demonstrates that the horizontal component of the down-slope directed pseudostatic force is responsible for the reduction of the available strength as well as for the increasing of the acting force along the sliding surface. On the contrary, the vertical component of the same force causes an increase of the force acting along the sliding surface but it also contributes to increase the available strength. Moreover, the shaking maps or the seismic hazard evaluation generally only provide the expected values of the horizontal peak ground acceleration (PGA). For all these reasons, the vertical component of the pseudostatic force is usually neglected in pseudostatic slope stability analysis. The most critical assumption at the basis of this approach is that the same equivalent seismic action is applied within the landslide mass; such an hypothesis is theoretically suitable in the case of landslide length lower or almost equal to the half-length of the seismic waves (for simplicity considered as a monofrequencial sinusoid). As discussed by Hutchinson (1987), in this last condition the landslide mass should be moved along the slope toward the same direction, i.e. the acceleration values remain positive or negative within the landslide mass. For the same criterion, the absolute value of the equivalent seismic action should not significantly change within the landslide mass only in the case of seismic waves whose half-length ($\lambda/2 = 1/2 \cdot V_s \cdot T_m$, where V_s is the shear wave velocity and T_m is the sine wave period) is much more higher that the landslide one. Due to the low V_s of the earth-slide masses this condition is generally satisfied for high to very high T_m.

On the other hand, the conventional pseudostatic approach is properly suitable for evaluating the SF of already existing small landslides or of first time rock-slides. In this last case, the assumptions at the basis of the pseudostatic approach are admissible due to the high V_s value of the rocks, that generally leads to significantly high λ values also for low T_m, as well as to the rigid behavior reliable for block sliding mechanisms of entire blocks, i.e. for which internal strains are negligible during the seismic shaking.

2.1 Unconventional Pseudostatic Analysis

An alternative to the pseudostatic solution for slope stability analysis under seismic action is provided by an unconventional pseudostatic analysis.

In this case, the restrictions imposed by the conventional pseudostatic analysis can be limited by considering several distributions of the horizontal pseudostatic coefficient k_x that can be obtained by varying the horizontal acceleration values within the landslide mass ($k_x(x)$) according to the sine wave. At this aim, the landslide mass is portioned in slices (i.e. delimited by vertical boundaries) and different k_x values are applied to each slice based on the spatial distribution of the horizontal acceleration values associated to the sine wave. To take into account the variability of the seismic actions, sine waves with different periods ($T_m = 1/f_m$ where f_m is the sine wave frequency) and phases (Φ, varying in the range 0–360°) are considered; the different Φ values are attributed by changing the sine wave position from the top of the landslide mass (crown area). Because the length (λ) of the seismic wave changes with the value of T_m, the ($k_x(x)$) distributions are characterized by decreasing spatial variation with increasing T_m. For an infinite T_m value (i.e. for a null frequency of the seismic wave) the assumed conditions are theoretically identical to the pseudostatic ones (k_x is equal for all the considered slices). Based on such an approach the SF value distribution can be reported as a function of f_m only ($SF_{(fm)}$) or of both f_m and Φ ($SF_{(fm,\Phi)}$).

The unconventional analyses was experienced in the case study of the Diezma landslide, in Spain (Delgado et al. 2015). The Diezma landslide is located 25 km from the city of Granada (Fig. 1); although the slope had repeatedly suffered small scale stability problems since the construction of the A-92 highway, a larger failure occurred on 18 March 2001 causing several damages (Rodríguez-Peces et al. 2011; Delgado et al. 2015). The landslide has a main translational mechanism and involves an estimated volume of 1.2 Mm^3, constituted of a chaotic deposit of silt and clay with heterometric blocks of limestones (Azañón et al. 2010).

Despite the 18 million Euros spent since 1999 on geotechnical investigations and stabilization solutions, the numerous reactivations that occurred untill 2013 demonstrate the persistent activity of the landslide.

Sine waves with maximum accelerations of 0.14 g were considered according to the available hazard maps and to the expected PGA for a return time of 475 years (Benito et al. 2010); moreover, the sine waves are characterized by values of T_m between 2 and 0.33 s (i.e., corresponding to seismic wavelengths within the Diezma landslide mass from 100 to 600 m).

Different possible phases of the sine wave were assumed for each considered value of PGA and T_m to obtain several spatial distributions of k_x within the

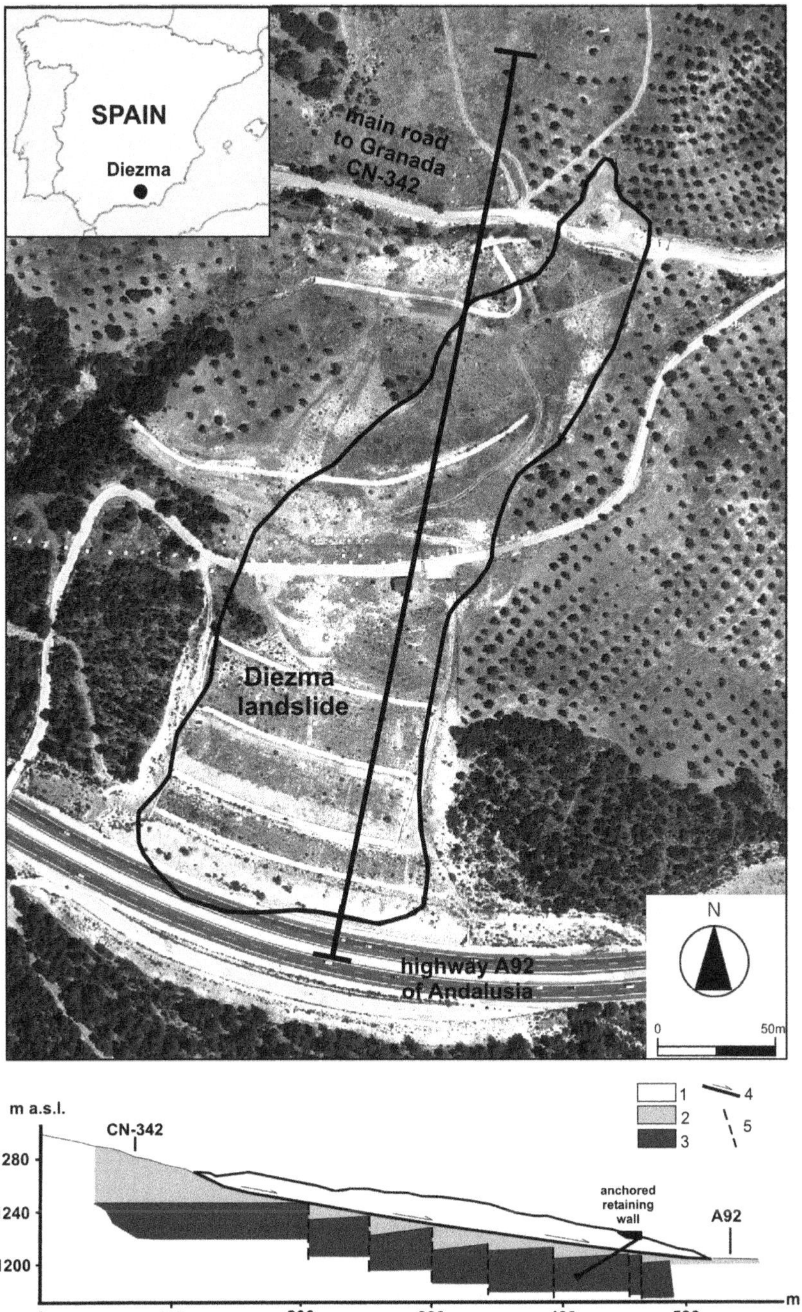

Fig. 1 Diezma landslide (Spain): GoogleEarth satellite view and geological cross section along the mapped trace. *1* landslide debris; *2* silty-clay with heterometric blocks of limestones of the Numidoide formation (Aquitanian–Burdigalian); *3* silts and clays of the Maláguide domain formations (Devonian–Triassic); *4* sliding surface; *5* fault

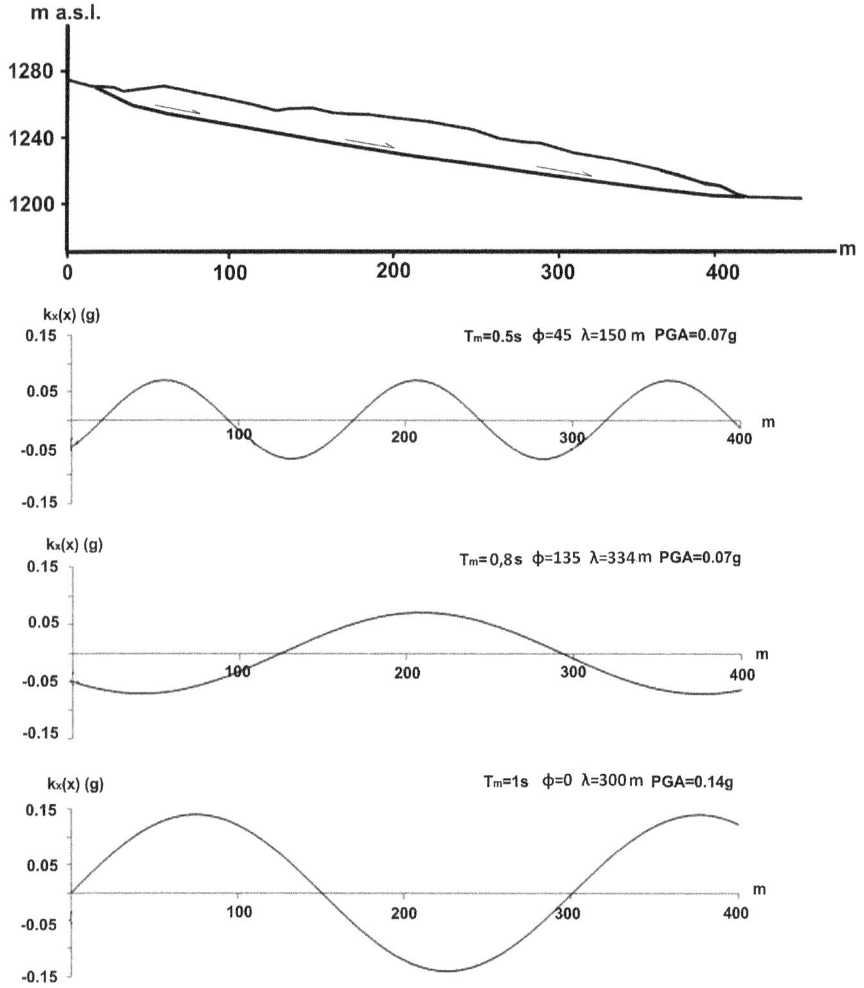

Fig. 2 $k_x(x)$ distributions within the Diezma landslide mass for different sine waves. The values of period (T_m), phase (Φ), wavelength (λ) and amplitude (PGA) of the sine waves are also reported

landslide mass (Fig. 2). Different distributions of water pressure were also considered by varying the Bishop coefficient r_u (Bishop 1955) from 0 to 0.3 (i.e. up to the maximum admissible value for the considered landslide).

$SF_{(fm,\Phi)}$ versus r_u distributions can be computed for each PGA value; moreover, an interesting distribution can be obtained for a fixed r_u by considering the minimum SF values ($SF(f_m)_{min}$) computed at a certain f_m for all the considered Φ; such a distribution highlights the critical frequency value (f_{m_cr}) which corresponds to a negligible variation of SF for a fixed r_u.

The results of this analysis (Fig. 3) show that the $SF_{(fm)min}$ values of the Diezma landslide significantly increase with increasing f_m from 0 (i.e., conventional pseudostatic conditions) to 1 Hz. In particular, for the considered PGA of 0.14 g, the f_{m_cr} corresponds to about 1 Hz while the critical stability conditions ($SF_{(fm)} = 1$) correspond to f_m values between 0.5 and 1.0 Hz.

The $SF_{(fm,\Phi)}$ distributions (Fig. 3) were obtained for the same range of admissible values of r_u (i.e. 0–0.3). They demonstrate that the stability of the Diezma landslide is strongly influenced by both Φ and f_m of the sine wave as: the $SF_{(fm,\Phi)}$ values decrease below 1 Hz; the minimum $SF_{(fm,\Phi)}$ values result for Φ varying in the range 0–90°; the maximum $SF_{(fm,\Phi)}$ values correspond to the Φ range 180°–270° (i.e. supplementary with respect to the previous one).

The computed $SF_{(fm,\Phi)}$ distributions show that the stability of the Diezma landslide slope depend on the Φ values more than on the f_m of the sine wave. Moreover, the obtained $SF_{(fm,\Phi)}$ distributions indicate that the probability of landslide failure for the same expected PGA and for the same f_m and r_u value is equal to

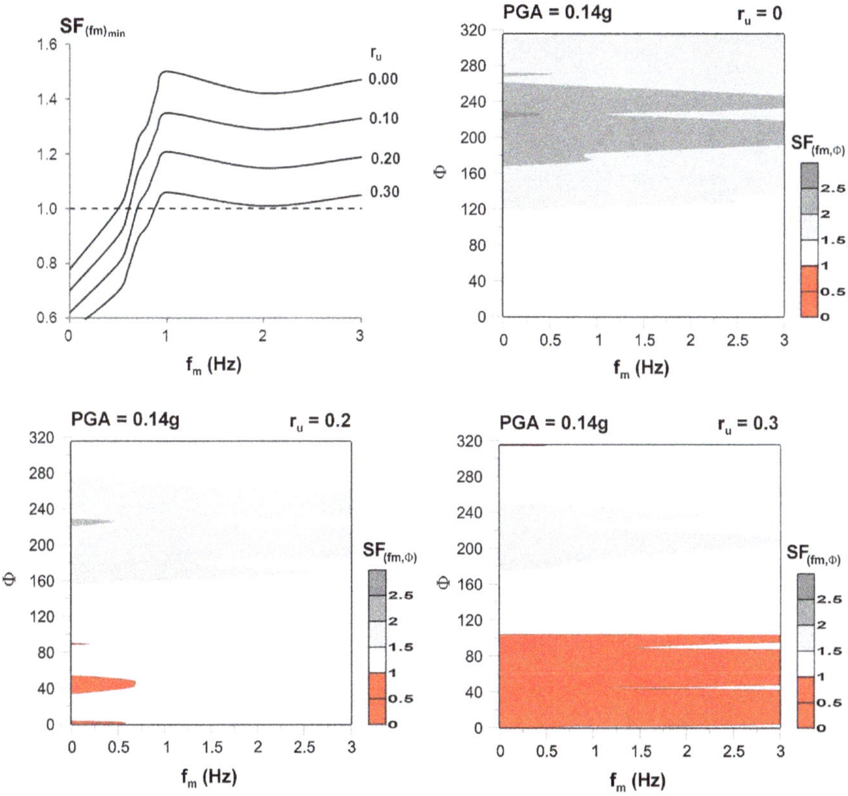

Fig. 3 Diezma landslide (Spain): SF $(f_m)_{min}$ and SF (Φ, f_m) distributions obtained from the unconventional pseudostatic analysis performed by considering a PGA value of 0.14 g and r_u values varying from 0 to 0.3

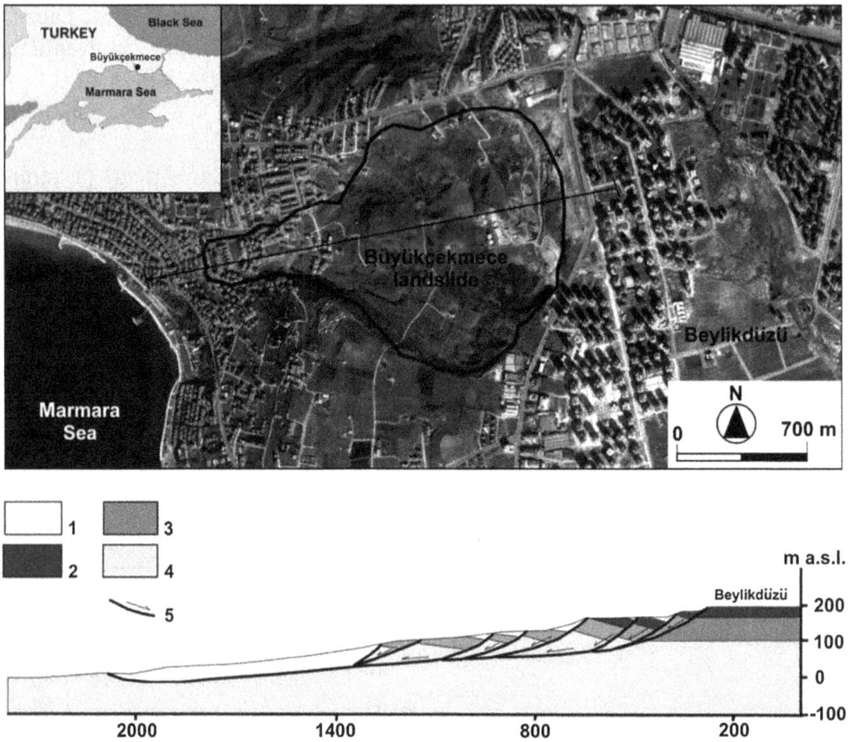

Fig. 4 Büyükçekmece landslide (Turkey): GoogleEarth satellite view and geological cross section along the mapped trace. *1* landslide debris; *2* calcarenites, conglomerates and sands of the Istanbul Formation (Upper Oligocene–Lower Miocene); *3* silty-clays, sands and tuffs of the Danisment Formation (Upper Oligocene); *4* clayey silts and clays of the Danisment Formation (Upper Oligocene); *5* sliding surface

almost 25 % by considering the ratio between the number of critical f_m (Φ) combinations (i.e. counted in the only range of Φ values that are suitable for the landslide reactivation) and all the considered ones.

The unconventional pseudostatic approach was also applied to the Büyükçekmece landslide case study, in Turkey (Fig. 4) as part of the European MARsite Project (Work Package 6—http://marsite.eu/). This landslide is located about 35 km SW from the center of Istanbul and about 15 km far from the Marmara Sea segment of the north-Anatolian seismogenetic fault (Kalkan et al. 2009).

The Büyükçekmece landslide involves sandy-silty clays ascribable to the Upper Oligocene—Lower Miocene succession (Dalgiç 2004; Duman et al. 2006) in a densely urbanized area that significantly extended after the M_w 7.4, 1999 Izmit earthquake. The landslide is characterized by a rototranslational mechanism (Cruden and Varnes 1996) and it is about 2.2 km of length and 1 km of width.

In this case, the $SF_{(fm,\Phi)}$ distributions resulting from the unconventional pseudostatic analysis (Fig. 5) show that, for an expected PGA of 0.35 g

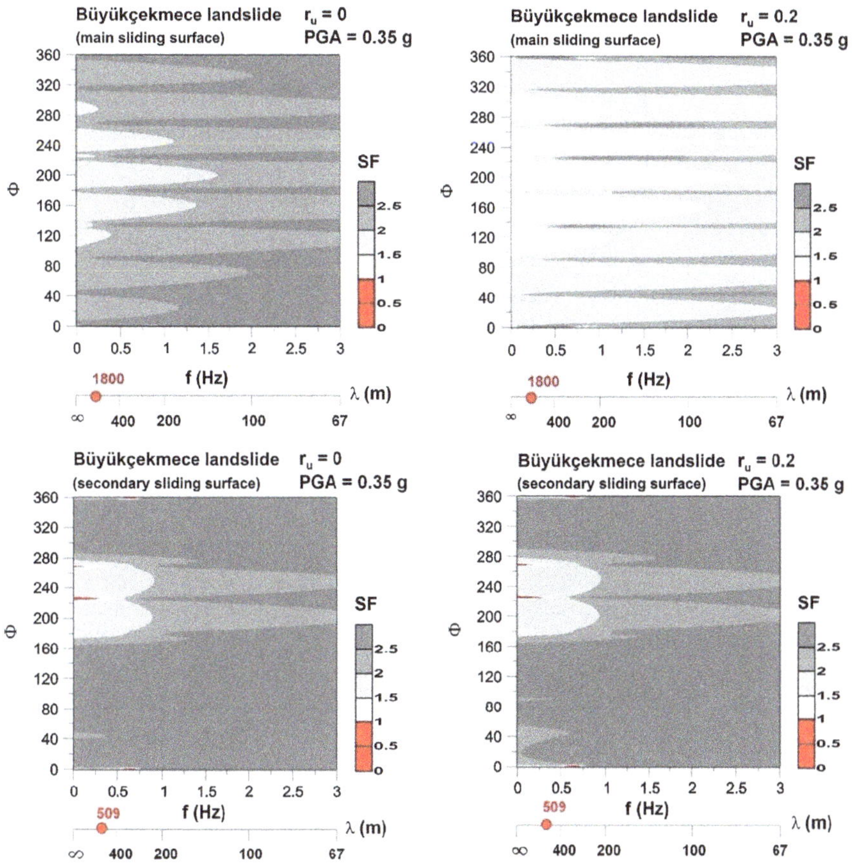

Fig. 5 Büyükçekmece landslide (Turkey): SF(Φ, f_m) distributions obtained from the unconventional pseudostatic analysis for a PGA of 0.35 g and r_u values varying from 0 to 0.2. The results are distinguished for the main sliding surface (*top*) and for one of the secondary sliding surfaces (*bottom*)

(corresponding to a return time of 475 years, according to Erdik et al. 1999 and Atakan et al. 2002) and for a severe distribution of pore water pressure along the main sliding surface ($r_u = 0.3$), the landslide stability along the main sliding surface becomes critical for f_m values lower to 1 Hz in case of 90° ≤ Φ ≤ 300° and for f_m values lower to 2 Hz in case of 140° ≤ Φ ≤ 260°. Based on the obtained results, the probability of landslide reactivation for the same PGA and r_u value is strongly influenced by the T_m as it varies from 33 % in case of 1 Hz ≤ f_m ≤ 2 Hz up to 58 % in case of f_m < 1 Hz. If one of the secondary sliding surface of the landslide is considered (Fig. 4), the $SF_{(fm,\Phi)}$ distributions show that the stability conditions becomes critical (1 ≤ SF < 1.5) for f_m values lower to 1 Hz in case of 180° ≤ Φ ≤ 280°, i.e. in this case the probability of landslide reactivation not exceed 28 % for the same PGA and r_u value.

3 Earthquake-Induced Displacements of Pre-existing Landslides

Earthquake-induced displacements can be referred to three different stages that are related to the seismic shaking (Ambraseys and Srbulov 1995): (1) co-seismic displacements which occur just during the seismic action (short duration effect) in undrained conditions; (2) post-seismic displacements, which follow immediately the seismic shaking (i.e. still undrained conditions) and depend on the disequilibrium conditions caused by the earthquake within the landslide mass; (3) post-seismic consolidation and creep displacements having a long to very long time duration after the seismic shaking, i.e. from undrained to drained conditions.

Limit equilibrium analyses for slope stability of seismically loaded slopes (Loukidis et al. 2003) assume a rigid landslide mass behavior and do not provide any information about displacements but only indicate if they can occur in case that seismic forces exceed the available strength along the sliding surface. A pseudostatic critical threshold can be computed that corresponds to the pseudostatic yielding coefficient (k_y). This threshold can be regarded as a trigger value for a rigid slide movement in case of co-seismic displacements (Jibson 1993, 2011). On the contrary, co-seismic and post-seismic displacements can be quantified via dynamic stress-strain numerical modeling by finite difference (FDM) or finite element (FEM) methods. These models avoid to assume a rigid movement of the sliding mass but it is necessary to define a rheological behavior as well as a dynamic input (i.e. accelerometric timehistory) that forces the numerical model. In this regard, the variability of the post-seismic displacements modeled via numerical approaches mainly depend on: (i) assumed rheology; (ii) attributed parameter values; (iii) reliability of the applied dynamic seismic action.

The assumed rheology has a significant influence on the numerical solutions in case of nonlinearity due to permanent displacements that are cumulated during and after the seismic action, as they depend on the effective stress paths of the involved soils (Sassa 1996). Uncertainty in parameter values can significantly affect the estimation of both co-seismic and post-seismic displacements as it is reflected in several studies that assume "order-of-magnitude" for predicting earthquake-induced landslide displacements or provide indexes for their evaluation (Strenk and Wartman 2011 and references therein). About the dynamic seismic action applied to the numerical models, the suitability and representativeness of the dynamic input play a fundamental role in the numerical computation of the earthquake-induced displacements (Murphy et al. 2002). The frequency resolution of the selected dynamic inputs applied in FDM and FEM models is related to both the grid geometry and the dynamic material properties (Kuhlemeyer and Lysmer 1973); moreover, the selected inputs should be representative for the seismic actions that are expected in the study area.

To perform parametric studies by numerical models, several inputs should be selected that respect the aforementioned properties; this explains the necessity to derive dynamic equivalent signals that are suitable for the numerical modeling and, at the same time, are characterized by a short-time duration to make it possible lots of modeling to be performed. In this regard, the recently proposed LEMA_DES (Leveled Energy Multifrequential Analysis for Deriving Equivalent Seismic inputs) approach (Lenti and Martino 2010) provides a procedure to derive dynamic signals that are equivalent to reference accelerograms in terms of frequency content, energy and PGA, i.e. the main characteristics which control the interactions between seismic waves and soils.

3.1 From the Rigid to the Flexible Sliding-Block Methods

Co-seismic displacements were traditionally computed from several decades by applying the Newmark's method (1965) which consider the existing landslide mass like a rigid sliding block whose displacements are cumulated during the seismic action only in case of PGA exceeding the k_y threshold. A double integration numerical solution provides the cumulative co-seismic displacements from the accelerometric timehistory of the earthquake. Due to the so resulting dependence of the earthquake-induced displacements from both mechanical properties and seismic input characteristics, several solutions were proposed to avoid using specific accelerograms for predicting earthquake-induced landslide displacements that are based on lots of numerical computations performed by using natural timehistories and by varying both the assumed k_y and the PGA of the seismic input (Jibson et al. 1998; Romeo 2000; Hsieh and Lee 2011; Hwang and Chen 2013). The use of natural timehistories is avoided by semi-empirical co-relations among PGA, k_y and earthquake-induced landslide displacements (Fig. 6a).

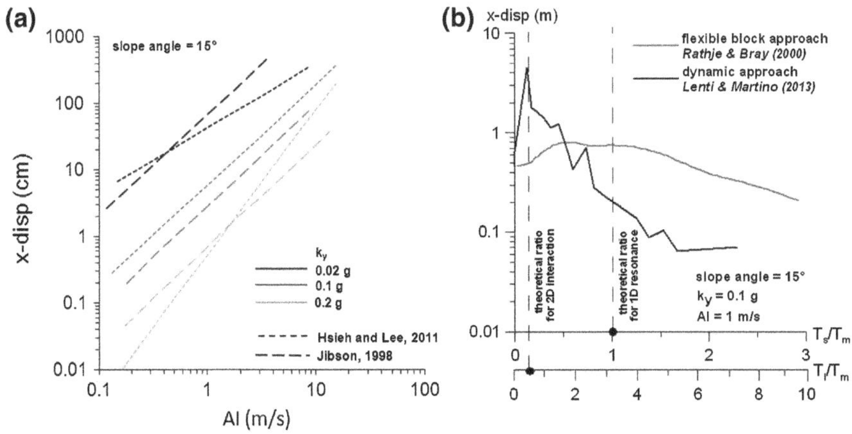

Fig. 6 Earthquake-induced landslide displacements computed by: **a** empirical co-relations based on the Newmark's method; **b** flexible block and dynamic approaches

Although the simplicity of the Newmark's approach and of its assumptions, several limits were pointed out on the representativeness of this method in case of not-rigid masses (i.e. in the most cases of earth-slides) due to the internal deformations produced during the seismic shaking which are responsible for amplification of the seismic motion. To take into account this feature, according to a 1D resonant column model, decoupled or coupled solutions were adopted (Makdisi and Seed 1978; Rathje and Bray 2000) that consider the simultaneous solution of the dynamic and of the sliding analysis. Such solutions were also implemented to take into account fully non-linear soil properties (Rathje and Bray 2000) as well as to consider the probabilistic variation of the seismic input properties (Bray and Travasarou 2007; Rathje and Antonakos 2010). As it results from these approaches, the maximum expected earthquake-induced displacement correspond to a resonant period due to the landslide mass thickness (T_s) equal to the characteristic period (Rathje et al. 1998) of the seismic input (T_m), that depends on the frequency content expressed by the Fast Fourier Transform (FFT). A characteristic period ratio T_s/T_m was introduced to describe such conditions, i.e. maximum displacement are theoretically expected for a T_s/T_m ratio equal to 1 (Fig. 6b).

The Newmark's method was also experienced to back-analyze the reactivation of the Calitri landslide (Southern Italy) which is about 700 m of length and 400 m of width, with an estimated volume of 23 Mm^3 and a sliding surface at about 90 m below the ground level (Hutchinson and Del Prete 1985 and references therein) (Fig. 7). The landslide involves stiff clays and sands and it was reactivated several times since 1964; the last historical reactivation occurred about 3 h after the mainshock of the 23rd November 1980 Irpinia earthquake.

This back-analysis demonstrated (Martino and Scarascia Mugnozza 2005 and references therein) that the computed earthquake-induced displacements by Newmark method are of about 10 cm and strongly underestimate the actually observed ones. Indeed, in correspondence to the landslide crown area, the observed displacements were of about 4 m and caused the complete destruction of the old Calitri village, located at the hill top.

The underestimation related to the computation by the Newmark method can be explained by the not suitable assumptions at the basis of the sliding block methods for evaluating co-seismic displacements in case of large earth-slide as well as by the neglected post-seismic displacements that were really observed 3 h after the triggering earthquake.

3.2 *Dynamic Approaches*

Stress-strain numerical models can be performed by FEM and FDM codes for dynamic analyses focused on evaluating landslide mass stability under seismic action and predicting both the co-seismic and the post-seismic earthquake-induced displacements.

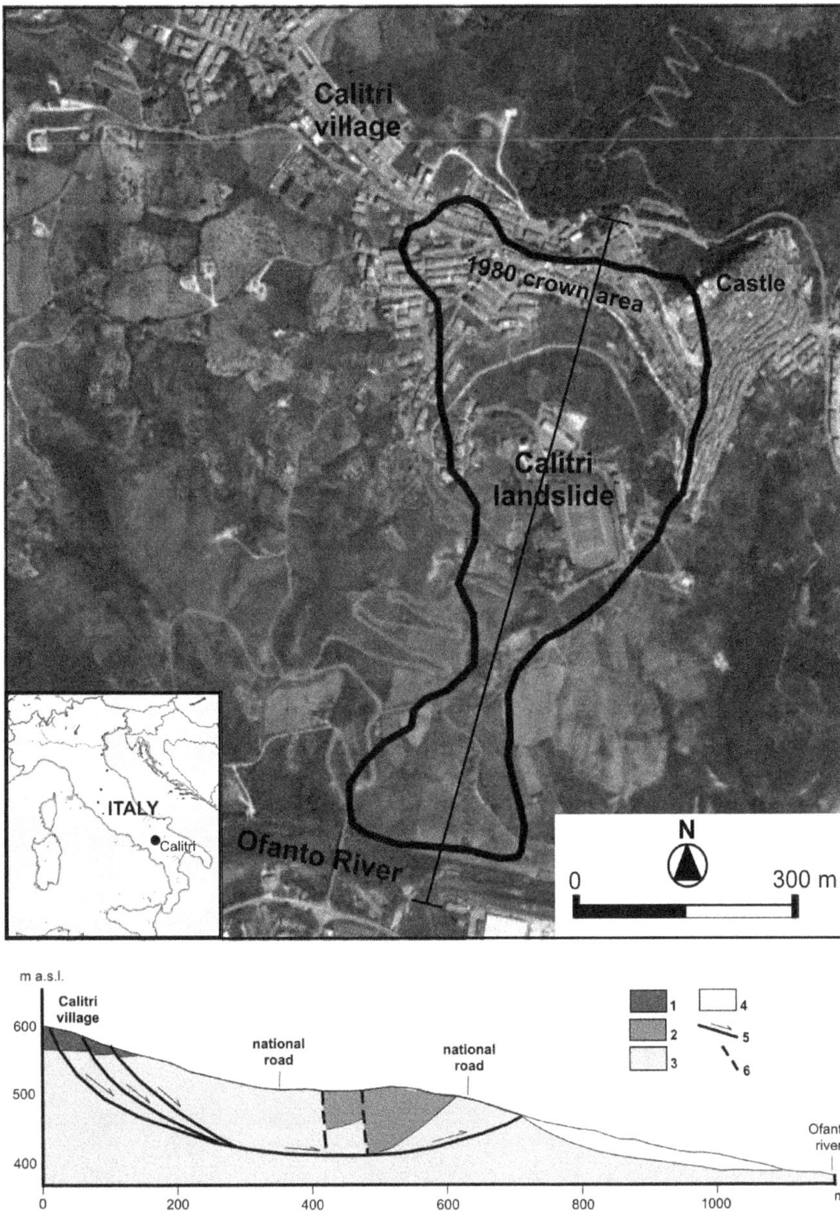

Fig. 7 Calitri landslide (Italy): GoogleEarth satellite view and geological cross section along the mapped trace. *1* sands and sandstones (Upper Pliocene–Lower Pleistocene); *2* sands (Pleistocene); *3* stiff clays (Upper Pliocene–Lower Pleistocene); *4* earth-flow debris; *5* sliding surface; *6* fault

These analyses allow to assume that deformations can occur within the landslide mass and to simulate more complex interactions between seismic waves and slope by considering 2D geometrical configurations. Such interactions have recently been analyzed in several case studies (Martino and Scarascia Mugnozza 2005; Sepulveda et al. 2005; Del Gaudio and Wasowsky 2007; Bourdeau and Havenith 2008; Danneels et al. 2008) whose results demonstrate that both landslide mechanisms and triggering conditions depend on seismic input properties such as energy, frequency content, directivity, and PGA. In particular, some case studies highlight the role of a "self-excitation" process (Bozzano et al. 2008b, c) and justify that seismic amplification effects can be responsible for triggering far-field and pre-existing large landslides, which represent outliers with respect to the proposed predictive curves (Keefer 1984; Rodriguez et al. 1999; Delgado et al. 2011).

Recent studies were also focused on the effects of step-like slope geometries on the local seismic response (Bouckovalas and Papadimitriou 2005; Lenti and Martino 2012, 2013) pointing out that earthquake-induced displacements of landslide masses significantly depend on the slope angle and on the landslide mechanism. Based on these studies not negligible differences exist among the Newmark's method displacements and the ones computed by dynamic stress-strain numerical models; more in particular, higher is the slope angle greater are the resulting differences for the same seismic input and for the same mechanical properties of the landslide mass (Fig. 8).

According to these results the role of the 2D geometry (i.e. slope angle and landslide mass length) could not be neglected for a better prediction of earthquake-induced landslide displacements. More in particular, as exemplified in

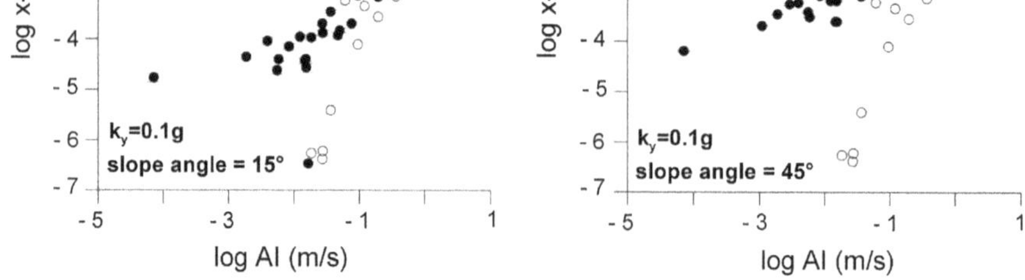

Fig. 8 Horizontal displacements (x-disp) versus Arias intensity (AI) obtained for a $k_y = 0.1$ g in the case of a 15° and 45° slope angle respectively. The values resulting from the dynamic numerical modeling (from Lenti and Martino 2013) are compared with those computed by the Newmark's method (according to Rathje and Antonakos 2010). The error bars, that show the values of standard deviation computed by averaging the x-disp predicted by numerical modeling, are negligible

Fig. 6b, the landslide mass mobility is theoretically favored by a characteristic period of the seismic input (T_m) which is double with respect to the period (T_l) associated to the length of the landslide mass itself (i.e. for a characteristic period ratio T_l/T_m equal to 0.5).

Lots of numerical models performed by using several dynamic inputs and different slope geometries (Lenti and Martino 2013) demonstrated that the values of expected earthquake-induced landslide displacements depend on a combination of 1D and 2D effects, these last ones related to more complex interactions between landslide mass and slope geometry (Fig. 6b). More in particular, for increasing energy of the input the 2D effects become more significant and the expected displacements are mainly related to the T_l/T_m ratio. On the other hand, the 1D resonance of the landslide mass is much more evident in case of more gentle slopes with respect to steeper ones.

Although a greater complexity can be taken into account by using dynamic stress-strain numerical models, the uncertainty related to the material rheology, to the associated mechanical parameters and to the assumed boundary condition for the dynamic modeling (Paraskevopoulos et al. 2010 and references therein) generally represent the main reasons for criticisms about these approaches.

3.2.1 The Giant Vokhchaberd Landslide (Armenia): A Deterministic Dynamic Analysis Using a Sine Wave Input

A numerical evaluation of the earthquake-induced displacements via stress-strain FDM numerical models was obtained (Martino et al. 2007) in the case of the giant Vokhchaberd landslide (Kotayk Province, Armenia) which involves the homonymous village (20 km east of Yerevan city) hosting about 1000 inhabitants (Fig. 9). The landslide bedrock is composed of Lower Oligocene clays, sandstones and tuff stones (Shorakhpiur Unit) overlaid by tuff breccias, conglomerates and clays of Lower Pliocene (Vokhchaberd Unit). The landslide, with an estimated volume of 450 Mm^3, is characterized by a rototranslational mechanism and it is about 3 km of length and 1.5 km of width. Historical and paleo-seismic studies (Balassanian et al. 1999a, b) show that M_w 7.0 earthquakes can occur along the Garni fault, very close to the Vokhchabertd landslide area. After the 1988 M_w 7.0 Spitak earthquake a landslide reactivation was testified by the village inhabitants; in the same period no significant change in the mean annual precipitation rate has been recorded (about 400 mm/year).

A dynamic numerical analysis was performed by a FDM code, taking into account the frequency content of earthquakes ($M_w > 5.0$) with epicentral distances within 100 km, recorded by a local USGS (Unites States Geological Survey) array (Fig. 10).

More in particular, a sine wave input was applied to the numerical models characterized by a 1.5 Hz frequency, a 0.1 g amplitude (derived as 65 % of the mean of the recorded PGA values) and a duration of 2.66 s (corresponding to 4 equivalent cycles according to the approach proposed by Seed 1979).

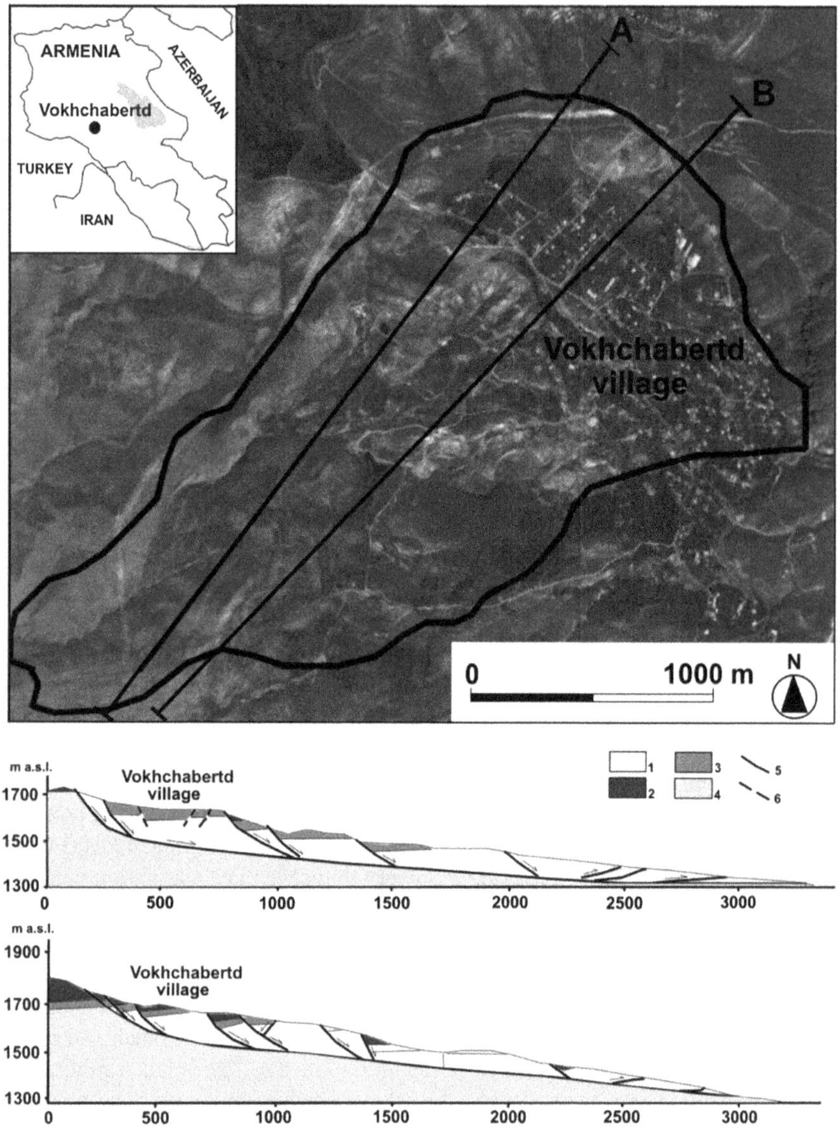

Fig. 9 Vokhchabertd landslide (Armenia): GoogleEarth satellite view and geological cross section. *1* landslide debris; *2* clays, sandstone and tuffstones (Shorakhpiur Unit, Lower Oligocene); *3* conglomerates and clays (Vokhchabertd Unit, Upper Oligocene–Lower Pliocene); *4* tuff breccias (Vokhchabertd Unit); *5* sliding surface; *6* fault

A nonlinear incremental analysis was performed to take into account the stiffness decay for increasing shear strain values. The obtained results provided earthquake-induced displacements up to 60 cm and pointed out a differential

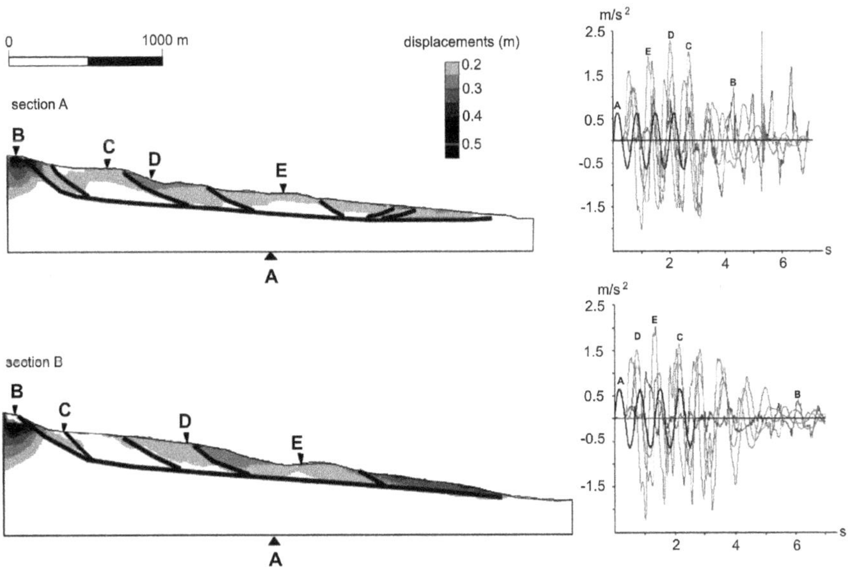

Fig. 10 Dynamic stress-strain FDM numerical modeling of the Vokhchabertd landslide: displacement distribution, sine wave input applied at the model base (*A*) and acceleration outputs at the model surface (*B*, *C*, *D*, *E*). The sliding surfaces of the landslide reported in Fig. 9 are also shown

kinematic behavior between the right and the left sectors of the landslide mass (see the geological cross sections A and B of Fig. 9 respectively).

These results are consistent with the observed landforms which show an almost uniform distribution of the displacements in the northern portion of the landslide mass with respect to the southern one. As the numerical models reach a new post-seismic equilibrium, a recurrent activity can be assumed for the Vokhchabertd landslide, that is consistent with a cumulative displacement field resulting after several earthquake occurrences.

A Newmark's analysis was previously performed in order to evaluate the possible earthquake-induced displacement for a PGA value of 0.55 g (corresponding to a maximum expected magnitude M_w 7.2); the obtained results indicate co-seismic displacements up to 63 cm. Since this value is the same resulting by the dynamic analysis performed assuming a 0.1 g PGA value for the sine wave input, also in this case the Newmark's method underestimates the expected earthquake-induced displacements.

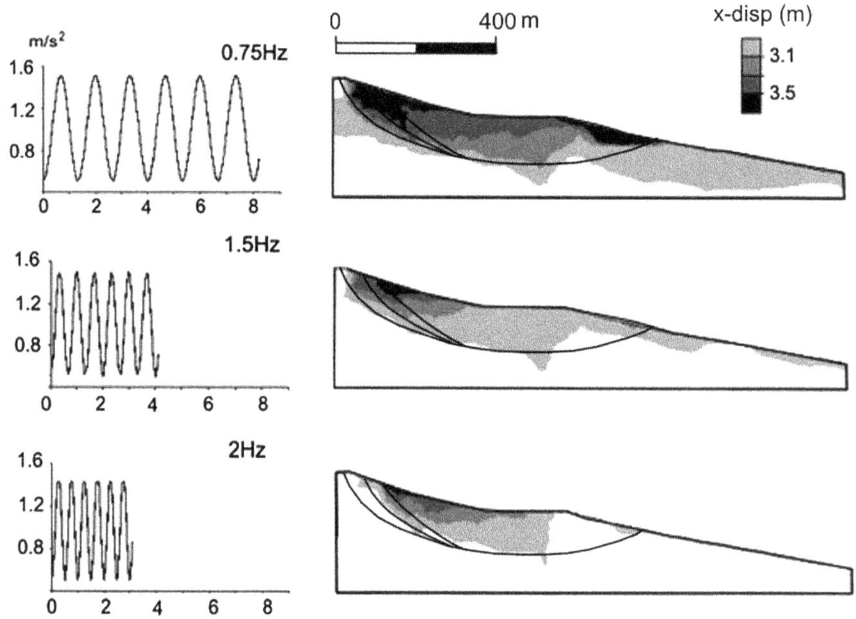

Fig. 11 Dynamic stress-strain FDM numerical modeling of the Calitri landslide reactivation occurred on 23rd November 1980: sine wave inputs applied to the model base (*left*) and corresponding horizontal displacement distributions (*right*). The sliding surfaces of Fig. 7 are also reported on the sections (*right*)

3.2.2 The Calitri Landslide (Italy): A Dynamic Sensitivity Analysis on the Frequency of Sine Wave Inputs

In the case of the Calitri landslide (previously introduced in the Sect. 3.1) a FDM dynamic numerical modeling was carried out (Martino and Scarascia Mugnozza 2005) to analyze the role of the characteristic frequency of the seismic input on the earthquake-induced landslide displacements by experiencing the use of different sine wave inputs.

Based on the 23rd November 1980, Irpinia earthquake record at the accelerometric station of Calitri (CTR—Italian Accelerometric Network), three sine wave inputs were derived, characterized by a PGA value of 0.1 g, 6 equivalent dynamic cycles (Seed 1979) and frequencies equal to 0.75–1.5–2 Hz, respectively (Fig. 11). The dynamic modeling pointed out a significant role of the sine wave frequency in the landslide mass movement. As it results from the numerical modeling, for decreasing frequency: (i) the value of both vertical and horizontal displacements increases up to the meter-order ones; (ii) the reactivation involves deeper portions of the landslide mass; (iii) the rototranslational mechanism of the landslide is much more evident.

3.2.3 The Cerda Landslide (Italy): A Dynamic Back-Analysis Using a Multifrequential Dynamic Input

The Cerda landslide (Sicily, Italy) was reactivated by the 6th September 2002 Palermo earthquake (M_w 5.9), about 50 km far from the epicenter (Bozzano et al. 2011a) (Fig. 12).

The landslide caused damage to farm houses, roads and aqueducts, close to the homonymous village, and involved about 40 Mm^3 of clay-shales; the first ground cracks due to the landslide movement formed about 30 min after the main shock.

A stress–strain dynamic numerical modeling, was performed by a FDM code (Bozzano et al. 2011a) to experience a back-analysis of the last earthquake-induced reactivation by using a multifrequential dynamic input (Fig. 13). Since accelerometric records of the triggering event were not available, a dynamic equivalent input was used for the numerical modeling that was derived according to the

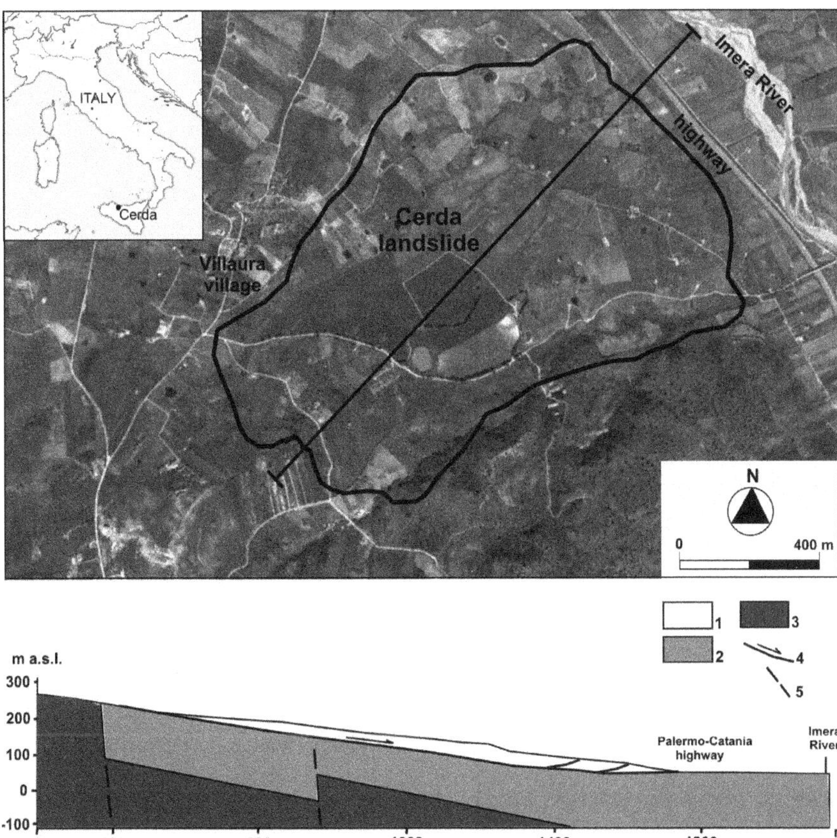

Fig. 12 Cerda landslide (Italy): GoogleEarth satellite view and geological cross section. *1* landslide debris; *2* scaly clays (Argille Varicolori Formation, Cretaceous–Eocene); *3* marls and calcarenites (Roccapalumba Formation, Triassic); *4* sliding surface; *5* fault

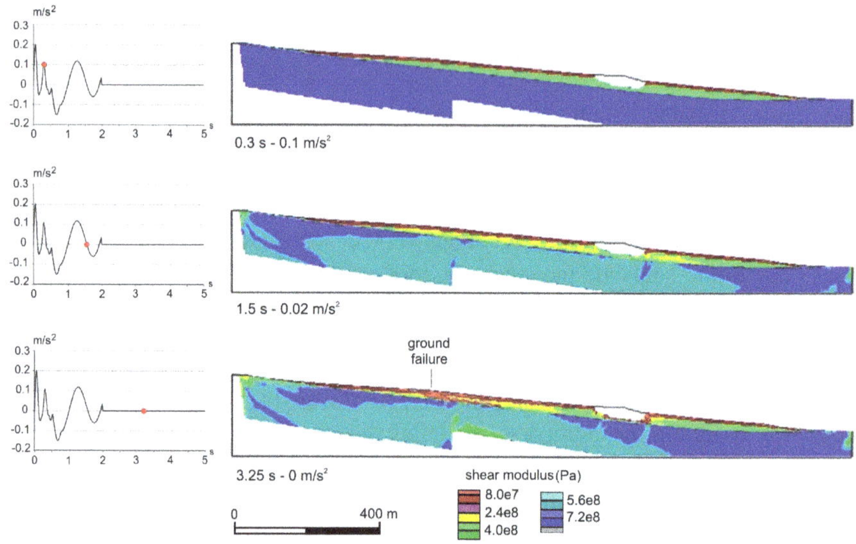

Fig. 13 Dynamic stress-strain FDM numerical modeling of the Cerda landslide reactivation on 6th September 2002: the resulting shear modulus distributions points out the induced decay due to the shear strain. The LEMA_DES derived equivalent input applied to the model base is also shown on the *left*

aforementioned LEMA_DES approach (Lenti and Martino 2010) by selecting accelerometric timehistories from the European Strong Motion Database (ESMD, Ambraseys et al. 2002) representative for the local ground shaking and with a PGA up to 0.02 g, i.e. the maximum expected in 475 years. The numerical modeling results support the idea that the combination of local geological setting of the seismic bedrock as well as the earthquake frequency content played a fundamental role in the landslide reactivation.

Before performing the dynamic numerical analysis, a k_y of 0.17 g was computed and an equivalent PGA of 0.006 g, acting within the Cerda landslide mass, was obtained by an unconventional pseudostatic analysis, that was performed by assuming a PGA of about 0.01 g for the Palermo earthquake. As this value is significantly lower than the computed k_y, the back-analysis of the Cerda landslide reactivation cannot be verified by performing a pseudostatic stability analysis. Moreover, co-seismic displacements cannot be computed by the Newmark's method in case of a PGA values lower than the k_y threshold. On the contrary, the stress–strain dynamic analysis, performed by applying dynamic equivalent inputs with PGA values ranging from 0.01 to 0.02 g, pointed out a reactivation of the Cerda landslide and justified the displacements up to 50 cm that were observed in the detachment area.

This represents a relevant result since the Cerda landslide can be regarded as an outlier event (Delgado et al. 2011), with respect to the empirical curves of expected maximum distances for landslide earthquake-induced reactivation (Keefer 1984; Rodriguez et al. 1999).

Nonetheless, as better discussed in the following Sect. 4.1.1, the landslide should not be re-triggered if the marly-limestone, corresponding to the local seismic bedrock, were not considered in the numerical model.

3.2.4 The Diezma Landslide (Spain): A Dynamic Sensitivity Analysis to the Frequency Content of Seismic Inputs

In the case of the Diezma landslide (introduced in the Sect. 2.1) a stress-strain numerical modeling was carried on by a FDM code to evaluate the role of the seismic input properties on earthquake-induced displacements. At this aim, 36 dynamic signals were derived by the LEMA_DES approach from accelerometric records of the European database integrated with records from K-NET and COSMOS databases in order to be representative for a wide range of characteristic ratios T_s/T_m and T_l/T_m referred to the Diezma landslide (i.e. $0.05 \leq T_s/T_m \leq 2.50$ and $0.33 \leq T_s/T_m \leq 16.63$) as well as for an energy value ranging within three orders of magnitude (i.e. Arias intensity, AI, varying from 0.01 to 1 m/s). As it resulted from the modeling, the earthquake-induced displacements significantly depend on both AI and characteristic periods of the selected dynamic signals (Fig. 14).

The horizontal displacements that resulted from the dynamic numerical modeling incorporate the contributions to the landslide mass movement, caused by both the 1D resonance and the 2D input-slope interaction. More in particular, at the lowest AI values the maximum earthquake-induced displacements result for T_m close to the landslide mass resonance period (i.e. to a T_s/T_m ratio almost equal to 1). On the other hand, for increasing AI the maximum modeled displacements result for long period signals, i.e. whose T_m is almost equal to $2T_l$. These findings output the relevance of dynamic input properties in evaluating earthquake-induced displacements of pre-existing landslide masses; moreover, as it results from the Diezma case study, higher is the energy content of the earthquake at the landslide site more relevant is the role of 2D interactions between landslide mass and seismic waves. As already demonstrated in the theoretical study by Lenti and Martino

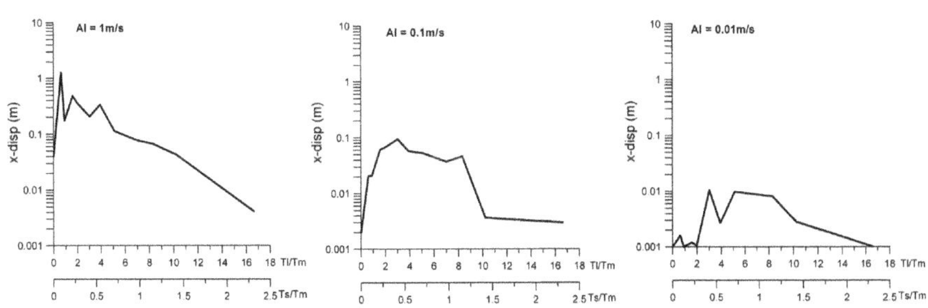

Fig. 14 Diezma landslide (Spain): T_s/T_m and T_l/T_m versus horizontal displacement (x-disp) obtained from dynamic stress-strain FDM numerical modeling for different Arias intensity (*AI*) values

(2013), at low T_s/T_m ratios and at high AI values (i.e., greater than 1 m/s) the traditional sliding block methods, which do not account for the bi-dimensional propagation of seismic waves, predict earthquake-induced displacements lower than those predicted by the dynamic numerical models. These differences are justified since the traditional methods cannot consider the increasing displacements at low T_l/T_m ratios, i.e. related to a seismic waves propagation within the landslide mass with a predominant half-period very close to the length of the landslide mass itself.

4 Seismic Response of Landslide Masses

Local site conditions responsible for the local seismic amplification (tectonic features, stratigraphic conditions, morphology) can be also regarded as responsible for a significant reduction of the slope stability due to the possible interactions between seismic input, local geological setting, slope geometry and pre-existing landslide mass.

Since the last decade several authors analyzed such effects by instrumental measurements (Gallipoli et al. 2000; Havenith et al. 2003a, b; Bozzano et al. 2004, 2008c; Del Gaudio and Wasowsky 2007; Méric et al. 2007; Delgado et al. 2015), laboratory experiments (Wang and Lin 2011 and references therein) as well as by numerical models focused on the interactions between seismic inputs and slopes (Bourdeau and Havenith 2008; Lenti and Martino 2012) or on the back-analysis of earthquake-triggered landslide reactivations (Bozzano et al. 2011a and references therein).

These studies pointed out that physical interactions between seismic waves and landslide masses are very complex since they result on several phenomena which might be often simultaneously invoked (Fig. 15): (i) stratigraphic resonance due to the impedance contrast between the landslide mass and its substratum; (ii) basin-like amplification effects related to the geological setting of the landslide slope as well as to the landside mass itself; (iii) seismic wave polarization within the landslide mass.

In this regard, for the Caramanico landslide case study (Italy) Del Gaudio and Wasowsky (2007) demonstrated an existing relation between site response directivity, topography and geological features by seismometric measurements. Nevertheless, the authors found that the observed directivity is caused by a directional re-distribution of energy, which is not necessarily associated to an overall site amplification.

In case that these site-dependent phenomena occur, the resulting effect on the earthquake-induced landslide displacements can be generally referred to a "self-excitation" process (Bozzano et al. 2011a) which implies a direct role of the landslide mass during the dynamic shaking.

Such an effect consists in modifying the seismic wave properties (i.e. PGA, frequency content, polarity) and favoring the landslide reactivation; the

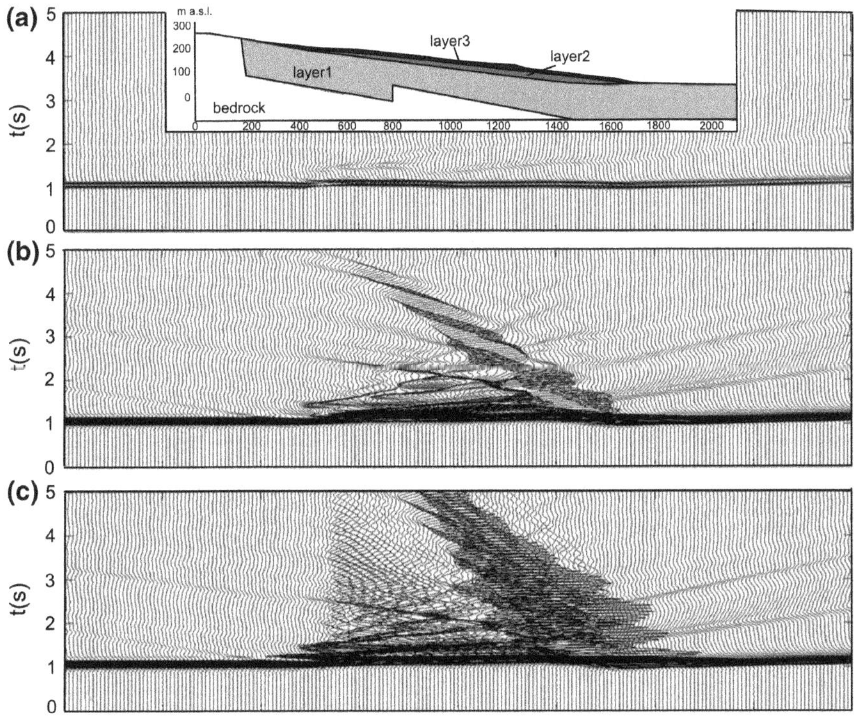

Fig. 15 Cerda landslide (Italy): S-wave propagation resulting from 2D numerical modeling performed by considering: **a** the bedrock and the layer 1 (scaly clays); **b** the bedrock, the layers 1 and 2 (landslide debris); **c** the bedrock, the layers 1, 2 and 3 (weathered landslide debris)

self-excitation process can also be invoked to explain far-field earthquake reactivation of large landslide (Delgado et al. 2011).

4.1 Case Studies: From Seismic Measurements to Numerical Models

4.1.1 Evidences of Local Seismic Amplification in the Diezma Landslide Slope (Spain)

In the Diezma landslide area (see Sect. 3.2.4) the local seismic response was studied (Delgado et al. 2015) by ambient noise measurements at 21 sites that were analyzed through the horizontal/vertical spectral ratio (HVSR) technique (Nakamura 1989) (Fig. 16).

The ambient noise was measured with a broadband seismometer Guralp CMG-6TD which has been processed according to the SESAME Working Group

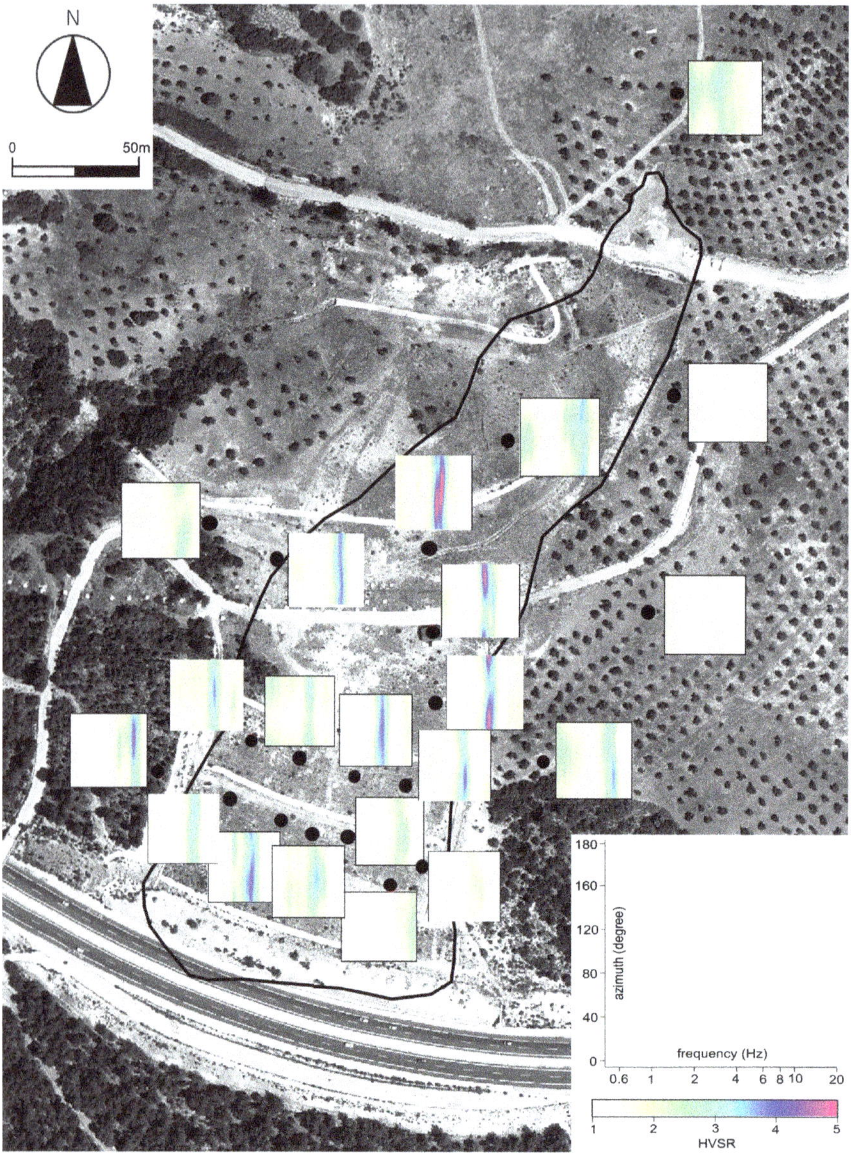

Fig. 16 Diezma landslide (Spain): map of the azimuthal distribution of the HVSRs obtained from the noise measurements, according to the HVSR technique

standards (2004). The results obtained at sites located outside of the landslide are characterized by flat HVSRs with amplitudes below 2.

On the other hand, it is noticeable that all the measurements performed on the landslide mass exhibit a clear peak occurs at frequencies of 4–5 Hz in the middle-low part of landslide mass, while the frequency varies for sites located in the upper parts of the landslide. The amplitude of the HVSR peak is usually higher than 4 but it decreases toward the landslide flanks and toe, where a large rock-fill exists. The low amplitude of the HVSRs and the corresponding lack of spectral peaks at sites located outside and around the landslide mass are a consequence of the direct outcrop of the seismic bedrock. Such outputs are in very good agreement with the stress-strain dynamic analysis, reported in the Sect. 3.2.4, as they demonstrate the significance of the landslide mass resonance in the local seismic response and justify the relevance of the T_s/T_m ratio for evaluating the expected earthquake-induced displacements, especially in case of low-energy seismic events.

4.1.2 Evidences of Local Seismic Amplification in Large-Landslide Slopes: The Salcito and Cerda Case Studies (Italy)

In addition to the Cerda landslide (introduced in the Sect. 3.2.3) another large landslide was reactivated in Italy on 2002 by the 31st October Molise earthquake (M_w 5.7), close to the Salcito village (Bozzano et al. 2004) (Fig. 17). The Cerda and the Salcito landslides show very similar features since: they involve volumes in the order of 20–40 × 10^6 m^3 of scaly-clays; they are about 1.5 km extended; they are typified by translational mechanisms with failure zones up to 50 m below ground level; they are characterized by measured S-wave velocity of the landslide mass of about 400 m/s; they occurred at an epicentral distance of about 50 km (i.e. significantly higher with respect to the co-relations between maximum landslide epicentral distance and earthquake magnitude by Keefer 1984).

Seismic noise measurements were performed by SS1-KINEMETRICS triaxially arranged seismometers (Bozzano et al. 2008c); the records were processed according to the HVSR technique (Nakamura 1989) and to the SESAME standard (SESAME 2004). The HVSR analysis of ambient noise measurements pointed out significant amplification effects, related to both the pre-existing landslide masses and the geological setting, that are responsible for HVSR peaks at about 1 Hz as well as at higher frequencies (i.e. in the range 2–6 Hz). Given the observed amplification effects and based on stress-strain dynamic numerical models, a self-excitation process was invoked for justifying the earthquake-induced reactivations of both the landslides (Bozzano et al. 2008c).

In the Cerda landslide area (Fig. 18), the HVSRs point out a wide amplifying zone in the range 0.5–1 Hz close to the crown, while frequencies in the range 2.5–4.5 Hz are amplified all along the landslide perimeter (Bozzano et al. 2011a). A temporary velocimetric array was also installed to record possible small-magnitude events (Bozzano et al. 2008a). The recorded earthquake accelerograms were FFT transformed to obtain the receiver functions (Field and Jacob 1995) in the related stations as well as spectral ratios to the reference station (Borcherdt 1994). The results are consistent with those obtained from the HVSRs from ambient noise.

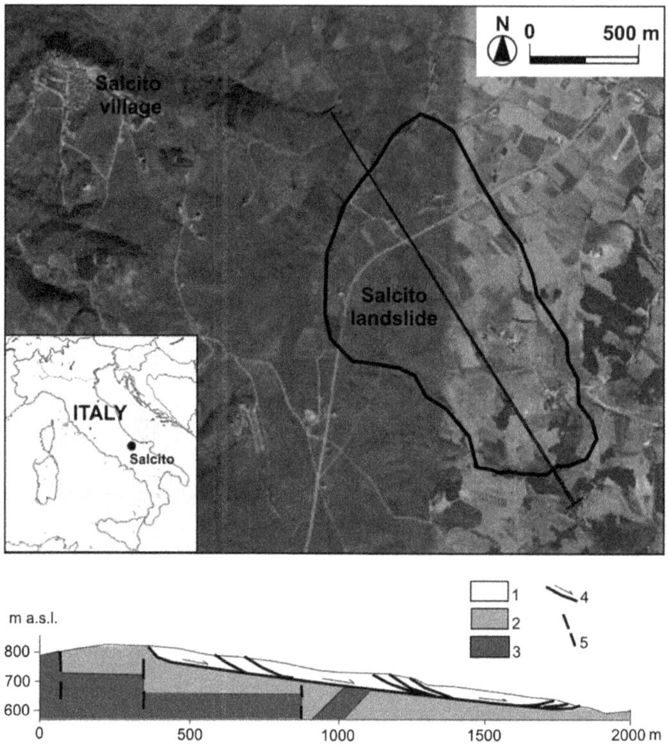

Fig. 17 Salcito landslide (Italy): GoogleEarth satellite view and geological cross section along the mapped trace. *1* landslide debris; *2* scaly clays (Sannio Unit, Oligocene–Miocene); *3* marls and calcarenites (Sannio Unit, Oligocene–Miocene); *4* sliding surface; *5* fault

In the Salcito landslide area (Fig. 18) the HVSRs point out an amplifying zone in the upper part of the slope, close to the left flank of the landslide; the main HVSR peaks correspond to frequency values in the range 1–2 Hz, while frequency values in the range 3.5–4.5 Hz are amplified within the landslide mass only (Bozzano et al. 2008b).

For both the Cerda and Salcito landslides, seismometric data as well as 2D numerical models (Bozzano et al. 2008b, 2011a) were used to define the contributions of different geological elements to the local seismic response and derive the numerical amplification function (A(f)). At this aim, the engineering-geological models were transposed into physical basin-like models where the landslide masses result part of a more complex geological system.

The obtained results (Fig. 18) show that after past activations the landslides became more suitable for seismically induced reactivations, since their softened masses can induce significant amplification effects at frequency values in the range 2–6 Hz, related to both the 1D seismic resonance and the 2D interactions mainly due to the superficial seismic waves. Such conditions are emphasized by a shallow debris layer over the landslide mass which is generated by weathering and remolding, also due to superficial earth-slides and soil creep processes.

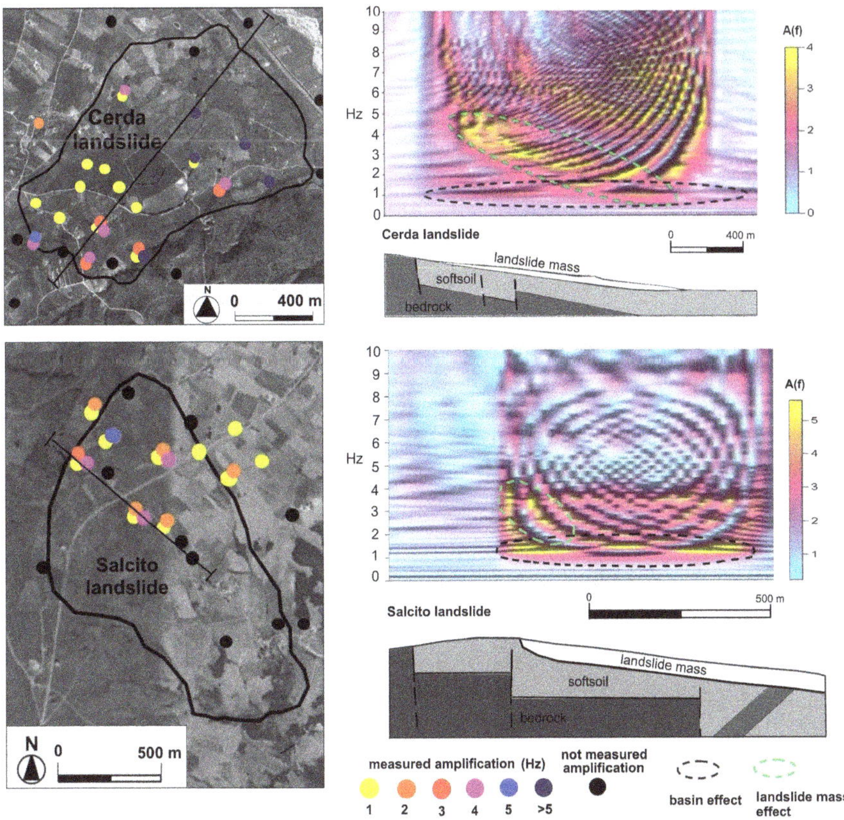

Fig. 18 Frequency peaks of HVSRs from noise measurements performed in the Cerda and Salcito landslide areas (*left*) and amplification function (A(f)) obtained from 2D numerical models showing the amplification effects related to the basin-like geological setting and to the softened landslide mass (*right*)

On the other hand, the amplification effect observed at lower frequency (<2 Hz), can be related to the local geological setting of the landslide areas (i.e. to the thick scaly clay layer above the local seismic bedrock). This amplification increases the amplitude of the long-period waves that are more suitable for interacting with the kilometer-extended masses of the two landslides.

The FDM numerical analyses performed for the Cerda and the Salcito landslides demonstrated that the low-frequency content (<2 Hz) of the triggering earthquakes is responsible for a relevant decay of the shear modulus along the sliding surface during the seismic shaking as well as for ground failures in post-seismic conditions (see Fig. 13 for the Cerda landslide case study).

Moreover, if the stiff bedrock is not reproduced in the models, the landslide reactivations do not occur highlighting the significant role played by the local geological setting for both the landslides.

5 Remarks and Perspectives

The evaluation of earthquake-induced landslide displacements represents a main goal for the prevision of damage due to seismic event occurrence. Although its complexity, this topic is a challenging field for multidisciplinary approaches as several theoretical and practical contributions are requested from geology, physic and engineering. During the last decades, several studies provided new insights on this regard, pointing out the main role of physical interactions among seismic waves, slope geometry, landslide mass and local geological setting. More in particular, thanks to the inventory of documented earthquake-induced landslides it was possible to output far-field outliers (respect to the literature co-relations between epicentral distance of landslide occurrence and earthquake magnitude), that gave the opportunity to back-analyze and explain their triggering conditions. In case of local seismic amplification the resulting effect on the earthquake-induced landslide displacements can be generally referred to a "self-excitation" process which implies a direct role of the landslide mass in its reactivation during the earthquake. The significant contribution of geophysical investigations (including seismic noise measurements and earthquake recording) supports more sophisticated methods of numerical analysis, such as FEM and FDM stress-strain dynamic models.

Some theoretical criteria were recently proposed to take into account the physical features of the seismic shaking in earthquake-induced landslide displacements. The "physical link" between the acting forces (earthquake) and the involved masses (landslide) is expressed by ratios of characteristic periods (i.e. T_s/T_m and T_l/T_m) which can be regarded as "markers" for evaluating if the seismic actions best tune the inertial forces of the landslide mass. Nonetheless, to correctly use these markers very detailed engineering-geological models are requested for the landslide slopes as well as very strong constraints to the local seismic response. In this regard, further studies are necessary to better understand the complex interactions among seismic waves and slopes observed and recorded so far in several case histories. More in particular, a greater attention should be devoted to relate strain effects induced by dynamic actions with 3D-models of landslide masses. Finally the presence of anthropic structures or infrastructures in landslide areas (i.e. buildings, bridge, tunnels, pipelines among others) is another element to be considered in future for completing the general scheme of interaction among seismic actions and landslide involved slopes.

Acknowledgments I wish to thank my colleague and friend Luca Lenti who provided the rigorous physical support to the researches carried on together; to Jose Delgado Marchal for the opportunity to discover and study several and very interesting case histories of earthquake-induced landslides in Spain; to Antonella Paciello for her precious contribution to the seismometric measurements in the Italian case studies; to Francesca Bozzano and Gabriele Scarascia Mugnozza, for their guide during several years of scientific co-operations; to my tutor Alberto Prestininzi who gave me the opportunity to approach and to carry on my research.

References

Alfaro, P., Delgado, J., García-Tortosa, F. J., Lenti, L., López, J. A., López-Casado, C., & Martino, S. (2012). Widespread landslides induced by the Mw 5.1 earthquake of 11 May 2011 in Lorca. *Engineering Geology, 137–138*, 40–52.

Ambraseys, N., Smit, P., Sigbjornsson, R., Suhadolc, P., & Margaris, B. (2002). Internet-site for European strong-motion data, European Commission, Research-Directorate General, Environment and Climate Programme. On-line web site: http://www.isesd.cv.ic.ac.uk/ESD/frameset.htm.

Ambraseys, N., & Srbulov, M. (1995). Earthquake induced displacements of slopes. *Soil Dynamics and Earthquake Engineering, 14*, 59–71.

Atakan, K., Ojeda, A., Meghraoui, M., Barka, A. A., Erdik, M., & Bodare, A. (2002). Seismic Hazard in Istanbul following the 17 August 1999 Izmit and 12 November 1999 Düzce Earthquakes. *Bulletin of the Seismological Society of America, 92*(1), 466–482.

Azañón, J. M., Azor, A., Yesares, J., Tsige, M., Mateos, R. M., Nieto, F., et al. (2010). Regional-scale high plasticity clay-bearing formation as controlling factor on landslides in Southeast Spain. *Geomorphology, 120*, 26–37.

Balassanian, S. Y. U., Martirosyan, A. H., Nazaretian, S. N., Arakelian, A. R., Avanessian, A. S., Igumnov, V. A., & Ruttener, E. (1999a). Seismic hazard assessment in Armenia. *Natural Hazards, 18*, 227–236.

Balassanian, S. Y. U., Melkoumian, M. G., Arakelian, A. R., & Azarian, A. R. (1999b). Seismic risk assessment for the territory of Armenia and strategy of its mitigation. *Natural Hazards, 20*, 43–55.

Benito, M. B., Navarro, M., Vidal, F., Gaspar-Escribano, J., Garcia-Rodriguez, M. L., & Martinez-Solares, J. M. (2010). Seismic hazard assessment in the region of Andalusia (Sothern Spain). *Bulletin of Earthquake Engineering, 8*, 739–766.

Bird, J. F., & Bommer, J. J. (2004). Earthquake losses due to ground failure. *Engineering Geology, 75*, 147–179.

Bishop, A. W. (1955). The use of the slip circle in the stability analysis of slopes. *Géotechnique, 5* (1), 7–17.

Borcherdt, R. D. (1994). Estimates of site-dependent response spectra for design (methodology and justification). *Earthquake Spectra, 10*, 617–653.

Bouckovalas, G. D., & Papadimitriou, A. G. (2005). Numerical evaluation of slope topography effects on seismic ground motion. *Soil Dynamics and Earthquake Engineering, 25*, 547–558.

Bourdeau, C., & Havenith, H. B. (2008). Site effects modeling applied to the slope affected by the Suusamyr earthquake (Kyrgyzstan, 1992). *Engineering Geology, 97*, 126–145.

Bozzano, F., Cardarelli, E., Cercato, M., Lenti, L., Martino, S., Paciello, A., & Scarascia Mugnozza, G. (2008a). Engineering-geology model of the seismically induced Cerda landslide. *Bollettino di Geofisica Teorica ed Applicata, 49*(2), 205–226.

Bozzano, F., Lenti, L., Martino, S., Paciello, A., & Scarascia Mugnozza, G. (2008b). Self-excitation process due to local seismic amplification responsible for the reactivation of the Salcito landslide (Italy) on 31 October 2002. *Journal Geophysical Research, 113*, B10312.

Bozzano, F., Lenti, L., Martino, S., Paciello, A., & Scarascia Mugnozza, G. (2008c). Self-excitation process due to local seismic amplification and earthquake-induced reactivations of large landslides. In Z. Chen, et al. (Eds.), *Landslides and engineered slopes* (pp. 1389–1395). London: Taylor and Francis Group. ISBN 978-0-415-41196-7.

Bozzano, F., Lenti, L., Martino, S., Paciello, A., & Scarascia Mugnozza, G. (2011a). Evidences of landslide earthquake triggering due to self-excitation process. *International Journal of Earth Sciences, 100*, 861–879.

Bozzano, F., Martino, S., Naso, G., Prestininzi, A., Romeo, R. W., & Scarascia Mugnozza, G. (2004). The large Salcito landslide triggered by the 31st October 2002, Molise earthquake. *Earthquake Spectra, 20*(2), 1–11.

Bozzano, F., Lenti, L., Martino, S., Montagna, A., & Paciello, A. (2011b). Earthquake triggering of landslides in highly jointed rock masses: Reconstruction of the 1783 Scilla rock avalanche (Italy). *Geomorphology, 129*, 294–308.

Bray, J. D., & Travasarou, T. (2007). Simplified procedure for estimating earthquake-induced deviatoric slope displacements. *Journal of Geotechnical and Geoenvironmental Engineering, 133*(4), 381–392.

Cruden, D. M., & Varnes, D. J. (1996). Landslide types and processes. In A. K. Turner & R. L. Schuster (Eds.), *Landslides: Investigation and Mitigation* (pp. 36–75). Transportation Research Board, Speciation Report 247, National Research Council, National Academy Press, Washington, D.C.

Dalgiç, S. (2004). Factors affecting the greater damage in the Avcılar area of Istanbul during the 17 August 1999 Izmit earthquake. *Bulletin of Engineering Geology and the Environment, 63*, 221–232.

Danneels, G., Bourdeau, C., Torgoev, I., & Havenith, H. B. (2008). Geophysical investigation and dynamic modeling of unstable slopes: Case-study of Kainama (Kyrgyzstan). *Geophysical Journal International, 175*(1), 17–34.

Del Gaudio, V., & Wasowski, J. (2007). Directivity of slope dynamic response to seismic shaking. *Geophysical Reseach Letters, 34*, L12301.

Delgado, J., Garrido, J., López-Casado, C., Martino, S., & Peláez, J. A. (2011). On far field occurrence of seismically induced landslides. *Engineering Geology, 123*, 204–213.

Delgado, J., Garrido, J., Lenti, L., López-Casado, C., Martino, S., & Javier Sierra, F. (2015). Unconventional pseudostatic stability analysis of the Diezma landslide (Granada, Spain) based on a high-resolution engineering-geological model. *Engineering Geology, 184*, 81–95.

Duman, T. Y., Can, T., Gokceoglu, C., Nefeslioglu, H. A., & Sonmez, H. (2006). Application of logistic regression for landslide susceptibility zoning of Cekmece Area, Istanbul, Turkey. *Environmental Geology, 51*, 241–256.

Erdik, M., Biro, Y. A., Onur, T., Sesetyan, K., & Birgoren, G. (1999). Assessment of earthquake hazard in Turkey and neighboring regions. *Annali di Geofisica, 42*, 1125–1138.

Evans, S. G., & Bent, A. L. (2004). The Las Colinas landslide, Santa Tecla: A highly destructive flowslide triggered by the January 13, 2001, El Salvador earthquake. *GSA Special Papers, 375*, 25–38.

Field, E. H., & Jacob, K. (1995). A comparison and test of various site response estimation techniques, including three that are non reference-site dependent. *Bulletin of the Seismological Society of America, 85*, 1127–1143.

Gallipoli, M., Lapenna, V., Lorenzo, P., Mucciarelli, M., Perrone, A., Piscitelli, S., & Sdao, F. (2000). Comparison of geological and geophysical prospecting techniques in the study of a landslide in southern Italy. *European Journal of Environmental and Engineering Geophysics, 4*, 117–128.

Havenith, H. B., Strom, A., Jongmans, D., Abdrakhmatov, K., Delvaux, D., & Tre′fois, P. (2003a). Seismic triggering of landslides, part A: Field evidence from the northern Tien Shan. *Natural Hazards and Earth Systems Sciences, 3*, 135–149.

Havenith, H. B., Vanini, M., Jongmans, D., & Faccioli, E. (2003b). Initiation of earthquake-induced slope failure: Influence of topographical and other site specific amplification effects. *Journal of Seismology, 7*, 397–412.

Hsieh, S. U., & Lee, C. T. (2011). Empirical estimation of the Newmark displacement from the Arias intensity and critical acceleration. *Engineering Geology, 122*, 34–42.

Hutchinson, J. N. (1987). Mechanism producing large displacements in landslides on pre-existing shears. *Memoir of the Geological Society of China, 9*, 175–200.

Hutchinson, J. N., & Del Prete, M. (1985). Landslide at Calitri, southern Apennines, reactivated by the earthquake of 23rd November 1980. *Geologia Applicata e Idrogeologia 1985, XX*(1), 9–38.

Hwang, G. S., & Chen, C. H. (2013). A study of the Newmark sliding block displacement functions. *Bulletin of Earthquake Engineering, 11*, 481–502.

Jibson, R. W. (1993). Predicting earthquake-induced landslide displacements using Newmark's sliding block analysis. *Transportation Research Record, 1411*, 9–17.

Jibson, R. W. (2011). Methods for assessing the stability of slopes during earthquakes—A retrospective. *Engineering Geology, 122*, 43–50.

Jibson, R. W., Harp, E. L., & Michael, J. M. (1998). A method for producing digital probabilistic seismic landslide hazard maps: An example from the Los Angeles, California area (pp. 98–113). U.S. Geological Survey Open-File Report.

Kalkan, E., Gülkan, P., Yilmaz, N., & Çelebi, M. (2009). Reassessment of probabilistic seismic hazard in the Marmara region. *BSSA, 99*, 2127–2146.

Keefer, D. K. (1984). Landslides caused by earthquakes. *Geological Society of America Bulletin, 95*, 406–421.

Kuhlemeyer, R. L., & Lysmer, J. (1973). Finite element method accuracy for wave propagation problems. *Journal of Soil Mechanics & Foundations Div ASCE, 99*(SM5), 421–427.

Lenti, L., & Martino, S. (2010). New procedure for deriving multifrequential dynamic equivalent signals (LEMA_DES): A test-study based on Italian accelerometric records. *Bulletin of Earthquake Engineering, 8*, 813–846.

Lenti, L., & Martino, S. (2012). The interaction of seismic waves with step-like slopes and its influence on landslide movements. *Engineering Geology, 126*, 19–36.

Lenti, L., & Martino, S. (2013). A parametric numerical study of the interaction between seismic waves and landslides for the evaluation of the susceptibility to seismically induced displacements. *BSSA, 103*(1), 33–56.

Loukidis, D., Bandini, P., & Salgado, R. (2003). Stability of seismically loaded slopes using limit analysis. *Geotechnique, 15*(2), 139–160.

Makdisi, F. I., & Seed, H. B. (1978). Simplified procedure for estimating dam and embankment earthquake-induced deformations. *Journal of Geotechnical and Geoenvironmental Engineering ASCE, 104*(7), 849–867.

Martino, S., Paciello, A., Sadoyan, T., & Scarascia Mugnozza, G. (2007). Dynamic numerical analysis of the giant Vokhchaberd landslide (Armenia). In *Proceedings of 4th International Conference on Earthquake Geotechnical Engineering, Tessaloniki (Greece) 25–28 June 2007*. ISBN: 9781402058929, Tessaloniki (Grecia), 25–28 Giugno 2007.

Martino, S., Prestininzi, A., & Romeo, W. R. (2014). Earthquake-induced ground failures in Italy from a reviewed database. *Natural Hazards and Earth System Sciences* (In press).

Martino, S., & Scarascia Mugnozza, G. (2005). The role of the seismic trigger in the Calitri landslide (Italy): Historical reconstruction and dynamic analysis. *Soil Dynamics and Earthquake Engineering, 25*, 933–950.

Mazzanti, P., & Bozzano, F. (2011). Revisiting the February 6th 1783 Scilla (Calabria, Italy) landslide and tsunami by numerical simulation. *Marine Geophysical Research, 32*(1–2), 273–286.

Méric, O., Garambois, S., Malet, J. P., Cadet, H., Guéguen, P., & Jongmans, D. (2007). Seismic noise-based methods for soft-rock landslide characterization. *Bulletin de la Societe Geologique de France, 178*(2), 137–148.

Murphy, W., Petley, D. N., Bommer, J. J., & Mankelow, J. M. (2002). Seismological uncertainty in the assessment of slope stability during earthquakes. *Quarterly Journal of Engineering Geology and Hydrogeology, 35*, 71–78.

Nakamura, Y. (1989). A method for dynamic characteristics estimation of subsurface using microtremor on the ground surface. *Quarterly Report of RTRI, 30*(1), 25–33.

Newmark, N. M. (1965). Effects of earthquakes on dams and embankments. *Geotechnique, 15*(2), 139–159.

Paraskevopoulos, E., Panagiotopoulos, C. G., & Manolis, G. D. (2010). Imposition of time-dependent boundary conditions in FEM formulations for elastodynamics: crytical assessments of penalty-type methods. *Computational Mechanics, 45*(2–3), 157–166.

Petley, D. (2012). Global patterns of loss of life from landslides. *Geology, 40*(10), 927–930.

Rathje, E. M., & Bray, J. D. (2000). Nonlinear coupled seismic sliding analysis of earth structures. *Journal of Geotechnical and Geoenvironmental Engineering ASCE, 126*(11), 1002–1014.

Rathje, E. M., Abrahamson, N. A., & Bray, J. D. (1998). Simplified frequency content estimates of earthquake ground motions. *Journal of Geotechnical and Geoenvironmental Engineering ASCE, 124*(2), 150–159.

Rathje, E. M., & Antonakos, G. (2010). Recent advances in predicting earthquake-induced sliding displacements of slopes. In *Proceedings of Fifth International Conference of Recent Advances in Geotechnical Earthquake Engineering and Soil Dynamics* (pp. 1–9), May 24–29, San Diego, California.

Rodriguez, C. E., Bommer, J. J., & Chandler, R. J. (1999). Earthquake-induced landslides: 1980–1997. *Soil Dynamics and Earthquake, 18*, 325–346.

Rodríguez-Peces, J., Azañón, J. M., García-Mayordomo, J., Yesares, J., Troncoso, E., & Tsige, M. (2011). The Diezma landslide (A-92 motorway, Southern Spain): History and potential for future reactivation. *Bulletin of Engineering Geology and the Environment, 70*, 681–689.

Romeo, R. (2000). Seismically induced landslide displacements: A predictive model. *Engineering Geology, 58*(3–4), 337–351.

Ruttener, E. (1999). Seismic hazard assessment in Armenia. *Natural Hazards, 18*, 227–236.

Sassa, K. (1996). Prediction of earthquake induced landslides. In K. Senneset (Ed.), *Landslides* (pp. 115–132). Rotterdam: Balkema.

Seed, H. B. (1979). Soil liquefaction and cyclic mobility evaluation for level ground during earthquakes. *Journal of the Geotechnical Engineering Division ASCE, 105*(GT2), 102–155.

Sepulveda, S. A., Murphy, W., Jibson, R. W., & Petley, D. N. (2005). Seismically induced rock slope failures resulting from topographic amplification of strong ground motions: The case of Pacoima Canyon, California. *Engineering Geology, 80*, 336–348.

SESAME Working Group. (2004). Guidelines for the implementation of the h/v spectral ratio tecnique on ambient vibration measurements, processing and interpretation. http://sesame-fp5.obs.ujf-grenoble.fr/Delivrables/Del-D23-HV_User_Guidelines.pdf.

Strenk, P. M., & Wartman, J. (2011). Uncertainty in seismic slope deformation model predictions. *Engineering Geology, 122*, 61–72.

Wang, K. L., & Lin, M. L. (2011). Initiation and displacement of landslide induced by earthquake —a study of shaking table model slope test. *Engineering Geology, 122*, 106–114.

Resilience, Vulnerability and Prevention Policies of Territorial Systems in Areas at High Seismic Risk

A. Teramo, C. Rafanelli, M. Poscolieri, F. Lo Castro, S. Iarossi, D. Termini, M. De Luca, A. Marino and F. Ruggiano

1 Introduction

Seismic vulnerability and resilience (McAslan 2010) of a territorial system are the elements of reference for the characterization of an effective prevention policy, and the identification of the multidimensional and interdisciplinary character of the urban system exposure (Carreño et al. 2007). The first one is referable to the damage susceptibility of buildings due to an earthquake in relation to their seismic characteristics; the second one is referable to the ability of the urban system to reconfigure its structure to cope with the effects of an earthquake.

The territorial systems of regions located in high seismic hazard areas (Fig. 1) should be therefore characterized in terms of specific territorial, seismic, medical, social and juridical indicators, on the basis of the results of targeted research activities in each urban context. Research of this type are in progress in the city of Messina (Italy) in the framework of two research projects (*PRISMA* and *DIONISO*) funded by the Italian Ministry of Education, University and Research (MIUR), under the program *Smart Cities and Communities and social innovation.*

Here following it will be shown how it is possible to assess the resilience of an urban system, also using procedures and devices aimed at emergency management after a strong earthquake:

- building monitoring through wireless sensor networks, which allows a remote assessment of: the safety level of buildings after the earthquake, the practica-

A. Teramo · D. Termini · A. Marino
Seismological Observatory of Messina University, Messina, Italy

C. Rafanelli · M. Poscolieri (✉) · F. Lo Castro · S. Iarossi · M. De Luca
CNR-Institute of Acoustics and Sensors "O. M. Corbino", Rome, Italy
e-mail: maurizio.poscolieri@idasc.cnr.it

F. Ruggiano
Messina University, Messina, Italy

S. D'Amico (ed.), *Earthquakes and Their Impact on Society*,
Springer Natural Hazards, DOI 10.1007/978-3-319-21753-6_11

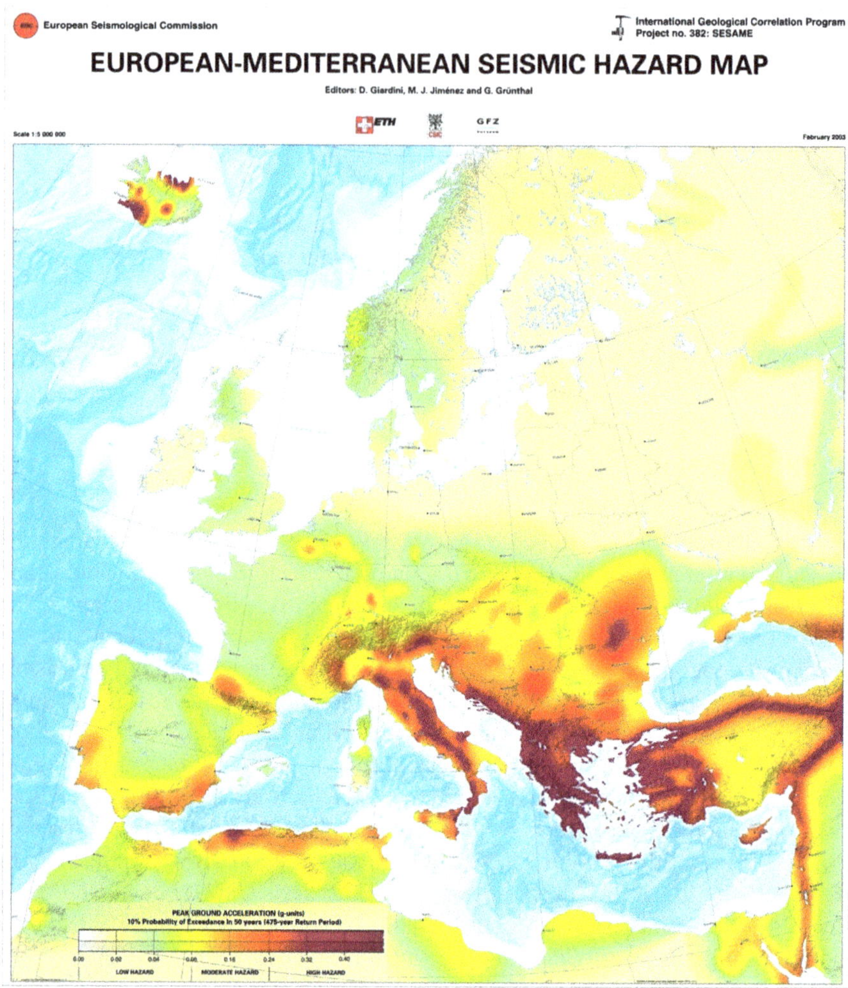

Fig. 1 European-Mediterranean seismic hazard map, after European Seismological Commission, UNESCO-IUGS International Geological Correlation Program Project no. 382 SESAME

bility of the main roads, the number of collapsed buildings and the number of involved people;

- development of innovative domotic devices, originally designed to improve the quality of life, and today, more consistently with the current needs of the community, oriented to relate comfort to safety and the right to safety.
- emergency management and rescue coordination, within the assessment of the resilience of the hospital system of Messina, on the basis of current research activities, as an expression of significant integration of interdisciplinary, scientific and entrepreneurial skills.

Through the characterization of the territorial system in its different dimensions of economic, environmental and anthropogenic type, the strong dependence of its vulnerability and resilience on the lack of effective policies for seismic safety, as well as on specific elements of the socio-economic fragility of the territory, is highlighted. The analysis of complex systems and the set of relative relationships, however, shows a goal-seeking behavior which is related to the activation of mechanisms of adaptation, learning and self-organization that can considerably reduce both the effects of a strong earthquake and the reactivation times of the conditions preceding the event disturbing the system equilibrium.

From proposed applications it is possible to infer the relationship between vulnerable systems, which do not have resilience, and those resilient which identify in the processes of self-organization an opportunity for sustainable economic development. A significant factor of territorial resilience is in fact related to the ability of the local business community to establish targeted interventions and activities, specifically configured on the promotion of innovation and technological entrepreneurship capable of supporting the economic and social development of the community, through processes of interaction and integration of the research system with the territory.

Finally, some legal aspects related to the right of safety are analyzed, together with legal macro-criticalities relating to the application of the existing rules on the territory safety, in the context of transparency, sharing, inclusion and participation criteria related to Smart Communities *e-government*, where the right to safety of citizens is the main element of reference.

2 Real Time Monitoring of Buildings for Seismic Damage Scenario Construction and Emergency Management

Not many years ago systems of building monitoring were based on wired grids of sensors, characterized by high costs, considerable size and poor flexibility. In recent years, the gradual development of wireless sensor networks (WSN) technology has given a significant innovation opportunity in building monitoring. A WSN node is able to communicate via a radio connection through devices, battery powered, that do not require any cable deployment, significantly reducing installation cost of the overall system. Structural health monitoring (SHM), however, may present some peculiar requirements, like a long-term operating lifetime (often over several years), and the deployment of sensor nodes in positions consistent with the structural geometry of building. One of the frequently adopted strategies for the development of WSN based SHM systems has been the use of general purpose communication platforms, suitably supplemented by specific sensing daughter boards (Federici et al. 2013). This strategy can simplify hardware and software development phase and speeds up system deployment, but it may nevertheless have some disadvantages. For example, general purpose platform are often not optimized in terms of energy consumption and it is not guaranteed that the addition of external hardware blocks will bring an overall optimized solution. Different typologies of monitoring actions—dynamic analysis oriented monitoring, seismic analysis oriented monitoring (Lo Castro et al. 2013), crack growth oriented monitoring, environmental or chemical oriented monitoring (Martinelli et al. 2007; Paolesse et al. 2008)—may have different or conflicting requirements (Krishnamurthy et al. 2008). In this regard, it is to be observed that dynamic analysis oriented monitoring systems must guarantee a precise measurement synchronization (Lynch et al. 2006), while this requirement may result less relevant when monitored quantities are slowly variable. Indeed, it is possible to identify the following general requirements for a SHM oriented WSN: reliability, reconfigurability, energy efficiency. Reliability is a critical requirement in building monitoring application considering that sensor nodes must be able to correctly report measured data, assuring a good communication. Reconfigurability and energy efficiency represent key requirements in relation to both the potential failures of any sensor node, and the sensor node battery lifetime.

An high energy efficiency is generally guaranteed by two different actions: (i) design or selection of low power consumption devices (i.e. devices with low active state power consumption and efficient stand-by and sleep modes); (ii) adoption of operational strategies oriented to the maximization of low power mode operating time for various blocks, using duty-cycling strategy (Federici et al. 2013).

A large number of structural health monitoring systems combine the use of wireless communication and innovative sensors such as fiber optic, piezoelectric or MEMS technology based accelerometers.

Over the last years, piezoelectric devices have been used as Acoustic Emission (AE) sensors in several applications due to its capability to detect crack growth, damage localizations in concrete structures, infrastructures in general and also, for example, in historical monuments (Carpinteri and Lacidogna 2006). In Italy, the sudden collapse of the Noto Cathedral in 2007 and the effects of the L'Aquila earthquake in April 2009, brought the problem of structural safety as a priority in the maintenance of civil structures and monuments. These recent events lead to the conclusion that a large number of structures need reliable, periodic, low cost monitoring and inspection procedures, easy to be implemented. During the last few years the AE technique has been used during long-term monitoring in order to analyze the time evolution of microcracking phenomena (Yoon et al. 2007; Ohtsu 1996; Pollock 1973; Brindley et al. 1973; Grosse et al. 2003; Carpinteri et al. 2005; Shah and Li 1994; Carpinteri et al. 2007; Niccolini et al. 2012). According to this technique, it is possible to detect the onset and the evolution of stress-induced cracks. Crack opening, in fact, is accompanied by the emission of elastic waves which propagate within the bulk of the material. These waves can be detected and recorded by transducers applied to the surface of the structural elements. AE monitoring is performed by means of piezoelectric sensors that give out signals when subjected to a mechanical stress (Gregori et al. 2007, 2012; Poscolieri et al. 2012). In this way, the AE technique makes it possible to estimate the amount of energy released during the fracture process, to obtain information on the criticality of the process underway and to localize the damage source locations. Due to the attenuation of acoustic waves and geometrical spreading in concrete structures, numerous sensors have to be applied to cover all critical parts. These circumstances make the traditional way of applying AE techniques too expensive. Monitoring systems for large structures should be based on a new kind of AE equipment using wireless transmission systems. In the new monitoring system, AE signals are detected by the sensor array, recorded in situ by a synchronization and storage unit, and, subsequently, they are sent via the GPRS/UMTS system to the central server for the elaboration phases.

The study of wireless sensor network for a structural health monitoring has been proposed by the research group of the authors, and it has been chosen to implement a low cost complete inertial system, for the measurement of accelerations, angular speed, rotations and inclinations. Its low cost will meet the need for widespread use of this WSN.

A possible implementation of the sensor node is shown in Fig. 2.

After a careful study of sensors cost evaluations, the attention has been focused on the products of ST Microelectronics and Kionix; both acceleration sensors produce low noise and high dynamic characteristics (±2 g ± 6 g F.S.; 45–50 μg noise).

The intrinsic noise, substantially stochastic, can be further lowered making an average of more accelerometers, configuration that is possible using low-cost MEMS. The measurement of the angular velocity is performed by low noise and high dynamic range ST Microelectronics three-axis MEMS gyroscope. Sensors can be calibrated at low frequencies using vibrational and centrifugal or Earth's Gravitational methods (ISO 16063-21:2003 2013; ISO 16063-15:2006 2010; Ferraris et al. 1994; Veldman 2006; Schuler et al. 1967; Lo Castro et al. 2007).

In order to obtain an inertial reference, the measurement of the rotations is carried out by a three-axis, high-sensitivity magnetometer from Freescale. The chosen MEMS sensor shows a scale of ±1000 μT and a sensitivity of 0.10 μT. The offset generated by the proximity of any ferrous objects can be compensated by software.

The complex management of the sensor system proposed requires the use of a high performance microcontroller provided with FPU (Floating Point Unit).

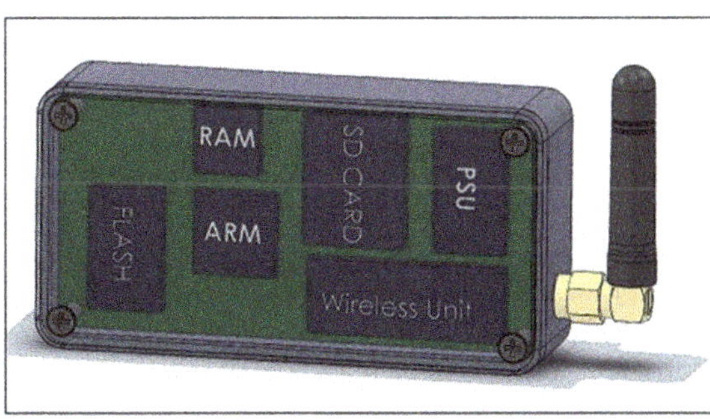

Fig. 2 Structural health monitoring sensor node (an example)

Systems based on core ARM 32-bit Cortex ensures high-speed computing, DSP functions and include all the necessary peripherals for sensor handle. Moreover, it is supplied with a SD Card interface to create a solid-state mass memory. For the management of the sensor system, it is necessary to use a real-time operating system as open source ChibiOS. The connectivity of the system is ensured by a high-power module based on ZIGBEE protocol, which allows easy deployment of the sensor network within the structures to be monitored.

The power is ensured by lithium polymer batteries that are charged by power supplies or small solar panels, depending on the position inside or outside the structure to be monitored.

The protection of people in buildings and the problem of managing them in case they are involved in a catastrophic event could find a valuable support in technologies developed in *domotics*. In this context, it can be considered that the new home technologies are trying to improve the everyday life of the inhabitants and particularly of elderly people who are most at risk in terms of health. Remote patient monitoring and remote assistance are now expanding also in Italy especially in the elderly accommodations, thanks to initiatives of public structures, in order to reduce home care costs of elderly and, generally, of the chronically ill patients.

The computer processing of digital images to reliably interpret human behavior in the home environment presents difficulties related to the presence of other people, under different light intensities, the movement of objects on the scene, etc. Nevertheless, it is reasonable to consider a number of objectives for interpreting these visual systems, which vary in complexity, from monitoring the occupancy of the room, to represent a detailed analysis of the forms of activity. These video monitoring capabilities can be used to improve information that is obtained from passive detectors wearable by people at home, especially when they are not put on because they can provide wrong data so causing false alarms. On the contrary, distributed sensors systems, installed inside the house, have the advantage of being independent from the people. Therefore, the most reliable system should be a combination of either sensor types. The sensors used in the current domestic environment often perform a single function. For example, the passive infrared, pressure sensors and door monitoring, respectively, provide occupancy time of a room or the presence in a particular area. Conversely, the video sensors are much more flexible, in fact they are applied for example in: controlling the frequency of presence in a room, detection of non-activity or reduced mobility of a person.

These kind of sensors could be used to better evaluate the vulnerability of a territory. Several approaches are used to define different types of vulnerability (social, economic, medical, etc.), strictly related to the level of the seismic vulnerability of a territory and to be assessed even through topological features (Agarwal et al. 2011).

The need to assess the seismic risk (Cornell 1968) and damage, also in terms of social and economic losses, has led in fact to develop many Earthquake Loss Estimation (ELE) software packages (Molina et al. 2010; Daniell 2011); some of these packages are open source, such as SELENA (Molina et al. 2007), whose core is based on the HAZUS®99 methodology (FEMA HAZUS®99 1999 and

FEMA-366 2001), DBELA (Crowley 2005) and others. They can be implemented and modified, in order to comply with the requirement of the proper damage assessment system (Crowley et al. 2009; Erduran et al. 2012).

The analysis typologies (probabilistic, deterministic, and based on real-time data) are related to the different ways of ground motion acquisition (probabilistic shaking maps, historical or user defined earthquakes and through real-time shake maps), the building inventory, the seismic vulnerability of the buildings in the urban context of a given city, the exposure in terms of dynamics of population distribution in the territory and the economic value of entrepreneurial activities.

In Fig. 3 an example of the output data provided by ELE software is shown: the physical damage, economic loss and human loss.

Physical or structural damage is expressed under a statistical point and divided into several damage levels as slight, moderate, extensive, and complete (Molina et al. 2007, FEMA HAZUS®-MH 2.1). In Table 1 damage is classified according to other methods, and additional classifications are provided in Rossetto and Elnashai (2003) and Dimova and Negro (2005); the human losses also are divided in severity levels: light injuries, moderate injuries, heavy injuries and death (FEMA HAZUS®-MH 2.1).

Considering that the main goal of a seismic prevention policy is to be related to an effective emergency management and rescue coordination, moreover coherent

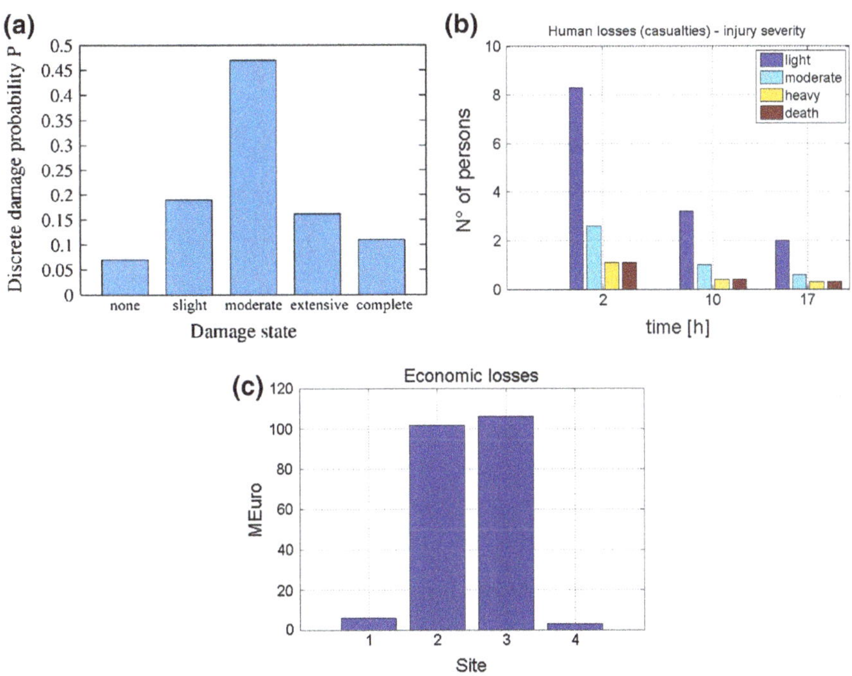

Fig. 3 Example of output graphics using Selena: **a** damage state, **b** human losses, **c** economic losses

Table 1 Approximate damage scale correlation state used in different ELE software

DI_{HRC}	HRC	HAZUS-MH 2.1	FEMA 273	SEAOC vision 2000	ATC-20	EMS-98	ISO 28841:2013
	Damage	Damage	Building performance level	Building performance level	Immediate post-earthquake inspection	Damage	Damage
0	–	–	No damage	No damage	Green tag	No damage	No damage
<50	Slight/light	Slight	Very light (operational level)	Negligible (fully operational)	Green tag	Grade 1	Insignificant
50–79	Moderate	Moderate	Light (immediate occupancy)	Light (operational)	Green tag	Grade 2	Slight
80–99	Extensive	Extensive	Moderate (life safety)	Moderate (life safety)	Yellow tag	Grade 3	Moderate
100	Collapse total/partial	Complete	Severe (collapse prevention)	Severe (near collapse)	Red tag	Grade 4	Serious
?	–	–	–	Collapse total/partial	–	Grade 5	Severe

HRC Homogenized reinforced concrete scale

DI_{HRC} The HRC-damage index

HAZUS Federal Emergency Management Agency. HAZUS user and technical manuals. Federal emergency management agency report: HAZUS 1999, Washington D.C., Vol. 7, 1999

SEAOC Structural Engineering Association of California (SEAOC). Vision 2000. Seismology Committee, Structural Engineering Association of California, CA, April 1995

FEMA 273 Federal Emergency Management Agency. NEHRP guidelines for seismic rehabilitation of buildings. Federal Emergency Management Agency Report: FEMA 273. Washington D.C., 1997

ATC-20 Applied technology council—Procedures for Post-earthquake Safety Evaluation of Buildings, 1989

EMS-98 Conseil de L'Europe. European macroseismic scale 1998 (EMS-98). Cahier du Centre Européen de Géodynamique et de Séismologie, G. Grünthal, editor. Luxemburg, Vol. 15, 1998

ISO 28841 2013 Guidelines for simplified seismic assessment and rehabilitation of concrete buildings

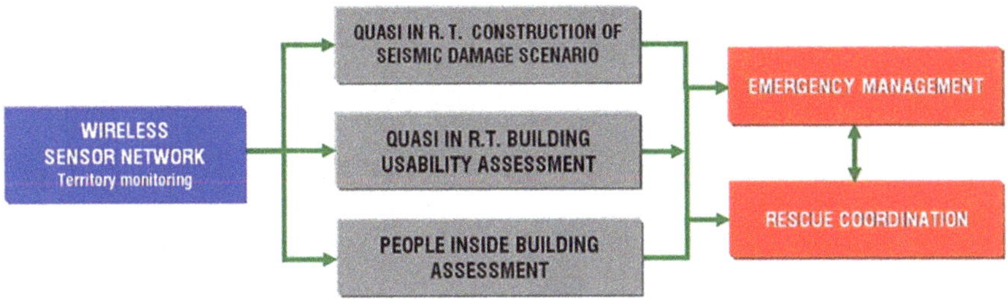

Fig. 4 Territory monitoring to improve the urban context resilience

with data acquired in real-time after a strong earthquake, an approach aimed at improving the territory resilience and based on real-time data is the most promising to achieve the best results for civil protection purposes (Pitilakis and Kakderi 2011; Erduran et al. 2012; Fiorini et al. 2012; Lin et al. 2012; Picozzi et al. 2013).

A wireless sensor network carried out at different scale (local, urban, and district level) can monitor the ground motion, the building storey accelerations and displacements (Kurata et al. 2004; Gattulli et al. 2013), in order to better assess the capacity curve (Blume et al. 1961; Freeman et al. 1975) after damage and to improve the estimation of the fragility function relative to the type of structure (Lin et al. 2012; Banerjee and Chi 2013; Bothara et al. 2010).

In this way it is possible to assess quasi in real-time the damage level and usability of the monitored buildings, after a strong earthquake, according to the procedure shown in Fig. 4 that highlights how the territory monitoring can improve the resilience of the urban context.

One of the main critical point of this system is the infrastructure network that must have a lower vulnerability than that of the buildings, in order to keep in service the communication system in case of earthquake, because it provides information on the critical scenarios that occur.

Berdica and Mattsson (2007) defined network vulnerability as a susceptibility to incidents that can result in a service ability loss. Vulnerability can be evaluated also in terms of connectivity loss (Barzel and Biham 2009; Albert and Barabási 2002; Gong et al. 2008; Erath et al. 2009; Nazarova 2009).

Figure 5 presents the flowchart of an hypothetical system where each building has a wireless sensor network able to pre-elaborate and store data on the building health and on the occupancy level. The recorded data are sent to the remote server to handle information and to provide the local earthquake loss estimation to the local and regional emergency agencies, in order to design and implement better risk management strategies (ReLuis PDC Project 2010–2013).

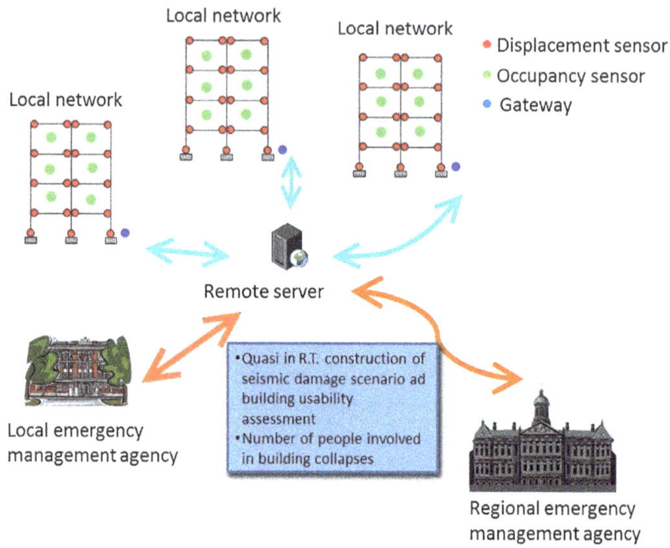

Fig. 5 Data flowchart from monitored buildings to emergency management agencies

3 On the Improving the Resilience of a Hospital in High Seismic Risk Area: The Case Study of the City of Messina (Italy)

Some of the elements of a study in progress are presented; they aim at the assessment of the resilience of the hospital system of the city of Messina (Italy) that assume, in this case, a particular significance, given the high level of seismic vulnerability of the urban context and the extent of the morphology of territory, as well as the number and location of existing health facilities.

The study is the result of an interaction between different disciplinary sectors in the context of emergency management and relief coordination after a strong earthquake.

The proposed methodological approach is based on analyses of territorial, seismic, demographic and medical type performed on the towns of Messina province, for a preliminary assessment of hospital functionality in case of mass casualty afflux due to an earthquake.

The hospital resilience, within the post-event emergency, has been checked on the basis of the seismic damage scenario of the city of Messina (Teramo et al. 2008a, b), referable to an earthquake with a magnitude equal to the 1908 event, as well as of a casualty number determined with reference to the seismic vulnerability of buildings, and related to an adjustment of the casualty treatment capabilities of each hospital due to a relevant medical care request (Teramo and Pinho 2008).

A preliminary evaluation of the resilience of the hospital system has been carried out, under the hypothesis that, after an earthquake, hospital structures do not cope with damage and accessibility of streets leading to hospitals. The goal is the assessment of the hospital response in terms of beds' availability within intra/extra-hospital areas, and a suitable number of doctors and nurses specifically trained.

The characteristics of the territory of the province of Messina are highlighted in a regional context with an indication of the main and secondary routes, ports, airports, hospitals, emergency centers, the so-called *118* (Fig. 6).

This territory is divided into several health districts (Fig. 7), which include 108 municipalities with a total population of about 632,000 inhabitants. Each health district coincides with the territory of one or more municipalities and characterizes the level of health care that can be provided, in relation to both the geographical distribution of health facilities and hospitals (Fig. 7), and the extent of medical care request of citizens, following a strong earthquake.

An assessment of the health vulnerability of the territory is carried out with reference both to the distribution of population by health district (Fig. 8) and the ordinary and disabling diseases that make people more vulnerable in case of an earthquake.

In such a context, it has been assessed per each health district, as significant element of reference, the actual ability of wounded treatment in 24 h (Fig. 9),

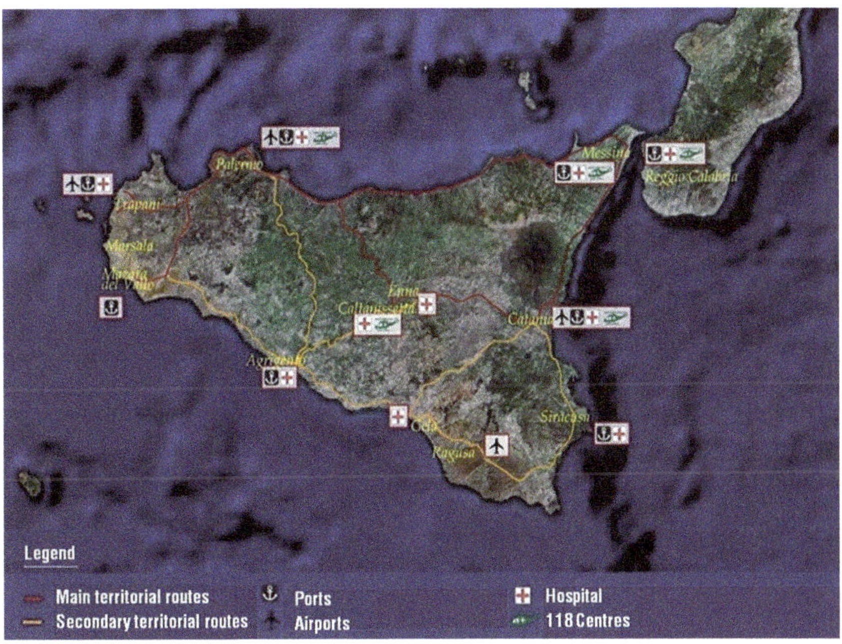

Fig. 6 Sicily region territory and main infrastructures

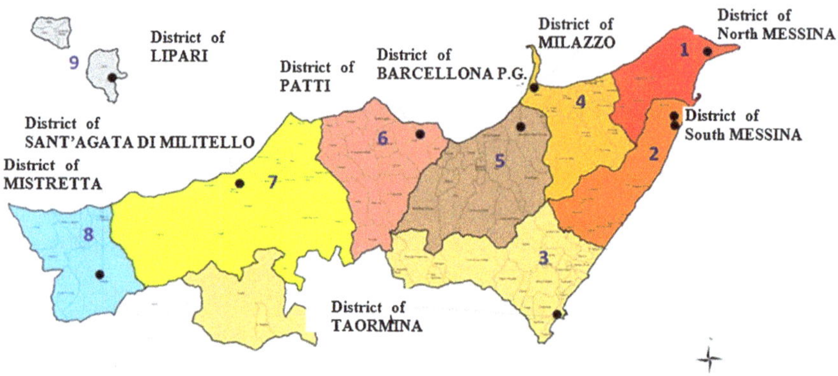

Fig. 7 Health districts of the Messina Province

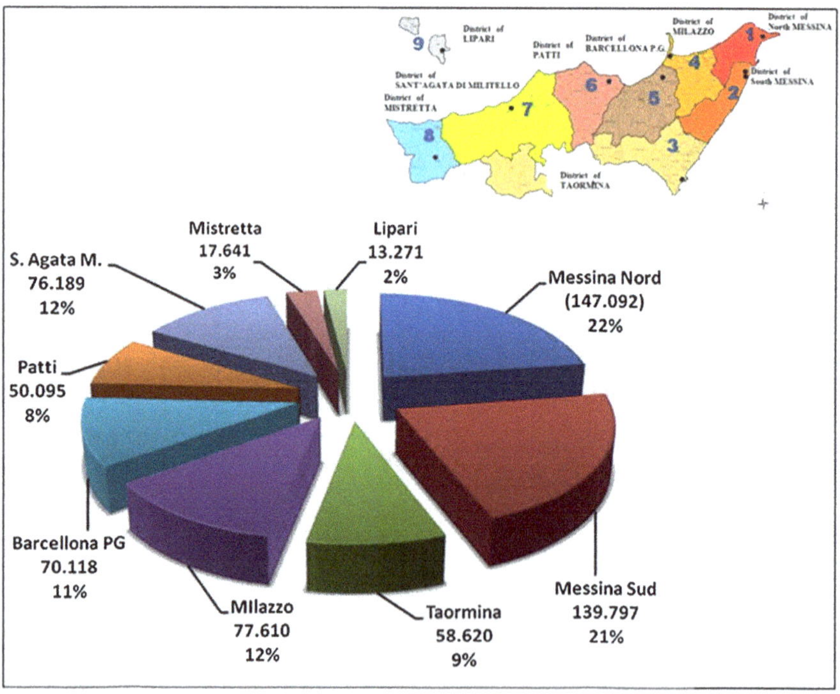

Fig. 8 Population per health district

depending on the number of available doctors and paramedics, as well as the number of beds available in emergency (Fig. 10).

The evaluation of the level of resilience of the hospital system of the Messina province, under the aforementioned working hypothesis of the absence of damage

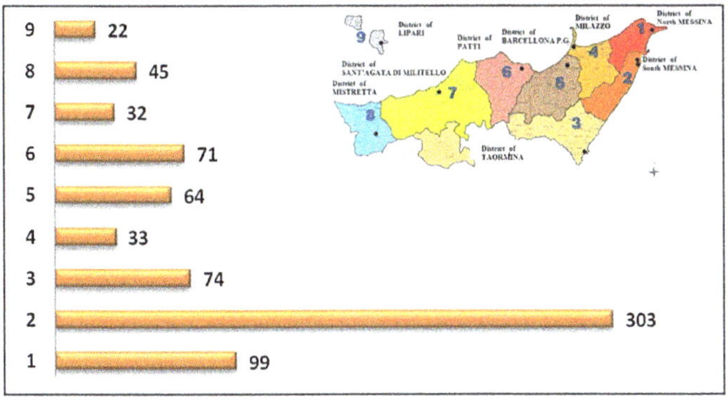

Fig. 9 Medical functional vulnerability per Health District: HTC (health treatment capacity)/24 by mass casualty afflux

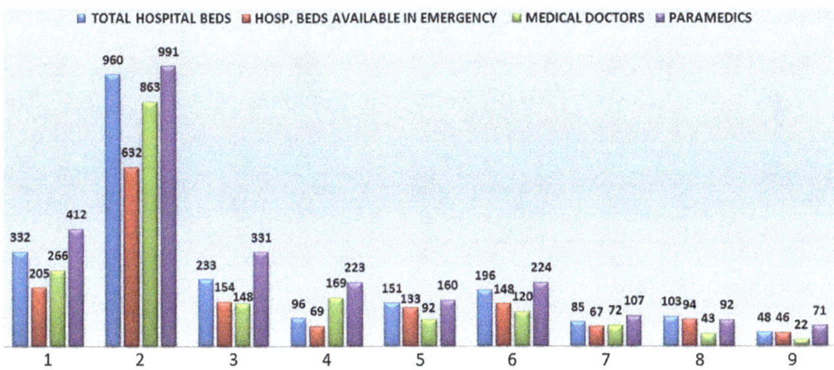

Fig. 10 Medical functional vulnerability per Health District: Number of available beds, doctors and paramedics

to hospital buildings and practicability of roads leading to hospitals after a strong earthquake, must be carried out on the basis of actual possibility of reconfiguration of the health facility, also in terms of locating equipped areas close to the existing hospitals, to be assigned to post-disaster health emergencies.

An analysis of the urban context of Messina, with reference to the areas close the *Piemonte Hospital* located in the centre of the city, has highlighted the availability of a wide military area (Fig. 11) scheduled for disposal (about 90,000 m^2 outdoor and 60,000 m^2 indoor). This area, after a specific assessment of its level of seismic vulnerability, could be suitably equipped to be used as multi-purpose structure to service the community and, throughout emergency after a strong earthquake, to cope with the mass casualty afflux.

The aforementioned example highlights just one of the aspects of the territorial character of the hospital resilience, which must be however expression of a suitable

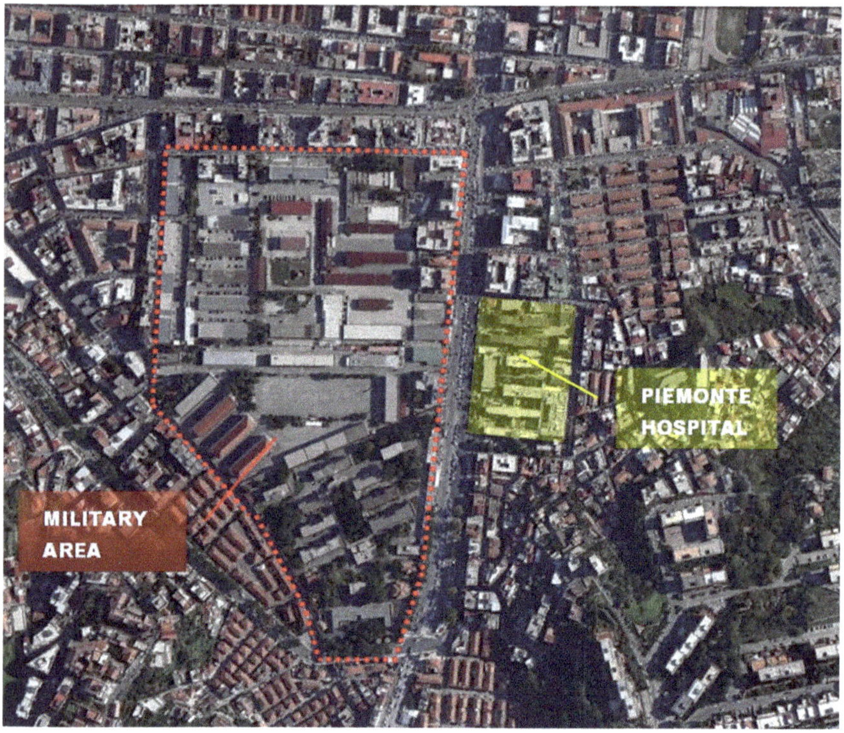

Fig. 11 Resilience of the Piemonte hospital in Messina: a possible location of extra-hospital areas for mass casualty afflux

policy of territory planning in emergency, given that to such extra hospital areas, characterized by low level of systemic vulnerability, can be associated a more timely and effective relief coordination.

It should be noted, finally, that the adoption of a suitable prevention policy must be based however on a preliminary testing of the seismic and systemic vulnerability of the urban context, ensuring the safety of the structures of hospitals and the practicability of the strategic routes, after a strong seismic event.

4 Seismic Safety of Territory: Dissemination, Education and Economic Growth

The success of a program designed to improve the living conditions of a community rests substantially on communication. Any program should, in fact, provide for a channel through which involve people in the activities and protocols carried out, in order to drive their willing participation with those activities.

Education at school, training, information, dissemination are key factors in the arousal of the interest of people for procedures which might appear uselessly complicated at first. Such procedures would be accepted and adopted only if the public benefit and the improvement of the quality of life, deriving from them, are clearly stated.

School is a crucial environment where to start spreading the principle of damage prevention connected to the seismic event. All the scientific disciplines, involved in the description and explanation of the event and in the organization of the human response to it, can concur to shape all-encompassing educational projects directed to kids. The exposition of young people to such topics should aim at raising their knowledge about earthquakes and awareness of the risk and then at making preparedness a part of common life. Educational programs should be designed in compliance with the kids' age and cover both the scientific part (plate tectonics, what is a seismic area, what is an earthquake and how it affects the surface world) and the prevention part (safety procedures in different environments). Becoming familiar with the mechanics of the event and with the right behaviors to adopt when it happens would also dramatically lower the negative psychological effects of the seism both during the event and in the aftermath.

At a higher level, school might become the ideal soil where to grow an innovative attitude related to damage prevention in the newer generations. The revolutionary principle to disseminate among young people, around which shaping a new way to think of the earthquake, is that safety is not a cause of limitations but an opportunity for economic growth. People should see and value the advantages arising from resilient organization of life: research and knowledge promotion, building protocols, materials and standards development, technological devices design and creation, urban and land planning reassessment, social resetting. Resilience could become the spring from which giving life to a more efficient and smarter community.

Education should be matched with an information and dissemination policy in the community about the culture of resilience. This is a harder goal to reach than raising more aware kids, because it deals with adults accustomed to think of the earthquake in fatalistic terms, as a remote event, not related to day by day life. Impossibility of predicting the earthquake is associated with impossibility of preventing the damages it may cause. Lack of preparation and planning is largely tolerated, in accordance with the dominant common view that men are defenseless against nature. The idea of risk prevention is commonly related only with rescuing people involved in the collapse of a building or in other possible scenarios.

A program aiming at the arousal of common awareness about disastrous events cannot be designed as a set of behaviors to be shifted onto people's shoulders. Such behaviors would hardly be integrated in the community day by day life, and would be quickly dropped (as they are as a matter of fact) as marginal details.

The way to effectively change the habits and beliefs of people, about prevention and preparation to a disastrous event, is to drive prevention in the economic fabric of the community, making an opportunity for growth out of it.

Communication should, then, follow two paths, education and information on one side, connection to the productive systems on the other.

Different productive sectors can be influenced and revitalized by the introduction of new protocols and devices related to prevention, safety, security, in a word: resilience. The sector of construction, restoration, urban and building planning is the one which would benefit more by the shift to a newer building conception, resting on energetic and safety efficiency. Businesses operating in this sector would have the chance to start over, rethink their procedures, enter totally new markets, where features, scarcely valued before, have become the heart of the matter, customers (educated, informed citizens) require all new items, materials, devices and appliances in their homes. These innovative solutions would not only meet the needs of the market through adequate business policies, but would also be inspired to environment sustainability, in compliance with the target of lowering seismic vulnerability levels, which is, in fact, related to resilience of the territorial system.

Construction businesses would then start cooperating with professionals newly formed and trained on this scopes, hiring them or requiring their advice. Geologists, engineers, architects and urban planners and technicians of different areas, including those able to certificate the resilience of a building, would broaden their market, supplying at the same time a really useful service.

Research in the field of materials, geotechnics, electronics and robotics would be boosted by such a new planning and building conception.

Designing and producing home and urban automation devices, sensors, robots would not be enough. Cloud platforms and software to manage data coming from the installed sensors would be needed, involving more professional figures: information technologists, software specialists, electronic technicians and so on.

For such a model to come true, all the components of a society (people, actors of the economic growth and institutions) must work as a network and cooperate with one another. Education of the newer generations is only one side of a program aiming at shaping a resilient community, not only ready to tackle the earthquake and its consequences, but also evolving and reshaping its rules, growth directions, relations among people and between men and nature around the idea of resilience. The control room of this profound renovation process can only be the institutional framework, which needs, then, to reform itself in compliance with the aforementioned principles. The program, in fact, needs to shift the principles which inspire the institutional interventions in the scope of prevention.

Institutions think of seismic events and their consequences in terms of building regulations and civil defense in different damage scenarios. They usually show a scarcely proactive attitude, considering land management a footnote in the scope of prevention, and giving no space to technologically innovative solutions in urban planning. In addition, they set building regulations as limitations to business and creativity, failing to stir, among the economic growth actors, the perception of regulations as opportunities to rethink their work procedures, resorting to research centers to find innovative ways to comply with rules, to reach for higher levels of safety and so to satisfy the needs of a market which is more aware than the past on the topic of resilience.

Land management and urban planning on one side, structures and infrastructures building regulations on the other are crucial topics among which driving the components of society and the economic growth actors together to cooperate. Innovating the relations among all these parts would favor the transformation of the government into *e-government*, and the evolution of thc community into a smart community. The *e-government* of a smart community is an integrated system that rests in communication and data sharing technologies to spread the responsibility in managing the territory and running an efficient community life.

Provided that the government promotes solutions and carries out services and social regulation within the scope of the territory's economic growth, its function joins improving the quality of life with solidarity, safety, research and a shared policy for land management. In addition to this usual function, *e-government* also favors the creation of technology clusters in different productive sectors, dealing with sensors to gather data on the running social processes and with the safety constructions industry, and drives the rise of a smart community concerning safety. This innovation contributes to make the general public economy grow as a consequence of the cooperation of Public Administration, companies and people in dealing with the common issues of planning and managing the community life and the territory.

The innovation in the sector of *e-government*, with a specific reference to the scope of territory safety, would be achieved, on one side, through the formation of a participative management model, extended to citizens and businesses, in a network logic, characterized by cooperation in the practices and processes sharing, on the other, through the development of prototypical *ICT* solutions running on cloud platforms to improve the networking of data processing and logistics systems.

E-government technological solutions and services may be adapted to communities on different scales, starting from the small towns and reaching a whole Country. The activities that would stir the change would be all the same: organizing prevention policies and post-earthquake emergencies in a network logic, drawing institutions, businesses, research centers and trained and informed citizens together to make decisions about rules, procedures, standards to adopt and live by.

5 Legal Aspects: The Right to Safety

The right to safety characterizes a new generation right, whose definition is closely related to technological progress within the disaster prevention field and whose realization would spell the end of inertia and delay on the part of public administrations, because they are those who should ensure the effectiveness of this right.

An international interdisciplinary research group has been formed in Messina to deal with the legal and technical aspects of giving legislative expression to this right, with reference both to the general right of citizen to an efficient administration and to real disaster predictability. The group consists of experts in different sectors (law, engineering, geophysics, seismology, medicine, economics, computer science, public administration), whose various skills, knowledge and cooperation are essential in defining the elements of reference of this right within the topic of seismic safety of territory.

At the same time, possible responsibilities of local administrations in this field, referable to an absence or shortage of disaster prevention instruments, will be characterized. The significance of these deficiencies depends, of course, on local risk typologies which vary from one region to another.

This is a challenging task which aims at involving the citizen in the realization of safe living space, by sharing policies of prevention and post-disaster emergency management. In particular, the most advanced international experiences in other countries, subjected to natural disasters, show how the approach to prevention and risk management, which is based on the direct involvement of local communities, provides significant benefits both in view of action effectiveness and cost efficiency.

This approach is based on the concept of *community resilience* that is the ability of a community to cope with critical events that challenge their physical environment and the social fabric.

In this context it is important the role of the *community preparedness* to the risk factor and its direct involvement in the definition of solutions and strategies to protect the safety of citizens. The models of prevention and management of risks from natural disasters are, in fact, based on the promotion of resilience factors in a perspective of *empowerment*, and *participatory risk management*.

An approach to planning seismic safety of the territory, which would set a risk prevention plan based on innovative models and best practices, will therefore:

- Identify the risk factors and territory resources;
- Engage and empower citizens as supervisors of their own safety;
- Select and train local social resources to respond effectively to natural disasters;
- Adopt a policy of seismic safety and prevention that is a prerequisite for the economic development of the territory, identifying in this area shared solutions for emergency management and relief coordination.

6 Concluding Remarks

The territorial and relational dimension of the exposure territorial systems to catastrophic events, the procedures for assessing the seismic vulnerability and resilience, in the more general context of the adoption of policies of prevention and seismic safety, have led to an increased interest in systemic approaches expressly arranged for the analysis of complex systems. This has encouraged the development of studies on the characterization of resilient urban systems that have shown that the extent of the response to a strong seismic event is an expression of a process that depends primarily on the level of preparedness of the community.

In this context, with reference to the mandatory nature of the rules of territory security, as well as to specific elements of accountability, transparency and inclusion, for which the right to safety is the main element of significance, the results of some research activities being carried out in the framework of research projects PRISMA and DIONISO were shown.

The expected impact of such research in the territory is referable to improved dialogue between the actors of economic development in the context of a shared project on territory safety. Moreover, it may contribute concretely to the institutional repositioning of Public Administration in the context of *e-government* of the territory, with the aim of fostering the adoption of a safety policy, consistent with the actual level of the seismic vulnerability of the selected area and the characteristics of resilience of the territorial system. Contextually, it could encourage the emergence of an innovative technological entrepreneurship, reference element for the competitive repositioning of the companies in their respective product segments.

References

Agarwal, J., Liu, M., & Blockley, D. (2011). A systems approach to vulnerability assessment. In *Proceedings of the 1st International Conference on Vulnerability and Risk Analysis and Management (ICVRAM 2011)*, pp 230–237.

Albert, R., & Barabási, A.-L. (2002). Statistical mechanics of complex networks. *Reviews of Modern Physics, 74*(1), 47–97.

Banerjee, S., & Chi, C. (2013). State-dependent fragility curves of bridges based on vibration measurements. *Probabilistic Engineering Mechanics, 33*, 116–125.

Barzel, B., & Biham, O. (2009). Quantifying the connectivity of a network: The network correlation function method. *Physical Review E 80*(4), 046104-1-11.

Berdica, K., & Mattsson, L.G. (2007). Vulnerability: a model-based case study of the road network in Stockholm. In, *Critical Infrastructure*, (pp. 81–106). Springer, Berlin.

Blume, J. A., Newmark, N. M., & Corning, L. M. (1961). *Design of multi-story reinforced concrete buildings for earthquake motions*. Chicago, USA: Portland Cement Association.

Bothara, J. K., Dhaka, R. P., & Mander, J. B. (2010). Seismic performance of an unreinforced masonry building: An experimental investigation. *Earthquake Engineering and Structural Dynamics, 39*(1), 45–68.

Brindley, B. J., Holt, J., & Palmer, I. G. (1973). Acoustic emission-3: The use of ring-down counting. *Non-Destructive Testing, 6*(6), 299–306.

Carpinteri, A., & Lacidogna, G. (2006). Damage monitoring of an historical masonry building by the acoustic emission technique. *Materials and Structures, 39*, 161–167.

Carpinteri, A., Lacidogna, G., & Pugno, N. (2005). Creep monitoring in concrete structures by the acoustic emission technique. In G. Pijaudier-Cabot, B. Gérard, & P. Acker (Eds.), *Creep, shrinkage and durability of concrete and concrete structures. Proceedings of the 7th CONCREEP Conference, Nantes, France, 2005* (pp. 51–56). London: Hermes Science Publishing.

Carpinteri, A., Lacidogna, G., & Niccolini, G. (2007). Acoustic emission monitoring of medieval towers considered as sensitive earthquake receptors. *Natural Hazards and Earth Systems Sciences, 7*, 251–261.

Carreño, M. L., Cardona, O., & Barbat, A. (2007). Urban seismic risk evaluation: A holistic approach. *Natural Hazards, 40*, 137–172.

Cornell, C. A. (1968). Engineering seismic risk analysis. *Bulletin of the Seismological Society of America, 58*(5), 1583–1606.

Crowley, H. (2005). DBELA: A new methodology for earthquake loss assessment. In *Proceedings of the SECED Young Engineers Conference* (pp. 1–6), 21–22 Mar 2005, Bath, UK.

Crowley, H., Colombi, M., Borzi, B., Faravelli, M., Onida, M., Ml, Lopez, et al. (2009). A comparison of seismic risk maps for Italy. *Bulletin of Earthquake Engineering, 7*(1), 149–180.

Daniell, J. E. (2011). Open source procedure for assessment of loss using global earthquake modelling software (OPAL). *Natural Hazards and Earth Systems Sciences, 11*, 1885–1900.

Dimova, S. L., & Negro, P. (2005). Seismic assessment of an industrial frame structure designed according to Eurocodes Part 2: Capacity and vulnerability. *Engineering Structures, 27*, 724–735.

Erath, A., Birdsall, J., Axhausen, K. W., & Hajdin, R. (2009). Vulnerability assessment methodology for swiss road network. *Transportation Research Record, 2137*, 118–126.

Erduran, E., Lang, D. H., Lindholm, C. D., Toma-Danila, D., Balan, S. F., Ionescu, V., Aldea, A., Vacareanu, R., & Neagu, C. (2012). Real-time earthquake damage assessment in the Romanian-Bulgarian border region. In *Proceedings of the 15th WCEE* 2012 (pp. 1–10).

Federici, F., Alesii, R., Colarieti, A., Graziosi, F., & Faccio, M. (2013). Design and validation of a wireless sensor node for long term structural health monitoring. In *Sensors, 2013 IEEE Conference* (pp. 1–4), 3–6 Nov 2013, Baltimore, MD, USA.

FEMA. (1999). *HAZUS®99: Earthquake loss estimation methodology: User's Manual*. Washington, D.C., USA: Federal Emergency Management Agency.

FEMA 366. (2001). HAZUS®99 Estimated annualized earthquake losses for the United States. Federal Emergency Management Agency, Washington, D.C., USA, Feb 2001.

FEMA Earthquake loss estimation methodology: HAZUS®MH 2.1—Advanced Engineering Building Module (AEBM)—Technical and User's Manual.

Ferraris, F., Gorini, I., Grimaldi, U., & Parvis, M. (1994). Calibration of three-axial rate gyros without angular velocity standards. *Sensors and Actuators A, 41–42*, 446–449.

Fiorini, E., Borzi, B., & Iaccino, R. (2012). Real Time damage scenario: case study for the L'Aquila earthquake. In *Proceedings of the 15th WCEE* 2012 (p. 9).

Freeman, S. A., Nicoletti, J. P., & Tyrell, J. V. (1975). Evaluations of existing buildings for seismic risk—A case study of Puget Sound Naval Shipyard, Bremerton, Washington. In *Proceedings of U.S. National Conference on Earthquake Engineering* (pp. 113–122), Berkeley, USA.

Gattulli, V., Graziosi, F., Federici, F., Potenza, F., Colarieti, A., & Lepidi, M. (2013). Structural health monitoring of the Basilica S. Maria di Collemaggio. In *Proceedings of the Fifth International Conference on Structural Engineering, Mechanics and Computation (SEMC 2013)* (p. 6), 2–4 Sept 2013, Cape Town, South Africa.

Gong, B., Liu, J., Huang, L., Yang, K., & Yang, L. (2008). Geographical constraints to range-based attacks on links in complex networks. *New Journal of Physics, 10*(1), 013030.

Gregori, G. P., Lupieri, M., Paparo, G., Poscolieri, M., Ventrice, G., & Zanini, A. (2007). Ultrasound monitoring of applied forcing, material ageing, and catastrophic yield of crustal structures. *Natural Hazards Earth System Science, 7*, 723–731.

Gregori, G. P., Paparo, G., Poscolieri, M., Rafanelli, C., & Ventrice, G. (2012). Acoustic emission (AE) for monitoring stress and ageing in materials, including either manmade or natural structures, and assessing paroxysmal phases precursors. In W. Sikorski (Ed.), *Acoustic emission* (pp. 365–398). Rijeka, Croatia: InTech Publisher. ISBN 978-953-51-0056-0.

Grosse, C. U., Reinhardt, H. W., & Finck, F. (2003). Signal based acoustic emission techniques in civil engineering. *ASCE Journal of Materials and Civil Engineering, 15*, 274–279.

ISO 16063-15:2006. (2010). Methods for the calibration of vibration and shock transducers—Part 15: Primary angular vibration calibration by laser interferometry (p. 42).

ISO 16063-21:2003. (2013). Methods for the calibration of vibration and shock transducers—Part 21: Vibration calibration by comparison to a reference transducer (p. 29).

Krishnamurthy, V., Fowler, K., & Sazanov, E. (2008). The effect of time synchronization of wireless sensor on the modal analysis of structures. *Smart Materials and Structures, 17*(5), 1–13.

Kurata, N., Spencer Jr, B. F., & Ruiz-Sandoval, M. (2004). Risk monitoring by ubiquitous sensor network for hazard mitigation. In *Proceedings of International Symposium on Network and*

Center-Based Research for Smart Structures Technologies and Earthquake Engineering (SE04) (pp. 1–6), 6–9 July 2004.

Lin, S.-L., Li, J., Elnashai, A. S., & Spencer, B. F, Jr. (2012). NEES integrated seismic risk assessment framework (NISRAF). *Soil Dynamics and Earthquake Engineering, 42*, 219–228.

Lo Castro, F., De Luca, M., Iarossi, S. (2007). MEMS accelerometers calibration at low frequencies. In *Sensors and Microsystems: Proceedings of the 12th Italian Conference, Napoli, Italy* (pp. 343–349), 12–14 Feb 2007. World Scientific Publishing Company.

Lo Castro, F., De Luca, M., & Iarossi, S. (2013). Localization, recognition and classification of a superficial seismic source in a inhomogeneous mean. In *Sensors and Microsystems—Proceedings of the 17th National Conference, Brescia, Italy* (pp. 47–51), 5–7 Feb 2013. Springer, Berlin.

Lynch, J. P., Loh, K. J., Lynch, J. P., & Loh, K. J. (2006). A summary review of wireless sensors and sensor networks for structural health monitoring. *The Shock and Vibration Digest, 38*(2), 91–128.

Martinelli, E., Zampetti, E., Pantalei, S., Lo Castro, F., Santonico, M., Pennazza, G., et al. (2007). Design and test of an electronic nose for monitoring the air quality in the international space station. *Microgravity Science and Technology, 19*(5–6), 60–64.

McAslan, A. (2010). *The concept of resilience—understanding its origins, meaning and utility* (pp. 1–13). Adelaide, Australia: Torrens Resilience Institute Paper.

Molina, S., Lang, D. H., & Lindholm, C. D. (2007). SELENA v2.0 User and Technical Manual v2.0, NORSAR2007.

Molina, S., Lang, D. H., & Lindholm, C. D. (2010). SELENA—An open-source tool for seismic risk and loss assessment using a logic tree computation procedure. *Computers & Geosciences, 36*(3), 257–269.

Nazarova, I. A. (2009). Solution methods for the vertex variant of the network system vulnerability analysis problem. *Journal of Computer and Systems Sciences International, 48*(4), 581–591.

Niccolini, G., Xu, J., Manuello, A., Lacidogna, G., & Carpinteri, A. (2012). Onset time determination of acoustic and electromagnetic emission during rock fracture. *Progress in Electromagnetics Research Letters, 35*, 51–62.

Ohtsu, M. (1996). The history and development of acoustic emission in concrete engineering. *Magazine of Concrete Research, 48*, 321–330.

Paolesse, R., Lvova, L., Nardis, S., Di Natale, C., D'Amico, A., & Lo Castro, F. (2008). Chemical images by porphyrin arrays of sensors. *Microchimica Acta, 163*(1–2), 103–112.

Picozzi, M., Bindi, D., Pittore, M., Kieling, K., & Parolai, S. (2013). Real-time risk assessment in seismic early warning and rapid response: A feasibility study in Bishkek (Kyrgyzstan). *Journal of Seismology, 17*, 485–505.

Pitilakis, K. D., & Kakderi, K. G. (2011). Seismic risk assessment and management of lifelines, utilities and infrastructures. In *5th International Conference on Earthquake Geotechnical Engineering, Santiago, Chile* (pp. 10–13), Jan 2011.

Pollock, A. A. (1973). Acoustic emission-2: Acoustic emission amplitudes. *Non-Destructive Testing, 6*, 264–269.

Poscolieri, M., Rafanelli, C., & Zimatore, G. (2012). Viable precursors of paroxysmal phenomena as detected by applying RQA to acoustic emission time series. *Journal of Acoustic Emission, 30*, 29–39.

ReLuis PDC Project 2010–2013. www.reluis.it.

Rossetto, T., & Elnashai, A. (2003). Derivation of vulnerability functions for European-type RC structures based on observational data. *Engineering Structures, 25*, 1241–1263.

Schuler, A. R., Grammatikos, A., & Fegley, K. A. (1967). Measuring rotational motion with linear accelerometers. *IEEE Transactions on Aerospace and Electronic Systems AES-3, 3*, 465–472.

Shah, S. P., & Li, Z. (1994). Localization of microcracking in concrete under uniaxial tension. *ACI Materials Journal, 91*, 372–381.

Teramo, M. S., & Pinho, R. (2008) Seismic vulnerability assessment of hospital buildings in the city of Messina. In *Proceedings of the 14th World Conference of Seismic Engineering*, 12–17 Oct 2008, Beijing, China.

Teramo, M. S., Teramo, A., Bottari, A., Termini, D., Crowley, H., Pinho, R., et al. (2008a). *From the Messina 1908 earthquake to scenario damage assessment in 2008* (p. 36). Pavia, Italy: Iuss Press. ISBN 978-88-6198-027-3.

Teramo, M. S., Crowley, H., Lopez, M., Pinho, R., Cultrera, G., Cirella, A., Cocco, M., Mai, M., & Teramo, A. (2008b). A damage scenario for the city of Messina, Italy, using displacement-based loss assessment. In *Proceedings of the 14th World Conference of Seismic Engineering* (pp. 1–8), 12–17 Oct 2008, Beijing, China.

Veldman, C. S. (2006). ISO 16063; A comprehensive set of vibration and shock calibration standards. In *Proceedings of the XVIII IMEKO World Congress Metrology for a Sustainable Development* (pp. 1–5), 17–22 Sept 2006, Rio de Janeiro, Brazil.

Yoon, D. J., Lee, S., Kim, C. Y., & Seo, D. C. (2007). Acoustic emission diagnosis system and wireless monitoring for damage assessment of concrete structures. In *Proceedings of NDT for Safety* (pp. 301–308), 7–9 Nov 2007, Prague, Czech Republic.

Numerical Study of the Seismic Response of a Mid-Rise RC Building Damaged by 2009 Tucacas Earthquake

Juan Carlos Vielma, Angely Barrios and Anny Alfaro

1 Introduction

Given the constant seismic hazard that the buildings in Venezuela are exposed to, it is necessary to design structures whose behaviour is appropriate to withstand to seismic actions, in other words, that these structures will be able to dissipate energy, have enough strength and adequate deformation capacity without reach the collapse. Although the current seismic code regulations establish minimum requirements to reduce the risk to the collapse of the buildings, these requirements are insufficient in order to assess the seismic behaviour of the whole structure and each of the members who compose it. For this reason in this paper it is proposed to evaluate the performance of this particular building, based on advanced non-linear analysis methods, taking into consideration mathematical models that include the calculation of the displacements that they can achieve when are subjected to strong motions, so providing more efficient tools to design structures which are likely to suffer lower damages during a strong motion.

As a highlight characteristic of the building subject of this study, is that it has been subjected to a seismic event in 2009 (see Fig. 1), which occurred in the Venezuelan coast with Mw = 6.4, causing light damage to non-structural elements (masonry and stucco-finishes). The main objective of this research is to evaluate the

J.C. Vielma (✉)
Universidad Lisandro Alvarado UCLA, Universidad de las Fuerzas Armadas ESPE, Av. La Salle entre Av. Las Industrias y Av. J.A. Benítez, Barquisimeto, Venezuela
e-mail: jcvielma@ucla.edu.ve; jcvielma@espe.edu.ec

A. Barrios · A. Alfaro
UCLA-CIMNE Classroom, Av. La Salle entre Av. Las Industrias y Av. J.A. Benítez, Barquisimeto, Venezuela
e-mail: angely.brrs@gmail.com

A. Alfaro
e-mail: annybalfaro@gmail.com

S. D'Amico (ed.), *Earthquakes and Their Impact on Society*,
Springer Natural Hazards, DOI 10.1007/978-3-319-21753-6_12

Fig. 1 Tourist facilities damaged by 2009 Tucacas earthquake

seismic response and therefore the vulnerability of the building to probable future seismic events; this main objective leads to increase knowledge on earthquake resistant behaviour of similar reinforced concrete framed buildings located in seismic-prone areas, thus enabling improved design procedures and managing to improve the safety of the population in less developed countries with similar structures.

In this paper, a new method for calculating fragility curves based on the analysis of the results of incremental dynamic analysis (IDA), formulated by Vamvatsikos and Cornell (2002), is also presented. The procedure was applied plotting the evolution of the interstorey drifts versus the roof drift of the building affected by a specific strong motion. Additionally, non-linear dynamic results provide useful values for the evaluation of the effect of pounding for the seismic performance of the entire structure.

2 Description of the Case Studied

The case studied is a tourist RC framed structure, which consists of four separate modules, three of which are part of a building of 8 levels: a central module and two symmetrical lateral modules. The building also has a one-level independent structure destined for parking use, see Fig. 2. The separation between modules in the building is 35 cm measured center to center of adjacent columns, resulting in a 3 cm total gap between them.

The adopted gravity load resisting-system was defined by means of ductile rein-forced concrete frames which represents an interesting case, because throughout the country similar buildings adopt this structural typology.

Lateral modules have 3 frames in x direction, and the central module has 4 frames. While in y direction, the lateral modules have 5 frames each, and the central

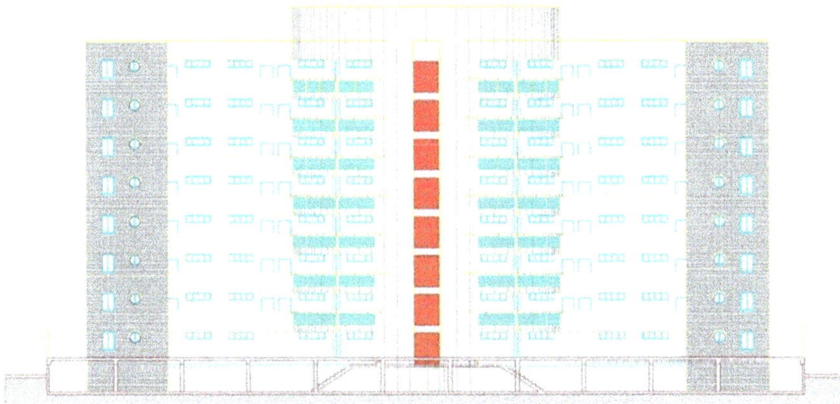

Fig. 2 Front view of the studied building

module has 4 frames. The story height is 3.10 m for both modules. The system of floor slabs and roof consists of one-way ribbed slabs of 25 cm thick, with ribs oriented according to the *y* axis. The stairs are at the core of circulation which is located within the central module, see Fig. 3.

The geometric properties of the reinforced concrete members are variable for each module, the columns dimensions vary from 40 × 40 cm to 40 × 90 cm, while the beams have dimensions ranging from 30 × 50 m and 30 × 60 cm. The foundation consists in a 50 cm thick RC mat for the main modules of the building, see Fig. 4.

In order to evaluate the behaviour of the structure designed according to current Venezuelan codes, a new structure which maintains the same original building

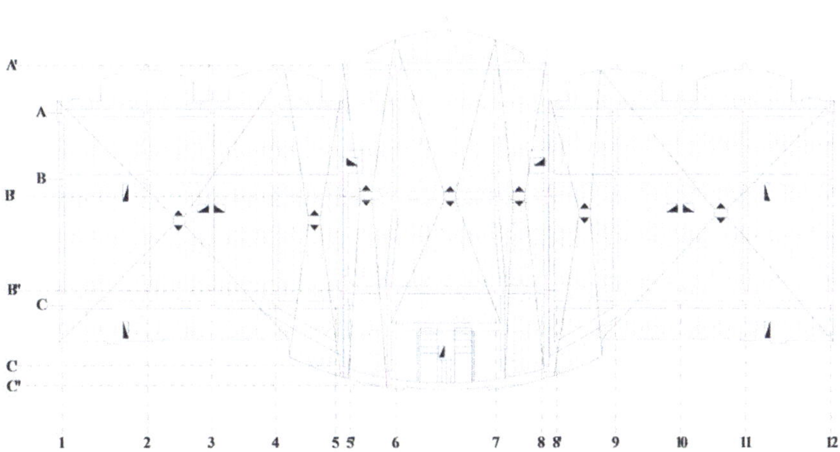

Fig. 3 Plan view of the studied building

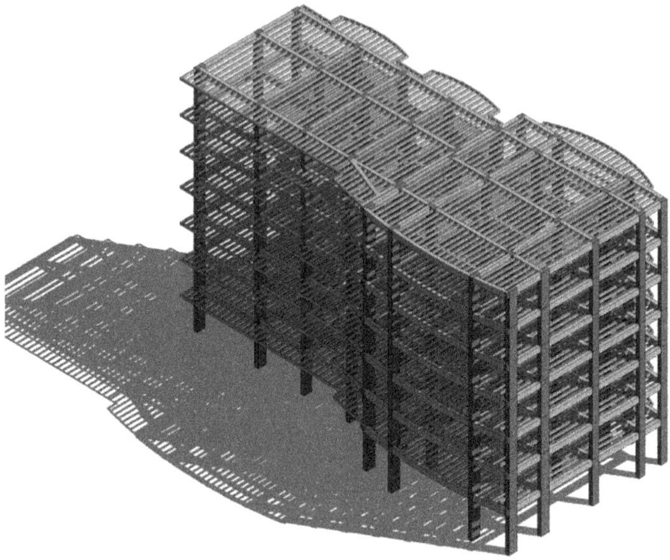

Fig. 4 Isometric view of the central and left modules of the building

structural characteristics and service load cases was designed, increasing the dimensions of beams and columns based on the principle of strong column-weak beam (Fardis 2009), and fulfilling the requirements of admissible interstorey drifts.

The general procedure includes the verification of the maximum interstorey drift obtained from the combination of the modal displacements. For this purpose the displacement should be amplified by the response reduction factor (R) used to calculate the inelastic design spectrum, with a displacement de-amplification factor equal to 0.8. Additionally, a second verification criterion for permissible maximum value corresponding to the interstorey drifts, even more demanding with the stiff-ness of the members of the structure, was applied. According to this criterion the displacements obtained in the elastic analysis increased by the displacement amplification factor equal to 1.5, which depends on the overstrength factor and structural redundancy (Vielma et al. 2011). Subsequently these inelastic displacements were used to calculate inelastic interstorey drifts and finally were compared with the maximum value pre-scribed by Covenin 1756-1:2001 seismic code ($\delta = 0.018$) (FONDONORMA: Norma venezolana Covenin 1756-1 2001).

Other additional criterion applied to the design and detailing of the columns was the verification of the maximum longitudinal reinforcement ratio ρ, which must not exceed 2.5 %, (Vielma et al. 2011). There were also considered the special confinement zones, where the presence of stirrups is augmented in order to ensure the ductile behaviour, pre-venting the fragile failure of the structural members near the beam-columns joints.

Table 1 Resulting sections and reinforcement of columns of the re-designed building

Member	Cross section	Longitudinal reinforcement	ρ (%)
Columns	60 × 90	22 Ø 7/8″	1.55
		20 Ø 7/8″	1.41
	60 × 80	20 Ø 1″	2.05
		18 Ø 1″	1.84
		20 Ø 7/8″	1.58
		18 Ø 7/8″	1.43
		18 Ø 3/4″	1.06
	60 × 70	20 Ø 7/8″	1.81
		18 Ø 7/8″	1.45
		16 Ø 7/8″	1.84
		16 Ø 3/4″	1.08
	60 × 60	16 Ø 3/4″	1.26
	50 × 50	14 Ø 5/8″	1.11

Table 2 Resulting sections and reinforcement of beams of the re-designed building

Member	Cross section	Longitudinal reinforcement (sup)	Longitudinal reinforcement (inf)
Beams	30 × 60	6 Ø 7/8″	6 Ø 7/8″
		5 Ø 7/8″	5 Ø 7/8″
	30 × 55	6 Ø 7/8″	6 Ø 7/8″
		4 Ø 7/8″	4 Ø 7/8″
		3 Ø 7/8″	3 Ø 7/8″
		4 Ø 3/4″	4 Ø 3/4″
		3 Ø 3/4″	3 Ø 3/4″
		2 Ø 3/4″	2 Ø 3/4″
	30 × 50	5 Ø 7/8″	5 Ø 7/8″
		4 Ø 7/8″	4 Ø 7/8″
		2 Ø 7/8″	2 Ø 7/8″
		4 Ø 3/4″	4 Ø 3/4″
		3 Ø 3/4″	3 Ø 3/4″
		2 Ø 3/4″	2 Ø 3/4″
		4 Ø 5/8″	4 Ø 5/8″

In Table 1 are show the geometrical characteristics of the columns resulting from the re-designed modules, and the corresponding longitudinal steel reinforcement ratio of the columns (in %).

In Table 2 are shown the geometrical characteristics and the resulting reinforcement of the beams.

3 Structural Analysis

Analytical methods have evolved gradually as advances in earthquake engineering have been occurred. So in order to evaluate the seismic response of the building there were possible to apply pseudo-static and dynamic non-linear analysis, which allow to find weaknesses that the elastic analysis cannot be able capture and also meet more closely the response of a building when is under a strong motion.

The first step of conventional pseudo-static analysis is the analysis with conventional incremental lateral forces, where the structure is subjected first of all to gravity forces, being the lateral loads equivalent to the effect of the earthquake, initialized to zero and increased sequentially along the building height. It is common to use a forces pattern distributed following an inverted triangle shape, which approximates the first mode of vibration of the whole structure. The software used in order to perform non-linear analysis was Zeus-NL (ZEUS NL 2010). This software is adapted to assess the seismic response of complex structures, as is the case studied. All structural members of the building were modelled considering the constitutive and geometric nonlinearities. The contribution of transverse reinforcement was obtained using the model developed by Mander et al. (1988).

The modelling process was performed using the technique of finite elements, where the columns and beams of the frames that belong to the structure were discretized into four elements per member, see Fig. 5. Thus, on each element two

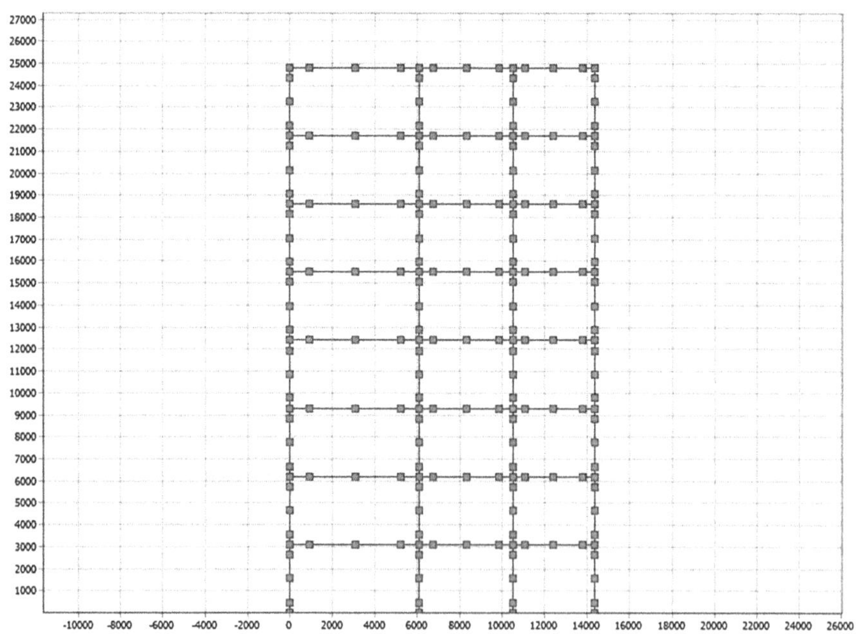

Fig. 5 Typical discretization of frames

zones were considered: a confined area, corresponding to the areas that are designed with special confinement (near beam-column joints) and ordinary-confined areas, corresponding to the central sections of beams and columns, as is prescribed by the Venezuelan reinforced concrete code 1753:2006 (FONDONORMA: Norma venezolana Fondonorma 1753 2006).

The results of the analysis were plotted showing the displacement of a control node located in the gravity centre on the roof level. This type of representation is known as the "pushover curve" of the frame or building.

From non-linear static analysis is possible to calculate a set of parameters that characterize the seismic response of the whole structure. These parameters are: the overall ductility (μ), over-strength (Ω) and the inherent response reduction factor (*Rinh*) (Vielma et al. 2011).

Figure 6 shows the computed values of the global ductility of the original and re-designed buildings' frames. Figure 7 shows the overstrength values and Fig. 8 shows the values of the inherent response reduction factor, computed as the product of the above mentioned values.

As an important feature, all frames must meet the basic requirement in seismic design, as it should have values of overstrength greater than 1.0. This condition is necessary but not sufficient to ensure that a structure has enough lateral strength in order to resist minimum code prescribed seismic forces. Other important parameter in order to evaluate the seismic performance is the global ductility, indicating what the failure mode of the structure, which shows that all values are greater than 1.0, reflecting the structure has the capacity to dissipate energy when the structure is subjected to a strong earthquake. It is important to note that the values obtained for the lateral resizing module show a slight increase of the overstrength values.

Due the massive stresses that an earthquake imposes to structures and the development of procedures of structural analysis, the non-linear response of the building was determined; all of the modules of the building were subjected to three

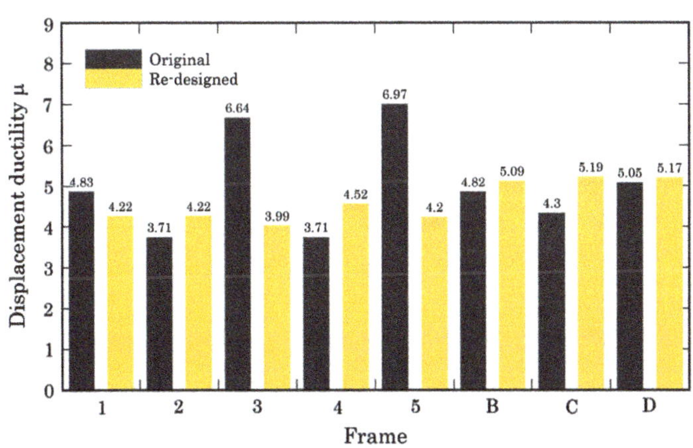

Fig. 6 Values of the global ductility μ

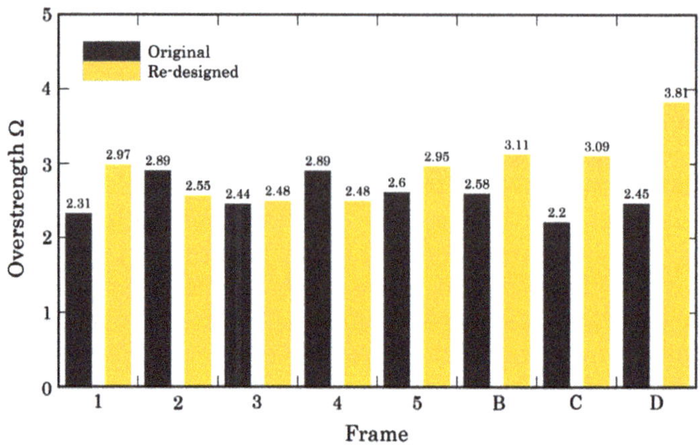

Fig. 7 Values of the overstrength Ω

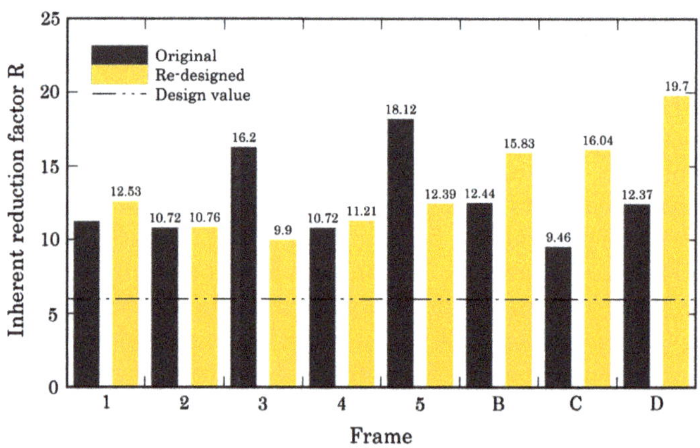

Fig. 8 Values of the inherent response reduction factors *Rinh*

synthetic accelerograms and the record of the Tucacas earthquake, amplified to meet three specific hazard levels associated with three Limit States, which are shown in Table 3 (Elnashai and Di Sarno 2008).

Each earthquake was scaled in order to reproduce three events of a different magnitude according to the procedure formulated in (Kappos and Stefanidou 2009):

- Frequent earthquake ($Tr = 95$ years): It has a de-amplification factor of 0.4 for a exceedance probability of 50 % in 50 years. The response of the structure to an earthquake of this magnitude allows verifying the serviceability Limit State (SLS). The interstorey drifts values must be kept below 0.5 %.
- Rare Earthquake ($Tr = 475$ years): Presents an amplification of 1.0 and a probability of exceedance of 10 % in 50 years. By applying this accelerogram it is

Table 3 Intensity levels of the applied accelerograms

Intensity	Limit state	Return period (years)	Exceedance probability (in 50 years) (%)
Frequent	Service	95	50
Rare	Repairable	475	10
Very rare	Collapse	2475	2

possible to evaluate the Repairable Damage Limit State (RDLS). In order to do not violate this Limit State the computed interstorey drifts must not exceed 1.5 %

- Very rare earthquake (*Tr* = 2475 years): Finally, this earthquake has an amplification factor of 2.0 and a probability of exceedance of 2 % in 50 years. The evaluation of the structure under this earthquake, to verify the Limit State of collapse prevention (LSCP), whit the maximum interstorey drifts must be under 3 % to satisfy this Limit State (Fardis 2009).

4 Results

Dynamic results are particularly important, first of all because they represent the response of structures against accidental loads like earthquakes and second, because the effect of the dynamic action usually impose actions more severe than the effect of the action considered in the pseudo-static analysis, since displacements of different sign occur, causing higher angular distortions in the elements.

Finally, using dynamic non-linear analysis the maximum inelastic displacements were computed in each of the frames of each defined module, in order to determine the values of global drifts and interstorey drifts. These values were determined according to:

$$\delta_{global} = \frac{D_{Roof} \times 100}{H_{Total}} \tag{1}$$

In this Equation δ_{Global} is the global drift, D_{Roof} is the roof displacement of the control node and H_{Total} is the roof height. The interstorey drift Δ_i is obtained from:

$$\Delta_i = \frac{|D_{i+1} - D_i|100}{H_{i+1} - H_i} \tag{2}$$

where D_{i+1} and D_i are the inelastic displacements of the i + 1 and i stories, respectively, and H_{i+1} and H_i are the heights of the i + 1 and i stories, respectively. Figure 9 shows the evolution of the global drift versus time, for the central module.

This figure represented global drifts obtained by applying three synthetic accelerograms scaled to meet a very rare event and the accelerogram obtained from Tucacas earthquake. It is important to note that the dynamic response is satisfactory,

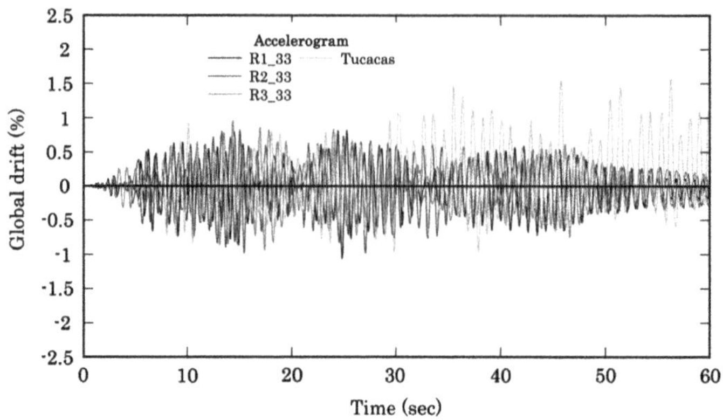

Fig. 9 Evolution of the global drift according to time computed from the synthetic accelerograms and the Tucacas accelerogram

since for each of the considered accelerograms the global drifts have not exceeded the threshold that mark the imminent collapse (2.5 %). Similar behaviour has been obtained for the other modules, and concluded that the earthquake resistant design is capable of producing structures that do not even reach the collapse even they are exposed to earthquakes of very rare occurrence.

The evolution of interstorey drifts of the central module computed for one of the considered synthetic accelerograms is shown in Fig. 10. Note that none of the interstorey drifts exceed the threshold associated with the Limit States of Table 3.

Note that none of the buildings exceed the thresholds that mark the Limit States of Table 3, so the global and local response of these buildings meet the main goal of the seismic design.

4.1 Fragility Curves

Fragility curves represent the margin between two different states of damage. They have been formulated based on numerical simulations or experimental test. Regardless of the method applied, it is essential to define a set of thresholds to calculate the state of damage. In this Chapter a new method based on the evolution of the relationship between the interstorey drifts and spectral accelerations is proposed.

This procedure involves the determination of the values of the accelerations (PGA) that correspond to specific values of interstorey drifts corresponding to 0.5, 1, 1.5, 2.25 and 3 %, calculated from the incremental dynamic analysis (IDA). This analysis was performed using a set of three synthetic accelerograms compatible with the design spectrum. This set is obtained from the elastic spectrum using the PACED program (Vielma 2009).

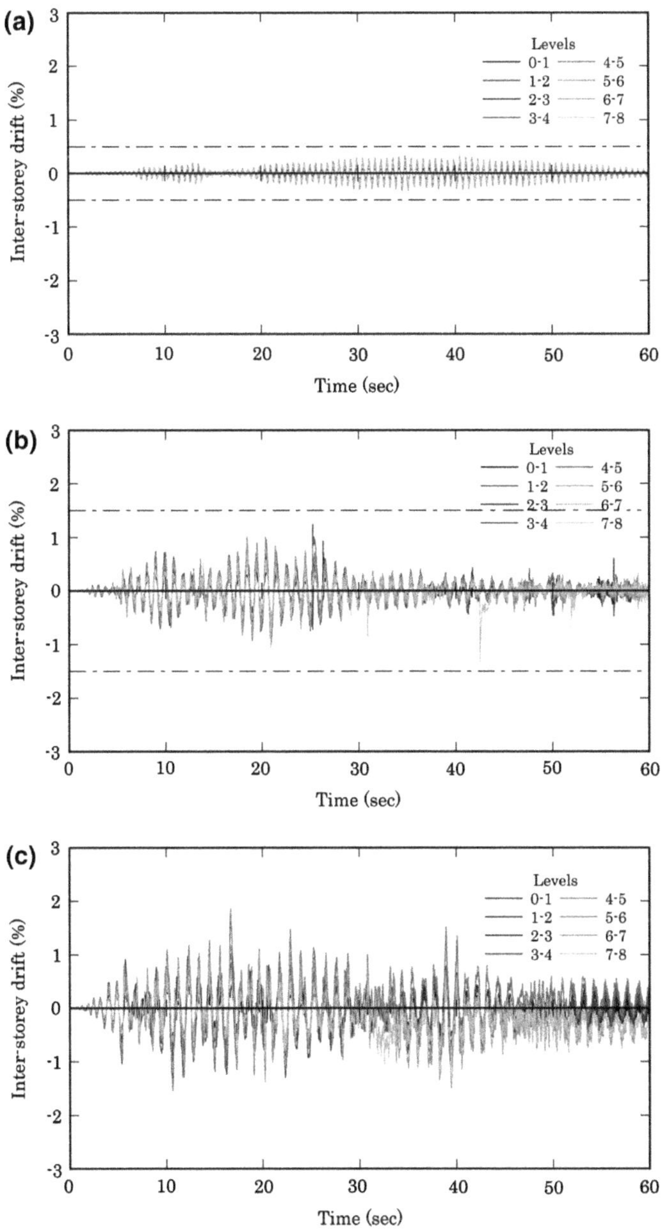

Fig. 10 Evolution of the interstorey drifts according to time, computed from the scaled accelerogram for three intensity levels: **a** frequent, **b** rare and **c** very rare

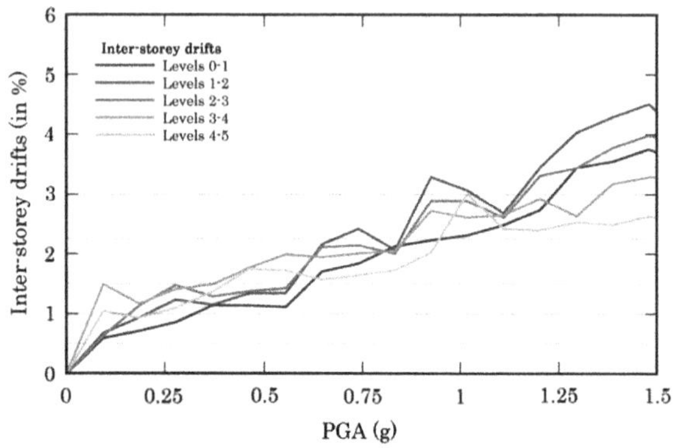

Fig. 11 General procedure for the thresholds of limit states determination

The values of the peak ground acceleration (PGA) corresponding to each threshold are obtained from the intersection of any curve, regardless of the level at which it occurs. A typical representation of the process is shown in Fig. 11.

Fragility curves are obtained using the given spectral accelerations which correspond to the above mentioned damage thresholds and considering a probability density function of the parameters that define the damage states demand corresponding to the lognormal distribution, (Pinto et al. 2006; Barbat et al. 2006, 2008; Lantada et al. 2009).

$$F(S_d) = \frac{1}{\beta_{ds} S_d \sqrt{2\pi}} \exp\left[-\frac{1}{2}\left(\frac{1}{\beta_{ds}} \ln \frac{S_d}{\bar{S}_{d,ds}} \right)^2 \right] \tag{3}$$

Fragility curves computed for lateral module of the original and redesigned building are shown in Fig. 12.

According to Fig. 12, the frames oriented in x direction of the Original Module have a high probability of reaching a state of light to moderate damage if they are subjected to an acceleration of 0.3 g, which is the acceleration prescribed by the Venezuelan seismic code for the location of the building. Moreover is seen that the lateral re-designed module achieves a state of slight damage when subjected to seismic action of 0.3 g. This indicates that these frames would reach repairable damage, indicating that the design is acceptable because it is within the safe zone (Vielma et al. 2010).

In Fig. 13, fragility curves of the frames oriented according to y direction of the original lateral module and the lateral re-designed module are shown. Fragility curves of this figure show a similar behaviour to the frames oriented according to x direction.

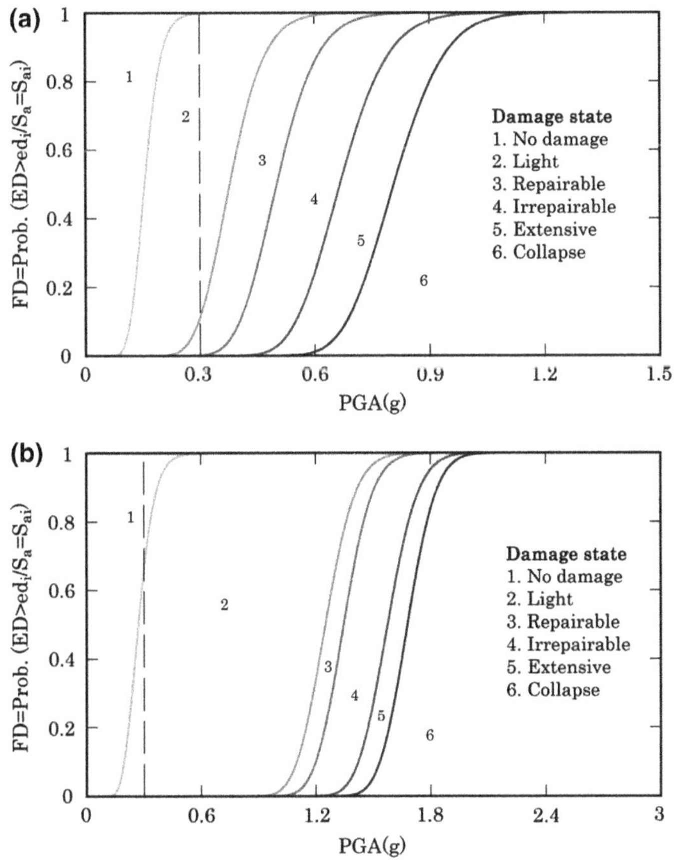

Fig. 12 Fragility curves of the frames of the **a** original lateral module and **b** re-designed lateral module oriented according to *x* direction

4.2 Pounding

The effects of pounding may be critical during a severe earthquake if there is not enough gap between adjacent buildings. As outlined previously, the studied building was splitted into several independent bodies, in order to avoid the effect of plan irregularity. This feature did not produce any adverse effect during Tucacas earthquake.

It is also important to note that the Venezuelan seismic code does not reflects in an explicit way the need to consider the effect of pounding, because there are only recommendations about the divisions of structures when dealing with plan irregularities, such “H” or “C” plan shapes. In order to quantify the damage that occurs in the building by the effect of pounding, in this study the non-linear dynamic response of adjacent bodies of the building have been plotted together, considering

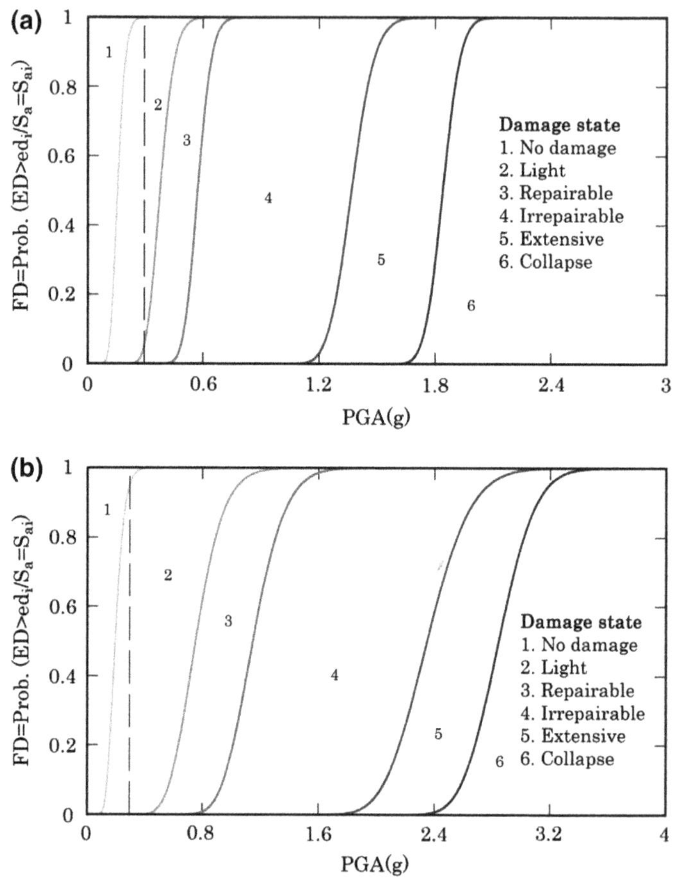

Fig. 13 Fragility curves of the frames of the **a** original lateral module and **b** re-designed lateral module oriented according to *y* direction

a shift equal to the gap between central and lateral module of the original design. Figure 14 shows that adjacent modules reach lateral displacements longer than the gap distance, therefore pounding could occurs for a seismic action defined by a synthetic design-spectrum compatible accelerogram.

5 Contribution to Reduce the Seismic Vulnerability of Society

The numerical results show the appropriate behaviour that the building has against an earthquake of low, medium and high intensity. This type of structure is similar to that found in many reinforced concrete buildings built recently in Venezuela and

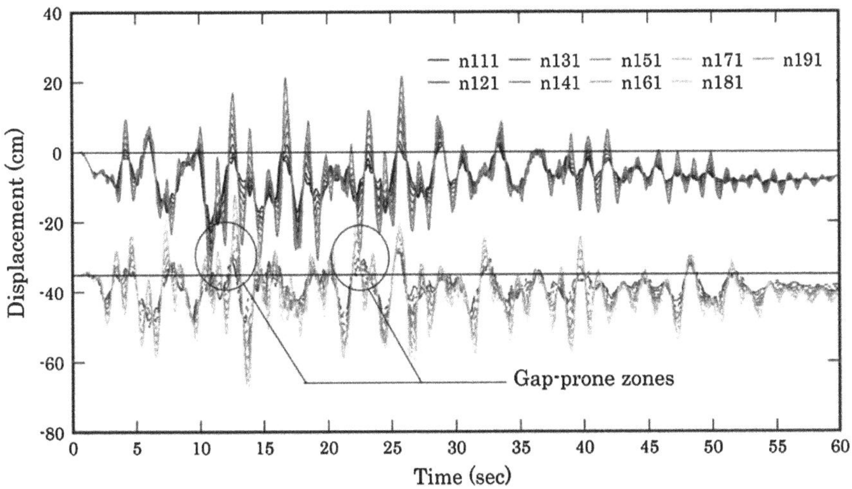

Fig. 14 Pounding between adjacent columns of lateral and central modules

other Latin-American countries, therefore it must be emphasized that it is a major achievement that the damage has not reached the structure, but only secondary elements during the Tucacas earthquake.

One of the main goals of research like this is to confirm that the observance of the code's prescriptions produce adequate response of structures. Numerical studies have the advantage to simulate the behaviour of very complex structural systems without performing very expansive laboratory tests.

While the behaviour of the building indicates that the lateral displacements caused no damage associated with pounding during the earthquake, the uncertainty associated with this phenomenon should motivate to perform researches which lead to improve the knowledge of the behaviour of structures under strong ground motions, to promote improvements to the earthquake-resistant design codes and in the build practice (Vielma et al. 2012), helping to reduce vulnerability of structures with such problems and thereby improving the safety of society in seismic-prone countries.

6 Conclusions

The analysis carried out through this Chapter allows studying the overall response of this type of structure which is common in Venezuela. Using advanced methods, analysis were performed using the computational tool Zeus NL, which allows performing non-linear static analysis, dynamic analysis and non-linear incremental dynamic analysis.

The results showed that the expected behaviour of the original structure satisfies the seismic requirements of the current version of the Venezuelan seismic code, however it is expected that the structure can be repairable structural damage when subjected to the action of an earthquake with a 10 % probability of exceedance in 50 years, which is the time for which this type of construction is designed.

A redesigned structure was modelled and analyzed in order to compare its behaviour of the same with the behaviour of the original structure. The response of this structure meets the overall objectives of performance-based design.

Fragility curves were calculated by using an innovative procedure based on incremental dynamic analysis, where it is clear that for effective peak ground acceleration of 0.25 g original building likely get to have repairable damage states, which are more significant compared with the probability of the redesigned building.

Finally, results indicate that the separation movement between modules used may be insufficient to prevent problems associated patter between adjacent modules. It is recommended to study this effect in buildings of different heights and with irregular shapes, within a comprehensive study in order to propose changes to the Venezuelan seismic code.

Acknowledgements The authors are especially grateful to the Research Council (CDCHT) of Lisandro Alvarado University. the first author also wishes to thank to Prometeo Program, under whose auspices this chapter has been written. We also express our gratitude to the Mid-American Earthquake Center and the National Science Foundation (award number EEC-9701785), the developers of Zeus NL software used in this research. Finally, we thank to Eng. Angel Delgado, for supplying the technical documents of the original building, which he designed and supervised.

References

Barbat, A. H., Pujades, L. G., & Lantada, N. (2006). Seismic damage evaluation in urban areas using the capacity spectrum method: Application to Barcelona. *Comput-Aided Civil Infrastructure Engineering, 21*, 573–593.

Barbat, A. H., Pujades, L. G., & Lantada, N. (2008). Seismic damage evaluation in urban areas using the capacity spectrum method: Application to Barcelona. *Soil Dynamics and Earthquake Engineering, 28*, 851–865.

Elnashai, A., & Di Sarno, L. (2008). *Fundamentals of earthquake engineering*. Chichester, UK: Wiley.

Fardis, N. M. (2009). *Seismic design, assessment and retrofitting of concrete buildings*. Heidelberg: Springer.

FONDONORMA: Norma venezolana Covenin 1756-1:2001. Edificaciones sismorresistentes, Parte 1. Fondonorma, Caracas, Venezuela (2001).

FONDONORMA: Norma venezolana Fondonorma 1753:2006, Proyecto y construcción de obras en concreto estructural. Fondonorma, Caracas, Venezuela (2006).

Kappos, A., & Stefanidou, A. (2009). Deformation-based seismic design method for 3d R/C irregular buildings using inelastic dynamic analysis. *Bulletin of Earthquake Engineering, 8*, 875–895.

Lantada, N., Pujades, L. G., & Barbat, A. H. (2009). Vulnerability index and capacity spectrum based methods for urban seismic risk evaluation. A comparison. *Natural Hazards, 51*, 501–524.

Mander, J. B., Priestley, M. J. N., & Park, R. (1988). Observed stress-strain behaviour of confined concrete. *Journal of Structural Engineering (ASCE), 114*, 1827–1849.

Pinto, P. E., Giannini, R., & Franchin, P. (2006). *Seismic reliability analysis of structures*. Pavia, Italy: IUSS Press.

Vamvatsikos, D., & Cornell, C. A. (2002). Incremental dynamic analysis. *Earthquake Engineering and Structure Dynamics, 31*, 491–514.

Vielma, J. C. (2009). *PACED: Program for the generation of spectrum compatible accelerograms*. Venezuela: Lisandro Alvarado University. Barquisimeto.

Vielma, J. C., Barbat, A. H., & Oller, S. (2010). Non linear structural analysis. Application to evaluating the seismic safety. In M. Camilleri (Ed.), *Structural analysis* (pp. 50–74). New York: Nova Science Publishers.

Vielma, J. C., Barbat, A. H., & Oller, S. (2011a). Dimensionado sísmico de edificios de hormigón armado mediante factores de amplificación de desplazamientos con base en el balance de energía. *Hormigón y acero, 63*, 83–96.

Vielma, J. C., Barbat, A. H., & Oller, S. (2011b). Seismic safety of RC framed buildings designed according modern codes. *Journal of Civil Engineering and Architecture, 5*, 567–575.

Vielma, J. C., Barbat, A. H., & Oller, S. (2012). The quadrants method: A procedure to evaluate the seismic performance of existing buildings. In *15 World Conference on Earthquake Engineering*. Lisbon, Portugal.

ZEUS NL: User manual. Version 1.8.9. Mid America Earthquake Center. Urbana, USA (2010).

Analysis of Seismic Vulnerability of Rural Houses in China

Jue Ji, Zening Xu and Xiaolu Gao

1 Introduction

Most seismic disasters in China have taken place in rural areas (Wang et al. 2005). So far in China, the research related to the potential vulnerability assessment has been made but most of them are limited to quake-affected areas (Yin 2001; Wang et al. 2005). On the other hand, knowledge about the vulnerability of rural areas in the whole country was poor and inadequate.

The damage of dwelling houses has been the biggest cause of casualties and asset loss, which cause enormous impact on rural societies and economic systems (Wang and Wang 2001). However, the development of rural areas had lagged behind, although China has achieved fast economic development in the past three decades. This situation is reflected in the quality of rural houses. National statistics for 2005 showed that by 2005, 20.1 % of rural houses were built in wood or other informal materials, and only 3 % of all rural families had spent 10 thousand US dollars or more for building or buying their houses. Actually, some 94.2 % of rural houses were self-built that ran out of the regulations of seismic design code (NBSC, 2006).

Accordingly, the challenge of protecting rural houses from disasters in China is serious. In 2005, a national effort named "movement of building new countryside" was initiated to speed up rural development, where the Chinese government contrived to improve the quality of rural houses by promoting the construction of new collective houses, which must comply with seismic design standards, but a huge number of unsafe rural houses still remain. The widespread existence of

J. Ji · Z. Xu (✉) · X. Gao
Key Laboratory of Regional Sustainable Development Modeling, Institute of Geographic Sciences and Natural Resources Research, CAS, Beijing 100101, China
e-mail: xuzening13@mails.ucas.ac.cn

Z. Xu
Graduate University of Chinese Academy of Sciences, Beijing 100049, China

S. D'Amico (ed.), *Earthquakes and Their Impact on Society*,
Springer Natural Hazards, DOI 10.1007/978-3-319-21753-6_13

poor-quality rural houses not only puts the lives and properties of millions of people in danger, it is also a risk for the whole society considering the huge recovery cost from disasters. It is quite necessary for the policy makers to assess the seismic vulnerability of rural houses in the whole country and adopt effective risk reduction strategies.

To do so, a holistic comprehension of the seismic vulnerability of rural houses in different areas in China is indispensable. Vulnerability, broadly defined as the potential for loss, is an essential concept in hazards research (Cutter 1996). Structural characteristics of housing, possible geological and geographical earthquake hazards and social economic characteristics were agreed as critical factors to calculate potential damage. With earthquake damage statistics in the past, empirical studies of damage probability matrices (DPM) (Whitman et al. 1973; Braga et al. 1982) and vulnerability index (Benedetti and Petrini 1984) or vulnerability curves of houses (Yin 1994, 1995) in many countries have been studied for years to solve this uncertainty problem since most vulnerable assessment focus on post-disaster assessment. Accordingly, based on empirical parameters, techniques or models have been developed for predicting the possible damages (Kohiyama and Yamazaki 2004; Schweier et al. 2004).

One of the main reasons for this phenomenon lies in the paucity of detailed data on rural houses at appropriate spatial scale. The Chinese administrative system is composed of five hierarchies: (1) the central government, (2) province, autonomous region, municipality directly under the central government, and special administrative region (SAR) (hereafter all referred to as 'province'), about 34 province-level administrative units, (3) prefecture, prefecture-level city, and autonomous prefecture (referred to as 'prefecture'), 333 prefecture-level ones (4) county, county-level city, autonomous county, and district of prefecture-level city (referred to as 'county'), about 2852 county-level administrative units altogether, and (5) township and town. In practice, county is the basic administrative unit responsible for making disaster prevention plans, conducting emergency rescues, and organizing post-disaster reconstructions. So studies at county-level are particularly important for understanding seismic risks and developing damage reduction strategies in rural areas.

2 Database of Housing Structures in Chinese Rural Areas

Among various quality indicators, the structure of rural houses is the most straightforward one concerning seismic vulnerability (Ge et al. 2008). At present, the national one percentage sampling survey conducted in 2005 (the 2005 census) is the most up-to-date data source of the structure of rural houses. From the perspective of catastrophe prevention, it is more urgent to conduct such a research in Chinese rural areas, because in urban areas, new buildings are all subject to the regulation of seismic design code, but most rural houses are beyond the regulation. That is why this study has taken this special issue for study.

According to the standard in national census, the structure of rural houses was divided into five types, i.e., reinforced concrete (RC), brick with concrete pillars and poles (mixed), brick, wood, and other informal structure (others). In general, the RC structure is the strongest, followed by the mixed, brick, wood and others. The 2005 census report gives the amount of houses in each structure in the country.

To build a good county-level database about the structure of rural houses is a challenge because the 2005 census data being published are incomplete. One of the problems is that the spatial scales of different provinces are inconsistent. We have confined the scope of study region to China's 31 provinces excluding Hong Kong SAR, Macau SAR, Taiwan province and the South China Sea islands and collected the census data in the 31 provinces. All have released the information of rural houses aggregated at provincial-level, however, 10 provinces (include 990 counties) have provided data of rural houses aggregated by prefectures, 8 provinces (include 636 counties) released rural houses data aggregated by provinces, and 2 other provinces (include 164 counties) provided data in county-level, but they are the summation of rural and urban houses. As a matter of fact, only 10 provinces (include 590 counties) have released rural housing structure data in county level, which satisfy the required spatial resolution.

To establish a complete database of rural houses in county-level, down-scaling models applying large scale information to estimate the information of specific units which encompasses a part of the original dataset were adopted to solve the problem (Hewitson and Crane 1996; Wilby et al. 1998; Carter et al. 2004; Gaffin et al. 2004).

With the method presented by Gao and Ji (2013), a set of down-scaling models were established (Table 1). The impact of climate, economic development level and ethnic minority culture factors on rural housing structure, as well as the spatial autocorrelation of neighboring spatial units were considered in different models.

The established models for estimating the county-level ratios of five structures from prefecture-level data explained about 60–70 % of the total variance, and those estimated from province-level sorting data explained about 25–55 %. The latter results are not so satisfactory and improvement should be further made, but still they implied that the variables being considered were important for capturing the influencing factors of rural housing structures. Subsequently, a database of county-level rural housing structure in China was established.

Houses in wooden and other informal structures are most vulnerable during seismic. According to housing damage matrices presented by Liu et al. (2011), when the intensity scale of earthquake is over IX, most wooden and informal structure houses are totally damaged or destroyed, while the probability of RC houses for being damaged is only 60 %. Figure 1 reveals that the ratios of wooden and other informal houses are relatively high in Shaanxi, Sichuan, and Yunnan provinces (in particular along the edge of Tibet upper land), Heilongjiang province, and a few in southern Zhejiang as well, where as much as 40 % of the rural houses in some counties were in wood structure.

We noticed that in economically advanced east China region such as the south of Zhejiang province, many traditional wooden houses have been preserved for their historical value and the quality of the houses are relatively well (Fig. 2a, b). But in

Table 1 Down-scaling and separation models of different scale to county-rural level

Estimation objectives	Description of down-scaling models
1. Down-scale from prefecture to county level	$y_{ck} = a \cdot y_d + \sum_j \beta_j x_{jk} + \varepsilon_k$
	x_{jk} is the *j*th variable that explains the spatial differences of housing structure in different counties. And variables represent architectural climate zone, architectural thermal zone, GDP, ethnic minority (=1 if it belongs to an autonomous region in China, =0 otherwise) were used
2. Down-scale from province to county level	$y^{ct} = g(y^{pv}, X, Y^*)$
	y^{ct} and y^{pv} represent the proportions of certain structures at county and province level, respectively, X, the factors influencing spatial differentiation for which the formerly defined Climate, thermal, GDP, ethic minority dummy were used. Besides, Y*, a vector of the average ratios of five structures in the counties of neighboring provinces
3. Split the statistics of rural houses and urban houses	$\sum_{i=1}^{k} H_i \times \mu_i + \mu^* \times (H - \sum_{j=1}^{t} H_j) = H_u$
	H is the total amount of houses in the prefecture
	H_u is the amount of urban houses (=population/average family size)
	H_j is the amount of houses in county j (for j = 1 to t)
	μ* is the ratio of urban families, which is assumed to be constant

other regions, the prevalence of wooden and informal houses results from poverty (Fig. 2c, d). They usually looked bad and are weak to natural disasters. In such areas, effective house reinforcement policies and disaster prevention plans should be made.

3 Estimation Methods for Seismic Vulnerability of Rural Houses in China

Vulnerability, broadly defined as the potential for loss, is an essential concept in hazards research. Most researches related to housing seismic vulnerability tried to evaluate from perspectives of seismic risk and loss (Steinbrugge 1969; Okada and Takai 2000; Schweier et al. 2004). It is generally agreed that housing seismic vulnerability has a close relationship with potential earthquake intensity of a region as well as housing fragilities.

In China, the national standard isoseismic maps give the seismic intensity of different areas. These maps show the horizontal peak accelerations and eigen-frequencies of ground mobility at a normal site which have a 10 % probability of being exceeded during a 50-years period. The most recent one is GB18306-2001 at the map scale of 1:4,000,000, which is revised in 2001 for the fourth time. Estimated from data relating to the frequency and intensity of past

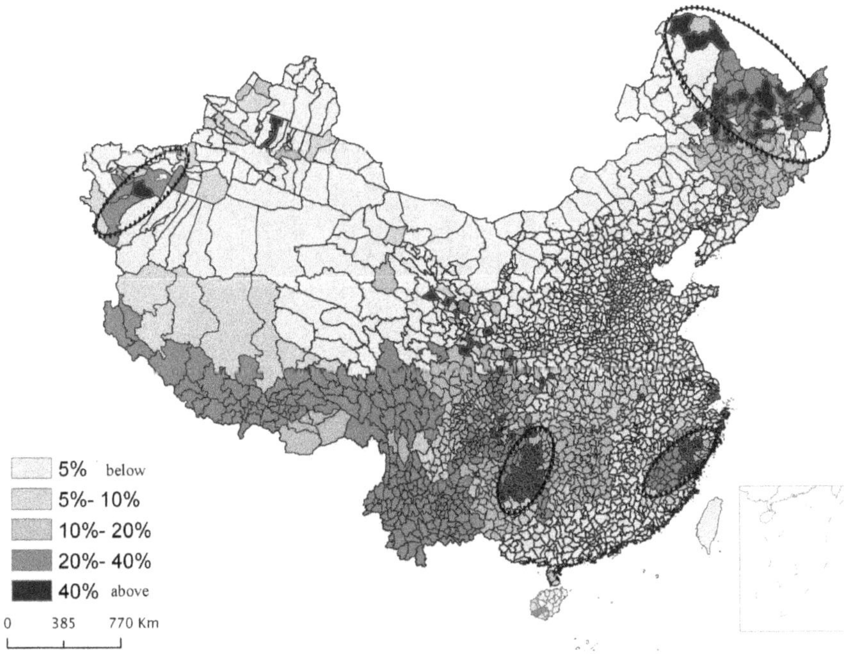

Fig. 1 Distribution of the ratio wood-structured rural houses by county in 2005

earthquakes, the national standard isoseismic maps describe the potential long-run seismic risk of different areas (IQP 2000) and act as the basis for land use planning and for setting buildings' reinforcement levels required by the seismic code in different regions. Generally speaking, the whole is divided into five zones by intensity scale: <VI (below Ms 6.0),VI (Ms 6.0–7.0),VII (Ms 7.0–8.0),VIII (Ms 8.0–9.0), and ≥IX (Ms 9.0 and above).

It is worth commenting that previous vulnerability assessment models in the literature have used post-disaster survey and building damage experiment data to determine the damage or risk parameters in the models, but due to spatial differentiation, variability of ground motion and uncertainties, the models being presented are varying a lot. With earthquake damage statistics in the past, empirical studies of damage probability matrices (DPM), vulnerability index or vulnerability curves of houses in many countries have been investigated for years to solve this problem. In China, Liu et al. (2011), with earthquake damage statistics from 1990 to 2008 and provincial data on the proportion of different housing structures, estimated the average rate/degree of damage of buildings with four different structures (wood and other structures were combined) with respect to the intensity of earthquakes (Table 2).

With the isoseismic maps and county level units being intersected, we first got small polygons with GIS applications. Secondly, the number of rural houses in different structures in each polygon was calculated. Thirdly, the expected number of

Fig. 2 Some example of wooden and informal structure houses in China. **a** Preserved wooden residence in Zhejiang province. **b** Xiangxi vernacular dwellings in the west of Hunan province. **c** Informal cave dwelling in Yan'an, Shan'xi Province. **d** Mud made residence in northern mountainous areas

Table 2 Building damage rates in earthquakes in China

Intensity scale	Seismic peak ground acceleration (g)	Average housing damage rate (reference range) (HDR), %			
		RC	Mixed	Brick	Wood and other
<VI	<0.05	0	0	0	0
VI–VII	0.05–0.15	5 (3.5–15)	8 (4.5–18.5)	10 (4.5–19.5)	15 (4.5–21)
VIII–IX	0.2–0.4	20 (20–50)	35 (30–57)	40 (36.5–58)	50 (39–61)
≥IX	≥0.4	60 (60–100)	90 (83–100)	95 (84–100)	100 (86–100)

damaged houses (A) was aggregated by county and the percentage of damaged rural houses in all (R) of each county was calculated (Fig. 3).

Both potential housing damage amount (A) and housing damage ratio (R) are frequently used indicators of potential loss. As indicator of the whole country, however, due to extensive variations of rural population and housing densities in China, neither may precisely reflect the vulnerability of a county alone. For example, in sparsely populated areas the total amount of damage is deemed more useful but in densely inhabited areas, the ratio of damaged houses is more important indicator of seismic vulnerability. To facilitate a comprehensive assessment of seismic vulnerability, an integrated index of A and R was anticipated. We have taken principal components analysis to generate such an evaluator.

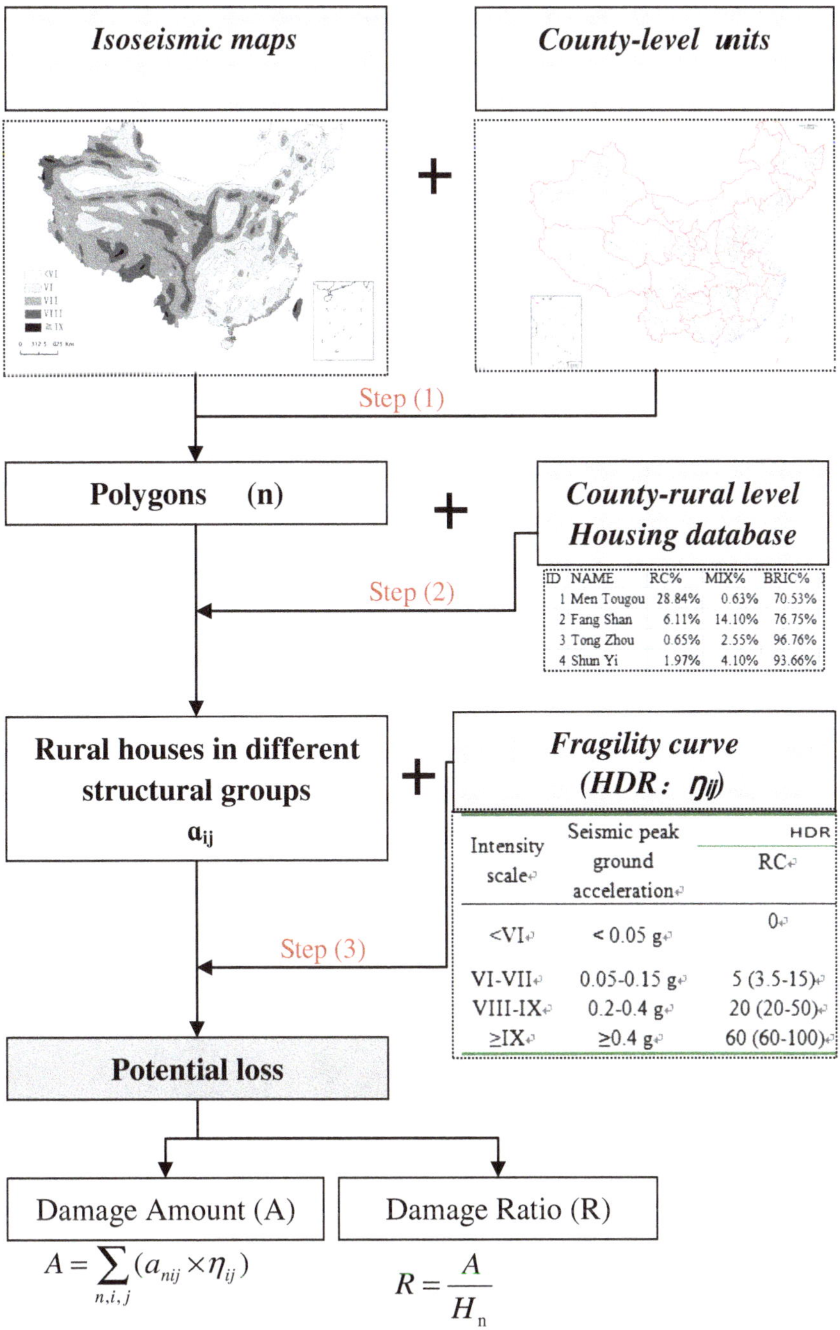

Fig. 3 Evaluation method of seismic vulnerability of rural houses

4 Evaluation of Seismic Vulnerability and Regional Strategies for Risk Reduction

Based on the evaluation method being introduced above, we systematically evaluated the seismic vulnerability of rural houses in China. The degree of seismic vulnerability for the 2403 counties in China in 2005 was represented by the general index of potential damage housing amount and potential damage housing ratio. As a result, four vulnerability levels were specified by standard deviation method: safe (<−0.5 std. dev.), relatively safe (between ±0.5 std. dev.), vulnerable (+0.5 to +1.5 std. dev.), and highly vulnerable (>+1.5 std. dev.) (Fig. 4). Details can be found in Gao and Ji (2013).

4.1 *Spatial Patterns of Seismic Vulnerability*

The evaluation results show that approximately 19.6 % of all counties had a low degree of vulnerability, 56.2 % were relatively vulnerable, 9.9 and 7.5 % were in high and extremely high classes. The spatial distribution in Fig. 4 shows that areas of high and extremely high vulnerability concentrated in the following areas: (a) the south of Xinjiang, (b) countries in northern Xinjiang, (c) the joint areas of Shaanxi,

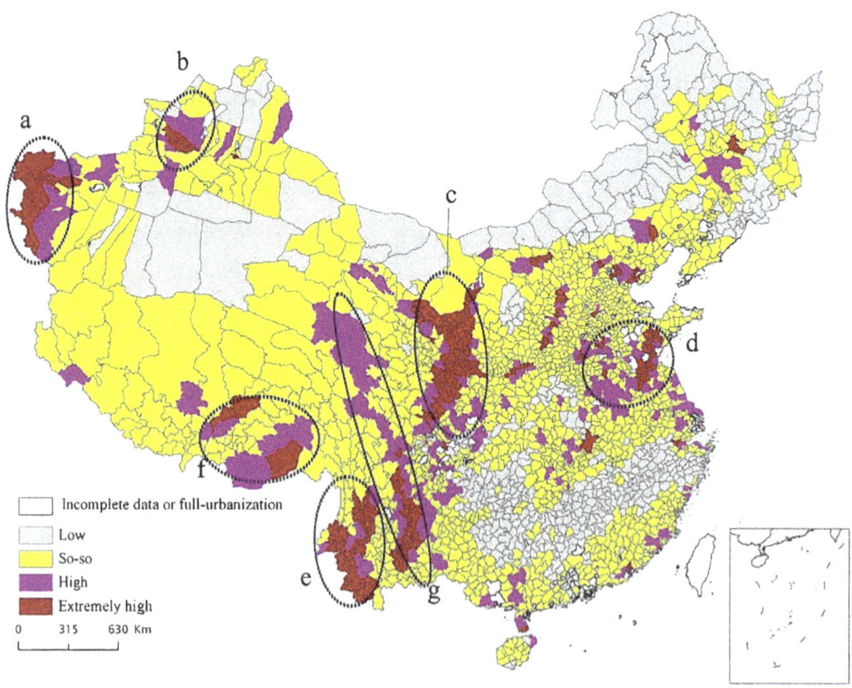

Fig. 4 Distribution of counties by four housing vulnerability levels (*Source* Gao and Ji 2013)

Gansu, and Ningxia, (d) Shandong, Jiangsu, and Anhui provinces, (e) the southwest part of Yunnan and Guizhou provinces, (f) southeastern Tibet, and (g) the eastern fringe of Qinghai and Tibetan Plateau.

4.2 *Risk Reduction Strategies*

Among the main reasons of seismic vulnerability in our research, poor housing structure, dense population, and potential high seismic intensity are major contributors. The spatial variations of these factors shed lights to spatial policies of risk reduction.

4.2.1 Improving Housing Quality

The majority of the deaths are caused by building collapse in earthquakes and the great majority have occurred in the developing world (Kenny 2009). Therefore, in the high potential seismic risk regions, it is an urgent task to rebuild or improve the structure of poor quality houses. In China, as Fig. 5 shows, most of these areas located in the southeast.

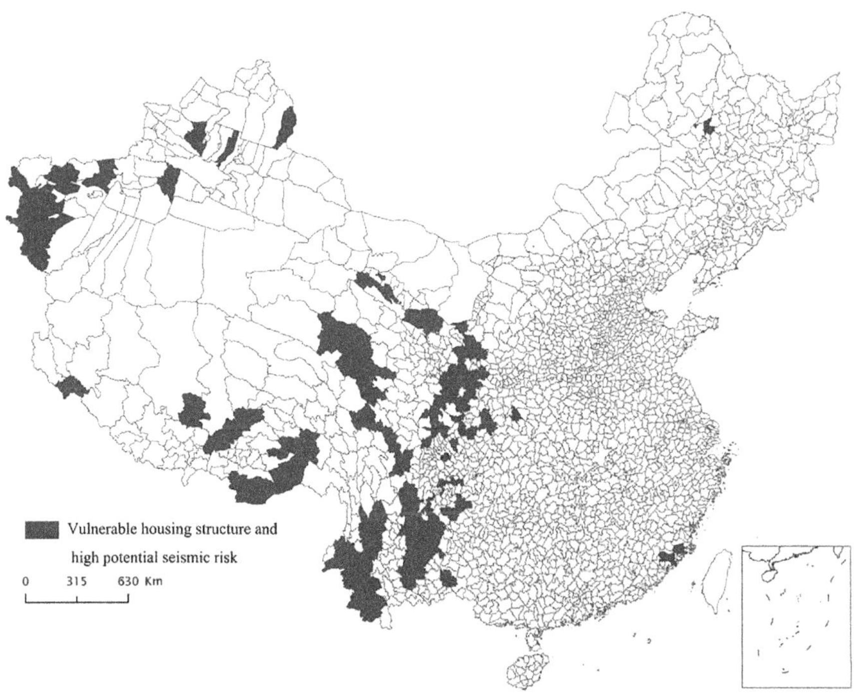

Fig. 5 Vulnerable houses in high seismic risk areas

Despite that engineering solutions can eliminate the risk of deaths caused by building collapse to a large degree. However, engineering solutions are both expensive and technically demanding, so that the benefit-cost ratio of such solutions is often unfavorable compared with other interventions designed to save lives in developing countries (Kenny 2009). This phenomenon is more obvious in under-developed regions where local governments and low-income residents governments are unable to implement housing reinforce projects because of limited economic powers. Wang et al.'s (2005) survey in poor villages in Yunnan province, China, for example, revealed that even though the villages had suffered severe damages in a recent earthquake, a big proportion of rural residents were reluctant to rebuilding houses, largely due to poor economy.

It is quite impossible to implement wholesome poor house renewal or reinforcement projects in all regions. Intensive public investment should be input to regions with highly concentrated poor houses, and a range of transparency and oversight mechanisms for public construction projects are indispensable. In short term, it is also significant to emphasize disaster education among rural residents.

4.2.2 Disaster education to rural residents

With the difficulty of improving housing structure in very short period, especially in poor economy counties, being understood, it is essential to instruct people in these areas about the risks of the surrounding environments. By improving people's awareness and escaping abilities of self protection through education and training, the adaptability and ability to protect themselves from disasters can be greatly enhanced.

Especially, in those highly populated areas, total damage will be greater than other things. The cost of reconstruction and damage costs will be huge once disaster happens. Moreover, large rural population increases the difficulties of escape. Besides, once seismic disasters happen, it is less easy to prevent diseases, fires, and sometimes riots. It is important to inform people about the risks and teach them how to escape from dangers and help each other in the case of disasters.

4.2.3 Enhancing Rural Planning

For villages or settlements located on active faults, local government should encourage people to relocate to safe places and to live together so that basic infrastructures and services can be provided at reasonable costs. Meanwhile, the relocation scheme should be carefully planned, which need more accurate spatial information of relocation regions, and pre-evaluation of its optimum population, households and industry.

In particular, sufficient emergency space and emergency roads should be provided to allow for quick evacuation and escape for people. Besides, as transportation and communication infrastructures play important roles in the case of disasters, it is important to secure lifelines and access.

5 Conclusion

In 2008, the MT 8.0 Great Wenchuan Earthquake claimed about 80 thousand lives, destroyed as many as 1.6 million houses and heavily damaged another 1.9 million houses. In the MT 7.1 Qinghai Yushu earthquake in 2010, nearly 2300 people died. In recent years, the Chinese government has put great emphases on quake-resistance and planning strategies designed to diminish the damage to dwellings in rural areas. From the perspective of catastrophe prevention, it is more urgent to conduct such a research in Chinese rural areas, because in urban areas, new buildings are all subject to the regulation of seismic design code, but most rural houses are beyond the regulation. That is why this study has taken this special issue for study.

In this paper, the seismic vulnerability of counties was assessed with regard to the structural features of rural houses, measured by the number and ratio of potentially damaged housing in the case of seismic disasters. Two main problems were discussed: (1) how to deal with the incomplete data problems with respect to rural houses, and to solve the problem, a set of downscaling models was built for estimating absent data at different accuracy levels and replenish the data of county units. This rural houses' database provided the basic information relating to the amount and ratio of houses with different structures in more than 2300 China's counties; (2) how to evaluate the seismic vulnerability of rural houses. The amount of houses in different structures, the standard scale of seismic intensity and the fragility of certain structures in the face of seismic disasters were taken as main factors in the evaluation model, and an integrated indicator of the amount and ratio of damaged dwellings was used to present the seismic vulnerability.

According to evaluation result, counties were classified by their vulnerability levels: safe, relatively safe, vulnerable, and highly vulnerable. The results revealed that a fairly large number of counties were subject to high risks, i.e., 17.4 % of them belong to the vulnerable and highly vulnerable groups. From the viewpoint of national disaster prevention policies, these counties should be given priority.

Upon further examination of the spatial distribution of housing structure, population density, and seismic intensity factors, as well as the environmental and social-economic characteristics of vulnerable counties, we drew a number of policy implications: (1) poor housing quality should be improved, especially those made of timber and other structures, attention should be paid to improving or reinforcing housing structures; (2) disaster education is generally essential for improving peoples' adaptability and ability to deal with seismic disasters, and in under-developed areas, it is even more significant; (3) rural planning should be enhanced. In those high seismic risk areas, the plans of relocating rural families to safe and flat places should be supported. Strategies to enhance of the provision of emergency space and evacuation roads, lifeline infrastructures, and the establishment of efficient disaster precaution mechanisms is also crucial in rural planning.

References

Benedetti D., Petrini V. Sulla vulnerabilità sismica degli edifici in muratura: proposta di unmetodo di valutazione. L'Industria delle Costruzioni n. 149, Rome (March 1984).

Braga, F., Dolce, M., & Liberatore, D. (1982). A statistical study on damaged buildings and an ensuing review of the MSK-76 scale. In *Proceedings of the Seventh European Conference on Earthquake Engineering*, Athens, Greece, pp. 431–450.

Carter, T. R., Fronzek, S., & Bärlund, I. (2004). A framework for developing consistent global change scenarios for Finland in the 21st century. *Boreal Environment Research, 9*(2), 91–107.

Cutter, S. L. (1996). Vulnerability to environmental hazards. *Progress in Human Geography, 20* (4), 529–539.

Gaffin, S. R., Rosenzweig, C. R., Xing, X., et al. (2004). Downscaling and geo-spatial girding of socio-economic projections from the IPCC special report on emissions scenarios (SRES). *Global Environmental Change, 14*(2), 105–123.

Gao, X. L., & Ji, J. (2013a). Analysis of the seismic vulnerability and the structural characteristics of houses in Chinese rural areas. *Natural Hazards, 70*(2), 1099–1114.

Gao, X. L., & Ji, J. (2013b). Estimate of rural housing structure and its vulnerability in China. *Journal of Geographical Sciences, 01*, 179–191.

Ge, Q. S., Zou, M., Zheng, J. Y. et al. (2008). Integrated assessment of natural disaster risks in China. Beijing: Science Press (in Chinese).

Hewitson, B. C., & Crane, R. G. (1996). Climate downscaling: Techniques and application. *Climate Research, 7*, 85–95.

IQP (Inspection and Quarantine of the People's Republic of China) (2000). Seismic ground motion parameter zonation map of China (GB18306–2001). Beijing: General Administration of Quality Supervision, Standardization Administration of the People's Republic of China.

Kenny, C. (2009). *Why do people die in earthquakes? The costs, benefits and institutions of disaster risk reduction in developing countries*. World Bank Policy Research Working Paper No. 4823. http://papers.ssrn.com/sol3/papers.cfm?abstract_id=1334526##

Kohiyama, M., & Yamazaki, F. (2004). *Vulnerability functions for wooden houses in Japan based on seismic diagnosis data*. Paper No. 130 presented on the 13th World Conference on Earthquake Engineering, Vancouver.

Liu, Y., Wu, S. H., Xu, Z. C., et al. (2011). Methodology for assessment and classification of natural disaster risk: A case study on seismic disaster in Shanxi Province. *China, Geographic Research, 30*(02), 195–208. (in Chinese).

National Bureau of Statistical of China (NBSC). (2006). *China population statistics yearbook*. Beijing: Chinese Statistics Press.

Okada, S., & Takai, N. (2000) Classifications of structural types and damage patterns of buildings for earthquake field investigation. In *Proceedings of the Twelfth World Conference on Earthquake Engineering*, Auckland, New Zealand, January 30–February 4, paper no. 0705.

Schweier, C., Markus, M., & Steinle, E. (2004). Simulation of earthquake caused building damages for the development of fast reconnaissance techniques. *Natural Hazards and Earth System Sciences, 04*, 1–9.

Steinbrugge, K. V. (1969). Studies of seismology and earthquake damage statistics. Washington DC: U.S. Department of Housing and Urban Department, pp. 1–100.

Wang, W. Y., & Wang, J. A. (2001). The distributive pattern of hail disasters based on three data sources in China. *Geographic Research, 20*(03), 380–387. (in Chinese).

Wang, Y., Shi, P. J., Wang, J. A. (2005). Impact of earthquake disaster on rural residents: a case study on Dayao County of Yunnan Province. *Journal of Natural Disasters 14*(06), 110–115 (in Chinese).

Whitman, R. V., Reed, J. W., & Hong, S.T. (1973). Earthquake damage probability matrices. In *Proceedings of the Fifth World Conference on Earthquake Engineering*, Rome, Italy (Vol. 2, pp. 2531–2540).

Wilby, R. L., Wigley, T. M. L., Conway, D., et al. (1998). Statistical downscaling of general circulation model output: A comparison of methods. *Water Resources Research, 34*, 2995–3008.

Yin, L. F., Li, M., & Song, L. J. (2001). Application of Bayes model to the prediction of buildings damages in earthquakes. *Platean Earthquake Research, 13*(04), 34–40. (in Chinese).

Yin, Z. Q. (1994). A dynamic model for predicting earthquake disaster losses. *Journal of Natural Disasters, 3*(02), 72–80. (in Chinese).

Yin, Z. Q. (1995) Prediction method of earthquake disaster and loss. Beijing: Earthquake Press (in Chinese).

Finite Element Modelling for Seismic Assessment of Historic Masonry Buildings

Michele Betti, Luciano Galano and Andrea Vignoli

1 Introduction

A large portion of the European cultural heritage is made of masonry buildings that have a growing economic and social value in many countries. Therefore, their preservation is considered an important issue in modern societies both for their historical interest and for the economic contribution in contexts where tourism has become a major industry (Bowitz and Ibenholt 2009). During past and recent earthquakes (Lucibello et al. 2013; Ceci et al. 2013; Brandonisio et al. 2013) these historic buildings have demonstrated to be particularly prone to damage, showing partial or total collapse. In many cases, this was due to restorations non-respectful of the original structural layout (Ramos and Lourenço 2004; Borri et al. 2000). Generally, masonry buildings are capable of carrying out vertical loads in a safe and stable way, while they are rather sensitive to horizontal loads such as the seismic ones. The high seismic vulnerability of these buildings is due both to their particular structural configuration and to the mechanical properties of the masonries. Open spaces, slender walls, lack of effective connections among the structural elements and the highly nonlinear behaviour with very small tensile strength are some examples of structural and material lacks.

In principle, prediction of the structural behaviour of a monumental building is similar to that of other constructed facilities. However, the analysis of an historic building is an even more challenging task (Del Piero 1984; Carpinteri et al. 2005;

M. Betti (✉) · L. Galano · A. Vignoli
Department of Civil and Environmental Engineering (DICeA),
University of Florence, Via di Santa Marta, 3, 50139 Florence, Italy
e-mail: mbetti@dicea.unifi.it

L. Galano
e-mail: luciano@dicea.unifi.it

A. Vignoli
e-mail: avignoli@dicea.unifi.it

S. D'Amico (ed.), *Earthquakes and Their Impact on Society*,
Springer Natural Hazards, DOI 10.1007/978-3-319-21753-6_14

Bartoli and Betti 2013) and, in some cases, train to extrapolate analytical procedures specifically developed for modern buildings is inadequate. A correct structural evaluation should be based on a deep knowledge of: (i) the history of the building and its evolution, (ii) the geometry, (iii) the structural details, (iv) the cracking pattern and the material damage map, (v) the masonry constructive technique, (vi) the material properties, and (vii) the global behaviour (Siviero et al. 1997; Leftheris et al. 2006). This knowledge can be reached through both in situ and laboratory experimental investigations (Corradi et al. 2002a, b; Binda et al. 2000) joined with structural analyses with appropriate models (Lourenço et al. 2007). Nevertheless, the restrictions set on inspections and on performing reliable quantitative strength evaluations results in limited information on the constructive system and properties of the materials.

These issues have been recently addressed by the Italian Guidelines for seismic vulnerability assessment of cultural heritage (DPCM 2011) that introduce the concept of knowledge level (KL) for monumental buildings specifying the confidence factor (CF) obtained through in situ tests and investigations. These Guidelines propose a methodology of analysis based on three different levels of evaluation, according to an increasing knowledge of the structure. The first level of analysis (Level 1, LV1) allows to evaluate the collapse acceleration of the structure by means of simplified models based on a limited number of geometrical and mechanical parameters (and qualitative tools such as visual inspections). The second level (Level 2, LV2) is based on a kinematic approach performed to analyse the local collapse mechanisms that can develop on several macro-elements. The identification of proper macro-elements is based on the knowledge of structural details of the building (cracking pattern, connections between the architectonic elements, etc.). The last level of evaluation (Level 3, LV3) requires a global analysis of the whole building under seismic loading by suitable numerical models.

Each level of evaluation should be appropriate with the achieved knowledge level. Usually, the LV1 level is a simplified territorial or urban scale seismic assessment performed on the whole building. The aim is to provide general guidelines to establish priority of interventions for the protection of historic monuments. The LV2 level provides the horizontal load multiplier that activates the local collapse of each macro-element considered. This level has to be considered when the structure needs of local interventions. The LV3 level is required in the case of global seismic assessment or interventions that modify the whole structural behaviour. This evaluation is usually performed by means of a finite element model of the structure in which the seismic action is considered performing static or dynamic nonlinear analyses. Compared to the previous two levels the LV3 is the most accurate, but it requires large amount of input data and great computational effort.

The above issues enlighten the attention that must be paid in modelling strategies of historic masonry constructions. Each monumental building is, by definition, a unique building characterised by its own history, often resulting in a composite mixture of added or replaced, strongly interacting, structural elements. Hence, engineers involved in the study of cultural heritage are called to have a specific care in the understanding of the historical process since modifications occurred through the building history produced several uncertainties in the model definition.

In this respect, the scientific literature presents a significant number of exemplary case studies that cover a wide range of applications around the European Community. Lourenço (2005) analyses the Church of Saint Christ in Outeiro (Bragança, North of Portugal). The papers show that sophisticated tools of structural analysis offer significant information for both understanding the existing damage and designing the strengthening of ancient structures. Romera et al. (2008a, b) analyse the Basilica of Pilar in Zaragoza, one of the most famous Spanish temples. The authors identify the actual structural state of the Church, its safety level and the relationship between structural behaviour and actual damage. A global numerical model of the Church was built and the masonry was simulated as a nonlinear material with brittle behaviour in tension and plastic properties in compression. The authors take in consideration the construction steps, including the reinforcement works added to the structure. The paper shows that suitable numerical models can offer effective information in both reproducing structural pathologies and check out the efficiency of historic restoration. Another example is presented by Lourenço et al. (2007). The authors investigate the structural safety of the Monastery of Jeronimos in Lisbon (Portugal). The paper advocates that the computational models can be used as a numerical laboratory, where the sensitivity of the results to the input material parameters and boundary conditions can be efficiently analysed, offering effective information in the design of in situ testing and structural monitoring. Betti and Vignoli (2008) analyse a Romanesque church, the Abbey of Farneta near Cortona (Italy). The authors build a 3D numerical model of the monument and develop linear and nonlinear analyses to assess the seismic vulnerability and the efficiency of traditional strengthening techniques. Taliercio and Binda (2008) analyse the Basilica of San Vitale in Ravenna (Italy), a Byzantine building which suffers diffused cracking and excessive deformations. The authors take into account the results of in situ topographical and mechanical investigations and build a global finite element model of the Basilica conceived as a first step toward the understanding of the structural behaviour. Ivorra et al. (2009) assess the seismic behaviour of the San Nicolas Bell Tower in Valencia (Spain). The finite element model of the Tower was first calibrated by means of in situ dynamic tests and subsequently used to evaluate the seismic response with respect to the seismic Spanish standards. The numerical simulations showed a satisfactory performance of the Tower. Del Coz Diaz et al. (2007) analyse the Palatine Chapel of San Salvador de Valdediós near Oviedo (Spain) and combine the finite element method with a frictional contact problem. The analyses were based on the application of the finite element technique to each stone block, and the blocks were assembled side by side using contact elements in order to reproduce the mechanical behaviour of the mortar. The authors show that sophisticated analysis tools provide a clear understanding of the structural behaviour. Betti et al. (2010) analyse the cracking pattern of a historic Italian palace and show a careful use of the finite element technique when dealing with practical engineering problems. The authors provide an interpretation of the manifested damage in the palace, and use the numerical results to design an extensive in situ investigation on the building. Ivorra et al. (2010) report a study carried out on the Bell Tower of the Church of Santas Justa and Rufina in Orihuela in Alicante (Spain). The model was first calibrated based on the dynamic characteristics in free vibration and then used to predict the evolution of the dynamic behaviour of the Bell Tower, considering the subsidence caused by variations in water table level.

Within this field of research the chapter focused the attention to some aspects of numerical modelling, to offer a contribution to the analysis of historic masonry buildings under seismic actions. The chapter is hence organized as follows. In the first part a numerical model used to replicate the nonlinear behaviour of the masonry is described. The model is quite general and is herein discussed with specific reference to the commercial finite element code ANSYS (1998). The identification of the required parameters is discussed using the results of experimental tests performed in old masonry buildings. In a second part the above approach is employed to build a finite element model of two relevant case studies and to analyse their seismic behaviour. The two examples illustrate the use of numerical analyses to face practical engineering problems in the field of seismic assessment of historic constructions.

2 The Numerical Model of Masonry

2.1 *Modelling Approach*

Masonry is an anisotropic non-homogeneous and nonlinear material composed by units (bricks and/or stones) and mortar joints. So, the modelling depends both on dimensions and arrangement of the units and on the size of the joints. According to the approaches proposed in literature, the numerical models can be performed with two different levels of detail (Zucchini and Lourenço 2002; Theodossopoulos and Sinha 2013). In the so-called micro-modelling approach, units and mortar joints are represented by continuous elements and are modelled separately. Discontinuous elements represent the interfaces between units and mortar. Since all material characteristics of the components are considered separately, this modelling strategy correctly reflects the actual behaviour of the masonry when experimental data are available for each component. A simplified version of the micro-modelling approach makes use of interface elements to account for both mortar joints and contacts between units and mortar. The units are still modelled as continuous elements (Gambarotta and Lagomarsino 1997; Da Porto et al. 2010). The micro-modelling approach (detailed or simplified) is usually employed to analyse specific problems of little-size since, it is hardly computational demanding in the application to large structures. For large-size models the so-called macro-modelling approach is the most commonly employed technique. Bricks, mortar joints and interfaces are globally represented by single continuous elements. The mechanical properties of the homogenous elements depend on those of the basic components.

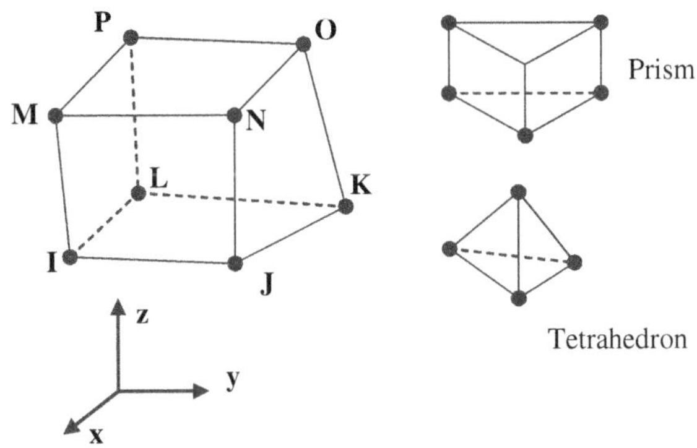

Fig. 1 Geometry, node locations and coordinate system for the *Solid65* element

Proper homogenisation techniques (Zucchini and Lourenço 2007) and experimental setup were accordingly proposed to determine these average parameters. This approach is appropriate to analyse large or complex structures, because the calculation is less demanding. Another field of application of the macro-modelling strategy is represented by the structural masonries with *opus incertum* texture.

Taking into account that the chapter aims at discussing the finite element (FE) modelling technique for the seismic assessment of historic masonry structures, the macro-modelling approach is considered. Specifically, 8-nodes isoparametric finite elements having three degrees of freedom at each node (*Solid65*) were employed to model the masonry assemblages (Fig. 1).

2.2 The Elastoplastic Model

In the follows it is considered a material with isotropic behaviour, according to the chaotic texture of the masonry in many existing historic buildings. The elastic behaviour of the equivalent continuum is ruled by the following classical equations:

$$\{\dot{\sigma}\} = \mathbf{E}\{\dot{\varepsilon}\}^{el} = \mathbf{E}\left(\{\dot{\varepsilon}\} - \{\dot{\varepsilon}\}^{pl}\right) \tag{1}$$

$$E_{ijkl} = \left(K - \frac{2}{3}G\right)\delta_{ij}\delta_{kl} + G\left(\delta_{ik}\delta_{jl} + \delta_{il}\delta_{jk}\right) \tag{2}$$

$$G = \frac{E}{2(1+v)}; \quad K = \frac{E}{3(1-2v)} \tag{3}$$

where E and G denote the longitudinal and the tangential modules of elasticity, K denotes the bulk modulus and v is the Poisson's ratio. $\{\sigma\}$ and $\{\varepsilon\}$ denote the stress and the total strain tensor, respectively. Dots indicate the incremental formulation of the law. To characterize the elastic stress-strain rule of a homogeneous and isotropic material only two constants are required. The plastic law, that characterizes the material behaviour over the elastic range, requires the definition of the following three conditions:

- a yield function that bounds the elastic domain (which identifies the condition at which the plastic flow begins);
- a rule of plastic flow, which correlates the increase of plastic deformation $\{\dot{\varepsilon}\}^{pl}$ to the current state of stress;
- a hardening rule that modifies the yield function during the plastic flow.

Taking into account the available material laws in the ANSYS code, these requirements are accomplished assuming the Drucker-Prager (DP) plasticity material model (ANSYS 1998; Drucker and Prager 1952). This is typically employed for pressure-dependent inelastic materials such as soils, rocks and concretes, and it is a modification of the Von Mises yield criterion that accounts for the hydrostatic stress component (the confinement pressure). The yield surface of the DP plasticity criterion depends on the first and the second invariant of the stress tensor and remains fixed in the stress space. Usually, the invariants considered to express the yield surface are the mean hydrostatic stress σ_m and the effective shear stress $\bar{\sigma}$:

$$\bar{\sigma}^2 = \frac{1}{2} s_{ij} s_{ij}; \quad \sigma_m = \frac{\sigma_{ii}}{3} \tag{4}$$

where s_{ij} are the deviatoric components of the stress tensor σ_{ij}. The DP yield condition is defined as follows:

$$F = 3\alpha\, \sigma_m + \bar{\sigma} - k = 0 \tag{5}$$

The constants α and k are two parameters related to the friction angle φ and to the cohesion c of the material, according to the following equations:

$$\alpha = \frac{2 \sin \varphi}{\sqrt{3}(3 - \sin \varphi)}; \quad k = \frac{6c \cos \varphi}{\sqrt{3}(3 - \sin \varphi)} \tag{6}$$

The two parameters α and k allow to evaluate the yield stresses in uniaxial tension and compression, f_{tDP} and f_{cDP} respectively, by:

$$f_{tDP} = \frac{k}{\frac{1}{\sqrt{3}} + \alpha}; \quad f_{cDP} = \frac{k}{\frac{1}{\sqrt{3}} - \alpha} \tag{7}$$

In case of elastic-perfectly plastic behaviour, the friction angle φ and the cohesion c are constant and do not depend on the plastic deformation. The normal to the yield surface is calculated as follows:

$$\mathbf{Q} = \frac{\partial F}{\partial \sigma} = \alpha \delta_{ij} + \frac{1}{2\bar{\sigma}} s_{ij} \tag{8}$$

The flow rule, that determines the direction of the plastic straining, is hence given as:

$$\{\dot{\varepsilon}\}^{pl} = \langle\lambda\rangle \mathbf{P}; \quad \mathbf{P} = \beta\delta_{ij} + \frac{1}{2\bar{\sigma}} s_{ij}; \quad \langle\lambda\rangle = \frac{1}{2}(\lambda + |\lambda|) \tag{9}$$

being **P** the plastic potential. If it is assumed $\alpha = \beta$ (then $\mathbf{P} = \mathbf{Q}$) the flow rule is called associated and the plastic straining occurs in direction normal to the yield surface. The experimental results available for soils and rocks show that the volumetric dilatation predicted by the associated DP flow rule is often larger than that obtained by the experiments. Therefore, a non-associated flow rule should be used through a proper definition of the plastic potential. In the present case a third parameter for the DP plasticity behaviour is introduced, called dilatancy angle δ. This parameter rules the flow degree of associativity. If $\delta = \varphi$ the flow is associated, whereas if $\delta = 0$ no plastic volumetric strains will be produced. In conclusion, the definition of the DP model requires three parameters: the friction angle φ that describes the slope of the yield surface (if $\varphi = 0$ there is no dependence on the hydrostatic pressure), the cohesion c (the yield strength at zero hydrostatic pressure) and the dilatancy angle δ.

The DP yield surface can be considered as a smooth version of the Mohr-Coulomb yield surface, and usually the parameters c and φ are introduced in such a way that the circular cone of DP corresponds to the outer vertex of the hexagonal Mohr-Coulomb yield surface. The resulting surface is a right-circular cone with apex at $\rho = k/\sqrt{3}\alpha$ (Fig. 2).

Depending on the parameter α and on the ratio f_{tDP}/f_{cDP}, the yield function, for the plane stress, has three conical forms: elliptic, parabolic and hyperbolic. These forms can be effectively analysed in the two-dimensional space ($\sigma_{III} = 0$), considering the cross-section of the DP cone in the plane (σ_I, σ_{II}). Taking into account that for masonry the ratio between the uniaxial compressive and tensile strengths is usually greater than 3, it is obtained $\alpha^2 > 1/12$ and the conical form of the intersection is parabolic (Fig. 3).

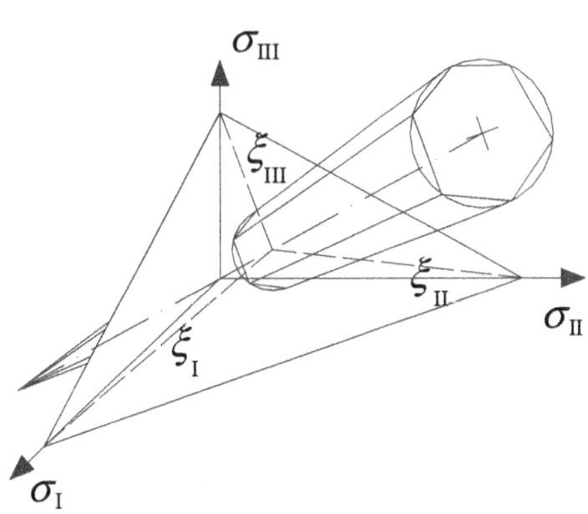

Fig. 2 Drucker-Prager yield surface in the Haigh-Westergaard stress space

Fig. 3 Cross-section of the Drucker-Prager cone with the plane (σ_{I}, σ_{II}) when $\alpha^2 > 1/12$

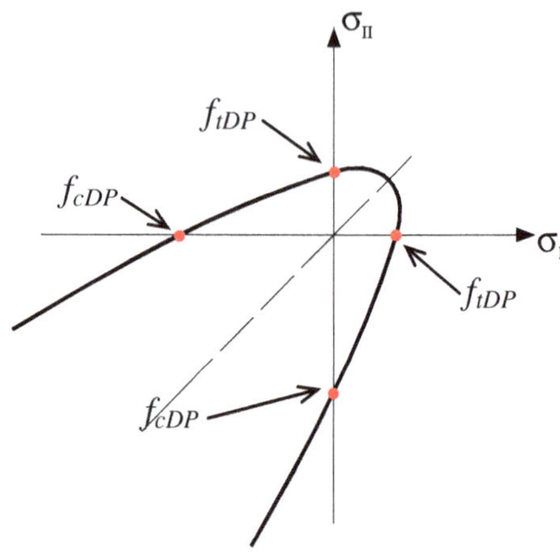

2.3 The Smeared Crack Model

A smeared crack model is introduced through the definition of a crushing and cracking rule. The model makes use of the failure surface employed in ANSYS for concrete and proposed by Willam and Warnke (WW) (William and Warnke 1975; Salari et al. 2004; Hansen et al. 2001). According to this criterion the element is capable of cracking in tension and crushing in compression. The failure surface shows an elliptic trace on the deviatoric sections in each sextant, and a parabolic trace in the meridian sections (Fig. 4).

The WW surface is defined by five parameters: the uniaxial compressive strength f_{cWW}, the uniaxial tensile strength f_{tWW}, the biaxial compressive strength f_{cb} and, two additional parameters ρ_1 and ρ_2. The last two parameters define the curvature of the parabolic traces in the meridian sections for high values of the hydrostatic compression, for anomalies $\zeta = 0°$ and $\zeta = 60°$. The failure surface is characterized by

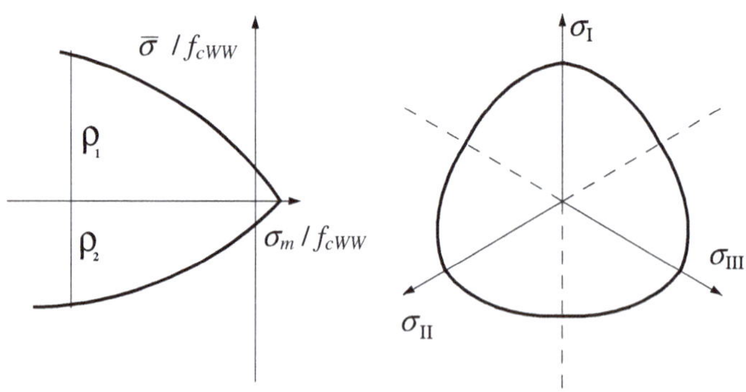

Fig. 4 Hydrostatic section of the yield surface (*left*) and section across a deviatoric plane (*right*)

proper expressions into four domains: CCC, CCT, CTT and TTT (C = compression, T = tension). In the CCC zone, f.i., this surface can be expressed as follows:

$$F' = \frac{1}{r(\sigma_m, \varsigma)}\sqrt{\frac{3}{5}\frac{\overline{\sigma}}{f_{cWW}}} - 1 = 0 \tag{10}$$

where r and ζ are the polar coordinates (radius vector and anomaly) of the representative point of the stress state in the deviatoric plane. The criterion accounts for both cracking and crushing failure modes through a smeared model and the crisis surface is completed by cut-off conditions. At each Gauss point cracking is permitted in three orthogonal directions and is modelled by modifying the material properties of the element introducing a plane of weakness normal to the crack face. It is worth nothing that the model is a fixed crack model.

Despite the need for five constants, in most practical cases (thereby when the hydrostatic stress is limited by $\sqrt{3}f_{cWW}$) the definition of the failure surface can be specified by means of only two constants, f_{tWW} and f_{cWW}, since the three other constants can be assumed as follows:

$$f_{cb} = 1.2 f_{cWW}, \quad \rho_1 = 1.45 f_{cWW}, \quad \rho_2 = 1.725 f_{cWW} \tag{11}$$

The model allows for the introduction of two additional coefficients, denoted as β_t and β_c, that account for a shear strength reduction of the stress producing sliding across the crack face for open (β_t) or re-closed cracks (β_c) (ANSYS 1998).

The WW failure criterion can be joined with the DP plasticity criterion and as a result, the material behaves as an isotropic medium with plastic deformation, cracking and crushing capabilities. The parameters required are the following:

- elastic parameters: E and ν;
- plastic parameters (DP): c, φ and δ;
- cracking and crushing parameters (WW): f_{cWW}, f_{tWW}, β_t and β_c.

The combination of the two models requires a careful definition of the above parameters. According to the experimental evidence such choice must comply with the following criteria:

- the tensile strength f_{tWW} must be smaller than the tensile strength derived from the plasticity model f_{tDP};
- the compressive strength f_{cWW} must be greater than the compressive strength derived from the plasticity model f_{cDP}, to ensure the correct plastic behaviour of the masonry in the mixed tensile-compression zone.

The proper combination of these parameters allows for an elastic-brittle behaviour in case of biaxial tensile stresses or biaxial tensile-compressive stresses with low compression level. On the contrary, the material is elastoplastic in case of biaxial compressive stresses or biaxial tensile-compressive stresses with high compression level.

Both the DP and the WW criteria have been extensively employed in the scientific literature to model the inelastic behaviour of masonry assemblages. Discussing the homogenisation approach for masonry Zucchini and Lourenço (2007) adopt the DP model for the simulation of the plastic deformation in masonry cells. They show that it is possible to account for the degradation of the masonry mechanical properties in compression. The DP criterion was adopted by Cerioni et al. (1995) to discuss the seismic behaviour of the Parma Cathedral Bell-Tower. Adam et al. (2009) used the WW criterion to model cracking and crushing phenomena, and the comparison between numerical and experimental results showed a good agreement. Chiostrini et al. (1998) combine the DP and the WW criteria to model the results of several diagonal tests on masonry panels, obtaining good agreement with the experimental results. Betti and Vignoli (2008) combine the two criteria to discuss the seismic vulnerability of a masonry church.

The assignment of the mechanical parameters required by the DP and the WW criteria requires a careful calibration (Chiostrini et al. 1998). The following section discusses the identification of these parameters through the results of experimental investigations.

3 Tuning of the Numerical Model

3.1 Experimental Tests

The constitutive parameters of the model should be evaluated on the basis of the in situ mechanical properties of the masonry walls. This requires to perform a set of experimental investigations. The paragraph discusses some of the tests that can be performed on masonry panels to evaluate the required parameters, and shows a calibration of the model based on these tests. It is worth noting that the difficulties in removing specimens in buildings of historic value suggest the calibration of the parameters using experimental results available from similar masonry textures. Herein, for illustrative purposes, the results of past experimental researches (Chiostrini and Vignoli 1992, 1994; Chiostrini et al. 2000, 2003) aimed to assess both strength and deformability of masonry walls of historic masonry buildings in Tuscany (Central Italy) are discussed.

The experimental researches comprehended laboratory tests on masonry samples and destructive in situ tests on masonry panels. The in situ tests were: direct shear tests (S), shear-compression tests (SC) and diagonal compression tests (DC). The first two types of test reproduce with good approximation the stress state of a masonry pier under seismic loading. The third test is useful for a direct evaluation of the masonry tensile strength. In a first experimental campaign (Chiostrini and Vignoli 1992, 1994) S and SC tests were performed on nine panels selected from different buildings: the S. Orsola Monastery in the historic centre of Florence (four panels: T1, T2, T3 and T4), an old residential building in Florence (three panels: COR1, COR2 and COR3) and two buildings in Pontremoli (Lunigiana, Central

Tuscany), the Istituto Belmesseri and the Town Hall (one panel for each case: BEL and COM). Common characteristics of these buildings were: bearing walls made with stone or mixed stone and brick masonry with chaotic textures and wood floor slabs with insufficient linkage between slabs and walls. Thickness of the panels varied from about 300 to 600 mm.

In a second experimental investigation (Chiostrini et al. 2000, 2003) three shear-compression tests and four diagonal compression tests were performed on seven masonry panels selected in five different rural buildings. Common characteristics of these buildings were: two or three stories height, bearing walls made by stone masonry with typical chaotic textures and wood or steel floor slabs.

Figures 5 and 6 show the experimental setup of the direct shear test (S) and the shear-compression test (SC). After having isolated the masonry panel under

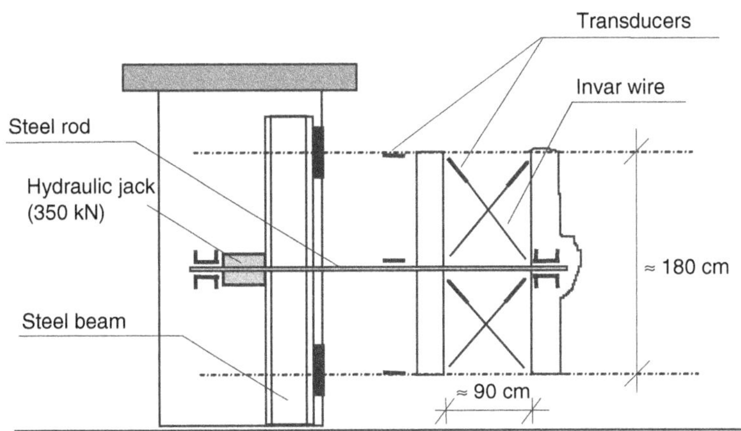

Fig. 5 Experimental setup of the direct shear test (S)

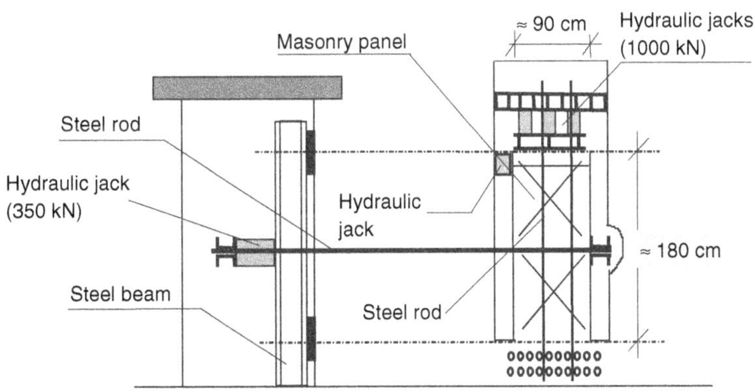

Fig. 6 Experimental setup of the shear-compression test (SC)

Table 1 In-situ direct shear and shear-compression tests on masonry panels

Panel ID	Location	Building	Panel cross-section (cm^2)	In-situ test	Masonry texture
T1	Florence	S. Orsola	4503	SC	SO1
T2	Florence	S. Orsola	4536	SC	SO2
T3	Florence	S. Orsola	4648	SC	SO2
T4	Florence	S. Orsola	3669	SC	SO1
COR1	Florence	Residential	2640	S	VC
COR2	Florence	Residential	2760	S	VC
COR3	Florence	Residential	2511	S	VC
BEL	Pontremoli	Ist. Belmesseri	4480	S	Bel
COM	Pontremoli	Town hall	2880	S	Com
A	Pieve Fosciana	Town hall	5765	SC	PF
B	Pieve Fosciana	Town hall	4798	SC	PF
E	Pognana	Residential	5400	SC	Po

SO1 = mixed stone and brick masonry with chaotic texture; *SO2* = mixed stone and brick masonry with regular texture; *VC* = mixed stone and brick masonry with chaotic texture; *Bel* = stone and brick infilled masonry with chaotic texture; *Com* = stone infilled masonry with chaotic texture; *PF* = two facings stone wall compact and interlocked; *Po* = two facings stone wall with irregular masonry texture infilled with packed mortar

investigation from the rest of the wall through two vertical cuts, the direct shear test is performed applying a horizontal force until the panel collapses. During the test the values of the deformations of the two main diagonals (the one in compression and the one in tension) of the panels are acquired. The deflection of the specimen in the middle section is also monitored. The shear-compression test is similar to the previous, but in this case a vertical pressure is firstly applied through hydraulic actuators (steel plates are positioned over the panel to apply a uniform compressive stress). Table 1 reports a summary of the performed S and SC tests. The table shows: an identification code of the panel (ID), the location, the type of building, the area of the cross-section of the panel, the type of test (S or SC) and a brief description of the masonry texture (thickness of the masonry panels was variable from about 300 to 600 mm).

Results of S and SC tests are summarised in Table 2 where the following data are collected: the vertical pressure acting on the panel σ_0, the characteristic shear strength τ_k and the value of the b shape factor employed for the calculation of τ_k. In addition, the table reports the values of the elastic tangential modulus G obtained through the shear-strain curve of the elastic range part of the tests, and the elastic longitudinal modulus E obtained in the preliminary tests. It is possible to observe the high variability of the mechanical properties of these masonries.

Table 2 Results of the experimental in-situ tests

Panel ID	σ_0 (N/mm^2)	τ_k (N/mm^2)	b	E (N/mm^2)	G (N/mm^2)
T1	0.800	0.114	1.37	/	200
T2	0.800	0.090	1.40	/	116
T3	0.400	0.109	1.29	/	274
T4	0.400	0.170	1.19	/	241
COR1	0.230	0.081	1.23	/	173
COR2	0.430	0.090	1.33	/	325
COR3	0.120	0.197	1.0	/	333
BEL	0.190	0.096	1.14	/	290
COM	0.130	0.203	1.0	/	249
A	0.378	0.234	1.0	1468	179
B	0.433	0.320	1.0	1333	435
E	0.165	0.065	1.21	250	102

3.2 *Identification of Model Parameters*

The direct shear tests on panels COR2, BEL and COM and the shear-compression tests on panels A and B were used to identify the model parameters. The setup of both tests was modelled using the finite element *Solid65* (Fig. 1) and two cases were considered. In the first case only the plasticity model was accounted for, in the second case both the plasticity and the failure models were considered. As an example, Fig. 7 shows the FE model used for the simulation of the SC test on panel

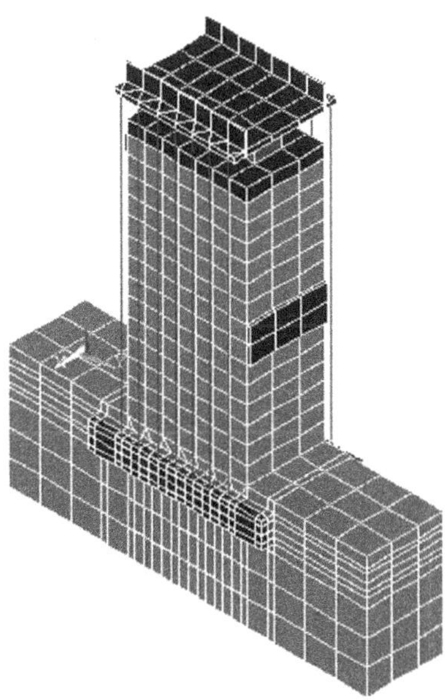

Fig. 7 FE model of the shear-compression test on panel B in Pieve Fosciana (Lunigiana)

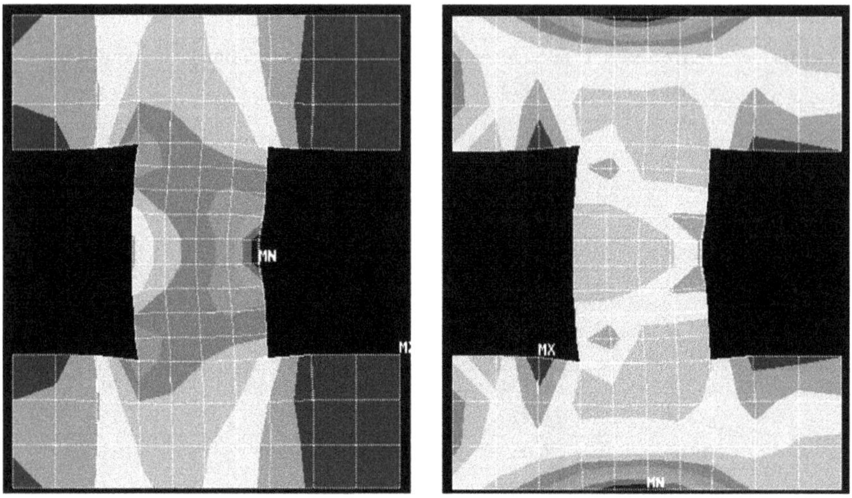

Fig. 8 Numerical modelling of shear test on panel COR2: principal compressive stresses (*left*) and principal tensile stresses (*right*)

B in Pieve Fosciana. The model includes a considerable portion of the underlying masonry, the steel plates on the head and the system of vertical bars that originate the vertical load. Similar FE models were built to simulate the remaining shear and shear-compression tests. The numerical simulations were performed in two steps according to the experiments. In a first time the vertical pressure was applied, subsequently a monotonically increasing horizontal shear force was applied under load control using the Newton-Raphson method to solve the nonlinear equations. Figure 8 shows the maps of the principal stresses in an intermediate step of the analysis for the panel COR2. It is easy to recognize the formation of the two diagonal struts, in the map of the principal tensile stresses.

For the same panel, Fig. 9 shows a comparison between the numerical and the experimental results for both material models (plasticity only and plasticity and smeared crack models). In both cases it possible to observe a good approximation of the collapse load. The adoption of the smeared crack model in combination with the plasticity model allows to reproduce with good accuracy even the collapse displacement observed during the test (Fig. 9-right). On the contrary, the adoption of the elastic perfectly plastic model alone does not obviously reproduce the collapse displacement, since the panel behaves like an elastoplastic continuum with no limits to deformation.

Figure 10 shows, in the plane of the principal stresses (biaxial state), the particular combination of the DP and WW domains obtained for the numerical tuning of the test (Fig. 9-right). Values of the cohesion c = 0.15 N/mm^2 and φ = 40° provide the masonry tensile strength f_{tDP} = 0.189 N/mm^2, which is cutted by the choice of the parameter f_{tWW} = 0.14 N/mm^2 adopted in the WW model. The choice f_{cWW} = 2.5 N/mm^2, greater than f_{cDP} = 0.643 N/mm^2 of the plasticity model, allows

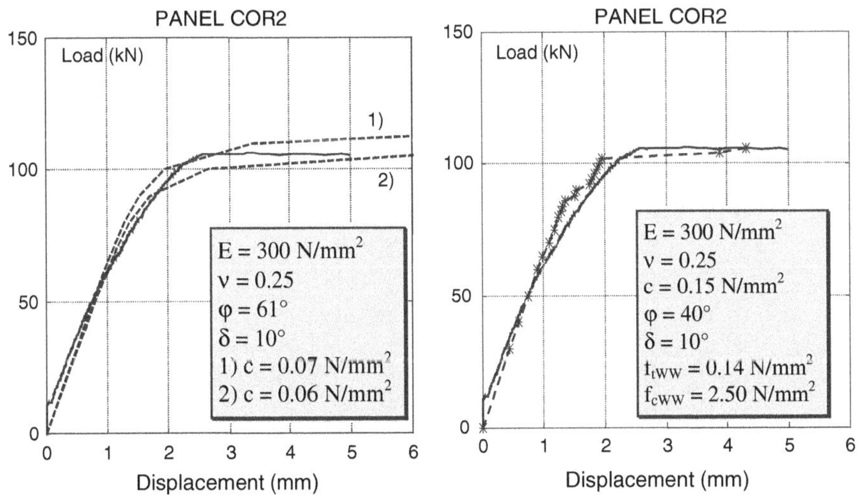

Fig. 9 Comparison between experimental and numerical results for shear test on panel COR2: plasticity model (*left*) and combined plasticity and failure models (*right*)

Fig. 10 Intersection between the plasticity and the failure domains (panel COR2)

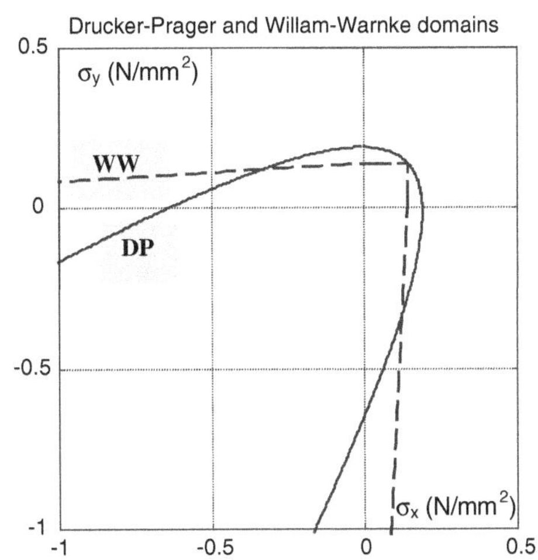

the correct simulation of the behaviour of the masonry in the mixed zone tension-compression. For the reproduction of the experimental results by using only the plasticity model, the friction parameter φ must be significantly higher than the correct choice for the combined models. The opposite happens for the cohesion c. Furthermore, the analysis highlighted the role of the dilatancy parameter δ which identification has always provided values lower than the friction angle. This confirms the impossibility of using the associated flow law. The values of β_t and β_c were assumed equal to 0.25 and 0.75 for all the analyses.

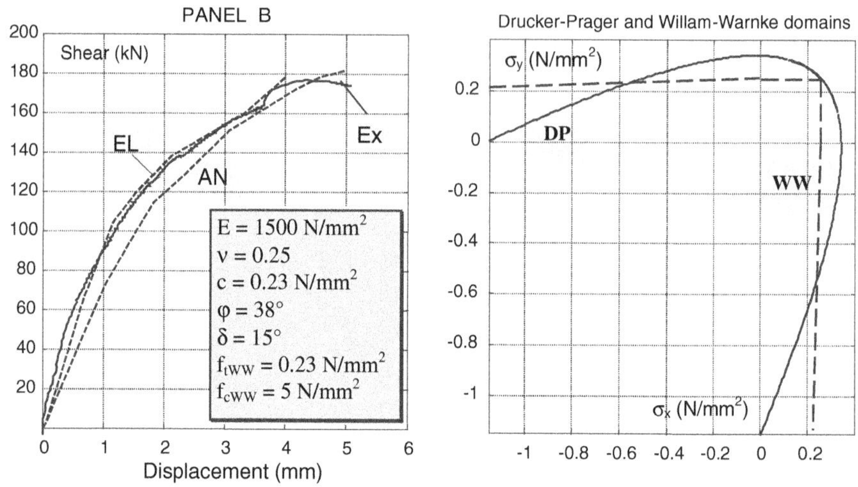

Fig. 11 Comparison between experimental and numerical results for shear compression test on panel B (*left*) and intersection between the plasticity and the failure domains (*right*)

As an additional example, Fig. 11-left shows the result of the comparison of the curves shear-displacement in the middle cross-section obtained for the simulation of the test on panel B. The comparison refers to the model with both plasticity and cracking failure. The two curves (EL and AN) refer to different numerical modelling of the horizontal constraint of the head of the panel. The tuned combination of the models is reported in Fig. 11-right. Comparing the results of the numerical simulations on panels B and COR2 it is possible to observe a significantly higher values of the elastic modulus E, cohesion c and tensile strength for B. This is in agreement with the experimental evidence because the masonry texture of panel B was more compact than that of panel COR2.

A summary of the constitutive parameters which allowed a good reproduction of the experimental results in terms of shear-displacement curves is shown in Tables 3 and 4. In particular, Table 3 resumes the parameters identified for the DP model, whereas Table 4 reports the parameters identified for the DP model used together with the WW model. The tables report:

Table 3 Identified model parameters (only plasticity model)

Panel ID	E (N/mm^2)	ν	c (N/mm^2)	φ (°)	δ (°)	f_{cDP} (N/mm^2)	f_{tDP} (N/mm^2)
COR2	300	0.25	0.065	61	10	0.503	0.049
COM	500	0.25	0.120	61	35	0.928	0.090
BEL	500	0.25	0.07	61	15	0.541	0.052
A	1500	0.25	0.28	40	15	1.201	0.353
B	1500	0.25	0.27	40	15	1.158	0.341

Table 4 Identified model parameters (plasticity model combined with the smeared cracking and crushing model)

Panel ID	c (N/mm^2)	φ (°)	δ (°)	f_{cWW} (N/mm^2)	f_{tWW} (N/mm^2)	f_{cDP} (N/mm^2)	f_{tDP} (N/mm^2)
COR2	0.150	40	10	2.5	0.14	0.643	0.189
COM	0.215	40	30	2.5	0.20	0.922	0.271
BEL	0.110	40	20	2.5	0.102	0.472	0.139
A	0.240	38	15	5.0	0.24	0.984	0.314
B	0.230	38	15	5.0	0.23	0.943	0.301

- the identified elastic parameters E and ν;
- the identified plastic parameters (DP): c, φ, δ (f_{cDP} and f_{tDP});
- the identified parameters of the cracking and crushing model (WW): f_{cWW}, f_{tWW}.

According to the experimental results in terms of shear strength τ_k, the panels can be divided into three classes (Chiostrini et al. 2000): (a) HS (high strength): panels A, B, COR3 and COM; (b) MS (medium strength): panels T1, T3, T4 and BEL; (c) LS (lower strength): panels T2, COR1, COR2 and E. HS corresponds to a good quality masonry; MS corresponds to a masonry with little internal voids, well filled by mortar and small dimension units; LS represents very poor assemblages of blocks and mortar, with many internal voids and facing walls weakly pinned. On the basis of such classification it is then possible to identify the ranges of variation of the parameters for the three classes of masonry (Table 5 for DP model and Table 6 for DP and WW models). Considering the difficulty to obtain specific results from monumental buildings, the reported values are typological reference data providing a reasonable estimation of these parameters (useful for the modelling of complex buildings whose walls textures are similar to those covered by the tests).

Table 5 Variability of the parameters of the plasticity model

Masonry	E (N/mm^2)	ν	c (N/mm^2)	φ (°)	δ (°)
HS	500–1500	0.25	0.12–0.28	40–61	15–35
MS	350–500	0.25	0.07–0.09	38–61	12.5–15
LS	100–300	0.25	0.065	38–61	10–12.5

Table 6 Variability of the parameters of the plasticity model combined with the failure criterion

Masonry	c (N/mm^2)	φ (°)	δ (°)	f_{cWW} (N/mm^2)	f_{tWW} (N/mm^2)
HS	0.22–0.24	38–40	15–30	2.5–5.0	0.20–0.24
MS	0.11	38–40	12.5–20	2.5–3.0	0.10–0.11
LS	0.08–0.15	38–40	10–12.5	2.5–3.0	0.08–0.14

4 Illustrative Case Studies

In this section the model previously described is employed to analyse the seismic behaviour of two relevant masonry structures: a basilica-type church and a historic residential building. The two case studies demonstrate the careful use of nonlinear numerical analyses to address practical engineering problems in the field of seismic assessment of historic constructions.

4.1 Masonry Church in Impruneta (Tuscany)

As a first application, the analysis of the 14th century Basilica of Santa Maria all'Impruneta near Florence is discussed (Fig. 12). The plan view of the church (Fig. 13) shows the typical layout with a single nave and a polygonal apse separated by a triumphal arch. A pronaos, composed of five wide arches, and surmounted by rectangular windows, was built in 1634 and covers almost totally the old façade (Fig. 12).

4.1.1 Geometry and Materials

An in situ survey of the church was made to obtain basic information on the geometry, the structural details and any irregularities. The investigation consisted in a geometrical relief, aimed at a check-up of wall-to-wall and wall-to-roof

Fig. 12 Front view of the Basilica of Santa Maria all'Impruneta (Tuscany)

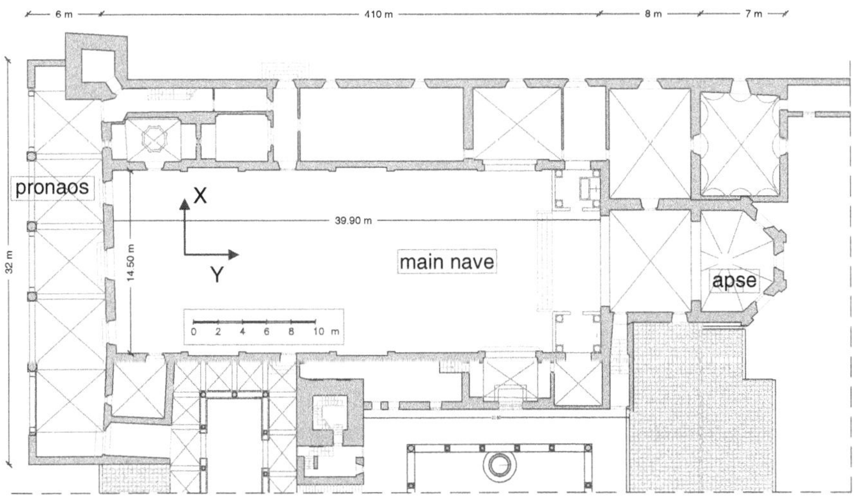

Fig. 13 Plan layout of the Basilica of Santa Maria all'Impruneta

Table 7 Elastic parameters of the main elements

	Nave	Apse	Columns
E (N/mm^2)	1400	1530	14,500
ν	0.2	0.2	0.2
γ (kg/m^3)	1900	2000	2300

connections, and useful for the characterization of the masonry texture. This step was accompanied by a historical survey of the building to determine the original shape and eventually consider the church modifications over the centuries. Main dimensions of the nave are the following: maximum length of about 39.90 m, maximum width of 14.50 m and walls with height of about 15.0 m. The thickness of the masonry walls ranges between 0.70 m (nave walls) and 0.80 m (apse walls) and the nave roof is made of a timber structure. The walls are single-leaf, with several types of masonry weaving, differing both in materials (stones, bricks, etc.) and textures (*opus incertum*, *opus mixtum*, etc.). However, despite these differences, the construction is mostly made of irregular sandstone masonry with thick lime mortar joints. Local stones are also used for windows, apse and doors jambs. For the mechanical characterization of the elastic properties of the masonry the Italian Recommendations were employed (OPCM 2003; NTC 2008), assuming the conservative values for historic masonry (Table 7, γ denotes the specific weight).

4.1.2 Numerical Modelling

The structural behaviour of the church was analysed through a complete 3D analysis using the FE technique and the macro-modelling strategy. This computational strategy, as reported above, is convenient for large scale models. Other strategies, suitable mainly for small size models, rely on micro-modelling approaches where units and mortar are modelled separately. An additional comprehensive recent discussion on these aspects is reported in Berto et al. (2005), Adam et al. (2010), Lourenço and Pina-Henriques (2006). The model of the whole *fabrica* was built by the code ANSYS (1998) to accurately reproduce the geometry of the structure, focusing on the variations in the wall thickness, on the geometrical and structural irregularities and on the wall connections.

Masonry walls were modelled by means of *Solid65* elements (Fig. 1); *Shell63* elements (two-dimensional isoparametric elements with four nodes) were used to model the main vault and the annexed buildings; *Beam44* elements (one-dimensional isoparametric elements with two nodes) were employed for modelling the queen truss on the timber roof of the main nave. The major openings in the building were reproduced, and the nonlinear analyses were performed assuming a rigid ground foundation (fixed base model). The final 3D model (Fig. 14) consists of 27,779 nodes, 76,895 3D *Solid65* elements, 1751 2D *Shell63* elements and 547 1D *Beam44* elements, that correspond to 81,021 degrees of freedom. The nonlinear behaviour of the masonry was reproduced by the proper combinations of the DP model with the WW failure surface. The conservative values for historic masonry were assumed, taking into account the Italian

Fig. 14 Finite element model of the Basilica

Table 8 Drucker-Prager yield criterion (main elements)

	Nave	Apse	Columns
c (N/mm^2)	0.1	0.1	0.5
φ (°)	38	38	38
δ (°)	15	15	15

Table 9 Willam-Warnke failure surface (main elements)

	Nave	Apse	Columns
f_{cWW} (N/mm^2)	7.5	8.5	40
f_{tWW} (N/mm^2)	0.15	0.15	3.5
β_c	0.75	0.75	0.75
β_t	0.15	0.15	0.15

Recommendations (OPCM 2003; NTC 2008) and the identification reported in the previous section. Tables 8 and 9 report the selected parameters of the model with respect to the main elements of the Basilica.

4.1.3 Static Analysis

Firstly, the static behaviour of the Basilica was analysed under the vertical loads deriving from the own weight of the walls and from the roof loads (1.1 kN/m^2 according to the Italian standards). The overall structure was analysed in the nonlinear range through the complete 3D model with the aim to obtain valuable information on the global behaviour and on the interaction among the single elementary parts. Finally, it was possible to identify the weak points of potential failure in the church. Results of the static analysis in terms of vertical stresses are reported in Figs. 15 and 16. In general, the stress state induced on the church by the static vertical loads is quite moderate. The average compressive stress on the nave walls is about 0.6 N/mm^2. Small values of tensile stresses appear on the top surface of the nave walls, due to the timber roof loads. Mainly, this is a local numerical effect depending on the punctual connection between the beams and the solid elements.

The maximum value of the compression is reached in the columns of the pronaos (Fig. 16) where the compressive stresses reach values of about 2.0–2.5 N/mm^2. Even if this stress level seems high when compared with the medium value of the compressive stresses it is lower than the crushing limit of the sandstone material of the columns. Anyway, among the various structural elements that compose the church, these columns (actually reinforced by steel collars) are the critical elements.

Results of the analysis substantially confirm that the church is adequate to withstand the vertical loads. This is a quite common result for this typology of buildings designed by skilled manufacturers to attempt very slender schemes. Results in terms of displacement show that the maximum value is reached close to the triumphal arch between the nave and the apse.

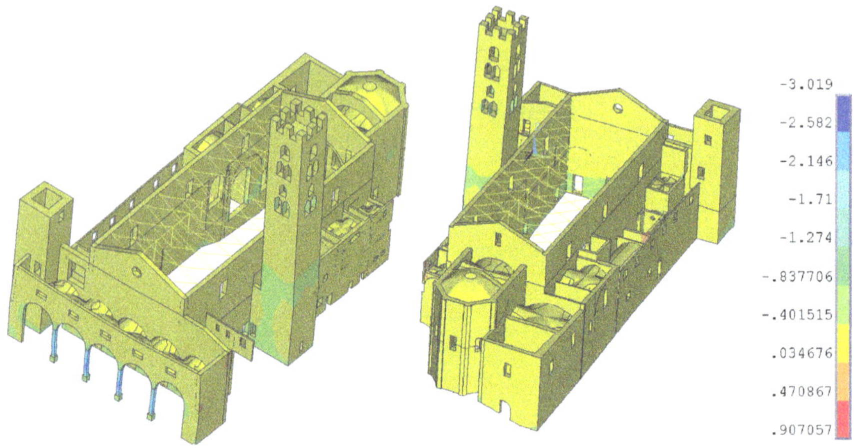

Fig. 15 Vertical stresses σ_z (N/mm^2): global view

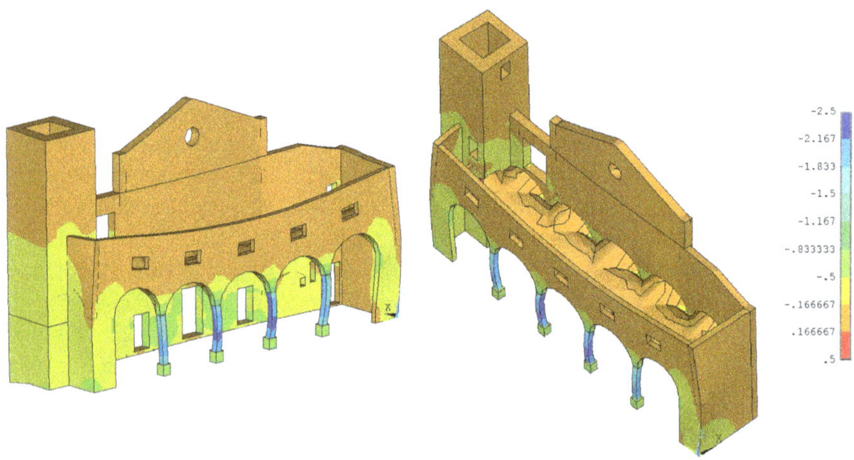

Fig. 16 Vertical stresses σ_z (N/mm^2): façade wall detail

4.1.4 Modal Analysis

The 3D model was also used to assess the modal behaviour of the Basilica. The first 100 modal shapes of the church were evaluated to assure that the total effective modal mass of the model was at least 90 % of the actual mass. The outcome was that 90 % of the total mass was accounted for by using the first 87 modes. Effective and cumulative masses of the first 10 global vibration modes in the transversal, longitudinal and vertical direction of the church are reported in Table 10. The first modal shape of the church involves the translation in the weakest transversal direction of the main nave, with significant out-of-plane deformation of the

Table 10 Church modal effective masses for transversal, longitudinal and vertical direction

Mode n °	Period (s)	X direction (transversal)		Y direction (longitudinal)		Z direction (vertical)	
		M_{eff} (%)	ΣM_{eff} (%)	M_{eff} (%)	ΣM_{eff} (%)	M_{eff} (%)	ΣM_{eff} (%)
1	0.585	14.11	23.95	0.03	10.58	0.00	0.24
2	0.480	0.01	23.96	2.92	13.50	0.00	0.24
3	0.404	0.72	24.68	0.02	13.53	0.00	0.25
4	0.385	1.28	25.96	8.62	22.15	0.00	0.25
5	0.334	1.13	27.68	0.93	27.57	0.01	0.36
6	0.298	18.38	46.06	0.19	27.76	0.26	0.62
7	0.297	5.31	51.37	0.02	27.78	2.86	3.48
8	0.289	0.67	55.22	0.73	28.53	0.00	8.43
9	0.261	0.00	55.22	6.10	34.64	0.28	8.71
10	0.242	5.66	62.87	0.29	35.11	0.12	8.88

orthogonal elements (Fig. 17). The higher modal shapes of the church are a combination of transversal vibration modes and torsional modes. The distribution of the modal shapes demonstrates that the church, though characterised by stiff structural elements on the perimeter, displays low transversal and torsional stiffnesses, with significant out-of-plane deformations of the elements. Furthermore, the deformed plan configuration confirms that the seismic loads acting along either longitudinal or transversal direction involve remarkable out-of-plane deformations of the orthogonal structural elements.

4.1.5 Seismic Vulnerability

The analysis of the seismic behaviour was performed by means of a pushover analysis (Falasco et al. 2006; Kim and D'Amore 1999). Monotonically increasing horizontal loads were applied under conditions of constant gravity loads. Based on this analysis method, the effects of the seismic loads were evaluated by applying two systems of horizontal forces perpendicular to one another. These forces, not acting simultaneously, were evaluated taking into account two load distributions. The first distribution was directly proportional to the masses of the church (uniform); the second distribution was proportional to the product of the masses by the displacements of the corresponding first modal shape. These two load distributions could be considered as representative of two limit states for the capacity of the building. The first distribution assumes that the horizontal loads are constant with respect to the height. This means that the displacements of the lower level of the church are overestimated, while the opposite happens for the displacement of the top level. On the contrary, the second distribution overestimates the displacement on the top level. Is it noteworthy to point out that a conventional pushover was performed in the study, i.e. loads applied on the building didn't change with the

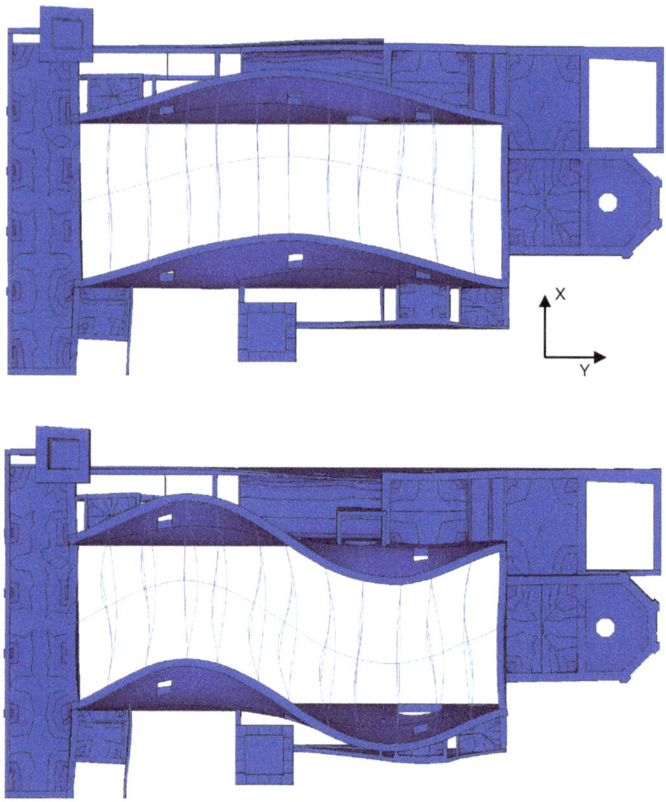

Fig. 17 First two transversal modal shapes of the church: T_1 = 0.585 s (f_1 = 1.704 Hz) and T_3 = 0.404 s (f_3 = 2.475 Hz)

progressive degradation occurring during the loading process (Antoniou and Pinho 2004; Chopra and Goel 2004).

The critical load distribution for the Basilica corresponds to the case of the load acting in the transversal direction, perpendicular to main nave. So, results are mainly next detailed with respect to this case. This behaviour was expected due to the fact that the transversal direction of the church involves remarkable out-of-plane deformations of the orthogonal structural elements. Figure 18 reports the displacements in the transversal direction at the end of the analysis and Fig. 19 reports the corresponding cracking pattern; it involves almost all the nave walls, together with the orthogonal ones. The cracking behaviour, compared with the overall deformative behaviour, shows a poor stiffness of the church in the transversal direction. Figures 20 and 21 report, respectively, the deformative behaviour and the cracking pattern that arises in the church with respect to the load acting in the longitudinal direction. The structural elements more vulnerable are the pronaos and the façade. The analyses stop at a level of the horizontal load of about the 18–20 % of the overall weight of the church.

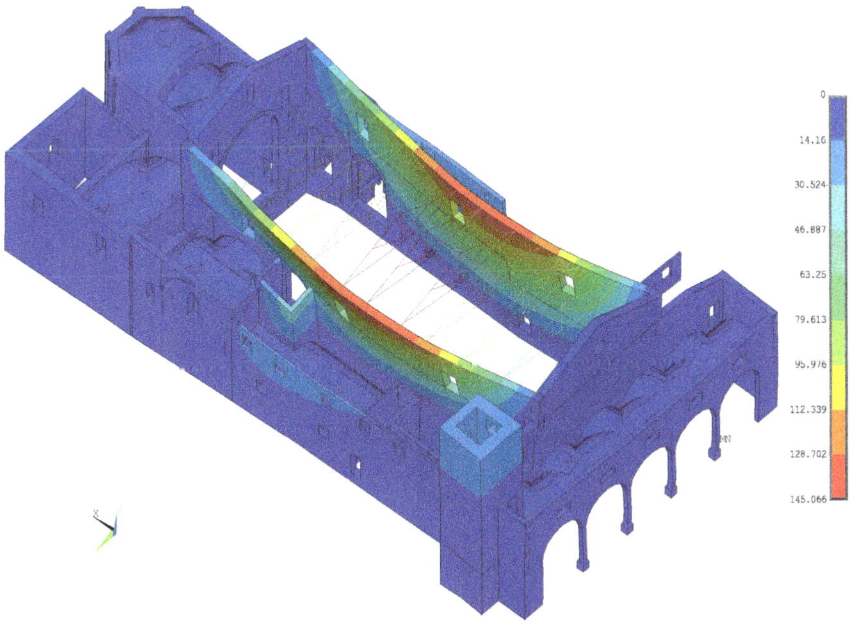

Fig. 18 Pushover analysis with uniform horizontal loads in the transversal direction: displacement (mm)

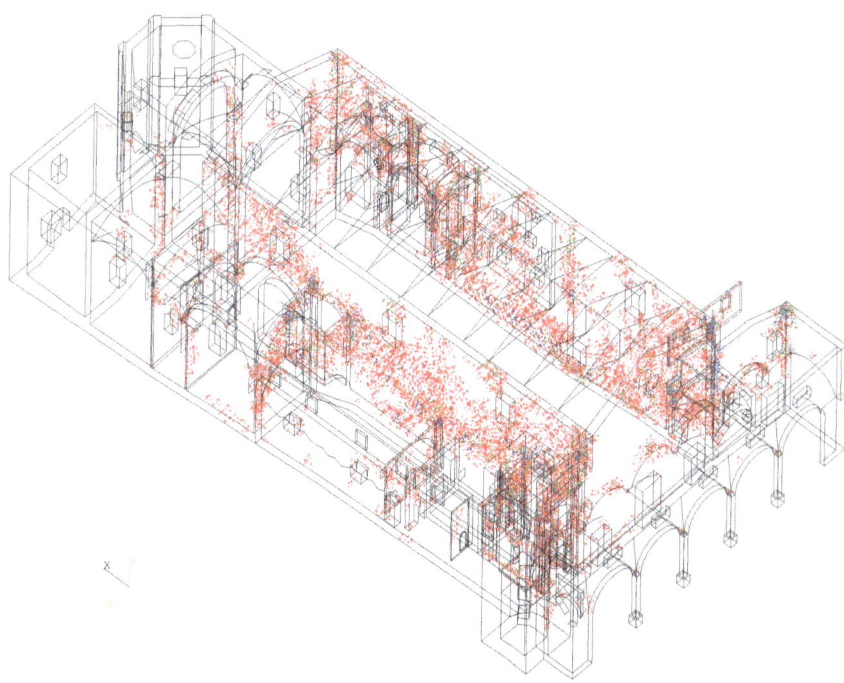

Fig. 19 Pushover analysis with uniform horizontal loads in the transversal direction: cracking pattern

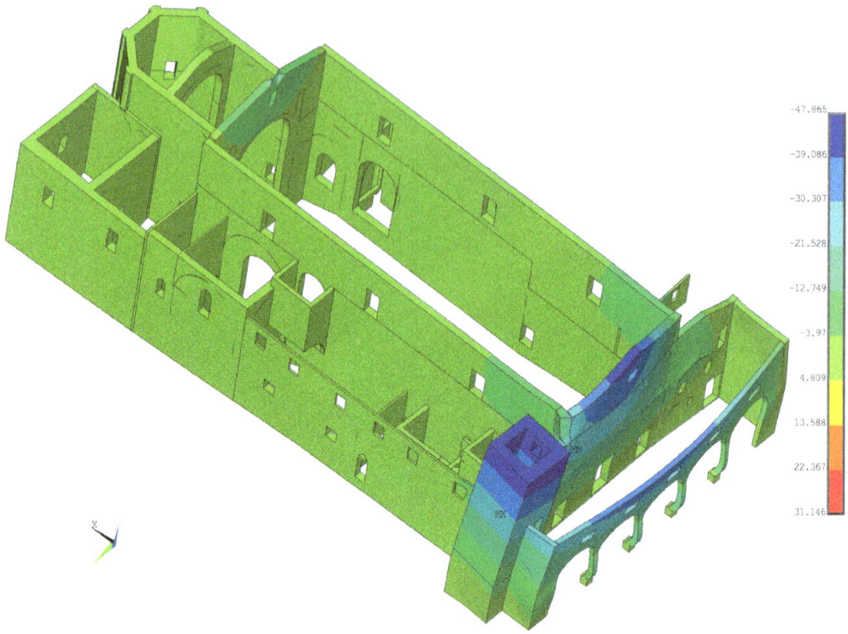

Fig. 20 Pushover analysis with uniform horizontal loads in the longitudinal direction: displacement (mm)

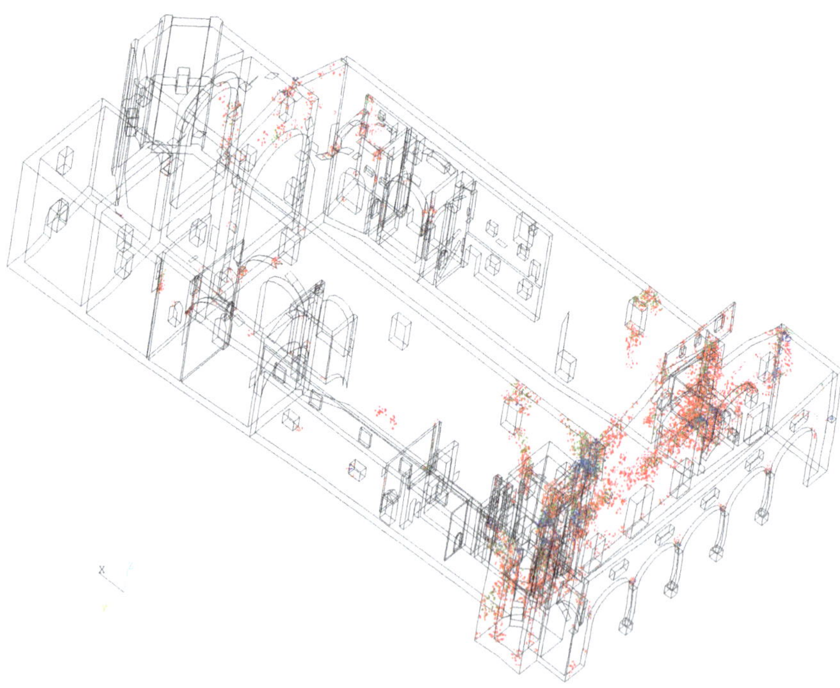

Fig. 21 Pushover analysis with uniform horizontal loads in the longitudinal direction: cracking pattern

The cracking pattern suggests the mechanism to be considered in the limit state analysis and offers an indication on the potential strengthening design (Lourenço and Oliveira 2007). The critical elements of the church are the lateral walls of the nave, which require to increase their out-of-plane strength. In order to generate a box-behaviour a global reinforcement of the queen roof can be performed by means of a system of horizontal counterbracing to be inserted between the extrados of the timber roof and the tile covering. The steel counterbracing, directly connected to the wooden beams of the queen roof, could be designed to create a top rigid floor that may ensure a box-behaviour of the nave. A local steel reinforcement of the connections between the timber structure and the masonry walls needs to be provided to avoid local failure. Additional local reinforcements may be inserted along the Basilica, in particular a steel tie could be placed along the apse perimeter.

4.2 Masonry Building in Fivizzano (Lunigiana)

The object of the second application is a residential building located in the historic city centre of Fivizzano in Lunigiana (Fig. 22). The building was damaged by an earthquake in October 1995 (4.7 magnitude on the Richter Scale), and subsequently retrofitted in December 1997. The building is an illustrative example, since it is representative of many typical masonry buildings of Central Italy, with architectural and structural features (as well as the seismic damages suffered during the earthquake) similar to those found in many other buildings. The moderate damages caused by the earthquake suggested the employment of traditional retrofitting techniques such as the insertion of steel chains at several levels. The numerical analyses aimed at assessing the effectiveness of the retrofitting by comparing the seismic behaviour of the building after and before the insertion of the steel chains.

4.2.1 Description

The construction is composed of two adjacent buildings with different heights, joined by a common wall (Fig. 22). Over time, the original structure of the building, that dates back to the eighteenth century, has been undergoing continuous structural and architectural changes, especially after the 1920 and 1967 earthquakes which led to the current configuration. As a part of the last restructuring in 1967, an additional floor was built with the creation of two units. Today the building has an irregular shape with two levels above ground and one basement.

The masonry walls are made of rubble stone masonry (an example of the chaotic texture is reported in Fig. 23) with the exception of the corners, where regular-shaped well-connected stones are visible. On the first floor there are two

Fig. 22 Front view (South-West)

Fig. 23 View of the masonry texture

heads brick walls, on the second floor a concrete wall was found, probably introduced during the last renovation in 1967. Lintels over doors and windows are made of hewn stone. The thickness of the stone walls varies from 0.55 to 1.00 m. Different types of floors are present. On the first floor there are stone vaults, the other levels are made of steel profiles (NP 140) and hollow flat blocks without concrete. The roof is made with wooden beams simply leaned on the perimeter walls. In the South-East portion of the building the roof is made by means of a reinforced concrete (RC) slab supported by rectangular concrete beams still leaned on the perimeter walls. Despite the last restructuring of the building dates back to 1967, there are no perimeter concrete beams.

During the 1995 earthquake the building was damaged and a series of cracks with a thickness of a few mm, formed on the first and the ground floor walls. Cracks passing through the wall thickness arose on North-West façade and in some internal walls. The seismic strengthening of the building, which ended in December 1997, was made through the introduction of steel chains at the level of the first and the second floor, and in correspondence of the first floor vault (Fig. 24).

The retrofitting was limited to the North-West area of the building, the most damaged during the 1995 earthquake and circular steel chains with diameter ranging from 18 to 24 mm were used. It is worth noting that the position and the number of the steel chains was decided without any provisional design, being selected only based on the building damage.

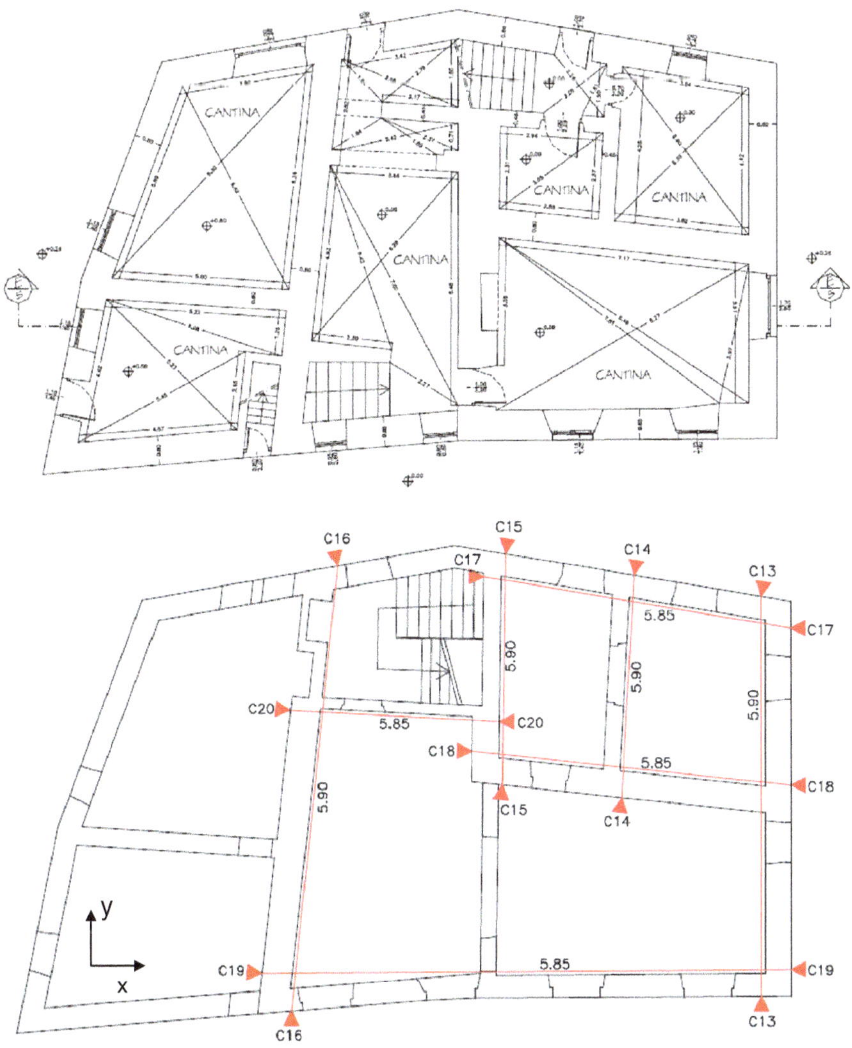

Fig. 24 First floor layout (*up*) and position of the steel chains (*down*)

4.2.2 Numerical Modelling

To analyse the seismic behaviour of the building before and after the retrofitting, a numerical model of the construction was built using the commercial code ANSY. The model includes the following structural elements: the masonry walls, the structural part of the stone vaults, the steel floors, the wooden cover roof, the reinforced concrete slab and the concrete wall (Fig. 25). Stairs were not considered in the model, since most of these are cantilever stairs without structural continuity with the confining walls (stairs were assumed as additional masses). The numerical model of the building before the retrofitting with steel chains comprised a total of 8404 elements, corresponding to 41,676 degrees of freedom (Fig. 25). In particular, the masonry walls were modelled by the elements *Solid65*, the steel floors by the elements *Beam8* and the masses by the elements *Mass21*. Finally, the RC slab was modelled through the elements *Shell63*.

The masonry nonlinear behaviour was defined by the combination of the DP plasticity model with the WW failure model. The mechanical characteristics of the in situ masonry were defined assuming the ones corresponding to the typology HS of Sect. 3 (Table 11). The elastic properties of the main elements are resumed in Table 12.

A preliminary static analysis under vertical dead and live loads was performed according to the Italian Recommendations in force at the time of retrofitting (1997) (DM96 1996). The maximum compression in the masonry at the ground level was found to be about 0.697 N/mm^2. The modal analysis provided the following results (Fig. 26). The first modal shape (f_1 = 5.34 Hz) involves the translation in the Y direction and mainly concern the bending deformation of the second floor. The second modal shape (f_2 = 6.74 Hz) is a torsional mode and again mainly involves deformations of the second floor. The third modal shape (f_3 = 7.49 Hz) is another torsional mode similar to the previous mode. Considering the first 20 modes, they activated more than 90 % of the total mass in the X and Y directions, only 59.7 % in the vertical direction.

4.2.3 Assessment of the Past Retrofitting

The structural behaviour of the building with and without the steel chains was analysed performing nonlinear seismic analyses, to assess the effectiveness of the post-earthquake strengthening. Two distributions of horizontal forces were preliminary considered. The first seismic equivalent load distribution was assumed according to the Italian Recommendations in force at the time of retrofitting (DM96 1996). The second load distribution was considered according to Eurocode 8 (1996).

The two codes differ in the manner in which the seismic actions are evaluated. The equivalent static analysis proposed by the DM96 (1996) considers a set of horizontal loads along the height of the building according to the distribution of the masses multiplied by the heights. The simplified dynamic analysis proposed by the

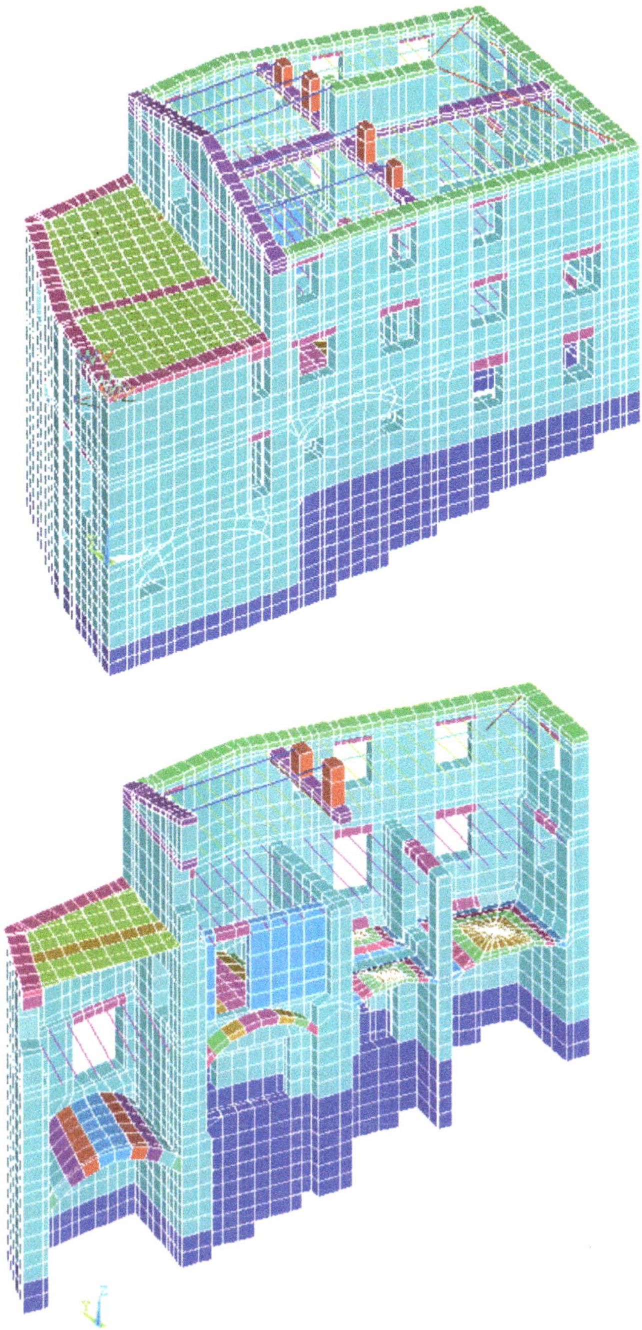

Fig. 25 Finite element model of the building in Fivizzano (Lunigiana)

Table 11 DP and WW model parameters

c (N/mm^2)	φ (°)	δ (°)	f_{cWW} (N/mm^2)	f_{tWW} (N/mm^2)	β_c	β_t
0.24	38	15	6.0	0.25	0.75	0.15

Table 12 Elastic parameters (main elements)

	Masonry walls	Stone jambs	Reinforced concrete	Wooden beams
E (N/mm^2)	1800	3000	30,000	11,500
ν	0.2	0.3	0.16	0.25
γ (kg/m^3)	2200	2200	2500	600

Eurocode 8 (1996) adopts instead a distribution of seismic loads based on the displacement of a modal shape, without any assumptions about the shape of this mode. The first mode of the building in Fivizzano is significantly different from the linear shape along the height assumed in DM96 (1996), which is therefore not fully reliable. This fact was indirectly confirmed by a preliminary analysis where quite good agreement with the state of the 1995 damage was obtained only with the simplified dynamic analysis performed using the modal load distribution proposed by the Eurocode 8 (1996). Therefore, the comparative nonlinear analyses were performed assuming the seismic forces distribution proposed by the Eurocode 8 (1996). The procedure consisted of two phases: the first phase in which vertical loads were applied (dead and live loads) and the second phase, divided into sub steps, in which the equivalent horizontal seismic actions where applied.

4.2.4 Nonlinear Analyses

In the second phase, the analysis of the building before retrofitting was stopped in the step corresponding to a value of the seismic loading of about 70 % of the full load (evaluated assuming a behaviour factor q = 1.5, a damping ratio ξ = 0.08, a peak ground acceleration a_g/g = 0.25, and a soil type A). The analysis of the results in terms of cracked area showed that the damage was localized at the top of the walls W1 and W3 (Fig. 27) with out-of-plane deformation. Other damages were observed in the wall W6, in the zone above the door and under the second floor window (Figs. 27 and 28).

To evaluate the improvement obtained with the insertion of the steel chains, three additional nonlinear analyses with different values of pretension of the steel chains were performed: pretension σ_f equal to 70 N/mm^2 (Case 1), 100 N/mm^2 (Case 2) and 120 N/mm^2 (Case 3). Higher valued were not considered to avoid local damages on the masonry. The chains were modelled with two-nodes linear elements (*Link8*), and the coupling of the link with the masonry (the steel plate) was modelled through shell elements (*Shell63*), to allow for a local distribution of the stresses. The failure modes predicted by the retrofitted model remain basically the

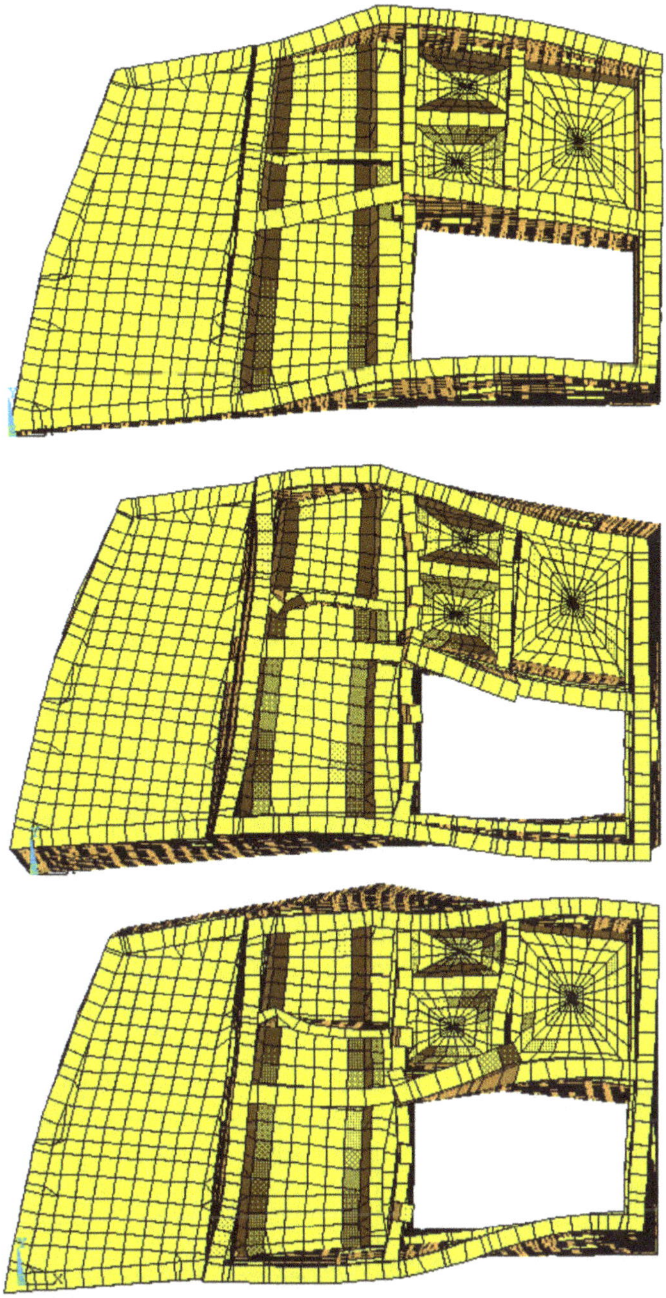

Fig. 26 First three modal shapes of the building: T_1 = 0.187 s (f_1 = 5.34 Hz), T_2 = 0.148 s (f_2 = 6.74 Hz) and T_3 = 0.133 s (f_3 = 7.49 Hz)

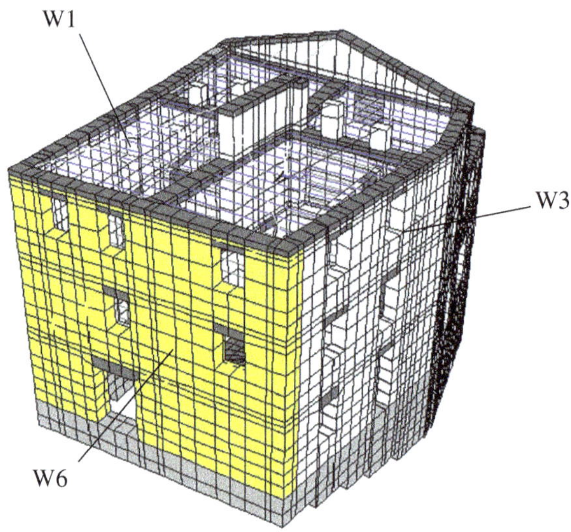

Fig. 27 Designation of the main walls

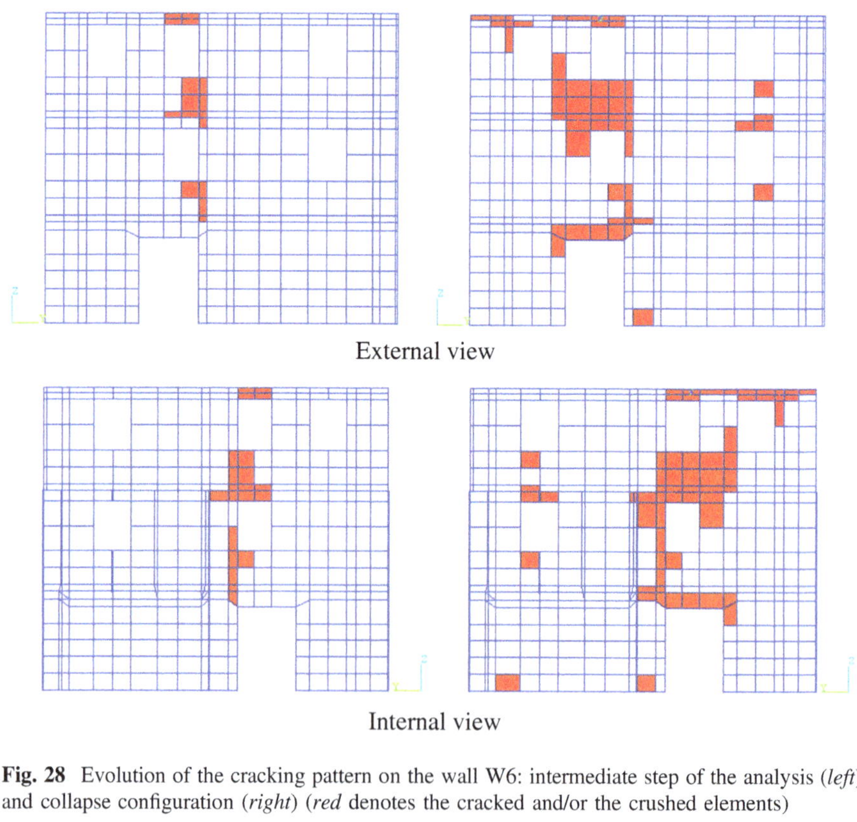

Fig. 28 Evolution of the cracking pattern on the wall W6: intermediate step of the analysis (*left*) and collapse configuration (*right*) (*red* denotes the cracked and/or the crushed elements)

Table 13 Benefits of the different levels of strengthening retrofitting

Nonlinear analysis	Chain pretension σ_f (N/mm^2)	Collapse loads (% of the seismic load)	Increment (%)
Without steel chains	0	70	/
Case 1	70	72	2.86
Case 2	100	76	9.29
Case 3	120	82	17.14

same, but an increment of the collapse load was observed (Table 13). The table reports also the improvement obtained increasing the stress in the chains. This benefit varies from about 2.8 % in the first case up to 17 % in the last case. The benefit obtained by the insertion of the chains is shown in Fig. 29 that reports the distribution of the cracked areas at collapse in the wall W6.

5 Conclusions

Modern societies consider preservation of built heritage, and passing it to future generations, a major issue (Fioravanti and Mecca 2011) since it contributes to consolidating a collective memory that creates a sense of belonging in citizens. In addition, from an economic point of view, and especially in contexts where tourism is becoming a major industry, accessibility to cultural heritage significantly contributes to the community's development (Bowitz and Ibenholt 2009).

Conservation of heritage buildings is an historical, cultural and engineering process (ICOMOS 2001) where safety evaluation should be correlated with proper principles of structural conservation (conserve as found, minimal intervention, like-for-like repairs, repairs reversible, etc.) according to a multidisciplinary and multicultural approach. From the specific engineering perspective, preservation calls for an interconnected series of operations aimed at obtaining a satisfactory broad-spectrum knowledge level of the building, where traditional in situ investigations must be performed in parallel with advanced numerical analyses such as the one discussed in the chapter. In fact, a clear understanding of the actual structural behaviour based on sophisticated numerical tools is an effective item of the path of knowledge that is needed for the proper design of a reliable strengthening that prevents invasive and inappropriate retrofitting.

With this focus the chapter discussed on the application of the finite element technique for seismic vulnerability assessment of historic masonry buildings. In a first part, two numerical models employed to reproduce the nonlinear masonry behaviour were explained, together with the identification of the required parameters. The models were a plasticity model and a combined plasticity and smeared cracking/crushing model. The tuning of the required parameters was performed based on the numerical reproduction of in situ experimental tests. So, a set of values

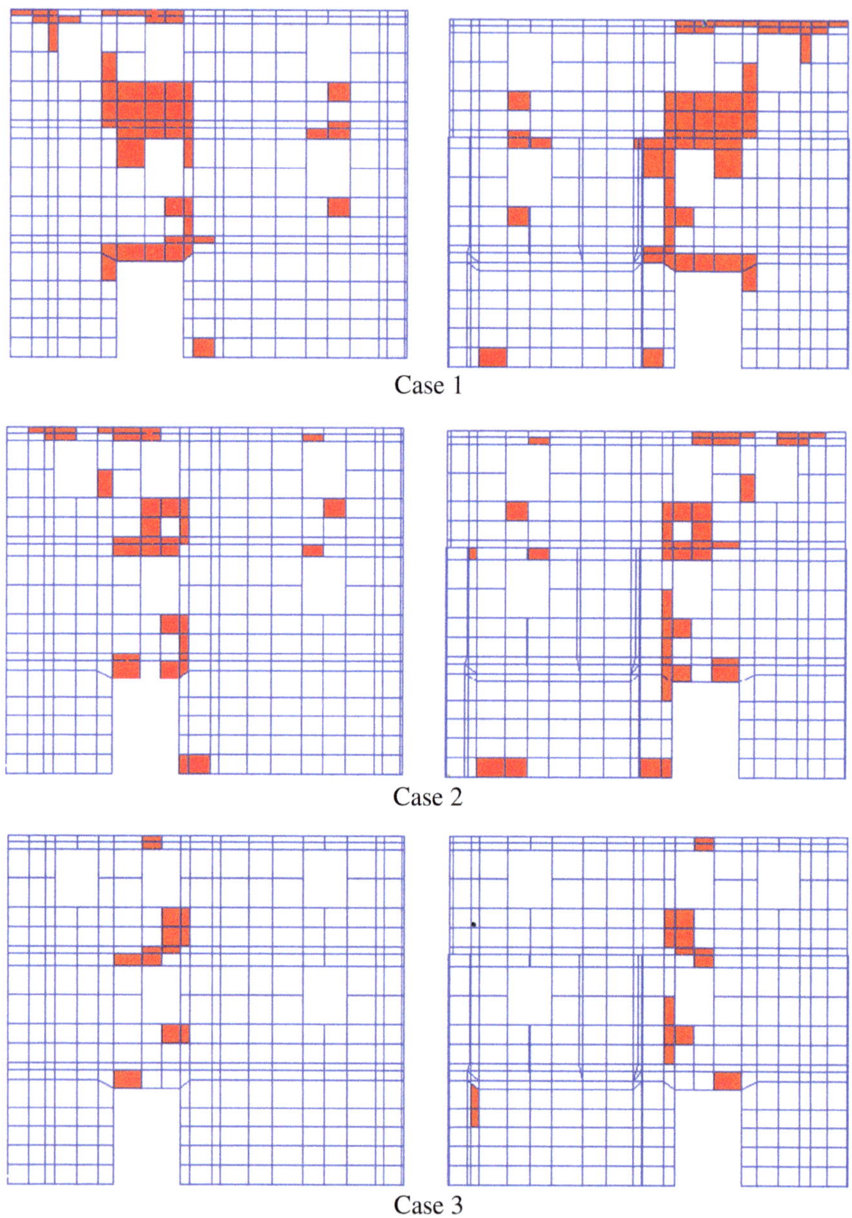

Fig. 29 Collapse configuration on wall W6: external view (*left*); internal view (*right*) (*red* denotes the cracked and/or the crushed elements)

were proposed for three typologies of historic masonries characterized by high, medium and low strength. Subsequently, in a second part, the models were employed to analyse the seismic behaviour of a monumental masonry church and a residential masonry building.

According to the presented numerical modelling, the chapter indicates that advanced computational analyses can significantly contribute to the understanding of the actual behaviour of historic buildings under seismic loading. Such knowledge, in its turn, allows to correctly design the strengthening interventions required for the safety and accessibility of the building.

References

Adam, J. M., Brencich, A., Hughes, T., & Jefferson, T. (2010). Micromodelling of eccentrically loaded brickwork: Study of masonry wallettes. *Engineering Structures, 32*(5), 1244–1251.

Adam, J. M., Ivorra, S., Pallarés, F. J., Giménez, E., & Calderón, P. A. (2009). Axially loaded RC columns strengthened by steel caging. Finite element modelling. *Construction and Building Materials, 23*(6), 2265–2276.

ANSYS Inc. (1998). *ANSYS manual.* USA: Southpoint.

Antoniou, S., & Pinho, R. (2004). Advantages and limitations of adaptive and non adaptive force-based pushover procedures. *Journal of Earthquake Engineering, 8*(4), 497–522.

Bartoli, G., & Betti, M. (2013). Cappella dei Principi in Firenze, Italy: Experimental analyses and numerical modeling for the investigation of a local failure. *ASCE's Journal of Performance of Constructed Facilities, 27*(1), 4–26.

Berto, L., Saetta, A., Scotta, R., & Vitaliani, R. (2005). Failure mechanism prism loaded in axial compression: Computational aspects. *Materials and Structures, 38*, 249–256.

Betti, M., Bartoli, G., & Orlando, M. (2010). Evaluation study on structural fault of a Renaissance Italian Palace. *Engineering Structures, 32*(7), 1801–1813.

Betti, M., & Vignoli, A. (2008). Modelling and analysis of a Romanesque church under earthquake loading: Assessment of seismic resistance. *Engineering Structures, 30*(2), 352–367.

Binda, L., Saisi, A., & Tiraboschi, C. (2000). Investigation procedures for the diagnosis of historic masonries. *Construction and Building Materials, 14*, 199–233.

Borri, A., Corradi, M., & Vignoli, A. (2000). Il comportamento strutturale della muratura nelle zone terremotate dell'Umbria: alcune sperimentazioni. *Ingegneria Sismica, XVII*(3), 23–33. (in Italian).

Bowitz, E., & Ibenholt, K. (2009). Economic impacts of cultural heritage—Research and perspectives. *Journal of Cultural Heritage, 10*(1), 1–8.

Brandonisio, G., Lucibello, G., Mele, E., & De Luca, A. (2013). Damage and performance evaluation of masonry churches in the 2009 L'Aquila earthquake. *Engineering Failure Analysis, 34*, 693–714.

Carpinteri, A., Invernizzi, S., & Lacidogna, G. (2005). In situ damage assessment and nonlinear modelling of a historical masonry tower. *Engineering Structures, 27*, 387–395.

Ceci, A. M., Contento, A., Fanale, L., Galeota, D., Gattulli, V., Lepidi, M., & Potenza, F. (2013). Structural performance of the historic and modern buildings of the University of L'Aquila during the seismic events of April 2009. *Engineering Structures, 32*(7), 1899–1924.

Cerioni, R., Brighenti, R., & Donida, G. (1995). Use of incompatible displacement modes in a finite element model to analyze the dynamic behavior of unreinforced masonry panels. *Computers & Structures, 57*(1), 47–57.

Chiostrini, S., Galano, L., & Vignoli, A. (1998). In situ tests and numerical simulations on structural behaviour of ancient masonry. In *Proceedings of Monument-98, Workshop on Seismic Performance of Monuments, Lisbon.*

Chiostrini, S., Galano, L., & Vignoli, A. (2000). On the determination of strength of ancient masonry walls via experimental tests. In *Proceedings of 12 WCEE, Auckland, New Zealand*, January 30–February 4, 2000.

Chiostrini, S., Galano, L., & Vignoli, A. (2003). In situ shear and compression tests in ancient stone masonry walls of Tuscany, Italy. *ASTM Journal of Testing and Evaluation, 31*(4), 289–304.

Chiostrini, S., & Vignoli, A. (1992). An experimental research program on the behavior of stone masonry structures. *ASTM Journal of Testing and Evaluation, 20*(3), 190–206.

Chiostrini, S., & Vignoli, A. (1994). In-situ determination of the strength properties of masonry walls by destructive shear and compression tests. *Masonry International, 7*(3), 87–96.

Chopra, A. K., & Goel, R. K. (2004). A modal pushover analysis procedure to estimate seismic demands for unsymmetric-plan buildings. *Earthquake Engineering and Structural Dynamics, 33*, 903–927.

Corradi, M., Borri, A., & Vignoli, A. (2002a). Strengthening techniques tested on masonry structures struck by the Umbria-Marche earthquake of 1997–1998. *Construction and Building Materials, 16*(4), 229–239.

Corradi, M., Borri, A., & Vignoli, A. (2002b). Experimental study on the determination of strength of masonry walls. *Construction and Building Materials, 17*(5), 325–337.

Da Porto, F., Guidi, G., Garbin, E., & Modena, C. (2010). In-plane behavior of clay masonry walls: Experimental testing and finite-element modeling. *Journal of Structural Engineering, 136*(11), 1379–1392.

Del Coz Díaz, J. J., García Nieto, P. J., Martínez-Luengas, A. L., & Álvarez Rabanal, F. P. (2007). Evaluation of the damage in the vault and portico of the pre-Romanesque chapel of San Salvador de Valdediós using frictional contacts and the finite-element method. *International Journal of Computer Mathematics, 84*(3), 377–393.

Del Piero, G. (1984). *Le costruzioni in muratura.* Berlin, Heidelberg: Springer. (in Italian).

DM96. (1996). Decreto Ministero dei Lavori Pubblici del 16 Gennaio 1996. Norme tecniche relative ai Criteri generali per la verifica di sicurezza delle costruzioni e dei carichi e sovraccarichi. G.U. 5/2/1996, No. 29 (in Italian).

DPCM. (2011). Direttiva del Presidente del Consiglio dei Ministri per la Valutazione e la riduzione del rischio sismico del patrimonio culturale con riferimento alle norme tecniche per le costruzioni di cui al decreto del Ministero delle infrastrutture e dei trasporti del 14 Gennaio 2008, G.U. 26/2/2011, No. 47 (in Italian).

Drucker, D., & Prager, W. (1952). Soil mechanics and plastic analysis or limit design. *Quarterly of Applied Mathematics, 10*(2), 157–165.

Eurocode 8 (1996) Design provisions for earthquake resistance of structures. Part 1–4: General rules—Strengthening and repair of buildings. ENV 1998-1-4: 1996. CEN, Brussels.

Falasco, A., Lagomarsino, S., & Penna, A. (2006). On the use of pushover analysis for existing masonry buildings. In *Proceeding of the First European Conference on Earthquake Engineering and Seismology, Geneva, Switzerland*, September 3–8, 2006.

Fioravanti, M., & Mecca, S. (Eds.). (2011). *The safeguard of cultural heritage: A challenge from the past for the Europe of tomorrow*. Florence: Firenze University Press.

Gambarotta, L., & Lagomarsino, S. (1997). Damage models for the seismic response of brick masonry shear walls. Part I: The mortar joint model and its applications. *Earthquake Engineering and Structural Dynamics, 26*(4), 423–439.

Hansen, E., William, K., & Carol, I. (2001). A two-surface anisotropic damage/plasticity model for plain concrete. In *Proceedings of Framcos-4 Conference 2001*.

ICOMOS (International Council on Monuments and Sites). (2001). *Recommendations for the analysis, conservation and structural restoration of architectural heritage.* International Scientific Committee for Analysis and Restoration of Structures of Architectural Heritage, Paris, 2001.

Ivorra, S., Pallares, F. J., & Adam, J. M. (2009). Experimental and numerical results from the seismic study of a masonry bell tower. *Advances in Structural Engineering, 12*(9), 287–293.
Ivorra, S., Pallares, F. J., Adam, J. M., & Tomás, R. (2010). An evaluation of the incidence of soil subsidence on the dynamic behaviour of a Gothic bell tower. *Engineering Structures, 32*(8), 2318–2325.
Kim, S., & D'Amore, E. (1999). Push over analysis procedures in earthquake engineering. *Earthquake Spectra, 15*(3), 417–434.
Leftheris, B. P., Stavroulaki, M. E., Sapounaki, A. C., & Stavroulakis, G. E. (2006). *Computational mechanics for heritage structures*. Southampton: WIT Press.
Lourenço, P. B. (2005). Assessment, diagnosis and strengthening of Outeiro Church, Portugal. *Construction and Building Materials, 19*(8), 634–645.
Lourenço, P. B., Krakowiak, K. J., Fernandes, F. M., & Ramos, L. F. (2007). Failure analysis of Monastery of Jero´nimos, Lisbon: How to learn from sophisticated numerical models. *Engineering Failure Analysis, 14*, 280–300.
Lourenço, P. B., & Oliveira, D. V. (2007). Improving the seismic resistance of masonry buildings: Concepts for cultural heritage and recent developments in structural analysis. In *Atti del XII Convegno Nazionale ANIDIS L'Ingegneria Sismica in Italia, Pisa, 2007*.
Lourenço, P. B., & Pina-Henriques, J. (2006). Validation of analytical and continuum numerical methods for estimating the compressive strength of masonry. *Computers & Structures, 84*, 1977–1989.
Lucibello, G., Brandonisio, G., Mele, E., & De Luca, A. (2013). Seismic damage and performance of Palazzo Centi after L'Aquila earthquake: A paradigmatic case study of effectiveness of mechanical steel ties. *Engineering Failure Analysis, 34*, 407–430.
NTC. (2008). Decreto Ministero delle Infrastrutture e dei Trasporti 14 Gennaio 2008. Nuove Norme Tecniche per le Costruzioni, G.U. 4/2/2008, No. 29 (In Italian).
OPCM. (2003). Ordinanza Presidente del Consiglio dei Ministri 3274/2003. Primi elementi in materia di criteri generali per la classificazione sismica del territorio nazionale e di normative tecniche per le costruzioni in zona sismica. G.U. 8/5/2003, No. 105 (In Italian).
Ramos, L. F., & Lourenço, P. B. (2004). Modeling and vulnerability of historical city centers in seismic areas: A case study in Lisbon. *Engineering Structures, 26*, 1295–1310.
Romera, L. E., Hernandez, S., & Reinosa, J. M. (2008a). Numerical characterization of the structural behaviour of the Basilica of Pilar in Zaragoza (Spain). Part 1: Global and local models. *Advances in Engineering Software, 39*, 301–314.
Romera, L. E., Hernandez, S., & Reinosa, J. M. (2008b). Numerical characterization of the structural behaviour of the Basilica of Pilar in Zaragoza (Spain). Part 2: Constructive process effects. *Advances in Engineering Software, 39*, 315–326.
Salari, M. R., Saeb, S., Willam, K. J., Patchet, S. J., & Carrasco, R. C. (2004). A coupled elasto-plastic damage model for geo-materials. *Computer Methods in Applied Mechanics and Engineering, 193*(27–29), 2625–2643.
Siviero, E., Barbieri, A., & Foraboschi, P. (1997). *Lettura strutturale delle costruzioni*. Milano: Città Studi Edizioni. (in Italian).
Taliercio, A., & Binda, L. (2008). The Basilica of San Vitale in Ravenna: Investigation on the current structural faults and their mid-term evolution. *Journal of Cultural Heritage, 8*, 99–118.
Theodossopoulos, D., & Sinha, B. (2013). A review of analytical methods in the current design processes and assessment of performance of masonry structures. *Construction and Building Materials, 41*, 990–1001.
William, K. J., & Warnke, E. D. (1975). Constitutive model for the triaxial behaviour of concrete. In *Proceeding of the International Association for Bridge and Structural Engineering, Bergamo, Italy, 1975*.
Zucchini, A., & Lourenco, P. (2007). Mechanics of masonry in compression: Results from a homogenisation approach. *Computers & Structures, 85*(3–4), 193–204.
Zucchini, A., & Lourenço, P. B. (2002). A micro-mechanical model for the homogenisation of masonry. *International Journal of Solids and Structures, 39*, 3233–3255.

Earthquake-Resistant and Thermo-Insulating Infill Panel with Recycled-Plastic Joints

Marco Vailati and Giorgio Monti

1 Introduction

The impellent need of endowing our cities with environment-friendly buildings that are also safe against destructive events, such as earthquakes, has been the main inspiration for this work.

In particular, the attention of the authors has been focused on the role of infill panels, which are in recent times gaining a widespread attention in the most advanced construction codes.

Thereby, they are now regarded as critical elements, both from the thermal insulation (Lgs, D. 311 2006) and from the earthquake resistant standpoints (NTC-08 2008; Circolare n.617 2009).

Infill panels are the key elements to ensure that the internal temperature of the building be kept constant, regardless of the external environment temperature.

Unfortunately, we are all aware that most of the heat (or cold) dispersion in buildings is due to the scarce insulating properties of infill panels.

In addition, their role as non-structural elements, not designed to sustain horizontal loads, becomes dramatically inadequate during exceptional events, such as medium-high intensity earthquakes, which always produce severe damage in these elements, if not complete collapse due to either in-plane shear or overturning.

These aspects have stimulated the scientific community towards a significant research effort to arrive at conceiving new types of external infill panels—and internal partition walls—that could satisfy both basic requirements: being eco-friendly from the thermal standpoint, while also resisting to severe earthquakes without endangering the people living and working in those buildings.

M. Vailati (✉) · G. Monti
Department of Structural and Geotechnical Engineering,
Sapienza University of Rome, Rome, Italy
e-mail: marco.vailati@uniroma1.it

G. Monti
e-mail: giorgio.monti@uniroma1.it

S. D'Amico (ed.), *Earthquakes and Their Impact on Society*,
Springer Natural Hazards, DOI 10.1007/978-3-319-21753-6_15

The literature on this subject has produced a wealth of results and interesting approaches so that current advanced structural codes incorporate specific procedures aimed at checking the safety of non-structural elements, as well.

For example, when the seismic action acts orthogonal to the infill panel, it may tend to overturn (Abrams et al. 1996) and it is now mandatory that this verification be carried out to ensure its stability.

Besides, several studies have pointed out that, in seismic conditions, infill panels become elements that contribute to the seismic response, by working in parallel with the main structural system and by changing its stiffness and strength (Bertero et al. 1981; Fardis and Panagiotakos 1998).

The approach currently followed to tackle this problem is as follows: if the interaction between infill masonry panels and reinforced concrete frame is not negligible, their contribution must be explicitly taken into account when modeling the structural system.

For the sake of exemplification, Fig. 1 shows some numerical examples on a simple non-linear frame with concentrated plastic hinges, without and with infill interaction (Vailati 2004).

One should readily notice that the presence of the infills dramatically changes the response.

In general, we observe: (a) a significant increase in the initial stiffness; and (b) an abrupt decrease of stiffness and strength (saw-tooth shape) corresponding to the in-plane collapse of the infills.

The interaction with the infills may also trigger partial failures in the main columns, due to the locally applied shear coming from the inclined strut that naturally develops within the infill panel.

These simplified analyses confirm the conclusions reached when observing the damage in buildings after many severe earthquakes: infills have in general too high an in-plane stiffness, which gives rise to in-plane shear collapses.

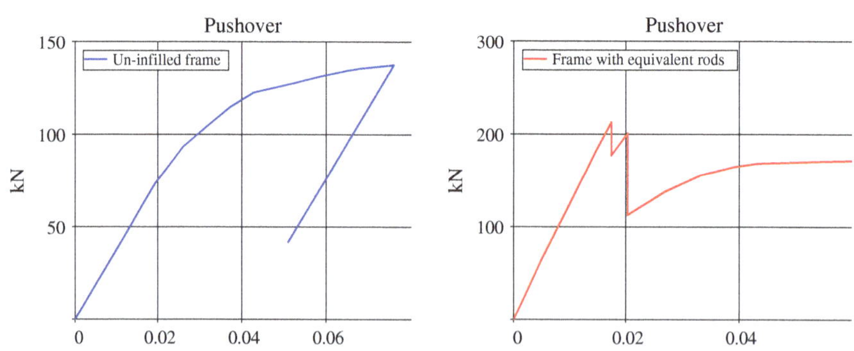

Fig. 1 Simplified force-displacement response of a multistory frame building. *Left* Without infills. *Right* With infills interacting with the structural system. The strength drop is due to the in-plane collapse of the infills

The main conclusion is therefore that such undesirable behavior is due to the exceedingly high stiffness of those elements. Why then not to try and solve the problem by following a completely different strategy? In other words: by pursuing flexibility rather than stiffness.

In an essay published in *An American Architecture* magazine (Kaufmann 1955), the great American architect F.L. Wright said:

> We solved the problem of the menace of the quake by concluding that rigidity could not be the answer, and that flexibility and resiliency must be the answer… Why fight the quake? Why not sympathize with it and outwit it?

This brief reflection, which 80 years ago anticipated one of the founding principles of modern seismic engineering, suggests a strategy to effectively reduce the vulnerability of infill panels subject to the devastating effects of earthquakes.

However, in many post-quake surveys, it has also been observed that infill panels often exhibit an out-of-plane instability, which gives rise to collapses due to partial/total overturning. Both behaviors are of course extremely dangerous for the safety of occupants. In these cases, it would be instead desirable that the infill panels be endowed with a high stiffness when subjected to out-of-plane forces.

These two requirements appear therefore to be contradicting. How can one develop a solution that ensures adequate flexibility in the in-plane response and, on the other hand, high stiffness in the out-of-plane behavior?

This unconventional problem stimulated the development of an innovative construction system of infill panels and partition walls having large in-plane displacement capacity and not interacting with the surrounding structural system. Moreover, these newly conceived infills also exhibit a remarkable stability with respect to overturning actions of any kind.

The key concept of the proposed solution was found in the horizontal connection system among the infill blocks: it consists of thermoformed recycled plastic joints that are dry-assembled and are meant to replace the traditional mortar joints.

Under the in-plane displacement imposed by the seismic action, the wall can be made to behave as an assemblage of blocks that slide relatively to each other along the plastic bed joints. This modifies the mechanics of the resisting system; the infill panel acts in fact as a series system consisting of rigid blocks, the bricks, and of flexible elastic interface springs, the plastic joints.

The horizontal displacement is localized at the joints, thus keeping the blocks essentially undeformed and undamaged. In a sense, the approach taken follows the line of the capacity design principle: failure of brittle elements, the bricks, is avoided by increasing the ductility of the deformable elements, the joints.

Under the overturning out-of-plane forces, the panels attain stability by means of plastic strips hidden in the vertical joints and connected to the horizontal plastic joints, which realize a continuous vertical reinforcement that prevents the panel from bending outwards. The technological details and the relevant design issues are dealt with in this paper.

2 Safety of Infill Panels Under Earthquakes

The most advanced construction codes are concerned with the safety of a construction as a whole and thus require the verification of both structural and non-structural elements.

Non-structural elements are generally defined as "those having stiffness, strength and mass such to influence significantly the structural response, and those, though not influencing the structural response, that may put the safety of people at risk".

Infill panels fall in this category and therefore their safety under seismic actions has to be verified. Both Eurocode 8 (EN 1998-1) and the current Italian Code NTC-08 (and also its more recent proposed revision of 2014) require that non-structural elements be verified for the in-plane response, under the Damage Limit State (DLS) seismic action, and for the out-of-plane response, under the Life Safety Limit State (LSLS) earthquake.

For DLS, it is required to check that under the design seismic action the non-structural elements are not damaged, so that the entire building can still be usable after the earthquake.

An implicit check is performed by limiting the interstory drift of structural elements to a certain percent of the story height, depending on the panel type.

By doing this, it is expected that the infill panels, which undergo a drift equal to that of the structural elements, can sustain low-intensity earthquakes without damage. As mentioned, the drift limits are given depending on the panel type: infills interacting with the structural elements require small interstory drifts (of the order of 0.005 the story height), while infills having limited or no interaction with the surrounding structural elements allow for a larger (double) interstory drift.

This suggests to follow an alternative approach to that pursuing stronger infills: these can be made to have no interaction with the structural elements so to increase the limit drift under frequent earthquakes.

As a matter of fact, it is interesting to note that, in case of infills purposely designed as collaborating with the earthquake-resisting structure, it is recommended that their design and construction be performed following widely accepted documents.

Nothing is said about infills that are purposely designed not to collaborate with the structure, which clearly reveals a lack of available solutions.

For LSLS, it should be underlined that the only verification performed is that of stability of the infills, that is, to check that panels are not going to overturn in case of medium-high-intensity earthquakes. Nothing is currently said about their in-plane behavior.

This amounts to saying that, when strong earthquakes hit the building, the main concern is to ensure that all structural elements perform adequately, while it is implicitly accepted that non-structural elements can actually fail in-plane, while they are not allowed to overturn outside the building.

To ensure this, verifications are carried out by comparing the overturning force with the corresponding capacity. The overturning force is determined as:

$$F_a = \frac{S_a \cdot W_a}{q_a} \tag{1}$$

in which: F_a = horizontal seismic force applied at the centroid of the non-structural element in the most unfavorable direction; W_a = weight of the element; q_a = behavior factor of the element (at most $q_a = 2$); S_a = spectral acceleration, in terms of gravity acceleration *g*, which can be determined from formulations of proven validity, such as, for example:

$$S_a = \alpha \cdot S \cdot \left[\frac{3 \cdot \left(1 + \frac{Z}{H}\right)}{1 + \left(1 - \frac{T_a}{T_1}\right)^2} - 0.5 \right] \geq \alpha \cdot S \tag{2}$$

where: α = peak Ground Acceleration, in terms of gravity acceleration *g*; S = coefficient accounting for soil type and topographical conditions; T_a = fundamental period of vibration of the non-structural element; T_1 = fundamental period of vibration of the building in the considered direction; Z = height of the overturning line of the non-structural element, measured from the foundation level; H = height of the building, measured from the foundation level.

The demand generated on the element by the force calculated in Eq. (1) is compared with the capacity of the system.

An interesting aspect of these modern codes is that they recognize the importance of a proper design of these non-structural elements and thus they identify also the corresponding areas of responsibility of the different actors involved in the design process.

Thus, responsibilities are defined as follows: "When the non-structural element is constructed on site, it is up to the designer of the structure to identify the demand and to design its capacity according to formulations of proven validity and it is the task of the project manager to follow their implementation; on the other hand, when the non-structural element is only assembled on site, it is up to the designer of the structure to identify the demand, while it is the duty of the supplier and/or installer to provide elements and connection systems of adequate capacity".

It is finally worth mentioning that codes also deal largely with the role of non-structural elements within the overall response of the building, and with the way of correctly modeling them.

It is said: "When setting up the structural analysis model, all non-structural elements (cladding and partitions), should be represented only in terms of mass, while their contribution to the stiffness and strength of the structural system should be considered only if their strength and stiffness can significantly change the behavior of the model".

However, despite the intentions of the Code, very rarely happens that designers consider explicitly the presence of infills within the structural model, even if they interact with the response of the structural system and modify it.

This is a matter of particular relevance, since in most cases the design is performed on a structural model without infills, whose response, both local and global, can be very different from that of the building that will be realized.

It is sufficient to think about the interaction with the structural elements at beam-column joints and about the natural period of oscillation, which may be significantly different.

Even in case of seismic assessment of existing buildings, neglecting the contribution of the infills to the strength and stiffness of the structure may result in an underestimation of the risk of exceeding the limit states considered.

Again, this brings us towards the choice of a solution that reduces the interactions between infills and structures to the least possible.

3 The "PlastiBloc®" System

All the consideration expressed in the previous sections found a natural outcome in a non-structural element that can actually solve all the concerns raised with respect to the seismic safety, with the additional desirable feature of being completely sustainable.

The developed infill panel is composed by the traditional (concrete or clay) hollow-core blocks, by an insulation layer, and by the recycled-plastic joints, which is the truly innovative part. The plastic joint is a 300 × 258 mm horizontal plane with a thickness of about 2 mm, having some thermoformed extruded hollow teeth on both sides. These are meant to be inserted into the holes of the blocks, which are then transformed into blocks ready to be dry-assembled by simply stacking them on top of each other. Figure 2 shows a graphical representation of the joint. The recycled-plastic joints are produced according to a controlled industrial process that guarantees stability of the physical-mechanical properties of each piece.

Since we had to comply with performance requirements of multidisciplinary nature, the block system was optimized to satisfy all of them; the double alignment of the blocks, represented in Fig. 2, right, provides the wall with excellent thermal properties (up to 0.29 W/m^2 K), thanks to the insertion of insulating elements in the air chamber between the blocks. Moreover, the teeth size is optimized so to have the needed deformation capacity and give the joint the desired sliding capability, which significantly reduces the horizontal stiffness of the wall. Consequently, the so-realized wall does not interact with the surrounding structural system, thus reducing the damage potential during earthquakes.

In addition to the above, the system was also endowed with another useful feature: in order to provide stability with respect to forces acting orthogonal to the wall face, a series of vertical plastic strips can be placed, if needed, in between the adjacent blocks, hidden in the vertical joints.

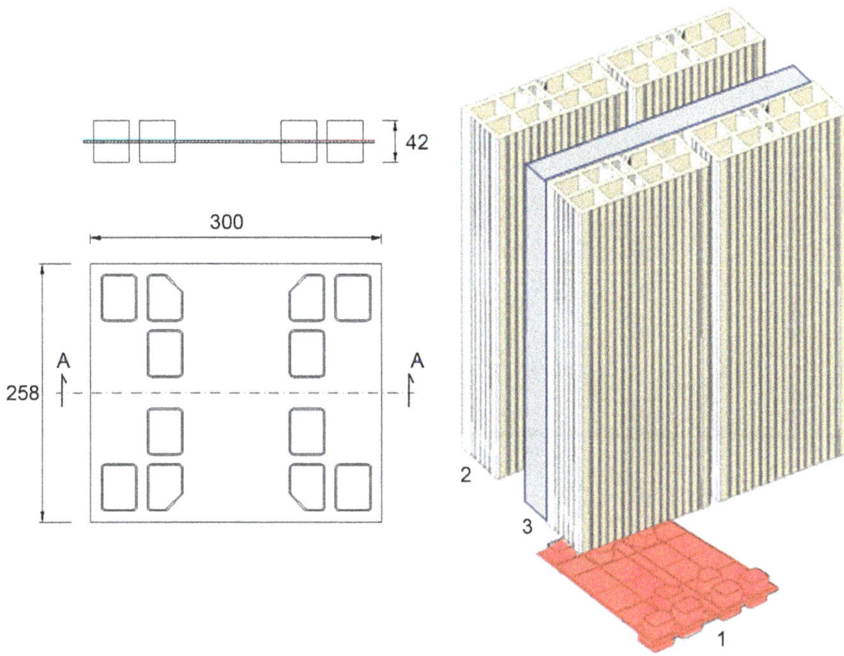

Fig. 2 *Left* Geometry of the recycled-plastic joint in mm. *Right* Assembled PlastiBloc®, system with horizontal joint (*red*, numbered as 1), blocks (*yellow*, numbered as 2), insulation layer (*grey*, numbered as 3)

These strips, shown in Fig. 3, are connected to the horizontal joints by means of special self-seizing connectors and can be regarded as a continuous reinforcement along the entire panel height that provides the necessary resistance when it tends to bend outwards.

Thus, the system so conceived complies with two apparently contradicting requirements: high flexibility under in-plane actions and high strength and stiffness under out-of-plane actions.

3.1 Comparison with Traditional Infills

The differences between the innovative system and traditional infill panels, made from either concrete or clay blocks connected with mortar bed joints, are several and all remarkable.

As a first instance, PlastiBloc® has a better constructability in that it is dry-assembled and thus it does not require preparation, application and curing of mortar joints as in the traditional infills. This significantly speeds up construction time, which is almost halved. Also, from the construction standpoint, there is no

Fig. 3 PlastiBloc® system with vertical reinforcing plastic strips between blocks (as shown by *arrows*)

need of shifting the vertical joints. It should as well be noted that the internal voids of the hollow core blocks, which are closed by the mortar in the traditional infills, are in this case all connected along the panel height and can be used to host piping and lines.

In addition, it is safer in case of earthquakes, both because of its already mentioned performance in the in-plane and in the out-of-plane behavior, but also because, being lighter that the traditional panels, it reduces the overall mass of the building and, consequently, the corresponding seismic forces.

On passing, being lighter than traditional infills makes transportation and handling at the construction site easier.

From the sustainability point of view, it should be remarked that, being dry-assembled, the entire infill panel can be disassembled and reused. In this sense, it can be considered as fully recyclable and environment-friendly.

It is even more eco-friendly if one considers that the plastic used to make the joints is recycled from production leftovers that otherwise would be wasted.

Its thermal insulation capability allows for a significant energy saving, both in cold and hot climates. Finally, the unit costs is lower than traditional infills.

3.2 Some Experimental Studies

The infill panel system has recently undergone an extensive program of experimental tests. The study, reported in Vailati et al. (2014), shows in great detail the results of tests performed on the constituent material, on the single joint, on small assemblages of blocks and joints, and on full-scale panel assemblages, under both in-plane and out-of-plane actions.

Figure 4 shows the force-displacement graph of a failure test conducted on a small assemblage of three blocks with two plastic joints interposed.

Note that, shifting the diagram by 5 mm, the joint can attain a maximum displacement of about 20 mm while still remaining elastic.

The tests performed on 3 × 3 m panels with opening, to represent real cases, gave very satisfactory results (Fig. 5): under an imposed drift of 100 mm, equivalent to 3 % of the interstory height, no constructive element of the panel, neither the plastic joints nor the blocks, showed any visible damage and remained perfectly functional.

All tests conducted on full-scale infills confirmed the remarkable displacement capacity of the infills with plastic joints, which reached much higher displacement levels than those usually experienced during seismic events.

The out-of-plane tests (not shown here) provided equally satisfactory outcomes: the infill panel was rotated by several degrees until the horizontal position, which represents an orthogonal acceleration of 1 g, without any sign of instability.

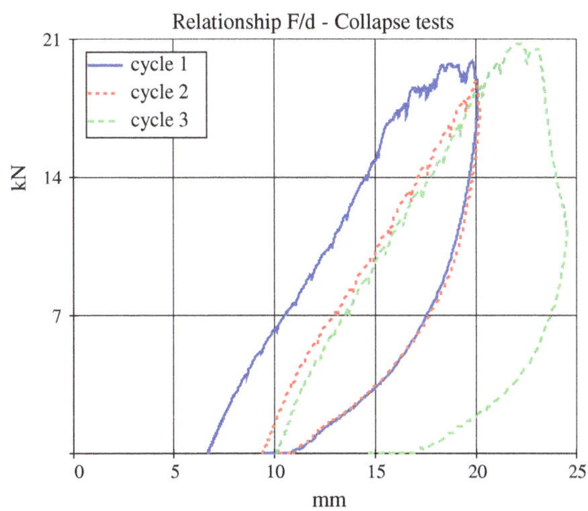

Fig. 4 Failure test of an assemblage of blocks with plastic joints. The Force-displacement graph shows the remarkable displacement capacity of the system

Fig. 5 Shear test on full-scale model of infill panel made by plastic joints

3.3 Some Modeling Studies

The out-of-plane behavior of the proposed system was studied using two models, one analytical and one numerical, both aiming at comparing the seismic demand on the infill panel and its outwards bending capacity.

In the analytical model, the seismic demand on the infill panel was evaluated by considering two types of horizontal force patterns: a point force at the panel centroid (mid-height), and a uniformly distributed force along the height of the panel. Their values were determined, for the former case, directly from Eq. (1), while for the latter, by smearing the force in Eq. (1) over the height h of the panel.

The capacity of the infill panel was evaluated by assuming a linear compressive stress distribution on the panel cross-section, and a constant tensile stress in the tension strips. These give rise to compression and tension resultants as shown in Fig. 6, having a lever arm equal to 0.8 d, where d is the overall thickness of the panel.

Figure 6 shows a generic cross-section of the panel and the corresponding stress distribution, used to compute the outward bending capacity.

In addition, the elastic stiffness of the system was computed under the hypothesis that the entire panel made of the stacked blocks be regarded as a simply supported beam.

The numerical model was developed by using a commercial nonlinear finite element software. The following considerations are worth recalling:

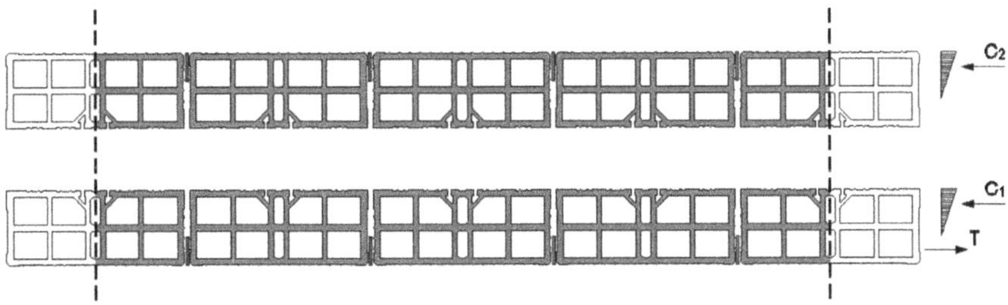

Fig. 6 Generic cross-section of the PlastiBloc® system of 1 m length. The reinforcing plastic strips (between blocks, in *red*), placed inside the vertical joints of the hollow-core blocks, provide the tensile resultant that equilibrates the compressive one on the blocks

Each block is assumed as undeformable;
The panel cross-section does not remain plane throughout, because the two block layers are actually coupled by the plastic joint, which is deformable;
The plastic joints do not offer tensile resistance;
The plastic strips do not offer compressive resistance;
The panel is hinged at both ends.

The tests were conducted in the nonlinear field. Sources of nonlinearity are to be found in the plastic material (constitutive laws are all non-symmetric) and in the geometry. In fact, the system geometry changes significantly under horizontal actions, because its stiffness is initially very low before the vertical strips are actually activated.

Analytical and numerical results were compared with respect to six parameters:

- 1st mode natural period of vibration of the panel;
- C_1 the compression force in the external layer of the panel;
- C_2 the compression force in the internal layer of the wall;
- T the tensile resultant in the strips;
- d the mid-height displacement under seismic action.

Table 1 shows a summary of the results obtained with the two models. It can be seen that the results obtained with the analytical model subjected to distributed load are closer to numerical model, where such load pattern naturally arises.

The results allow concluding that the analytical model can be effectively used for designing panels made with the proposed system.

Table 1 Comparison between the results obtained with analytical and numerical models

Model	1st mode (s)	C_1 (kN)	C_2 (kN)	T (kN)	d (mm)
Analytical (point load)	0.28	15.4	15.4	13.2	5.3
Analytical (distributed load)	0.28	7.7	7.7	6.6	5.3
Numerical	0.11	5.4	4.0	2.3	7.2

4 An Application

The PlastiBloc® system has been recently adopted in the construction of the infill walls of the expansion of the Faculty of Law, within the University Campus of Rome, Italy.

A two-story steel frame building was built on top of the existing masonry building. Figure 7 highlights the newly constructed portion. The first interstory had a net height of almost 6 m, while the second reached almost 5 m. The infill panels installed on the perimeter had to ensure adequate thermal insulation and at the same time, notwithstanding their slenderness, a high resistance against possible overturning, due to either seismic actions or accidental thrusting forces.

The extremely unusual slenderness of the walls required a dedicated study that finally led to the development of an additional vertical strengthening, made of the special plastic strips, shown above in Fig. 3. As mentioned previously, these strips are designed to act as surface tensile reinforcement, suitably arranged inside the vertical joints of the blocks and connected to the horizontal plastic joints by means of special connectors, so to form a continuous plastic strip along the entire height of the infill, which provides the required lateral resistance against out-of-plane actions.

Figure 8 shows two phases of the assemblage of the infill panels. One should note the layering of the system that is composed of two adjacent blocks, made of clay in this case, with the insulating element placed in between, to provide the required thermal resistance. In the picture, the horizontal plastic joints are clearly visible. The walls were assembled without using any mortar, with the only exception of that needed to place the bottom joints over the bottom slab. Right

Fig. 7 *Center* Super-elevated portion of the building where the PlastiBloc® system was adopted (two last levels). *Left* First of two levels with height of almost 6 m. *Right* Top level with height of almost 5 m

Fig. 8 Two phases of the assemblage of the PlastiBloc® system. In evidence, at *right*, the plastic strips

picture in Fig. 8 also shows some vertical plastic strips. Also, it should be noted that the holes in correspondence of the teeth of the plastic joint, allow to easily install pipes and lines through the wall thickness.

Finally, Fig. 9 shows a completed infill wall, having height close to 6 m. Construction details were devised in discontinuity zones, to allow assembling the blocks near openings, at corners, and at the beams soffit. At the construction site, it was estimated that this system allowed assembling all the walls in a quarter of the time that would have been taken by traditional infills.

5 Impact on Society

The innovative system for infill panels here presented can be defined, by several points of view, as a sustainable and eco-friendly product. It brings several advantages at the social and economical level, by introducing a series of changes in the usual production process and in the quality of the resulting product. The following is a, certainly not exhaustive, list:

Fig. 9 A completed infill wall of almost 6 m in height, with and without openings

1. Being made from recycled plastic waste, the joints material can be fully recycled for the production of new artifacts, thus becoming a product with close to zero CO_2 emissions (ISO/TS 14067 2013);
2. The assembly process is fully reversible so that all the basic elements: blocks, insulation, joints and strips can be reused;
3. The dry assembly of the plastic joints significantly reduces construction time, eliminating not only the time necessary for mixing and curing of mortar, but also the very realization of laying surfaces for the blocks;
4. A weight reduction between 30 and 60 %, which implies less transportation costs and less related emissions;
5. High safety against horizontal forces, either due to earthquakes or accidental impact;
6. Self-extinguishing, thus avoiding problems associated with possible propagation of a fire.

6 Conclusion

An innovative solution for infill panels has been presented, which have interesting and promising properties for as regards, both, thermal insulation and seismic resistance. The basic idea is that of replacing the traditional mortar bed joints with recycled plastic joints that are thermoformed. These are placed on top of the traditional (concrete or clay) blocks for infills and make them into blocks that are ready to be dry-assembled on top of each other.

These joints are designed to deform and to allow stacked layers of blocks to slide with respect to each other, thus offering the least possible strength and stiffness contribution to the main structural system. In this way, the interaction between the main structural elements and the non-structural elements is reduced to a minimum.

During a low intensity earthquake, this has the beneficial effect that the infill panels can easily accommodate the interstory drifts of the structural elements without suffering any damage, while during a medium-high-intensity earthquake, the panel can suffer some light damage, but they do not interact with the structural elements, which then can exploit the performance levels they were designed for.

This technology does not require avoiding alignment of the vertical joints. On the contrary, the vertical alignment of the joints allows for the insertion of plastic vertical strips that can be helpful in those cases where tensile reinforcement is need in too slender walls. In this way, the panel can effectively resist horizontal forces acting orthogonal to its plane, be they of accidental nature or due to earthquakes. This prevents the panels from overturning during medium-intensity earthquakes.

Some recent experimental and modeling studies are also presented (Vailati et al. 2014), which have highlighted the potentialities of the system and have given some insight in its mechanics.

The system is therefore to be seen as a first step towards a way of designing and realizing buildings that comply with different requirements, be they related to structural safety or to sustainability. It is also a further step towards a structural design that looks at the building as a compound ensemble of structural and non-structural elements, each of them worth being protected in case of exceptional events.

The technology proposed lends itself for a smooth introduction in the construction industry, since it effectively replaces a traditional way of realizing infills, without any abrupt change, rather, by using the same basic components and by replacing mortar with an element, the plastic joint, which has the advantage of producing more eco-friendly panels, with less construction time and less cost.

Acknowledgments The authors thank the probe! company in Rome, Italy, patent holder of the plastibloc® system. Additional information about the system can be find at www.probeitalia.com.

References

Abrams, D. P., Angel, R., & Uzarski, J. (1996). Out of plane strength of unreinforced masonry infill panels. *Earthquake Spectra, 12*(4), 825–844.

Brokken, S., Steven, T., Bertero, V. V., & Vitelmo, V. (1981). *Studies on effects of infills in seismic resistant R/C construction*. Berkeley: University of California (Report UCB/EERC 81-12).

Circolare n. 617 C.S.LL.PP. (2009). Istruzioni per l'applicazione delle nuove norme tecniche per le costruzioni di cui al decreto ministeriale 14 gennaio 2008. Supplemento ordinario n. 27 alla Gazzetta Ufficiale 26 febbraio 2009.

NTC-08. (2008). Nuove norme tecniche per le costruzioni. Gazzetta Ufficiale della Repubblica Italiana 4 febbraio 2008.

Fardis, M. N., & Panagiotakos, T. B. (1998). Seismic response and design of masonry infilled reinforced concrete buildings. In N.K. Srivastava (Ed.), *Proceedings of the Structural Engineering World Congress*. San Francisco.

ISO/TS 14067. (2013). Greenhouse gases. Carbon footprint of products. Requirements and guidelines for quantification and communication.

Kaufmann, E. (1955). *An American architecture*. New York: Horizon press.

Lgs, D. 311. (2006). Disposizioni correttive ed integrative al decreto legislativo 19 agosto 2005, n. 192, recante attuazione della direttiva 2002/91/CE, relativa al rendimento energetico nell'edilizia. Gazzetta Ufficiale della Repubblica Italiana 1 febbraio 2007.

Vailati, M. (2004). Contenuti dell'ordinanza n.3274 in materia di costruzioni antisismiche, risvolti applicativi e confronti con altre normative. 2nd Level Master Degree Thesis, University of Rome.

Vailati, M., Caluisi, A., & Monti, G. (2014). Environmentally-friendly joints for seismic resistant infill panels. Atti delle giornate AICAP 2014-Strutture nel tessuto urbano, Bergamo, 1998.

Base Isolation and Translation of a Strategic Building Under a Preservation Order

Giorgio Monti, Marco Vailati and Roberto Marnetto

1 Introduction

The office building object of the retrofitting intervention is one of the headquarters of the Italian Highways Company. The construction of the building started at the end of the 1950s, when earthquake engineering was still in its immature phase and seismic codes were far away to come. The owner, given the strategic importance of the building, deemed necessary to perform an assessment of its seismic vulnerability and eventually verify the need for a retrofitting intervention.

After a preliminary yet detailed study, it was assessed that the building was not able to withstand the earthquake intensity corresponding to the Life Safety Limit State (LSLS), as prescribed in the Italian code NTC-08 (2008) and its explanatory note (circolare n. 617 2009).

Among the many requirements requested by the owner, the need of maintaining the office building operational during construction works has proved particularly challenging and has determined the choice of the type of intervention.

Eventually, base isolation was chosen as the more appropriate, especially for the possibility offered by the particular geometry of the building to easily create an isolation interface at the ground level. The intervention saw a preliminary phase in the shear-strengthening of the ground level columns, which was carried out through the so-called CAM system (prestressed stainless steel strips), with the aim of

G. Monti · M. Vailati (✉)
Department of Structural and Geotechnical Engineering, Sapienza University of Rome, Rome, Italy
e-mail: marco.vailati@uniroma1.it

G. Monti
e-mail: giorgio.monti@uniroma1.it

R. Marnetto
De.La.Be.Ch. Costruzioni Srl, Rome, Italy
e-mail: r.marnetto@libero.it

S. D'Amico (ed.), *Earthquakes and Their Impact on Society*,
Springer Natural Hazards, DOI 10.1007/978-3-319-21753-6_16

improving their low shear capacity. In fact, during the survey phase, it was observed that the transverse reinforcement of the columns was largely insufficient and, most of all, heavily corroded.

As mentioned, the building is made of a central entrance hall, with stairs and elevators, and of two symmetrical side "wings". Since all the working activities were carried out in the two "wings", the project was organized in phases, taking advantage of the symmetry, each working phase being carried out firstly on a wing and then, after its completion, on the other wing.

The most interesting and innovative aspect was the approach followed to cope with the large displacements expected in the base-isolated structure under the design earthquake. In general, base isolation is an effective seismic protection system provided pounding to adjacent buildings be prevented. In the case at hand, this called for a widening of the insufficient existing joints, which was obtained by pushing the two wings away from the central hall building. The initial solution foresaw using low-friction sliders under the elastomeric isolators and pushing the wings with hydraulic jacks contrasting against the central hall building. This approach, though straightforward, proved more expensive than that eventually selected. In fact, a non-conventional solution was finally devised that proved extremely effective in terms of cost and practical advantages: installing pre-deformed isolating devices. The details of this solutions are given in the following.

The building geometry allowed realizing a seismic isolation intervention under conditions that can be defined as ideal: the isolation devices were placed under the ground floor columns of the two "wings", where no stairs were present, thus avoiding the usual problems related to isolation of stairs or elevators. In addition, during the translation phase, the central hall building was used as a reference point to control the symmetric displacements of the two symmetric wings.

2 Description of the Building

A view of the building is in Fig. 1. It consists of three buildings, as shown in Fig. 2. The retrofitting operations described hereafter refer to the two lateral buildings, in gray, called "wings", which host the offices where the strategic activities are carried out. The left one is represented in a larger scale in Fig. 3.

The wings have rectangular shape with sides of 11.50 × 32 m; they are about 12.80 m high above ground. The ground floor is of the "pilotis" type and is 5.00 m high. The two upper floors are used as offices with height of 3.90 m each; there is also a basement area 3.40 m high, where equipment and pipings are accommodated. The central building provides entrance to the entire building and the vertical connections, hosting stairs and elevators. The structure is made of reinforced concrete frames, obtained by connecting precast columns.

A vertical cross-section of the building is in Fig. 4, while a detail of a wing with its columns at ground floor is shown at a bigger scale in Fig. 5.

Fig. 1 Aerial view of the building

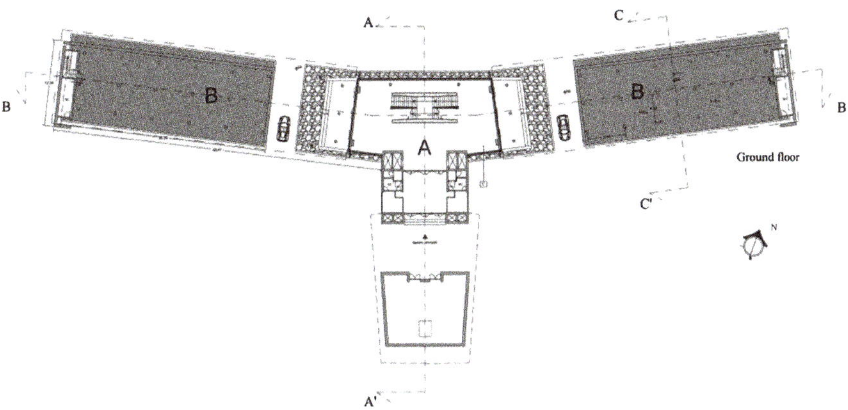

Fig. 2 Ground floor of the building. *A* Main hall, *B* Wings

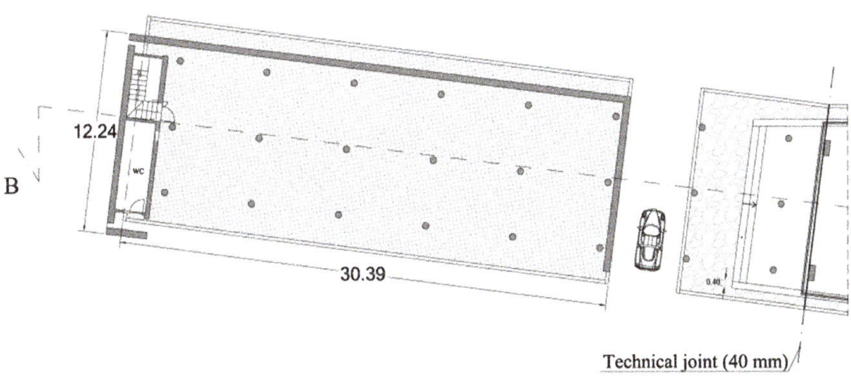

Fig. 3 West wing of the building. At *right*, technical joint with main hall (measures in meters)

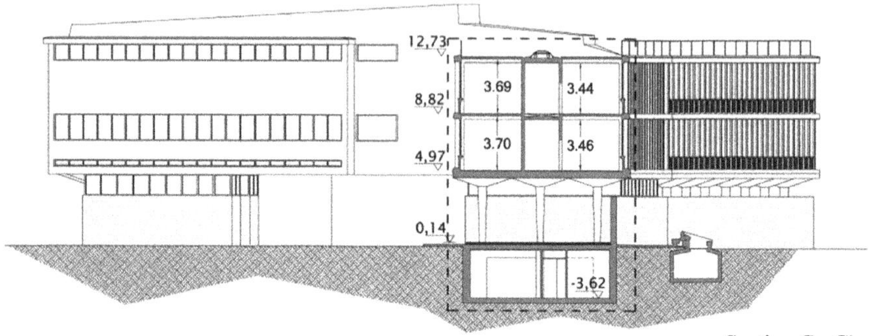

Fig. 4 Vertical section C-C of Fig. 2. The *dashed box* highlights the east wing (measures in meters)

Fig. 5 Detail of vertical section C-C of the east wing (measures in meters)

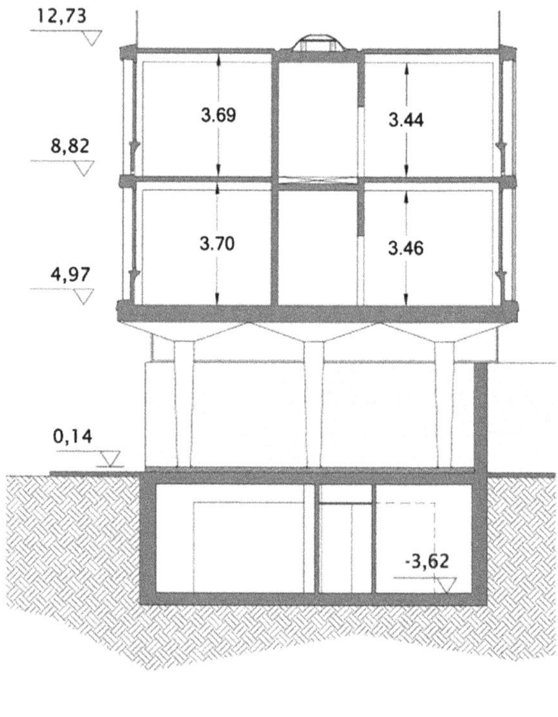

At the ground level of each wing there are 24 conic columns having a hollow-core circular section. Their diameter varies along the height from 400 mm at the bottom to 600 mm at the top; the top is connected to the first floor slab with a conic portion, as shown in Fig. 6, with outer diameter ranging from 600 mm up to 4.00 m. Overall, each column has a characteristic "mushroom" shape.

The hollow core has a constant diameter of 150 mm and hosts the drainpipes. It is worth noticing that the internal hole produces a significant reduction of the

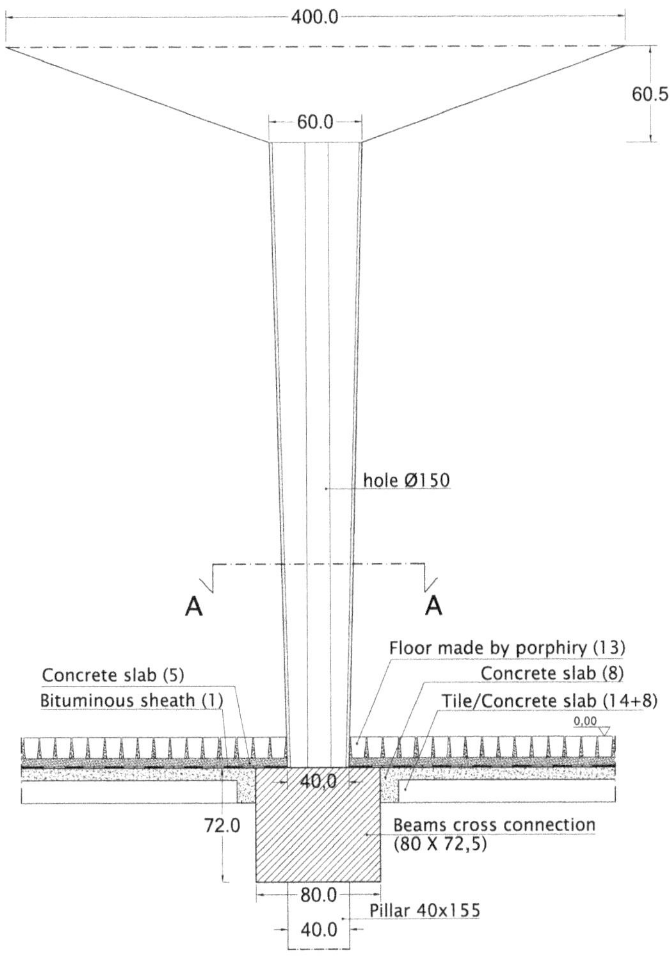

Fig. 6 Mushroom-shaped column. Note the hollow-core circular section inside it (measures in cm)

column shear capacity, especially in the bottom portion, where the thickness of the circular crown is only 125 mm.

The first level floor is made of a concrete slab with depth of 0.50 m, while those at second and roof floor are made of concrete and masonry, with depth of 0.12 m and joists with width of 0.40 m.

Note that the feasibility of the base isolation solution adopted was ensured by the presence of a basement area, with large columns placed right under the mushroom-shaped columns. This favored the displacement of the latter, while still maintaining the eccentricity of the vertical load to acceptable levels in the basement columns below.

Moreover, the top side of each basement column is rigidly connected to beams having depth of about 0.70 m and width of 0.80 m.

Fig. 7 Shows an archive picture taken during construction of the basement area; the reinforced concrete columns that today support the isolating devices are clearly visible

Therefore, the structure of the basement area provides a continuous bearing surface, in the neighborhood of each column, where it was possible to place the sliding devices and to insert the seismic isolators under the columns (Fig. 7).

3 Construction Details and Materials

The size of the structural elements and their reinforcement were found from the original drawings (dated 1959) and then verified directly on site. The columns are reinforced, from design, with 27ϕ24 longitudinal bars; transversal reinforcement is made by stirrup of ϕ10/80 mm. As it often happens, the actual reinforcement found on site was quite different: the vertical bars are instead 24ϕ22 and the transverse reinforcement is made of spirals ϕ5/50 mm, which are very corroded and ill-distributed, as seen in Fig. 8.

With such inadequate transverse reinforcement, the column shear capacity is provided by non-conventional mechanisms, including the "dowel action" guaranteed by the almost continuous longitudinal reinforcement cage. Regarding the concrete strength, it has been observed that the very construction technique of these hollow columns, made by centrifugation, has produced the segregation of the

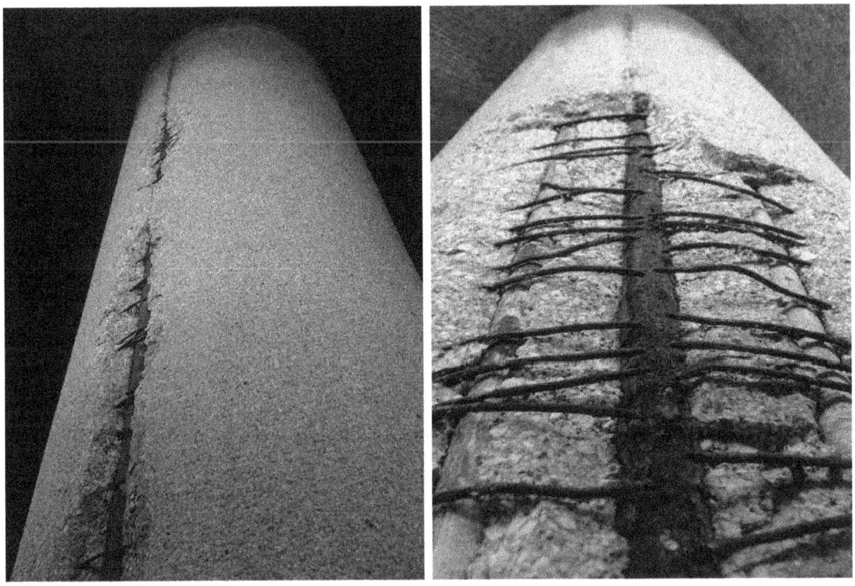

Fig. 8 *Left* mushroom-shaped column damaged by corrosion of bars. *Right* detail of the corroded longitudinal bars and transverse ties

heaviest aggregates towards the external surface, so that the concrete properties change through the thickness of the column cross-section. Thus, in the seismic assessment carried out, the actual reinforcement found during the survey was adopted and a reduced concrete strength was considered.

4 Seismic Assessment and Retrofit Strategy

It is well known that buildings with *pilotis* story often show poor performance under earthquakes: the upper infilled part remains rigid, so that the displacement demand localizes on the columns of the ground floor. In this situation, plastic hinges can only develop at the base of the columns, where the ability to dissipate the earthquake energy input is minimal, both because of the axial load that makes them less ductile and because of P-delta effect.

The seismic assessment has been carried out by evaluating the demand on the structural elements, using modal analysis with linear elastic response spectrum at the Life Safety Limit State (LSLS); the seismic hazard is determined from the local parameters for (stiff) soil class A. For a 10 % probability of exceedance in 75 years (reference period for strategic buildings), that is, a return period of 712 years, the PGA is equal to 0.154 g.

As shown in the next section, linear analysis with response spectrum was adopted because the building clearly exhibits elastic-brittle behavior. In these conditions, a low dissipation behavior is expected, which implies a low behavior factor q.

The Capacity/Demand ratio obtained from the assessment has highlighted that, as largely expected, the critical elements are the columns at the ground floor. Particularly, it was observed that the shear capacity, determined by the method of the variable-angle truss, was lower than the flexural capacity.

Therefore, the columns would collapse in shear before developing a plastic hinge, thus showing a brittle behavior.

This reflects on the entire building, which shows an unacceptable elastic-brittle behavior overall.

For this reason, it was decided to take action and improve the building seismic performance by retrofitting the columns to meet the code requirements pertaining to new constructions.

Regardless of the design strategy chosen, practical and formal requirements needed to be fulfilled:

1. the office activities on the upper floors had to continue without interruption during construction operations,
2. the peculiar "mushroom" shape of the columns could not be changed: the retrofitting measures had to preserve their shape and their architectural value.

The initial solution meant to increase the columns capacity by concrete-jacketing them. This intervention would have greatly hampered the architectural value of the columns.

Therefore, an alternative solution was searched, oriented to reducing the earthquake demand, rather than to increasing the structural elements capacity.

Base isolation immediately appeared as the best solution for this case, with the insertion of isolation devices under the existing mushroom-shaped columns; in this way, they are protected without changing their appearance (Marnetto et al. 2004).

This strategy proved to be optimal under all points of view—technical, architectural, functional, and economical—and therefore it was adopted for the seismic retrofitting of the building.

5 Implementation of the Retrofitting Intervention

5.1 Construction Phases

The retrofitting intervention consisted of three phases in sequence, whereby each phase is organized in sequential operations, as described below:

1. Strengthening of columns:
    - S1. Restoration of the external surface of the columns and passivation of corroded steel bars,
    - S2. Application of pre-tensioned confinement strips (CAM system) to all columns and surface finishing with structural plasterwork.
2. Base isolation of the wings:
    - I1. Laying of equipment and carpentry for load transfer from the columns,
    - I2. Cut at the base of the columns,
    - I3. Laying of isolators and sliders.
3. Translation of the wings:
    - T1. Laying of the anchorage devices for translation,
    - T2. Laying of the hydraulic jacks for the power-assisted release of the isolators,
    - T3. Release of the jacks and translation of the building with continuous monitoring,
    - T4. Protection of the devices with covering steel cases.

5.1.1 Phase 1: Strengthening of the Mushroom-Shaped Columns

The first phase consists in the restoration of the outer layer of concrete and in the replacement of the stirrups with high resistance metal strips. This technique is commonly known as CAM (Italian acronym for Active Confinement of Members) that consists of the application of pre-tensioned stainless steel strips (Dolce et al. 2001).

The steel strips have thickness of 0.9–1.0 mm and width 19 mm; the design strength is 532 MPa.

The lower part of the columns, except those with drain pipe, has been filled with expansive mortar, to increase the effectiveness of the confinement supplied by CAM, by changing the section from hollow to full. Figure 9 shows a column after the strengthening.

5.1.2 Phase 2: Seismic Isolation at *Pilotis* Floor

The second phase consists in the insertion of high-damping elastomeric isolation devices (HDRB) under the existing mushroom-shaped columns. Figure 10 shown this first type of devices.

In some columns, it has been necessary to adopt sliders, characterized by low values of friction coefficient (generally <0.003). Such devices were named "tripods" because they are made from an assemblage of three smaller sliders as shown in Fig. 11. Figure 12 shows their placement under a column, after removing its lower

Fig. 9 View of the columns after shear strengthening with CAM system. *Left* west wing from main hall. *Right* bottom view

Fig. 10 HDRB devices. *Left* devices ready to install. *Right* laboratory test at Sapienza University of Rome, Valle Giulia

Fig. 11 *Left* support points at the base of "tripods". *Right* the internal flexible joint

portion. They are installed under the columns that contain drainpipes to allow drainage to the main sewage system.

Fig. 12 A "tripod" during the installation phases

For this reason, the tripods were conceived with three support points at the vertices of a triangle around the column base section.

5.1.3 Phase 3: Translation of Side "Wings" Buildings

The current Italian code NTC-08 (2008) allows, among the various intervention techniques, also the widening of technical joints. In this case, this operation is carried out by pushing the two side "wings" buildings away from the central one, after installing the isolator devices, blocked in the deformed configuration with specific steel locking devices. Figure 13 shows the assembled system.

After removing the locking devices, the isolators tend to return elastically to their undeformed position, thus translating the building above. The translation speed is controlled by a system of tie rods placed on two alignments, shows in Fig. 14. After the vertical position is reached, the last operation is the installation of protection cases around the devices and the surface finishing of the columns.

Fig. 13 Isolator device with its locking system. The isolator is deformed to its design displacement

Fig. 14 The two alignments of the displacement system control. *Left* across the main hall, from east to west wing. *Right* outside the building, from west to east wing

Fig. 15 Two of the elastomeric isolators in the deformed position, before returning to the vertical position. *Left* starting position. *Right* nearly final position

Figure 15 shows a view of two elastomeric isolators at removal of the locking devices. In this position, the isolators are at their design displacement; this already represents an on-site test on the deformation capacity of the system and its stability against the vertical loads.

Figure 16 shows the lower portion of a mushroom-shaped column, in which the elastomeric isolator has finally returned to the undeformed position: the building above has therefore shifted by the same amount.

Finally, Fig. 17 shows the visual impact of the intervention at the end of work. Note that all isolator devices have been covered with the same metal case.

Fig. 16 Final position of elastomeric isolators

Fig. 17 Final phase of the work. The isolator devices are covered by metal protection cases. *Left* columns close to the driveway. *Right* columns at the lateral entrance of main hall

6 Impact on Society

Seismic isolation is oddly enough one of the most effective retrofitting techniques and yet one of the least used by professionals. It has many well-known advantages, such as, the total protection of the building against even destructive earthquakes, or the possibility of working only at the foundation level without intrusive works in the upper stories. Ideally, one may think of only working at the isolation level: once the seismic capacity of the building is assessed, the isolation system is designed to filter the seismic action to a level compatible with that capacity. In this manner, the top portion of the building can be left untouched, with a significant advantage for the occupants or for the activities thereby carried out. From the social standpoint, this implies not having to displace the occupants to different locations while working on the building. This entails, apart from the additional cost involved, also the psychological distress related to misplacement, especially for old people. From the economic standpoint, it is also well accepted that base isolation is a competitive strategy, especially when applied to buildings with a number of stories between 3 and 6. In this particular case, the great advantage was of cultural nature: the building at hand is considered part of the architectural heritage of that region and base isolation was the only possible strategy that could respect its architectonic value.

7 Conclusions

The seismic retrofitting of a strategic building owned by the Italian Highway Company, built in the 50ies, was presented.

The building is endowed with mushroom-shaped columns with hollow section at the *pilotis* floor, which makes them particularly vulnerable to shear.

It was therefore decided to design a seismic isolation system at the base of the building, by inserting the isolation devices under the existing columns at the *pilotis* floor.

The main benefit of this solution is the possibility to operate exclusively at the ground level, without interrupting the work activities at the upper floors. In fact, the installation of the devices, the positioning in the final configuration of the lateral buildings, the covering of the technical joints between the three buildings and the re-connection of the water drainage system with flexible joints at the ground floor, have been all performed during normal working time.

The isolation system is made of 13 sliding devices (called tripods) under those columns containing drainpipes, and 11 elastomeric devices under all other columns.

The isolation system was designed so to have a first period of the isolated system equal to about 2.4 s, while the equivalent viscous damping ratio has been taken equal to 16 %. The efficiency of the solution was verified by modal analysis with response spectrum, from which it was seen that the shear demand on the columns is reduced to about 1/3 with respect to the original situation, thus remaining well below the corresponding capacity.

Overall, base isolation has proved to be an effective measure to reduce the demand on the ground floor columns. Only minimal strengthening interventions by the so-called CAM system, to increase the columns capacity, had to be carried out.

The efficiency of the isolation system during the earthquake depends on the width of the joints to the adjacent building, which must allow for the displacement in seismic conditions. However, the existing joints were found to have a small width, insufficient to accommodate the design displacements.

Therefore, the project involved also the widening of the joints, as indeed allowed in the current Italian code NTC-08 (2008). This is achieved by slowly translating the two lateral buildings away from the central one, using hydraulic systems and special tendons anchored at the intrados of the first floor.

This operation is followed by the removal of the locking devices used to re-deform the elastomeric isolators. At translation completed, protection cases are installed to cover the isolation devices.

The preliminary studies allowed assessing the intervention feasibility, by gradually reducing the many uncertainties (and doubts!) that initially accompanied the proposal, to finally make it possible. The chosen solution proved to be eventually the only one that could meet the requirements of achieving total protection with no disruption of the normal strategic activities and without diminishing the architectural value of the building.

References

Circolare n. 617 C.S.LL.PP. (2009). Istruzioni per l'applicazione delle nuove norme tecniche per le costruzioni di cui al decreto ministeriale 14 gennaio 2008. Supplemento ordinario n. 27 alla Gazzetta Ufficiale 26 febbraio 2009.

NTC-08. (2008). Nuove norme tecniche per le costruzioni. Gazzetta Ufficiale della Repubblica Italiana 4 febbraio 2008.

Dolce, M., Gigliotti, R., Laterza, M., Nigro, D., & Marnetto, R. (2001). Il Rafforzamento dei Pilastri in C.A. Mediante il Sistema CAM. In: atti del 10° Convegno Nazionale L'ingegneria Sismica in Italia, Potenza-Matera, 2001.

Marnetto. R., Massa, L., Vailati, M. (2004). Progetto sismico di strutture nuove in cemento armato ai sensi dell'ord. 3274 del 08/05/2003 e successive integrazioni n.3316. Edizioni Kappa, Roma.

Lessons from the Wenchuan Earthquake

Yunsheng Wang, Shuihe Cao and Xin Zhang

1 Introduction

The May 12, 2008 Great Wenchuan Earthquake has resulted in 69,185 deaths, 18,467 missing, 374,171 injured and material losses in the hundreds of billions RMB. In which most losses were concentrated in the towns or villages along the active fracture zone (Li et al. 2013; Wang 2008). After the Wenchuan earthquake, several extreme rainstorms hit the meizoseismal area in 2009, 2010, 2012, leading to debris flow and avalanches that buried towns, villages and roads: The highway between Dujiangyan and Wenchuan was interrupted by debris flows and slope collapses. The bridge between Taoguan tunnel and Futang tunnel collapsed; a dammed lake was formed in Yingxiu town; a large number of residential buildings were covered by flood; dozens of people died or disappeared. People began to wonder: "Are the mountain towns in Longmenshan area still safe? What we have done in reconstruction, is that effective? What lessons did we learn from this earthquake? Therefore, it is high time that we analyze the disaster chain after a big earthquake. In this paper, we draw some lessons in hazard chain effects in the Longmenshan area after the Wenchuan earthquake.

Y. Wang (✉) · S. Cao · X. Zhang
State Key Lab of Geohazard Prevention and Environment Protection,
Chengdu University of Technology, Chengdu 610059, Sichuan, China
e-mail: wangys60@163.com

S. D'Amico (ed.), *Earthquakes and Their Impact on Society*,
Springer Natural Hazards, DOI 10.1007/978-3-319-21753-6_17

2 Location of Towns and Villages Concentrated Along the Active Earthquake Fault Zone

As the frequency of high magnitude earthquakes is low and the living space is limited in mountainous areas, our ancestors chose the valleys along active earthquake faults valleys as the location for towns and villages (Jiang et al. 2005), because the valleys provide drinking water, and relatively wide river terrace space for living and tilling. The main towns along the Wenchuan earthquake central fracture are Yingxiu, Longchi, Qingping, Beichuan and Chengjiaba etc.; the main towns along the back range fracture are Genda, Wenchuan, Maoxian, Qingchuan; the main towns along the front range fracture are Dujiangyan, Xiang'e, Hongbai and Hanwang etc. (Fig. 1). But the inhabitants did not realize that their homes were situated in a potentially dangerous location. The losses in those towns and villages were more serious than in other settlements in the Wenchuan earthquake zone. For example, the whole town ruined and nearly 10,000 people were killed in Yingxiu, 20,000 were killed in Beichuan, and 780 were killed in Donghekou, Guanzhuang, Qingchuan county.

According to "Sichuan earthquake data base" (Zhang 2008; Deng 2013), earthquake magnitudes over Ms ≥ 7.0 happened once every 30 or 50 years in the west of Sichuan Province in the past, such as the Diexi earthquake in 1933, the Songpan-Pingwu earthquake in 1976 and the Wenchuan earthquake in 2008. Although a number of strong earthquakes happened in recent history in Sichuan

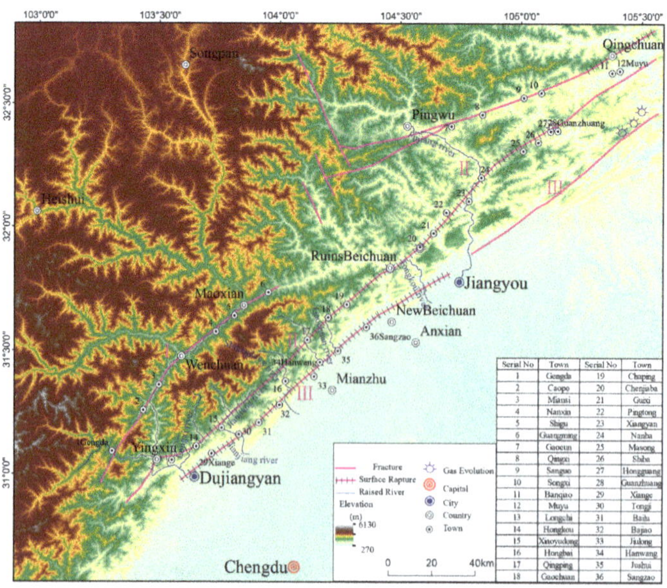

Fig. 1 Location of earthquake hit towns in the Longmenshan area. *I* Back range fracture; *II* Central range fracture; *III* front range fracture

Province, a 30–40 years interval between strong earthquakes let people forget the pain of the last earthquake, and the younger generation does not realize that they are in a potentially dangerous situation. On the other hand, as living space is limited, they almost have no other choice, so the local residents (mainly Qiang Minority people) are still living in the same fault valleys or mountain planation surface areas.

As we know, the old wooden structure houses had a reasonable resistance against seismic loading, but in the 80s and 90s of last century, people were not anymore satisfied with the old wooden structure houses, Brick or brick-concrete houses were in fashion, but unfortunately most houses that were built at that time in the meso-seismal zones have no or a poor anti-seismic design. Those houses were seriously damaged during the Wenchuan earthquake, such as 90 % of the houses in Hongkou, Yingchanggou (Fig. 2a), Yingxiu, Hongbai, and Bailu towns, collapsed or were seriously damaged. Even some brick-concrete structure buildings with precast concrete slabs collapsed in Hanwan and Chengjiaba town, because of the poor connections between the slabs (Fig. 2b).

During the Wenchuan earthquake, one third of the deaths were caused by the secondary disasters in the three earthquake fault zones (the central range fracture, the back range fracture and the front range fracture). As these three fractures activated during the Quaternary, rivers and small streams developed along the

Fig. 2 **a** House collapse in Yingchanggou where the central earthquake fracture passes nearby. **b** Teaching building collapse at Xuankou middle school, Yingxiu. **c** The Shibantan landslide buried a village and dammed the Qingzhu River. **d** The Wangjiayan landslide buried the old district of Beichuan

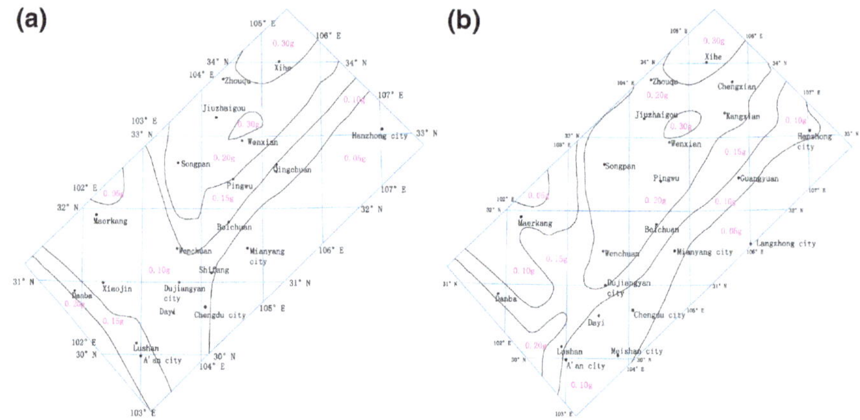

Fig. 3 **a** The seismic intensity prediction map before the Wenchuan earthquake. **b** The modified seismic intensity map after the Wenchuan earthquake

fractures, forming NE striking valleys. The fault valley width ranges from several hundred meters to 2 km, with high and steep slopes, especially at the hanging wall side. The length of these valleys ranges from several kilometers to tens of kilometers. Slope failure on both sides of the valleys destroyed the towns and villages on the bottom of the valleys or at the foot of the slopes (Fig. 2c, d). In addition to the surface rupture, the seismic intensity is higher along the fracture zone than in the adjacent areas. Considering the safety distances suggested by Zhou et al. (2008), most towns and villages in the Longmenshan area cannot avoid geological (earthquake) disasters.

Since 2000, the construction in mountain towns began to implement a-seismatic design, but unfortunately, due to under-estimation of the seismic intensity (Fig. 3a), the seismic reinforcement criteria for some new buildings was too low. They could not resist the Wenchuan earthquake. The zonation map of seismic intensity in this area has been modified after the big earthquake (Fig. 3b). In the event of a great earthquake in the future, a large number of casualties and property loss is still inevitable.

3 Consequences of the Large Amount of Loose Material Remaining on the Slopes After the Earthquake

3.1 Under-Estimation of the Loose Material Sources in the Gully Catchments

The total amount of loose material left on the slopes after the earthquake is over 28×10^8 m^3 in the Longmenshan area as estimated by Cui et al. (2010), which is considerably higher than any previous estimation. Before the Wenchuan earthquake, the estimation of the loose material sources of the gullies was correct on the whole, however, the estimation method of the loose material sources did not reflect the real situation after the Wenchuan earthquake, especially the difference between estimation of the active material sources and real runoff (Table 1). In fact, the occurrence and scale of debris flows is much larger than that what would be expected from the estimation of the increased amount of material sources. As a consequence, the pre-earthquake debris flow control works are not suitable for the new estimations of runoff (Xu 2010; Gan et al. 2012; Hu et al. 2011). For example, check dams in Niumian gully was filled up by debris flow in 2013 (Fig. 4a), the drainage canal in Chediguan gully was covered by the lateral slope failure (Fig. 4b), the debris flow of Wenjia gully, Qingping on 12th, August, 2010 buried 400 houses, 1000 m of road,

Table 1 Comparison of predicted and observed debris runoff

Gully	Runoff of debris flow ($\times 10^4$ m^3)	
	Prophetical (estimated)	Practical (real) amount
Hongchun gully	1.7	65 (2013.7.10)
Qipan gully	1.6	85 (2013.7.10)
Taoguan gully	3	110 (2013.7.10)
Wenjia gully	224	450 (2010.8.12)

Fig. 4 **a** Check dam in Niumiangou was filled by a debris flow on 10th, July, 2013. **b** Drainage channel of the Chediguan gully was destroyed by a debris flow on 10th, July, 2013. **c** Qingping town buried by Wenjia gully debris flow on August 12, 2010. **d** Resettlement area buried by the Qipan gully debris flow on July 10, 2013

destroyed a bridge, killed 5 people. The run-out deposit is 1000 m long, 300–400 m wide, 5–15 m thick, with a volume of about 450 × 10^4 m^3. During this event, all geological hazard control works in Wenjia gully were destroyed.

Some town reconstruction areas have encroached the debris flow run-out zones. The Wenjia gully debris flow triggered by a heavy concentrated rainfall on August 13, 2010 resulted in flooding of the whole Qingping town (Fig. 4c). The Qipan gully debris flow, also triggered by a heavy concentrated rainfall on July 13, 2013 lead to flooding of the new residential site (Fig. 4d).

3.2 Landslides Form the Material Sources for the Debris Flows

In the Subao river catchment in Beichuan country as example, debris flows had never been recorded of before the Wenchuan earthquake, but it has become one of the most serious areas of debris flows since the earthquake due to the large amounts of loose material on the slopes due to the strong shock. On the basis of mud level investigation (morphology investigation method) and rain-flood method, the outcome of runoff calculations for Guanmenzi gully, Subao gully and Da'anshan gully had increased by about 50–100 % after the Wenchuan earthquake. The characteristic trend for the development of debris flows is as follows: after a first period of high activity a more quiet period will follow. The first active period will last at least for 15 years (Wei and Xie 2002).

3.3 Rising of the River Bed Base-Level Caused by Debris Deposited into the River

After the earthquake, large numbers of debris flows occur every year in the earthquake affected area, which leads to a notable rise of the river (Fig. 5) bed base level (Table 2). For example, the mean river bed rise of Tongkou river since the Wenchuan earthquake is 10 m, and the river bed rise of Mianyuan river has increased by 20 m. The mean river bed rise of Yuzixi river in Wenchuan 10–20 m (Table 2).

During the design of the relocation of new roads a possible rise of the floodplain was hardly taken into account. The valley bottom of the Yuzixi (Fig. 5c), Minjiang mainstream, Shitingjiang, Mianyuan River (Fig. 5a, b) and Tongkou River was quickly filled up by the debris flows of Sept. 24, 2009, Aug. 13th, 2010 and Aug. 17, 2010. Due to the rapid rise of river beds, a number of relocated roads along rivers had to be abandoned because of *conservative design* without considering river erosion at a higher base level. Residential buildings on both sides of the river on terraces have been covered by flooding; some of bridges had to be redesigned with an increased construction height (Fig. 5d).

Fig. 5 **a** The river bed near Xiaomuling power station on 14th, May, 2008. **b** Rise of the river bed at the Xiaomuling power station on 9th, July, 2013. **c** Abandoned shed-tunnel along Yuzixi River due to the rise of the river bed. **d** Redesign of bridge Pier. **e** New houses buried by a debris flow in Qingping. **f** Apartment building in the outflow area destroyed by debris flows

Table 2 River bed rise since Wenchuan earthquake

River	Rise of river bed (m)	
	Average	Range
Yuzixi	5	10–20
Main stream of Minjiang	7	6–8
Longxi	9	8–10
Baishui	6	6–7
Shitingjiang	7	6–8
Mianyuan	16	15–20
Gaochuan	8	8–10
Jianjiang	10	10–12
Fujiang	7	7
Tongkou	10	10–20

4 Occupation of Flood Discharge Space in Reconstruction Activities

Cui et al. (2008) pointed out that mountain town planning and construction after Wenchuan earthquake have not always been appropriate. Due to potential slope stability problems, many dispersed houses have been removed to the Terrace I, higher flood plain or to the mouth of the outflow area (debris flow fan), which has reduced the debris flow discharge space, leading to a reduction of the gully discharge capacity such as happened in the Qingping, Qipan gully, Taoguan gully, Futang gully, Mianchi gully, Mianyuan River, Gaochuan river and Chaping river, etc. More than half of the discharge space has been occupied in some resettlement areas. These residents are confronted with serious debris flow threats such as in the Qingping town (Fig. 5e), Longchi town, and Qipan gullies (Fig. 5f).

5 Geochemical Anomaly After the Earthquake

A large number of geochemical anomalies were discovered along the central fault zone after the Wenchuan Earthquake located in Chenjiaba town, Shikan village, and in Donghekou respectively. One spot is located in Jianfeng, Jiangyou (on the front range fracture). The main features of these spots are that (hot) gas escapes from ground fissures (Fig. 6a). The temperatures can reach 200 °C (Fig. 6b), and the environmental effects are still on evaluation. The residents worry about the local environment ground water, and air could be contaminated, and even some fear the possibility of explosions in the future. Fears of the population may last a long time.

Fig. 6 **a** Gas escaping in Chenjiaba. **b** Gas with high temperature escaping in Chenjiaba

6 Conclusions

Huge personal and material losses since the Wenchuan earthquake woke up the population to review our town reconstruction and relocation planning, disaster evaluation method and geological hazard control thinking. The main lessons from Wenchuan earthquake can be summarized as follows:

1. The town planning along the active fault valleys results in towns and villages with serious seismic hazards. Reconstruction and relocation must pay more attention to avoid surface rupture traces and follow a-seismic design.
2. The estimation methods to determine gully loose material should be revised after a high magnitude earthquake. A correct estimation is the basis for a sound disaster control.
3. The resettlement planning after a high magnitude earthquake should not block parts of the flood discharging space, as large scale debris flows could then wipe out the resettlement area.
4. Reconstruction and relocation of roads, bridges, and tunnels should take the rise of river beds into account. This is also valid for the ground level of the resettlement areas.

Acknowledgments This study is financially supported by the National Foundation for Natural Science of China (41072231) and the China Geological Survey Bureau (Grant No. 12120113010100). The authors would like to thank Niek Rengers for revising the article and his many helpful suggestions.

References

Cui, P., Wei, F.-Q., Chen, X.-Q., et al. (2008). Geo-hazard in Wenchuan earthquake area and counter measures for disaster reduction. *Bulletin of Chinese Academy of Sciences, 23*(4), 317–323.

Cui, P., Zhuang, J., Chen, X., et al. (2010). Characteristics and countermeasures of debris flow in Wenchuan area after the earthquake. *Journal of Sichuan University (Engineering Science Edition), 42*(5), 10–19.

Deng, S. (2013). Earthquake on the Longmen Mt-Minshan fault zone. Journal of Xihua University (Philosophy and Social Sciences), (2), 18–22.

Gan, J., Sun, H., Huang, R., et al. (2012). Study on mechanism formation and river blocking of Hongchungou giant debris flow at Yingxiu of Wenchuan Country. *Catastrophology, 27*(1), 5–9.

Hu, H., Cui, P., You, Y., et al. (2011). Influence of debris supply on the activity of post-quake debris flows. *The Chinese Journal of Geological Hazard and Control, 22*(1), 1–6.

Jiang, W., Wang, S., Wang, Y., et al. (2005). Active tectonic system and its control of seismic activity in the east part of the northwest fault block of Sichuan, China. *Journal of Chengdu University of Technology (Social Sciences), 30*(4), 340–344.

Li, W., Huang, R., Xu, Q., et al. (2013). Rapid prediction of co-seismic landslides triggered by Lushan earthquake, Sichuan, China. *Journal of Chengdu University of Technology (Social Sciences), 40*(3), 264–274.

Wang, Z. (2008). A preliminary report of the great Wenchuan earthquake. *Earthquake Engineering and Engineering Vibration (Earthq Eng And Eng Vib), 7*(2), 225–234.

Wei, F., & Xie, H. (2002). Debris flow and its mitigation in the construction of mountain towns in western China. *The Chinese Journal of Geological Hazard and Control, 13*(4), 23–28.

Xu, Q. (2010). The 13 August 2010 catastrophic debris flows in Sichuan province: Characteristics, genetic mechanism and suggestions. *Journal of Engineering Geology, 18*(5), 596–608

Zhang, L. (2008). Earthquake on the Longmen mountain fault zone. *Journal of Literature and History, 4*, 66–68

Zhou, Q., Xu, X., Yu, G., et al. (2008). Investigation on widths of surface rupture zones of the M 8.0 Wenchuan earthquake, Sichuan, China. *Seismology and Geology, 30*(3), 778–788

Lessons Learned from the Recent Earthquakes in Iran

Mohammad Ashtari Jafari

1 Introduction

During 1984–2003 about four milliard people were influenced by extreme natural events and between 1990 and 1999 the costs of natural disasters were about 15 times higher than the period 1950–1959. It seems that the basic causes behind rising economic losses are changes in land use and increases in the concentration of people and capital in high risk areas (Amendola et al. 2008). For example, in Iran, the capital Tehran and some other important urban areas have been exposed to disastrous earthquakes (e.g. Tabriz, Roudbar, Bam …). Historical seismicity of Iran confirms that even medium sized earthquakes had left catastrophic results so learning from the past catastrophic events would be useful at least in considering land use and future urban area developments.

The Persian Plateau is under the effect of the multiple convergences between African, Indian and Eurasian Plates which resulted in high mountain ranges extended along the northern (Alborz), northeastern (Kopeh-Dagh) and southwestern (Zagros) boundaries of this plateau and lower rugged mountains (East Iran) along the eastern margins. The Azarbaijan region which is located along the northwestern of the Persian Plateau can be considered as the transition between the Turkish Plateau and Central Iran. The Makran in the southern part is an area where the oceanic crust is subducting beneath the Persian Plateau (Ashtari Jafari 2014). The seismicity patterns (Fig. 1) show nonuniform distribution that are concentrated along the active fold-thrust mountain belts surrounding the relatively aseismic undeformed rigid and stable blocks (Berberian 2005). Catastrophic earthquakes are one of the features of this plateau. Several urban and industrial complexes has been developed near mountain foothills that are usually located near the bordering of

M. Ashtari Jafari (✉)
Institute of Geophysics, University of Tehran, Iran P.O. BOX 16765-533, Tehran 16, Iran
e-mail: muhammad@ut.ac.ir

S. D'Amico (ed.), *Earthquakes and Their Impact on Society*,
Springer Natural Hazards, DOI 10.1007/978-3-319-21753-6_18

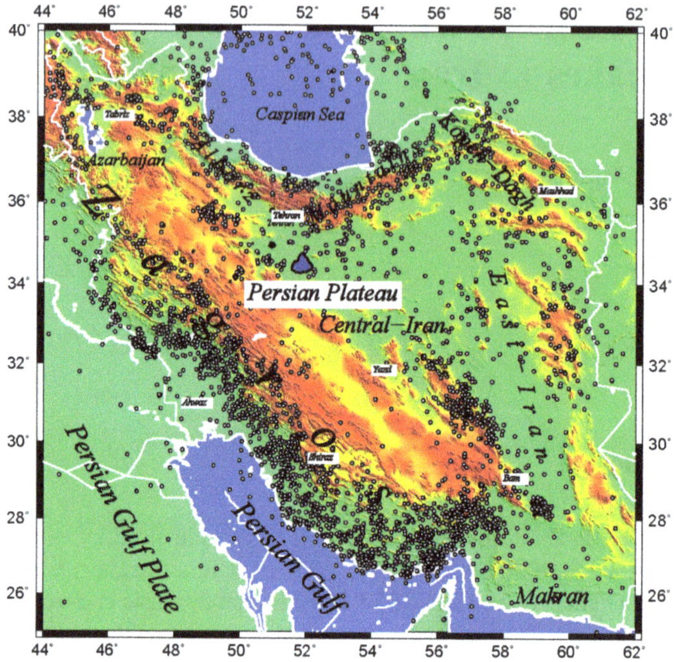

Fig. 1 Epicentral distribution of the earthquakes (Mn ≥ 3.5; depth ≥ 50 km) in Iran from early 1996 to early 2013 (epicenters from Iranian Seismological Center, irsc.ut.ac.ir)

active faults, so great earthquakes along such faults has been destroyed many cities during historical and recent times while interaction between reverse and strike-slip faults has been revealed by several clusters of events, also (Ambraseys and Melville 1982; Berberian 1995; Berberian and Yeats 1999; Ashtari Jafari 2013a). It seems that seismicity is of shallow type and takes place around active fault zones that surround nearly stable microplates. Coseismic rupture studies are mainly realized by teleseismic modeling that usually use segmentation of rupture in time and space (e.g. studies on Rudbar, Golbaf and Zirkuh earthquakes).

2 Seismicity and Seismotectonics

Seismicity of Iran has been studied by different authors (Wilson 1930; Niazi and Basford 1968; Nowroozi 1976; McKenzie 1972; Kaila et al. 1974; Berberian 1995; Shoja-Taheri and Niazi 1981; Ambraseys and Melville 1982; Ahmadi et al. 1989, Engdahl et al. 2006; Ashtari Jafari 2013a). The Persian Plateau which is located within the continental collision zone between the Persian Gulf and the Eurasian plates include recent volcanics, high mountain belts and active faults. This region has gone under different mountain building phases and is characterized by concurrent magmatism and metamorphism especially within early Paleozoic, middle Triassic, early Jurassic and early Cretaceous (Berberian 1981; Sengor et al. 1988). The approximate rate of convergence between Persian Gulf-Eurasia has been already estimated to be about 30 mm/yr at longitude 50°E and 40 mm/yr at longitude 60°E (Jackson 1992; DeMets et al. 1994; Chu and Gordon 1998) but the present GPS measurements has estimated lower values for deformation in Iran. Vernant et al. (2004) suggested deformation is about 20 mm/yr for the last Ma (Zagros: 10 mm/yr, Alborz: 8 mm/yr and eastern-Iran: 16 mm/yr). The regional convergence is partitioned into shortening perpendicular to the strike of the folds and the faults while active deformation is not uniformly distributed and consists of shortening, thickening, reverse and strike-slip faulting plus subduction of the oceanic crust. On the other hand a great part of the shortening is taken up in mountain building (Alborz, Zagros and Kopeh-Dagh), earthquakes and some parts are accommodated in Central-Iran. One can roughly divide the active faults of Iran into reverse and strike-slip faults. Active faults in Iran are mainly short length compressive faults located within active fold-thrust border zones plus the strike-slip faults seen over narrow zones accompanied with reverse sub-faults. This is an example of the complex tectonic behavior of the reverse and strike slip faults, accordingly the main features revealed by CMT solutions are consistent with strike-slip and compressive motions (Fig. 2). Around Zagros the reverse solutions

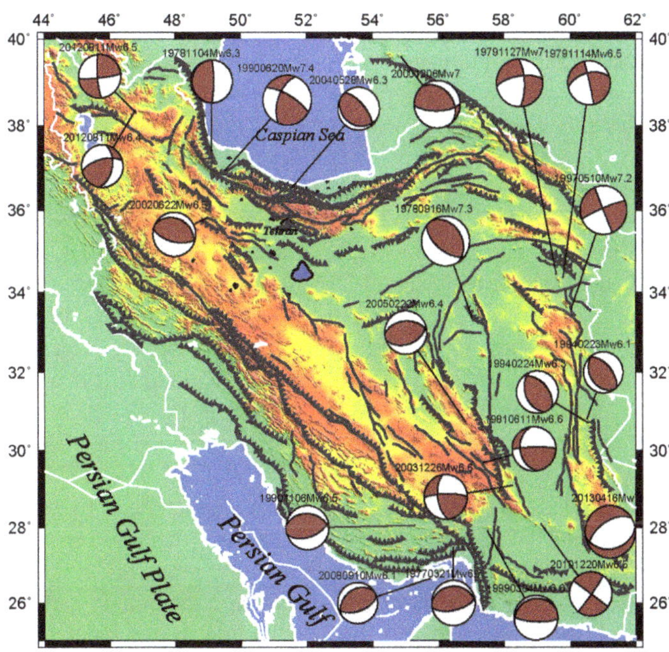

Fig. 2 Distribution of the CMT solutions (Mw ≥ 6.0) in Iran (CMT solutions from GlobalCMT, www.globalcmt.org)

are dominant while in Alborz, Kopeh-Dagh and Central-Iran strike-slip solutions start to grow in number. Focal depth of earthquakes in Iran are mainly within 5–20 km (upper continental crust), although in the regions associated with subduction (e.g. Makran) one would expect for deeper earthquakes (Engdahl et al. 2006).

With a V-shaped structure the Alborz Mountains is about 600 × 100 km range along the southern part of the Caspian Sea including active deformation within the African–Eurasia collision zone that presents coeval strike-slip and compressional deformation. It can be considered as an example for inactive fold and thrust belts which involve a component of oblique shortening. Deformation in Alborz is partitioned along range-parallel thrusts and left lateral strike-slip faults. Recent GPS studies showed that N-S shortening across the Alborz occurs at 5 ± 2 mm/yr and that the left-lateral shear along the overall range has a rate of 4 ± 2 mm/yr (Vernant et al. 2004). Other features of Alborz have been discussed elsewhere (e.g.: Allen et al. 2003; Ashtari Jafari 2007a, b, 2009, 2010; Ashtari Jafari et al. 2005).

In Azarbaijan the faults show a northward bending tectonic arc where deformation is characterized by E–W striking reverse and thrust faults, oblique strike–slip faults, as well as N–S striking normal faults. Several aspects of these faults including geometry, slip rates and kinematics are not well documented. Historical and paleo-seismological data show that several large earthquakes occurred in this region which was accompanied by volcanic eruptions, landslides, river migrations and other natural structures. Right-lateral strike-slip faulting is also present around the Chaldran fault and the Gailatu SiahChesmeh-Khoy (GSCK) fault system; however there are evidences for oblique normal faulting on the Serow normal faults. Based on recent GPS measurements regional crustal shortening is about 10 ± 2 mm/year, meanwhile the right lateral motion on the Tabriz fault has been estimated to be around 8 mm/yr (Ambraseys and Melville 1982; Ambraseys and Jackson 1998; Ashtari Jafari 2013b; Berberian 1997; Karakhanian et al. 2004; Reilinger and Barka 1997; Trifonov et al. 1996).

The Central Iranian Block is constructed from different micro-continental blocks that detached from Gondwana from the Permian to early-Triassic and accreted into Eurasia along the Alborz and Kopeh-Dagh during the late Triassic closure of Palaeo-Tethys. It consists of the Tabas, the Lut, the Posht-Badam and the Yazd blocks. The borders of these blocks are strike-slip faults that are subject to rotation. Epicenters of the most destructive earthquakes show good coincidence with these tectonic ties. Central Iran is bounded by Alborz and Kopeh-Dagh in the north, Zagros in the south/south-west, Makran in the south-east and East Iran in the east. The block is laterally trapped between the Persian Gulf plate in the south and southwest and the Indian Plate in the east that is characterized by coherent plate motion with internal deformation of about 2 mm/yr. The low seismicity of this region suggests that the Central Iranian Block is rigid, meanwhile it is characterized by discontinuous seismic activity with shallow large magnitude earthquakes and some earthquakes are associated with surface faulting in this block (Bachmanova et al. 2004; Vernant et al. 2004; Ashtari Jafari 2012a).

The East-Iran includes conjugate strike-slip faulting. In the south there are N–S right-lateral faults of the Sistan Suture Zone and in the north the dominant tectonic features are the E–W left-lateral Dasht-e Bayaz and Doruneh faults. These fault systems accommodate N–S right-lateral shear between Central Iran and the E–W left-lateral Dasht-e Bayaz and Doruneh faults by rotating clockwise about a vertical axis. Vertical axis fault rotation seems to be a significant tectonic process in eastern Iran. The N-NNE movement of central Iran relative to Afghanistan results in right lateral motion of about 15 mm/yr in eastern Iran (Allen et al. 2006; Berberian and Yeats 1999; Berberian et al. 2000; Freund 1970; Jackson and McKenzie 1984; Vernant et al. 2004).

The Kopeh-Dagh extends for about 700 km across northeast Iran and Turkmenistan between the Caspian Sea and Afghanistan. The right-lateral Ashgabat fault lies at the northern margin of this range which has a component of thrust motion to the north. The Kopeh-Dagh is separated in the south from the eastern Alborz Mountains by several reverse faults. The west central part reaches the Caspian Sea by the Atrak River and the southeastern part drains to the Kashaf-Rud depression (Allen et al. 2004; Nowrouzi et al. 2007; Priestley et al. 1994).

The Makran has been formed by the subduction of the oceanic crust beneath Eurasia. Subduction may probably be started during Paleocene and accretion started during Eocene and it has developed since Late Miocene that is propagating seaward at a rate of about 10 mm/yr yet. It is bounded to the west by continent-continent collision between Africa and Eurasia and to the east by continent-continent collision between Indian and Eurasian plates. Important features of this subduction include the high sediment thickness over the oceanic crust and the low dip angle of subduction. The eastern Makran has been experienced large historical earthquakes and currently presents low seismic activity, however the western Makran seem to be quiet, meanwhile it shows a steeper slab in the west compared to the east (Byrne et al. 1992; Kopp et al. 2000; Platt et al. 1988; Ashtari Jafari 2012b; Ashtari Jafari 2014).

The Zagros is actively deforming due to the shortening between the Persian Gulf and Eurasian plates. The main structural trend of Zagros is NW–SE and the convergence direction happen at 45° relative to the range axis. Different measurements estimate the motion of Afro-Eurasia to be about 20 mm/yr with a northward direction at the longitude of the Persian Gulf. Some of this convergence is accommodated almost completely in Zagros Mountains (6–9 mm/yr). Based on seismicity and GPS data, Central Iran acts like a backstop for this belt. The main recent fault (MRF) has restricted the tectonic activity of NW-Zagros where no extensive series of large earthquakes is occurred along it. Most of the seismic deformation in the south of the main recent fault (MRF) is related to the thrusting events including slip vectors normal to the main structural trend, so it may be considered that most of the parallel component of oblique convergence is taken by the MRF. According to the transfer of motion induced by the Kazeroon fault, tectonic activity of the SE Zagros seems to be different. The Main Zagros Thrust (MZT) is the principal tectonic feature bordering the Zagros to the NE (Allen et al. 2003; Ambraseys and Melville 1982; Bachmanova et al. 2004; Berberian 1995; Vernant et al. 2004; Vernant and Chery 2006).

3 Damage from Recent Earthquakes in Iran

The global annual earthquake casualty rate is 8.895 ± 30 deaths/year, which indicates that 0.8895 million deaths will occur in the 21st century even if no disastrous earthquakes occur. As destructive earthquakes follow a non-stationary Poisson process in which their rate is proportional to world population, there will be 8.7 ± 3.3 disastrous earthquakes in this century assuming medium-fertility population projections by the United Nations (UN), so a minimum of 0.87 ± 0.33 million people will die in destructive earthquakes. The combination of these two estimates denotes minimum world earthquake casualties in this century will be 1.76 ± 0.33 million. According to the average post-1900 death toll of disastrous earthquakes (193,000), world earthquake casualties in 21st century will be 2.57 ± 0.64 million (Holzer and Savage 2012). Holzer and Savage (2012) also forecast that the number of earthquakes with death tolls greater than 50,000 will increase in the 21st century.

Direct and indirect damages caused by earthquakes have been recorded globally. Several historical and recent devastating events have destroyed different regions in Iran (Table 1). According to the seismic macrozonation hazard map of Iran provided by Building and Housing Research Center (BHRC), many cities are located within the high and very high relative risk areas. As is shown by previous earthquakes man made constructions will collapse even under shakings caused by

Table 1 Earthquakes caused more than 1000 deaths in Iran from 1900–2015

Date	Location	Magnitude	Comments
1990/06/20	Roudbar 37.0° N, 49.4°E	7.4	Felt in most of northwestern Iran, including Arak, Bakhtaran and Tabriz. Nearly all buildings were destroyed in the Rudbar-Manjil area. Substantial damage occurred at Khalkhal and Now Shahr and slight damage occurred at Tehran. Estimated 40,000 to 50,000 people killed, more than 60,000 injured, 400,000 or more homeless and extensive damage and landslides in the Rasht-Qazvin-Zanjan area, Iran
2003/12/26	Bam 28.99°N 58.31°E	6.8	Believed to be the largest earthquake in this area in more than 2000 years About 31,000 people killed, 30,000 injured, 75,600 homeless and 85 % of buildings damaged or destroyed in the Bam area. Landslides occurred in the epicentral area
1978/09/16	Tabas 33.2°N 57.4°E	7.8	About 15,000 people killed, many were injured. Tabas, Dehesk and Kurit had the highest death toll the remainder of the deaths were in surrounding areas

(continued)

Table 1 (continued)

Date	Location	Magnitude	Comments
1962/09/01	Buein Zahra 35.6°N 49.9°E	7.1	About 12,225 people killed and felt at Tehran, Tabriz, Esfahan and Yazd. Based on damage to old structures, this was probably the largest earthquake in this immediate area since at least 1630. Some landslides and sandblows occurred. Earthquake lights (a red to orange glow) from the Rudak area were observed prior to the event by various people
1968/08/31	Dasht-e Bayaz 33.9°N 59.02°E	7.3	About 12000 people killed. A strong aftershock on Sep 01 destroyed the town of Ferdows (see next event). In all, more than 175 villages were destroyed or damaged in this rather sparsely populated area of Khorasan Province. Most buildings in the area were built of adobe with very were destroyed, the few steel-frame or brick-and-mortar structures in the area generally survived with only minor to moderate damage. The maximum strike-slip (horizontal) offset was about 4.5 m near Dasht-e Bayaz with a vertical offset of about 2 m. Extensive ground ruptures and sandblows occurred in the Nimbluk Valley east of Salayan, south of the main fault trace
1909/01/23	Silakhor 33.4°N 49.1°E	7.3	About 6000 people killed. Many villages destroyed or severely damaged. Casualties occurred in 130 villages. Over 40 km of surface rupture was seen on the Dorud Fault
1972/04/10	Ghir 28.4°N 52.8°E	7.1	About 5054 people killed and injured 1700. In Ghir, 67 % of the population of 5,000 were killed, and 80 % of the buildings were leveled. Many of the victims were women and children, as the men had departed for the fields. Landslides blocked roads hampering rescue work
1929/05/01	Koppeh Dagh 37.85°N 57.75° E	7.2	More than 3,250 people were killed and 88 villages destroyed or damaged in the Baghan-Gifan area, Iran. Damage also occurred at Bojnurd. Nearly all buildings were destroyed at Garmab-Turkmenistan. About 50 km of surface faulting was observed on the Baghan-Germab fault

(continued)

Table 1 (continued)

Date	Location	Magnitude	Comments
1981/06/11	Golbaf 29.9°N 57.7°E	6.9	About 3000 people killed, many injured, and extensive damage in Kerman Province
1930/05/06	Salmas 38.15°N 44.70°E	7.2	About 2500 people killed and 60 villages destroyed in the Salmas Plain and surrounding mountains. The town of Dilman was completely destroyed, but there were only 1,100 deaths because a magnitude 5.4 foreshock had occurred at 07:03 UTC. Although the foreshock killed 25 people, it probably saved thousands of lives since many people chose to sleep outdoors that night
1923/05/25	Torbat-e Heydariyeh 35.2°N 59.2°E	5.7	About 2200 people killed and 5 villages completely destroyed southwest of Torbat-e Heydariyeh
1997/05/10	Qaen 33.9°N 59.7°E	7.3	More than 1567 people killed, 2300 injured, 50,000 homeless, 10,533 houses destroyed, 5474 houses damaged and landslides in the Birjand-Qaen area. Felt in the Kerman, Khorasan, Semnan, Sistan va Baluchestan and Yazd regions of Iran. This earthquake occurred on the Abiz fault. This fault is north of the collision zone between the Arabian and Eurasian plates. The region of the Abiz fault is comprised of several microplates and is tectonically very active
1981/07/28	Sirch 30.0°N 57.8°E	7.3	About 1500 people killed, 1000 injured, 50,000 homeless and extensive damage in the region
1957/07/02	Sang Chal 36.14°N 52.70°E	7.1	About 1200 people killed. Nearly all villages destroyed in the Ab-e Garm-Mangol-Zirab area on the north side of the Alburz Mountains. Many landslides and rockslides blocked the Amol-Tehran road and caused nearly as much damage in some villages as had been caused by shaking. It was felt strongly at Tehran
1957/12/13	Sahneh 34.35°N 47.67°E	7.1	About 1130 people killed, 900 injured and 211 villages destroyed or severely damaged in the Sahneh-Songor-Asadabad area in Kermanshahan and Hamadan Provinces

Data from USGS (earthquake.usgs.gov/earthquakes/world/world_deaths.php)

Table 2 Losses caused by natural hazard in Iran from 1900–2015

Disaster type	Disaster subtype	# of events	Killed	Total affected	Damage (000USD)
Drought	Drought	2	–	37,625,000	3,300,000
	Average per event		–	18,812,500	1,650,000
Earthquake	Earthquake	106	147,475	2,691,488	11,826,628
	Average per event		1391	25,391	111,571.96
Epidemic	Unspecified	1	76	–	–
	Average per event		76	–	–
	Bacterial infectious diseases	2	296	2500	–
	Average per event		148	1250	–
Extreme temperature	Heat wave	1	158	–	–
	Average per event		158	–	–
Flood	Unspecified	37	4177	1,745,305	470,840
	Average per event		112.9	47,170.4	12,725.4
	Flash flood	14	2689	1,291,066	253,700
	Average per event		192.1	92,219	18121.4
	General flood	32	1269	1,076,948	6,990,528
	Average per event		39.7	33,654.6	218,454
Mass movement wet	Avalanche	3	73	44	–
	Average per event		24.3	14.7	–
	Landslide	1	43	100	–
	Average per event		43	100	–
Storm	Unspecified	8	248	19,785	13,540
	Average per event		31	2473.1	1692.5
	Local storm	3	88	–	15000
	Average per event		29.3	–	5000
	Tropical cyclone	1	12	160,009	–
	Average per event		12	160,009	–
Wildfire	Scrub/grassland fire	1	–	–	–
	Average per event		–	–	–

Data from EM-DAT (www.em-dat.net)

moderate earthquakes, because the construction quality especially in small villages and towns is usually poor (most buildings are built of adobe with thick arched roofs). The residences and other constructions are generally built without considering seismic design regulations, so they are highly vulnerable (Daniell et al. 2011; Ashtari Jafari 2008). Not only earthquakes but also other natural hazards hit Iran, time by time, and leave considerable damage (Table 2).

Another way to discuss the economic losses is to consider other indices like GDP (Gross domestic product). Precise application of country-adjusted CPI (consumer price index), GDP and salary, is necessary to convert earthquake losses into present monetary indices (e.g. today's dollar). Daniell et al. (2012a) used the concept of HNDECI (Hybrid Natural Disaster Economic Conversion Index) to estimate worldwide economic losses from 1900–2012. They estimated the over 2.9 trillion USD loss attributable to earthquake in this period that imposed serious impacts in Central Asia and Pacific rim. They employed direct and indirect losses and they disaggregated secondary effect economic losses also. Their most important conclusion was that losses are increasing worldwide.

From Daniell et al. (2011, 2012a, b) we consider three economic parameters: GDP, net capital stock and indirect losses. Annual average loss caused by earthquakes in Iran based on GDP (PPP: purchasing power parity) is estimated to be about 0.02–0.03 % per year. Cumulative net capital stock loss from 1900–2012 in Iran is about 4–6 %. Meanwhile indirect damages seem to be lower in Iran (in comparison with countries like China, Japan and USA which are the countries that suffer periodic great earthquake similar to Iran).

4 Society and Recent Earthquakes in Iran

Earthquakes leave significant damage to man-made structures, lifeline infrastructures, such as hospitals, cultural centers, the water supply systems, power lines, educational buildings etc..... Comparing yearly earthquake death rates among Iran, Japan and USA during three different periods revealed that while Japan and USA have been reduced their yearly rates; Iran's status has been worsening (Spence 2007). The interaction of society in different aspects will be a great help to improve the situation of Iran, in this regard. Based on the in hand experiences from recent earthquakes in Iran, in this section society and its relation with several important aspects of pre and post earthquake phases including: public awareness, medical aspects, survey and reconstruction and economical impacts will be discussed (Ashtari Jafari 2008).

4.1 Public Awareness

Public education can play an important role to improve the society role and its better engagement with the subject of earthquakes as a whole, because the public awareness of the earthquake risk is still low in Iran (Tierney et al. 2005). Four models developed by McEntire et al. (2002) i.e.: the disaster-resistant society, the disaster-resilient society, sustainable development/sustainable hazards mitigation, invulnerable development can be employed as a basis for community interaction with earthquakes. After the Bam 2003 destructive earthquake Bamdad (2005) suggested the disaster knowledgeable society model (by involving the external and internal extents of knowledge that relies on the distribution of the knowledge among the societies in order to enhance community engagement). His model identified four post disaster waves following the Bam earthquake: physical, social, economic, political, and medical waves. He proposed that the principal concern for the Iranian society to cope with earthquakes is the public education.

4.2 Medical Aspects

Earthquake occurrence will perturb the common medical and non-medical infrastructures. So emergency supports involving: urgent treatment centers, evacuation of the injured people, supporting temporary public health, disease surveillance, supplying safe food and water … etc., will be disturbed. As the great earthquakes leave significant overload on the existing medical and health capacities so the community involvement in: search and rescue, triage and initial stabilization and definitive medical care is very important. Search and rescue assistance from neighboring societies (e.g.: neighboring cities, provinces and even countries) is essential. Triage could be managed: on-site, medical and during evacuation. Principally medical services should reach the effected area rapidly after the quake in order to be effective. Meanwhile rescue personal must be self-sufficient and should have suitable capabilities (Abolghasemi et al. 2006).

4.3 Survey and Reconstruction

Recent researches in Iran especially after the Bam 2003 earthquake (Mehrabian and Achintya 2005) show that adobe and unreinforced masonry constructions are very susceptible to earthquake shakings and could not be used in seismically active zones. The existing constructions of these kinds must be retrofitted by available technologies. As a result of inappropriate choice of material and the poor quality of workmanship even reinforced concrete and steel constructions did not perform as expected, so improved design, appropriate use of materials and proper inspection is essential. In reconstruction phase the community opinions such as: preserving the area identity in urban design, reinforcing the residences and active participation of community in different aspects of physical, environmental, social and economic issues, should be considered. Supplying construction material and financial aid to the local people are important factors for a successful reconstruction program. Educational services covering building safety techniques is required for reconstruction and mitigation programs in order to improve the overall local construction techniques. However effective construction inspection for private housing and small projects should be taken into consideration (Fallahi 2007; Manafpour 2005).

4.4 Economic Impacts

Fundamental problems of development are among the problems that contribute to the regional vulnerability and turn natural events into a human and economic disaster. The essential causes of regional vulnerability involve fast and uncontrolled urbanization, the persistence of extensive urban and rural poverty, the depreciation of the region's environment resulting from the mismanagement of natural resources, ineffective public policies, and lagging and misguided investments in infrastructures. Development and disaster-related policies have largely focused on emergency response, leaving a serious underinvestment in natural hazard prevention and mitigation. In comparison to previous decades the frequency of extreme natural disasters has increased by a factor of three and the direct economic costs have increased by a factor of nine. The per capita impact on the developing or transition countries is nearly 20 times as great as in the developed world (Iranian Studies Group at MIT 2004).

Direct and indirect economical damages left by earthquakes in Iran (Table 2) are considerable. For example crude estimates of economic loss from the 2003 Bam earthquake are about 1.5 milliard USD, including direct damage of 1.2 milliard USD and indirect damage 0.3 milliard USD. Official statements consider between 2 and 3 years for the reconstruction of the area, while the previous experiences have shown that it might take even a decade or more to complete the reconstruction (Manafpour 2005).

5 Lessons Learned and Discussions

During recent decades, the understanding of the reasons and impacts of natural disasters has improved. Models to compute the frequency and severity of disastrous earthquakes have been coupled with techniques to measure the vulnerability of capital stock to catastrophe losses. This has not only improved the probability estimates of natural disasters, but also of the potential impacts of these events. Although sophisticated modeling of both the nature and costs of natural disasters exists for developed countries, little reliable data on the risks of natural disasters has been developed for developing and transition countries (Iranian Studies Group at MIT 2004). Meanwhile, deaths due to earthquakes have been continuously decreasing relative to the population of the world, which is due to the factors like: employing better building practices, applying seismic resistant codes, early warning systems, earthquake risk awareness, non-megacity affecting earthquakes and a wealth increase. However, when compared to the global death rate from 1900 to 2012 a slight increase can be seen in earthquake deaths as a percentage of total deaths. This means that perhaps greater efforts have been made to reduce human death tolls in other fields when compared to earthquakes (Daniell et al. 2012a).

The earthquake fatalities have been primarily occurred in developing countries (e.g.: Iran) and one of the principal causes of death is buildings collapse. Many destructive quakes during recent and historical times have been located in Iran. The country seems to be powerless in considering the earthquake risk and it appears that the seismic damage has been increased in place of decrease for recent decades (Spence 2007). High death tolls and financial losses are among the main characteristics of the Iranian earthquakes. Four days before the Bam earthquake a similar magnitude event in California/USA left just a few of the killed people. Then the knowledge of protecting human life against earthquakes exists and the international society could provide this knowledge to high seismic risk countries of the world in order to minimize and mitigate the risks (Manafpour 2005). Memories of the catastrophic Buien-Zahra, Dasht-e-Bayaz, Tabas, Golbaf and Manjil-Rubar earthquakes have not been forgotten but the obvious lessons of these earthquakes are being ignored. The periodical nature of the destructive earthquakes should not impose the culture of ignorance and acceptance of such earthquakes and their consequences and preventing the responsibility of society. The recent disastrous earthquakes indicates that the most important lesson are: the fundamental earthquake hazard reduction needs to engage national consciousness at all levels of society, public education, solving the problems in the natural disaster preparedness system and the deficiencies in current construction practice.

References

Abolghasemi, H., Radfar, M., Khatami, N., SaghafiNia, N., Amid, A., & Briggs, S. M. (2006). International medical response to a natural disaster: Lessons learned from the Bam earthquake experience. *Prehospital and Disaster Medicine, 21*, 141–147.

Ahmadi, G., Mostaghel, N., Nowroozi, A. A. (1989). Earthquake risk analysis of Iran: Probabilistic seismic risk for various peak ground accelerations. *Iranian Journal of Science and Technology B*, 115–156.

Allen, M. B., Ghassemi, M. R., Shahrabi, M., & Qorashi, M. (2003). Accommodation of late Cenozoic oblique shortening in the Alborz range northern Iran. *Journal of Structural Geology, 25*, 659–672.

Allen, M., Jackson, J., & Walker, R. (2004). Late Cenozoic reorganization of the Arabia-Eurasia collision and the comparison of short-term and long-term deformation rates. *Tectonics, 23*, TC2008.

Allen, M. B., Walker, R., Jackson, J., Blanc, J. P., Talebian, M., & Ghassemi, M. (2006). Contrasting styles of convergence in the Arabia-Eurasia collision: Why escape tectonics does not occur in Iran. *Memoir Geological Society of America Special Papers, 409*, 579–589.

Ambraseys, N. N., & Jackson, J. A. (1998). Faulting associated with historical and recent earthquakes in the Eastern Mediterranean region. *Geophysical Journal International, 133*, 390–406.

Ambraseys, N. N., & Melville, C. P. (1982). *A history of Persian earthquakes*. UK: Cambridge University Press.

Amendola, A., Linnerooth-Bayer, J., Okada, N., & Shi, P. (2008). Towards integrated disaster risk management: Case studies and trends from Asia. *Natural Hazards, 44*, 163–168.

Ashtari Jafari, M. (2007a). Seismicity characteristics of Central Alborz. *Journal of the Earth, 2*, 51–64.

Ashtari Jafari, M. (2007b). Time independent seismic hazard analysis in Alborz and surrounding area. *Natural Hazards 42*, 237–252.

Ashtari Jafari, M. (2008). *Lessons learned from the Bam Urban earthquake* (pp. 12–17). Paper presented at the 14th World Conference on Earthquake Engineering, Beijing, China.

Ashtari Jafari, M. (2009). Short period fluctuations of seismicity around Tehran inferred from "a" and "b" values. *Journal of Earth Space Physics, 35*, 45–57.

Ashtari Jafari, M. (2010). Statistical prediction of the next great earthquake around Tehran, Iran. *Journal of Geodynamics, 49*, 14–18.

Ashtari Jafari, M. (2012a). Seismicity anomalies of the 2003 Bam, Iran, earthquake. *Journal of Asian Earth Sciences, 56*, 212–217.

Ashtari Jafari, M. (2012b). Teleseismic source parameters of the Rigan county earthquakes and evidence for a new earthquake fault. *Pure and Applied Geophysics, 169*, 1655–1661.

Ashtari Jafari, M. (2013a). Spatial distribution of seismicity parameters in the Persian Plateau. *Earth Planets Space, 65*, 863–869.

Ashtari Jafari, M. (2013b). Combination of double couple and non-double couple events during the Van, Turkey, 2011 earthquake sequence. *Journal of Asian Earth Sciences, 67–68*, 63–75.

Ashtari Jafari, M. (2014). The 16 April 2013 Mw7.8 Ghosht, Iran earthquake. *Journal of Asian Earth Sciences, 87*, 26–36.

Ashtari Jafari, M., Hatzfeld, D., & Kamalian, N. (2005). Microseismicity in the region of Tehran. *Tectonophysics 395*, 193–208.

Bachmanova, D. M., Trofonova, V. G., Hessami, K. T., Kozhurina, A. L., Ivanovac, T. P., Rogozhind, E. A., et al. (2004). Active faults in the Zagros and central Iran. *Tectonophysics, 380*, 221–241.

Bamdad, N. (2005). *The role of community knowledge in disaster management: The Bam earthquake lesson in Iran*. Iran: Institute of Management and Planning Studies.

Berberian, M. (1981). Active faulting and tectonics of Iran. In H. K. Gupta (Ed.), *Zagros-Hindu Kush-Himalaya Geodynamic Evolution, AGU Geodynamics Series*, pp. 33–69.

Berberian, M. (1995). *Natural hazards and the first earthquake catalogue of Iran*. IIEES-UNESCO.

Berberian, M. (1997). Seismic sources of the Transcaucasian historical earthquakes. In D. Giardini & S. Balassanian (Eds.), *Historical and prehistorical earthquakes in the Caucasus* (pp. 233–311). Dordrecht: Kluwer Academic Publishing.

Berberian, M. (2005). The 2003 Bam urban earthquake: A predictable seismotectonic pattern along the western margin of the rigid Lut block, southeast Iran. *Earthquake Spectra, 21*(S3), s35–s99.

Berberian, M., Jackson, J. A., Qorashi, M., Talebian, M., Khatib, M. M., & Priestley, K. (2000). The 1994 Sefidabeh earthquakes in eastern Iran: Blind thrusting and bedding-plane slip on a growing anticline, and active tectonics of the Sistan suture zone. *Geophysical Journal International, 142*, 283–299.

Berberian, M., & Yeats, R. S. (1999). Patterns of historical earthquake rupture in the Iranian Plateau. *Bulletin of the Seismological Society of America, 89*, 120–139.

Byrne, D. E., Sykes, L. R., & Davis, D. M. (1992). Great thrust earthquakes and aseismic slip along the plate boundary of the Makran subduction zone. *Journal Geophysical Research, 97*, 449–478.

Chu, D., & Gordon, R. G. (1998). Current plate motions across the red sea. *Geophysical Journal International, 135*, 313–328.

Daniell, J. E., Khazai, B., Wenzel, F., & Vervaeck, A. (2011). The CATDAT damaging earthquakes database. *Natural Hazards and Earth Systems Sciences, 11*, 2235–2251.

Daniell, J. E., Khazai, B., Wenzel, F., & Vervaeck, A. (2012a). *Worldwide CATDAT damaging earthquakes database in conjunction with Earthquake-report.com–presenting past and present socio-economic earthquake datapp* (pp. 24–28). Paper presented at the 15th World Conference on Earthquake Engineering, Lisbon, Portugal.

Daniell, J.E., Khazai, B., Wenzel, F., & Vervaeck, A. (2012b). *The worldwide economic impact of historic earthquakes* (pp. 24–28). Paper presented at the 15th World Conference on Earthquake Engineering, Lisbon, Portugal.

DeMets, C., Gordon, R. G., Argus, D. F., & Stein, S. (1994). Effect of recent revisions to the geomagnetic reversal time scale on estimates of current plate motions. *Geophysical Research Letters, 21*, 2191–2194.

Engdahl, E. R., Jackson, J. A., Myers, S. C., Bergman, E. A., & Priestley, K. (2006). Relocation and assessment of seismicity in the Iran region. *Geophysical Journal International, 167*, 761–778.

Fallahi, A. (2007). Lessons learned from the housing reconstruction following the Bam earthquake in Iran. *Australian Journal of Emergency Management, 22*, 26–35.

Freund, R. (1970). Rotation of strike slip faults in Sistan, southeast Iran. *Journal of Geology, 78*, 188–200.

Holzer, T. L., & Savage, J. C. (2012). Global earthquake fatalities and population. *Earthquake Spectra, 29*, 155–175.

Iranian Studies Group at MIT. (2004). Earthquake Management in Iran, A compilation of literature on earthquake Management.

Jackson, J. (1992). Partitioning of strike slip and convergent motion between Eurasia and Arabia in Eastern Turkey and the Caucasus. *Journal of Geophysical Research, 97*, 12471–12479.

Jackson, J., & McKenzie, D. (1984). Active tectonics of the Alpine-Himalayan Belt between western Turkey and Pakistan. *Geophysical Journal Royal Astronomical Society, 77*, 185–264.

Kaila, K. L., Rao, N. M., & Narain, H. (1974). Seismotectonic maps of southwest Asia region comprising eastern Turkey, Caucasus, Persian Plateau, Afghanistan and Hindukush. *Bulletin of the Seismological Society of America, 64*, 657–669.

Karakhanian, A. S., Trifonov, V. G., Philip, H., Avagyan, A., Hessami, K., Jamali, F., et al. (2004). Active faulting and natural hazards in Armenia, eastern Turkey and northwestern Iran. *Tectonophysics, 380*, 189–219.

Kopp, C., Fruehn, J., Flueh, E. R., Reichert, C., Kukowski, N., Bialas, J., & Klaeschen, D. (2000). Structure of the Makran Subduction zone from wide angle and reflection seismic data. *Tectonophysics, 329*, 171–191.

Manafpour, A. (2005). The Bam, Iran Earthquake of 26 December 2003. *Field Investigation Report*. Halcrow Group Limited.

McEntire, D. A., Fuller, C., Johnston, C. W., & Weber, R. (2002). A comparison of disaster paradigms: The search for a holistic policy guide. *Public Administration Review, 62*, 276–91.

McKenzie, D. (1972). Active tectonics of the Mediterranean region. *Geophysics Journal of Royal Astronomical Society, 30*, 109–185.

Mehrabian, A., & Achintya, H. (2005). Some lessons learned from post-earthquake damage survey of structures in Bam, Iran earthquake of 2003. *Structural Survey, 23*, 180–192.

Niazi, M., & Basford, J. R. (1968). Seismicity of Iranian Plateau and Hindukush region. *Bulletin of the Seismological Society of America, 58*, 417–426.

Nowroozi, A. A. (1976). Seismotectonic provinces of Iran. *Bulletin of the Seismological Society of America, 66*, 1249–1276.

Nowrouzi, G., Priestley, K., Ghafory-Ashtiany, M., Javan-Doloei, G., & Rham, R. (2007). Crustal velocity structure in Iranian Kopeh-Dagh, from analysis of P-waveform receiver functions. *Journal of Seismological Earthquake Engineering, 8*, 187–194.

Platt, J. P., Leggett, J. K., & Alam, S. (1988). Slip vectors and fault mechanics in the Makran accretionary wedge, southwest Pakistan. *Journal Geophysical Research, 93*, 7955–7973.

Priestley, K., Baker, C., & Jackson, J. (1994). Implications of earthquake focal mechanism data for the active tectonics of the South Caspian Basin and surrounding regions. *Geophysical Journal International, 118*, 111–141.

Reilinger, R., & Barka, A. (1997). GPS constraints on fault slip rates in the Africa-Eurasia plate collision zone: Implications for earthquake recurrence times. In D. Giardini & S. Balassanian (Eds.), *Historical and Prehistorical earthquakes in the Caucasus* (pp. 91–108). Dordrecht: Kluwer Academic Publishing.

Sengor, A. M. C., Altiner, D., Cin, A., Ustaomer, T., & Hsu, K. J. (1988). Origin and assembly of the Tethyside orogenic collage at the expense of Gondwana Land. In: Hallam (Eds.) *Gondwana and Tethys* (Vol. 37, pp. 119–181). Oxford: Oxford University Press.

Shoja-Taheri, J., & Niazi, M. (1981). Seismicity of the Iranian plateau and bordering regions. *Bulletin of the Seismological Society of America, 71*, 477–489.

Spence, R. (2007). Saving lives in earthquakes: Successes and failures in seismic protection since 1960. *Bulletin of Earthquake Engineering, 5*, 139–251.

Tierney, K., Khazai, B., Tobin, T., & Krimgold, F. (2005). Social and policy issues following the 2003 Bam Iran earthquake. *Earthquake Spectra, 21*, 513–534.

Trifonov, V. G., Karakhanian, A. S., Berberian, M., & Ivanova, T. P. (1996). Active faults of the Arabian Plate bounds, Caucasus and Middle East. *Journal of Earthquake Predict Research, 5*, 363–374.

Vernant, P., & Chery, J. (2006). Mechanical modeling of oblique convergence in the Zagros, Iran. *Geophysical Journal International, 165*, 991–1002.

Vernant, P., Nilforushan, F., Hatzfeld, D., Abassi, M., Vigney, C., Mason, F., et al. (2004). Present day crustal deformation and plate kinematics in Middle East constrained by GPS measurements in Iran and north Oman. *Geophysical Journal International, 157*, 381–398.

Wilson, A. T. (1930). Earthquakes in Persia. *Bulletin of the School of Oriental and African Studies, 6*, 103–131.

The Impact of the Great 1950 Assam Earthquake on the Frontal Regions of the Northeast Himalaya

R.K. Mrinalinee Devi and Pabon K. Bora

1 Introduction

The Himalaya not only constitutes the world's highest mountain range, but they are also the locus of the world's greatest intra-continental earthquakes (Molnar 1990). The study of contemporary and recent earthquakes represents perhaps the major contribution to the understanding of tectonic processes active in the world today (Allen 1986). The tectonically active Northeast/Eastern Himalayan mountain belt experienced two great earthquakes of 1897—Shillong (M = 8.7) and 1950—Assam (M = 8.6). One of the biggest earthquakes of the twentieth century, the great 1950 Assam earthquake, was recorded at about 7.40 pm (Indian Standard Time) on the 15th August, 1950. It caused widespread devastation throughout the Upper Assam, particularly in the frontier tribal districts of the Mishmi and Abor hills and parts of the Lakhimpur and Sibsagar districts (Poddar 1950).

Though this earthquake is known as the Great Assam Earthquake, its epicenter was located in the territory near the junction of India and China, adjacent to the Mishmi Hills of Arunachal Pradesh of India (then NEFA), in higher reaches of the Lohit river (Nandy 2001) near Rima, Tibet. It was the 10th largest earthquake of the 20th century (USGS 2011). This earthquake was caused by the convergence of two continental plates of India and Eurasia. The earthquake shock lasted for a period ranging from 4 to 8 min within the severely affected area and it was felt throughout the Eastern India. In badly affected areas, the ground cracked and fissured, water and sands spouted through the fissures, road and railway tracks were broken up and twisted. Many bridges were destroyed and river beds silted up. Immediately after the shock several tributaries of the Brahmaputra River, particularly the Subansiri, Dihang, Dibang and Tiding, were blocked by landslips caused by the violent

R.K. Mrinalinee Devi (✉) · P.K. Bora
Geoscience Division, North-East Institute of Science and Technology, Jorhat 785006, India
e-mail: mrinalineerk@rediffmail.com

S. D'Amico (ed.), *Earthquakes and Their Impact on Society*,
Springer Natural Hazards, DOI 10.1007/978-3-319-21753-6_19

shaking of the earthquake. Coarse grey sand and silt, together with water, spouted from vents during the shocks over large areas nearby the rivers (Poddar 1950).

Paleoseismological studies in the Himalaya were undertaken due to lack of recognized surface rupture associated with the great earthquakes and the quest to understand the relationship between strain accumulation and strain release associated with the largest continental thrust system. Such events pose a great threat to the population in the Himalayan and adjoining regions (Bilham et al. 2001). Earthquakes in India are caused by the release of elastic strain energy created and replenished by the stresses resulting from India's collision with Asia (Bilham and Gaur 2000). The top surface of the basement of the Indian Shield lies at a depth of about 5 km beneath the Ganga Basin (Karunakaran and Rao 1979; Sastri et al. 1971). The 1950 Assam earthquake occurred on the detachment surface between the Indian shield and Himalaya which caused landslide on enormous scale (Satyabala 2002) near the Eastern Syntaxis of the Himalaya. Satyabala (2002) has given the location of the 1950 Assam-Arunachal earthquake at 96.76°E longitudes and 28.38°N latitude with maximum intensity of X (ten). The meizoseismal area covered 99,840 km^2 and the felt area measured 2,892,000 km^2 with 1526 death (500 due to subsequent flood); 40–50 % wild lives perished in the hilly tract of the meizoseismal area; widespread collapse of buildings took place in Upper Assam. Estimated loss of the Assam Oil Company at Digboi was Rs. 11,000.00 (1950 price index) and damage to road, railways and bridges were more than Rs. 5,000,000.00 (1950 price index). 126 acres of tea garden in Upper Assam area were covered with thick silts and sand. Sadiya, a flourishing Upper Assam town, lying in between the confluences of Dibang and Lohit rivers with the Brahmaputra, was totally devastated in this great earthquake of the Eastern Syntaxis of the Northeast Himalaya. From Dibrugarh to Saikhoaghat, roads and railway lines as well as bridges were severely damaged and at some places roads were buckled and railway lines were hanging in the air (Nandy 2001). Ni and York (1978), Ben-Menahem et al. (1974) suggested from the geological and tectonic setting of the epicentral tract and using focal mechanism solution of the event that the great earthquake occurred due to right lateral shear movement.

This great earthquake, destructive in Assam and Tibet, has a calculated magnitude of 8.6 and it was the most important earthquake event since the introduction of seismological observing stations. Alterations of relief were brought about by many rock falls in the Mishmi Hills and destruction of forest areas. In the Arbor Hills 70 villages were destroyed with 156 casualties due to landslides. Dykes blocked the tributaries of the Brahmaputra; that in the Dibang valley broke without causing damage, but that at Subansiri opened after an interval of 8 days and the wave, 7 m high, submerged several villages and killed 532 persons (ISC Bulletin, The International Seismological Summary). According to an observer in Digboi, which was under intensity VIII zone, two flashes of light were seen to the north which was not due to lighting. This pre-earthquake lighting was perhaps due to stress experienced by the quartz rocks before an earthquake which may produce as much as 500 million volts at a distance of the order of a seismic wavelength, as reported by the U.S. Geological Survey (Chouhan et al. 1974).

2 Meizoseismal Area and Earthquake Effects

The meizoseismal area of an earthquake is the area of maximum damage (Bolt 2005). The meizoseismal area of the Great 1950 Assam Earthquake (after Poddar 1950) is shown (Fig. 1). From the intensity map, syntaxial region of Pasighat, Likhabali, Ledum and Sadiya area are found to be in the high intensity zone of intensity X. It has also been said that Siji/Gai and Siang River changed their river courses since the 1950 Assam Earthquake. The Eastern parts of Papumpare district, North Lakhimpur and Saikhowaghat region fall under the zone of intensity IX,

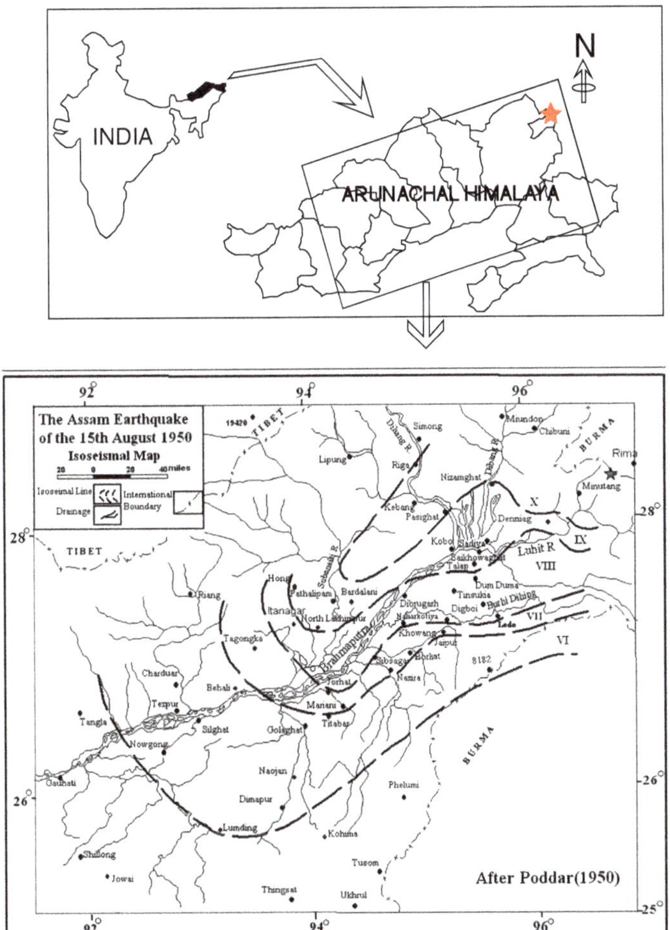

Fig. 1 The Meizoseismal map of the Great 1950 Assam Earthquake with epicentre (*red star*)

while Itanagar, capital city of Arunachal Pradesh, Dibrugarh, Tinsukia, Dumduma, Digboi falls under the zone of intensity VIII. Eyewitness accounts say that the Siji River changed its flow afterwards of this great earthquake. Geologic evidences also indicated recent shifting of the river channel. Present-day marketplace on the northern side of the Likhabali Township were observed to be the past river flow site and the river shifted its flow towards north. Active tectonic activities uplifting Quaternary unconsolidated gravelly sediments for 40 m nearby the petrol pump in Likhabali area were also observed. The sequence of uplift event is yet to be accounted after proper dating of the event. Devastation was complete in the Sadiya Township during the great Earthquake. Stream channels were filled with huge amounts of sand and silt flows. Large ripple waves of sand and silt particles were recorded in the sandy terraces of Jiya stream river channel. Once prosperous, Kundil Bazar, in Sadiya town, marked in the map before the earthquake, is not available at present. When we match the longitude-latitudes of the area with our present GPS reading, the difference in locations is quite visible. The landscape changed terribly in the region after the 1950 earthquake. Likewise, most of the buildings in North Lakhimpur were either destroyed or damaged. Fields sunk many feet below level, portions of road disappeared and communications disrupted. Shipping on the Brahmaputra River was dislocated by a tide of tens of thousands of uprooted trees and the bodies of wild animals down the river from the earthquake epicentral area while up in the mountains, the river had been dammed by landslides.

The shock of the earthquake was more damaging in the Assam region than the great Shillong earthquake (M-8.7) in terms of property loss. Widespread flooding and siltation had taken place in the downstream flows of the Subansiri and Siang River channels and their tributaries, with the blocking of the river channels in the upstream parts due to the co-seismic events of landslides and blockages of river channels. Pilots flying over the meizoseismal area reported great changes in topography which was largely due to enormous slides. Alterations of relief were brought about by many rockfalls in the Mishmi Hills and destruction of forest areas. 1526 deaths were recorded, out of which 600 were from Lahimpur and Sibsagar districts of Assam alone. In Abor Hills 70 villages were destroyed with 156 casualties due to landslides F. Kingdon-Ward (1955), a botanical explorer who was at Rima during the earthquake confirmed violent shaking, extensive slides and rise of the streams. Many rock falls altered the relief of the Mishmi Hills and forest areas were destroyed. Many tributaries of the Brahmaputra River, like the Dihang, Dihing and Subansiri were blocked by huge landslides. Dykes blocked the Dibang River, but it was broken without causing damage. The Subansiri River dried up after the shock due to damming for four days and opened only on 19th August 1950 and the wave, 7 m high, submerged several villages and killed 532 persons. Mathur (1953) concluded that at least 5×10^{10} m^3 of material was involved in the sliding. This is about 30 times of the average load of detritus carried by the river Brahmaputra annually. Some worst liquefaction damage was reported from the area where the river enters the plain.

Estimated loss of the Assam Oil Company at Digboi was Rs. 11,000.00 (1950 price index) and damage to road, railways and bridges was more than Rs. 50,00,000.00

(1950 price index). 126 acres of tea garden in Upper Assam area were covered with thick silts and sands (Nandy 2001). Watery sandy and silty materials sprouted from the vents during the shock over a large area along the bank of the Brahmaputra River. Isoseist X covered the lower reaches of the Subansiri, Siang, Dibang and Lohit rivers.

This earthquake was supposed to have caused due to slip on the Jiali and Po Chu Faults in southern Xixang, along the border with the Northeast India. The source of the Assam earthquake of August 15, 1950 is revealed from amplitude observations of surface and body waves at Pasadena, Tokyo and Bergen. And it is believed that the earthquake was caused by a motion of the Asian plate relative to the eastern flank of the Indian plate where the northeast India block has imparted a tendency of rotation with fracture lines being developed along its periphery (Ben-Menahem et al. 1974). Sapkota et al. (2013) challenged the popular belief that along the Himalayas, the source sizes and recurrence times of large seismic events are particularly uncertain, since no surface signatures were found for those that shook the range in the twentieth century. They significantly suggested the traces of the rupture zone of the *M*w 8.2 Bihar–Nepal earthquake on 15 January 1934, along at least 150 km of the Main Frontal Thrust fault in Nepal, between 85′50° and 87′20°E and from their findings it was implied that surface ruptures of other reputedly blind great Himalayan events might exist. Till now, there are no surface rupture zone findings for the great 1950 Assam Earthquake, even though it was suggested to have occurred along the Po-Chu Fault in Tibet, China (p. 161; Nandy 2001). The earthquake was followed by a large number of aftershocks, most of which were of magnitude 6.0 or greater. These were very frequent following the main shock and continued for many years. After the heavy toil, the Assam Rifles were able to open all important tracks in the Abor and Mishmi Hills, even though they themselves struggle for existence. They rendered invaluable help in alleviating the sufferings of the civil population. Kvale (1955) effectively used the term seiche to describe oscillations of lake levels in Norway and England caused by the earthquake. Widespread damage and devastations due to the great earthquake, occurred in the Indo-China border were reported from Upper Assam (Fig. 2). Huge amount of dried and seasoned tree trunks were also observed on the left bank of Siang River opposite to Ranaghat area in Pasighat till recently as an evidence of the aftermath of the great earthquake after sixty years of the devastation which carried the logs and got deposited in the point bar region of the Siang River, where it takes a rectangular turning due to the presence of Ranaghat fault which uplifted a tilted Quaternary terrace deposit. Uncountable destruction and damage were witnessed due to the earthquake and was less investigated. Older villagers and tribal dwellers had their own some kind of fearful stories running till now when asked about their experiences of the Great earthquake. We had been told by some tribal dwellers that original Pasighat town was situated north of the present location midway of the Siang River, and the river got shifted its channel after the 1950 earthquake. Lots of dating of the events, for confirmations, is needed even though those loose sandy samples were washed annually by heavy floods.

Fig. 2 **a** Epicenter of the great 1950 earthquake, 20 miles northwest of Rima, Tibet. **b** Broken RCC bridge over the Ranganadi River in 1950 Assam EQ. **c** Cracks developed at the Upper Assam Trunk road at Khowang, Assam. **d** Rail link broken during the 1950 Assam earthquake at Saikhoaghat, Assam. **e** A temple got destroyed after the 1950 Assam earthquake at Dibrugarh, Assam. **f** Active erosion started after the 1950 Assam EQ. in Majuli river island in Brahmaputra River. G. Sangester Lake formed after the Great 1950 EQ. in Tawang, Arunachal Pradesh

3 Geological Setting

The great 1950 Assam earthquake occurred in the Eastern Syntaxis (Fig. 2a) which is also known as Namche Barwa Syntaxis. The Himalayan orogenic belt is a 2000 km long, east–west trending feature arcing the collision of the Eurasian and

Indian plates. At the western and eastern ends, the belt terminates at the Nanga Parbat and Namche Barwa syntaxes, areas that are characterized by exhumation rates of up to 10 mm year^{-1} (Zeitler et al. 1993; Burg et al. 1998). The Syntaxis is characterized by a northeast-plunging antiform and is bounded by two northeast striking strike-slip shear zones: a left-slip shear zone on the western side and a right-slip shear zone on the eastern side. These strike-slip shear zones are linked by east-west trending thrusts and served either as (1) a roof thrust to a large duplex system or (2) transfer faults to a south-directed thrust system that accommodated northward indentation of a folded Indian plate (Ding, et al. 2001). From the finding of ~8 Ma $^{40}Ar/^{39}Ar$ age on hornblende from a metadiorite within the core of the antiform, they suggested that the Namche Barwa Syntaxis has been characterized by rapid cooling and exhumation since at least Late Miocene time. The southerly extension of the Namche Barwa syntaxis is known as Eastern Syntaxial Bend or Siang Antiform, a regional scale NNW-SSE trending antiformal structure (Singh 1993). In the Eastern Syntaxial Bend, the major tectonic units show a bend in their regional strike from ENE-WSW to NW-SE trend.

The Himalayan Orogen has been dominated by north-south contraction during the Cenozoic Indo-Asian collision (Le Fort 1996; Yin and Harrison 2000). However, at times, this phase of deformation has also been coeval with north-south and east-west extension (Burchfiel et al. 1992; Harrison et al. 1995; Hodge 2000). In contrast to dominantly dip-slip faulting along the Himalayan orogen, strike-slip tectonics prevails east and west of the two Himalayan syntaxes (Ding et al., 2001). Right-slip dominate along the Indo-Burman ranges and along the northern edge of the Indochina block, east of the Namche Barwa Syntaxis (Ni et al. 1989; Wang and Burchfield 1997). The overall structure of the Namche Barwa syntaxis is a large north-plunging antiform (Burg et al. 1998; Zhang et al. 2004). This crustal-scale fold is outlined by a south facing U-shaped shear zone system that consists of multiple strands of ductile thrusts and strike-slip shear zones. West of Namche Barwa, the Indus-Yalu suture separates the Cretacious-Tertiary Gangdese batholiths of the Asian plate in the north from the Tethyan Himalayan sequence of the Indian plate in the south (Ding et al. 2001). Within the Namche Barwa antiform, a granulite-bearing complex containing ultramafic fragments has been regarded as a deep crustal exposure of the Indus-Yalu Suture (Zheng and Chang 1979). Gneisses within the granulite complex are interpreted to be of Indian origin while the mafic boudins within the gneiss, contain inherited. Jurassic—Cretaceous zircons suggesting an affinity to the Gangdese arc (Burg et al. 1998; Ding et al. 2001).

Holt et al. (1991) indicated that focal mechanisms of Eastern Himalayan events show oblique thrust consistent with the N–NE directed movement of the Indian plate as it under thrusts a boundary that strikes at an oblique angle to the direction of convergence. Moreover, movement of India past southeast Asia may be causing distributed deformation within Burma and southwest China (Le Dain et al. 1984). The Eastern Syntaxis can be viewed as a complex triple junction that joins the Indian and Eurasian plates with the northern end of the Burma platelet (Curray 1989). The great 1950 Assam earthquake occurred within this region. Ben-Menahem et al. (1974)

obtained nearly pure strike-slip mechanisms from amplitude observations for this earthquake event while Chen and Molnar (1977) maintained that first-motion data can be described by a shallow north-dipping thrust much like other thrusts along the Himalayas. The aftershock distribution may well suggest a complicated event involving more than one plane, i.e., both strike-slip and shallow underthrusting (Chen and Molnar 1977). Molnar and Pandey (1989) found that nearly all the relocated 1950 aftershocks, recorded by 50 or more stations, lie east of the main shock beneath the Himalayas, in a 250-km-long by 100-km-wide zone, suggesting the 1950 Assam earthquake occurred on a gently north-northeast dipping thrust fault. However, Armijo et al. (1989) prefer the view that the 1950 Assam Earthquake involved right-lateral slip, consistent with their hypothesis that right-lateral slip on the Jiali—Po Qu fault zones in southeast Tibet wrap around the Eastern Syntaxis and connect with the right-lateral strike-slip motion on the Sagaing fault. It has been suggested that a discontinuation of the ophiolite in the region of the Eastern Syntaxis is evidence of 450 km of right-lateral displacement on faults that wrap around the syntaxis (Mitchell 1981; Armijo et al. 1989). Choudhuri et al. (2009) provided a briefly compiled geological map of the Eastern Sytaxis (Fig. 3) by using published works of Singh (1993), Kumar (1997), Gururajan and Choudhuri (2003, 2007) and Quanru et al. (2006) with field inputs of their studies on geology and structural evolution of the eastern Himalayan syntaxis.

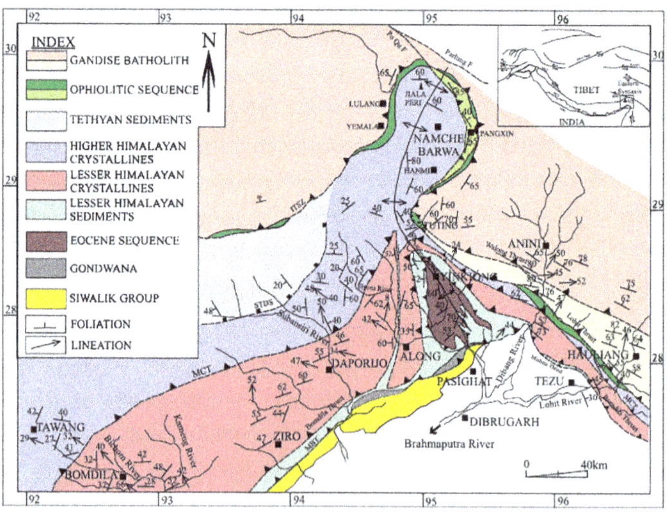

Fig. 3 Compiled geological map of Eastern Himalayan syntaxis (Choudhuri et al. 2009, compiled after Singh 1993; Kumar 1997; Gururajan and Choudhuri 2003, 2007; Quanru et al. 2006)

4 Stress Pattern and Analysis

Stress makes geologic processes happen, and geologic processes make stress. Since stress is the continuum equivalent of force, it is expected that any mechanical process in the solid or fluid parts of the Earth must involve stress (Ruff 2002). Accurate estimates of the background stress field and perturbations caused by smaller groups of events contribute to the understanding of kinematics and seismotectonics of the region. The regional stress patterns are obtained from focal mechanism solutions of the region. Focal mechanism of an earthquake depends on the tectonic stress field prevalent in the vicinity of the hypocenter. In a rock mass under stress, slip can occur on pre-existing zones of weakness with a variety of orientations relative to the principal stresses. In this case, focal mechanism solutions may differ appreciably under a single stress field. In addition, local and regional stress parameters may vary significantly (Yeh et al. 1991). Alam and Hayashi (2003), in their study of stress distribution and seismic faulting in the Nepal Himalaya with Focal mechanism solutions of earthquakes in the Himalayan region suggested the existence of thrust faults stretching along E-W with one plane dipping gently north beneath the Himalaya. Baruah et al. (2013) estimated the tectonic stress pattern for the frontal Northeast Himalaya through two different techniques namely, Michael's (1984, 1987) method and Gauss method (Zalohar and Vrabec 2007). Michael's (1984) stress tensor inversion method is derived in order to invert slickenside data for the stress field that caused the faulting episode. However, Michael's (1987) method uses the focal mechanism solutions for determination of stress field. The inversion scheme tries to minimize the differences between the slip direction, computed for stress tensor and observed slip on each plane of the focal mechanisms. The Gauss method involves the concept of the best-fit stress tensor. The compatibility function, the Gaussian function, depends on the compatibility measure, taking into account both the angular misfit between the resolved shear stress and the actual direction of movement on the fault plane, and the ratio between the normal and shear stress on the fault plane. Both the compatibility function and the compatibility measure represent a measure of correspondence between some trial stress tensor and fault datum. Stress analysis of the three major regions viz, REG 1 (Arunachal Himalaya, including the ESB), REG 2 (Mishmi Thrust Region) and REG 3 (Eastern Indo Myanmar Region), of the Northeast/Arunachal Himalaya, shows variation of stress directions (Fig. 4).

Region1 (Arunachal Himalaya, including ESB), shows N-S compression approximately perpendicular to the major thrusts i.e., MCT and MBT, indicates the continued northward compressive stress in the Himalaya due to collision tectonics between the Indian and Eurasian plates. Region 2 (Mishmi Thrust), shows NNE-SSW compression; an extension along the intermediate axis is also observed, that causes strike-slip movement which reflect slips of the Himalaya mega thrusts, due to stress build up, caused by inter-continental collision. Holt et al. (1991) and Kayal (2010) emphasized that the syntaxis zone is dominated by strike-slip faulting earthquakes. The fault plane solution of the 1950 great earthquake (Ms 8.7) also

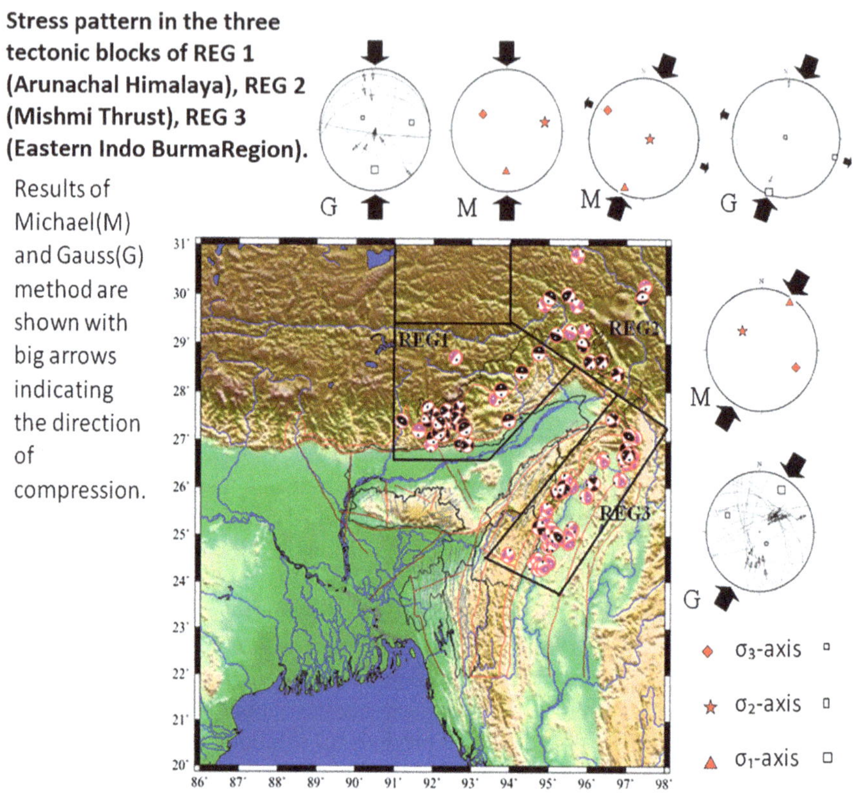

Fig. 4 Stress pattern in three tectonic blocks of Arunachal Himalaya (Reg. 1), Mishmi Thrust (Reg. 2) and Eastern Indo-Burma region (Reg. 3) (modified after Baruah et al. 2013)

shows a preferred strike-slip mechanism (Ben-Menahem et al. 1974). In Region 3 (Eastern Indo-Burma Region) also, the NE-SW compression prevails, and this may be explained by decoupling of the overriding Burmese plate over the Indian plate and might be due to oblique plate convergence. This can be explained by the northward mechanical dragging of the overriding plate along with the Indian plate (e.g. Le Dain et al. 1984).

5 Urbanization Activities Vis-á-Vis Active Tectonics in the Northeast Himalaya

The Eastern Himalaya extends in Bhutan, Sikkim, Arunachal Pradesh and Assam, in which the largest portion lies in Arunachal Pradesh, the largest state in the northeastern region of India, with its virgin forest and pristine mountains which is fast approaching towards urbanization. Developmental activities with increasing population and expanding civilization are also increasing. The demands of enlarging population with unsystematic build-up activities and constructions are making hazardous implications for the tectonically delicate region. The northeastern portion of the Indian landmass is seismically one of the most active regions of the world. Earthquakes and landslides are the major natural calamities, prevalent to this state. Landslide and erosion of land masses frequently cause lost of human lives and properties and blockage of lifeline roads. Unplanned random rock cutting and disturbance of tectonically disturbed region for urbanization, causing landslide and erosion, is on the increase in the state. So, now it would be proper for the planners to understand properly, the critical tectonic condition of the region, while taking up developmental schemes and work programmes. So that, time, money, machinery and labour can be saved by considering the structural elements of lineaments and faults, which happens to be the prime factor for land sliding. By understanding active tectonics concerning active faults, which usually gives rise to active landslides, we can always try not to disturb these areas as much as possible. Identification of Active Faults becomes the need of the hour so as to save life and property and also not to waste time and money. After the Great 1950 Earthquake, once prosperous Sadiya region had gone into oblivion and active erosion of the Kundil River had started playing havoc with the township till now (Fig. 5) by shifting its channel frequently during the recent geological time period.

The study area is located in the seismically active northeast India region. The dynamically active frontal Arunachal Himalaya is undergoing compression tectonics, as well as adjustment along the transverse neotectonic trends of N-S, NNW-SSE and NW-SE directions, resulting into continued fault activity which gave rise to heavy landslides occasionally being aggravated by the heavy rainfall climatic conditions during the summer season. Geologically, stacking of the major Himalayan thrust faults, Main Central Thrust (MCT), Main Boundary Thrust (MBT) and Himalayan Frontal Thrust (HFT) exhibits the compressive nature of the India-Eurasia plate collision zone. Stress tensor inversion results indicate that N-S compression is represented in the three regions of Northeast Himalaya.

As many landslides are triggered due to human activities, largely due to rock cutting in unsystematic and in unplanned manner, we need to observe thoroughly and take precautionary measures while taking up developmental programmes in this part of the Sub-Himalaya. Active faults play a very important role in causing landslides in the unconsolidated terrain of the Sub-Himalaya with heavy rainfall factor. Network of roads, metalled and unmetalled, are the lifeline for the far and remote areas of this emerging state. Landslides cause heavy loss to the nation and many landslides cannot be checked fully, due to its association with tectonic activities. By providing the information about the seismotectonic conditions of the areas concerned, to the planners, we can help in saving life and properties. By knowing proper ground truths, sufficient precautionary measures can be taken up beforehand.

Fig. 5 **a** Sand bed tilted 12° NW direction on the right bank of Kundil River. **b** Unpaired Terraces on the bank of Kundil R. and incised terrace due to channel shifting. **c** Overall view of Kundil R. with unpaired terraces in Sadiya region. **d** Abandoned channel of Jia stream, a tributary of Kundil river, due to channel shifting. **e** One pond of the series of ponds along the abandoned channel of the Jiya stream. **f** Collapse of houses due to the tectonically active nature of the sub-surface rocks along the left bank of Siji river, which was reported to have shifted its channel after the 1950 Earthquake

6 Conclusion

The great 1950 Assam earthquakes had inflicted catastrophic damage to life and property. The economic impact of the major earthquake that causes damage to structures, loss of life and dwellings over a wide area is enormous. A realistic seismic hazard assessment essentially requires the identification of seismic source, evaluation of maximum possible earthquake and frequency-magnitude relation for individual source zone governed by the characteristics of the capable and potential active faults. So far, less attempts has been made for identification of exact source and rupture area of the earthquake based on the ground truth of complex tectonics of the region. Paleoseismological and stress studies were undertaken due to lack of recognized surface rupture associated with the great earthquakes and the quest to understand the relationship between strain accumulation and strain release associated with the largest continental thrust system.

Acknowledgements This paper is an outcome of the project no. SR/WOS-A/ES-04/2011, sponsored by the Department of Science and Technology, Government of India, New Delhi. The authors are grateful to Dr. D. Ramaiah, Director, North East Institute of Science and Technology, Jorhat, Assam, for providing the necessary facilities, encouragement and support.

References

Alam, Md. M., & Hayashi, D. (2003). Stress distribution and seismic faulting in the Nepal Himalaya :insights from finite element modeling. *Japanese Journal of Structural Geology, 47*, 37–48.

Allen, C. R. (1986). Seismological and paleoseismological techniques of research in active tectonics. In R. E. Wallace Chairman (Ed.), *Active tectonics: Studies in geophysics* (pp. 148–154). Washington, DC: National Academy Press.

Armijo, R., Tapponnier, P., & Han, T. (1989). Late Cenozoic right-lateral strike-slip faulting in Southern Tibet. *Journal of Geophysical Research, 94*, 2787–2838.

Baruah, S., Baruah, S., & Kayal, J. R. (2013). State of stress in Northeast India and adjoining South Asia region: An appraisal. *Bulletin of the Seismological Society of America, 103*(2A), 894–910. doi:10.1785/0120110354.

Ben-Menahem, A., Aboodi, E., & Schild, R. (1974). The source of great Assam earthquake-an interplate wedge motion. *Physics of the Earth and Planetary Interiors, 9*, 265–289.

Bilham, R., & Gaur, V. K. (2000). Geodetic contributions to the study of seismotectonics in India. *Current Science 79*(9, 10), 1259–1269.

Bilham, R., Gaur, V. K., & Molnar, P. (2001). Himalayan seismic hazard. *Science, 293*(5534), 1442–1444.

Bolt, B. (2005). Earthquakes: 2006 centennial update-the 1906 big one (5th ed., p. 6). W.H. Freeman and Company. ISBN 978–0716775485.

Burchfiel, B. C., Chen, Z., Hodges, K. V., Liu, Y., Royden, L. H., Deng, C., & Xu, J. (1992). The South Tibetan detachment system, Himalayan orogen: Extension contemporaneous with and parallel to shortening in a collisional mountain belt. *Geological Society of America Special Paper, 269*, 1–41.

Burg, J.-P., Nievergelt, P., Oberli, F., Seward, D., Davy, P., Maurin, J.-C., et al. (1998). The Namche Barwa syntaxis: evidence for exhumation related to compressional crustal folding. *Journal of Asian Earth Sciences, 16*, 239–252.

Chen, W. P., & Molnar, P. (1977). Seismic moments of major earthquakes and the average rate of slip in Central Asia. *Journal of Geophysical Research, 82*, 2945–2969.

Choudhuri, B. K., Gururajan, N. S., & Singh, R. K. B. (2009). Geology and structural evolution of the Eastern Himalayan Syntaxis. *Himalayan Geology, 30*(1), 17–34.

Chouhan, R. K. S., Gaur, V. K., & Singh, J. (1974). Investigations on the aftershock sequence of the great Assam earthquake of August 15, 1950. *Annals of Geophysics*, 245–266.

Curray, J. R. (1989). The Sunda arc: A model for oblique plate convergence, Netherlands. *Journal of Sea Research, 24*, 131–140.

Ding, L., Zhong, D., Yin, A., Kapp, P., & Harrison, T. M. (2001). Cenozoic structural and metamorphic evolution of the eastern Himalayan syntaxis (Namche Barwa). *Earth and Planetary Science Letters, 192*, 423–438.

Gururajan, N. S., & Choudhuri, B. K. (2003). Geology and tectonic history of the Lohit Valley, Eastern Arunachal Pradesh, India. *Journal of Asian Earth Sciences, 21*, 731–741.

Gururajan, N. S., & Choudhuri, B. K. (2007). Geochemistry and tectonic implications of the trans-Himalayan Lohit Plutonic Complex, eastern Arunachal Pradesh. *Journal Geological Society of India, 70*, 17–33.

Harrison, T. M., Copeland, P., Kidd, W. S. F., & Lovera, O. M. (1995). Activation of the Nyainqentanghla shear zone: Implications for uplift of the southern Tibetan Plateau. *Tectonics, 14*, 658–676.

Hodges, K. V. (2000). Tectonics of the Himalaya and southern Tibet from two perspectives. *Geological Society of America Bulletin, 112*(3), 324–350.

Holt, W. E., Ni, J. F., Wallace, T. C., & Haines, A. J. (1991). The active tectonics of the eastern Himalayan Syntaxis and surrounding regions. *Journal of Geophysical Research: Solid Earth, 96*(B9), 14595–14632.

Karunakaran, C., & Rao, R. (1979). Status of hydrocarbon in the Himalayan regions. Contributions to stratigraphy and structure. *Miscellaneous Publications of Geological Survey of India, 41*, 1–66.

Kayal, J. R. (2010) Himalayan tectonic model and the great earthquakes: an appraisal. *Geomatics, Natural Hazards and Risk, 1*(1), 51–67.

Kingdon-Ward, F. (1955). Aftermath of the Assam Earthquake of 1950. *The Geographical Journal, 121*(3), 290–303.

Kumar, G. (1997). *Geology of Arunachal Pradesh* (p. 217p). Bangalore: Geological Society of India.

Kvale, A. (1955). Seismic seiches in Norway and England during the Assam earthquake of August 15 1950. *Bulletin of the Seismological Society of America, 45*, 93–113.

Le Dain, A. Y., Tapponnier, P., & Molnar, P. (1984). Active faulting and tectonics of Burma and surrounding regions. *Journal of Geophysical Research: Solid Earth, 89*(B1), 453–472.

Le Fort, P. (1996). Evolution of the Himalaya. In A. Yin & T. M. Harrison (Eds.), *The tectonic evolution of Asia* (pp. 95–109). New York: Cambridge University Press.

Mathur, L. P. (1953). Assam earthquake of 15th August 1950—A short note on factual observations. In M. Ramachandra Rao, A compilation of papers on the Assam earthquake of August 15, 1950, Central Board of Ceophysics, Government of India, pp. 56–60.

Michael, A. J. (1984). Determination of stress from slip data: Faults and folds. *Journal Geophysical Research, 89*, 11517–11526.

Michael, A. J. (1987). Stress rotation during the Coalinga aftershock sequence. *Journal Geophysical Research, 92*, 7963–7979.

Mitchell, A. H. J. (1981). Phenerozoic plate boundaried in mainland SE Asia, the Himalaya and Tibet. *Journal of the Geological Society, London, 138*, 109–122.

Molnar, P. (1990). A review of the seismicity and the rates of active underthrusting and deformation at the Himalaya. *Journal of Himalayan Geology, 1*, 131–154.

Molnar, P., & Pandey, M. R. (1989). Rupture zones of great earthquakes in the Himalaya region. *Earth and Planetary Sciences, 98*, 61–70.

Nandy, D.R. (2001). Geodynamics of northeastern India and the Adjoining region (209 p.). Kolkata: ACB Publication.

Ni, J. F., Guzman-Speziale, M., Bevis, M., Holt, W. E., Wallace, T. C., & Seager, W. R. (1989). Accretionary tectonics of Burma and the three-dimensional geometry of the Burma subduction zone. *Geology, 17*, 68–71.

Ni, J., & York, J. E. (1978). Late Cenozoic tectonics of the Tibetan Plateau. *Journal of Geophysical Research, 83*(B11), 5377–5384.

Poddar, M. C. (1950). Preliminary report of the Assam earthquake of 15th August 1950. *Bulletin of geological survey of India Series B, 2*, 1–40.
Sapkota, S. N., Bollinger, L., Klinger, Y., Tapponnier, P., Gaudemer, Y., & Tiwari, D. (2013). Primary surface ruptures of the great Himalayan earthquakes in 1934 and 1255. *Nature Geoscience, 6*, 71–76 (published online Dec.2012).
Quanru, G., Guitang, P., Zheng, L., Chen, Z., Fisher, R. D., Sun, Z., et al. (2006). The Eastern Himalayan syntaxis: Major tectonic domains, ophiolitic mélanges and geological evolution. *Journal of Asian Earth Sciences, 27*, 265–285.
Ruff, L. J. (2002). State of stress within the Earth, international handbook of earthquake and engineering seismology, part A (pp. 539). London: Academic Press.
Sastri, V. V., Bhandari, L. L., Raju, A. T. R., & Datta, A. K. (1971). Tectonic framework and subsurface stratigraphy of the Ganga basin. *Journal of Geological Society of India, 12*, 223–233.
Satyabala, S. P. (2002). Historical earthquakes of India. In W. H. K. Lee, H. Kanamori, P. C. Jennings & C. Kisslinger (Eds.), In the international handbook of earthquake and engineering seismology (Chap. 48.3, Vol. 81A). USA: Academic Press (Elsevier Science).
Singh, S. (1993). Geology and tectonics of the Eastern syntaxial bend, Arunachal Himalaya. *Journal of Himalayan Geology, 4*, 149–163.
United States Geological Survey. (2011). Largest Earthquakes in the world since 1900.
Wang, E., & Burchfiel, B. C. (1997). Interpretation of Cenozoic tectonics in the right-lateral accommodation zone between the Ailao Shan Shear Zone and the eastern Himalayan Syntaxis. *International Geology Review, 39*, 191–219.
Yeh, H–Y., Barrier, E., Lin, C. H., & Angelier, J. (1991). Stress tensor analysis in the Taiwan area from focal mechanisms of earthquakes. *Tectonophysics, 200*, 267–280.
Yin, A., & Harrison, T. M. (2000). Geologic evolution of the Himalaya-Tibetan orogen. *Annual Review of Earth and Planetary Sciences, 28*, 211–280.
Zalohar, J., & Vrabec, M. (2007). Paleostress analysis of heterogeneous fault-slip data: The Gauss method. *Journal of Structural Geology, 29*, 1798–1810.
Zeitler, P. K., Chamberlain, C. P., & Smith, H. A. (1993). Synchronous anatexis, metamorphism, and rapid denudation at Nanga Parbat (Pakistan Himalaya). *Geology, 21*, 347–350.
Zhang, P., Shen, Z., Wang, M., Gan, W., Bürgmann, R., & Molnar, P. (2004). Continuous deformation of the Tibetan Plateau from global positioning system data. *Geology, 32*, 809–812.
Zheng, X., & Chang, C. (1979). A preliminary note on the tectonic features of the lower Yalu-Tsangpo river region. *Scientia Geologica Sinica, 2*, 116–126.

Archaeoseismology in Sicily: Past Earthquakes and Effects on Ancient Society

Carla Bottari

1 Introduction

In historical and prehistoric times, strong earthquakes occurred in Italy that heavily damaged many of the buildings belonging to the historical and archaeological heritage of our country. Historical descriptions of these destructive events have come down to us, some mentioned in the historical accounts of ancient historians (e.g. Pliny the Elder), while others in epigraphs. For instance, the destruction of the Selinunte temples and the rotation of the drums of the column of Marcus Aurelius in Rome have been related to earthquakes (Boschi et al. 1995; Guidoboni et al. 2002; Bottari et al. 2009), although the passing of time tends to mask the effects of earthquakes. The nearly perfectly aligned toppled columns of 'temple C' at Selinunte (Fig. 1) suggest an earthquake as the causing event (Fig. 2), although the proposed correlation of the column orientation and the ground motion direction has not yet been verified. In some cases, the recognition of earthquake-related damage in previously excavated ruins is a challenging task. Deformed and collapsed structures observed in archaeological sites affecting man-made constructions as well as historical buildings, monuments, pavements and defensive structures may have different causes: seismic ground motion, slope process, unstable soils, differential overburden, building decay or acts of war. All of these possibilities need a systematic identification and classification of trigger mechanism on archeological damaged ruins, to prove if indeed the cause was seismic or not. Several papers have focused on classifying archeological and stratigraphic markers of earthquakes (Nikonov 1988, 1995; Stiros 1996; Galadini and Galli 2004; Hinzen 2005; Galadini et al. 2006; Rodrìguez-Pascua et al. 2011). Recently, quantitative methods have also begun to be used in archaeoseismology. These procedures are used to validate a

C. Bottari (✉)
Istituto Nazionale di Geofisica e Vulcanologia, Sezione di Geomagnetismo, Aeronomia e Geofisica Ambientale, via di Vigna Murata, 605, 00143 Rome, Italy
e-mail: carla.bottari@ingv.it

S. D'Amico (ed.), *Earthquakes and Their Impact on Society*,
Springer Natural Hazards, DOI 10.1007/978-3-319-21753-6_20

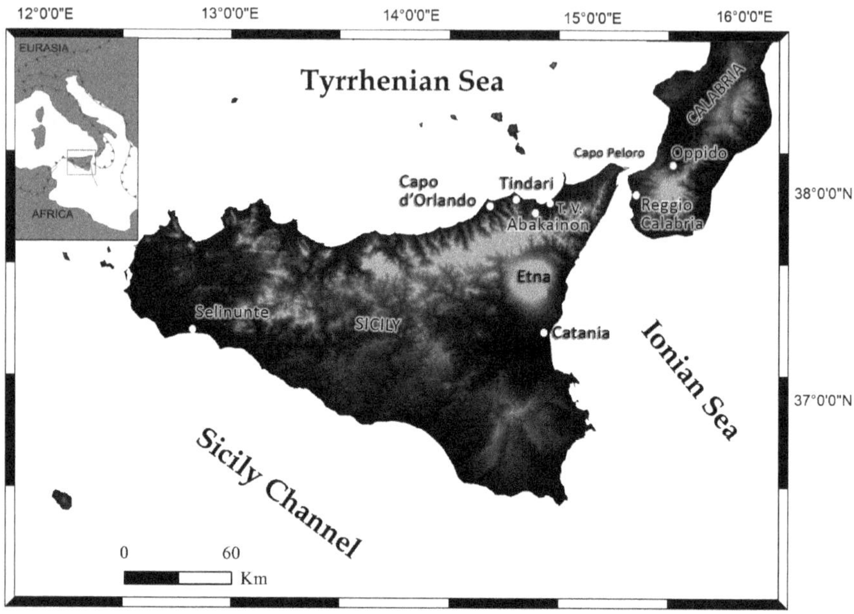

Fig. 1 Location map of the described archaeological sites in Sicily (*see inset for location of the area in the Italian peninsula*): Tindari, Abakainon, Reggio Calabria, Oppido and probably also Terme Vigliatore (TV) were affected by one or more earthquake in the 1st century AD; Selinunte was shaken by two seismic events in the 4th century BC and 4th–6th century AD; Capo d'Orlando was heavily damaged by the mid 7th century AD earthquake; Catania was damaged by the 122 BC eruption, which was probably preceded by an uncertain and poorly documented seismic event

Fig. 2 The northern colonnade of Temple C (Selinunte), which collapsed as a consequence of an earthquake occurring between the 4th and the 6th centuries AD, with the characteristic domino-style arrangement of its drums (Hulot and Fougères 1910; Bottari et al. 2009)

seismogenic hypothesis which is usually based on an archaeoseismological investigation. Advanced surveying techniques such as laser scanning and photogrammetry, finite and discrete element models of archaeological objects, geotechnical models of the subsurface of archaeological sites as well as synthetic site-specific strong motion seismograms, are all employed to prove that evidence is seismo-induced (Hinzen et al. 2011).

This short paper reviews some evidence of strong earthquakes occurring in Sicily (Fig. 1), showing how archaeoseismic data can be effectively used to improve the knowledge of past historical seismicity. Earthquakes were not simply catastrophic events, but in some cases played a catalyzing role in the evolution of urban and architectural style giving the opportunity to build new houses, palaces and churches.

2 Archaeoseismology

Archaeoseismology is a new discipline involving diverse specialists from human sciences (e.g. archaeology, history, anthropology, sociology) to geological sciences (e.g. seismology, geophysics) and engineering (e.g. architecture), in order to study past earthquakes from archaeological remains and evidence in geoarchaeological records. A review on the methodology (Galadini et al. 2006) summarized a complete archaeoseismic study in a flowchart, emphasizing quantitative models as crucial tools to validate or eliminate an earthquake hypothesis, which usually forms the basis for an archaeoseismic investigation. The main questions to be answered by archaeoseismological investigations are: (1) how probable are seismically induced ground motion or secondary effects as the cause of damage observed in ancient man-made structures? (2) when did the damaging ground motion occur? and (3) what can be deduced about the nature of the causing earthquake? When all the questions are successfully answered, archaeoseismology helps to extend the earthquake record of a region and eventually improve the hazard estimate. Nevertheless, part of the scientific community remains skeptical whether cultural and archaeological material data—such as destruction layers, structural damage to man-made constructions, displaced structures, indications of repair and abandonment and inscriptions—can reliably be used as earthquake indicators at all. Consequently, it remains outside of formal seismic hazards analysis procedure.

Sometimes archaeoseismology is used to confirm the evidence of earthquake effects reported by written sources to avoid circular reasoning (Rucker and Niemi 2010). These events in historical sources range from a treatment of how writers in the Greek and Roman worlds depicted seismic events and their impact on society (for example, what were the social and economic consequences of the earthquake of 62 AD that destroyed Naples, Pompeii and Herculaneum); to broad coverage of methods and cases studies (see Guidoboni 1989) and catalogues of ancient earthquakes (Guidoboni et al. 1994). In spite of the diversity of the approaches, there is a growing consensus in the seismological community that archaeological sources can contribute to a better understanding of earthquake history of a given area.

Archaeoseismology is often seen as the only tool that can constrain the repeat time of major earthquakes sources in areas where written records are limited and the geological expression of large faults is unknown or difficult to interpret.

3 Ancient Earthquakes in Sicily: Case Studies

In the past strong earthquakes have caused widespread damage and fatalities over large areas, with sociopolitical consequences, though this was sometimes not recognized at the time. The point is that each case has to be examined in its own historical framework.

3.1 The 1st Century AD Earthquake

The 1st century AD earthquake is one good example and seems to have had a tectonic expression with a source in southern Italy (Ferranti et al. 2008) and a tsunami that swamped the coastal areas of NE Sicily (Pantosti et al. 2008) and southern Calabria (Fig. 1). This event caused the destruction of a large part of the ancient city of Tindari in NE Sicily, and heavily damaged the necropolis of Abakainon and probably also the Roman Villa of Terme Vigliatore (Bottari et al. 2009, 2013), for which archaeologists document a major reconstruction in the first century AD (Tigano et al. 2008). The only seismic evidence still preserved in situ at the site of Abakainon is related to the oriented collapse of the columns of funerary monuments, the displacement, fractures and dipping broken corners in the basements of the tombs, and tilting of the tomb basements (Fig. 3). Tindari, instead, was entirely rebuilt and flourished again, though a later earthquake in the 4th–5th century AD ruined the city forever. On the Calabrian shore, the archaeoseismic evidence indicates that the Roman settlement near the ancient village of Oppido, was abruptly destroyed, with its collapsed walls lying directly on the road surface (Galli and Bosi 2002). This is in agreement with historical accounts (e.g. Pliny the Elder, Phlegon Trallensis) reporting that an earthquake struck Sicily and Calabria causing damage in many cities in Sicily and in the area around Reggio during the Tiberius reign (14-37 AD).

3.2 The 4th Century BC and the 4th–6th Century AD Earthquakes at Selinunte

Selinunte, in SW Sicily, was completely devastated by the 4th century BC earthquake which destroyed the temples of the western complex (e.g. temple M and Triolo temple; Fig. 4). Some remains of temple M were also found lying on the

Fig. 3 Archaeological evidence of the 1st century AD earthquake in the necropolis of Abakainon (after Bottari et al. 2013): **a** oriented collapse of the columns of the funerary monument; **b** example of a broken stele, the top is still in situ; **c** displacement, fractures and dipping broken corners in the basements of the tombs; **d** tilting of the tomb basement; **e** fractures in the tomb basement due to the impact of fallen columns and steles

smooth slope behind the western edge of the building. These blocks belonged to the upper triangular part of the western wall, which on falling preserved their original arrangement. This part of the wall is not parallel to its initial position (i.e., to the western foundations) but has rotated about 30° along a vertical axis. The style of collapse of the temple (rotation and absence of deformations in the foundations) led us to assume that a dynamic effect was responsible for at least one phase of destruction of this building (Bottari et al. 2009 and reference therein). At approximately 400 m from temple M, below a thick layer of sand, the remains of the Triolo temple were uncovered during the excavations (Tusa et al. 1984, 1986). The northern wall collapsed outside the building parallel to the foundation wall, while the blocks of the southern wall fell inside the building. The style of collapse leaves

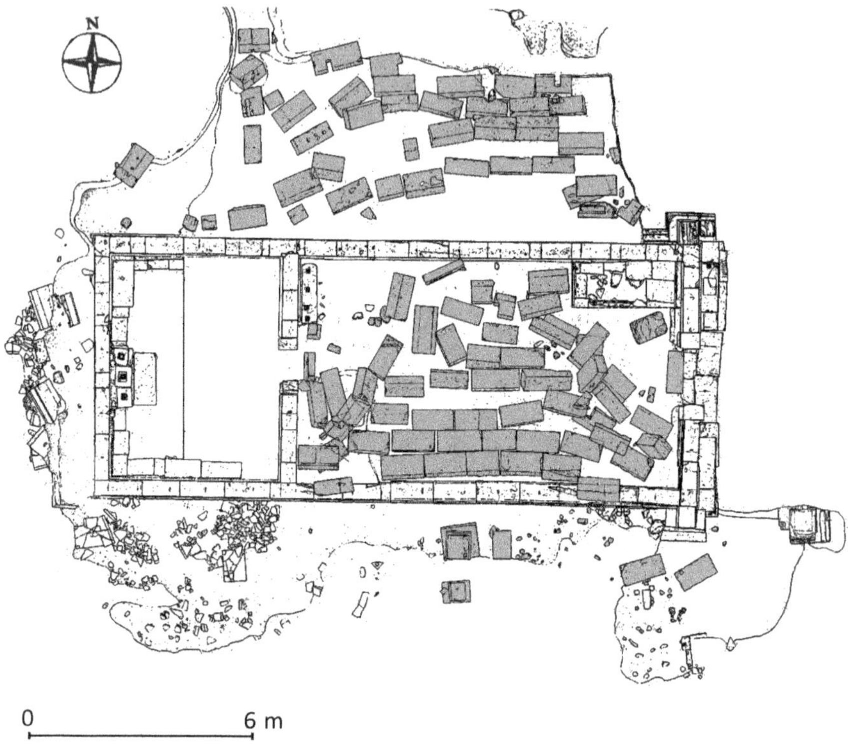

Fig. 4 Selinunte. Plan of the Triolo building showing the collapsed blocks of the southern and northern walls. Its destruction was caused by the 4th century BC earthquake (ca. 390–300 BC)

little doubt that the destruction was due to an earthquake. Indeed, the overturning of both walls in the same direction is typical of earthquake damage produced by a dynamic oscillation (Fig. 4).

The wealthy commercial Greek and post-Greek city of Selinunte was hit not only by the 4th century BC event but also by a major event in the 4th–6th century AD, destroying the large temples of the Acropolis (e.g. Temple C). Archaeologists were surprised to observe that the superstructure seemed to have tilted from north to south; the roof collapsed, followed by the columns which fell parallel to each other (Hulot and Fougères 1910). At the end of 19th century, systematic archaeological excavations uncovered the collapsed remains of Temple C, with all the columns having fallen in a domino-style arrangement. Two main directions of collapse of the columns were identified: the northern and southern colonnade were found lying on the ground in a N-S direction, with their drums in a domino-style arrangement (Fig. 2; Bottari et al. 2009 and reference therein), while the eastern colonnade fell in an ESE direction and the western colonnade collapsed in a different direction from all the others, in a non-systematic direction. A few columns of the northern side fell on top of a Roman house built on the steps of the temple (Fig. 5). Shortly after this

Fig. 5 Selinunte. Detail of the Roman house built on the steps of temple C. The letters A, B, and C represent the architectonical elements of the second, third, and fourth columns which fell onto the Roman house (D) during the 4th–6th AD earthquake (Bottari et al. 2009)

discovery, the northern colonnade of the temple was re-erected, mostly using the fallen blocks while the other three sides were left in their collapsed position.

However, the interest not only lies in the mechanism of destruction but also in the social implication or response to the calamities. For temples M and Triolo, there was no attempt at reconstruction after the 4th century BC event: most of the collapsed stone blocks were reused for other purposes. The same occurred for the late 4th–6th century AD earthquake, a much stronger event compared to the previous one, considering the dimension and height of the temple C and the large volume to fall toward the north and east. Selinunte retained its prosperity but the wealthy class evidently decided not to fund the rebuilding of major civic buildings. The collapse of the temples and the pagan cult centres of the city hastened the end of the late antique city, a factor that led to the transformation of Selinunte into a Roman city after the 4th century BC event and to a Byzantine city after the late Roman seismic event (4th–6th century AD).

3.3 Is the Middle 7th AD Earthquake a False Earthquake?

Despite late medieval sources documenting an earthquake around the middle of the 7th century AD, there is no general consensus between ancient and modern authors about the event. It was reported by Zonara, a scholar of the 11th–12th century AD and later also by Bardi, 16th century AD (see Guidoboni 1989 and reference

Fig. 6 Capo d'Orlando. *Photo* of room 4 of the Roman villa, showing the cracks across the floor in an E–W direction and in the internal walls of the room (after Bottari et al. 2008)

therein). Bardi, based on Zonara's account, dates the seismic event to 659 AD and locates it in Sicily. A recent re-examination of the historical sources led Guidoboni et al. (1989) to interpret it as a "false earthquake". The following questions thus arise: (1) did the earthquake really occur? (2) when did it occur? (3) what type of damage did it produce?

A detailed archeoseismic investigation (Bottari et al. 2008) identifies evidence of an earthquake among the ancient ruins of the Roman Villa of Bagnoli at Capo d'Orlando (NE Sicily). The seismic event produced extensive damage to the ancient building towards the sea. In detail, the complex suffered the collapse of the bath, the tilting of parallel dry masonry walls in the same direction and the cracking of the bath floor (Fig. 6). The high degree of destruction indicates a strong shaking caused its destruction shortly before 644-647 AD. Could the seismic event documented at the site of Capo d'Orlando be the same one reported by historical sources or were there perhaps two seismic events in the middle of the 7th century AD? The archaeoseismic investigation has not identified other archeological sites (apart from the Roman Villa) damaged by an earthquake in this period. It may therefore be deduced that the event reported by historical sources is the same as the one documented at Capo d'Orlando.

3.4 *Catania- the Case of the 122 BC Earthquake*

Historical sources report a Plinian eruption in 122 BC at Etna, an active basaltic volcano. This eruption was characterized by bombs and lapilli fallout on the south-eastern flank of the volcano down to Catania. It caused fires, widespread roof collapse and obscured the sun behind an ash cloud for days. The damage was so severe that the inhabitants were exempted from paying taxes to Rome for ten years. This eruption in 122 BC produced a Plinian column which is typical of explosive volcanoes. Coltelli et al. (1998) suggested that during the emplacement of the 122 BC magma within the volcanic pile, a nonelastic deformation due to the powerful buoyancy of this >1 km^3 large magma body produced a major permanent displacement of the eastern upper flank of the volcano, as was the case of both 1986 and 2002 eruptions when an earthquake swarm preceded/accompanied these eruptions. 122 BC earthquake, not documented in the historical sources, is instead reported by Mercalli (1883) who, based on a Diodoro Siculo letter (see Mercalli 1883), describes the earthquake effects at Catania. The author reports that the fortified walls collapsed, the Cibele temple was ruined, the Cerere temple was damaged in the upper part, and the ancient harbor of Catania was covered by the 122 BC lava. An accurate re-examination of Diodoro text by Pietrasanta (2005) reveals that the description of earthquake damage is misleading. The coeval sources (*Orosius* 5, 13, 3; *Augustinus* 3, 31; *Cicero* 2, 96) described the eruption in detail but failed to mention any earthquake. It is possible that the historians did not lend much importance to an earthquake that probably preceded the large Plinian eruption and decided not to document it. However, archaeological reports note the collapse of walls and roofs of ancient buildings buried by a thick layer of ash (Branciforti 2010), but it is not clear if the collapse is related to the ash weight or to earthquake shaking. A future archaeoseismic investigation could help solve the problem of the eruption mechanism and also document the seismic effects on the city.

3.5 Historical Data Versus Archaeological Data

The interpretation of historical sources is often troubling, and the cases reported in the last paragraphs offer two different examples. In the first, the historical text provides information that can only be validated with archaeological data; the second instead concerns a potential earthquake that is missing in the historical source. Both disciplines (history and archeology) should always be integrated in order to confirm or not data and avoid false or missing earthquakes. Earthquakes may be represented by one or more testimony such as historical texts, epigraphs, damaged archaeological structures and fault ruptures. Clearly, if evidence for the earthquake is found in all these then the case becomes much stronger. The correlation among data should be real and not contrived to avoid circular reasoning.

4 Earthquake Impact on the Ancient Society

Earthquakes have often been associated with ancient societal disasters and their impact has frequently been considered secondary for two reasons: erroneous archaeological interpretation of excavated destruction layers and misconceptions about patterns of seismicity.

The indiscriminate use of earthquakes as the 'deus ex machina' to explain otherwise inexplicable evidence of abandon and destruction was common practice in the past. Often historians and archaeologists added drama to a site history by using earthquakes as the only possible explanation for the observed destruction. Such interpretations led to multiple seismic events or gave rise to universal catastrophic earthquakes such as the 365 AD earthquake.

However, a better understanding of the irregularities of the time–space patterns of large earthquakes suggests that they were probably responsible for some of the great and enigmatic catastrophes in ancient times. The most relevant aspect of seismicity is the episodic time-space clustering of earthquakes occurring in the 4th century AD in the eastern Mediterranean area. The archaeological sites of Crete, Cyprus, Libya, Egypt and southern Italy were all affected by earthquakes around 365 AD (Table 1). According to Di Vita (1990), a 'universal' earthquake swept across all the Mediterranean coasts from Algeria to Syria in AD 365; other authors propose a seismic sequence of disasters which occurred between AD 361 and 450 (Guidoboni et al. 1994) or limit the effects of the 365 earthquake to Crete and the Nile Delta (Jacques and Bousquet 1984). The lack of consensus among the scholars is mainly due to the decline of Roman Empire and a consequent period of struggle between the 4th and the 5th centuries AD. Therefore, the historical information for this dark period is not reliable.

The clustering of earthquakes, probably triggered one after the other, was interpreted by historians as a 'universal' catastrophe that affected the entire Mediterranean area "…an earthquake occurred throughout the world, the sea swamped the coastline and destroyed countless nations and cities of Sicily and of many [other] islands- Jerome's continuation of the Chronicles of Eusebius, referring to the second year of the reign of Valentian in 365 or 366 AD (see also Jensen 1985; Guidoboni et al. 1994; Stiros 2001). By contrast, the 365 AD earthquakes with different epicentral localizations caused major destruction in the Mediterranean area, and hence correspond to events of different scales (Fig. 7).

4.1 A Social Perspective

Although archaeoseismology benefits from collaboration between earthquake geologists, historians and archaeologists decoding the earthquake parameters etc. a larger role should be given to the impact of earthquakes on ancient society within the cultural history of a site. A better understanding of the complex dynamics by

Table 1 Ancient sites damaged by earthquakes around AD 365

Region/site	*Dating of earthquake*	*Sources*
Crete		
Kisamos	Shortly after AD 355-361	Stiros and Papgeorgiou (2001)
Gorthyn	10–15 years before AD 383	Stiros (2001)
Eleuterna	Shortly after AD 351-361	Themelis (1988)
Cyprus		
Paphos	Between AD 364 and 365	Jensen (1985)
Kourion	Between AD 364 and 365	Soren and Davis (1985), Soren and Leonard (1989)
Libya		
Ptoleimas	Shortly before AD 364-365	Pesce (1950), Bacchielli (1995)
Balagrae	Shortly after AD 364	Bacchielli (1995)
Cyrene	Shortly after AD 361	Bacchielli (1995)
Leptis Magna	Shortly after AD 364-367	Di Vita (1995)
Sabratha	Shortly after AD 364-367 and shortly before AD 378	Di Vita (1995), Lepelley (1984)
Italy		
Selinunte	After AD 330	Bottari et al. (2009)
Agrigento	Shortly after AD 364-367	Bottari et al. (2009)
Reggio Calabria	10 years before AD 374	Bottari et al. (2009)
Egypt		
Alexandria	July 21st, AD 365	Jacques and Bousquet (1984)

Dating of earthquakes is based on numismatic and historical data as well as epigraphic evidence
After Drakos and Stiros (2001), Bottari et al. (2009), Stiros (2010)

which ancient cultures dealt with and responded to damaging earthquakes (e.g. antiseismic measures), could shed light on the resilience of past societies and their relative capacity to withstand seismic events.

The question whether ancient societies were aware of antiseismic techniques is matter of debate between specialists. According to Kirikov (1992), ancient builders did not consider the earthquake resistance of megalithic monuments. Stiros (1995) argued this debate on the existence or lack of antiseismic construction is due to the inaccuracy and fragmentary knowledge of building style and its evolution over the centuries, as well as the manner of building by ancient architects and engineers. In the Greek and Roman world it is difficult to say what they did or did not know. However, we cannot exclude the possibility that in some cases the empirical methods of ancient builders, who adopted erroneous procedures due to the lack of mathematical calculations, produced at least some positive results in relation to earthquakes (e.g. the buildings did not collapse).

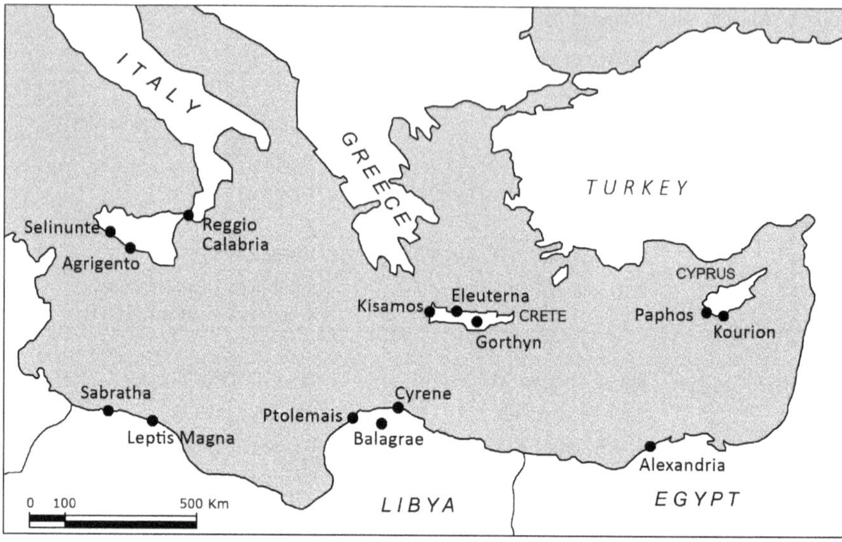

Fig. 7 Map of the Central-Eastern Mediterranean Sea with the location of the cities damaged by earthquakes around 4th century AD

5 Conclusive Remarks

Archaeoseismology may be considered a new tool for analyzing past earthquakes, by integrating, checking, validating or rejecting the descriptions of the historical sources. In other cases, archaeoseismology may reveal the occurrence of vaguely reported or uncatalogued ancient earthquakes. Ancient societies lived with earthquakes, and the relationship between historical, social development and changes due to natural seismic disasters also falls within the focus of archaeoseismology. Whatever the case, this growing discipline is opening the opportunity to view already excavated archaeological remains as ancient "seismoscopes". This will allow testing site specific earthquake effects by cataloguing and parameterizing archaeological earthquake evidence, contributing greatly to a better characterization of poorly known ancient earthquakes by producing a more realistic definition of earthquake scenarios. Integrating the historical seismic catalogues with archaeoseismic data can play a key role in establishing a local seismic culture of a city/region/country, on one hand maximizing earthquake awareness and on the other hand minimizing the seismic risks of an area.

References

Augustinus of Hippo (5th century AD) 1928–1929. In B. Dombart & A. Kalb (Ed.), *De civitate Dei, Leipzig*.

Bacchielli, L. (1995). A Cyrenaica earthquake post 364 AD: Written sources and archeological evidence. *Annals of Geophysics-Italy, 38*(5–6), 977–982.

Boschi, E., Caserta, A., Conti, G., Di Bona, M., Funicello, R., Malagnini, L., et al. (1995). Resonance of subsurface sediments: an unforeseen complication for designers of roman columns. *Bulletin of the Seismological Society of America, 85*, 320–324.

Bottari, C., Barbano, M. S., Pirrotta, C., Azzaro, R., Ristuccia, G., & Gueli, A. (2013). Archaeological evidence for a possible first century AD earthquake in the necropolis of Abakainon (NE Sicily). *Quaternary International, 316*, 190–199. doi:10.1016/j.quaint.2013.10.004.

Bottari, C., Bottari, A., Carveni, P., Saccà, C., Spigo, U., & Teramo, A. (2008). Evidence of seismic deformation of the paved floor of the decumanus at Tindari (NE, Sicily). *Geophysical Journal International, 174*, 213–222.

Bottari, C., Stiros, S. C., & Teramo, A. (2009). Archaeological evidence for destructive earthquakes in sicily between 400 BC and AD 600. *Geoarchaeology, 24*, 147–175.

Branciforti, M. G. (2010). Da Katane a Catina. In M. G. Branciforti & V. La Rosa (Eds.), *Tra lava e mare, contributi all'archaiologhia di Catania* (pp. 135–258). Catania: Le Nove Muse.

Cicero (1st century BC) 1955–1958. In A. S. Pease (Ed.), *De natura deorum*. Cambridge (Mass).

Coltelli, M., Del Carlo, P., & Vezzoli, L. (1998). Discovery of a Plinian basaltic eruption of Roman age at Etna volcano, Italy. *Geology, 26*(12), 1095–1098.

Di Vita, A. (1995). Archaeologists and earthquakes: The case of the 365 AD. *Annals Geophysics-Italy, 38*, 971–976.

Di Vita, A. (1990). Sismi, urbanistica e cronologia assoluta. Terremoti e urbanistica nella città di Tripolitania tra il I secolo a.C. ed il IV d.C: in L'Afrique dans l'Occident romain (Ier siècle av. J.-C. - IVe siècle ap. J.-C.) (pp. 426–494). Rome: École Française de Rome.

Drakos, A., & Stiros, S. (2001). The AD 365 earthquake. From legend to modelling. *Bulletin of the Geological Society of Greece, 24* (5), 1417–1424.

Ferranti, L., Monaco, C., Dorelli, D., Antonioli, F., & Maschio, L. (2008). Holocene activity of the scilla fault, Southern Calabria: Insights from coastal morphological and structural investigations. *Tectonophysics, 453*, 74–93.

Galadini, F., & Galli, P. (2004). The 346 AD earthquake (central-southern Italy): An archaeoseismological approach. *Annals Geophysics-Italy, 47*, 885–905.

Galadini, F., Hinzen, K. G., & Stiros, S. C. (2006). Archaeoseismology: Methodological issues and procedure. *Journal of Seismology, 10*, 395–414.

Galli, P., & Bosi, V. (2002). Paleoseismology along the Cittanova fault: Implications for seismotectonics and earthquake recurrence in Calabria (southern Italy). *Journal Geophysical Research, 107*, 1–19.

Guidoboni, E. (1989). *I terremoti prima del Mille in Italia e nell'area mediterranea*. Storia, Archeologia, Sismologia: INGV-SGA, Bologna.

Guidoboni, E., Comastri, A., & Traina, G. (1994). *Catalogue of ancient earthquakes in the Mediterranean area up to 10th century*. Roma: ING.

Guidoboni, E., Muggia, A., Marconi, C., & Boschi, E. (2002). A case study in archaeoseismology. The collapses of the Selinunte Temples (Southwestern Sicily): Two earthquakes identified. *Bulletin of the Seismological Society of America, 92*, 2961–2982.

Hinzen, K. G. (2005). The use of engineering seismological models to interprete archaeoseismological findings in Toblacium, Germany: A case study. *Bulletin of the Seismological Society of America, 95*, 521–539.

Hinzen, K. G., Fleischer, C., Reamer, S. K., Schreiber, S., Scütte, S., & Yerli, B. (2011). Quantitative methods in archaeoseismology. *Quaternary International, 241*(1), 31–41.

Hulot, J., Fougères, G. (1910). *Sélinonte*. La Ville, l'acropole et les temples. Paris.

Jacques, F., & Bousquet, B. (1984). Le raz de marée du 21 juillet 365—Du cataclysme local à la catastrophe cosmique. *MEFRA, 96*(1), 423–461.

Jensen, R. (1985). *The Kourion earthquake: Some possible literary evidence* (pp. 307–311). Cyprus: Report of the Department of Antiquities.

Kirikov, B. (1992). *Earthquake resistance of structures: From antiquity to our time*. Moscow: MIR Publisher. 240.

Lepelley, C. (1984). L'Afrique du Nord et le prétendu séisme universel du 21 juillet 365. *MEFRA, 96*(1), 463–491.

Mercalli, G. (1883). *Vulcani e fenomeni vulcanici in Italia*. Forni.

Nikonov, A. A. (1988). On the methodology of archaeoseismic research into historical monuments. In P. Marnos & G. Koukis (Eds.), *The engineering geology of ancient works, monuments and historical sites: preservation and protection* (pp. 1315–1320). Rotterdarm: Balkema.

Nikonov, A. A. (1995). The stratigraphic method in the study of large earthquakes. *Quaternary International, 25*, 47–55.

Orosius (4th–5th century AD) 1976. In A. Lippold (Ed.), *Historiarum adversus paganos libri septem*, Milano.

Pantosti, D., Barbano, M. S., Smedile, A., De Martini, P. M., & Tigano, G. (2008). Geological Evidence of Paleotsunamis at Torre degli Inglesi (northeast Sicily). *Geophysical Research Letters, 35*, L05311. doi:10.1029/2007GL032935.

Pesce, G. (1950). Il "*palazzo delle colonne*" in Tolemaide di Cirenaica. L'Erma di Bretschneider Roma.

Pietrasanta, D. (2005). *Le epistole di Diodoro Siciliano* (p. 166). Laruffa: Catania.

Pliny the Elder (1st century) Natural History, Book 2 (Chapter 94). http://penelope.uchicago.edu/Thayer/L/Roman/Texts/Pliny_the_Elder/2%2A.html.

Rodríguez-Pascua, M. A., Pérez-López, R., Giner-Robles, J. L., Silva, P. G., Garduño-Monroy, V. H., & Reicherter, K. (2011). A comprehensive classification of earthquake archaeological effects (EAE) in archaeoseismology: Application to ancient remains of Roman and Mesoamerican culture. *Quaternary International, 242*, 20–30.

Rucker, J. D., & Niemi, T. M. (2010). Historical earthquake catalogues and archaeological data: Achieving synthesis without circular reasoning. In M. Sintubin, I. S. Stewart, T. N, Niemi & E. Altunel (Eds.), *Ancient Earthquakes, Geological Society of America Special Paper* (471, 97–106). Boulder: Geological Society of America.

Soren, D., & Davis, T. (1985). Seismic archaeology at Kourion: The 1984 campaign, Report of the Department of Antiquities (pp. 293–301), Cyprus, pl. LVI.

Soren, D., Leonard, J. R. (1989). Archeologia sismica a Kourion: un approccio multidisciplinare in azione per un terremoto del IV secolo d.C. In E. Guidoboni (Ed.), *I terremoti prima del Mille in Italia e nell'area mediterranea*. Storia, Archeologia, Sismologia. SGA, Bologna.

Stiros, S. C. (1995). Archaeological evidence of antiseismic constructions in antiquity. *Ann Geophysics-Italy, 38*, 725–736.

Stiros, S. C., & Papageorgiou, S. (2001). Seismicity of Western Crete and the destruction of the town of Kisamos at AD 365: Archaeological evidence. *Journal of Seismology, 5*, 381–397.

Stiros, S. C. (1996). Identification of earthquakes from archaeological data: methodology, criteria and limitations. In S. Stiros & R. Jones (Eds.), *Archaeoseismology, British School at Athens, Fitch Laboratory Occasional Paper* (Vol. 7, pp. 129–152). Amsterdam: Elsevier.

Stiros, S. C. (2001). The AD 365 Crete earthquake and possible clustering during the fourth to sixth centuries in the Eastern Mediterranean: A review of historical and archaeological data. *Journal of Structural Geology, 23*, 545–562.

Stiros, S. C. (2010). The 8.5+magnitude, AD365 earthquake in Crete: coastal uplift, topography changes, archaeological an historical signature. *Quaternary International, 216*, 54–63.

Themelis, P. (1988). *Eleutherna. Kretiki Estia, 2*, 298–302. (in Greek).

Tigano, G., Borrello, L., & Lionetti, A. L. (2008). Terme Vigliatore-S. Biagio: Villa romana, Palermo: Regione siciliana, Assessorato dei beni culturali, ambientali e della pubblica istruzione, 44 pp.

Tusa, V., Ferruzza, L., Fanfara, G., Parisi Presicce, C., De Wailly, M., Delh, C., et al. (1986). Selinunte–Malophoros: Rapporto preliminare sulla II campagna di scavi. *Sicilia Archeologica, 19*, 13–88.

Tusa, V., De Wailly, M., Gregari, B., Parisi Presicce, C., Valente, I., Pacci, M., et al. (1984). Selinunte–Malophoros: Rapporto preliminare della prima campagna di scavi—1982. *Sicilia Archeologica, 17*, 17–58.

The Earthquakes of Southern Italy from the 18th to the 20th Centuries

Andrea Catalani

1 The Earthquake in the Italian Environmental Historiography

In May 1906, in the «Bollettino della Reale Società Geografica Italiana», the geographer and father of historical seismology, Mario Baratta, when commenting the nth seismic event in Calabria in 1905, said:

> [In Calabria] man – who lives in strange, unspeakable isolation – is in a continuous war atmosphere. He is forced to fight against the lie and nature of land, which obstruct communications, the exchange and trade with other regions and among the various towns and villages high up in the mountains [...] he fights the floods, which, left to run, flow down hurriedly from the crags because of the *wild deforestation*. They drag in their floods large amounts of branches and wood devastating and destroying the cultivated lands together with the settlements. Man must fight *malaria*, which depopulates wide and fertile territories, and not only turns away men from their daily work, but also takes away easy and suitable shelters. It inoculates the fatal germ, which wears out life and opens up the door to the tomb prematurely. He is forced to fight drought, which lashes the farmer's daily efforts; [he fights] against the *earthquake* which, violently and frequently, takes a heavy toll [...]; he must always be ready to fight the most dirty usury, which dries up the fruit of the harshest work; and let's say that, he must face powerful and domineering people who want to trample on his rights and want him poor, illiterate and submissive (Baratta 1906, p. 434).

The earthquake was identified by this geographer as one of the environmental factors that had influenced the life of the Calabrian communities throughout history. A well-known politician and expert on issues concerning the south of Italy Francesco Saverio Nitti adopted these remarks the following years. In fact he

A. Catalani (✉)
Roma, Italy
e-mail: andrea_catalani@virgilio.it

S. D'Amico (ed.), *Earthquakes and Their Impact on Society*,
Springer Natural Hazards, DOI 10.1007/978-3-319-21753-6_21

included the seismic events among the most serious modifying and "delaying" causes of development in Basilicata and Calabria together with migration, malaria, and hydrogeological instability; the latter due to deforestation (Nitti 1919).

The above observations originated from a deep awareness of those environmental and social phenomena that did not characterize only Calabria and Basilicata, but they could also be associated with the whole area of the South of Italy. Malaria, migration, deforestation and the imbalance between populations and resources, to varying degrees, had characterised the history of Southern Italy in the nineteenth century and before. Certainly, earthquakes had played an important role with their negative consequences on the demography and in general on socio-economic aspects and the environment.

We will examine here the historical period between the beginning of the 18th and the 20th centuries. In this period the high density of population in Southern Italy, the characteristics of the territory (mainly mountainous with malarial planes along the coast), the economic system based on the land, had pushed the southern communities to adapt to their territory, finding forms of appropriate use and exploitation of the land itself. Even the building industry had adapted to the territory (Tino 2007). However, these people had also learnt how to cope with the worst effects of earthquakes. In fact, we have to consider that, between 1501 and 1929 there were about 481 devastating earthquakes in the Mediterranean basin, of which 188 occurred in Italy (Bevilacqua 1996, p. 74). Most of them occurred between the 18th and 20th centuries, at least this is the period from which we have most of the documentation. Even if this figure has probably been rounded down, what is most striking, is that they were more frequent in the South of Italy. In 1915, Baratta stated that there had been 17 strong earthquakes during the 19th century, meaning those, which had caused the death of more than a hundred people each. Over the century, there were 23,687 victims altogether, which is an average of 237 deaths per year. Furthermore Baratta added that "the provinces in the South had been, unfortunately, the worst hit and among these Calabria Citra had been the most frequently, if not the most violently hit by the earth tremors" (Baratta 1915, pp. 4–5).

Nevertheless, it is not possible to understand the importance of earthquakes in Italy completely and in particular in the Southern regions, if only we consider them as natural phenomena. Earthquakes have surely contributed to the physical alteration of the landscape, just like other natural forces (such as wind, rain and changes in temperature) together with human activities. However, over the centuries the earthquake has been marked as a dynamic element of the "cultural landscape" (Guidoboni 2005). This, because all catastrophic events, therefore even earthquakes, mark the life of communities and become part of their historical memory. Communities build settlements, organize their economy and establish their social relationships upon which these disastrous events then take effects.

What is surprising about earthquakes is that, in spite of their relevance to people's lives, especially in the South, they have been ignored by Italian historiography. They are not part of the "Great History"; therefore, we have partly lost memory of them even though Italy is a seismic country, with 23 million people living in areas at high risk.[1]

There are many reasons for this scarce attention to natural disasters, earthquakes included, with all their tragic consequences: one of the most important is a certain "progressive view" of the past, for which events are studied according to growth and progress, conceiving progress as "slow, but safe and straight-line progress" (Bevilacqua 1996, pp. 73–91). This historical approach has confined earthquakes to the rank of simple physical phenomena, therefore an issue that only concerns natural sciences, in spite of their relevant effect on people and society. Yet the seismic event has occurred with its own specific characteristics, which make it distinct from other disasters in Europe during the *ancien régime*. Its destructive power, in fact, is not limited to demographic depletion. Epidemics and famine were also part of those mechanisms of the *forbice maltusiana* (*Malthusian scissor*) (Macry 1995, pp. 77–103), which contributed to "decimate part of the active population", to restore the balance between resources and people which tended to occur after a period of excessive in demographic growth. After a natural disaster, in fact, the depleted communities not only regained their possessions, but, in many cases, they could enjoy a surplus of resources. Instead, the consequences of an earthquake were more similar to those of a war; indeed a particularly violent earthquake added the destruction of goods and properties to the death of people. In fact, in addition to the collapse of buildings and the heavy damage to lifelines, a very destructive earthquake could add famine and epidemics, spoiling the agricultural production of the year and the food supply stock (wheat, oil and wine) of the community. In the economic system of the ancient regime, which was closely tied to the natural rhythm of the land, with quite slow growth, this implied a paralysis of production. There was something more. As stated recently by Piero Bevilacqua, "Beyond men, houses, goods and properties, it was the *time* itself to be hit. In fact, earthquakes often swept away the technical instruments that mediated the relationship between men and nature. Earthquakes took the communities back to semi-primitive conditions, forcing them to retrace outdated steps from previous generations" (Bevilacqua 1996, pp. 76–77).

In spite of the enormous influence of earthquakes on the destabilization of society and economy, historical research has tended to ignore seismic events, except in the last two decades, when there have been studies about this issue. The reasons for such an attitude are the same that characterised "the neglected study of the territory in Italian historiography until not long ago". This extended blindness depends on the leading cultures, which, mainly in the 20th century, were philosophical-literary ones, together with political ideologies. On the one hand, they have contributed to link the history of Italy to the history of Europe; on the other hand, they have left in the shadows, or nearly removed, from the national scene,

[1]It is important to notice that according to the census of the Italian population, which was taken in 2001, the population was 57 million; therefore, the ones living in the territory at high seismicity (23 million) were 48 %. According to the 2011 census, the population was 59 million, but the data on population living in territories of high seismicity was almost unchanged.

technical and academic cultures and the folk wisdom, which are the most connected to the territory on which the community live (Bevilacqua 2005, p. 12).

To justify the removal of environmental dynamics, and therefore of earthquakes from historical events, historians have often used anti-determinist motivations. Vico himself had stated that while men make history, Nature is God's work. Therefore, "the rejection of determinism has denied every natural determination of social behaviour"; that means we have forgotten that man's action is often the answer, through the means of culture, to the environmental situation that he has to face (Bevilacqua 1996, p. 79).

Another element of Italian historiography that has contributed to this removal is certainly the lack of a "partnership" between history and geography. A union which has in fact occurred in France, where it has been at the base of the fortune and originality of French historiography in the 20th century, embodied in the "Annales" by Marc Bloch and Lucien Febvre and afterwards by Fernand Braudel.[2]

Concerning this Guidoboni (2005) pays attention to the responsibilities, of historians and geographers in dropping the studies of those historians who had given appreciable and precious contributions to the history of the territory in general, and to seismic history in particular. According to Guidoboni, the awareness of this issue has been assigned only to the mass media which often talk about it but only superficially and sometimes not properly, while the issue should have found more room in academic studies. Guidoboni goes even further in her criticism of historiography and, introducing earthquakes among the elements of the territorial dimension, she writes:

> If the consistent economic and social damages and the victims that [earthquakes] have caused to Italian society had been inflicted by a human, there is no doubt that our history books, and very many essays, would be filled up with these references. Every war and peace treaty would be remembered with every little detail, historians would have wasted no time in giving theories, analysis and suggestions, and in making generations of students aware of this danger and threat; sociologists and journalists would have spread their analysis through the mass media, up to a time when a rooted culture of awareness would generate either a revenge attitude to fight the enemy or a wide spread ambition to smooth out the threat peacefully: in short, a shared social and cultural reaction would emerge to mitigate such a danger (Guidoboni 2005, p. 21).

Instead, Italian historiography has considered the earthquake as distant and unfamiliar to the history of society and culture. Of course, there have been exceptions, for example the work by Augusto Placanica on the Calabrian earthquakes in 1783,[3] and the pioneering work by the geographer Mario Baratta. At the

[2]About the historiography of "Annales" see also Burke (1995).

[3]Among Placanica's works about this item, see also volumes about the privatization of Calabria ecclesiastical properties (Placanica 1970, 1979), and the ones about a study on the social and cultural impact of a seismic event (Placanica 1972).

beginning of the 20th century, with his great catalogue[4] on the earthquakes in Italy, Baratta laid the foundations for research which had been in the shadows for a long time, but which was going to lead to the publication of richer and more up-to-date catalogues in Italy and abroad in the last twenty years of the century.[5]

Despite the partial silence and neglect shown by historiography until very recently, the earthquake can be considered an influential element in history; therefore, it can be the object of historical analysis with its own significant complexity and importance. As Guidoboni wrote "From the social point of view, with all the burden of reconstruction and the risks of the desertion, the history of seismic disasters is, in every respect, the history of economic and social costs, of modified settlements, of broken or only partly reactivated cultural habits. Moreover, it is also the history of great loss of life: it is not only the death of people, but also the loss of daily habits and marginalized cultural identities, besides social and psychological unease due to the loss of connection with one's own roots, dispersion of people and affection, broken links and daily customs" (Guidoboni 2005, p. 33).

2 Destructive Earthquakes from 1700 to 1950 and Their Geographical Location

At the beginning of the 20th century Baratta had already referred to the southern Italian regions as the worse part of the peninsula affected by earthquakes between the 18th and the 20th centuries (Baratta 1901). However, as Baratta stated the entire peninsula had faced seismic risks and suffered their consequences (Figs. 1 and 2). According to legislation until 2003, 48 % of Italian people lived in a seismic area, and 2960 municipalities (35 %) were classified as seismic. These were divided into four categories whose decreasing levels, from 1 to 4, corresponded to suitable building regulations. However, 2004 law defined 8101 Italian municipalities as seismic areas. (Guidoboni 2005, p. 19) They are divided into the following:

716 municipalities are located in "area one" (the most dangerous);
2324 are located in "area two";
1634 are situated in "area three",
3427 are situated in "area four", where residential building regulations are more flexible and the earthquake proof rules must be followed mainly for "strategic" public buildings (schools, hospitals, etc.) (Guidoboni 2005, p. 19).

[4]Baratta (1901). This work is divided into 3 parts: the first one is about the chronicle of the most destructive earthquakes in Italy; the second one is a survey of seismicity of each region; the third one is a valuable bibliography.

[5]Guidoboni (1989a), Boschi and Guidoboni (1997).

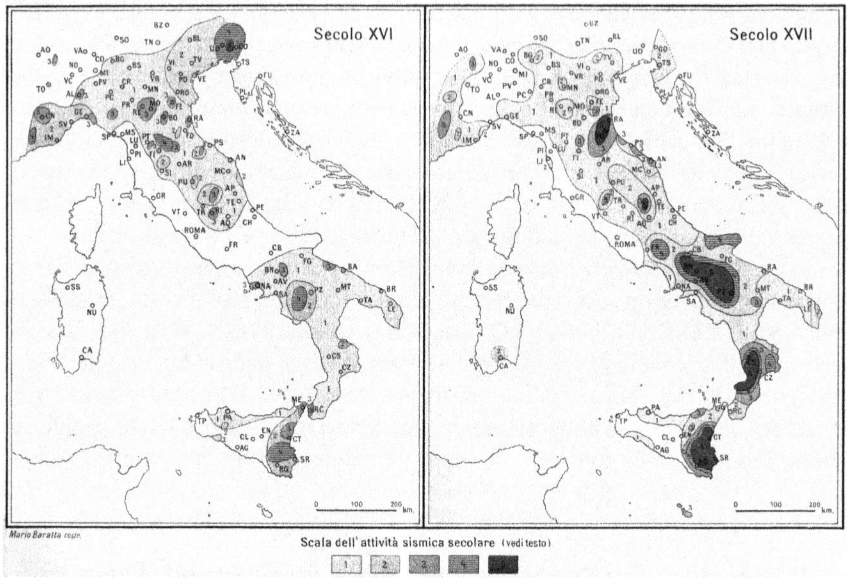

Fig. 1 Baratta's maps on seismic activity in Italy (Baratta 1936)

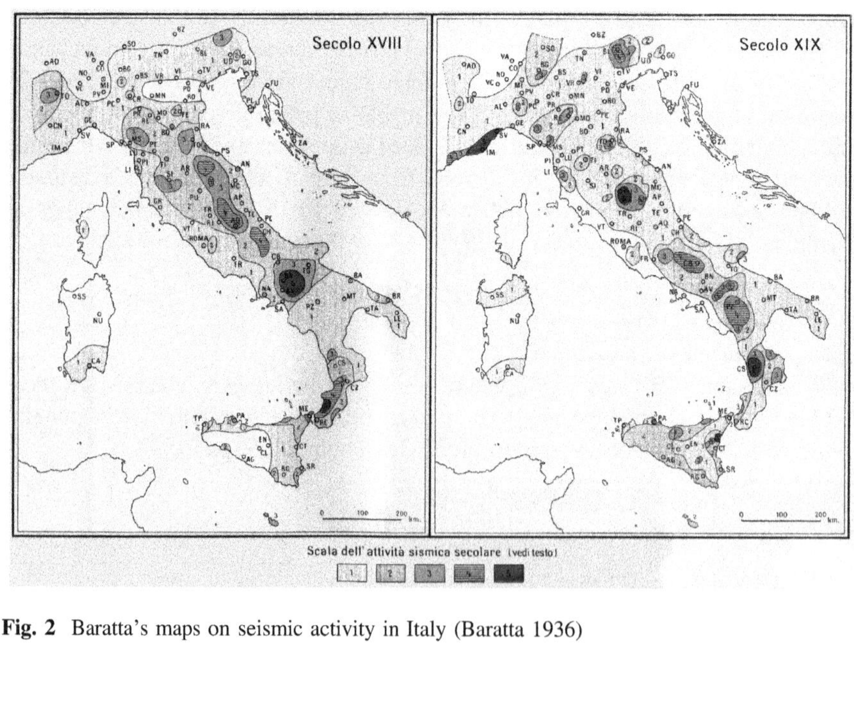

Fig. 2 Baratta's maps on seismic activity in Italy (Baratta 1936)

The 2004 law has highlighted the territorial complexity of the whole peninsula from the point of view of seismicity (Fig. 3). Next to areas at high seismic risk such as Calabria, Sicily, Friuli and many areas in the central Apennines there are others, including Sardinia, where there is no memory of earthquakes, either in ancient or modern times. Therefore, there is no mention of these areas in the *Catalogo dei forti terremoti in Italia* (*Catalogue of Severe Italian Earthquakes*), which is the main source for historical seismicity studies in our country. However, this diversification

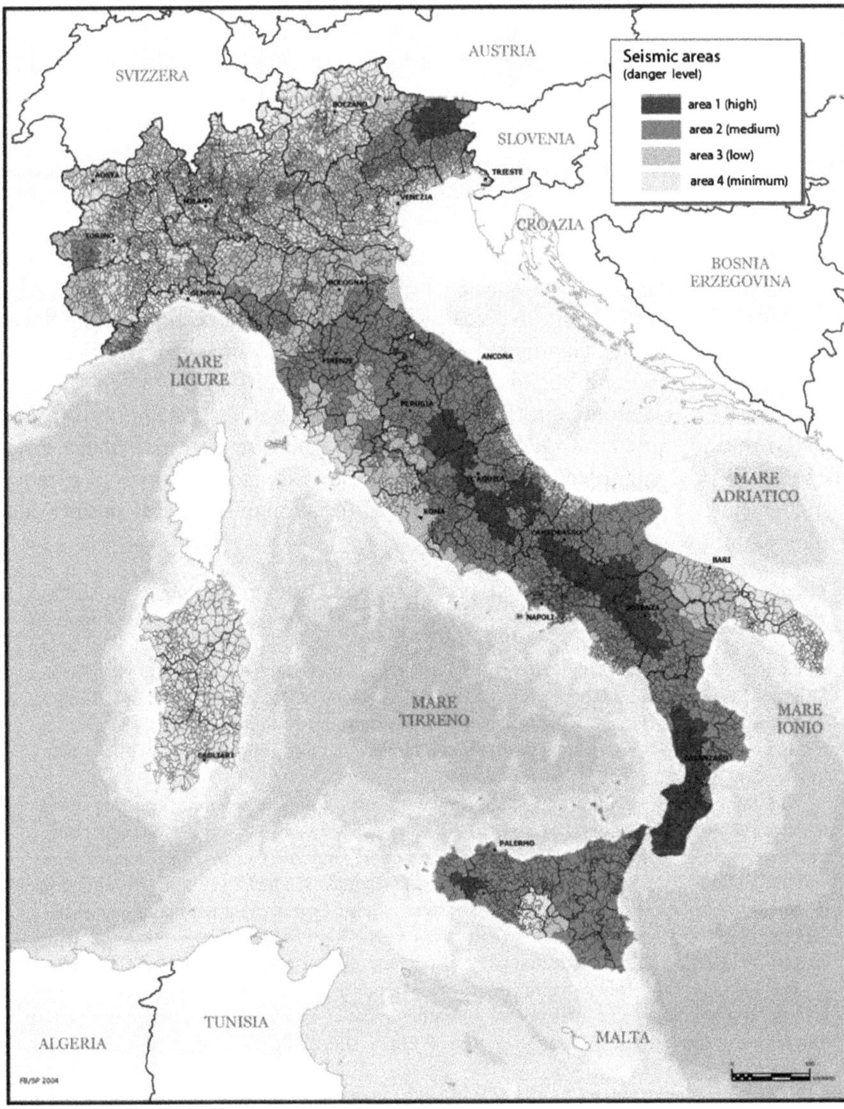

Fig. 3 Map of seismic areas in Italy according to 2004 regulation

is also reflected in the wider perspective of the Mediterranean basin. In fact, there are areas which are frequently struck by high magnitude earthquakes such as Greece and Turkey; but there are also countries that are considered of medium seismicity both for frequency and intensity, like Italy and the Maghreb countries. Finally, the Iberian coast and the other north-African coasts have much lower risk levels than the two previous areas. From this point of view Italy differs from the northern countries of Europe which do not have a significant seismic threat. However, it is with these countries that Italy has had a shared political, economic and social history over the last five centuries. Somehow the history of seismicity in Italy is an element that links it to the history of the Mediterranean world (Guidoboni 1989b, pp. 15–16). This area was also brought together by the ancient administration union of the Roman Empire, as well as commercial trade with the East in the Middle Ages and part of the modern era. This is particularly true of the southern regions of Italy.

All along the Italian peninsula, from the 5th century B.C. to the 20th century, there have been 364 earthquakes, considering those higher than VII degree (MCS). If we only consider those of high intensity—higher than IX degree MCS—there have been 97 destructive events in the last four centuries, that is, on average, a strong earthquake nearly every five years. Fifty-three of these occurred in the South of Italy, in particular in Calabria and Sicily, followed by Campania, Basilicata and then Abruzzo, Molise and Puglia (Guidoboni 2005, pp. 19–21).

As for their occurence in Italy, considering Baratta's catalogue, Argerio Filangeri[6] has identified, for the South of Italy, the epicentre zones, as "those zones whose subsoil had their own tremor and were not just affected by waves of external seismic events". He describes the territorial manifestation of seismicity in the South of Italy as follows:

> The most important [epicentre zones], proceeding from South to North, are: in Abruzzo where the orographic systems of the Gran Sasso and Maiella are the two most important zones. The former extends along Aquila's valleys from Antrodoco to San Pio delle Camere and the latter from Torre dei Passeri to Conca di Castel di Sangro across Maiella and Sulmona's valley. Smaller zones are located in the towns of Amatrice, Ancarano, Teramo, Atri and Orsorgna. Then we move on to the area around Matera – between Molise and Campania – whose tremors have hit the inhabitants of Isernia's and Boiano's valleys many times.
>
> In Campania, Benevento is right in the centre of a complex seismic zone that affects most of the province area and extends to Alife on one side and Ariano on the other. In the province of Salerno the wide hollow to the north of Alburni and Cervati, the Tanagro valley and the Vallo di Diano are the most affected by earthquakes. They have some ramifications in Basilicata towards Balvano and Vietri and in the South towards Piaggine. Still in Campania, a third zone crosses the plain from Santa Maria Capua Vetere to Nola. Other smaller zones are centred respectively on: Roccamonfina, Ischia, Salerno and Caposele.
>
> Basilicata has various epicentre zones: one extends around the area of Potenza, a second is in the triangle between Marsiconuovo, Grumento Nova and Tramutola, a third zone is on the Vulture mountains, a fourth on the Pollino, along the line between Episcopia,

[6]The author mentions Filangeri as in the years after the Second World War, he is one of the first to consider the socio-economic aspects of the earthquake together with Plancanica and Gambi.

Castelluccio and Laino, and a fifth is in the territory between Matera and Ferrandina. There are isolated centres in Tito, Craco, Maratea and Lagonegro.

In Puglia the hollow near Gargano is the most seismic zone where we can identify a few epicentres: one between Larino and Termoli, one between San Severo and Torremaggiore and a third which includes the towns to the west of Gargano, from Vico to San Marco in Lamis. In the province of Bari a long, wider zone lies between Barletta and Bari; there are other isolated centres in Foggia, Mattinata, Ascoli Satriano, Cerignola and Spinazzola. Furthermore south the land of Otranto is one of the quietest; very few tremors were reported in Brindisi, Nardò and Massafra.

Opposite Otranto is the seismic zone of Calabria: where we can identify five zones of high seismic activity: the valley of Cosenza, the hollow of Polia-Vibo Valentia, the hollow of Polistena-Palmi, the coast of Reggio and the plain of Cutro. There are also more isolated seismic centres than in other regions, including Castrovillari, Bisignano, Rossano, San Giovanni in Fiore, Nicastro, Girifalco, Africo, Gerace, Paulonia and Stilo (Filangieri 1979, pp. 73-87).

Recent studies of historical seismicity, recorded by the INGV[7] in the *Catalogo dei forti terremoti*, confirm this occurence of earthquakes: towns like Benevento, Foggia and the areas around Aquila and Cosenza are often recorded among the zones with more frequent earthquakes. Using the information concerning seismic events in that Catalogue, we have produced the following chart (Chart 1) reporting the earthquakes of an intensity higher than VIII MCS that occurred in southern Italy in the period from 1700 to 2000:

This chart examines the southern regions according to their present administrative borders, which corresponds approximately to what was formerly the Bourbons' kingdom.[8]

The data shows that only 94 earthquakes out of 259, in the whole peninsula, were located in Southern Italy, despite the fact that the South is regarded, by the seismological community, as the most seismic part of Italy. Actually if we consider the above definition of the destructive earthquake, we have a different vision. The data has been taken from the records quoted in the catalogue.[9] This list includes all the seismic events, from II to XI MCS, which occurred in Italy from 461 B.C. to 1990, while we have indicated as destructive those earthquakes with an intensity higher than VIII MCS.

[7]INGV stands for *Istituto Nazionale di Geofisica e Vulcanologia*. For further information see www.ingv.it.

[8]The Regions, as administrative entities, were recognized in Italy after the Second World War; but some territories of Abruzzo and Campania (for example Benevento) were papal properties before the unification of Italy.

[9]See Boschi and Guidoboni (1997). For the earthquakes which occurred from 1990–2000, we referred to data available on the web site of the INGV where it is also possible to find the catalogue updated to 2000: http://storing.ingv.it/cfti4med/.

Chart 1 Earthquakes in the southern regions and in Italy from 1700 to 2000

Years	Abruzzo-Molise	Campania	Puglia	Basilicata	Calabria	Sicilia	Total in Southern Italy	Total in Italy
1700–1750	2	5	2	–	–	3	12	23
1751–1800	2	1	–	–	7	3	13	45
1801–1850	2	1	–	2	3	5	13	38
1850–1900	3	3	1	4	5	4	20	54
1901–1950	8	2	4	1	5	3	23	61
1951–2000	3	3[a]	–	–	1	6	13	35
Tot.	20	15	7	7	21	24	94	259

[a]The earthquake in Irpinia in 1980 (the official number of victims assessed was 2914; the number of injured was about 9000; the evacuated were about 394,000) affected mainly the territories in Campania, that is why it is considered among the data in the chart. However, this seismic event also struck some territories of the Appennino Lucano heavily

Having clarified that, and always taking this list into account, I have produced the following two charts: the first one of the "destructive" seismic events in southern Italy from 1700 to 2000 (Chart 2), the second one concerning the same category of events but in the central-northern regions over the same period (Chart 3). In both charts, the earthquakes are those measuring VIII grade MCS upwards and therefore the ones labelled "destructive".

Comparing the data in the charts we can see that between Charts 1 and 2, for the southern regions the number of seismic events does not change much: 5 fewer in Sicily, 4 fewer in Campania and the same number in Puglia, 3 fewer for Abruzzo and Molise, and only 1 fewer in Calabria and Basilicata. In total there were 18 fewer earthquakes in the south. It is important to stress, that Calabria, compared to the first chart, is a seismic region with a destructive earthquake every 15 years, a very sad primacy even worse than Sicily with its 19 destructive seismic events. In fact, as said before, Calabria has always been considered the worst affected seismic region together with Sicily. Furthermore, if we consider the total number of earthquakes in Italy in the three centuries, out of 259 significant events, which affected the whole country (nearly one per year), 137 were destructive and more than half of them occurred in the South.

The seismic nature of the south is far stronger than the seismicity of the centre-north. This area too is highly affected by earthquakes with 18 destructive earthquakes in Umbria and Marche, exceeding the 8 and 10 of the larger regions of Tuscany and Emilia Romagna. This part of the peninsula is often affected by tremors, which in general turn out to be of low intensity, while destructive earthquakes are less frequent (37) compared to 76 in the south. While it is true that we are considering a smaller area (68,692 km^2 in the centre-north, 98,973 km^2 in the south), which does not include another critical area of our country, Friuli Venezia Giulia. It is also true that with 37 seismic events in the centre-north, the remaining 24 occurred in the rest of Italy, we are below the total number of seismic events which involved the south tragically.

To sum up, the data confirms that the south is more frequently struck by seismic destructive events, even though it only accounts for 32 % of the national territory (Graph 1). Furthermore, it is important to notice that 76 destructive earthquakes in 300 years correspond to an average of one earthquake every three or four years. This is some important information, highlighting a certain time frequency for seismic events that directly influence not only the economy but also the psychology of the involved communities. This frequency also affected, to various degrees, government expenditures during those periods, with the financing of first aid and reconstruction.

Chart 2 Earthquakes in the southern regions and in Italy from 1700 to 2000 (from the VIII grade M.C.S. onwards)

Years	Abruzzo-Molise	Campania	Puglia	Basilicata	Calabria	Sicilia	Total in Southern Italy	Total in Italy
1700–1750	2	2	2	–	–	3	9	16
1751–1800	2	1	–	–	7	1	11	23
1801–1850	2	1	–	2	3	5	13	22
1850–1900	3	3	1	3	4	4	18	26
1901–1950	6	2	–	1	5	3	17	32
1951–2000	2	2	–	–	1	3	8	18
Total	17	11	3	6	20	19	76	137

Chart 3 Earthquakes in the central-northern regions from 1700 to 2000

Years	Toscana	Umbria-Marche	Emilia Romagna	Liguria	Central-Northern Italy	Total in Italy
1700–1750	2	4	1	–	7	16
1751–1800	–	4	4	–	8	23
1801–1850	1	1	–	–	2	22
1850–1900	1	1	2	1	5	26
1901–1950	4	2	2	–	8	32
1951–2000	–	6	1	–	7	18
Total	8	18	10	1	37	137

There were a total of 24 earthquakes, I > VIII MCS, in the other Italian regions during the period in question

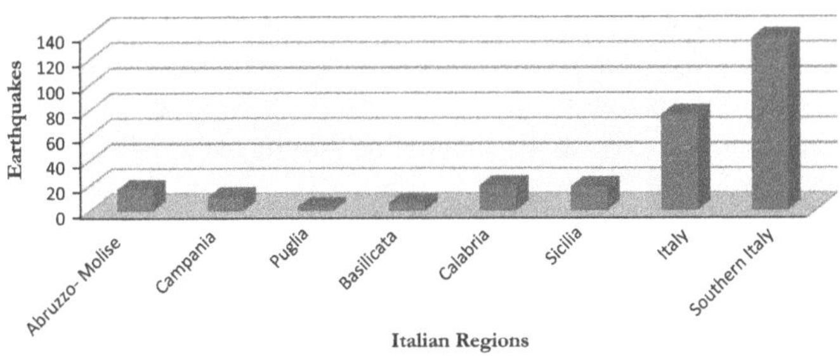

Graph 1 The destructive earhquakes in the southern regions and in Italy from 1700 to 2000

3 The Demographic, Social and Environmental Effects of Earthquakes Between 1700 and 1900 in Southern Italy

3.1 *The Demographic Consequences*

We have considered the frequency of earthquakes in the southern regions of Italy, and where they occurred more frequently.

Despite the severe impact on properties and loss of resources, the demographic consequences of earthquakes in the South between the 18th and the beginning of the 20th centuries has been negligible. For example, Filangieri says, “unlike epidemics which severely affected the population and influenced the development curve, the *seismic events* do not significantly affect demographic development, […] [as] being anticipated by subtle early warning signs which allowed people to escape their houses, this reducing the number of victims” (Filangeri 1979, p. 73).

Actually, such a remark is not entirely correct for two reasons. The first is that the analysis of the early-warning signs of an earthquake—including the inexplicable anxiety of animals before a tremor—is still theories. The desire to foresee an earthquake is mainly the willingness of the human mind to control the event by finding out in advance when it is going to happen. We are still far from being able to say with certainty and precision the day and place of the next event, but thanks to the research of historical seismology and to technical and scientific development we can identify the most dangerous areas. So that we can try to prevent damages to the territory and learn to protect ourselves and therefore to deal with earthquakes. It is not possible to foresee an earthquake, but it is possible to prevent its effects.

The second reason is more specific. Actually, the ability to detect the danger in advance, either because of small tremors, animals' behaviour, or special unexpected physical phenomena such as the tide and volcanic activity, can occur but only in certain situations. In 1893, for instance, in spite of the violence of the earthquake, which struck the Gargano in northern Puglia either destroying or damaging most of the buildings, only four people died. Local people had been frightened by some light tremors the days before the main devastating shake at 20:52 on 10th August and had taken refuge in makeshift shelters in the countryside.[10] But nothing like this happened in Val di Noto in 1693, when at 1 p.m. on 9th January a strong preliminary strong shake was sensed in Catania, Augusta and Lentini. It actually measured VIII grade on the M.C.S. scale and people left their houses for just one day, because of the folk belief that strong earthquakes recurred within twenty-four hours. So when on 11th January at around 9 p.m. the main tremor of this long (about two years long) seismic event hit, most of the people in Val di Noto were back home. 45 towns and villages were totally destroyed and 54,000 people died, according to the estimate of the time. In January 1915 in Abruzzo (Marsica area) an earthquake suddenly struck at 7:53 in the morning, without any early-warning sign, and most of the people were still in their houses. Therefore, this earthquake caused the death of 32,610 people.(Margottini et al. 1999, p. 50).

The key to limiting the number of victims are building techniques as well as the awareness and knowledge of simple seismic rules, how to behave in dangerous situations and how far the population, especially the most vulnerable ones, has adopted these rules. However, not all of this is applicable to the historical period that we are discussing in this work. In fact, in the 18th and 19th centuries, people had superstitious and fatalist belief about earthquakes that prevented any rational reaction and caused collective panic and fear; moreover, the buildings had been built using inadequate techniques and poor quality materials.

In contrast to Filangeri's assertions, mainly in the period from the beginning of the 1700s to the end of 1800s, seismic events did affect the demographic characteristics of a territory. This was not only because of the event itself, but also for the many deaths caused by the consequent epidemics, famine and the migration of the population elsewhere. These factors became obstacles to the reconstruction of

---

[10]See Boschi and Guidoboni (1997). Vol II (cd-rom, sisma 1983 Gargano).

buildings and the economy in the affected areas. For example, the above-mentioned earthquake, which occurred in Campania and Basilicata in 1857, was followed by all these circumstances. To the initial 19,000 victims of the earthquake, which accounted for between 10 and 75 % of the population in some towns, we have to add the death toll caused by cold weather (the earthquake struck at the beginning of winter), and malaria, which spread rapidly among the people, also because the earthquake enlarged the marshy areas. Furthermore, neither the Bourbon authorities nor the Italian ones adopted adequate economic and building measures in a territory that was already suffering a crisis before the earthquake. That caused a long demographic decline due in great part to emigration. From the data of the two population censuses, in unified Italy, in 1861 and 1881 emerged that the Vulture area and its surroundings lost 23 % of its inhabitants (Boschi and Guidoboni 1997). On the contrary, the violent earthquake that struck Eastern Sicily in 1693 did not stop the economic growth of the area. 54,000 people died, but the demographic crisis was overcome thanks to the reconstruction process started soon afterward, which attracted a large flow of people and workers from outside (Guidoboni and Ferrari 1997, p. 101). The same occurred after the 1908 earthquake, which claimed about 80,000 victims. Despite the low number of 4000 survivors who decided to remain in Messina after the earthquake, there was rapid urbanization with immigration of people from the nearby villages and other Italian regions hoping to make their fortune in the reconstruction. However, in this case the lack of financial resources, the trade inexperience and the strong rural traditions of the new inhabitants, were, according to the geographer Lucio Gambi (1960), the main reasons for the lack of a social, cultural and economic rebirth of the town. Before the earthquake in fact, Messina was one of the most important trade centres in the country, a role that was greatly reduced after the dramatic event of the earthquake (Oteri 2005, p. 22).

If we analyse the 76 seismic events in the South of Italy that occurred between the 18th and 20th centuries from the demographic point of view (Chart 4[11] and Graph 2), we can note that the three regions, previously defined as the most affected by destructive seismic events, were also the ones that suffered the highest death tolls. In about three centuries, Calabria lost 76,999 inhabitants; Sicily had 60,904 victims, and then Abruzzo and Molise had 47,969 dead. The data referring to Basilicata is rather striking as, in spite of having fewer earthquakes than Campania (six vs. eleven), from 1700 to 2000 Basilicata lost 20,575 inhabitants. This is mainly because of the destructive 1857 earthquake, which occurred in the Campania-Basilicata area, where 19,000 people died. It is the same if we consider

[11]We have completed this chart, just like the three previous ones, elaborating the information concerning the seismic events in the Catalogue. The number of victims in Campania and Basilicata of the 1900s were elaborated comparing the total amount of victims of different seismic events to the number of victims of each single province that had been struck. The considered earthquakes were the Irpino-Lucano earthquake in 1910 (50 victims), the 1930 earthquake in Irpinia (1404 victims) and the Irpino-Lucano earthquake in 1980 (2914 victims). This same procedure was used to calculate the victims of the earthquake of Reggio and Messina in 1908 (80,000 victims).

Chart 4 The number of the earthquake victims in the South of Italy from 1700 to 2000

Years	Abruzzo-Molise	Campania	Puglia	Basilicata	Calabria	Sicilia	South
1700–1800	9761	2331	3880	–	35,185	250	51,407
1801–1900	5583	2377	4	20,019[a]	1086	195	29,264
1901–2000	32,625[b]	3721	108	556	40,728	60,459	138,197
Total	47,969	8429	3992	20,575	76,999	60,904	218,868

[a]The number of victims of the earthquake which occurred in the Vulture area in 1857 referred to the unofficial data, which counts 19,000 victims. The official data, in fact, does not consider the victims of the countryside

[b]The data referred mainly to the 1915 earthquake in Marsica area where the official number of victims was 32,610. Several authors stated that the number exceeded 33,000 victims

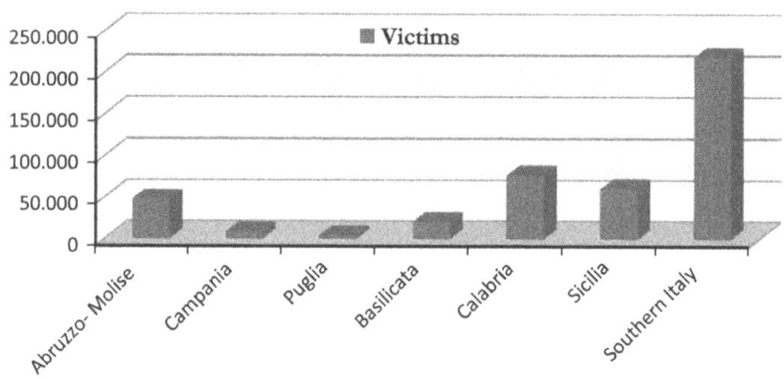

Graph 2 The number of the earthquake victims in the South of Italy from 1700 to 2000

the data of Calabria where two main seismic events played a seriously destructive role, the 1783 earthquake and the 1908 Calabria-Sicilian earthquake. The latter contributed to the increased total number of victims in Sicily. Even the number of dead in Abruzzo and Molise was highly influenced by the catastrophic seismic event in the Marsica area in 1915.

If we look at the estimated number of deaths in the last two centuries considered, it is between 1901 and 2000 that the number increases significantly in the three regions, with tragic primacy. This happens mainly in the first twenty years of the 1900s when the two destructive earthquakes of 1908 and 1915 occurred. However, this is not sufficient to justify the very high number of victims compared to the two previous centuries. Because we have more sources for this historical period, we can have a more complete overview of the precise number of deaths. However, the higher number of victims may be linked to the population increase in the south of Italy and in the rest of Europe and the consequent effects of urbanization from the second half of the 18th century onward. Concerning this matter, Filangieri writes: "In 1793 there were just 80,000 people living in municipalities of more than 20,000 inhabitants (except for Naples), but in 1901 that number had increased to 1,348,000 of which 773,000 in Puglia and 354,000 in Campania [...] In the other regions, the population increase was limited only to the provincial capitals" (Filangieri 1979, p. 149).

This population increase in major towns led to a haphazard urbanization both at horizontal and vertical level. Horizontally there was the enlarging of the town toward the countryside; vertically there was the raising of the buildings with the consequent increase of risk in areas at high seismic danger. In fact, the highest buildings in historical centres were the first to collapse, blocking the narrow roads where people tried to escape during an earthquake. This actually happened in Messina and Reggio Calabria in 1908 and afterwards in Marsica (Abruzzo) in 1915. This problem became a much discussed issue among scientists and politicians soon after the two earthquakes and some forerunners wrote that in Italy it would not be possible to build skyscrapers like in Chicago and New York, as they would be the first buildings to collapse (Riccò 1915, pp. 8–9).

Baratta underlined that the increase of the death toll in the earthquake in 1908 was due to a cultural problem: "forgetfulness" (Baratta 1910, pp. 26–27). In fact, the effects of the earthquake in 1783 had been forgotten, as had the building rules adopted by the Bourbons in the reconstruction of Messina and Calabria. Those rules forbade the raising of buildings: the height of a building was linked to the width of the road, which it overlooked. They had been adopted after Placanica had studied the destructive seismic event of 1783. However, according to Baratta (1910, pp. 26–27), "forgetfulness" together with the very bad quality of the building materials played a part in the negative effects of the national seismic tragedies (Otieri 2005, pp. 13–64; Maniaci and Stellino 2005, pp. 89–110).

Another element to remember is that "up to some decades after the unification of Italy (1861), people still lived mainly in the mountains and hill territory just like in the 18th century. Only in the last century [1900], with the reclaiming of the land and the disappearance of malaria have the hills undergone a slight change in the importance of farming (52 %): while the mountains gets to 16 % only of their population; the plain reaches 32 %" (Filangieri 1979, pp. 148–149; Tino 2002, pp. 15–63). The high population density in the montains and hills is therefore the factor explaining the high number of victims up to the beginning of the 20th century in montainous Calabria for the 1783 earthquake and in hilly Basilicata for the 1853 and 1857 earthquake.

The following chart (Chart 5) is also in Filangeri's book and shows the permanent increase in population in Italy between the 18th and the 20th centuries, with particular reference to the south of Italy. What emerges is the increase in the South,

Chart 5 Development of the Italian population by geographical areas from the 18th to the 20th centuries[a] (thousand inhabitants)

Years	**Centre-north**	**South**	**Islands**	**Total**
1771	10,800	4150	1900	16,850
1801	11,350	4950	2200	18,500
1861	16,122	6530	2981	25,633
1901	20,751	8099	4322	33,172
1931	25,788	9652	4870	40,310

[a]The data of the chart is extracted from Filangieri (1979) p. 150

which goes from 4,150,000 inhabitants in 1771 to 9,652,000 in 1931 and reaches 12,435,000 people in 1961. By claiming 218,868 victims in 300 years the destructive earthquakes certainly did not stop this gradual growth entirely, but they did slow it down. Filangieri himself admitted that the "anomaly" of the demographic figures showing the decrease in the South's population in 1783, 1784 and 1793 is due to the earthquake in Calabria in 1783. Together with migrations, earthquakes contributed to limiting demographic growth in the South especially in the 19th and the 20th centuries. In some cases, the effect of earthquakes at a regional level has led to a demographic crisis as for the mentioned earthquake in Campania and Basilicata in 1857.

When analyzing the earthquake data from the South, we must surely consider that the number and the type of victims tended to change according to the season and the time of day the earthquake struck. If the earthquake suddenly struck in the evening or in winter, it certainly caused more casualties than if it happened during the day and in summer, when most of the people were out in the fields working. Generally, it affected the people living in towns more than people in the countryside. Of course, the latter lived in hovels or shacks therefore they risked less than people living in brick houses which had generally been built with poor quality or inappropriate materials. Furthermore, in those areas affected by seasonal migration, as most of the southern regions, where the victims were more often women and children. Mercalli himself, when talking about the demographic influence of the seismic event in 1908 confirmed what has been said so far:

> [...] The causes of the number of dead [much higher in large inhabited places], must be found somewhere else. In my opinion the causes are: taller houses, too many tenants, being close to high buildings already collapsed onto smaller ones, finally narrow and winding streets in the old parts of the town. It is natural that rural houses, which are usually on one floor above the ground floor, allow the tenants to get out quite easily before the collapse. Furthermore, in little towns inhabited by many farmers, at 5.20 they were already up and some had already left to go and work on the land or at the spinning factories.[...] At Cannitello I was told that most of the dead were women and children; as the men were in America at the time. I mention it for the comparison with another similar,[...] where you can read that the destructive earthquake on 27 March 1638 "about 70 people died, they were mostly women and children, the only victims among men were the priest Don G.B. Galletta and the churchman Nicola Barberio, who died under the collapse of the Church Santa Maria Maggiore". (Mercalli 1910, p. 334)

Things changed radically after the Second World War when more efficient building methods and the increase in prevention led to fewer earthquake victims and the number of injured and evacuated was limited, too. Therefore, earthquakes do not bear on demography as in the past, but there has still been a certain impact with victims: we can remember the 231 victims in Belice (1968) and the 939 victims in Friuli (1976), or the 2914 who died in Irpinia (1980). More recently, the 308 victims of the earthquake on 6th April 2009 in the Aquila area show the challenge that the earthquake brings not only to the local communities, who are periodically involved, but also to the population of the whole country. Therefore, we need a better level of prevention, not only to limit further the demographic effects of seismic events. It is necessary to face the consequences of the not always legitimate housing speculations of the last thirty years of the 20th century, driven by excessive urbanization of the land, which has led to uncontrolled building development in high-risk areas even because of "Forgetfulness". This uncontrolled development, which is often condoned by law, could represent a major economic cost for the safety of the involved territory also for future generations.

3.2 *Aspects of the Intervention Policy*

3.2.1 Economic Aspects

"In this last almost ignored region of Italy, Basilicata, a huge disaster has set civilization back 50 years, losing the significant yet little known economic progress that people can achieve in half a century". This is how Francesco Saverio Nitti, quoting the Basilicata historian Giacomo Raccioppi, introduced the economic and social effects of earthquakes in Calabria and Basilicata in the parliamentary inquiry on the farmers' conditions (Nitti 1919, pp. 60–61).

In fact, as in the past the survivors of a destructive seismic event experience first the evidence of having survived and then the fact of being taken back, to have lost everything and have nothing left to carry on.

The feeling of having lost everything, which characterizes the "psychology of the disaster", is a distinguishing factor of the communities struck by an earthquake. The narration of the destruction and the loss that earthquakes produce goes beyond rhetoric, which is actually often utilized by the sources. Many areas of the South of Italy between the 18th and the 20th centuries had to pay the cost of what Guidoboni calls "a permanent social challenge" (Guidoboni 2005, pp. 23–25). With this definition Guidoboni tries to highlight the answers given by the communities involved in earthquakes to defend themselves from the seismic threat. In fact, it is possible to understand whether the earthquake has contributed to the decline of the community or there is the opportunity to grow from the tragic event. For example, after the 1857 earthquake, Basilicata went through a long economic and demographic crisis for two decades, made worse by the earthquake. On the contrary, in the Etna area, hit by numerous seismic events caused by the volcano during the 19th century, these events did not stop the positive economic phase, which characterized this area up to the beginning of the 20th century. The reconstruction contributed to its growth. (Boschi et al. 1997)

In studying the communities' responses to protect themselves from the seismic threat, the government's intervention plays an important role. For the three centuries we are considering, in the analysis of the policies adopted by the governments in the management of the emergency and the reconstruction, we must bear in mind two peculiar aspects. Until the unification of Italy, the first aspect deals with the various institutional methods of intervention after an earthquake that depended on the political and administrative divisions of the peninsula: in fact, each state, before the union of Italy, had its own rules and procedures. That caused different "seismic disaster cultures" all along the territory and the unified country had to face these differences. Still today, there are differences at regional levels (Guidoboni and Ferrari 1997, pp. 99–101). A second important element in the reaction to seismic events is that the national authorities tend to minimize the gravity of the earthquake's consequences, trying to take advantage of it to strengthen their image. That has often occured either by minimizing the effects of the earthquake in order not to seem weak to other countries; or by publicising the actions of the government to strengthen their efficiency in terms of the central control of the territory. For instance, in 1570 the Este family tried to circumscribe the news of the earthquake in Ferrara to protect the fame of the powerful ducal family and not to lay themselves open to criticism and possible attacks. The same happened in fascist Italy in the 1930s, just before the Second World War; the Italian press accurately avoided divulging details of the damaged places and the victims of the earthquakes in Irpinia in 1930 and then in Abruzzo in 1933. They only wrote about the seismic events in terms of the efficiency of the regime's intervention (Margottini et al. 1999, p. 11). In both cases the news of the earthquake became harmful for the image of the State. Another meaningful example is the 1832 earthquake in the Modena Dukedom, where Duke Francesco IV in his proclamation of the administrative measures to deal with the damages, defined the earthquake as the divine punishment against the "iniquity of the time". He said the earthquake should be considered a warning, given that in the Este family possessions, instead, there had been no victims (Fulci 1916, pp. 88–89). For political reasons, the Duke tried to play on his people's fear of the natural disaster. In fact, the previous year his power had been undermined by Ciro Menotti's plot.[12] This example can be used also to underline another aspect, which is how religious or scientific theories could be used to justify measures taken by central authorities.

We should ask what the institutional interventions might have been like after a seismic event in Italy before the unification. In spite of the differences we have mentioned, there are three categories of interventions to face the emergency and the reconstruction: economic incentives, tax cuts; the distribution of funds through a one-off earthquake fee plus tax cut.

> The first kind of policy of intervention was bound to the variability of available money, money which was given in the form of a loan at a very low rate and over a long period. That money came from the state coffers. This was the administration policy in Grand Duchy of Tuscany, by Medici and Lorena. The tax cut policy consisted instead in the exemption of taxes from one to ten years; that was used in the Kingdom of Naples (except for the exceptional case of the reconstruction in 1783), in the Kingdom of Sicily during the Spanish and Bourbon time [...] and partly in the Republic of Venice. Finally, the fund distribution policy with non-recurring tax plus tax cut was a sort of "mixed" policy: first a general tax was applied, then, once the regime of tax exemption had been defined, it was decided how the funds were distributed. This was the administration policy in the Papal States starting from the first part of the 1600s. The redistribution of the funds was usually

---

[12]The plot is part of the risings of 1831. At the beginning, the Duke himself supported that uprising, trying to expand his Dukedom. Then he changed side and he condemned to death Menotti and many others among its allies.

left to the parish priests and it was slow and farraginous. (Guidoboni and Ferrari 1997, pp. 99–101)

It is necessary to underline that these three kinds of intervention could often be used at the same time in an affected zone. This often happened in bordering areas such as in Irpinia, which was part of the territory of the Kingdom of Naples, but included Benevento, a papal dominion. That meant different reconstruction policies depending on where the policy came from and how it was funded. No matter which the nature of the territory was; if the intervention was careless the town could be rebuilt in a less seismic area, but the damage could be equally disastrous because of the bad rebuilding intervention. "In some areas, the rebuilding could be rushed, temporary, or even in appearance only; while in other areas on the contrary, there could be more long-sighted and intelligent policies: which generated different results for the future" (Guidoboni and Ferrari 1997, pp. 99–101). This statement confirms that, even if the affected town was rebuilt on a less seismic territory, the poor quality and rushed reconstruction could still cause bad damage.

Concerning the three typologies of administrative reactions to earthquakes, we must add that economic incentives, in favourable conditions, boosted "an enterprising view of reconstruction". In other words, through loans and subsidized loans, commitment to the recovery was encouraged and the link to the central government increased through the bond of the debt to pay back. Loyalty to the territory was favoured among the people who acted out of love for their land. The second kind of intervention, which actually mainly concerns the Kingdom of Naples, was the exemption from taxes of the towns involved for a number of years in proportion to the damages. The Chamber of Sommaria, a financial institution, supervised this intervention policy. Through its offices, this institution of the Bourbon Kingdom dealt with fiscal census, tax collection, and exemption requests from the local communities and of the feudatories, in the wake of destructive earthquakes too. The Chamber sent royal officers to write about the people's conditions after the earthquake. Their reports were then analysed and the exemption measures were adopted together with the connected procedures. In this case, except for the enlightened intervention after the 1783 earthquake, the burden of the reconstruction was integrally charged on the local communities, therefore this procedure often further damaged the economy to subsistence level. An example was Basilicata in the first half of the 1800s, which was hit by four destructive earthquakes: 1826, 1836, 1851 and 1857. Although the latter two were the most disastrous, it is true that when they occurred, most of the population was still waiting for the reconstruction support, which had never come. The only adopted measures from the central authority had been the tax cut in an area that was already depressed. Things did not change even after the 1851 earthquakes in the Vulture and the 1857 earthquake in Basilicata and Campania. It is true that even after King Ferdinando II visited the affected areas establishing the measures for their reconstruction, they were never implemented because of lack of money. That probably influenced the economic and demographic crisis in Basilicata in the second half of the 1800s.

The policy of funds distribution through an earthquake fee plus tax cut adopted by the Papal States, is "a standard procedure exposed to much partisanship, which did not solve the reconstruction problems and, in many cases, discouraged the craftsmen from acting and many were forced to migrate" (Guidoboni and Ferrari 1997, pp. 99–101). Only in exceptional cases did the Papal Chamber directly commit to the reconstruction for Benevento's territory, an enclave of the Kingdom of Naples. In fact, after the many seismic events in the area (1688, 1701 and 1732) non-intervention would have meant the mass abandonment of the productive artisan class (Guidoboni and Ferrari 1997, pp. 99–101).

So far, we have talked about the different measures of intervention, which characterized, to a greater or lesser degree, the entire peninsula before the unification of Italy. In order to compare how the unified country reacted to seismic events, we can look at the first destructive earthquake faced by the Italian Kingdom in Casamicciola Terme in 1883.[13] This earthquake had its epicentre in this town in the centre of Ischia, which was known all around Europe for its spa and where there were already foreign tourists, especially during summer. Anyway, the economy of the island was mainly based on fishing and agriculture, specifically viticulture. The earthquake caused a great deal of damage to real estate and contents and of course to agriculture. In fact, landslides and mudslides badly damaged production. Of 2333 dead 625 were foreign tourists and this attracted attention not only at a national level, but there was international interest too. After the earthquake, some first aid was sent in the form of 1200 soldiers, and the government set up a committee to organize aid, which was led by the prefect of Naples. This central committee had the duty to raise and administer the funds sent by spontaneous solidarity committees set up all over Italy and in many European capital cities. To avoid epidemics, the authorities ordered the ruins to be covered with phenic acid and quick-lime. This measure generated debates in the national press, which reported insufficient help. Cases of typhus occurred and the infected were confined to isolated temporary homes away from the 800 built for the evacuated.

For the first time the great national and international outcry over the event attracted the attention of the Italian government to the seismic problem and the rebuilding systems in the areas at risk. The Minister of Public Works, in fact, set up a special committee of civil engineers to write down the building regulations for the island. The central committee handled the first emergency period, completing its function with the detection and assessment of the damage and the following allocation of funds for the reconstruction. The regulation for the allocation proposed by the committee came into effect on 3rd September 1883 when the government approved the related Ministerial Decree. The committee went on with additional activities. They required the building contractors to employ inhabitants of the island to reduce beggary. Maps of the highest seismic risk areas were prepared to identify the location of the areas for reconstruction. They also financed the private citizens

[13]For further information about the earthquake in Casamicciola Terme see Boschi and Guidoboni (1997). Vol. II (cd-rom, sisma 1883 Casamicciola Terme). See also De Marco (1998).

who rebuilt their properties, and the reconstruction of damaged public buildings including schools, police stations, post offices, town halls, roads and aqueducts.

Therefore, it can be seen that the central government of the united country played a role in the financing of immediate relief to the survivors, and appointed local authorities to directly oversee the reconstruction. They kept the local authorities under their control through the emergency committee, which was coordinated and monitored by the prefects or state representatives. The committee also administered the funds from national and other sources of solidarity. These funds played an important role in financing the reconstruction, especially in the absence of the more substantial contribution of the state. This method of intervention, as we will analyze later on, was generally used by the government of the united country for most of the earthquakes that followed. Some important exceptions were in the aftermath of the 1908 and 1915 earthquakes, when the political-legislative debate tried to overcome the system of special laws for natural disasters in order to prepare a consolidation act, which stated the terms of the state intervention[14] (Fulci 1916).

3.2.2 Reconstruction Aspects

To complete the issue of state intervention in the case of destructive earthquakes, before and after Italian unification, we must look more closely at the phase of reconstruction after each seismic event (Principe 2001).

In southern Italy between the 18th and 20th centuries, post-earthquake reconstruction has often meant investments in many projects for the relaunching of economic and urban planning. Unfortunately, most of them have never been completely carried out because of poor political management of the reconstruction, due to both local and central administration. The reconstruction procedures have acquired an important role as they influenced the social balances and the various interests beyond them together with the dynamics and logics that ruled the interests themselves. The financial procedures also influenced the relationship between the central administration of the state and the local authorities in the areas struck by the earthquake. "As a macro phenomenon which involves houses, a reconstruction involves heavy economic and social costs and it activates administrations and bureaucracy revealing the heart of a territory's political administration" (Guidoboni 2000, p. 253).

We must bear in mind that it is actually the reconstruction policy, as the administration's answer to the seismic event, which makes the earthquake an opportunity for a turning point in a territory. It can be either positive or negative, beyond the destructive dimension of the earthquake itself. It is in this answer that

[14]For further information M.F. Stati 1999 Regulation and legislation following the earthquake of 13th January 1915 in S. Castenetto and F. Galadini "13th January the earthquake in Marsica".

we can have the economic, social and cultural relaunching of the area if the establishment decides to see the earthquake in these terms.

From this point of view the 1783 earthquakes, which involved Calabria and Sicily, can be seen as the turning point in Italian seismic history. This depends not only on the tragic figures—consider that in Calabria 384 small towns were involved at various levels of destruction—1 million people were left homeless and about 50,000 dead both because of the tremors and the correlated effects such as landslides, fires, floods, tsunami and endemic fevers (Placanica 1972, pp. 16–22). Actually, it was also the first time that, either in the Kingdom of Naples or at international level, there was a heated cultural debate on seismic effects and the most appropriate procedures for reconstruction. Such a debate influenced the emergency plan of intervention by the Naples government, which at that time was considered among the most "enlightened". The most important element to be underlined in this case is the logic of intervention aimed at relaunching an area that was economically and socially backward. The setting up of *Cassa Sacra* was done purposely. It was divided into two main offices. The Council of Cassa Sacra with its seat in Catanzaro was at the centre of the provinces[15] struck by the earthquake. It had the task to expropriate and to administrate the ecclesiastical properties. The second office was the Council of Corrispondenza of the Cassa Sacra, with its seat in Naples, which had the task of controlling and appealing to the Council of Cassa Sacra.

The socio-economic objective at the base of this institution was the expropriations of ecclesiastic properties to assure land to the peasants for their subsistence and to use the sale profit of expropriations for the reconstruction of the villages and towns struck by the earthquake (Placanica 1979). Actually, these objectives were partly disregarded, especially the first as most of the expropriated properties went to the emerging lower middle class of entrepreneurs. Instead of reinforcing their entrepreneurial activities, they got hold of properties in order to grow rich rapidly.

In 1783 as in many other cases, one problem to be faced during the reconstruction was the difficulty of getting people back where possible to replace the work force, which was extremely necessary for the reconstruction. That was not all: actually, together with labour, financial resources were needed to sustain basic requirements after the earthquake. Therefore, another huge problem was to convince nobles and notables to go back to town from their properties in the countryside to meet the expenses for the restoration of their palaces in town. They would employ workers and restart the housing business again. Other times earls and barons had to be pushed away from the court or other palaces of the establishment to go back to their properties and cooperate in the rescue and reconstruction activities. That was the case in the Calabrian territory struck by the 1783

[15]At that time, Calabria was divided into 2 provinces: *Calabria Ultra*, with Catanzaro as provincial capital and Calabria Citra, with Cosenza as provincial capital. In the territory, there were also other small royal towns and each of them ruled over some hamlets. The most important royal towns among them were Reggio Calabria, Tropea and Amantea. All the other towns in Calabria were under the authority of landowners. See Maniaci and Stellino (2005).

earthquake, when the barons at the court in Naples were urged to go back to their properties as quickly as possible.[16]

Now as in the past, the administrations should avoid production stoppage after an earthquake. In the past, it was often the reconstruction policy itself that became an important driving force in facilitating the economic re-launch of the affected area. Earthquakes devastated infrastructures for economic life—from roads to watermills, granaries and stables—but it also influenced the emotional mood state of the fearful laborers. For the cases we are analyzing, such a production stoppage often occurred: sometimes it was temporary, lasting a few months or years. Sometimes it became chronic and for many people the only way to escape the already existing poverty was to emigrate.[17]

From the point of view of urbanization, the measures adopted have always been important. They were often made real through important decisions for the development of the affected towns. That was the case when adopting technology to improve the quality of the building industry in a town or change the layout of the town or even better to rebuilt it in a new position, which was perhaps more advantageous from the economic point of view than before. For many inhabited centres in the South of Italy between the 18th and the 20th centuries, earthquakes led to great changes in the settlements in some valleys of the central and southern Apennines as in eastern Sicily (Guidoboni and Ferrari 1997, pp. 103–105).

In the South of Italy, between 1700 and the first decades of the 1900s, after a destructive earthquake, the permanence of the original site struck by the seismic event was very often doubtful. The habit of leaving the settlement after an earthquake has ancient roots and it has also characterized the last three centuries. In fact, even in 1968 it was decided to evacuate the towns in Belice valley and the territories of Sicily struck by the earthquake.[18] Generally speaking, we can define three main different kinds of answers that oriented the reconstruction of settlements in the South of Italy between the 18th and the 20th centuries: reconstruction in "situ"; the withdrawal and disappearance of the original settlement; the variation of the site and complete reconstruction. In the first case, "The criteria which defined this choice were linked to economic and communication advantages. Therefore, even very seriously damaged sites could be reconstructed or repaired on the same site. That happened in Catania, after severe damages in 1693: the town changed its layout although it remained on the same site […]. The average time for a reconstruction, for the period between the 12th and the 19th centuries was between 10 and 40 years. Even when the previous urban plan was maintained, the

[16]See *Catalogo dei forti terremoti in Italia.* Vol II (cd-rom, sisma 1783 Calabria-Messina).

[17]Once again, we refer to the most emblematic case of the earthquake that struck the Vulture area in 1857. Instead, an example of good administration of the reconstruction which had avoided production stoppage, is that of the Habsburg government during the earthquake which struck Foggia and the surroundings in 1731 (The Kingdom of Naples came under the influence of the Austrian Habsburgs from 1707 to 1738) See *Catalogo dei forti terremoti in Italia.* Vol II (cd-rom, sisma 1731 Foggiano).

[18]See *Catalogo dei forti terremoti in Italia.* Vol II (cd-rom, sisma 1968 Valle del Belice).

reconstruction changed the use of still existing buildings in a restoration that modified the previous structures and included the construction of new buildings" (Guidoboni and Ferrari 1997, pp. 103–105). In the second case: "when the destructions were very large, especially in small villages, and on a territory which had no resources, the "preference" was to leave the place. The people moved to other sites or they migrated. Some small villages in central-southern Italy have completely disappeared because of seismic events for which there were no reconstruction resources" (Guidoboni and Ferrari 1997, pp. 103–105). One example is Sperone, a hamlet in Gioia dei Marsi, which was destroyed by the 1915 earthquake. The reconstruction was actually started with many difficulties on a site 300 m away from the original, but the new urbanization was completely abandoned within a few decades as its inhabitants moved to Gioia dei Marsi. In the third case, "the site was moved between 300 m (e.g. reconstruction of Basilicata in 1857) and 12–15 km (e.g. reconstruction of Sicily after 1693). In modern and contemporary times, the changes of the site followed the direction of new communication infrastructures, with a delocalization of the previous centres, towards the plain away from the mountains, peaks and rocks" (Guidoboni and Ferrari 1997, pp. 103–105). This also occurred in Abruzzo for the 1915 earthquake.

However, the reconstruction did not only mean buildings, it also had a social aspect.

> Without any doubt – Nitti wrote about the barren land of Calabria struck by the earthquake in 1908 – everything can be renewed. Death and life are two phenomena of the same nature; tomorrow the orange tree and bergamot will bloom where yesterday there were cries of death. However, the psychology of the people in these areas must be influenced by profound changes. After terrible disasters, those people often acquire a form of indifference to evil, an inability to dare. What is human work, if a little violence of nature can destroy it at once? (Nitti 1919, p. 62)

It was to this question and the deep feeling of alienation caused by earthquakes that religion gave an answer, of interest to communities hit by such catastrophic events. Such demonstrations, either for whose "miracle" was said to have saved the village while others in the neighborhood had been struck[19]—or the cult of San Emidio, patron saint of the earthquake,[20] had an important role in reassuring the

[19]An example of such gratitude is the case of Naro in Sicily. The towns in the valley of Noto and the ones along the west coastline expressed their thanks to their patron San Calogero, who had saved their villages from the earthquake of 1693. Each year in the date of the earthquake (11th January) they make a procession carrying the painting of the miracle. This painting is preserved over the high altar (Lombardo 1993, P. 10).

[20]The long seismic sequence in Umbria in 1730, Foggia in 1731 and Irpinia in 1732, had caused a great emotional impact also on the non-directly involved communities of the Papal States and of the kingdom of Naples. Therefore, during the 18th century, the Church re-proposed the cult of San Emidio, thaumaturge and protector from sickness and calamities including earthquakes. There are many images, dating from that period, that represent the Saint speaking to God trying to "control" the earthquake by divine virtue. The cult for the Saint spread quickly all along the Italian territory, but it became particularly important in those areas of the central and southern Apennines (Varrasso 1989).

communities and recomposing the broken social links. These demonstrations, based on a medioeval conception of the earthquake as the scourge of God, were carried out even in the 1900s, together with an increasingly scientific interpretation of seismic phenomena (Walter 2009).

References

Baratta, M. (1901). *I terremoti d'Italia*. Bocca, Torino: Saggio di storia geografia e bibliografia sismica italiana.

Baratta, M. (1906). *I terremoti in Calabria*. Bollettino della Reale Società Geografica Italiana, Maggio.

Baratta, M. (1915). *Difendiamoci dai terremoti: a proposito del recente disastro sismico della Marsica*. Nuova Antologia, Maggio.

Baratta, M. (1910). *La catastrofe sismica calabro messinese*. 28 dicembre 1908. Reale Società Geografica Italiana, Roma.

Baratta, M. (1936). *I terremoti in Italia*. Firenze: Le Monnier.

Bevilacqua, P. (1996). *La natura imprevedibile e l'umana imprevidenza*. Terremoti e disboscamenti. In P. Bevilacqua (Ed.), Tra natura e storia. Ambiente, economie, risorse in Italia (pp. 73–91). Donzelli, Roma.

Bevilacqua, P. (2005). Sulla impopolarità della storia del territorio in Italia, in Natura e società. Studi in memoria di Augusto Placanica. In P. Bevilacqua & P. Tino (Eds.), *Natura e società. Studi in memoria di Augusto Placanica. Meridiana Libri* (pp. 7–16). Donizzelli, Roma.

Boschi, E., & Guidoboni E. (1997). Catalogo dei forti terremoti in Italia dal 461 a.C. al 1997. English edition: Boschi E. Guidoboni E. Ferrari G. Mariotti D & Valensise G. (1997). *Catalogue of strong Italian earthquake from 461 b.C. to 1997*. Bologna: Istituto Nazionale di Geofisica & SGA—Storia Geofisica Ambiente.

Burke, P. (1995) *Una rivoluzione storiografica la scuola delle Annales. 1929–1989*. La Terza, Roma, Bari.

De Marco, R. (1998). *Il terremoto del 28 luglio 1883 a Casamicciola nell'isola di Ischia. Presidenza del Consiglio dei Ministri, Dipartimento per i servizi tecnici nazionali, Servizio Sismico Nazionale*. Roma: Istituto Poligrafico e Zecca dello Stato.

Ferrari, G. & Guidoboni, E. (1997). Scenari sismici e stime d'intensità: alcune costanti nell'applicazione della scala MCS. In E. Boschi & E. Guidoboni (1997) (Eds.), *Catalogo dei forti terremoti in Italia dal 461 a.C. al 1997* (vol. 1, pp. 78–82). Istituto Nazionale di Geofisica & SGA—Storia Geofisica Ambiente.

Filangieri, A. (1979). *Territorio e popolazione nell'Italia Meridionale*. Franco Angeli, Milano: Evoluzione storica.

Fulci, L. (1916). Le leggi speciali italiane in conseguenza di terremoti. Esposizione e commento. In *Enciclopedia giuridica italiana*, Soc. Ed. Libraria, Milano.

Gambi, L. (1960). La più recente e meridionale conurbazione italiana. In *Quaderni di geografia Umana per la Sicilia e la Calabria* (V, pp. 3–7). Messina.

Guidoboni, E. (1989a). *I terremoti prima del Mille in Italia e nell'area mediterranea*. SGA—Storia Geofisica Ambiente, Bologna.

Guidoboni, E. (1989b). Sismicità naturale e disastri sismici prima del Mille: il lungo periodo e i punti di vista. In E. Guidoboni (Ed.), *I terremoti prima del Mille in Italia e nell'area mediterranea* (pp. 15–16) Bologna: SGA—Storia Geofisica Ambiente.

Guidoboni, E. (2000). Un'antirisorsa del Sud: disastri sismici nella sfida economica. In P. Bevilacqua & G. Corona (Eds.), *Ambiente e risorse nel mezzogiorno contemporaneo* (pp. 245–261). Corigliano Calabro: Meridiana libri-Donzelli Editore.

Guidoboni, E. (2005). Dimenticare i terremoti? I segni dell'attività sismica nel paesaggio culturale e naturale in Italia, In P. Bevilacqua & P. Tino (Eds.), *Natura e società. Studi in memoria di Augusto Placanica, Meridiana Libri—Donizzelli* (pp. 17–36). Roma.

Guidoboni, E. & Ferrari, G. (1997). Scenari sismici urbani del passato. In E. Boschi & E. Guidoboni (Eds.), *Catalogo dei forti terremoti in Italia dal 461 a.C. al 1997* (vol. II, pp. 91–109). Bologna: Istituto Nazionale di Geofisica & SGA—Storia Geofisica Ambiente

Lombardo, L. (1993). *Catastrofi e storie di popolo. Terremoti ed eruzioni nella cultura popolare.* Terzopiano Edizioni, Siracusa.

Macry, P. (1995). *La società contemporanea.* Il mulino, Bologna: Una introduzione storica.

Maniaci, A., & Stellino, A. (2005), *La Calabria e il terremoto del 1783. Memorie dei danni e disegno della ricostruzione.* Storia urbana. N. 106/107, gennaio-giugno (pp. 89–110).

Margottini, C., Kozak, J., & Ceroni A. (1999). *Terremoti in Italia dal 62 A.D. al 1908. Frammenti di testimonianze storiche e iconografiche tratti dalla banca dati EVA dell'ENEA sulle catastrofi naturali in Italia.* Roma: ENEA.

Mercalli, G. (1910). *I danni prodotti dai terremoti nella Basilicata e nelle Calabrie. Inchiesta parlamentare sulle condizioni dei contadini nelle province meridionali e nella Sicilia*, (Vol. 5). Tomo 3 Relazione della sotto giunta parlamentare, Tipografia Nazionale, Roma.

Nitti, F. S. (1919). Inchiesta sulle condizioni dei contadini in Basilicata e in Calabria. In A. Massafra & P. Villani (Eds.), (1968), *Scritti sulla questione meridionale* (Vol. IV–I, pp. 57–62). Laterza, Bari.

Oteri, A. M. (2005). *Memorie e trasformazioni nel processo di ricostruzione di Messina dopo il terremoto del 1908.* Storia urbana. N. 106/107, gennaio-giugno (pp.13–64).

Placanica, A. (1970). *Cassa Sacra e beni della Chiesa nella Calabria del Settecento.* Napoli: Poligrafia & cartevalori.

Placanica, A. (1972). *Il filosofo e la catastrofe.* Torino: Einaudi.

Placanica, A. (1979). *Alle origini dell'egemonia borghese in Calabria: privatizzazione delle terre ecclesiastiche.* 1784–1815. Società editrice meridionale, Salerno-Catanzaro.

Principe, I. (2001). *Città nuove in Calabria nel tardo Settecento.* Roma: Gangemi.

Riccò, A. (1915). *Il terremoto di Avezzano* (pp. 1–11). Febbraio: Bollettino dell'Accademia Gioenia di Scienze Naturali in Catania.

Tino, P. (2002). Da centro a periferia. Popolazione e risorse nell'Appennino meridionale nei secoli XIX e XX. Meridiana. *Rivista di storia e scienze sociali, 44*, 15–63.

Tino, P. (2007). Territorio, popolazione, risorse. Sui caratteri originali della storia ambientale italiana. I frutti di Demetra. *Bollettino di storia e ambiente, 13*, 5–22.

Varrasso, A. A. (1989). *I terremoti e il culto di Sant'Emidio.* Chieti: Vecchio Faggio Editore.

Walter, F. (2009). *Catastrofi.* Angelo Colla Editore, Costabissara (Vicenza): Una storia culturale.

Earthquake and People: The Maltese Experience of the 1908 Messina Earthquake

Ruben Paul Borg, Sebastiano D'Amico and Pauline Galea

1 Introduction

On December 28, 1908 at 5:20 a.m. local time, a devastating earthquake (Mw = 7.2) struck Southern Italy along the Messina Strait (Fig. 1). This event caused severe ground shaking throughout the region and triggered a local tsunami. As result the cities of Messina along Sicily's coast and Reggio di Calabria were completely destroyed (Baratta 1910) causing more than 120,000 fatalities and with many left without shelter.

The effects of the earthquake were felt within a 300-km radius. Rescuers searched through the rubble for weeks, and people were still being pulled out alive days later, but thousands remained buried there. The 1908 earthquake had a significant impact on buildings and people and local communities which were displaced. The Maltese experience of the Messina 1908 earthquake relied on communication which reached Malta after the event. The assessment of the Maltese experience of the Messina Earthquake has so far been carried out with reference to published newspaper reports and other brief accounts including Herbert Ganado's *Rajt Malta Tinbidel*. Alfons Maria Galea a Maltese author and filanthropist published a book in Maltese on the earthquake and its devastating effects in the popular educational series *il-Kotba tal-Mogħdija taż-Żmien* just a few weeks after the event. The book is a vivid account of the destruction caused by the earthquake, the suffering of the survivors and the reaction of the population in reviving the city. The document presents first-hand accounts of the events in sufficient detail to give a clear picture of the severity of the event, extents of the damage and impact on the population.

R.P. Borg (✉)
Faculty for the Built Environment, University of Malta, Msida, Malta
e-mail: ruben.p.borg@um.edu.mt

S. D'Amico · P. Galea
Department of Geosciences, University of Malta, Msida, Malta
e-mail: sebastiano.damico@um.edu.mt

S. D'Amico (ed.), *Earthquakes and Their Impact on Society*,
Springer Natural Hazards, DOI 10.1007/978-3-319-21753-6_22

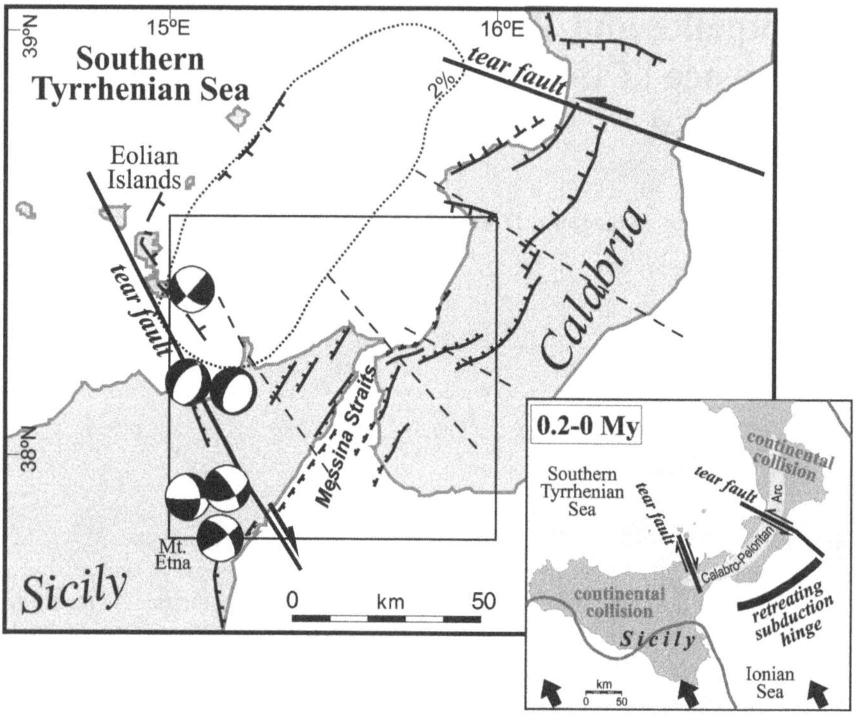

Fig. 1 The Messina Straits area investigated in the present study is indicated by a *square* in the *main plate* of this figure, while the *smaller frame* at the *lower right* (modified from Neri et al. 2009; D'Amico et al. 2010) shows the Calabro-Peloritan Arc region in southern Italy. This *smaller frame* indicates that while the central portion of the Arc corresponds approximately to the southeast-ward retreating subduction hinge, its northern and southwestern edges lie in continental collision zones which developed after local detachment of the subduction system. The northwestward *trending arrows* in the bottom of the same frame indicate the Nubia–Europe convergence direction (Nocquet and Calais 2004). The *main frame* shows the principal fault systems: *hatched*, normal faulting; *half arrow*, strike-slip component; *dashed*, presumed strike-slip but with different interpretations in literature (Fabbri et al. 1980; Finetti and Del Ben 1986, 2005; Monaco and Tortorici 2000). The *dotted curve* in the southern Tyrrhenian Sea marks the high velocity anomaly found by seismic tomography at a depth of 150 km (Neri et al. 2009) indicating the only part of the subduction system where the slab is still continuous (i.e. detachment has not yet occurred). The figure also shows the CMT focal mechanism solutions

It is mostly based on accounts received by Galea from persons in institutions including religious orders in Sicily who he knew. Newspaper reports in Malta and other countries together with Galea's book present clear first-hand accounts of this event and provide information on the building deficiencies and damage, limitations of communication infrastructure during that period, limits to timely emergency response to support the population and emergency action at the beginning of the 20th century.

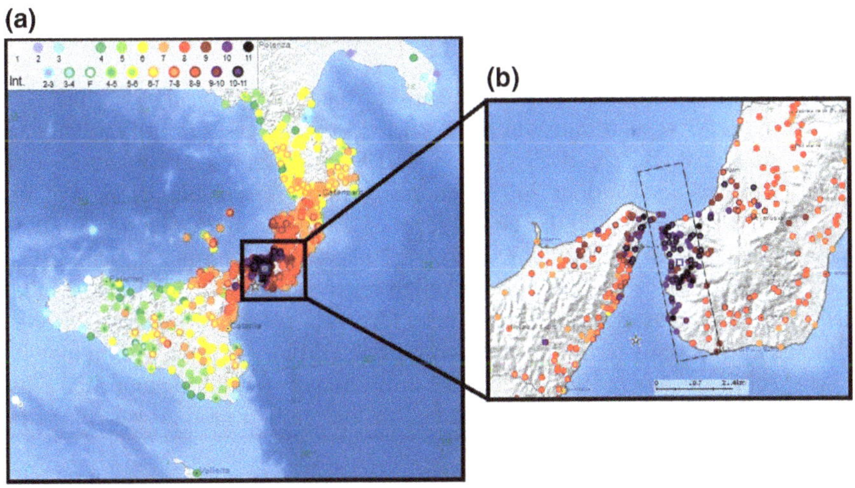

Fig. 2 Intensity map of the 1908 earthquake

2 Seismicity of the Region

2.1 Introduction

Southern Italy is one of the most earthquake-prone areas in the Mediterranean region and it has a long history of catastrophic events. The area lies on the Mediterranean-Alpine region which forms part of a complex boundary zone between the Eurasian, African, and Arabian plates (Fig. 2). The convergence of the Eurasian and African plates across the region has resulted in a wide zone of collisional tectonics and the Messina strait area is also affected by the subduction of the African Plate along the Calabrian Arc (D'Amico et al. 2010, 2011a). The subduction zone is characterized by an intermediate and deep seismicity clustered and aligned along a narrow (less than 200 km) and steep (about 70°) Wadati-Benioff zone striking NE-SW and dipping towards the NW down to 500 km depth (Piromallo and Morelli 2003; Neri et al. 2009).

2.2 The 1783 Seismic Event

The region of Calabria in the South of Italy, part of the Kingdom of Naples was hit by a sequence of 5 strong earthquakes in 1783. The first two earthquakes produced significant tsunamis. The epicenters of the earthquakes form a clear alignment extending nearly 100 km from the Straits of Messina to about 18 km SSW of Catanzaro. The epicenter of the first earthquake occurred in the plain of Palmi and the earthquakes occurred over a period of nearly two months, all with estimated magnitudes of 5.9 or greater. It is estimated that the deaths lie in the range 32,000–50,000.

2.3 The 1908 Earthquake

The 1908 earthquake may have been caused by the activation of a normal fault located in the Messina Straits, but the exact location and geometry of the fault are still controversial (Boschi et al. 1989; Bottari et al. 1989; Valensise and Pantosti 1992; Monaco and Tortorici 2000, Amoruso et al. 2002, 2006; DISS Working Group 2007; Pino et al. 2009). The low level of seismicity in the Messina Straits in the last few decades (Neri et al. 2003, 2004, 2008) has not permitted seismologists to define the locations and mechanisms of the seismogenic faults in this area. In the last thirty years, crustal seismicity has been recorded in a low-to-moderate activity with just a few events having magnitude above 5 (ISIDE at http://iside.rm.ingv.it).

The 28th December, 1908 earthquake was strongly felt throughout Sicily and southern Italy. Historical records indicate also that ground motion was felt as far away as Malta, Albania and Greece (Fig. 2).

The area of near-complete destruction by the ground shaking was concentrated in the cities of Messina and Reggio Calabira, with the geographic extent of the devastated areas more widespread in Calabria than in Sicily (Mulargia and Boschi 1983). In Sicily, damage was greatest in the Messina area, while Catania, the other major city about 80 km south of Messina did not suffer any significant damage from the ground shaking. Describing the damage in the city of Messina, Omori (1894) wrote: "*The enormity of the destruction of Messina is really beyond one's imagination. All the buildings in the city were, with a very few exceptions, considerably cracked or absolutely reduced to masses of ruin…*". In the major cities of Messina and Reggio Calabria more than 90 % of buildings were destroyed (Barbano et al. 2005). Messina suffered the most damage in the case of buildings constructed on soft soil (Barbano et al. 2005) while other localities sustained less intense damage because buildings were built on hard rock (e.g. Matagrifone and Gonzaga Castles). In several parts of the city the collapse of buildings caused obstructions covered by rubble and debris up to several meters thick (Fig. 3). Portions of the coast were also lost and also the submarine telephone cables were severed in several places and fires were observed in several locations (Comerci et al. 2008).

Furthermore the tsunami impact on the areas already heavily damaged by the ground shaking contributed to the devastation. In fact a few minutes after the largest event, a tsunami impacted the coastlines on either side of the Straits of Messina, with waves exceeding 6 m in several locations. Today the source mechanism of the tsunami is still debated within the scientific community. More recent research has suggested that the tsunami was generated by a seismically triggered submarine landslide (Billi et al. 2008). However, damage from the tsunami waves was most severe in the city of Messina and on the Calabrian coast and the tidal wave was also reported to have reached the Maltese islands. The large number of damaged buildings can be attributed to the type of building stock in Messina at the time, making the area very vulnerable. Buildings constructed with better quality materials or practices were less prone to collapse during the earthquake (Barbano et al. 2005; Luiggi 1909).

Fig. 3 **a** Structural damage of masonry buildings in Messina (raafyawan.wordpress.com). **b** Structural damage of masonry buildings in Messina (www.telegraph.co.uk). **c** Structural damage of masonry buildings in Messina. **d** Refugees awaiting transportation at Messina (www.ibblio.org)

Several fires also spread around the city following the earthquake. In the years following 1908, precautions were taken when reconstruction began, constructing buildings that would be able to withstand earthquakes of variable magnitude. As a result of the earthquake many people were left without a home and were relocated to various parts of Italy while others were forced to emigrate particularly to America.

2.4 Effects of the 1908 Earthquake in Malta

The 1908 earthquake led to a tsunami which affected the surrounding region in the central Mediterranean. The earthquake itself appears to have produced very mild shaking on the Maltese islands with no reports of structural damage in the contemporary newspapers (except for some alleged cracking in Marsaxlokk "cottages" mentioned in the Malta Herald of 28/12, which are difficult to substantiate). A local intensity of IV (EMS-98) may be assigned (Galea 2007). Unfortunately, the seismogram recorded by the Malta University seismograph installed at the time in Valletta remains missing from the local seismogram archive, although the Daily Malta Chronicle of the 29/12 reports that the seismograph was "thrown out of gear by the violence of its own action" (Malta Herald 1908).

The tsunami waves, on the other hand, appear to have caused more damage. There are inconsistencies between reports of different newspapers regarding the timing. The Malta Herald of the same day (28/12) reports the first wave arriving about 1 h 59 min after the earthquake, while the Daily Malta Chronicle of the 29/12 reports a 3 h interval. The Malta Herald reports initial wave heights up to 5 feet (1.7 m), with the sea receding and transgressing several times during the day at Marsamxett harbour. Boats in the Grand Harbour "*broke adrift*" and houses were inundated at Sliema, Msida and Pieta on the northern coast of Malta. Serious flooding was also reported in Marsaxlokk Bay, on the southeastern coast, where fishing boats on land were heavily damaged, and general panic ensued among the population (Malta Herald 1908; Daily Malta Chronicle 1908; Savona Ventura 2005).

The Daily Malta Chronicle of December 29, 1908 reported the event as follows:

> At about a quarter to eight (yesterday) the sea became strangely agitated. Thinking of the earthquake which had occurred some three hours before, one was inclined to conclude that the convulsion of the earth had been submarine and not far distant from us. The seabed appeared to be casting violently off the superincumbent mass of water and driving it to the shore. The Grand Harbour is protected by the breakwater; the tidal wave rushed unchecked into the Marsamuscetto harbour. In the creeks the agitation was great. In Misida creek the waters dashed right over the confining barriers and rushed up to, and into, the houses and shops by the shores. From the early morning it continued until after 4 p.m. People trembled at first to witness that which was taking place. After rising over the land, the waters receded and left the seabed bare near the shore, fish was picked up wriggling in the sand seeking to get back to their own element. This tsunami was caused by the earthquake that hit the Straits of Messina on December 28 and that levelled Messina and Reggio di Calabria. Because of the seismic activity in the area, Sicily and Calabria are referred to as la terra ballerina - the dancing land. The death toll was high reaching close to 200,000 because of two main factors: the extent of the quake - 7.5 according to today's Richter scale and the fact that it happened at about 5.20 a.m. when most people were indoors. The resulting tsunami triggered 40-foot waves. The intensity of the earthquake was felt in Malta. The seismograph at the university (of Malta) was thrown out of gear by the violence of its (the earthquake's) own action. The trouble of the earth lasted an hour-and-a-half. Maltese doctors, nurses and priests went to the stricken area with the assistance of the Royal Navy.

In *Rajt Malta Tinbidel*, Herbert Ganado reports on the first earthquake shock at 5.20 in the morning of the 28th of December 1908 and how the ground and the walls of houses in Malta were shaking. He states that the earthquake shocks were felt in Malta, reporting that furniture shifted from its place and that people were terrified and panicked. Ganado had heard first hand accounts of what happened. He reports that people ran out of their houses from their beds and writes about an aftershock and another 8 shocks at intervals felt in Malta, with the last one reported at 6.15 in the morning. Ganado reports that people ran to the open spaces and the countryside. He further writes about the rise in sea level at about 7.45 in the morning noting that the effect was mostly felt in Marsamxett Harbour since the new breakwater recently constructed protected the Grand Harbour. He reports on water reaching houses in Sliema, Msida and Pieta' (Ganado 1977).

3 Earthquake and People

3.1 *Introduction*

Knowledge on both the 1783 and 1908 seismic events in Messina and Calabria can be retrieved from a number of historic documents. Nevertheless the experiences of people in nearby communities and their reaction is relatively limited. In both earthquakes the Maltese experience including the emergency action, can be traced to few specific documents and is of particular interest given the proximity of the Maltese Archipelago to the affected region.

3.2 *The 1783 Calabria Earthquake*

Sutherland in his book The Achievements of the Knights of Malta, (Sutherland 1846) refers to the earthquake during the eight years of de Rohan's reign in Malta, and to the role that Order of the Knights of Malta had in emergency action in the region affected by the event. The Knights were engaged in assisting the survivors immediately after news of the event reached Malta. Under the direction of the French Frelon de la Frelonniere, the Knights left Malta with ships full of medicine, beds and tents. Galea, reporting from the source, refers to the shock and sorrow in Europe whilst the Knights of Malta were obliged to act and assist, the reason for belonging to the Order of St John (Sutherland 1846).

They reached Reggio, finding the devastation the earthquake and tsunami left behind. It is reported that earth tremors could still be felt. They left half the stocks in Reggio and proceeded to Messina, finding the city devastated, with collapsed buildings and destruction all around. The normally busy shoreline was deserted, with people running around on the building rubble heaps and reported to be disheartened and spread in different places around and in the fields. It is reported that the Neapolitan commander however had a safe and comfortable quarter set up for him including a band for entertainment, which contrasted deeply with the despair of the population and the dead and sick survivors around. He was not reported to be very welcoming to the Knights who offered to set up a hospital in Messina. The Knights were only allowed to distribute goods food and medicines to those in need and to care for the sick. Their help was however not welcomed by the Messina Commander and they left the city, stopping again in Reggio where they left their remaining supplies before returning to Malta (Sutherland 1846).

3.3 The 1908 Earthquake in Messina and Calabria

The knowledge of the 1908 earthquake was communicated and diffused to surrounding regions through the available means of communication of the time, including letters and postcards and newspaper reports. The accounts of the earthquake were reported in Malta mostly in the newspapers of the time which included the Malta Herald and the Daily Malta Chronicle. The concerns and fears of the community were communicated through letters and postcards sent by various personalities who had first-hand experience of the earthquake. Members of various religious orders in Messina sent letters to friends in Malta including Alfons Maria Galea, a Maltese filanthropist. The letters included reports of the earthquake, extents of the damage and impact on society. One relevant account is a book published by Galea in Maltese a few weeks after the event. Herbert Ganado also reported briefly about the earthquake and the experience in Malta, in the first volume of *Rajt Malta Tinbidel, L-ewwel Ktieb (1900–1933).*

4 A Maltese Account of the 1908 Earthquake

4.1 Alfons Maria Galea

Alfons Maria Galea was born in Valletta in 1861. He studied under the famous writer Annibale Preca in Malta, with the Jesuits in Gozo and eventually in Marseilles. He is remembered as a filanthropist having set up important religious convents and supported various religious institutions in Malta. He was senator between 1921 and 1927 and is best remembered for his publications in Maltese intended to educate the people who did not get proper education in School and the poor. In September 1899 Galea started publishing a new series of books in Maltese called *il-Kotba tal-Mogħdija taż-Żmien*. The series was well received and publication continued until 1915. The book *It-Theżhiża ta'Messina* written in Maltese, was published in the series during March 1909 (Fig. 4). The book includes an account of the earthquake, based on the letters which are reproduced in the same book and with additional comments and observations by the author. Galea's introduction to his book refers to the *Candlora* (2-2-1909), therefore indicating that the book was written and published just a few weeks after the earthquake. Furthermore in the introductory chapter, the book refers to the Calabria 1783 earthquake with Galea quoting from the following source: *The achievements of the Knights of Malta* by Alexander Sutherland, Volume 2, Chapter VIII (Galea 1909).

In the context of his Series of Publications in Maltese *il-Kotba tal-Mogħdija taż-Żmien,* the book is intended to communicate to the Maltese about this event and to inform about the suffering of the people of Sicily. This is further supported through his publication of the book in a relatively short time. In addition Galea published correspondence in newspapers at the time, asking for support to the population and also to thank those providing support.

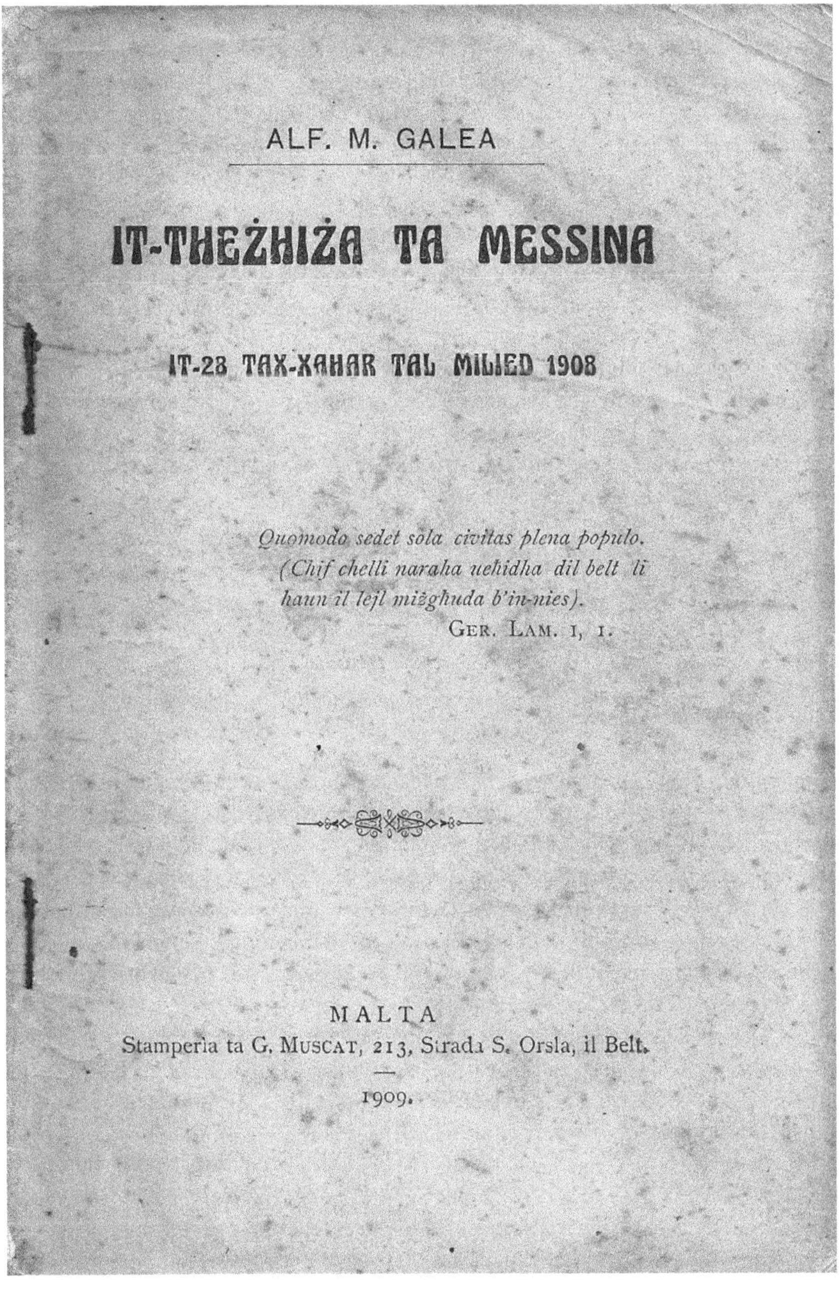

ALF. M. GALEA

IT-THEŻHIŻA TA MESSINA

IT-28 TAX-XAHAR TAL MILIED 1908

Quomodo sedet sola civitas plena populo.
(Chif chelli naraha uehidha dil belt li haun il lejl miżghuda b'in-nies).
GER. LAM. I, I.

MALTA
Stamperia ta G. MUSCAT, 213, Strada S. Orsla, il Belt.

1909.

Fig. 4 *It-Theżhiża ta'Messina*, it-28 tax-xahar tal-Milied, 1908, by Alf. M. Galea. Published in 1909 in Malta at Stamperija G. Muscat, 213, Strada S. Orsla, il-Belt (St Ursola Street, Valletta)

Galea's book refers to first-hand accounts of the earthquake from named sources presented in sufficient detail. Galea was familiar with the various buildings mentioned and had visited the specific places mentioned a short period before the event, having known well some of the survivors who provided the first-hand account, and in many instances those who lost their lives. At the beginning of the 20th century, trade relied on sea transport, with Sicily being the closest land to the Maltese Archipelago.

Few copies of the book survive as is generally the case with most of the volumes in the *Kotba tal-Mogħdija taż-Żmien* series. It is written in the Maltese language, but before new regulations were adopted for the written language years later. These two considerations have seriously limited access to the book, which explains why, in spite of being a relevant and important source of information, it has been ignored by scholars so far and has largely gone unnoticed. The document is a valuable reference in interpreting the impact of the earthquake for the local community in Messina but also in understanding the event as perceived in the Maltese Archipelago by the local population where communication was limited to printed media, in particular newspapers.

4.2 *An Account of the 1908 Earthquake:* It-Theżhiża ta'Messina

4.2.1 An Appreciation

In his introduction Galea presents to the readers an interesting account of an earlier earthquake reported by Sutherland, which occurred 128 years earlier and the services rendered by the order of the Knights of St John who sent help from Malta. He sets the scene for what is to follow in a book which presents the accounts of the 1908 earthquake, as reported by people most of whom he knew well. He makes a clear statement regarding the damage caused by the combined effect of the earthquake but also as a result of the tsunami hitting Messina, one of the brightest cities of Sicily, on the 28th December 1908. He even presents an overview about Messina and its history to the readers in the first chapter.

Galea in general reports on the earthquake on the basis of accounts based on the following: letters of Dun Ang Lovisolo, Dun Michiel Rua, Sister Ildegarda Zmiglio, Father Liborio Ferrara and father F.P. Nalbone. He reports on the *Sorijiet tax-Xjuħ* (Little Sisters of the Poor), the *Sorijiet tal-Bon Pastur* (Sisters of the Good Shepherd), the *Patrijiet ta' Ġieżu* (Franciscan Minors), the *Kabbuċċini* (Franciscan Capuchins), the *Franġiskani Konventwali* (Conventual Franciscans) and the *Sorijiet il-Blu* (Sisters of the Little Company of Mary also known as Blue Sisters). In the book he further refers to the *Sależjani* (Salesian Brothers of Don Bosco), the *Sorijiet ta' San Viċenz* (Sisters of Charity) and the *Ġiżwiti* (Jesuits, Society of Jesus).

Galea further reports on the organisation of help in Malta for the people in Messina both on personal experiences and on reports in the newspapers of the time including the Malta Herald of the 11th February 1909. He goes on to report on the

British and Russian aid, and on Maltese doctors. He further reports on the survival of people including children following the earthquake and on the criticism directed at priests, whom he defends in his book. He discusses the role of the Pope and bishops in the aftermath of the earthquake. He dedicates a chapter to Father Urso, who lost his life during the earthquake, and who he knew very well, and also presents two letters from Dun C Gatti and John Simeon, on Father Urso. He concludes the book with a chapter entitled "Messina is not dead". The book is clearly presented as an expression of deep sorrow of the prominent Maltese personality of the time, on what happened to Messina, a city he was very familiar with. It is an appreciation presented immediately in the aftermath of the earthquake, to the people he respected and knew well in his activities.

In his chapter dedicated to Dun Urso, Galea recounts his visit to Messina on the 11th November 1908, a few weeks before the earthquake. He recounts how Dun Urso wished he could be sent back to Sliema in Malta from Messina and about his wish to build in Malta a chapel for the *oratorju festiv* since this was needed due to the increasing number of children. Galea further reports that at the time of writing the chapel was under construction thanks to donations provided by rich and poor people. The photo of Dun Urso reproduced by Galea in the book, was taken in Messina on the 12th November 1908. Galea reports that once the news of the earthquake arrived in Malta they sent a telegram asking about Dun Urso and did not receive a reply until two or three days later, probably because the telegraph was not functional. The message received finally from Catania announced his death. A letter from his father on the 31st December 1908 from Belpassu and again on the 17th January 1909 informed that he had been removed from the temporary burial in Messina when he died and buried in his home town. Galea recounts the sorrow of the Maltese community at the *Juventus Domus* who gathered at the chapel of St Patrick's Institute in Sliema.

Galea concludes the book with a clear expression of sorrow but makes a statement in the last chapter *Messina m'hix mejta*, "Messina is not dead". He quotes from *Malta letterarja* in the conclusion with a statement of hope, with the opening again of the roads, reconstruction of buildings and churches, and people coming back to a city full of life and activity.

4.2.2 The Earthquake Experience

Galea presents first hand accounts as the one by Dun Lovisano from the Salesian Brothers' College. In his letter to Galea, Dun Ang Lovisano describes his experience of the earthquake at 05:20 of the 28th December 1908 when a loud sound like thunder was followed by a tremor with the house collapsing suddenly into a rubble mound. He ended squashed between the bed and the wall, with a timber beam falling onto him together with stone dust and render. He moved out, meeting others in the corridor and moved on to the dormitory on top of the church looking for the children, finding it to have collapsed down into the lower floor. The dormitory in the first floor overlying the school and the refectory in the lower floor collapsed. The external wall collapsed into the garden, and all, except five, were buried under stone. The same happened to the church and another dormitory some distance away. Dun Lovisano could hear crying from down below. The staircase had collapsed so they helped each other slide down to the ground floor. However the church's door could not be opened due to the rubble within the church. He continues to describe the action in getting people including the children out of the rubble, some dying immediately after, others surviving and joining in to help. He describes the chaos with survivors scarcely dressed and without shoes in the dark looking for people under the rubble of the collapsed church. He further describes the situation at day break with corpses of dead children all around, all wet with blood and rain. People were crying while trying to help, others asking for help. In the morning light, he could better appreciate the damage with collapsed classrooms and bedrooms. The fact that the earthquake struck at night, in adverse weather conditions, complicated the early action resulting in difficulties for survivors in appreciating the extents of the damages immediately and making it even more difficult to help those in distress and needing immediate help. Only at day break could the full extent of the damage and the extensive loss of life be well appreciated.

Dun Ang Lovisano describes how they called the names of the missing priests and that after an hour they could hear calls for help from under the rubble. He describes how very tired, under heavy rain and with adjacent walls at risk of collapse, they excavated without adequate tools to save Dun Urso. He recounts how professor Bertu Boeris came to help. Yet Father Urso could not be released from under the rubble because of serious injuries and died later. Eventually he recounts how they dug a hole and buried in it Father Urso. He further describes how children trapped in upper floors tied bed sheets one to the other to form a rope and could therefore slide down to ground level. Some priests went to look for help and returned later with some soldiers with tools to be able to help children trapped in the upper-most floors of the parts of the building which had not collapsed. He describes how difficult it was to get help since others were also stopping and asking the soldiers to help while on their way. He even describes how people became aggressive in the circumstances and the frustration of not being able to help trapped children in upper floors who needed urgent help. In addition he describes how, when asking for help from people passing by, these were helpless and desperate, asking for help themselves for their own relatives. For hours no help was available. Dun Ang Lovisano further describes how a fire was noted approaching the ruins of the college.

4.2.3 Damaged Buildings

Through the accounts, Galea describes the damage caused to the buildings ending up as piles of rubble (*ibrag mazcan*) with the occasional parts of walls standing after the earthquake. He further reproduces photos of damaged buildings including a number of the Salesian College (Fig. 5). The roofs which appear to be made from timber collapsed when walls overturned. Dun Ang Lovisano, describes how masonry walls overturned like the pages of a book. He indirectly described the overturning mechanisms of walls without adequate restraint and ties at floors, with the upper floors collapsing down to the lower levels. He describes how people ended up under rubble as a result of the collapsing walls and how people were thrown out of an upper second floor down into the street at a distance. In examining both the text and descriptions of the damage and sequence of failure and also the images of the damages in particular those for the Salesian College, it can be noted that most masonry wall structures consisted in sack masonry, and with timber roofs.

The corner of buildings, normally consisting of more regular stone blocks reinforcing the masonry structure, were noted to have survived in some instances. However the remaining masonry, in particular sack masonry type walls failed during the earthquake with expulsion out of the plane of the façade and in many cases collapse of the wall. Other mechanisms reported imply the out of plane rotation of the wall indicating also the absence of adequate ties. Collapse mechanisms related to this type of masonry consisting of irregular stone block walls, is reported in other similar non-reinforced masonry structures including the historic centre of L'Aquila following the 2009 Abruzzo earthquake more than 100 years later (Indirli et al. 2013; Borg et al. 2008, 2010, Borg 2010a, b)

Dun Lovisano describes the danger due to walls which though standing, were unstable and at risk of collapse on survivors. He further describes the damage on their route from the college to the coast and the calls for help from people buried under the rubble. He describes the harbour structures damaged and under water, on route to the train station and many people at the station with little clothing in the cold waiting to get into one of the train wagons. He also reports on people in distress holding onto planks at sea whilst travelling between Galati and Scaletta.

Father Liboriu Ferrara sj, wrote a letter to Father E. Scio sj on the 4th January 1909 and described how large timber beams and stones from walls fell on him, referring to the masonry walls and timber roofs. He further describes how he held to the bed sheets not to fall into the street below and how eventually he could slide down a timber beam.

From a letter of Father M.L.M. Tonna Barthet of the 26th January 1909, the Conventual Franciscans reported that they also lost their old church of the Immaculate, dating from the time of Pope Alexander IV. St Anthony of Padua is reported to have lived in the convent in 1219. With the loss of the church, the order lost important items including a statue of St. Anthony. This goes to show the significant loss of cultural heritage as a result of the earthquake.

4.2.4 Communication

Communication relied on printed documents and newspaper reports. Accounts of the earthquake reached other regions through postcards sent out by the survivors. The photographs published in the book accompanying the chapter based on Dun A. Lovisano's experience, show the damage caused to the Salesians College. Galea reports also that Dun A. Lovisano by the time of writing and printing the book, had arrived in Malta and therefore he could also rely on first-hand accounts of what

(a)
(b)
(c)
(d)
(e)
(f)
(g)
(h)

◀ **Fig. 5** **a** Surviving part of a masonry wall building with timber floors after the earthquake. **b** Façade of a church with bell tower standing after the earthquake with adjacent collapsed structures. **c** Remains of ta' Zacra Church near the Messina Cemetry. Surviving corner of a masonry building, with remaining walls collapsed. **d** The surviving façade of the Salesian college, with the collapsed upper floor and the remaining fissured wall structure. **e** The street where the Post office was located. View showing damage after the earthquake. Note the damage with expulsion of sack masonry. **f** Collapsed timber roof elements. **g** First floor bedroom in the Salesian College in Messina, and the recreation hall. Note the timber floors. **h** Salesian College in Messina, with the Dormitory at first floor collapsed onto the ground floor school and refectory. Note the collapsed back wall together the timber beams and free standing external walls. (Galea 1909)

happened and on the experiences of the survivors. Father Rua, the Salesian General reports on the ship Washington arriving in Catania from Messina bringing news of the destruction of the earthquake and also reports that the news of the earthquake damage was spread in Italy through the newspapers since the telegraph was not functional. It took days to get news about survivors. It is clear that Galea himself relied a lot on letters received from friends and people he knew in religious orders, together with newspaper accounts, in order to write the book. Galea makes an interesting reference to the fact that people had occasion to view the damages in cinemas (*ċinematografu*).

It is interesting to note the role of academia in Malta with Galea reporting on two conferences organised in Malta, concerning the earthquake. Galea reports that Prof. A. Bartoli organised a conference on Saturday the 13th February on Messina and Calabria, on how these places were before and after the earthquake. He refers to the use of a *Lanterna Magica*, a projector, to show to the people who attended this event the cities before the earthquake and then after the event. Another conference was organised by Prof. D. Fallon about 15 days later on the British activity and emergency support in the city of Catona in Calabria.

4.2.5 Transport

The access routes were interrupted due to the damage, with resulting difficulties in reaching the affected areas. Since the telegraph was interrupted due to the earthquake, communication to the surrounding areas was difficult. Yet Dun Lovisano describes how they managed to reach the train station and get to Catania by train indicating that routes were in part functional. He also describes how the harbour structures were destroyed and the shore line under the water level, yet reports on the use of the Russian ship for the transport of survivors to Catania. Father Rua describes how the train could not reach Messina and how the Provincial travelled to Messina trying to get as close as possible to the city and then continue on foot. The Director of the Sisters of Charity Ildegarda Zmiglio, reports from Rome on the 7th January 1909. She describes how difficult it was to care for and provide food for survivors for four days in Messina. She further describes how they carried the babies even in drawers and carried them under heavy rain to get the train from Geraci after walking 12 miles. She describes how sisters who survived in the women's hospital left to Palermo by train, once they could hear no more calls for help and then moved on to Naples. Father Liboriu Ferrara sj, reports how he could not travel to Gazzi since roads were full of rubble and had to travel through the hills of Peloritani till he got a train from Divietu near Barzola, to Palermo.

4.2.6 Post-event Support Organisation

Earthquake survivors were left homeless without shelter, during the darkness of the cold winter December night under heavy rainfall. Whilst many lost their lives under the rubble of the masonry walls, many others were left without shelter and refuge, in adverse weather conditions. Galea states that all of a sudden the rich and the poor shared the same experiences, irrespective of social standing before the earthquake. He argues that faith and religion were important for the survivors of the earthquake looking towards a future, with every family experiencing losses of dear ones, but also all possessions and homes. In addition Christian charity had provided for the needs of the people, many sending clothes, money and medicine, according to Galea. Others spontaneously left their home and went to the areas hit by the earthquake to help in the emergency action, in spite of aftershocks and earth tremors as described by Galea. Survivors, in spite of losing family and friends helped in recovering those trapped under the rubble whenever calls for help could be heard. Since an organised Civil Protection that could intervene in such catastrophic events did not exist, other organisations including religious orders and groups of survivors took the initiative of coordinating and providing assistance to survivors. Further, religious orders provided a reference point in a society where the Roman Catholic Church had an important and central role even before the earthquake. People found refuge in priests and members of religious orders as clearly indicated by Galea, even though these had suffered serious losses within their own communities, not only of members of the orders themselves but also many under their care including children and elderly. Dun A. Lovisano describes his role in providing moral support after the event to families who lost their dear ones, whilst facing losses within his own community including the loss of the Salesian College which had collapsed. Galea reports that 200,000 Liri (British Pounds referred to Liri in Malta at the time) were sent to Pope Pius X to be distributed to those in need. Galea further reports that besides the money sent to the pope, an additional 800,000 Liri were collected from around the world for the people of Messina and Calabria. He stated that the British sent 237,000 Liri and that the United States sent a ship full of goods.

Dun A. Lovisano describes how temporary shelters were set up with sheets. Lovisano once in Malta reports to Galea that after the earthquake and whilst travelling, they suffered hunger, relying on nuts and bread which they got with difficulty. He further describes how different people who survived shared and provided the little food they could get to the injured under the tents, including bread, boiled rice and goat's milk. He further describes how more and more tents were set up on the wet muddy ground and how they lit bonfires with pieces of timber to keep warm. The Russian soldiers assisted in the rescue of children trapped in upper floors using ladders and moving to the upper floors by forming openings into the roofs to reach the children. Lovisano describes how 15 Russian soldiers lost their lives helping the people during rescue operations. The rescued children were taken to the College in Catania on board the Russian ship.

Father Rua describes how the Provincial travelled from Catania to Messina to help. He further reports how the injured reaching Catania with Father Lovisano, were taken to the Garibaldi hospital. He describes how the Director of the Palermo college sent Fathers and young men to Messina to help with medicine and food and how they saved many people. This goes to show how the clergy provided assistance and emergency action to the survivors from the other cities which were not affected by the earthquake. The survivors were hosted in hospitals in Palermo. Further Father Rua informed the Mayors of Messina and Reggio via telegraph that he would welcome orphans in his Institute.

Father F.P. Nalbone director of the Society of Jesus wrote to the Jesuits in Sicily from Acireale on the 6th January 1909. He reports that when the Fathers and children realised that nobody could help, they started working with the help of a soldier to rescue other children. He describes how survivors put an effort to help those still under the rubble and how children who were orphaned now needed to be taken care of.

Galea reports that according to a letter from Father Luigi Attard, the Franciscan Minors, in spite of the serious damages in their convents in both Messina and Reggio and loss of lives within their communities, helped those who survived working with the little clothing they had to assist the injured. Galea further reports that the Franciscan Capuchins also lost their churches and convents of Messina and in Calabria, besides members of their communities.

The Director of the Blue Sisters sent four of her sisters to Messina to assist in helping the injured while she asked the Pope to send to her hospital next to St Stephen's church five of the injured people. Galea further reports that nine people were recovering in the hospital.

Sister Mary de Sales reported how the injured, full of blood, were hosted under tents with the Bishop visiting them from one tent to the next. In Palmi she describes a single pit with 665 dead bodies and writes on the bad smell. She reports that at least the ground tremors had ceased after nine days and that many people could find shelter in small spaces. She reports on the help of the priests who would do different tasks in assisting the injured including helping with the cooking.

4.2.7 Relocation of Survivors

Father Lovisano reports on discussions whether they should move on to Naples or reach other Fathers in Catania. In spite of the tragedy hitting the cities of Messina and Reggio and surrounding areas, many people considered moving out of the area while many others were reluctant and would not leave their land. Refugees also arrived and were hosted in Malta. The Little Sisters of the Poor helped get survivors to their other homes in Acireale, Catania, Modica, Naples and Rome to be cared for. Two sisters were sent to Ħamrun in Malta to rest and recover even though Galea reports that they were eager to get back and help.

Galea also reports on the Good Shepherd sisters whose convent he had been to a month before the earthquake. He refers to a sister who had witnessed the earthquake and with whom he spoke at Ħal Balzan in Malta. She reports on how the injured were transferred to Ħal Balzan in Malta after suffering the rain and the cold weather for four days with little clothing and after having to walk for many hours. Galea wrote a letter on the 8th January 1909 which appeared in both the Malta Chronicle and the Malta Herald, to thank on behalf of the Provincial sister of the Good Shepherd, the captain Evelyn Le Marchant of the ship Sutlej, for helping the sisters during their transfer to Malta. Further, Galea refers to the Archbishop of Syracuse who wrote to Galea to inform him that he thanked the Admiral for the help provided by his men, by the sisters and by Maltese doctors, to the people of Messina and Calabria.

When Galea reports on the goods collected by the Maltese and distributed to those in need, he also notes that 17 women who survived the earthquake travelled with the sisters from Messina and stayed at the Convent of the Good Shepherd in Ħal Balzan, Malta. He further reports on 40 Sicilians who came over to Malta following the earthquake.

4.2.8 Maltese Support

Maltese doctors left Malta on the British ships to assist the injured in Messina and Calabria. Galea reports on how this had also happened 128 years earlier, when the Knights of the Order of St John provided assistance in Messina and Reggio.

Once news of the earthquake reached the British ships, these joined the Russian ships in providing assistance. They carried large quantities of sheets, tents and medicine and all that was necessary for the injured. The Maltese and the British sent all they could to assist the people of Messina and in spite of the difficult situation in Malta, the rich and the poor collected what they could for the people of Messina. The Maltese collected clothes, sheets and money to help the people of Messina. According to Malta Herald of the 11th February 1909, Lawyer A.H. Stilon and Mr. R.G. Vassallo informed friends that they intended to send material to Messina and people started sending to them all sorts of goods. The material was distributed between various people who could pass them on to those in need. Galea reports that material was provided to the Convent of the Good Shepherd at Ħal Balzan for the 17 women who arrived there with the sisters from Messina, with additional material for the 40 Sicilians who came over to Malta after the earthquake. Austrian and Italian ships transferred the goods free of charge from Malta to Siracusa and Catania. Many Maltese sent goods also to the Malta Chronicle to be sent to Sicily as reported in this newspaper on the 7th, 9th and 14th January 1909.

People sent considerable sums of money to the Archbishops of Syracuse, Catania and also Acireale to pass them on to the Archbishop of Messina. The money was to be given to those surviving the earthquake. The Archbishop of Malta Mons. P. Pace and the Bishop Mons G.M. Camilleri expressed their wish to priests in Malta and Gozo for the collection of money to be sent to the Pope to be distributed to the survivors of the earthquake. Galea states that the Maltese contributed significantly both in terms of goods and money, depositing material with priests and also sending goods to the newspapers. In addition he states that the Maltese government spent 266 Liri and the Lord Mayor of London sent 1000 Liri to the Maltese government, to be used in the assistance provided to the people of Messina and Reggio. Galea reports that the Duke of Connaught left on the Aboukir on the 10th January to visit the hospital which the British set up in Reggio.

A funeral mass was held in St John's co-cathedral for the dead, celebrated by the Archbishop Pace. Galea reports that other funeral masses were held also in Gozo with flags flying at half mast. Galea expresses satisfaction at the generosity of the Maltese given that in those days the Maltese collected two thousand Liri for the people of Messina and Calabria, even though they had collected significant sums for other causes.

4.2.9 The British, The Russians and The Maltese

Galea states that the British and the Russians provided significant assistance to the people of Messina and refers in this regard to the British newspaper the Telegraph, and Italian Newspapers who had words of praise in their regard. A British Doctor on the ship Philomel wrote in the Telegraph that on the 30th December at 9.30 in the evening, Malta received orders from England that doctors on British ships were to go and assist in Messina. The Philomel left Malta 12 h later on the morning of the 31st December with 8 British and 13 Maltese doctors, with the Maltese doctors carried along as much medicine as they could get during the night. He reports that the sea was very rough and that they arrived in Messina at dawn of the 1st January 1909. The doctor gives an account of what they found in Messina, with building rubble all over the place with not a single house standing, the roads blocked with rubble and fire spreading all over. He notes that there were people under the rubble and notes that the survivors were not helping at all. He states that the prison collapsed and describes people going around the city to steal. He notes that a thousand doctors would have been necessary in such a disaster. He gives an account of the silence, the collapsed building throughout, with fire and the mountains covered in snow in the distance. He reports that they landed in Canatello, and that they had with them Italian guard. They set up 12 tents and 2 large shelters to form a hospital and that people gathered around asking for food. The language barrier during emergency activity is a relevant issue. The British doctor reports that the Maltese doctors could communicate in Italian with the local people. The doctor informs that only 500 people survived out of 3000. He notes that they had guards protecting the camp to prevent people from entering since they were hungry and had nothing left since their homes collapsed. He reports that the sailors worked to get the people out from under the rubble while the doctors worked in the temporary hospital set up. He reports on the many injured people requiring assistance including those who had been buried for 5 days. The doctor writes that the injured were in a terrible state, most dying whilst being assisted and in most cases little could be done to save people. He reports, that by midnight on the day they had assisted 200 people in the hospital and that on the 3rd January many people were still under the rubble with the dead lying here and there. They returned to Malta on the 5th January and he praises the selfless sailors who worked for 18 h at a go to assist the people in Canatello. Galea further reports on an account appearing in the Graphic of the 6th February, concerning the captain and sailors of the ship Afowen arriving in Cardiff, describing their activity in saving 10 children from the upper floors of a damaged building using ladders.

Galea reports that Pope Pius X expressed his gratitude for the support of British sailors, with British students who went to visit him on the 12th January. Galea reports that while the Pope expressed his gratitude also for the help of the Russians, he notes that the British could help more since Malta is so close to the affected area. Galea goes on to agree with this and refers to the goods transferred from Malta to Sicily from the Maltese drydocks and referring to the newspaper The Chronicle which lists the thousands of pounds of meat, flour and biscuits, medicine and sheets. The Pope made reference to the fact that the British and the Russians reached the area first, followed by the Italians who he states arrived late and lost time, though they made up for this later. Galea reports that then even French, Greek and Germans also assisted, as did many other countries.

The Maltese doctors who went to Messina are mentioned by name in Galea's book and he further states that on the 12th January 1909 they were invited by the Governor of Malta Sir Henry Fane Gant to be thanked personally for their work in Sicily. They were once more invited by the Duke of Connaught on the 16th January who informed them that he would inform King Eduard VII about their actions. Galea reports that he received a letter from the Archbishop of Syracuse thanking him for the help received from Malta, for the help of the British, Russians and the Maltese and also the Maltese doctors and sisters who assisted the injured people.

Galea presents also an account of the rescue of three children from under the rubble, 18 days after the event. He described how they were trapped in the cellar of the building and survived on the little food they could get (wine and onions) whilst they dug with their hands. They were taken on the ship Savoia to be assisted.

4.2.10 The Role of the Religious Orders

Galea presents clearly the role of the religious orders during the earthquake and defends their contribution in saving people, in spite of criticism directed towards these at the time. Galea had close ties with religious institutions who provided the main sources of the book. He dedicates a section of the book "The Priests escaped from Messina" to describe the role of the priests after the event. Referring to the newspaper Roma, he quotes Admiral Mirabelli who had stated that the priests were nowhere to be seen and that the Archbishop just sent his secretary to greet the King and the Queen. Galea then criticises Mirabelli and states that he was unable to handle the situation and that someone else and better than him should have been in charge of the emergency activity. Galea suggests that he should have worked more to find workmen to save people from beneath the rubble, and that he should have stopped ships sailing in the Straits to get help and should have sought food and medicine for the survivors.

Galea quotes from il Giornale d'Italia, referring to the parish priest of Mazacuva who shared the goods he could retrieve from his house which collapsed and though with little clothing, inspired the survivors in helping extract those under the rubble. The bishop of Mileto gathered members of the clergy and organised help by visiting villages to assist the injured, bury the dead and providing support to survivors. He wrote in newspapers to disseminate the news of the event so that people could help. Galea refers to il Giornale di Sicilia and reports how the Archbishop of Messina set up a hospital in the part of his house which had not collapsed and provided assistance to the injured. Galea reports from an account in the Civilta' Cattolica, which refers to another report in The Corriere della Sera concerning a mass and the people's sorrow, noting it as being even more important since those writing in the Corriere della Sera are not close to the church. Galea reports from a letter by a monk in the Corriere della Sera of the 22nd January 1909, who explains that priests could not be seen in Messina because they were not dressed as priests but bits and pieces of clothing while helping those in distress. Galea reports further how these priests put their own life at risk.

Galea further reports that Pope Pius X had asked what he could do to assist the priests of Calabria and Messina and that he started receiving money from all over the world to distribute to those in need. This is an important consideration when considering the financing of post earthquake emergency activity. Galea reports that at the time of publication the Pope had received 5 million francs or 200,000 Liri. Galea writes that the Archbishop of Catania reported that 20,000 people had arrived in Catania and that in spite of the priest's efforts, it was difficult to manage so many thousands of displaced people. Galea reports how the Archbishop of Syracuse hosted in his own home about 60 people and even gave up his own bed. Reporting from the Civilta' Cattolica of the 20th February 1909, Galea writes that the Pope was provided with the ship Catalonja by the Spanish Marchese de Carmillas, and this was used to transfer 180 orphans and about 20 injured and some priests who eventually arrived in Rome with the Hospital train—*Salib Ta'Malta* (Maltese Cross). Galea reports that the Pope had prepared to host 300 people. Galea also reports how a French priest named Santol offered the Pope to host 1000 orphans but was not allowed by the Italian government out of fear that these children will lose their Italian identity. He further states that ironically the Italian Government was allowing orphans to be taken by those who were not priests. Such references hint at the political situation at the time and Church-State relationship in Italy. Galea further notes that Father Rua had offered to welcome all orphans which could be sent to the Salesian brothers and argues that it is a pity if such an opportunity were lost. Galea emphasises that the Archbishop of Messina who was initially thought to be dead, would not leave the city as long as there were people still there and worked hard for those injured. He reports that in March 1909, at the time of writing the book, the Archbishop of Messina was in Rome meeting the Pope.

Galea further reports that an old general stated that in case of disaster those helping most are always the priests, the soldiers and the people of good will. Galea clearly makes a statement in this chapter to stress the importance of the role of the priests was in the emergency action.

5 Newspaper and Other Accounts in Malta

Printed media in particular newspapers served as the important means of communication at the time to get the news to the wider audience.

The Daily Malta Chronicle of December 29, 1908 reported on the local effects of the earthquake and tsunami. Local newspapers, for example the Malta Herald, continued to publish extensive and detailed reports about the Messina-Reggio disaster on a daily basis at least till the 11 January 1909. The reports were mostly based on eyewitness accounts from Maltese or Italian citizens who sent letters and telegrams to Malta, which were published very rapidly. Many of these reports give very vivid descriptions and personal accounts of individual human stories of survival and tragedy, as well as information about relief missions setting sail from Malta to Sicily, fund-raising activities, letters of gratitude etc. It is interesting to note the rapidity and efficiency with which aid and fund-raising activities were organised and coordinated. Three days after the earthquake, an extra theatrical performance of the opera La Boheme was announced to raise funds and a well-coordinated activity for collection and shipping of clothing items was quickly organised. By the 9th January, the sum of 281 pound sterling had been collected from local benefactors, besides the sum of 200 pound donated personally by Mr. and Mrs. A.M. Galea and a Miss Asphar.

Reporting on local newspapers was also relatively efficient. On the 29th December, the day after the earthquake, a letter by a presumably Maltese citizen, signed J.B., gives the first impressions of the earthquake (Fig. 6; Malta Herald, 30/12) whereas, the next day, a detailed account by a Sig. Diani from Reggio, on his arrival in Palermo, describes graphically the destruction in Reggio, and his slow progress to reach the harbour among dead bodies, walking wounded, broken furniture, telegraph and electric cables on the ground (Malta Herald 31/12). A Maltese gentleman, Joseph Said, was granted passage on board HMS Minerva, and vividly recounts the on-board assistance to the wounded on the way to Messina and Reggio and back to Malta (Malta Herald 04/01/1909). On the 4th January 1909, it was reported that strong aftershocks were still being felt in Messina and Reggio, and resulted in the final collapse of many damaged buildings. There were also accounts of how looters and marauders were being shot on site.

The Chronicle reported also that besides the Ambulance Corps, a section of a field bakery for the baking of bread on the spot was despatched by sea together with 5000 bags of flour (Daily Malta Chronicle 1908).

In his book, Galea himself presents not only the first hand accounts of people he knew and their letters to him and to people he know well. He refers also to the

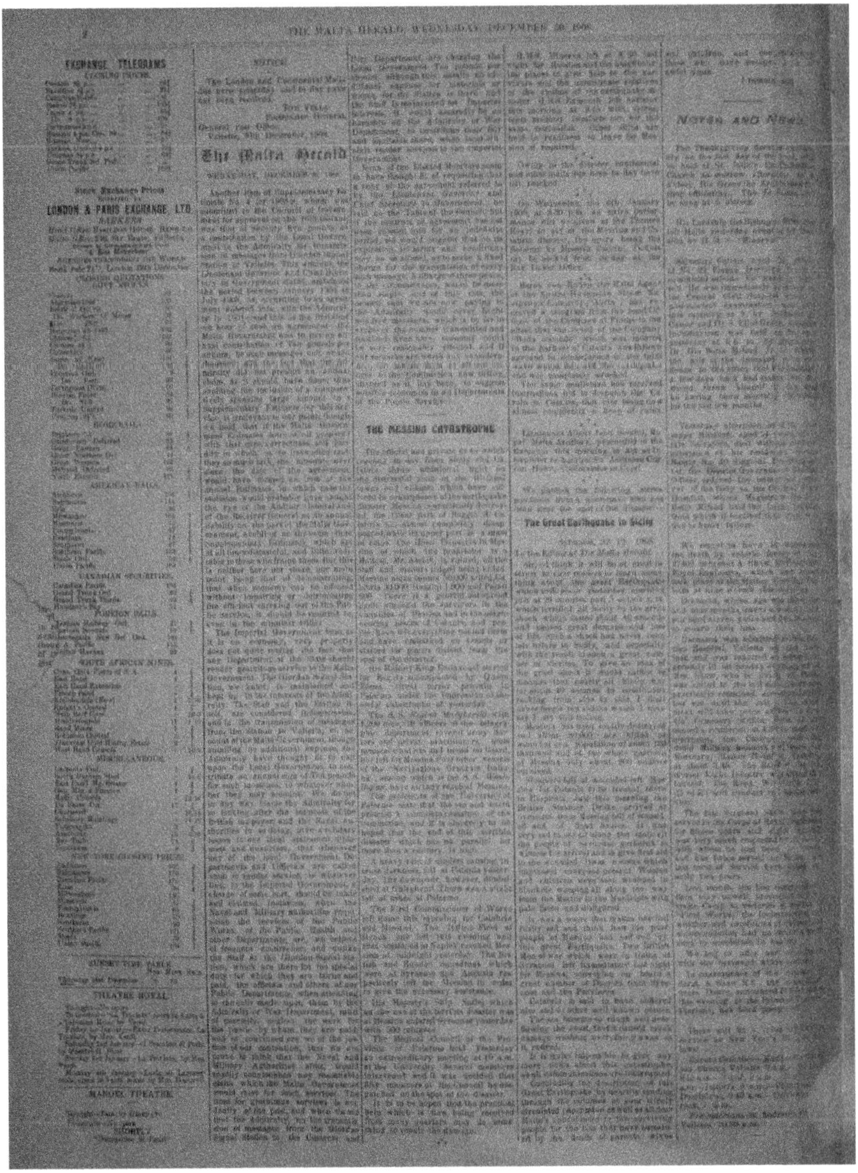

THE MALTA HERALD, WEDNESDAY, DECEMBER 30, 1908.

EXCHANGE TELEGRAMS

LONDON & PARIS EXCHANGE, LTD.

The Malta Herald

THE MESSINA CATASTROPHE

The Great Earthquake in Sicily

NOTES AND NEWS

Fig. 6 Extract of the Malta Herald of the 30th December, 1908 (Malta Herald 1908)

newspaper extracts which in particular include accounts on the help organised by the Maltese for the people in Messina, and letters in newspapers including one he sent himself, to thank others on behalf of the Good Shepherd sisters (Galea 1908).

Herbert Ganado writes briefly about the earthquake and the experience in Malta, in the first volume of *Rajt Malta Tinbidel, L-ewwel Ktieb (1900–1933)*. While he states that he was too young at the time, he refers to some first hand accounts of people who remembered the event. Ganado reports that his uncle, Robert Randon, was one of the five doctors, the other four being British, who were sent together with more than 50 nurses to assist in Messina by the Royal Army Medical Corps (RAMC). He reports that another 13 doctors also left Malta to assist. Ganado writes that his uncle's knowledge of the Italian language was an asset. The Admiral of the British Fleet in the Mediterranean ordered that the entire fleet ships goods to the affected area and Ganado reports that the Maltese people contributed. He further notes that Russian warships which were nearer, were the first to reach Messina. Ganado reports on the account of the captain of the ship Washington who reported seeing the collapse of buildings. The Italian Government presented Malta with a silver medal and a diploma in recognition of the help provided, while the King of Italy had honoured the 14 Maltese doctors as Knights of Italy (Ganado 1977).

Ganado reports that the Italian families living in Malta, were very anxious in the morning, awaiting news of their friends and families in Italy. He reports that the Carola, used to sail regularly between Malta and Sicily and that just some years earlier Maltese University students had travelled there and had friends there. Ganado reports that the telegraph messages arriving in the evening reported the bad news about the large number of dead and injured and that the cities of Messina and Calabria were in ruins. He reports on survivors who moved to the shore away from the ruins but were swept away by the tsunami. Ganado also mentions the winter cold and snow, and the fires. He reports that Sicily and Italy were asking for help.

6 Discussion

In 1909 Malta was still a British Colony. It is located geographically close to the area affected by the earthquake. Furthermore the earthquake was felt well in Malta and the effects of the tsunami were also reported in the coastal harbour areas. The Maltese and Italian families living in Malta had close ties with Sicily. Therefore it is not surprising that the British and the Maltese were among the first to assist the people of Messina and Calabria.

The accounts presented by Galea in his book provide a valuable source of information on the Maltese experience of the earthquake which has largely been overlooked during the past century. The reasons for this are related to language since the book is written in Maltese before new regulations were adopted later, and because the book is not commonly found. The book presents a unique and detailed account, with first hand accounts of the experience and the role of the Maltese in the emergency activity after the earthquake.

The book was published with the aim of informing the public about the event just a few weeks after the earthquake making it even more relevant. The book has to be understood also in the social context of the time, with the Catholic Church having a central role in society. Galea was a prominent person in Maltese Society and very close to the religious communities in Malta and in Sicily. The book presents in particular the experiences of religious communities, as is also demonstrated in his assessment of the role of priests and religious orders during the emergency activity.

Galea presents an account based largely on first hand experiences of survivors he knew well, based on letters sent to him and to the various religious orders by people who had a first hand experience of the events. He also refers to letters in newspapers which were the main source of information at the time and also makes reference to various activities in Malta which he was informed on. The newspaper accounts he refers to get an even more important significance since he refers to them at the same time when these were appearing and reaching the public. He therefore includes his personal experiences and mentions experiences which he was involved in himself or which he knew about through first hand knowledge at the time. This is demonstrated also by the fact that the book was published close to the date of the event, and as noted by the author himself, includes some mistakes which there was no time to remedy except by the insertion of a note within the book at the end.

In some instances the book is a collection of letters, reproduced by the author and probably edited and translated into Maltese. While one has to see this document in the light of other international sources which present various aspects of the earthquake, Galea's account is sufficiently detailed presenting the Maltese experience. The book also includes references to the role of different groups in society including the role of the priests and emergency action. Further the book presents the people's response in the immediate days following the earthquake. Whilst there was no organised civil protection in place at the time, the book reports on the confusion resulting after the event but also on the spontaneous support and help arriving from surrounding areas, also thanks to the religious orders in surrounding cities who were trying to establish contact with their communities in Messina and Reggio. The role of religious orders to get help through their already established networks is significant in this context, in the absence of an organised civil protection at the time.

It makes reference to the confusion resulting from the disaster, people losing all their belongings and their relatives and therefore not wanting to leave their homes even if there was nothing left but rubble heaps. It refers to the collapse of the prison and stealing from properties. The book makes reference to the damages to buildings which can be indirectly understood from the descriptions of collapse mechanisms expressed in basic terms by the survivors, the photographs of the damages and additional experiences of other earthquakes in Italy which affected similar constructions. The building damage mechanisms are presented through first hand accounts of the earthquake and its effect on buildings.

Galea presents the important role of the Russians and the British in assisting after the earthquake. It is relevant to note the interruption of communication links as a result of the non-functional telegraph service. In spite of the geographic proximity of Sicily and Malta, still Messina was perceived to be distant and far away. The report by Galea shows clearly the concern of the Maltese towards a community in a neighbouring country and the influence of the earthquake on the local population in Malta at a time when means of transport and communication technology were limited.

Galea's account shows that in spite of the perceived distance from the area affected by the event, the local community in Malta together with the British present in Malta still reacted strongly. The book presents a first hand account of personal relationships with people who died in Messina and the sorrow of the Maltese community. Galea makes clear reference to the newspapers and also to the already well established strong networks of the religious orders before the event, as having a key role in the emergency activity. The role of newspapers in coordinating the collection of goods from the population of Malta is also significant especially when one notes the important role of newspaper as the main source of information reaching the population. The support of the British and Maltese doctors in setting up temporary shelters and hospitals in the affected areas, confirm the important role of Malta and the British present on the Islands in organising help in the affected zone. The hosting in Malta of people from the affected areas also demonstrates the important role of the local Maltese community.

Galea's accounts are supported by Ganado who in his book *Rajt Malta Tinbidel*, makes clear reference to the sorrow and anxiety of the Maltese and Italian families living in Malta at the time, awaiting news of the earthquake.

In the context of an organised Civil Protection today, emergency action is expected to be significantly different, with better lines of communication both on the Italian territory but also with the external world. The SIMIT Project (2015) for example covering the Sicily-Malta territory results in a collaborative framework for civil protection and building vulnerability in the Sicily-Malta Region. Other considerations today relate to the building typologies and vulnerabilities of the built environment in view of new constructions, transport links and improved means of communication, exposure and population size.

7 Conclusion

The Maltese experience of the 1909 Messina Earthquake is based on information mostly presented in newspapers which were published after the earthquake and some other brief accounts. In addition Galea's book, *It-Theżhiża ta'Messina* presents a unique account of the experience of the people of Malta. Galea presents in particular various accounts from persons in religious orders, who had experiences the earthquake and who had shared their experiences. The accounts presented show the important role of the network of various religious orders and communities in different towns and countries, and that of the Russians, the British and the Maltese, in providing for the spontaneous post-event emergency action. The account also shows the serious limitations in communication and transport links. It presents the effects of a lack of an organised civil protection, and the relevant role of religious orders in providing assistance and attracting support from external sources. It further presents the collection of goods in Malta to support the people affected by the earthquake, the help of the Maltesein the field and also the hosting of people in Malta. Galea informs on the means of communicating the earthquake to the public at the time, through newspapers, letters and postcards and the telegraph, but also the cinema and conferences organised by the university. Galea's book in the series *Kotba tal-Mogħdija taż-Żmien* is itself a unique document, presented shortly after the earthquake and intended to reach a wide audience.

The account presents not only a vivid description of the disaster and the effect on the population but also the damage to buildings and building collapse mechanisms. The performance of masonry construction typologies present in the area, is noted to be similar to that of buildings in other regions in Italy following seismic events during the 20th century and the beginning of the 21st century.

The devastation of the earthquake had a significant impact on the people of Messina and Reggio, with long lasting effects. However the accounts of despair of the people affected had a significant effect on the people of surrounding regions including the Maltese. The Maltese experience as presented mostly in Galea's accounts is one of solidarity and deep sorrow expressed through generosity towards a desperate population which nevertheless shows courage at the face of tragedy. At the end of the book, Galea describes Messina as a city which shall be rising back to be full of life and activity once more. Galea concludes his book stating that Messina is not dead.

References

Amoruso, A., Crescentini, L., Scarpa, R. (2002). Source parameters of the 1908 Messina Straits, Italy, earthquake from geodetic and seismic data. *Journal of Geophysical Research, 107*, B4, doi:10.1029/2001JB000434.

Amoruso, A., Crescentini, L., Neri, G., Orecchio, B., Scarpa, R. (2006). Spatial relation between the 1908 Messina Straits earthquake slip and recent earthquake distribution. *Geophysical Research Letters, 33*, doi:10.1029/2006GL027227.

Baratta, M. (1910). *La Catastrofe Sismica Calabro-Messinese (28 dicembre 1908)*. Roma: Società Geografica Italiana.

Barbano, M. S., Azzaro, R., & Grasso, D. E. (2005). Earthquake damage scenarios and seismic hazard of Messina, North-eastern Sicily (Italy) as inferred from historical data. *Journal of Earthquake Engineering, 9*(6), 805–830.

Billi, A., Minelli, L., Orecchio, B., & Presti, D. (2008). Constraints to the cause of three historical Tsunamis (1908, 1783, and 1693) in the Messina Straits Region, Sicily, Southern Italy. *Seismological Research Letters, 81*, 907–915.

Borg, R. P. (2010a). Seismic damage assessment of structures; L'Aquila earthquake, April 2009. In *Seismicity and Earthquake Engineering: L'Aquila Earthquake of April 2009*. Valletta, Malta: Kamra Tal-Periti. ISBN-978-99932-0-879-2 (April 2010).

Borg, R. P. (Ed.). (2010b). Seismicity and earthquake engineering: L'Aquila earthquake of April 2009. Valletta, Malta: Kamra Tal-Periti. ISBN-978-99932-0-879-2 (April 2010).

Borg, R. P., Borg, R. C., & Borg Axisa, G. (2008). The seismic risk of buildings in Malta. In F. Mazzolani et al. (Eds.), *Urban habitat constructions Under catastrophic events*. Malta: University of Malta.

Borg, R. P., Indirli, M., Rossetto, T., & Kouris, L. A. (2010). Damage assessment methodologies: L'Aquila earthquake field investigations. In *Proceeding of the International Conference COST Action C26 Urban Habitat Constructions under Catastrophic events, Naples, Italy,* September 16–18, 2010. London: CRC Press, Taylor & Francis. ISBN 9780415606851.

Boschi, E., Pantosti, D., Valensise, G. (1989). Modello di sorgente per il terremoto di Messina del 1908 ed evoluzione recente dell'area dello Stretto. In *Proceedings VIII Meeting Gruppo Nazionale di Geofisica della Terra Solida, Rome* (pp. 245–258).

Bottari, A., Capuano, P., De Natale, G., Gasparini, P., Neri, G., Pingue, F., Scarpa, R. (1989). Source parameters of earthquakes in the Strait of Messina, Italy, duringthis century. *Tectonophysics, 166*, 221–234.

Comerci, V., Blumetti, A. M., Brustia, E., Di Manna, P., Esposito, E., Fiorenza, D., et al. (2008). One century after the 1908 Southern Calabria—Messina earthquake (southern Italy): A review of the geological effects. *Geophysical Research Abstracts* 10.

D'Amico, S., Orecchio, B., Presti, D., Gervasi, A., Guerra, I., et al. (2011a). Testing the stability of moment tensor solutions for small and moderate earthquakes in the Calabrian-Peloritan arc region. *Bollettino di Geofisica Teorica ed Applicata, 52*, 283–298 doi:10.4430/bgta0009

D'Amico, S., Galea, P., Borg, P. R., & Lotteri, A. (2011b). Earthquake ground-motion scenario: Case study for the Xemjia Bay (Malta) area. In *Proceedings of the International Conference of Europen Council of Civil Engineers* (pp. 149–160).

D'Amico, S., Orecchio, B., Presti, D., Zhu, L., Herrmann, R. B., & Neri, G. (2010). Broadband waveform inversion of moderate earthquakes in the Messina straits, Southern Italy. *Physics of Earth and Planetary Interiors, 179*, 97–106. doi10.1016/j.pepi.2010.01.012

DISS Working Group. (2007). Database of Individual Seismogenic Sources (DISS) Version 3.0.4: A Compilation of Potential Sources for Earthquakes Larger than M 5.5 in Italy and Surrounding Areas. ©INGV 2007—Istituto Nazionale di Geofisica eVulcanologia, Roma, http://www.ingv.it/DISS

Fabbri, A., Ghisetti, F., & Vezzani, L. (1980). The Peloritani-Calabria range and the Gioia basin in the Calabrian Arc (south Italy): Relationships between land and marine data. *Geologica Romana, 19*, 131–150.

Finetti, I. R., & Del Ben, A. (1986). Geophysical study of the Tyrrhenian opening. *Bollettino di Geofisica Teorica ed Applicata, 110*, 75–156.

Finetti, I. R., & Del Ben, A. (2005). Ionian thetys lithosphere roll-back sinking and backarc Tyrrhenian opening. In I. R. Finetti (Ed.), *CROP PROJECT: Deep seismic exploration of the Central Mediterranean and Italy* (pp. 483–503). Amsterdam: Elsevier.

Galea, A. M. (1909). It-Theżhiża ta'Messina; it-28 tax-Xahar tal Milied 1908, Il-Kotba tal-Moghdija taż-Żmien.

Galea, P. (2007). Seismic history of the Maltese islands and considerations on seismic risk. *Annals of Geophysics, 50*, 725–740.

Ganado, H. (1977). Rajt Malta Tinbidel, Malta. L-ewwel ktieb, 1900–1933.

Indirli, M., Kouris, L. A., Formisano, A., Borg, R. P., & Mazzolani, F. M. (2013). Seismic damage assessment of unreinforced masonry structures after the Abruzzo 2009 earthquake: The case study of the historical centres of L'Aquila and Castelvecchio Subequo. *International Journal of Architectural Heritage, 7*(7), 536–578. doi:10.1080/15583058.2011.654050

Luiggi, L. (1909). Quel che ho veduto all'alba del terremoto calabro -siculo del 1908, in Nuova Antologia di Lettere Scienze ed Arti, (Maggio-Giugno 1909), serie V(141): 297–317.

Malta 28/12/1908, 29/12/1908, 30/12/1908 (Newspaper)

Monaco, C., & Tortorici, L. (2000). Active faulting in the Calabrian arc and eastern Sicily. *Journal of Geodynamics, 29*, 407–424.

Mulargia, F., & Boschi, E. (1983). The 1908 Messina Earthquake and related seismicity. In: H. Kanamori & E. Boschi (Eds.), *Earthquakes: Observation, theory and interpretation, Proceedings of the International School of Physics, Enrico Fermi* (pp. 493–518).

Neri, G., Barberi, G., Orecchio, B., Mostaccio, A. (2003). Seismic strain and seismogenic stress regimes in the crust of the southern Tyrrhenian region. *Earth and Planetary Science Letters, 213*, 97–112.

Neri, G., Barberi, G., Oliva, G., Orecchio, B. (2004). Tectonic stress and seismogenic faulting in the area of the 1908 Messina earthquake, south Italy. *Geophysical Research Letters, 31*, pp. L10602-1-L10602-5.

Neri, G., Orecchio, B., Totaro, C., Falcone, G., & Presti, D. (2009). Subduction beneath southern Italy close the ending: Results from seismic tomography. *Seismological Research Letters, 80*, 63–70.

Nocquet, J.-M., & Calais, E. (2004). Geodetic measurements of crustal deformation in the Western Mediterranean and Europe. *Pure and Applied Geophysics, 161*, 661–681.

Omori, F. (1894). On the aftershocks of earthquakes. *Journal of the College of Science*, Imperial University of Tokyo *7*, 111–200.

Pino, N. A., Piatanesi, A., Valensise, G., & Boschi, E. (2009). The 28 December 1908 Messina Straits Earthquake (Mw 7.1): A great earthquake throughout a century of seismology. *Seismological Research Letters, 80*(2), 243–259. doi:10.1785/gssrl.80.2.243. http://srl.geoscienceworld.org/content/80/2/243.full

Piromallo, C., & Morelli, A. (2003). P wave tomography of the mantle under the Alpine-Mediterranean area. *Journal of Geophysical Research 108*(B2), 2065, 23 pp. doi:10.1029/2002JB001757.

Santini, A., & Moraci, N. (Ed). (2008). In *2008 Seismic Engineering Conference: Commemorating the 1908 Messina and Reggio Calabria Earthquake (AIP Conference Proceedings)*. American Institute of Physics. ISBN 978-0-7354-0542-4.

Savona Ventura, C. (2005). Tsunami events in Malta. *The Sunday Times of Malta*, January 9, 2005. http://www.timesofmalta.com/articles/view/20050109/letters/tsunami-events-in-malta.102636. Accessed 24 May 2014.

SIMIT. (2015). *SIMIT Project report. Italia-Malta operational programme: Integrated system for transboundary Italo-Maltese Civil Protection* (European Fund for the Regional Development. European Territorial Cooperation 2007–2013). Project Code: B1-2.19/11.

Sutherland, A. (1846). *The Achievements of the Knights of Malta* (Vol. 2). , Philadelphia: Carey and Hunt.

The Malta Herald 28/12, 29/12, 30/12, 31/12/1908 and 02/01, 04/01, 05/01, 07/01, 08/01, 09/01, 11/01/1909 (Newspaper)

The Daily Malta Chronicle 30/12/1908, 31/12/1908 (Newspaper)

Valensise, G., Pantosti, D. (1992). A 125 Kyr-long geological record of seismic source repeatability: The Messina Straits (southern Italy) and the 1908 earthquake (MS 71/2). *Terra Nova*, *4*, 472–483.

A Web Application Prototype for the Multiscale Modelling of Seismic Input

Franco Vaccari

1 Introduction

In the framework of the cooperation Project "Definition of seismic hazard scenarios and microzoning by means of Indo-European e-infrastructures", funded by Regione autonoma Friuli Venezia Giulia (Italy), a web application prototype has been developed, that enables scientists to compute a wide set of synthetic seismograms, dealing efficiently with the variety and complexity of the potential earthquake sources and of the medium travelled by the seismic waves.

The computational engine of the web application is based on the neo-deterministic seismic hazard assessment (NDSHA) methodologies (Panza et al. 2001, 2012) for the generation of synthetic seismograms. It allows for a rapid definition of seismic and tsunami hazard scenarios for a given event, at local or regional scales. Neo-deterministic means scenario-based methods for seismic hazard analysis, where realistic and duly validated synthetic time series, accounting for source, propagation, and site effects, are used to construct the earthquake scenarios. The user interface has been designed so to hide the intricacy of the underlying computational engine, yet it allows power users to act even on the deeper aspects of the model parameterisation.

Due to the employment of highly optimised computational codes, the web application is designed to run even locally on a properly configured laptop computer. For massive parametric tests, or single tasks targeted at modelling ground shaking scenarios at very large scale, it can be interfaced with different computational

F. Vaccari (✉)
Department of Mathematics and Geosciences, University of Trieste, Trieste, Italy
e-mail: vaccari@units.it

F. Vaccari
The Abdus Salam International Centre for Theoretical Physics,
SAND Group, Trieste, Italy

S. D'Amico (ed.), *Earthquakes and Their Impact on Society*,
Springer Natural Hazards, DOI 10.1007/978-3-319-21753-6_23

platforms, ranging from Grid computing infrastructures to HPC dedicated clusters up to Cloud computing.

The web application is addressed to engineers, urbanists, administrators, insurance companies and stake-holders who are interested in reliable territorial planning and in the design and construction of buildings and infrastructures in seismic areas. It also constitutes a powerful educational tool for seismologists and seismic engineers willing to better understand the earthquake phenomenon, since quick parametric studies can be easily performed for determining the influence of each model element on the resulting ground shaking scenarios.

2 The Computational Engine

The basic requirement for the generation of synthetic seismograms is a proper modelling of the earthquake source properties, and of wave propagation in the medium between the source and the site of interest.

2.1 *Source Models*

The source is introduced in the medium as a discontinuity in the displacement and shear stress fields. The source function, describing the discontinuity of the displacement across the fault, is approximated with a step in time and a point in space. Imposing the continuity of the normal stress across the fault, for the representation theorem the equivalent body force in an unfaulted medium is a double-couple with null total moment (Maruyama 1963; Burridge and Knopoff 1964). The simplest source model adopted for the computation of the synthetic seismograms is a double-couple size scaled point source (SSPS) (Gusev 1983). Roughly speaking, this approximation is acceptable when modelling the ground motion at large epicentral distances due to moderate magnitude events, neglecting the contribution of frequencies higher than 1 Hz.

The system is now being updated so that more realistic source models can be considered, taking into account source finiteness and the details of the rupturing process that lead to the directivity effects often observed at the sites around the fault. The most comprehensive approach being implemented is based on the extended source model (ES) described by Gusev and Pavlov (2006) and Gusev (2011), where the full characteristics of the slip distribution, and the rupturing velocity across the fault area, are taken into account. In a simplified representation, namely the size and time scaled point source model (STSPS), a combination of extended (ES) and point sources is used, with the purpose of reducing the computational times while still keeping into consideration the main characteristics of the rupturing process. In this approximation, the subsource time functions (i.e. the ones described for the ES model) are summed to obtain the equivalent single source representative of the entire space and time structure of the ES and the related Green's function. The resulting function is then convoluted with the unscaled seismogram obtained in the point source approximation.

2.2 Laterally Homogeneous Layered Models

The computational engine upon which the web application (WA hereafter) is based has its roots in the middle '80s, when the modal summation technique for the generation of P-SV synthetic seismograms (radial and vertical component of motion) was developed (Panza 1985). Equation (1) describes the asymptotic expression of the Fourier transform of the displacement for the radial (u_x) and the vertical (u_z) components of motion at a distance r from the source.

$$
\begin{aligned}
\mathrm{u}_x^R(r,z,\omega) &= \sum_{m=1}^{\infty} \frac{\mathrm{e}^{-\mathrm{i}3\pi/4}}{\sqrt{8\pi\omega}} \frac{\mathrm{e}^{-\mathrm{i}k_m^R \mathrm{r}}}{\sqrt{\mathrm{r}}} \frac{(\chi_m^R(\mathrm{h}_s,\omega))}{\sqrt{\mathrm{c}_m^R \mathrm{v}_m^R \mathrm{I}_m^R}} \frac{(\mathrm{F}_x^R(z,\omega))}{\sqrt{\mathrm{v}_m^R \mathrm{I}_m^R}} \\
\mathrm{u}_z^R(r,z,\omega) &= \sum_{m=1}^{\infty} \frac{\mathrm{e}^{-\mathrm{i}\pi/4}}{\sqrt{8\pi\omega}} \frac{\mathrm{e}^{-\mathrm{i}k_m^R \mathrm{r}}}{\sqrt{\mathrm{r}}} \frac{(\chi_m^R(\mathrm{h}_s,\omega))}{\sqrt{\mathrm{c}_m^R \mathrm{v}_m^R \mathrm{I}_m^R}} \frac{(\mathrm{F}_z^R(z,\omega))}{\sqrt{\mathrm{v}_m^R \mathrm{I}_m^R}}
\end{aligned}
\tag{1}
$$

The equivalent codes for the generation of the SH seismograms [transverse component of motion u_y, as shown in Eq. (2)] were created by Florsch et al. (1991) and finally allowed for the generation of three-component seismograms.

$$
\mathrm{u}_y^L(r,z,\omega) = \sum_{m=1}^{\infty} \frac{\mathrm{e}^{-\mathrm{i}3\pi/4}}{\sqrt{8\pi\omega}} \frac{\mathrm{e}^{-\mathrm{i}k_m^L \mathrm{r}}}{\sqrt{\mathrm{r}}} \frac{(\chi_m^L(\mathrm{h}_s,\omega))}{\sqrt{\mathrm{c}_m^L \mathrm{v}_m^L \mathrm{I}_m^L}} \frac{(\mathrm{F}_y^L(z,\omega))}{\sqrt{\mathrm{v}_m^L \mathrm{I}_m^L}} \tag{2}
$$

At that time, given the limited (when compared to nowadays…) computational facilities available to the researchers, algorithm efficiency and proper computer code implementation was of uttermost importance. The beauty of the modal summation technique computer codes for layered, inelastic structural models, based on the original paper by Panza, lies in the separation of the calculation of the spectral quantities related with the medium properties [phase velocity c, group velocity v, energy integral I and complex (due to anelasticity) wavenumber k appearing in Eqs. (1) and (2)], from those that describe the source and its position with respect to the sites of interest (radiation pattern χ, epicentral distance r, rupturing model etc). In such a way, once the medium is described by defining each layer's thickness, density, *Vp* and *Vs* wave velocities and the *Qp* and *Qs* factors that define attenuation, the "lengthy" computation of the spectral quantities that do not depend on the source position can be performed just once, and the obtained quantities can simply be stored and reused for as many seismograms have to be computed.

The "lengthy" adjective is really rooted in the '80s. At that time, as reported by Panza (1985), the computation of the spectral quantities associated with a layered model required about one hour of CPU time on an IBM 370/168 computer. In the author's memories, it was even longer than that. Due to job queueing and a less powerful mainframe computer, jobs had to be submitted in the evening and results were obtained, with some luck, the day after. Now, the same results can be obtained on a modern laptop in a couple of seconds. Most obviously because of the improved performance of modern computers, and partly due to code optimisation applied since then, and I/O rearrangements allowed by the increased memory now available to programs.

A further reduction in the computational times is achieved when a sequence of seismograms is generated keeping constant the hypocentral depth. Thus, the terms describing the eigenfunctions calculated at the depth of the source [terms F_x and F_z in Eq. (1) and F_y in Eq. (2)] can be computed for the first seismogram only, and reused for the following ones. Therefore, even in the '80s, after the spectral quantities were obtained, the modal summation approach allowed for a "quick" generation of the time series: 300 s for the first seismogram, and 30 s for the others, when keeping constant the hypocentral depth. Come back onboard of a modern laptop, and a thousand seismograms can be generated in a couple of seconds.

While the layered model approximation cannot be considered satisfactory for every scenario in the world, nevertheless the modal summation technique is an unbelievably powerful tool to explore in near real-time the influence of single model (source and layer) parameters on the generated ground shaking scenarios. It can help to quickly constrain part of the model parameterisation for the optimised planning of more complicated and computationally intensive tasks. It is also at the base of the seismic hazard computations at regional scale, as discussed in detail by Panza et al. (2001, 2012).

2.3 Laterally Heterogeneous Structural Models

More sophisticated computational approaches can be used when the geological setting of the studied area can hardly be approximated by a flat layered model (Panza et al. 2001). For the WA, the hybrid technique developed originally by Fäh and Panza (1994) is the engine adopted for the computation of synthetic seismograms along 2D heterogeneous profiles. A laterally homogeneous inelastic layered model is defined to represent the average lithospheric properties along the path from the source to the vicinity of the local, heterogeneous structure of interest. In this part of the model wave propagation is modelled by the modal summation technique, according to what described in the previous section. So there is no time penalty in this part of the model, associated with the length of the path. The generated wavefield is then introduced in the mesh that defines the local heterogeneous area characterising the site of interest, where it is propagated according to the finite difference scheme. With this hybrid approach, source, path, and site effects are all taken into account, and detailed ground shaking scenarios can be efficiently evaluated along the 2D profile even at large distances from the epicentre.

The computational time spent for the modelling depends primarily on the size of the finite difference mesh, and on the duration requested for the seismograms. Again, with respect to the middle '90s, when the theory and codes were developed, execution time has greatly improved. Results that required days of mainframe computer CPU time, can now be obtained in a matter of tens of minutes to hours, depending on the model characteristics.

The hybrid method has demonstrated its validity as a tool to perform the seismic microzonation of urban areas (e.g. Panza et al. 2002; Alvarez et al. 2005; Harbi et al. 2007; Zuccolo et al. 2008; Amponsah et al. 2009; Mohanty et al. 2013). Given its widespread application, quite some resources have been dedicated to the development of helper applications that permit an easy and nearly error-proof construction of the finite difference mesh. Still, we are quite far from reaching the level of speedy interactivity allowed by the modal summation technique for laterally homogeneous models. Furthermore, some training is required for the users to self-certify the quality of the results obtained. It may happen yet, although not very often, that the model parameterisation has to be finely tuned by the user, when the default input configuration fails to produce acceptable results.

Other computational techniques based on the modal summation theory may simplify and speed up the generation of ground shaking scenarios in laterally heterogeneous media. Their field of applicability is mostly driven by the characteristics of the medium, above all the geometric properties of the heterogeneities, and the acoustic impedance contrast across the interfaces. For instance, when the lateral heterogeneity can be reasonably approximated by a sharp vertical boundary separating two layered quarter-spaces, the modal coupling technique (Vaccari et al. 1989; Romanelli et al. 1997) can be efficiently used. In this approach, the energy carried by the incoming modes, excited by the source, is transmitted across the boundary and redistributed in the modes that characterise the second structure. This technique remains valid also in the presence of a sequence of vertical boundaries, noting that its effectiveness is inversely proportional to the number of considered interfaces: when their number increases, the ramification of the cross-coupling paths between modes of different order may grow dramatically, as well as the need to consider multiple reflections between the boundaries.

When considering smoothly varying media, a different approach can be taken for the computation of the synthetic seismograms. A good choice is the so called WKBJ approximation (acronym of the names Wentzel, Kramers, Brillouin and Jeffreys). Here, the heterogeneity can be seen as a perturbation of an initial lateral homogeneous model and, if such a perturbation is small within a wavelength, a procedure based on the ray method can be used to construct an approximate solution corresponding to the wave field (see e.g. Woodhouse 1974; Yanovskaya 1989).

The computational package based on the programs by Panza (1985) and Florsch et al. (1991) has been already expanded and integrated with new codes to permit the computation of synthetic seismograms in 2D and 3D heterogeneous media

(La Mura 2009; La Mura et al. 2011; Panza et al. 2012). The interfacing with the WA is in progress, as it requires the development of some additional routines not yet available, mostly aimed at the simplification of the model preparation; the same is true for the mode coupling approach.

2.4 *Future Developments*

In general, the WA is so structured, that it poses no big problems in implementing new computational techniques, should they became available, capable of producing ground shaking scenarios better and faster for specific classes of structural and source models. The computational engine is well separated from the user interface (written in the Xojo cross-platform, object-oriented language), so in principle any software developed in any programming language could be installed on the dedicated number-crunching hardware. When adding a new modelling tool, the extra work implied is limited to the modification of an existing panel or the addition of a new one to the WA user interface, properly designed to allow an easy definition of the input model, and the visualisation of the computational results. This generally requires the development of some middleware that improves and simplify the process of input preparation, and validates the data provided by the user.

3 The Web Application User Interface

3.1 *Login Page*

The WA requires the user to pass through a login page, shown in Fig. 1. Each of the user's input data and computational results are stored into well separated areas of the file system. In no way a logged-in user can access other users' data navigating through the WA interface panels. The administrator of the WA can create new accounts, lock or delete users and clean their stored computations when needed, authorise access to a subset of functionalities on a per-user basis, and check the system status.

After entering the system, a tabbed interface keeps the different computational tasks well separated. In the following, each tab functionality is described, with some screenshots shown to better highlight the way the user interacts with the system. Each panel has a help page for the basic explanation of the available functionalities, with a list of bibliographic references useful for those interested in a deeper understanding of the physical phenomenon and of the computational tools used for its modelling.

Fig. 1 Login page of the web application

3.2 *Laterally Homogeneous Structural Models*

The first panel available to the user after logging into the system is the one dedicated to the computation of the frequency domain quantities associated with the layered structural models (Fig. 2).

By default, some standard models are made available to the user upon the first login, so that one can immediately start experimenting with the WA. The models correspond to classes of ground types mentioned in seismic design rules like Eurocode 8 (EN 1998-1, 2004), ranging from hard rocks of type "A" to soft soils of type "E". For each class of ground, two variations are prepared with different quality factors $Qs = 1000$ for slowly attenuating models, and $Qs = 100$ for structures characterised by higher attenuation. Qp is taken equal to $2.2 \cdot Qs$.

The user must prepare a simple text file with the description of the layers as shown in Fig. 3, and then upload the file through the dedicated control. After that, the model appears in the list of available structures and can be selected to generate its Rayleigh and Love modes with the choice of the cutoff frequency for the computations: 1 Hz for modelling the ground motion very far from the source, or 10 Hz for obtaining ground shaking scenarios at shorter distances.

The underlying computational engine takes care of splitting the physical layers defined by the user into thinner layers that satisfy the requirements of the algorithms at the base of the modal summation technique, to avoid potential overflow conditions during the evaluation of the matrix elements for any given layer (Panza 1985). The inclusion of low-velocity layers is also permitted by the methodology.

While not too long, still the user has to wait some time before the computation get finalised, and the modes are shown in the web application panel of Fig. 2.

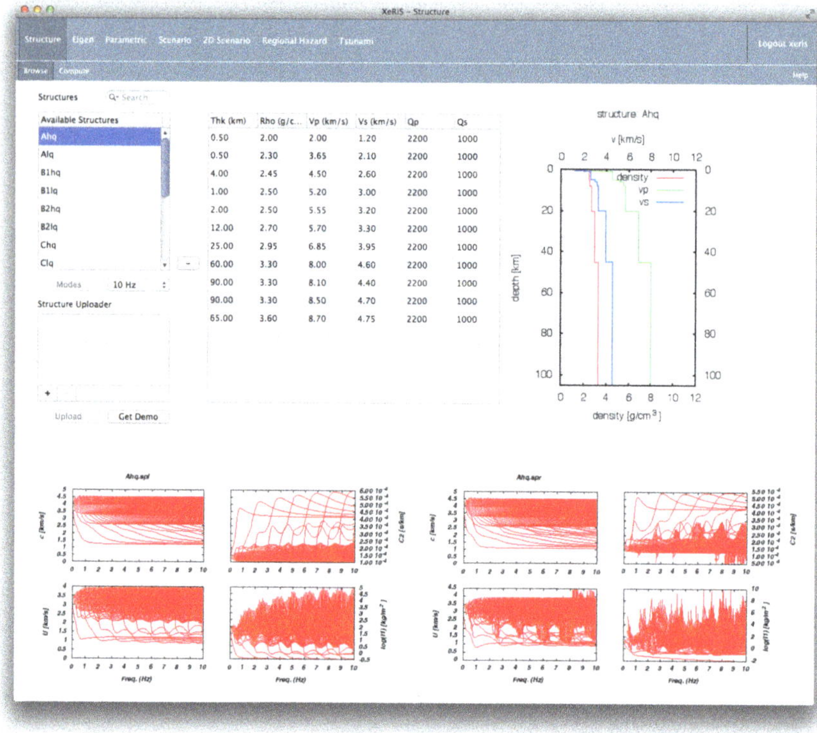

Fig. 2 Panel dedicated to the definition of the laterally homogeneous layered models, and the related frequency domain quantities required to compute the synthetic seismograms by the modal summation technique. *Top,* From *left* to *right* the list of structural models available to the user, with the uploader that allows the addition of more, user-defined structures; the properties associated with the layers of the selected structure; the plot of the model with density, *Vp* and *Vs* depicted. *Bottom* The spectral quantities calculated for Love and Rayleigh modes

The amount of this delay mostly depends on the system load at the time of the job submission. In general, it should not exceed a couple of minutes. Once the computational request has been submitted to the system, the user interface signals the job status through the appearance of an icon in the toolbar bar. The colour of the icon is associated with the job status: grey if the job is queued for execution; yellow after the execution has started, green when the modes are finally available and can be used for the generation of the synthetic seismograms. While the computations are taking place, there is no lock in the user interface. Other panels can still be accessed to play with the separate functionalities of the WA.

It may sometimes happen that for whatever reason, usually associated with typing errors in the file that defines the layer properties, or with very peculiar characteristics of the structural model, the mode generation fails. This condition is signalled by a red colour in the toolbar icon. In such a case, the user should first check if some errors have been made in the preparation of the structural model, or

```
   thk    rho   Vp    Vs    Qp   Qs
 2.000 2.20 4.67 2.70  220  100
 3.000 2.40 5.54 3.20  220  100
 3.000 2.50 5.89 3.40  220  100
 7.000 2.70 5.97 3.45  330  150
 4.000 2.75 6.23 3.60  330  150
 9.000 2.80 6.40 3.70  330  150
12.000 2.90 6.92 4.00  330  150
 5.000 3.10 7.09 4.10  330  150
11.000 3.24 7.27 4.20  330  150
 9.000 3.25 7.44 4.30  330  150
 9.000 3.26 7.61 4.40  330  150
18.000 3.29 7.79 4.50  330  150
```

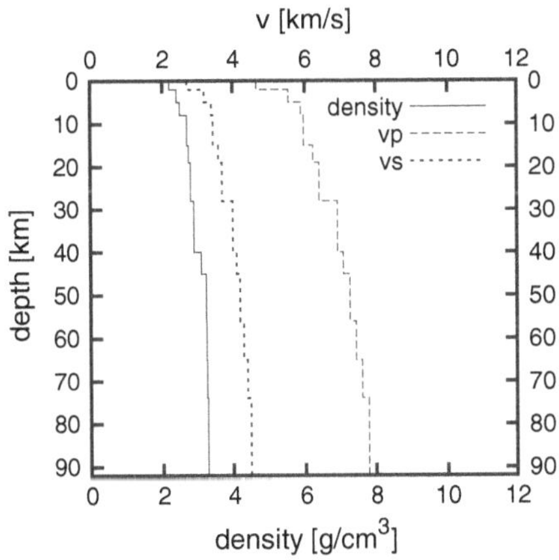

Fig. 3 Example of the content of a text file (*left*) the user should prepare for uploading a layered structural model (*right*) to the web application. Thickness (*thk*) is given in km, density (rho) in g/cm^3, layer velocities (*Vp* and *Vs*) in km/s while attenuations for *P* and *S* velocity (*Qp* and *Qs* respectively) are adimensional quantities. In this examples, the *Vp* value has been obtained from the *Vs* in the hypothesis of a Poissonian medium, but there is actually no constrain on the *Vp*/*Vs* ratio that can be adopted for each layer

modify slightly the layer properties to try to overcome the problem. The management of error conditions has not yet been thoroughly considered at this prototypal stage, but will definitely get the attention it deserves as the WA approaches a more mature state.

3.3 Eigenfunction Visualisation

This panel has to be taken mostly as an educational tool, rather than something designed or required for the definition of a ground shaking scenario. It allows the user to explore the distribution with depth of the eigenfunctions (displacements and stresses) associated with a given layered model. The user can choose the structural model for which the eigenfunctions should be generated, and visualise their depth distribution for selected Love and Rayleigh modes, at a given frequency. In such a way, one can understand which modes and which frequencies will be better excited by a source placed at a given depth, therefore contributing to the construction of the synthetic seismogram. For instance, the example of Fig. 4 shows that in order to excite mode no. 4 at the frequency of 2 Hz a source should be placed no deeper than 6 km.

The evaluation of the eigenfunctions is truly instantaneous, so the user can compute and visualise them in a really interactive way, quickly getting familiar with their behaviour as a function of the structure characteristics, the mode index or the frequency considered. Contrary to all other panels, due to the speedy computation,

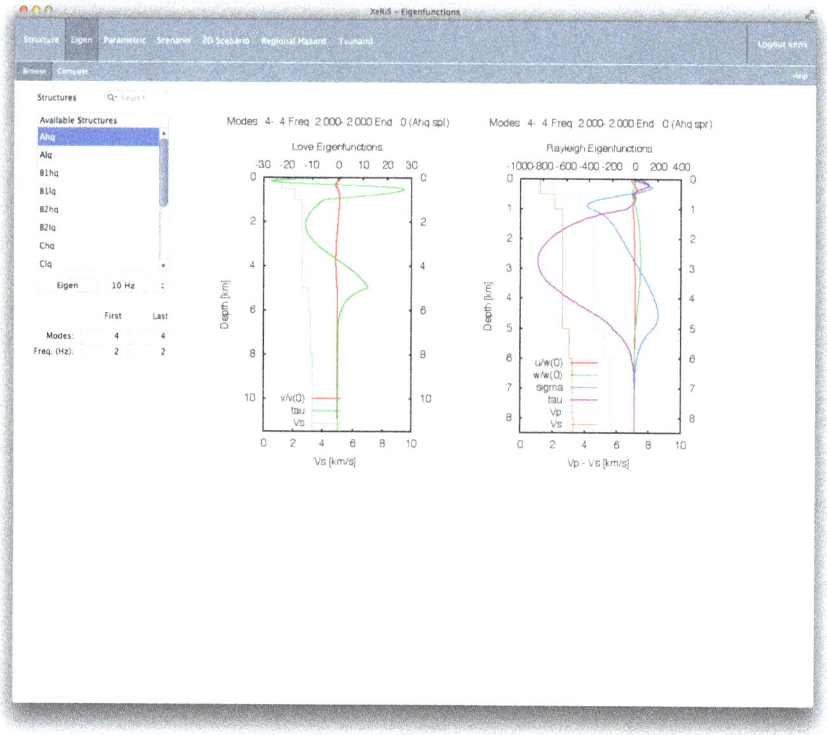

Fig. 4 Panel dedicated to the computation of the eigenfunctions (displacement and stresses) associated with a structural model for a given mode and frequency

the job is not queued upon submittal but is executed immediately at the push of the dedicated button. Therefore no toolbar icon is required in this panel to signal the job status. Only the structures for which the modes have been already generated appear in the list of the selectable models.

3.4 Parametric Tests

This panel is dedicated to the quick execution of parametric tests, where a set of synthetic seismograms is generated based on the variation of a single source parameter. The user can choose to variate in a loop one of the following quantities: event magnitude, epicentral distance, hypocentral depth, fault dip and rake, and the azimuthal position of the observation point with respect to the fault strike. The time series (radial, transverse and vertical component of motion) are generated in a matter of seconds, with the choice of visualising ground displacement, velocity or acceleration. A summarising graph showing the peak values taken from the seismograms as a function of the varied parameter is also given, to intuitively demonstrate the dependency of the ground motion amplitude on the single source characteristic being analysed. The example of epicentral distance variation is shown in Fig. 5.

A list of the structures for which the modes have been already computed is presented in the panel. This allows to easily explore the dependency of the ground shaking also on the layer properties. Each experiment can be quickly repeated selecting the desired structural model from the list. The results obtained with each simulation are properly labelled and stored for later retrieval and comparison with other experiments. Stored results can of course be deleted by the user when they are no longer needed.

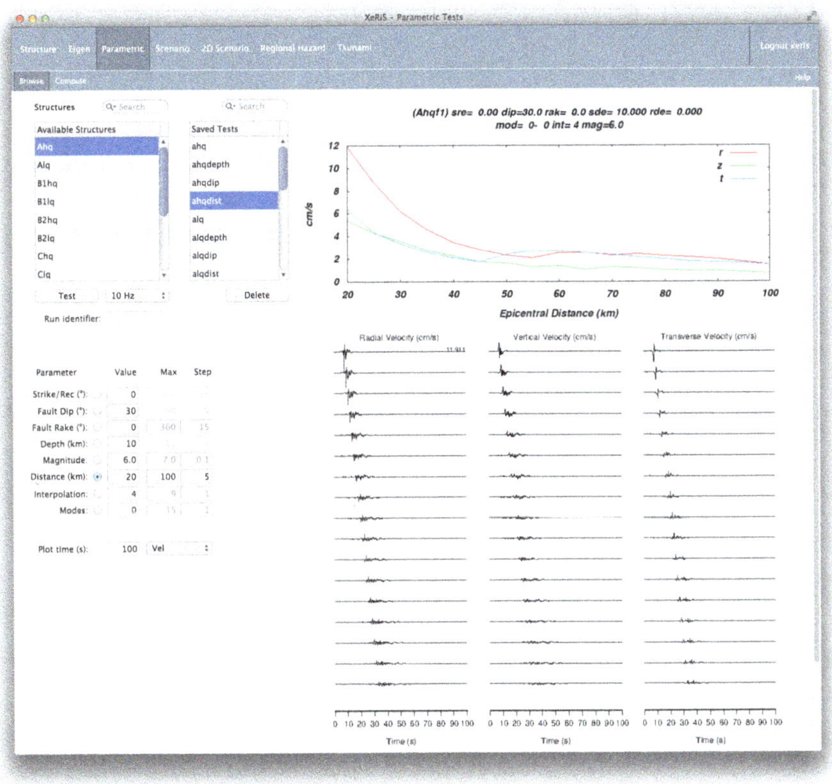

Fig. 5 Panel dedicated to the computation of a set of three-component synthetic seismograms obtained varying a single model parameter, to grasp its influence on the obtained ground shaking. *Top*, From *left* to *right* the list of available structural models; the list of stored parametric tests; the *graph* summarising the dependency of the peak ground motion on the selected parameter (epicentral distance in the shown example). *Bottom* The definition of the source model, with the selected parameter and its range of variation; the plot of the three component seismograms

For a given source parameterisation, and mostly for educational purposes, it is also possible to visualise a set of time series, each generated for a single mode of the considered structure. For instance, after configuring a loop over modes running from index 1–10, the contributions coming from the fundamental mode (index = 1) up to the ninth higher mode (index = 10) will be separately shown in the seismogram plot. Furthermore, if the starting index of the mode loop is set to 0 instead of 1, the first seismogram shown will be the "complete" one, that is the one obtained summing up all the modes associated with the considered structure.

3.5 Earthquake Scenarios

The next functionality of the WA is targeted at the generation of full earthquake scenarios, by calculating synthetic seismograms all around the epicentre in a user-defined range of epicentral distances and with properly discretised azimuthal and distance steps. In this modelling, the computational time depends on the number of seismograms that must be generated. But unless the user requires the investigation of a very broad area at very small discretisation steps, it is still a matter of seconds to obtain the requested scenario (Fig. 6).

As in the case of the computation of modes and parametric tests, already described, also for the generation of earthquake scenarios a toolbar icon appears after a computational job has been submitted, with the icon's colour signalling the job status.

The computed scenarios remain available to the user, and can be retrieved at any moment by selecting its associated label in the dedicated listbox.

3.6 Earthquake Scenarios Along Heterogeneous Profiles

The hybrid technique used for the computation of ground shaking scenarios along laterally heterogeneous profiles requires a much larger amount of CPU time with respect to the modelling procedures described up to now. It is definitely not the kind of process that can be done interactively. To simplify the operations, and to keep the user interface clean, the dedicated panel has been further organised in sub-panels.

The first one is dedicated to the definition of the model, and to the submission of new jobs (Fig. 7). The user can choose the layered structural model, among those for which the modes have been already computed, that represents the average lithospheric properties along the path from the source to the vicinity of the sites of interest. After that, the characteristics of the heterogeneous profile that includes those sites must be selected.

Finally, in the "Basic" configuration mode, the source parameterisation must be done. An "Advanced" mode is eventually made available to the power users, that allows for a fine tuning of the finite difference model. Tweaking to the mesh

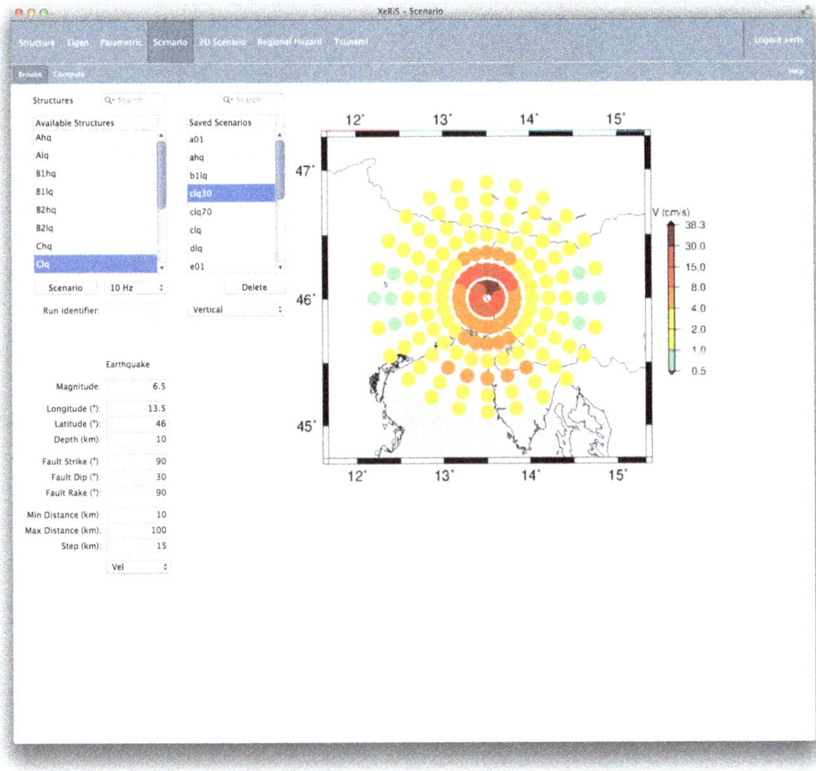

Fig. 6 Panel dedicated to the computation of ground shaking scenarios for a single earthquake. *Top*, From *left* to *right* the list of available structural models for the selected cutoff frequency; the list of stored scenarios with the popup menu for the selection of the component to be visualised; the map of ground motion shaking (vertical PGV in the shown example). *Bottom left* The parameters adopted to define the scenario

discretisation, and modifications to other rather obscure technical parameters, usually kept out of reach to the unexperienced users, can be applied here.

After all the above has been properly defined, either in "Basic" or "Advanced" mode, the modelling can be started at the push of a button.

The second panel (Figs. 8 and 9) is organised for the retrieval of the scenarios that have been already computed. No need for the user to remain logged in while the computational task proceeds. When a job finishes, its associated label is automatically added to the dedicated listbox. At the next login, if the job has completed, the user will be able to select the related entry in the list of available scenarios, and visualise the results produced. The choices are between the three components synthetic seismograms (in acceleration, velocity or displacement), shown in Fig. 8, and the spectral amplifications computed with respect to the average lithospheric model, taken as reference, as shown in Fig. 9.

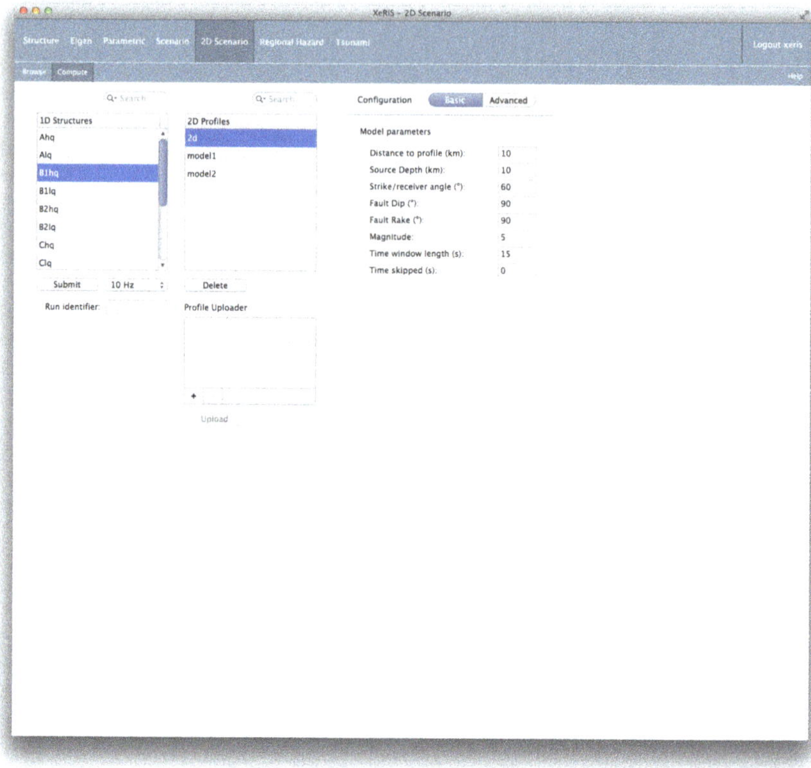

Fig. 7 Panel dedicated to the computation of ground shaking scenarios along a laterally heterogeneous profile. On the *left* is the list of available structural models used to define the average lithospheric properties along the path from the source to the site of interest, for which the modes have been computed already. In the middle is the list of available heterogeneous profiles already loaded into the WA. On the *right* there are further parameters, mostly source-related, that the user must adjust, in “Basic” or “Advanced” mode, to properly configure the experiment

A little more must be told about the preparation of the laterally heterogeneous part of the model. This step once required the manual editing of a quite complicated file, in which the lateral heterogeneities were defined through the superposition of rectangular “patches”, each covering a part of the layered model representing the average lithospheric structure. Every patch is characterised by its own set of density, P and S waves velocities and attenuation values. For complicated geometries the editing process was lengthy, convoluted and extremely error-prone. Most often, a stupid mistake in the preparation of this file resulted in wrong scenarios produced after days of wasted CPU time. Such a computational nightmare is now relegated into the past. A dedicated desktop application has been developed (Fig. 10), that allows the user to draw the model using the standard tools available in painting programs (straight and freehand pencils, colour picker, filler, magnifier lens etc.), plus some dedicated tools for the calibration of the model size, the digitisation of points and polygons

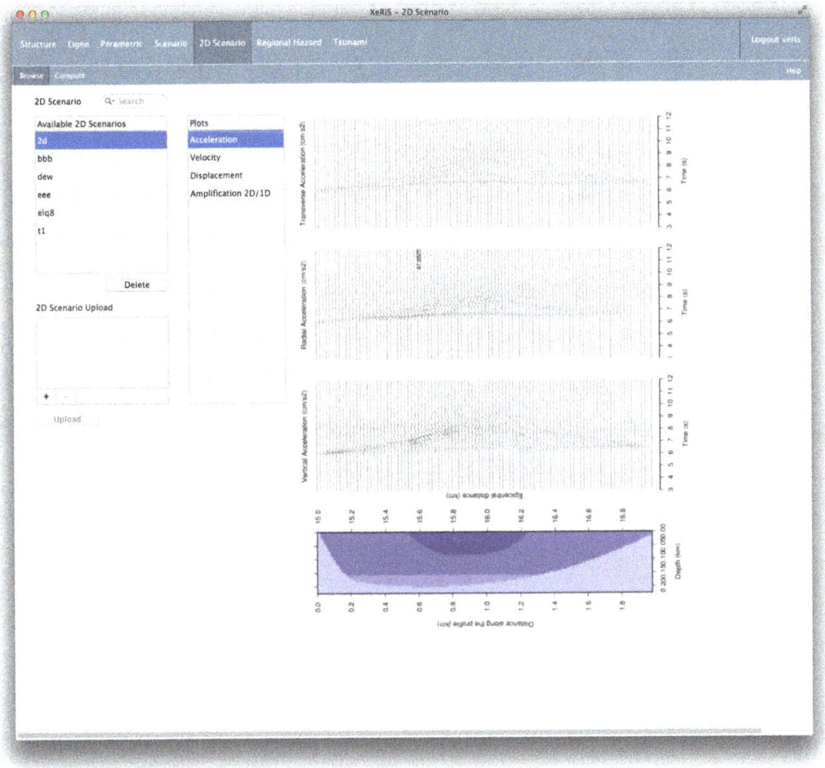

Fig. 8 Three component synthetic seismograms computed along the heterogeneous profile

coordinates, and for the chromatic correction of the model if an image acquired with a scanner is used as a reference when designing the profile. Each colour used in drawing the model is associated with a unique set of layer properties, that can be stored in a library for later reuse. After the drawing is complete, a rasterisation algorithm produces the rectangular patches required by the computational engine, and exports them in a properly formatted file, as needed by the finite difference program.

3.7 *Seismic Hazard Scenarios at Regional Scale*

The definition of seismic hazard with the neo-deterministic approach, as fully described in Panza et al. (2001, 2012), requires the preparation of several input datasets. They are needed for the definition of the characteristics (magnitude and focal mechanism) of the potential earthquake sources distributed within the active seismogenic zones, and for the description of the average lithospheric properties of the Earth in the considered region.

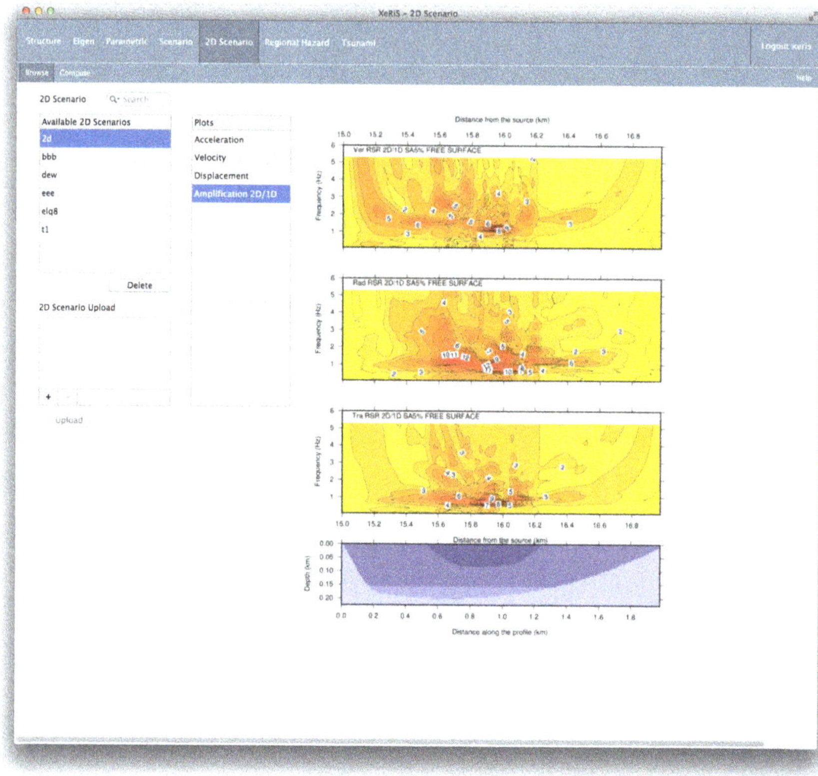

Fig. 9 Spectral amplifications computed along the heterogeneous profile

Once the input files are prepared, the computer package performs the following tasks, with no further user intervention:

- discretisation into cells of size 0.2° × 0.2° of the seismicity data taken from the historical earthquake catalogue, assigning to each cell the maximum magnitude obtained into each cell after a smoothing window has been applied to the original data;
- definition of a representative focal mechanism to be associated to all the sources belonging to each seismogenic zone;
- definition of a grid of sites evenly distributed in the considered region;
- generation of the paths between sources and sites, properly sorted by structural model and then by source depth, in view of an optimised computation of the synthetic seismograms;
- generation of the three-components synthetic seismograms (displacements, velocities, accelerations) for all the paths previously identified, with the possibility to adopt the SSPS or STSPS models;
- extraction and mapping of the peak values on the grid of sites.

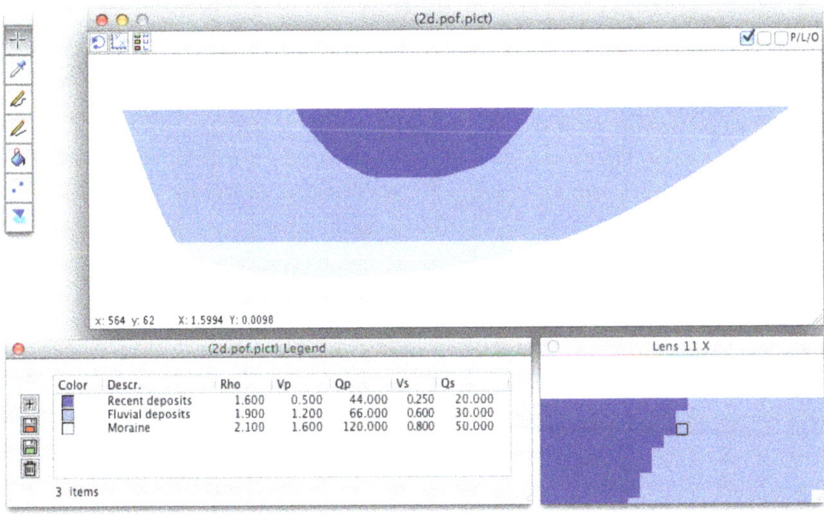

Fig. 10 The main elements of the desktop application developed for the preparation of the laterally heterogeneous profiles. From the *top–left corner*, *clockwise* the tools palette, the main window, the magnified view around the cursor position and the legend with the layer properties

The computational engine is based on the modal summation technique. Depending on the extension of the area being studied, and on the range of epicentral distances, requested by the user or automatically chosen by the system according to the magnitude of the considered sources, it might be necessary to compute a huge amount of seismograms. Typical numbers may span from a few hundred (for small areas defined ad hoc for quick preliminary parametric tests), to millions of seismograms for large sub-continental areas. For the cited extremes, the computational time required varies roughly between few seconds and about a day.

At the current stage of development, the "Regional Hazard" panel is fully functional in its browsing capabilities, as shown in Figs. 11 and 12. In that sub-panel, the user can upload the compressed archives generated automatically by the package at the end of the execution. Each archive contains both the input data and the output maps representing the hazard scenario.

For the generation of new scenarios from within the web application, the dedicated sub-panel layout is almost ready, but not all the planned functionalities are currently implemented. At present, it permits the upload of the required input datasets, including the optional ones that may be considered by the package and not yet mentioned here: the seismogenic nodes obtained through the morphostructural analysis as defined by Gorshkov et al. (2002, 2004, 2009), and the alerted areas declared by the CN or M8 algorithms (Peresan et al. 2005). What is currently being developed, and must be considered a truly essential improvement, is a set of helper applications aimed at guaranteeing that the uploaded input files are, at the very least, properly formatted according to the guidelines of the package. Better yet, a

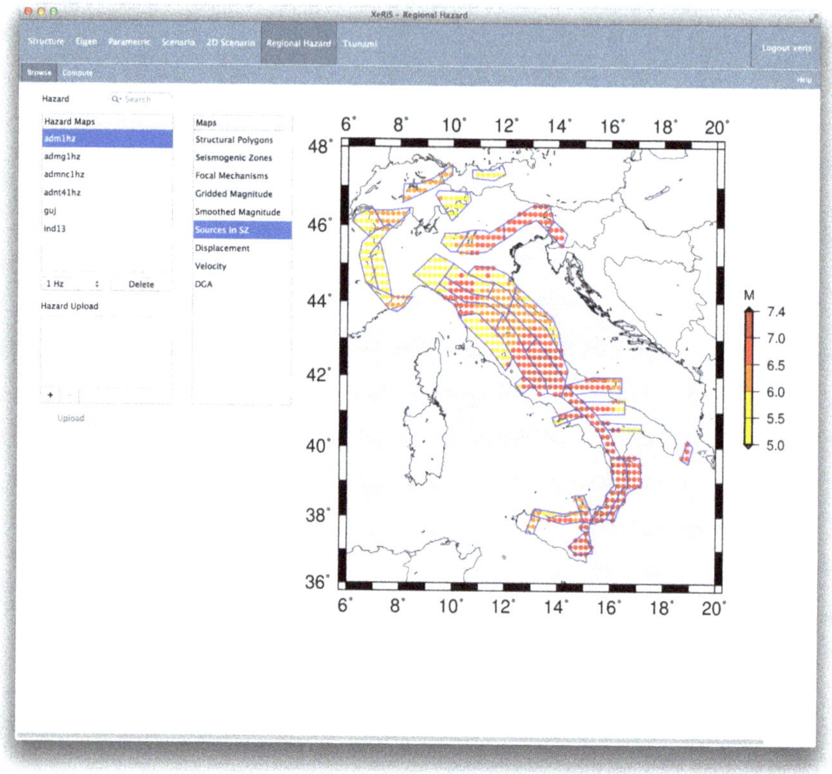

Fig. 11 Panel of the WA dedicated to the browsing of the input data and of the corresponding seismic hazard scenarios. From *left* to *right* the list of available computational experiments, with the control that permits to upload more items; the list of available maps for the selected experiment; the map. Here the distribution of sources inside the seismogenic zones, and their magnitude, is shown

further feature is already planned, that will allow a quality check of the data, targeted at the identification of unreasonable values mistakenly introduced in the datasets.

3.8 *Tsunami Hazard Scenarios*

This panel is the latest addition to the WA. Similarly to the seismic hazard panel, it currently allows the browsing of the uploaded results of tsunami modelling (Fig. 13). The computational engine upon which the modelling is based is the one described in Panza et al. (2000). The capability of generating new scenarios through a dedicated and user friendly web interface is currently being implemented, and will be shortly made available to the beta testers of the WA.

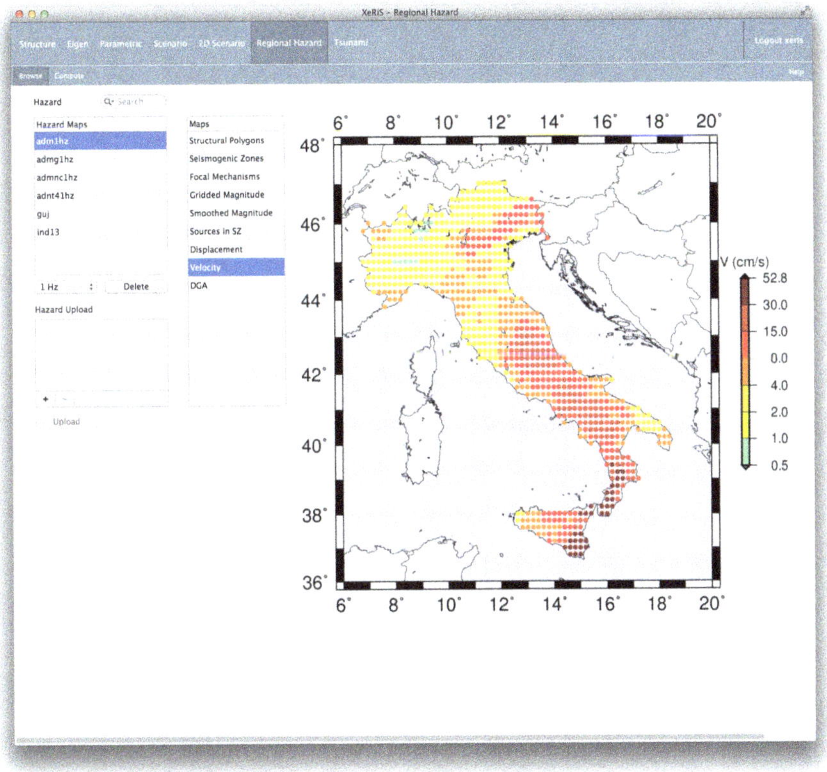

Fig. 12 In the same panel shown in Fig. 11, the obtained PGV distribution is visualised here

4 Final Remarks

It is a matter of fact that the seismic regulations currently adopted worldwide are mostly based on probabilistic estimates of the seismic hazard. This is very likely due to the lack of adequate tools that might have provided viable alternatives at the time the regulations were formulated.

The web application prototype here described aims at demonstrating that something new can finally be done to improve the preparedness against future events. With the underlying NDSHA computational engine, the end-user can be kept well isolated from the complexity of the modelling tools, and there is no steep learning curve for him to fight against, before he can generate his first ground shaking scenario. This is particularly true for the design and execution of quick parametric tests in laterally homogeneous layered models, where the source or layer properties are varied, and the effects (or lack thereof) of their variation on the ground shaking can be immediately verified in the obtained scenarios. Under these

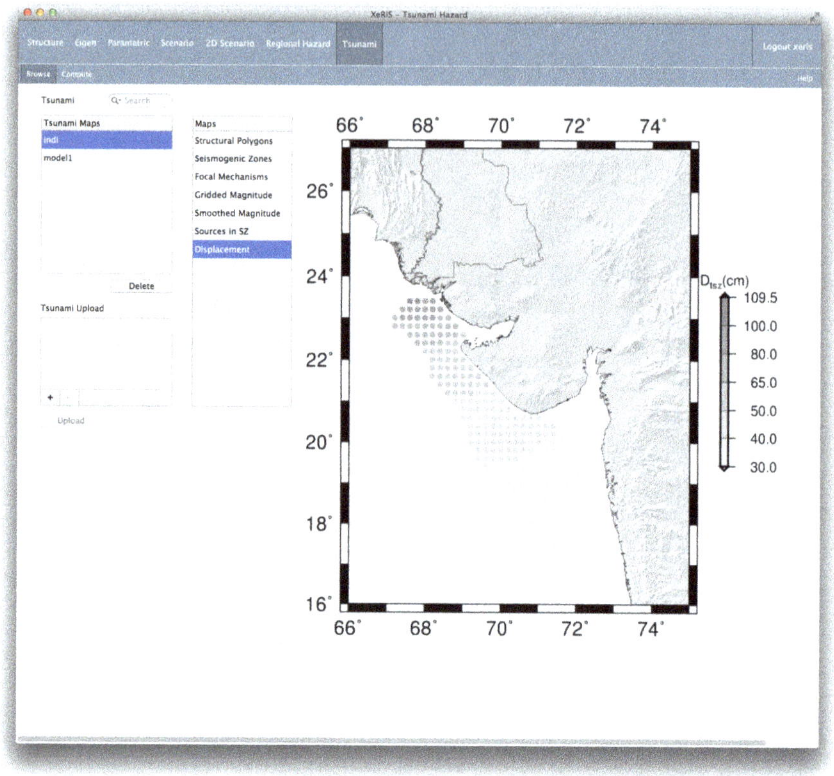

Fig. 13 Panel dedicated to the management of tsunami hazard calculations. Here the height of the modelled tsunami wave is shown

conditions, it is really a matter of minutes for civil engineers or city planners to get a first-order estimate of the seismic input expected for the next earthquake.

Conversely, the execution of massive experiments configured to explore the whole model parameter space, within the limits of the current knowledge, or the adoption of more realistic heterogeneous models, can be programmed taking advantage of modern computer architectures, based on the notions of Grid and Cloud computing.

The future development of this web application prototype is planned in such a way that the choice of the computational hardware on which the modelling will be executed should be almost transparent to the user. It has to become as simple as checking a radio button on the graphical user interface.

The benefits of the methodology upon which the web application is based have been recently proved in the city of Trieste (Italy). Commissioned by the Provincia di Trieste Authorities, ground shaking scenarios have been generated in the city along several 2D profiles. The amplifications obtained due to the combined interaction of source, path and site effects have lead to spectral accelerations higher than those predicted by the official hazard maps, based on probabilistic studies and currently adopted by the law. The seismic input specifically computed at selected sites of interest for the Provincia di Trieste Authorities has been used to verify the behaviour of some relevant buildings under the seismic load, and to plan the retrofitting that will hopefully avoid heavy damaging, or even the collapse of the structures when the earthquake will hit.

Acknowledgments First and foremost, this web application couldn't have seen the light without the underlying computational engine, based on the pioneering work by Prof. Panza. Giuliano's energy, and his continuous push and encouragement have played a decisive role in shaping up and transforming the original codes into the friendly tools they now are.

The web application development saw the light within the project "Definition of seismic hazard scenarios and microzoning by means of Indo-European e-infrastructures" funded by Regione autonoma Friuli Venezia Giulia in the framework of the interventions aimed at promoting, at regional and local level, the cooperation activities for development and international partnership —"Progetti Quadro ai sensi della Legge regionale n. 19 del 30 ottobre 2000". I'm truly grateful to Antonella for her impeccable project management.

The recent, strong evolution of the underlying computational engine would have never been possible without the dedicated work by the friends who made my life better here at the University: Fabio, Elisa, Andrea, Cristina and Davide, to name just those more deeply involved in the coding. And of course Enrico, for he knows what…

Stefano and Francesco have introduced me into the new universe of grid and cloud computing. I really enjoyed it, and I do hope our collaboration will continue and expand in the future.

Finally, a deep gratitude goes to my beloved family. I wish all the computational time spared through algorithm and code optimisation might be magically converted into time spent with them. Apparently I'm not too good in doing so, but don't loose your faith, I'll keep trying… :-)

References

Alvarez, L., et al. (2005). Seismic microzoning from synthetic ground motion parameters: Case study, Santiago de Cuba. *Soil Dynamics and Earthquake Engineering, 25*, 383–401.

Amponsah, P. E., et al. (2009). Deterministic seismic ground motion modeling of the greater Accra Metropolitan area, Southeastern Ghana. *South Africa Journal of Geology, 112*, 317–328.

Burridge, R., & Knopoff, L. (1964). Body force equivalents for seismic dislocations. *Bulletin of the Seismological Society of America, 54*, 1875–1878.

EN 1998-1. (2004). Eurocode 8: Design of structures for earthquake resistance—Part 1: General rules, seismic actions and rules for buildings [Authority: The European Union Per Regulation 305/2011, Directive 98/34/EC, Directive 2004/18/EC].

Fäh, D., & Panza, G. F. (1994). Realistic modelling of observed seismic motion in complex sedimentary basins. *Annali di Geofisica, 3*, 1771–1797.

Florsch, N., et al. (1991). Complete synthetic seismograms for high-frequency multimode SH-waves. *Pure and Applied Geophysics, 136*, 529–560.

Gorshkov, A. I., et al. (2002). Morphostructural zonation and preliminary recognition of seismogenic nodes around the adria margin in peninsular Italy and Sicily. *Journal of Seismology and Earthquake Engineering, 4*, 1–24.

Gorshkov, A. I., et al. (2004). Identification of seismogenic nodes in the Alps and Dinarides. *Bollettino della Società Geologica Italiana, 123*, 3–18.

Gorshkov, A. I., et al. (2009). Delineation of the geometry of the nodes in the Alps-Dinarides hinge zone and recognition of seismogenic nodes (M $\geq$ 6). *Terra Nova, 21*, 257–264.

Gusev, A. A. (1983). Descriptive statistical model of earthquake source radiation and its application to an estimation of short period strong motion. *Geophysical Journal of the Royal Astronomical Society, 74*, 787–800.

Gusev, A. A. (2011). Broadband kinematic stochastic simulation of an earthquake source: A refined procedure for application in seismic hazard studies. *Pure and Applied Geophysics, 168*, 155–200.

Gusev. A. A., & Pavlov, V. (2006). Wideband simulation of earthquake ground motion by a spectrum-matching, multiple-pulse technique. In *First European Conference on Earthquake Engineering and Seismology (a joint event of the 13th ECEE and 30th General Assembly of the ESC)*, Geneva, Switzerland, 3–8 Sept 2006. Paper Number: 408.

Harbi, A., et al. (2007). Seismicity, seismic input and site effects in the Sahel—Algiers region (North Algeria). *Soil Dynamics and Earthquake Engineering, 27*, 427–447.

La Mura, C. (2009). Wave propagation in three-dimensional anelastic media: The modal summation method in the WKBJ approximation, Ph.D. thesis, http://hdl.handle.net/10077/3141.

La Mura, C., et al. (2011). Three-dimensional seismic wave propagation by modal summation: Method and validation. *Pure and Applied Geophysics, 168*, 201–216.

Maruyama, T. (1963). On the force equivalents of dynamical elastic dislocations with reference to the earthquake mechanism. *Bulletin of the Earthquake Research Institute, 41*, 467–486.

Mohanty, W. K., et al. (2013). Influence of epicentral distance on local seismic response in Kolkata City, India. *Journal of Earth Science, 122*, 321–338.

Panza, G. F. (1985). Synthetic seismograms: The Rayleigh waves modal summation. *Journal of Geophysics, 58*, 125–145.

Panza, G. F., et al. (2000). Synthetic tsunami mareograms for realistic oceanic models. *Geophysical Journal International, 141*, 498–508.

Panza, G. F., et al. (2001). Seismic wave propagation in laterally heterogeneous anelastic media: Theory and applications to seismic zonation. *Advances in Geophysics, 43*, 1–95.

Panza, G. F., et al. (2002). Realistic modeling of seismic input for megacities and large urban areas (the UNESCO/ IUGS/IGCP project 414). *Episodes, 25*, 160–184.

Panza, et al. (2012). Seismic hazard scenarios as preventive tools for a disaster resilient society. *Advances in Geophysics, 53*, 94–165.

Peresan, A., et al. (2005). Intermediate-term middle-range earthquake predictions in Italy: A review. *Earth-Science Reviews, 69*, 97–132.

Romanelli, F., et al. (1997). Analytical computation of coupling coefficients in non-poissonian media. *Geophysical Journal International, 129*, 205–208.

Vaccari, F., et al. (1989). Synthetic seismograms in laterally heterogeneous, anelastic media by modal summation of P-SV waves. *Geophysical Journal International, 99*, 285–295.

Woodhouse, J. H. (1974). Surface waves in laterally varying layered structure. *Geophysical Journal of the Royal Astronomical Society, 37*, 461–490.

Yanovskaya, T. B. (1989). Surface waves in media with weak lateral inhomogeneity. In V. I. Keilis-Borok (Ed.), *Seismic surface waves in a laterally inhomogeneous earth* (pp. 35–69). Dordrecht, The Netherlands: Kluwer Academic Publishers.

Zuccolo, E., et al. (2008). Neo-deterministic definition of seismic input for residential seismically isolated buildings. *Engineering Geology, 101*, 89–95.

Rapid Response to the Earthquake Emergencies in Italy: Temporary Seismic Networks Coordinated Deployments in the Last Five Years

Milena Moretti, Lucia Margheriti and Aladino Govoni

1 Introduction

The National Institute of Geophysics and Volcanology (Italian: Istituto Nazionale di Geofisica e Vulcanologia, INGV) is in charge of the seismic monitoring and surveillance of the Italian territory. To this purpose it has installed and manages more than 300 permanent seismic stations included in the Italian National Seismic Network (Italian: Rete Sismica Nazionale, RSN; Amato and Mele 2008). Every time an earthquake that is likely to affect people (i.e. with magnitude above a certain threshold) occurs on the Italian territory the INGV surveillance room alerts the Department of Civil Protection (Italian: Dipartimento di Protezione Civile, DPC) providing detailed information about the location and the magnitude of the seismic event.

The earthquake hypocenter location accuracy is strongly dependent on the geometry of the network, the distance between the seismic stations and the type of seismic sensor. The RSN guarantees good locations for events of $M_L > 2$. The temporary seismic networks are deployed to improve the detection performance of the permanent monitoring systems. Knowing the number, the magnitude and the spatio-temporal distribution of small aftershocks can be useful for decision makers to assess the current situation during seismic crises (Dolce and Di Bucci 2014). The improvement in earthquake detection and location accuracy obtained installing a real time temporary dense seismic network is very significant. Defining the clustering of seismic events can help in foreseeing the characteristics of future seismic sequences in the same tectonic or volcanic environment (IAVCEI Subcommitee for Crises protocol 1999) and provide invaluable data for scientific studies related to hazard, tectonics and earthquake physics. Rapid-response seismic networks are a

M. Moretti (✉) · L. Margheriti · A. Govoni
Istituto Nazionale di Geofisica e Vulcanologia, via di Vigna Murata, 605, 00143 Rome, Italy
e-mail: milena.moretti@ingv.it

S. D'Amico (ed.), *Earthquakes and Their Impact on Society*,
Springer Natural Hazards, DOI 10.1007/978-3-319-21753-6_24

valuable element of the response to seismic crises. Temporary station sites are chosen to optimize network geometry for earthquake locations. The real time data contribute to the monitoring of the seismicity in the epicentral area, detecting the evolution of aftershocks in space and time. The off line analysis of the recorded seismograms allows the imaging of the fault system geometry and kinematics.

The INGV manages a Portable Seismic Network of about one hundred seismic stations some of them transmit data in real time to the seismic surveillance system of INGV. In the following paragraphs we describe in brief the INGV Portable Seismic Network, its history and the current coordination projects with other Italian and international Institutes. Moreover we describe INGV rapid response organization activities and emergency operations in the last five years, starting from the lesson learned in the 2009 L'Aquila emergency.

1.1 The INGV Portable Seismic Network in the Past

The INGV has been managing an emergency structure ready to face the occurrence of damaging earthquakes since 1990. The INGV Emergency Seismic Network was developed and implemented for scientific reasons but its importance during seismic crises raised the interest of DPC that decided to co-finance it. The Emergency Seismic Network has been usually deployed for $M_L \geq 5.0$ crustal earthquakes occurring in the Italian territory; it has been an important monitoring tool during the major seismic sequences of the last 25 years in Italy beginning with the Carlentini (Sicily) seismic sequence in 1990, up to Umbria-Marche (1997–1998), Forlì (Emila Romagna-2000), and San Giuliano di Puglia (Molise-2002), just to name some of the most important seismic crises when it was successfully deployed. Figure 1 shows the mobile acquisition center installed during the Umbria-Marche seismic sequence in Cofiorito: up to 6 stations could be received via UHF radio links by this center and their signals could be locally analyzed to determine hypocentral locations; later, off line, the data could be integrated into the RSN data and used to

Fig. 1 The rapid response real time network in the 90s, in the picture is shown the mobile acquisition center installed in Cofiorito during the Umbria-Marche seismic sequence (1997–1998)

produce refined locations together with RSN and seismic signals recorded by other stations installed for the emergenciy and recording on local memory supports (i.e. Selvaggi et al. 2001).

In time the emergency seismic network changed, following the technological evolution of seismic data logger, accelerometer and velocimeter sensors and improved its capability of acquiring and archiving data. In fact at the beginning of the 90s events were recorded in trigger mode on floppy disks and only in the middle of that decade we started recording seismic signals of temporary stations on hard disks in continuous mode.

The goal of transmitting the seismic data in real time directly to the seismic surveillance system of INGV was accomplished in 2008 when a new emergency structure was co-founded by INGV and the DPC (see 2007–2009, 2010–2012 and 2012–2021 Agreements between the DPC and the INGV; http://istituto.ingv.it/l-ingv/progetti/Convenzione-Quadro%20tra%20INGV%20e%20DPC). The new structure, developed and managed by the National Earthquake Center (Italian: Centro Nazionale Terremoti, CNT) a department of INGV, was tested for the first time during a regional seismic emergency simulation organized by the Civil Protection of the Marche Region. This structure is composed of two operational units: a temporary seismic network that transmit by satellite in real time to the seismic surveillance system of INGV; and a stand-alone temporary seismic network recording on local memory supports. The data in real time are archived together with the RSN data in the European Integrated Data Archive (EIDA, http://eida.rm.ingv.it/) in Standard for the Exchange of Earthquake Data (SEED) format (Mazza et al. 2012).

Moreover in 2008 a new module of the INGV emergency structure was developed: the Seismic Emergency Operational Centre (Italian: Centro Operativo Emergenza Sismica, COES; Govoni et al. 2008). It is a mobile office equipped with a satellite internet connection that can be rapidly installed in the disaster area. The COES supports the INGV staff operative needs and helps in the coordination with other national and international researchers working in the epicentral area. It is devoted to seismic emergency management (Moretti et al. 2009, 2010a) and cooperates with the DPC providing updated information 24 h a day, contributing to the decision-making stages during an emergency (Persico 2013). Furthermore, the COES may serve as a reference information center for the people involved in the crisis management providing important practical, and also psychological, support (La Longa and Crescimbene 2010) to the rescuers and to the population affected by an earthquake.

The response to the occurrence of the disastrous L'Aquila earthquake (central Italy) in April 2009 provides an example of spontaneous and collaborative rapid response seismic network coordination in a European framework. The INGV portable network deployed at the time was the one presented in 2008 but different rapid-response seismic networks from Italy, France and Germany (Margheriti et al. 2011) were deployed during the sequence. This deployment acquired the largest dataset ever recorded during a normal fault seismic sequence (Chiaraluce et al. 2011). A seismic network of more than 60 stations more than 60,000 earthquakes

(Valoroso et al. 2013). The data from most of the temporary stations were archived in EIDA and made available to the scientific community.

The experiences gained during this emergency emphasized the necessity to improve communication to facilitate rapid information exchange among the Italian and European rapid response network community. The experience of the L'Aquila earthquake in 2009 taught us that coordination between national and international Institutes in the deploying of seismic stations is fundamental to optimize instruments management, human resources and scientific results (Margheriti et al. 2011).

1.2 The INGV Emergency Seismic Network Today: The Co-Ordination at National and European Level

Today the INGV structure for rapid response to seismic emergency (Moretti et al. 2010b) is based on the one that operated during the L'Aquila earthquake, but with an increased number of temporary seismic stations. In 2011 was initiated "**Sismiko**", an INGV project to coordinate the activities during the emergencies and to establish common procedures for stations deployment, network maintenance and data archiving. The different sets of temporary seismic stations (networks) that belong to different INGV centers (Fig. 2) can either be independently deployed or coordinated into a unique network. This integrated seismic network is composed of: 1. real time stations equipped with radio-satellite, universal mobile telecommunications system transmission (UMTS) or WiFi transmission, which can become part of the RSN. This ensures substantial improvements in the real time monitoring systems. 2. Stand-alone recording stations that guarantee the acquisition of high resolution data devoted to the improvement of our scientific knowledge (e.g., seismic hazard, seismotectonics, earthquake physics, site response, wave propagation studies). The INGV seismic stations and the people involved in the emergency deployment are distributed in different centres in the Italian territory (Fig. 2: Ancona, Arezzo, Bologna, Catania, Gibilmanna, Irpinia, L'Aquila, Milan, Naples, Pisa and Rome).

After the occurrence of a damaging earthquake an alarm message is sent to all Sismiko participants, the deployment is coordinated by one central office that designs the possible network geometry and assigns the installation sites to the different groups in the field, according to the instrument availability and the achievable group reaction time. Station installation and maintenance can be performed by teams composed of staff coming from different centers, to guarantee the knowledge of the area and of the instruments. All the data recorded is archived in SEED data format and made available through the EIDA platform to the scientific community. While the real time data is immediately available, the data from stand-alone stations is periodically gathered from the field and converted.

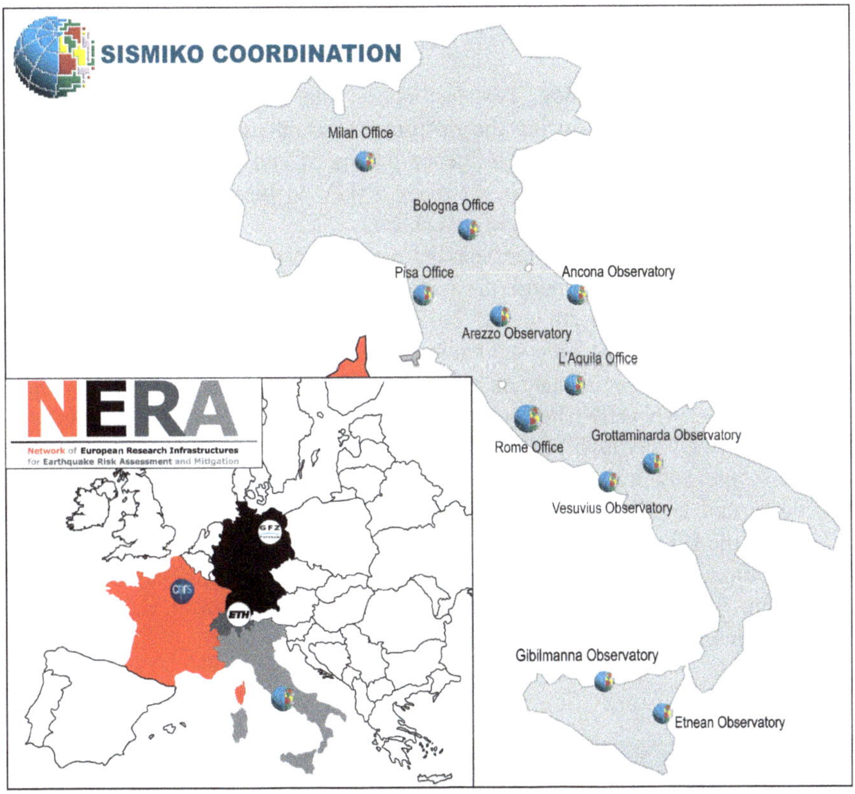

Fig. 2 Map of the INGV offices participating to the SISMIKO coordination group and map of the Institutes involved in the European Rapid Response seismic Network (project NERA)

At the European level, within the framework of the "Network of European Research Infrastructures for Earthquake Risk Assessment and Mitigation (NERA)" European project, a European Rapid Response Network (ERN) coordination project is being developed. Four European Institutes, INGV, Italy; GFZ, German Research Centre for Geosciences, Germany; ETH, Eidgenössische Technische Hochschule, Switzerland; CNRS, Centre National de la Recherche Scientifique, France (Fig. 2) are combining their efforts to set up a common European framework for the coordination of future post-seismic initiatives. This on-going project has facilitated the exchange of information at the European level and the coordination of station deployment within European institutions (http://www.nera-eu.org/index.htm?page=1207).

The Sismiko and the NERA coordination efforts have resulted, in the case of the May-June 2012 damaging **Emilia seismic sequence**, in a much faster reaction than in the past of all INGV components, in an easier management of the intervention and in a better coordination with other INGV, Italian and European emergency groups.

1.3 Seismic Emergency Preparedness

During a seismic emergency, key factors are the staff preparedness and the instrument efficiency. Enhancing the instruments maintenance plans and the constant training of the staff greatly influence the speed and the effectiveness of the action (Haddow et al. 2014). To get ready INGV in these last years organized several field simulations and training workshops.

In particular in the frame of the Sismiko and NERA projects INGV performed a simulation test of a rapid-response network deployment in Emilia Romagna on 26–30 September 2011 (Moretti et al. 2013). In the area of Montefeltro, already affected by a moderate seismic sequence in the previous months, a total of 8 real time seismic stations were planned and installed using either UMTS or a combination of HiperLAN (HIgh Performance Radio LAN) radios and satellite internet connections. The test was successful and allowed the different teams from Bologna, Milano, Rome and Irpinia to practice the different seismic stations installation procedures and to share expertize. The whole real time continuous data set was archived in the EIDA (station codes IV_T05** and IV_T06**) data bank. Figure 3 shows pictures of stations and people involved in the field simulation; the same teams were involved in the emergency deployment, which followed the Emilia 2012 seismic crises.

Fig. 3 Picture of the simulation test of a rapid response network deployment in Emilia Romagna on September 26th and 30th 2011

Moreover on 28–29 May 2013 a two day workshop on portable seismic networks was organized at INGV in Rome; the training course was devoted to staff updates and seismic instrumentation testing (see blog INGVTerremoti: http://ingvterremoti.wordpress.com/2013/05/28/inizia-il-corso-rete-sismica-mobile-allingv-28-29-maggio-2013/).

More than 70 researchers, technologists and technicians from INGV, but also from the National Institute of Oceanography and Experimental Geophysics (Italian: Istituto Nazionale di Oceanografia e Geofisica Sperimentale—OGS), the University of Genova—Department of Earth, Environment and Life Sciences (Italian: Dipartimento di Scienze della Terra dell'Ambiente e della Vita, DISTAV), from the Prato Ricerche Institute, gathered together in Rome for training on instrumentation usage and on real time data transmission and archiving. The course included a first theoretical session with several lectures on the technical aspects of the data acquisition and transmission. During the practical part people was divided in heterogeneous teams and sent to the field (actually the Institute garden) to install units of the CNT Emergency Network equipped with seismometers and accelerometers. All participants installed and configured a complete seismic station, discussed the problems related to acquire and download the data from different seismic stations. The training course was very important to strengthen the relations between INGV and DISTAV that collaborated in the Lunigiana emergency that occurred in June 2013 (Margheriti et al. 2014). Figure 4 shows pictures of the different phases of the course (YouTube: http://www.youtube.com/watch?v=SYmPM40Bh5E&feature=c4overview&list=UUWcylY2YDfioFmDAULj3vgA).

1.4 Main Seismic Emergencies in 2010-2014

In the last 5 year, starting after the L'Aquila seismic emergency (Margheriti et al. 2011), the emergency networks were installed 7 times during seismic swarms and after moderate main-shocks in the Italian territory (seismic sequence near Frosinone in 2009, seismic sequence near Fermo in 2010, seismic sequence in Montefeltro in 2011, seismic sequence near the Pollino mountain in 2011–2014, seismic sequence in Emilia region in 2012, seismic sequence in Lunigiana region in 2013 and seismic sequence in Matese region in 2013–2014). The larger events occurred in this time frame was in the Po-Plain, the Emilia seismic sequence in 2012, the longer swarm is the one occurring at the Calabria Lucania boundary started in 2011 and still active. The data recorded during the emergencies is archived in EIDA (Fig. 5).

1.4.1 Rapid Response to the Earthquake Emergency of May 2012 in the Po Plain: The Emilia Seismic Sequence

On Sunday, May 20, 2012 at 02:03 UTC, a M_L 5.9 earthquake of hit northern Italy (44.89°N, 11.23°E, 6.3 km in depth). After few minutes, the mainshock was followed by a M_L 5.1 event, and after a few hours, by a second M_L 5.1 earthquake. In

Fig. 4 The May 28 and 29 2013 portable seismic network Workshop: **a** one of the lectures on the data logger maintenance, **b** practice with an UMTS real time unit in the "field"

the following 72 h (20–22 May 2012) about 500 events occurred (data source: 2005–2012 ISIDe Italian seismological instrumental and parametric database, ISIDe Working Group 2010, Scognamiglio et al. 2012): 86 events had $M_L \geq 3.0$, of these 18 events had $M_L \geq 4.0$). After the mainshock struck, contacts among the different INGV centers started immediately and early in the morning of May 20, 2012, the first INGV emergency rapid response group was heading to the epicentral area. The first real time station (UMTS) started operating at about 10:30 UTC on May 20, 2012. The INGV groups from Ancona, Arezzo, Bologna, Milan, Irpinia, Pisa and Rome installed 16 seismic stations (8 in real time: 4 UMTS-4 radio-satellite) within the first 48 h.

In the meantime, the INGV started the coordination with other Italian and foreign Institutes (OGS, DPC, NERA project participants). In the following days the rate of seismicity remained high, with more than 100 events per day. On May 29, 2012, at 07:00 UTC, a new M_L 5.8 earthquake struck the area at the western edge of the ongoing seismic sequence (44.85°N, 11.09°E, 10.2 km in depth). This was followed by several earthquakes, including M_L 5.2 and M_L 5.3 events, at 10:55 UTC and 11:00 UTC, respectively, of May 29, 2012; finally a M_L 5.1 event occurred on June 3, 2012, at 19:20 UTC. After this earthquake the seismicity was characterized by a progressive decrease in the seismic rate and the seismic moment release. During the first week of June 2012 the emergency seismic network

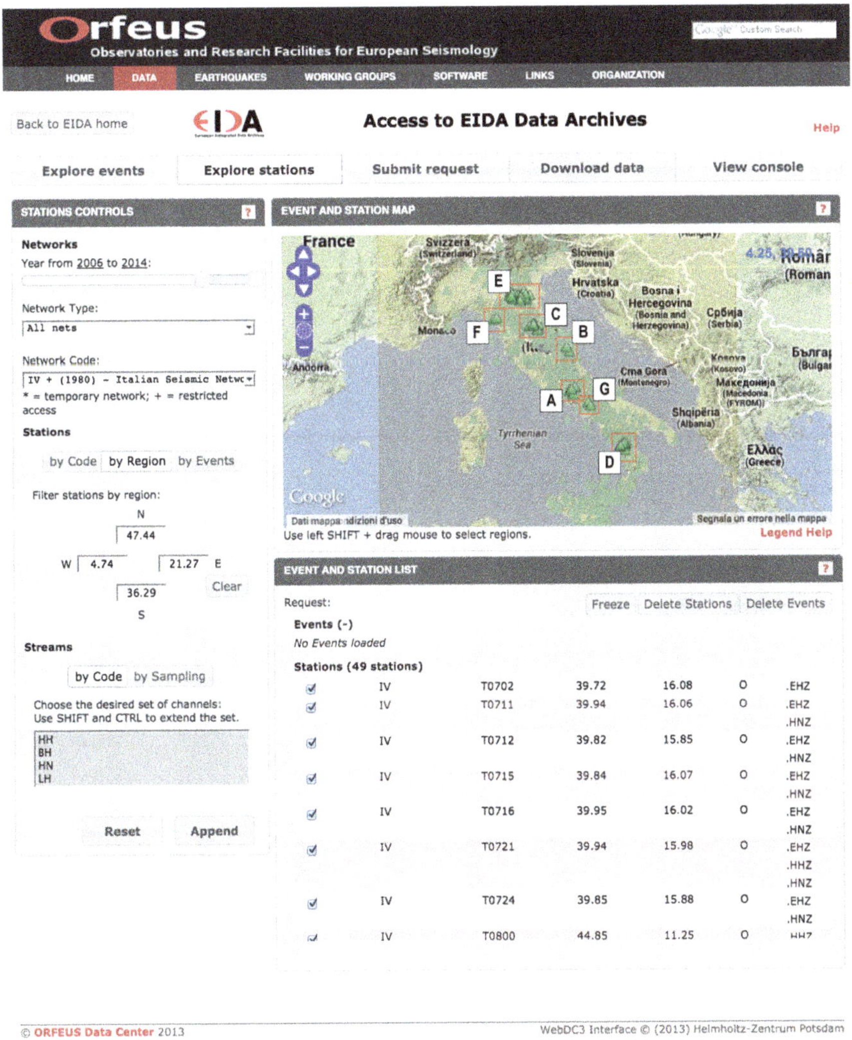

Fig. 5 Data in EIDA from the 7 emergencies in last 5 years in the Italian territory. Only stations open to the scientific community are shown as *green triangles*; the *light green* stations in the background belong to the RSN. Seismic sequences recorded and archived: A. "Frosinone 2009", B. "Fermo 2010", C. "Montefeltro 2011", D. Pollino 2011–2014", E. "Emilia 2012", F. "Lunigiana 2013", G. "Matese 2013–2014"

continued to improve under the central coordination of the INGV. As soon as further stations became available, the network was expanded, following the spatial evolution of the seismic sequence (Fig. 6b). On June 3rd about 70 seismic stations were operating for the emergency, including stations managed by French groups just deployed (see also Moretti et al. 2012). Data from 22 INGV stations, named as T0800-T0828 are available to the scientific community on EIDA; real time stations

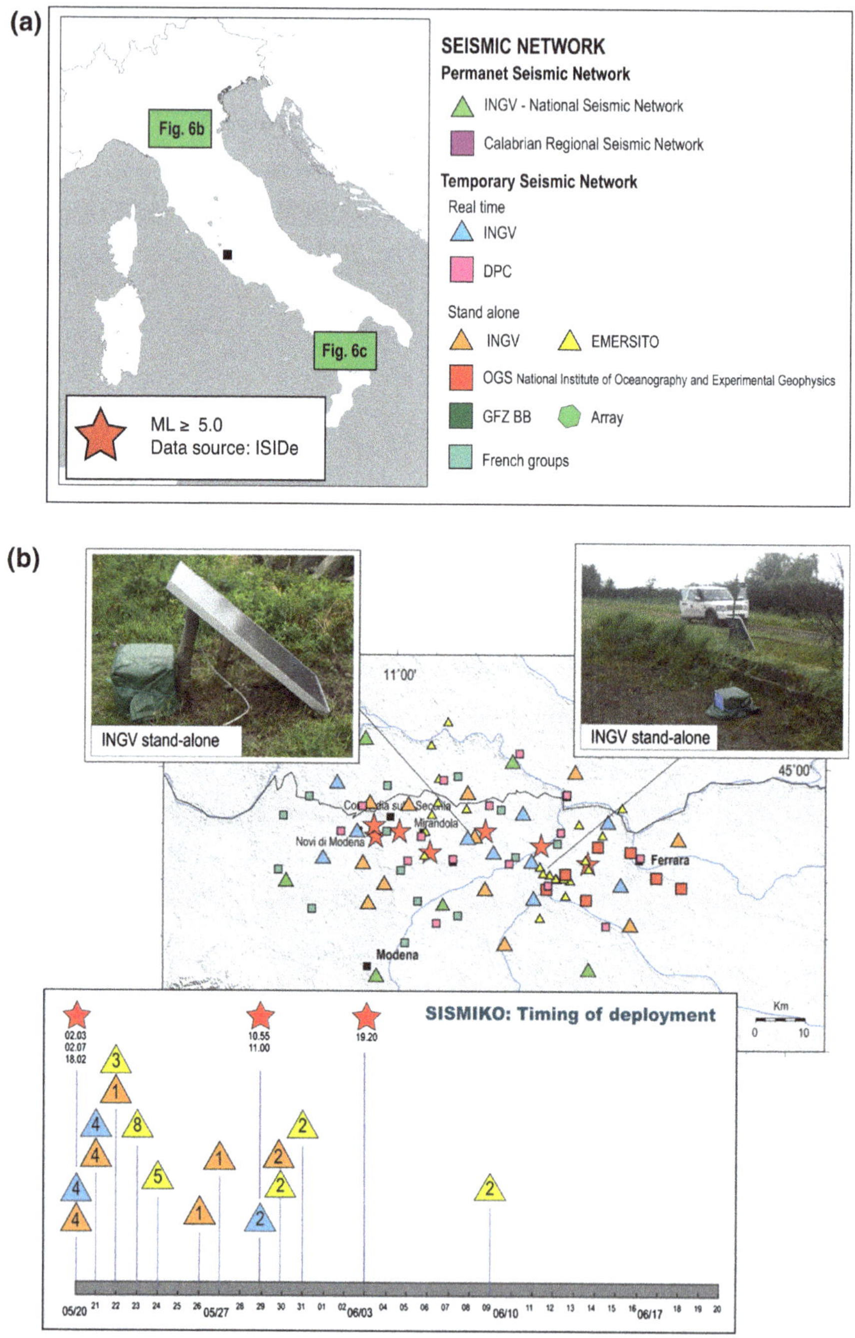

Fig. 6 **a**. Legend of the station types deployed during the "Emilia 2012" (see Fig. 6b) and the "Pollino 2011–2014" (see Fig. 6c) seismic sequences. **b** Map of the seismic stations deployed after May 20th. See the legend to Fig. 6a for the symbols. **c** Map of the seismic stations deployed in the Pollino region from November 2011 to the present. See the legend to Fig. 6a of the symbols

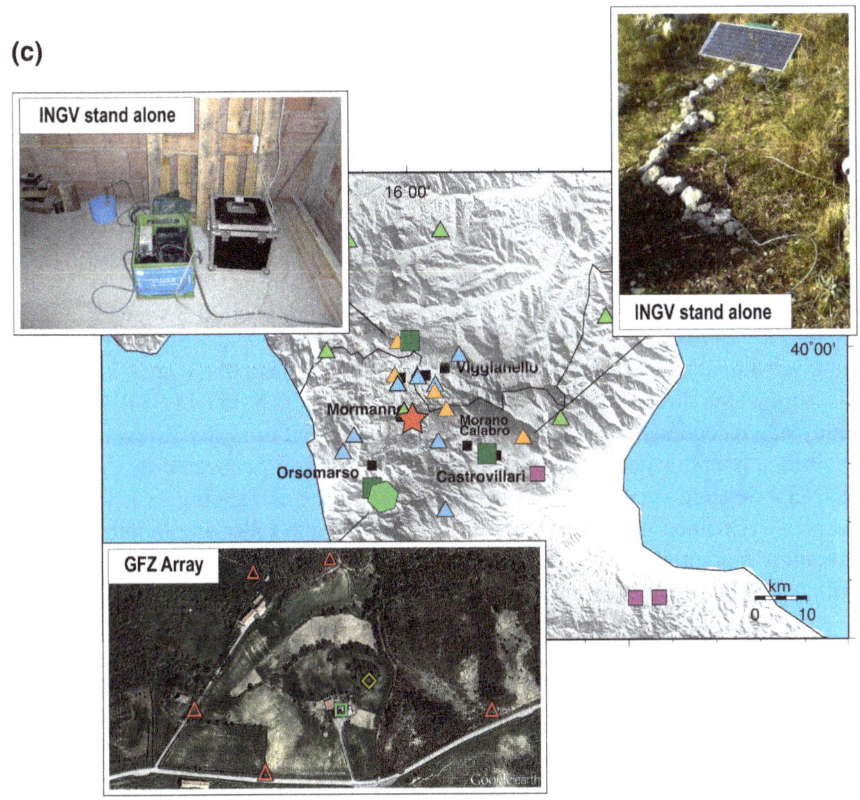

Fig. 6 (continued)

were available in real time; the other stations were archived in July 2012; other from different INGV groups were added later and some of them are restricted data and will become available in the next years. The continuous data are archived in SEED format and can be requested by the scientific community in SEED or SAC format. Data of the main events of the sequence are available in the accelerometric data base ITACA (Italian ACcelerometric Archive, Luzi et al. 2008; Massa et al. 2012).

The 10 stations in real time together with the RSN contributed to the accuracy of the hypocenter locations that were timely provided to DPC for civil protection purposes. Overall the real time stations operation scores were satisfactory: by the end of June, the real time stations had contributed to the location of about 2000 earthquakes, providing more than 5000 P-wave and more than 4000 S-wave pickings, allowing thus for enhanced completeness and accuracy of the earthquake list distributed by the INGV through ISIDe.

The about 60 seismic stations, recording stand-alone, from INGV and other Institutions (DPC, OGS, Università di Ferrara and the French colleagues coordinated by National Institute for earth sciences and astronomy—INSU), provide an excellent data-set for scientific investigations and allowed and will allow the scientific community to perform studies related to hazard (Cultrera et al. 2014), seismotectonics (Govoni et al. 2014), earthquake physics, site response and wave propagation (Emersito group; Bordoni et al. 2012; Milana et al. 2013).

1.4.2 Rapid Response to the Pollino Seismic Swarm 2011–2014

In November 2011 in the area around the Calabria Lucania boundary, close to the Pollino mountain range, the rate of seismicity increased reaching tens of events/day several earthquakes were felt by the population. Consequently INGV undertook initiatives aimed at upgrading the seismic monitoring system of the area, installing a new permanent station in Cetraro (CET2 lat 39.528756°N lon 15.954618°E) and deploying, in collaboration with the Dipartimento di Fisica dell'Università della Calabria, a temporary seismic network of five stations (3 connected in real time to the INGV Seismic Surveillance System; T0711, T012 e T0715; Fig. 6c). The rate of seismicity remained higher than the background for the following months, but the maximum magnitude did not reach M_L 4.0 until May 28th when a M_L = 4.3 earthquake occurred about 5 km to the west of the ongoing seismic swarm. Before the end of May 2012, during the climax of the Emilia seismic sequence, to optimize the network geometry in the Pollino area two other temporary stations in real time were deployed (T0701 e T0702; Fig. 6c).

The seismic activity remained concentrated in the M_L 4.3 source region and in the middle of July 2012, considering the seismicity decrease, the network at the Calabria Lucania boundary was reduced leaving only the T0701 T0702 real time temporary stations. These continued to make an important contribution to the localization in the Seismic Surveillance System of INGV headquarter in Rome. In early August seismicity jumped back westward to the previous area, with several earthquakes of magnitude larger than 3. On October 25, 2012 (23.05 h UTC), a M_L 5.0 event (39 881°N, 16.009°E Depth 6.3 km) hit the area. In the two days following the earthquake an INGV team installed 5 stand-alone stations T0721-T0725, all equipped with a Lennartz LE-3D1s velocimeter and an Episensor accelerometer. On November the 15th the T0724 station was equipped with an UMTS router and began to transmit real time data. At the beginning of November 2012 the international research team of the NERA ERN group (GFZ and INGV; Govoni et al. 2013) installed other seismic stations in the Pollino range to improve the detection capabilities; 3 stations are equipped with broadband sensors from the European Union's Cold Cases in Magma Propagation—Physics of Magma Propagation: A Multi-Methodological Investigation project (CCMP- POMPEI; http://www.gfz-potsdam.de/en/section/physics-of-earthquakes-and-volcanoes/projects/ccmp-pompei/), and 6 were deployed as a small-aperture seismic array. The seismic rate remained high for some months, but aftershock magnitudes did not exceed magnitude 3.7. The seismic rate then decreased at the beginning of 2013 and stayed quite low for the rest of the year up to the beginning of 2014, nevertheless the swarm is still active with a seismic rate higher than the background.

Figure 6c shows the complete deployment of the temporary network after the M_L 5.0 earthquake in the Pollino area (the deployment history starting from the end of 2011 is described in Margheriti et al. 2013).

The data acquired in real time by the temporary stations are available on EIDA, data from stand-alone stations were analyzed by INGV and the results of the refined locations are going to be described in De Gori et al. (2014).

1.5 Conclusive Remarks

Portable seismic networks are important tools in the rapid response to the earthquake emergencies. The INGV is constantly working to improve the performance of its emergency network both improving the instrumentation pools and training the staff. More over coordination projects both at national and European level have been developed in the last five years.

Every intervention during the seismic crises and every training event in the last five years improved our preparedness and influenced the speed and the effectiveness of subsequent actions. The cooperation and coordination between different groups is a difficult task to achieve but is very important because force the sharing of data and techniques improving the scientific results.

On January 26th 2014 a $M_L > 6$ stroke the Island of Kefalonia, Greece; the NERA ERN group was alerted and the Sismiko network decided to join the international efforts to improve the monitoring in the Island. Six stations from INGV were installed in the area which was the place of a site response seismic experiment in 2011 (NERA research activities). Hopefully this experience will push the Greek to join the ERN group in NERA.

Acknowledgments We would like to thank all the people involved in the Portable seismic Network deployments of the last 5 years especially the staff of Sismiko coordination project and Nera ERN project. This study has partially benefited from funding provided by the Italian Presidenza del Consiglio dei Ministri, Dipartimento della Protezione Civile (DPC). Scientific reports funded by the DPC do not represent its official opinion and policies.

References

Amato, A., & Mele, F. M. (2008). Performance of the INGV national seismic network from 1997 to 2007. *Annals of Geophysics, 51*(2–3), 417–431.

Bordoni, P., Azzara, R. M., Cara, F., Cogliano, R., Cultrera, G., Di Giulio, G., et al. (2012). Preliminary results from EMERSITO, a rapid response network for site-effect studies. *Annals of Geophysics 55*(4). doi:10.4401/ag-6153.

Chiaraluce, L., Valoroso, L., Piccinini, D., Di Stefano, R., & De Gori, P. (2011). The anatomy of the 2009 L'Aquila normal fault system (central Italy) imaged by high-resolution foreshock and aftershock locations. *Journal of Geophysical Research, 116*, B12311. doi:10.1029/2011JB008352.

Cultrera, G., Faenza, L., Meletti, C., Amico, V. D'., Michelini, A., & Amato, A. (2014). Shakemaps uncertainties and their effects in the post-seismic actions for the 2012 Emilia (Italy) earthquakes. *Bulletin of Earthquake Engineering*. published on-line 30 January 2014. doi:10.1007/s10518-013-9577-6.

De Gori P., Margheriti, L., Lucente, F. P., Govoni, A., Moretti, M., Pastori, M., et al. (2014). Seismic activity images the activated fault system. In *The Pollino area, at the Apennines-Calabrian Arc Boundary Region*. 33° Gruppo Nazionale di Geofisica della Terra Solida (Bologna 2014).

Dolce, M., Di Bucci, D. (2014). National civil protection organization and technical activities in the 2012 Emilia earthquakes (Italy). *Bulletin of Earthquake Engineering*. published on-line 25 February 2014. doi:10.1007/s10518-014-9597-x.

Govoni, A., Abruzzese, L., Amato, A., Basili, A., Cattaneo, M., Chiarabba, C., et al. (2008). Sequenze sismiche: La nuova struttura di Pronto Intervento dell'Istituto Nazionale di Geofisica e Vulcanologia. 27° Convegno Nazionale GNGTS—Trieste 6–8 ottobre 2008.

Govoni, A., Passarelli, L., Braun, T., Maccaferri, F., Moretti, M., Lucente, F. P., et al. (2013). Investigating the origin of seismic swarms. An international project addresses the question gaining insight on the seismicity in the Pollino range (Southern Italy). *Eos, Transactions American Geophysical Union, 94*(41), 361–362. doi:10.1002/2013EO410001.

Govoni, A., Marchetti, A., De Gori, P., Di Bona, M., Lucente, F. P., Improta, L., et al. (2014). The 2012 Emilia seismic sequence (Northern Italy): Imaging the thrust fault system by accurate aftershocks location. *Tectonophysics*. http://dx.doi.org/10.1016/j.tecto.2014.02.013.

IAVCEI Subcommitee for Crises protocol. (1999). Professional conduct of scientists during volcanic crises. *Bulletin of Volcanology, 60*, 323–334. doi:10.1007/PL00008908.

ISIDe Working Group (2010). *Italian seismological instrumental and parametric database*. http://iside.rm.ingv.it.

Haddow, G., Bullock, J., & Coppola, D. P. (2014). *Introduction to emergency management* (5th ed.). Oxford: Butterworth-Heinemann.

La Longa, F., & Crescimbene, M. (2010). La dimensione psicologica del terremoto che ha colpito l'Abruzzo. http://www.earth-prints.org/handle/2122/5869.

Luzi, L., Sabetta, F., Hailemikael, S., Bindi, D., Pacor, F., & Mele, F. M. (2008). ITACA (ITalian ACcelerometric Archive): A web portal for the dissemination of Italian strong motion data. *Seismological Research Letters, 79*(5), 717–723.

Margheriti, L., Chiaraluce, L., Voisin, C., Cultrera, G., Govoni, A., Moretti, M., et al. (2011). Rapid response seismic networks in Europe: Lessons learnt from the L'Aquila earthquake emergency. *Annals of Geophysics, 54*(4), 392–399.

Margheriti L., Moretti M., Pasta M., Chiaraluce C., Frepoli A., Piccinini D., et al. (2014). Il terremoto del 21 giugno 2013 in Lunigiana. Le attività del coordinamento Sismiko. Rapporti Tecnici INGV, 268.

Massa, M., Lovati, S., Sudati, D., Franceschina, G., Russo, E., Puglia, R., et al. (2012). INGV strong-motion data web-portal: a focus on the Emilia seismic sequence of May-June 2012. *Annals of Geophysics 55*(4). doi:10.4401/ag-6120.

Mazza, S., Basili, A., Bono, A., Lauciani, V., Mandiello, A. G., Marcocci, C., et al. (2012). AIDA—Seismic data acquisition, processing, storage and distribution at the National Earthquake Center, INGV. *Annals of Geophysics 55*(4). doi:10.4401/ag-6145.

Milana, G., Bordoni, P., Cara, F., Di Giulio, G., Hailemikael, S., & Rovelli, A. (2013). 1D velocity structure of the Po River plain (Northern Italy) assessed by combining strong motion and ambient noise data. *Bulletin of Earthquake Engineering*. published on-line July 2013. doi:10.1007/s10518-013-9483-y.

Moretti, M., Govoni, A., Nostro, C, La Longa, F., Crescimbene, M., Pignone, M., et al. (2009). The new emergency structure of the IstitutoNazionale di Geofisica e Vulcanologia during the L'Aquila 2009 seismic sequence: The contribution of the COES (Seismological Emergency Operation Center—Centro Operativo Emergenza Sismica). AGU Fall Meeting, 14–18 December, San Francisco, California, USA.

Moretti, M., Govoni, A., Basili, A., Amato, A., Doumaz, F., Vinci, S., et al. (2010a). Progettazione e realizzazione del Centro Operativa Emergenza Sismica (COES). *Rapporti Tecnici INGV, 172*, 19 pp.

Moretti, M, Govoni, A., Colasanti, G., Silvestri, M., Giandomenico, E., Silvestri, S., et al. (2010b). La Rete Sismica Mobile del Centro Nazionale Terremoti. *Rapporti Tecnici INGV, 137*, 66 pp.

Moretti, M., et al. (2012). Rapid-response to the earthquake emergency of May 2012 in the Po Plain, Northern Italy. In M. Anzidei, A. Maramai & P. Montone (Eds.), *The Emilia (northern Italy) seismic sequence of May-June, 2012: Preliminary data and results, Annals of geophysics*, Vol. 55, no. 4, pp. 583–590. doi:10.4401/ag-6152.

Moretti, M., Cattaneo, M., Pondrelli, S., Margheriti, L., Govoni, A., Nostro, C., et al. (2013). *Pianificazione e preparazione dell'emergenza. L'esercitazione a Santa Sofia (FC) – 26–30 settembre 2011: un esempio di gestione di una crisi sismica*. In pubblicazione: Quaderni di Geofisica, 108, 27 pp. ISSN 1590-2595.

Persico, F. (2013). La gestione degli eventi estremi: il coordinamento nei network organizzativi complessi. Tesi di dottorato, Università degli Studi di Milano-Bicocca, 2013. pp. 169. http://hdl.handle.net/10281/40099.

Scognamiglio, L., Margheriti, L., Mele, F. M., Tinti, E., Bono, A., De Gori, P., et al. (2012). The 2012 Pianura Padana Emiliana seimic sequence: Locations, moment tensors and magnitudes. *Annals of Geophysics 55*(4). doi:10.4401/ag-6159.

Selvaggi, G., Ferulano, F., Di Bona, M., Frepoli, A., Azzara, R., Basili, A., et al. (2001). The Mw 5.4 Reggio Emilia 1996 earthquake: Active compressional tectonics in the Po Plain. *Italy Geophysical Journal International, 144*, 1–13.

Valoroso, L., Chiaraluce, L., Piccinini, D., Di Stefano, R., Schaff, D., & Waldhauser, F. (2013). Radiography of a normal fault system by 64,000 high-precision earthquake locations: The 2009 L'Aquila (central Italy) case study. *Journal of Geophysical Research: Solid Earth, 118*(3), 1156–1176. doi:10.1002/jgrb.50130.

The Key Role of Eyewitnesses in Rapid Impact Assessment of Global Earthquakes

Rémy Bossu, Robert Steed, Gilles Mazet-Roux, Fréderic Roussel, Caroline Etivant, Laurent Frobert and Stéphanie Godey

1 Introduction

Rapid assessment of a global earthquake's impact, focusing on damage caused by ground shaking (rather than secondary effects such as fires, tsunamis, landslides…), relies first on the spatial distribution of earthquake shaking as estimated by ground-motions prediction equations and on the building stock inventory and related vulnerability (Erdik et al. 2011). However, the variability of the ground-motion predictions model is significant (e.g., Atik et al. 2010), and at the global scale, building stock inventory and related vulnerability are difficult to evaluate with sufficient accuracy and spatial resolution (Porter et al. 2008) leading to uncertainties in impact assessments.

Impact assessments can prove to be uncertain even when building stock and vulnerability are relatively well constrained. Shaking level is, as a first estimate, a function of magnitude and of distance to the fault rupture. For small to moderate magnitude earthquakes, source rupture can be approximated by a point. Variations of a few kilometres of epicentral location (i.e. within typical uncertainties of real-time location estimates) in relation to centres of population can lead to significant changes in impact scenario. In this type of situation, it can be difficult to rapidly identify the scope of the disaster without in situ information. A typical example is the 1999 M 5.9 Athens, Greece, earthquake. Located at about 18 km from the historical centre of the city, it caused 143 casualties (Papadopoulos et al. 2000). Undoubtedly, the death toll would have been significantly higher if, other things being equal, the epicentre had been closer to the city by 5 or 10 km. Figure 1

R. Bossu (✉) · R. Steed · G. Mazet-Roux · F. Roussel · C. Etivant · L. Frobert · S. Godey
European-Mediterranean Seismological Centre, Bruyères-le-Châtel, France
e-mail: bossu@emsc-csem.org

R. Bossu
CEA, DAM, DIF, F91297 Arpajon, France

S. D'Amico (ed.), *Earthquakes and Their Impact on Society*,
Springer Natural Hazards, DOI 10.1007/978-3-319-21753-6_25

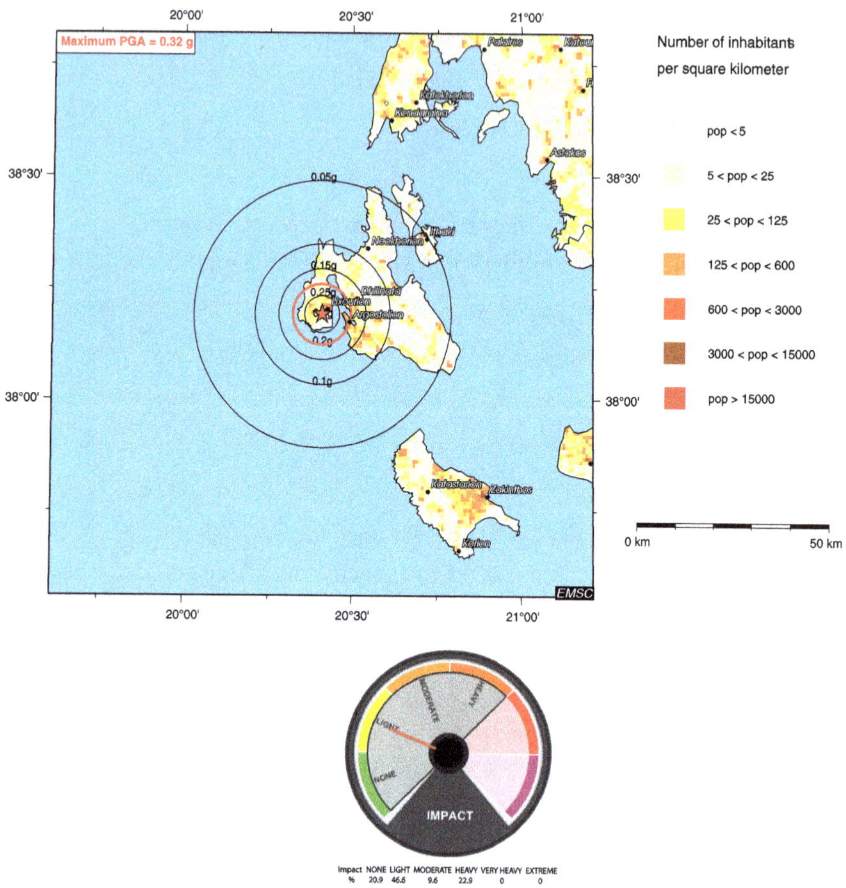

Fig. 1 Qualitative impact assessment for the M 6.1 Cephalonia earthquake, January 26, 2014. The map shows peak ground motion at bedrock through region-specific ground motion prediction equations (GMPEs) and the number of inhabitants (*shaded color*). The exploration of the location and magnitude uncertainty domains leads to different scenario (*lower figure*), the *red line* indicates the most likely one. The method named EQIA (Earthquake Qualitative Impact Assessment) based on the NERIES-ELER methodology (Erdik et al. 2011), has been calibrated on past global earthquakes (Bossu et al. 2009) and has been in operation since 2009, results being restricted to EMSC members. The scenario is very sensitive to the slightest change of epicentral location in relation to centres of population

shows the uncertainties on the qualitative impact scenario as automatically computed by EMSC for another Greek earthquake, Cephalonia, M 6.1 event of January 26, 2014 and its sensitivity to the slightest change in location and magnitude.

For large magnitude earthquakes, a point source approximation is no longer valid and fault rupture parameters (position, orientation, length) which are unknown or only partially determined in the immediate aftermath of an earthquake occurrence play an important role in the spatial distribution of ground shaking and hence on the distribution of damage. When striking a highly vulnerable and densely-populated region, like the 2010 M 7.0 Haiti earthquake, (e.g., Bilham 2010) the extent of the losses may not dramatically change with rupture parameters, but in situ observations are still helpful to ascertain locations of damaged areas and target search and rescue efforts.

We present in this article the strategy and methods implemented at the European-Mediterranean Seismological Centre (EMSC) for rapidly collecting in situ observations on earthquake effects from eyewitnesses to reduce uncertainties in rapid impact assessment of global earthquakes. We show how Internet and communication technologies are creating new potential for rapid and massive public involvement by both active and passive means. We underline the importance of merging results from different methods to improve performance and reliability. We then explore what could be the next technical development phase, by observing that the pervasive use of smartphones changes the way rapid earthquake information is accessed. Finally, we discuss how these approaches not only augment data collection on earthquake phenomenon at little cost but also how they change the way that we, as scientists, interface with eyewitnesses and how it pushes us to better understand and respond to the public's demands and expectations in the immediate aftermath of earthquakes through improved information services.

2 Indirect Eyewitnesses' Contributions

Two automatic real-time methods are currently in operation at EMSC to indirectly collect information from eyewitnesses on earthquakes' effects. One, named Twitter Earthquake Detection (TED), developed and operated by the US Geological Survey (USGS), is based on the social networking site Twitter which people use to send public 140-characters Twitter messages (tweets) soon after feeling shaking. It applies place, time, and key word filtering on tweets to rapidly detect shaking events through increases in the number of tweets related to felt experiences (Earle et al. 2010, 2011). The other method, named *flashsourcing* has been developed and is operated at EMSC. It analyses traffic patterns on its website, a popular rapid earthquake information website (www.emsccsem.org). The first step of *flashsourcing* detects the web traffic surges that are often observed after felt earthquakes (Wald and Schwarz 2000; Schwarz 2004; Bossu et al. 2007, 2008) that are caused by eyewitnesses converging on its website to find out the cause of their shaking experience (Bossu et al. 2014), a detection named Internet Earthquake Detection (IED).

Both detection methods are independent of seismic monitoring systems and based on real-time statistical analysis of Internet-based information generated by the reaction of the public to the shaking. A comparison of the data used by TED and IED is presented Fig. 2 for a 33 h window around the M 5.1, La Habra earthquake which shook Los Angeles area on March, 28 2014 (http://www.scsn.org/2014lahabra.html). Detections of felt earthquakes are typically within 2 min for both methods, i.e. considerably faster than seismographic detections in poorly instrumented regions of the world (Bossu et al. 2011a; Earle et al. 2011). They provide an efficient heads-up

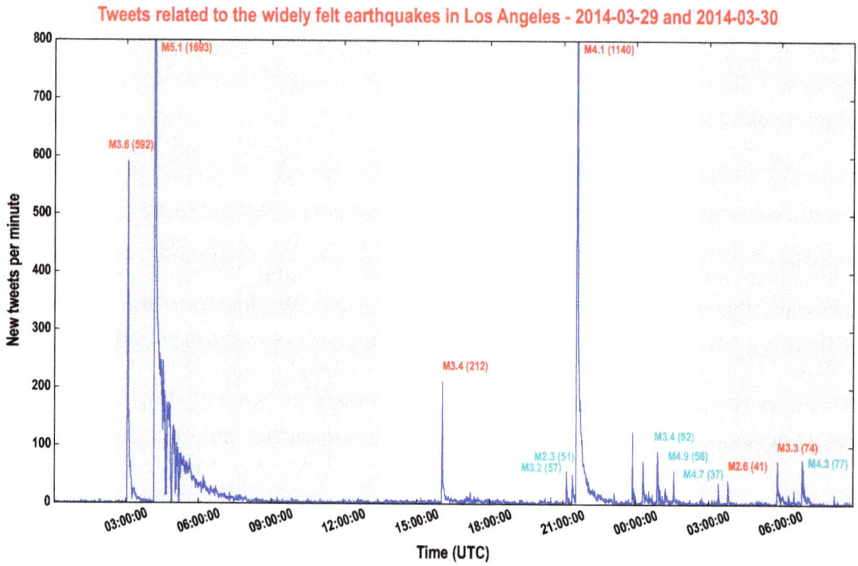
Tweets related to the widely felt earthquakes in Los Angeles - 2014-03-29 and 2014-03-30
New tweets per minute
Time (UTC)
M3.6 (592)
M5.1 (1693)
M3.4 (212)
M4.1 (1140)
M2.3 (51)
M3.2 (57)
M3.4 (92)
M4.9 (58)
M4.7 (37)
M2.6 (41)
M3.3 (74)
M4.3 (77)

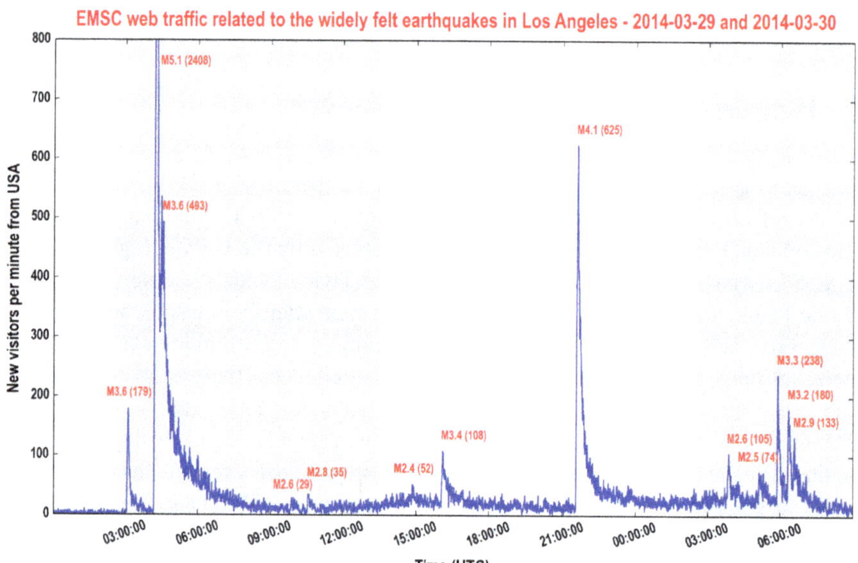
EMSC web traffic related to the widely felt earthquakes in Los Angeles - 2014-03-29 and 2014-03-30
New visitors per minute from USA
Time (UTC)
M5.1 (2408)
M3.6 (493)
M3.6 (179)
M2.6 (29)
M2.8 (35)
M2.4 (52)
M3.4 (108)
M4.1 (625)
M2.6 (105)
M2.5 (74)
M3.3 (238)
M3.2 (180)
M2.9 (133)

◀ **Fig. 2** Comparison of the data used for Twitter Earthquake Detection (TED, above) and Internet Earthquake Detection (IED, below) over the same time period of March 29–30, 2014 during which the La Habra earthquake shook Los Angeles area at 04:09 UTC. TED uses the number of published tweets per minute containing the keyword "earthquake" in various languages. IED is based on the number of new visitors per minute from a given country (the USA in this example) hitting EMSC website. A new visitor is defined as an IP address launching an HTTP request and which had no activity on the website during the previous 30 min. When identified, the magnitude of the causing earthquake is indicated along each surge, the number in brackets indicates the maximum number of new visitors. Text in *red* corresponds to earthquakes of the La Habra sequence, in *blue* from other regions and countries. The maximum value of the ordinate axis is truncated for the figure to display smaller surges. Contrary to IED, there is no a priori geographical filtering for TED which is based on a unique curve. IED is based on one curve per country which makes it easier to detect events close in time. In this case, the mainshock was followed 21 min later by a felt M 3.6 aftershock which was automatically detected by IED and not by TED. Several reasons could explain this difference. Firstly, there can be unrecoverable errors in the collection of data for the TED curve (P. Earle, personal communication, 2014) which can be observed in the hour following the mainshock. There is also an issue with the duration of exchanges after a felt earthquake which often last several tens of minutes on Twitter

for felt and potentially damaging earthquakes that are likely to attract public and/or media attention regardless their magnitude (Bossu et al. 2011b; Earle et al. 2011).

These two methods are complementary since they are based on 2 different types of Internet use that might occur after an earthquake occurrence. IED capacity depends on the notoriety of EMSC website in the shaken area while TED depends on the popularity of Twitter. As an illustration in the Euro-Med region, so far TED capacity has proved more efficient in Turkey, while IED better detects felt earthquakes in Northern Balkan countries. However, a specificity of IED is to exhibit near instantaneous improvement of detection capacity by a simple mention in a locally popular media, either traditional, online or social network; this being enough to permanently increase EMSC website visibility in the area. This point will be discussed further in the paragraph on the importance of social networks.

Once detected, a surge is automatically associated with the causing earthquake. The association is performed through time and spatial analysis using the geocoded tweets for TED (Earle et al. 2010) and IP (Internet Protocol) locations of website visitors for IED (Bossu et al. 2008, 2011b). Once associated, TED are published on Twitter under the username @USGSted for earthquakes worldwide with magnitudes greater than M 5.5 (http://earthquake.usgs.gov/earthquakes/ted/); all IED are published on Twitter under the username @LastQuake both before and after the association with the causing event (Bossu et al. 2011a). TED have recently been shared by USGS with EMSC which performs the association for earthquakes lower than 5.5 in magnitude and publishes them on its twitter account.

Flashsourcing is not limited to detection of felt earthquakes. It also provides rapid information (within 5 min) on the local effects of earthquakes (see Bossu et al. 2011a, 2014 for details). More precisely, it can automatically map the area where shaking was felt by plotting the geographical locations of statistically significant increases in traffic, a result that compares well with macroseismic maps (Bossu et al. 2011a, b). It can also discriminate localities affected by alarming shaking levels through a higher ratio of the number of new visitors to the number of

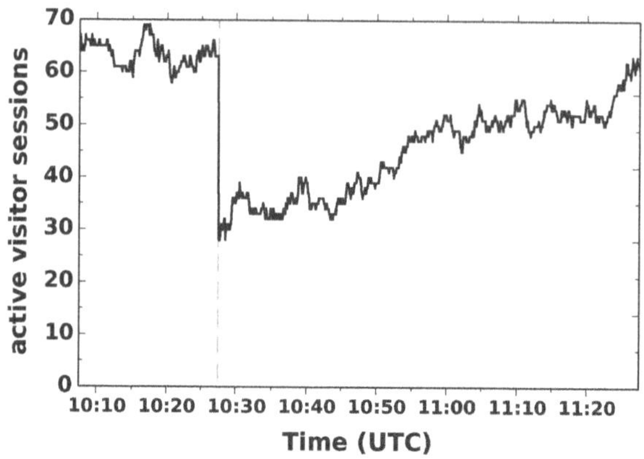

Fig. 3 Time evolution of the number of Internet sessions on the EMSC website and originating from Santiago del Chile (Chile) on March 6th, 2014. A partial and concomitant loss of session was observed at 10:27 UTC. There was no associated earthquake and no cause was identified. This is the type of pattern which is expected in cases of local damage when simultaneous with an earthquake occurrence

inhabitants (Bossu et al. 2014), and in some cases it can detect and map areas affected by severe damage or network disruption through the concomitant loss of Internet sessions originating from the impacted region (Bossu et al. 2011a). An example of a detected power cut is presented in Fig. 3. Previous testing failed to map areas where people evacuate buildings by discriminating indoor from outdoor Internet connections at city level (Bossu et al. 2014); however, in order to improve characterisation of earthquake effects, the initial system analysing webserver log files has now been fully switched to a traffic analysis system of web sessions based on websockets. The detection of a power failure (Fig. 3) is the first step towards a new system capable of detecting Internet interruptions or localised infrastructure damage.

The original system's statistics were performed on clicks i.e. on visitors' activity, while the new one monitors all visitors present on the website whether or not they click, i.e. whether or not they initiate a new HTTP (Hypertext Transfer Protocol) request. This change improves detection of new visitors by reducing background noise and allows us to detect areas where many sessions are suddenly lost in which case more information should be actively sought. A locality in which all visitors are lost at the time of the earthquake and no traffic is observed in the following minutes could have suffered severe damage. Contrastingly, a locality for which no visitors are lost and where traffic strongly increases after the earthquake's occurrence, shows that though the event was felt, the locality is known to have not suffered significant damage. The different theoretical cases and their interpretations are presented Fig. 4.

Flashsourced information alone does not provide a full description of earthquake impact, but within a few minutes, independently of any seismic data, and, at little cost, it can exclude a number of possible damage scenarios, identify localities

Time Evolution of the number of visitors after an earthquake occurring at T_0	Interpretations/Inferences*	Remarks
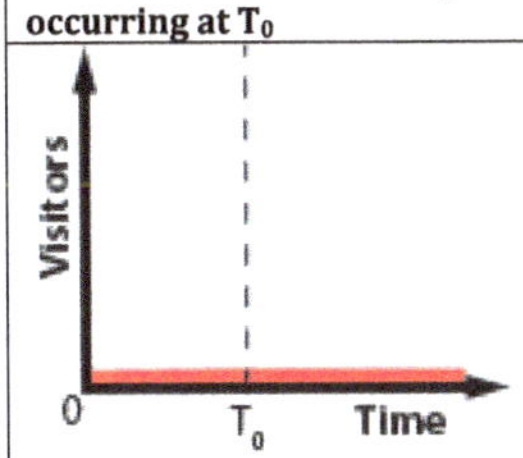	Not felt, or severe loss of Internet connectivity, or significant widespread damage, or severe perception of danger	Only valid in areas where, based upon previous data, the website is well known. Interpretations must be based on a cluster of IP locations rather than an isolated point
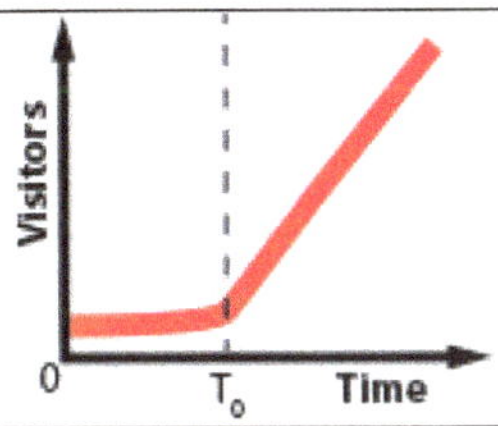	Felt and no significant loss of Internet connectivity and no significant damage	
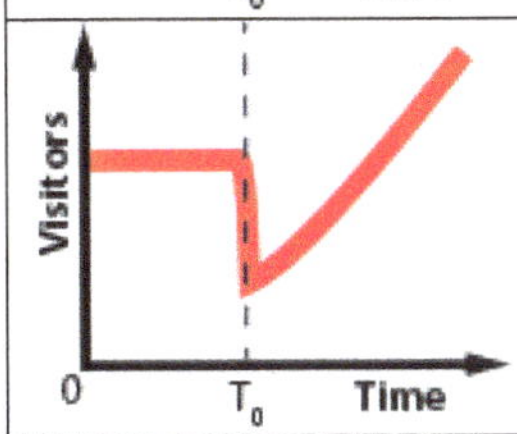	Felt and, local or temporary, loss of Internet connectivity, or, locally significant damage or, locally severe perception of danger	Capacity to discriminate between local or widespread depends on the statistical significance of the number of website visitors before the earthquake
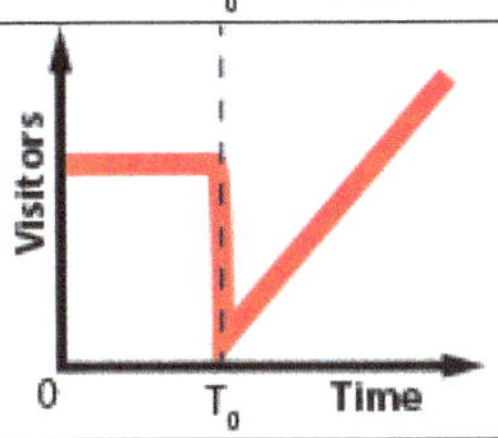	Felt and, local or temporary, loss of Internet connectivity, or, locally or widespread significant damage or, locally or widespread severe perception of danger	
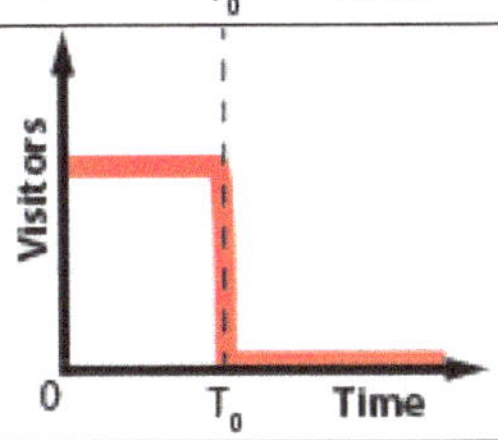	Significant loss of Internet connectivity, or significant damage, or severe perception of danger.	

Fig. 4 Interpretations of the different possible time evolutions of the number of Internet sessions following an earthquake occurrence at T_0 in a region where EMSC website is well identified. When the perception of danger is significant, eyewitnesses are more likely first to flee to safety rather than immediately browsing the Internet for information. The cases' various conclusions are non-exclusive

where no significant damage has occurred and others where damage cannot be excluded. *Flashsourcing* is the fastest tool for identifying localities which should be the focus of immediate and further investigations in order to determine the actual impact of an earthquake and is the first step of the strategy developed by EMSC on this matter.

3 Crowdsourcing or Direct Eyewitnesses' Contributions

Rapid earthquake information websites being natural convergence points of eyewitnesses after an earthquake occurrence, they are also a logical focal point for collecting eyewitnesses' observations on earthquake effects. The US Geological Survey was a pioneer in what is now known as "crowdsourcing" by initiating the "Did you feel it?" (DYFI) system, an automated approach for rapidly collecting macroseismic intensity data from Internet users, well before the term was even coined (Wald et al. 1999a, 2012). Shaking and damage reports are today routinely collected online by many seismological institutes around the world.

The questionnaire in use at EMSC is available in 32 languages to favor data collection at global scale, by removing linguistic hurdles and thus reducing border effects in data collection (Fig. 5). Location is collected through a Google map interface, making each questionnaire location potentially unique. Questionnaires' locations are automatically clustered before intensity assignment to reduce error; the higher the number of collected questionnaires, the larger the number of clusters and the better the spatial resolution of the macroseismic map. By default, macroseismic maps published by EMSC are expressed in EMS98 scale (Gruntal et al. 1998) but can also be expressed in MMI.

Macroseismic maps generally provide a more detailed picture of earthquake effects than *flashsourcing* for intensities up to VIII; for intensity VIII and above communication is significantly hampered and the public escapes to safety which restricts rapid questionnaire collection. The pace of questionnaire collection has shown a dramatic increase over the last few years: in 2009, only 30 % of the questionnaires were collected in the first 2 h (Bossu et al. 2011b) compared to 75 % for events between January 2013 and May 2014 (Fig. 6). Still, today, flashsourcing remains much faster for collecting information on earthquake effects with only 0.5 % of the macroseismic questionnaires being collected within the first 5 min of earthquake's occurrence i.e. the maximum time window used for the flashsourcing analysis (Fig. 6).

Thumbnails (Fig. 7) have replaced questionnaires on the EMSC website for mobile devices (http://m.emsc.eu) and in its LastQuake smartphone application (released during the preparation of this manuscript). This move is in response to the difficulty of filling online questionnaires on a small screen and without a proper keyboard, two common features of many mobile devices. It also solves all linguistic barriers. Thumbnails have been available since mid 2011 when the website for mobile devices was put online but more than half of the current collection of

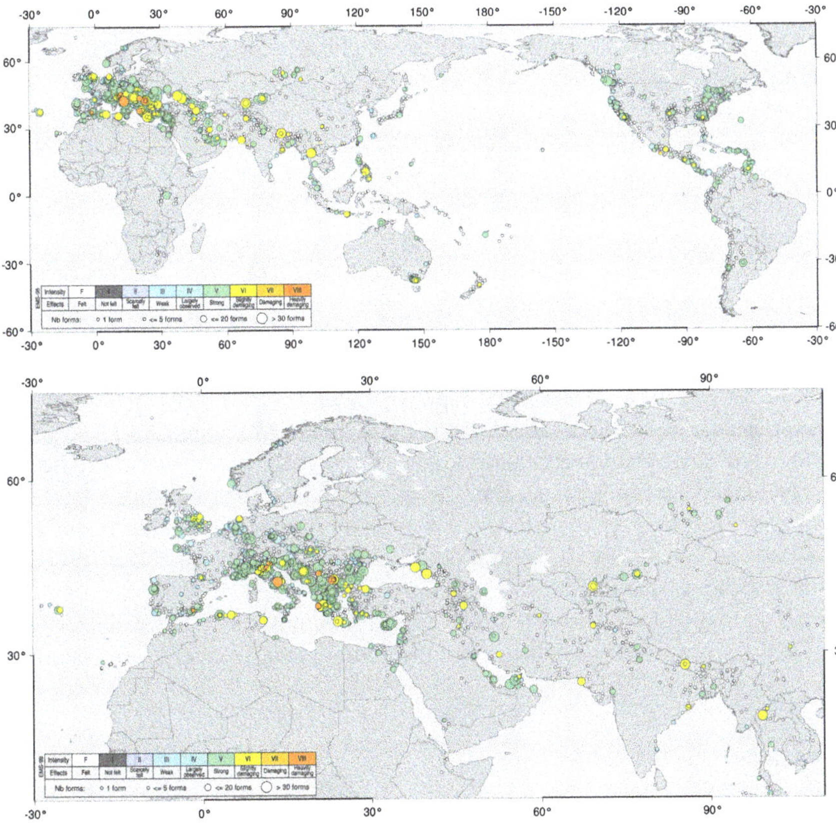

Fig. 5 Composite macroseismic maps at global and within Eurasia scales. The pace of collection has been increasing fast: 60 % of the 41,000 questionnaires collected between 2008 and May 2014 were collected between January 2013 and May 2014. A minimum of 5 questionnaires per cluster is required to assign an intensity value

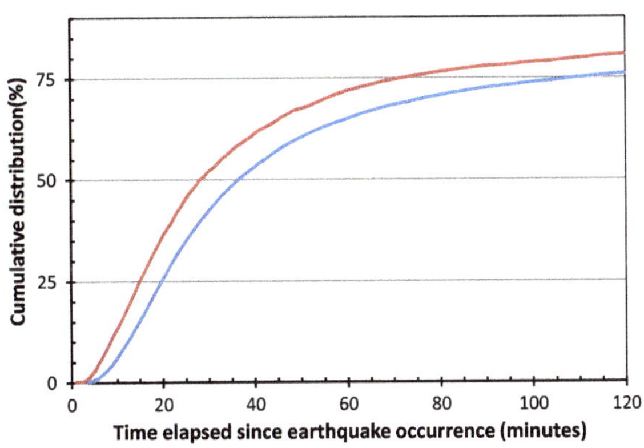

Fig. 6 Percentage of responses to the macroseismic questionnaire (*blue*) and thumbnails (*red*) with respect to time since earthquake occurrence. Cumulative curves for all earthquakes from January 2013 to May 2014. During this period 24,285 questionnaires and 6421 thumbnails were collected. Collection of thumbnails is significantly faster in the first 30 min

responses were collected in the last 12 months only (Fig. 8). This probably reflects the massive adoption of mobile devices at the global scale and if so, this trend is likely to continue in the future.

Thumbnails are not currently used in macroseismic maps. Initial comparisons between macroseismic data collected through questionnaires and thumbnails show a good correlation (the detailed analysis will be the subject of a separate study). Together with the rapid adoption of mobile devices, it opens the way for thumbnails to ultimately replace questionnaires on all EMSC crowdsourcing platforms. Such a move is likely to increase the number of collected testimonies: we already observe that for the same number of visits on the mobile website and traditional one there are 25 % more collected thumbnails than questionnaires. Furthermore, collection of thumbnails is faster especially in the first 30 min (Fig. 6). This may help to constrain an earthquake's effects more rapidly during the period when any in situ observation can be valuable to initiate the first phase of earthquake response. It should however be mentioned, that if implemented, this replacement would have a cost: thumbnails are not as accurate as questionnaires. For example, the exact situation of the observer (which floor, whether he was sleeping etc.) is not integrated in the thumbnails. Therefore their usefulness for detailed macroseismic studies may be reduced.

A second type of crowdsourced information is comments that eyewitnesses can leave in their own language at the end of the questionnaire or after having chosen a

Fig. 7 Example of thumbnails (intensity III and VII) used on EMSC website for mobile devices and in its LastQuake smartphone application to collect information on earthquake effects. One thumbnail has been prepared to illustrate each of the 12 level of the EMS98 macroseismic scale

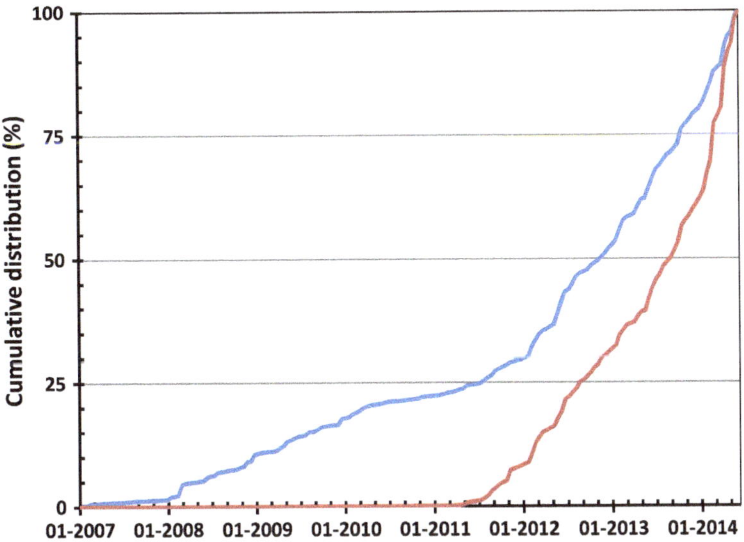

Fig. 8 Percentage of thumbnails (*red*) and questionnaires (*blue*) with repect to time. The popularity of thumbnails has been increasing fast over the last year: more than 50 % of them were collected over this period

thumbnail. About one third of the testimonies have an associated comment. They are validated to exclude inappropriate ones. Once validated, they appear on the website automatically translated in the language of the visitor browser. They contribute to catharsis for people who have been through a frightful experience.

Comments are not used to evaluate earthquake impact. They help us to better understand public questions, expectations and mood and, can also reveal some surprising misconceptions. They are then essential for adapting communication to meet public needs, particularly for social networks.

Two examples can be given. When several earthquakes are felt in the same area over a period of a few days or a few weeks, comments reveal a growing anxiety and the number of questions about the possibility of an imminent larger event multiply. The M3.8 earthquake which shook Skopje (FYR of Macedonia) on May 25, 2014 revealed that its inhabitants, or at least some of them, had associated earthquake occurrence with massive rainfalls: the 1963 earthquake which destroyed the city (Petrovsky 2004) occurred after floods and the May 2014 struck immediately after a strong storm suggesting a possible causative link between earthquake and rainfalls. Rainfall having been significant in the region, comments revealed that some people were afraid about the possible evolution of local seismic activity. In this type of situations, EMSC provides explanations on Twitter or Facebook to correct misconceptions and remind the populace of the importance of being prepared.

The last type of crowdsourced information is geo-located photos and videos of earthquake effects uploaded to the website (Fig. 9). Note that no photo or video is harvested from social networks. The number of collected videos is less than 10 and

Fig. 9 Example of geo-located photos shared by Robert Michael Poole an eyewitness of the 2013 Bohol, Philippines M 7.2 earthquake. Photos are efficient to assess the scale of damage in a given area

so we focus here on the images. Users are required to provide the location of observation, and to select the causing event. Photos are validated before website publication on 4 different criteria: they must respect the privacy and dignity of people, as well as copyright and be consistent with seismological data. Validation is based on a visual check, the inspection of the metadata contained in the EXIF (Exchangeable image file format) for possible inconsistencies of date or location, a web search to check whether the image is published on other websites, and finally the level of damage visible on the photo is compared with the estimated value of the peak ground acceleration (PGA). For example, a photo portraying flattened buildings will not be validated if the estimated local PGA value is only 5 % of g. This validation process is subjective but has proved to be efficient.

The geolocated photos are by definition related to the small fraction of earthquakes causing permanent effects and damage. 576 images have been validated so far for 63 different earthquakes; among them 122 are related to the M 6.3 2009 l'Aquila earthquake alone. One of the goals of our Last-Quake smartphone application (mentioned above) is to increase the number of collected pictures. Uploading a file from a smartphone to a website is indeed not straightforward and we believe that by simplifying the file transfer to EMSC servers, the application, if successful, will improve our rate of photo collection.

4 The Key Roles of Social Networks

Crowdsourcing and particularly flashsourcing are based on the natural convergence of eyewitnesses to our websites and their efficiency depends strongly on how easily eyewitnesses can find them when looking for earthquake information on the web. This is where, as illustrated in 2 different cases, social networks can play a significant role to rapidly guide them to our websites.

EMSC is present on Twitter, Facebook, Google+, Pinterest, Youtube and Linkedin; Twitter and Facebook being by far the most strategic ones for engaging with citizens. The EMSC Facebook page (www.facebook.com/EMSC.CSEM) offers general earthquake information, details of EMSC services and is the main platform for exchanging with the public and for providing explanations. Twitter @LastQuake focuses on automatic publication of real time information on felt and damaging earthquakes. It automatically merges internal and external information on felt earthquakes to generate and publish tweets in a time window ranging from 1 to 90 min after occurrence. Collated data feeds include TED, IED, macroseismic maps, automatic earthquake impact assessment (Fig. 1), or tsunami information released by the Pacific Tsunami Warning Centre. 28 different types of tweets are generated ranging from early detection, epicentral and macroseismic maps to revision of earthquake parameters. The full description of this Twitter information service, sometimes called *Quakebot* is beyond the scope of this paper.

The impact of social networks can be significant in regions of low seismic hazard where the populace are unlikely to be regular visitors to websites such as EMSC's. This is the case for Budapest, the capital city of Hungary, which was shaken on April 22, 2013 by a M 4.5 earthquake. Figure 10 shows that the web traffic increase originating from Hungary was unusually slow for such a widely felt event (see Fig. 11 for comparison). It took more than 10 min for the EMSC site to exhibit a strong increase and the vast majority of the Hungarian visitors came from Facebook pages and not necessarily the EMSC one. Although this cannot be proven, our presence on this social network has probably improved rapid website access. It is even more difficult to prove whether our presence on Facebook helped to increase the amount of crowdsourced questionnaires. One can only compare the 334 questionnaires collected at EMSC for this event with the 25 questionnaires collected by the US Geological Survey DYFI system (http://comcat.cr.usgs.gov/earthquakes/eventpage/usb000gdw3#dyfi), which remains by far the best referenced earthquake website in the world. Although this is not a proof, it is an indication that, at least in

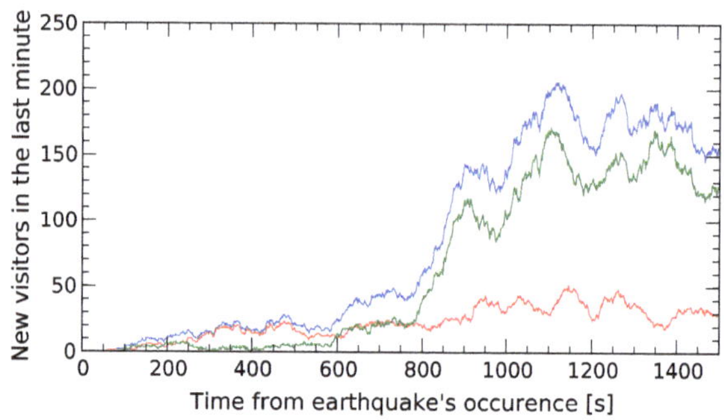

Fig. 10 Number of new Hungarian visitors per minute hitting EMSC website following the M 4.5, April 23, 2013 earthquake (for more details see Fig. 2 caption). The *blue curve* represents all the new Hungarian visitors, the *green curve* the ones coming from a Facebook page. The *red curve* represents the difference between the 2 previous curves, i.e. the EMSC website traffic which did not come from Facebook

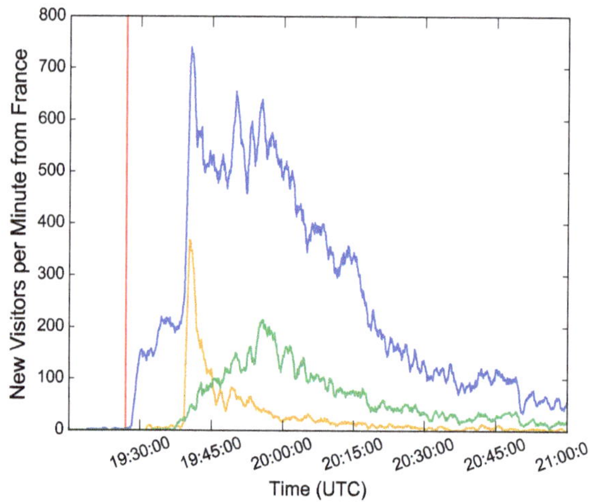

Fig. 11 Number of new French visitors per minute hitting EMSC website following the M 4.9, April 7, 2014 earthquake (for more details see Fig. 2 caption). The *blue curve* represents all the new French visitors, the *orange curve* the ones coming from Twitter, and the green curve the ones coming from a Facebook page. One can see the difference of dynamic between Facebook and Twitter generated traffics

certain cases, a presence on Facebook can be an efficient way to rapidly attract the attention of an emerging online community of eyewitnesses.

The second example shows the potential impact of Twitter. On April 7, 2014 a M 4.9 earthquake in SE France was detected in 96s through the traffic surge it generated (Fig. 11). A second massive surge occurred 10 min later and was caused by the retweet of one of our automatic tweets by the @LesNews account which counted at this time 286 thousand followers (compared to 5 thousand at the end of May 2014 for @LastQuake). Facebook also had a non-negligible impact (Fig. 11) but with a very different dynamic. Jointly, Facebook and Twitter were the source of a significant proportion of the eyewitnesses visiting the EMSC website (rapidly) after this earthquake.

5 Discussion and Next Developments

The EMSC has been developing a strategy to collect direct and indirect information about earthquake effects from eyewitnesses in order to rapidly constrain the impact assessment of global earthquakes and fill the distinct information gap that exists for the first hours after an earthquake occurrence This is possible thanks to its rapid earthquake information services which attracts eyewitnesses, allowing them to firstly make an indirect data contribution, secondly to answer some of their questions and needs, and finally to properly engage with them in active data collection. Our strategy is not restricted to crowdsourcing tools, it also comprises real time earthquake information services, Internet marketing and the use of different support websites, smartphones applications, social networks.

Flashsourcing and crowdsourcing are both based on the natural convergence of eyewitnesses to EMSC websites, a convergence which is based upon the good visibility of EMSC websites on search engines, websites, forums and social networks. This creates a virtuous circle: better information services bringing additional eyewitnesses and improved data collection which, in turn, further improve information services. By considering eyewitnesses not only as endusers but also as contributors, this strategy, named "citizen seismology" (Bossu et al. 2011b), transforms earthquake detection (Allen 2012; Young et al. 2013) and advantageously pushes scientists to better understand and respond to public demands and expectations after disasters.

Concerning the crowdsourced information, there are very few examples of intentional misuse, errors in selecting the causing earthquake are much more frequent. In most cases, misuses are comments which try to be funny. Interestingly, none of these comments has ever been collected in the first 30 min, so the rapidity of the crowdsourcing mechanisms probably helps to limits the frequency of misuses. Basic statistics are also used to identify outliers in the macroseismic questionnaires and the reliability of the crowdsourced data benefits from consistency checks with flashsourced and earthquake information (PGA).

Citizen seismology is not the only possible strategy to reduce the uncertainties of rapid impact assessment. Dense real time accelerometric networks are an alternative, but a much more expensive one. Whereas the main problem EMSC has been facing with citizen seismology is for its web infrastructure to cope with massive traffic surges. This is why it is essential, in addition to the adaptation of the website to mobile devices, to also develop push information services such as a smartphone application and automatic publication on social networks.

In the next steps, we plan to further exploit the collected data. For example, the comments probably contain more information than what is perceived at first sight. In a remarkable study based on 20,000 comments collected by the USGS, Celsi et al. (2005) showed that individuals in California tend to underestimate the potential consequences of earthquakes because they relate their felt earthquake experience to the reported event magnitude rather than their local shaking level by giving insufficient consideration to seismic wave attenuation. Collected comments could identify other important features and improves our understanding of earthquake perception, for example: what are possible cultural influences? or, how the frequency of felt earthquakes can modify the reaction of eyewitnesses?

A second development is to bring in new data. EMSC has initiated deployments of lowcost community-run Quake Catcher Network (QCN) (Cochran et al. 2009) sensors in Patrai and Thessaloniki (Greece), to set up dense urban ground motion networks. Whether shakemaps (Wald et al. 1999b) derived from such low cost networks can offer additional constraints to earthquake damage scenarios, has to be evaluated.

The most important evolution will be to automatically merge earthquake data, flashsourced and crowdsourced information into a single time evolving situation map to get a complete view of the available information. This will not be able to assess all situations and may miss individual accidents due, for example, to sub standards building but it could be real tool for decision support for the seismologist.

6 Conclusion

New technologies are transforming the way the seismological community collects data and interfaces with the public and authorities. These advances are creating tremendous potential for rapid and massive public involvement by both active and passive means. The strategy developed and implemented at EMSC aims to harness the collective power of eyewitnesses to determine the impact of global earthquakes and fill the information gap which follows the occurrence of a damaging earthquake. By doing so, it additionally favors interaction with the public and helps us to understand how best to communicate and improve seismic risk awareness. In short, it helps seismologists to do a better job for society.

Acknowledgments There are many people to be thanked for their contributions to this work. The first ones are obviously the citizens who are the actors of this research. We also want to thank the EMSC members, our host the LDG for its continued support and Digital Elements for providing IP-location software. We thank individuals for their fruitful collaborations: Elizabeth Cochran, Paul Earle, Jesse Lawrence, Christian Mendelewski, Zafeiria Roumelioti and Efthimios Sokos. The development of the smartphone application LastQuake has been partially funded by the Fondation MAIF. The work presented in this article has been partially funded by the EC Project REAKT, FP7-282862.

References

Allen, R. M. (2012). Transforming earthquake detection? *Science, 335*, 297–298.

Atik, A. L., Abrahamson, N., Bommer, J. J., Scherbaum, F., Cotton, F., & Kuehn, N. (2010). The variability of ground-motion prediction models and its components. *Seismological Research Letters, 81*(5), 794–801.

Bilham, R. (2010). Lessons from the Haiti earthquake. *Nature, 463*(7283), 878–879.

Bossu, R., Douet, V., Godey, S., Mazet-Roux, G., & Rives, S. (2007). On the use of Internet to rapidly collect earthquake impact information. *EMSC Newsletter, 22*, 31–34.

Bossu, R., Mazet-Roux, G., Douet, V., Rives, S., Marin, S., & Aupetit, M. (2008). Internet users as seismic sensors for improved earthquake response. *EOS, Transactions, 89*(25), 225–226.

Bossu, R., Merrer, S., & Mazet-Roux, G. (2009). *Applications and utilization of ELER software. Evaluation of the level 0 approach. NERIES Report* (p. 48). http://www neries429eu.org/main.php/JRA3_D5EMSCImpactEstimationJRA3_V3.pdf?fileitem=10272797.

Bossu, R., Gilles, S., Mazet-Roux, G., Roussel, F., Frobert, L., & Kamb, L. (2011a). Flash sourcing, or rapid detection and characterization of earthquake effects through website traffic analysis. *Annals of Geophysics, 54*(6), 716–727. doi:10.4401/ag-5265.

Bossu, R., Gilles, S., Mazet-Roux, G., & Roussel, F. (2011b). Citizen seismology or how to involve the public in earthquake response. In D. M. Miller & J. Rivera (Eds.), *Comparative emergency management: Examining global and regional responses to disasters* (pp. 237–259). Boca Raton: Auerbach/Taylor and Francis Publishers.

Bossu, R., Lefebvre, S., Cansi, Y., & Mazet-Roux, G. (2014). Characterization of the 2011 mineral, virginia, earthquake effects and epicenter from website traffic analysis. *Seismological Research Letters, 85*(1), 91–97.

Celsi, R., Wolfinbarger, M., & Wald, D. (2005). The effects of earthquake measurement concepts and magnitude anchoring on individuals' perceptions of earthquake risk. *Earthquake Spectra, 21*(4), 987–1008.

Cochran, E. S., Lawrence, J. F., Christensen, C., & Jakka, R. S. (2009). The quake catcher network: Citizen science expanding seismic horizons. *Seismological Research Letters, 80*(1), 26–30.

Earle, P., Guy, M., Buckmaster, R., Ostrum, C., Horvath, S., & Vaughan, A. (2010). OMG earthquake! Can Twitter improve earthquake response? *Seismological Research Letters, 81*, 246–251.

Earle, P. S., Bowden, D. C., & Guy, M. (2011). Twitter earthquake detection: Earthquake monitoring in a social world. *Annals of Geophysics, 54*(6), 708–715. doi:10.4401/ag-5364.

Erdik, M., Şeşetyan, K., Demircioğlu, M. B., Hancılar, U., & Zülfikar, C. (2011). Rapid earthquake loss assessment after damaging earthquakes. *Soil Dynamics and Earthquake Engineering, 31*(2), 247–266.

Grunthal, G. (1998). European Macroseismic Scale EMS98. Cahier du Centre Européen de Géodynamique et de Séismologie, 15.

Papadopoulos, G. A., Drakatos, G., Papanastassiou, D., Kalogeras, I., & Stavrakakis, G. (2000). Preliminary results about the catastrophic earthquake of 7 September 1999 in Athens. *Greece Seismological Research Letters, 71*(3), 318–329.

Petrovski, J. T. (2004). Damaging effects of July 26, 1963 Skopje earthquake. In *Middle East Seismological Forum, Cyber Journal of Geoscience Volume 2.*

Porter, K. A., Jaiswal, K., Wald, D. J., Greene, M., & Comartin, C. (2008). WHEPAGER Project: A new initiative in estimating global building inventory and its seismic vulnerability. In *Proceedings of the 14th World Conference on Earthquake Engineering*, Beijing, China.

Schwarz, S. (2004). Cyberseismology and teachable moments. *Seismological Research Letters, 75* (6), 749.

Wald, D. J., Quitoriano, V., Dengler, L., & Dewey, J. W. (1999a). Utilization of the internet for rapid community intensity maps. *Seismological Research Letters, 70*(6), 680–697.

Wald, D. J., Quitoriano, V., Heaton, T. H., Kanamori, H., Scrivner, C. W., & Worden, C. B. (1999b). TriNet "ShakeMaps": Rapid generation of peak ground motion and intensity maps for earthquakes in southern California. *Earthquake Spectra, 15*(3), 537–555.

Wald, D., Quitoriano, V., Worden, C., Hopper, M., & Dewey, J. (2012). USGS "Did You Feel It?" Internet-based macroseismic intensity maps. *Annals of Geophysics, 54*(6). doi:10.4401/ag-5354.

Wald, L., & Schwarz, S. (2000). The 1999 southern california network bulletin. *Seismological Research Letters, 71*(4), 401–422.

Young, J. C., Wald, D. J., Earle, P. S., & Shanley, L. A. (2013). *Transforming earthquake detection and science through citizen seismology. Case study series 2*. Wilson Center, Commons Lab. http://www.wilsoncenter.org/publication/transformingearthquakedetectionandsciencethroug-hcitizenseismology. Last accessed October 2013.

Real-Time Mapping of Earthquake Perception Areas in the Italian Region from Twitter Streams Analysis

Luca D'Auria and Vincenzo Convertito

1 Introduction

Since the definition of the earliest macroseismic scales (Mercalli 1902), the importance of information provided by citizens about the earthquake perception was already acknowledged. Technological developments in the last decades have led to a renewed interest in macroseismic surveys. In fact, the use of web-based technologies led to a quick supply of surveys allowing a rapid mapping of a macroseismic intensity field (Wald et al. 1999; Sbarra et al. 2009). While this technique relies on the voluntary compilation of macroseismic questionnaires, other approaches are based on the data mining of information provided by people, which usually are not aware of their usage. Bossu et al. (2008) showed that the analysis of the spatial and temporal pattern of users accessing the European-Mediterranean Seismological Centre's (EMSC) Web site (http://www.emsc-csem.org) can be successfully used to quickly assess the map of the area where an earthquake has been felt. This goal can be achieved by determining the location of users accessing the web searching for information about the event, by pairing their IP addresses to their geographical locations. The analysis of web accesses has been subsequently applied also to the detection and location of earthquakes, as well as to a speedy determination of macroseismic intensities (Bossu et al. 2011, 2014).

Earle et al. (2010) showed, for the first time, as data mining from a social network can be exploited to detect and determine the felt area of an earthquake. This approach was subsequently applied extensively leading to the so-called Twitter Earthquake Detector (TED, http://earthquake.usgs.gov/earthquakes/ted) (Earle et al. 2011).

L. D'Auria (✉) · V. Convertito
Istituto Nazionale di Geofisica e Vulcanologia, Sezione di Napoli "Osservatorio Vesuviano" via Diocleziano 328, 80124 Napoli, Italy
e-mail: luca.dauria@ingv.it

V. Convertito
e-mail: vincenzo.convertito@ingv.it

S. D'Amico (ed.), *Earthquakes and Their Impact on Society*,
Springer Natural Hazards, DOI 10.1007/978-3-319-21753-6_26

The use of social network for the purpose of an early detection of earthquakes at a global scale represents a radical change in a basic seismological paradigm (Allen 2012). The information carried by social networks travels much faster than seismic waves, allowing a fast and reliable detection within few minutes after the origin time (Sakaki et al. 2010; Earle et al. 2011).

Currently, the most useful social network for the purpose of earthquake detection is Twitter.

Twitter is a microblogging service operating since 2006 and currently is the leader among this type of sites (Sysomos 2014). It was designed to allow people to post and share very quickly short messages with the rest of the world. The content of the messages (called tweets) usually consists in sharing of opinions and feelings from common people, or in news and advertisements from commercial and organizations. Its software platform allows registered users to connect to its database using an open source API (Application Programming Interface) (https://dev.twitter.com/docs/faq). Among the possible types of APIs, one of the most useful is the "Streaming", which allows a permanent connection to Twitter, gathering in real-time all the tweets matching some specific request. For instance it is possible to collect the tweets containing one or more specific keywords. For this reason Twitter has been extensively used to map geographic trends of a specific topic (Thelwall et al. 2011).

Since November 2012, we have made operative a software system named TwiFelt (http://twifelt.ov.ingv.it). The aim of TwiFelt is to provide real-time earthquake perception maps from the analysis of Twitter streams. The system collects in real-timegeotagged tweets (i.e., tweets associated to a geographic position). The technical details of TwiFelt can be found in D'Auria and Giudicepietro (2013). Currently TwiFelt is limited to the study of the Italian region, collecting tweets containing keywords as earthquake, quake and their Italian equivalents and limited only to those falling within the Italian territory.

From Nov. 2012 to Dec. 2013 we have collected about 10,500 tweets from the Twitter stream. In Fig. 1a we show the location of geotagged tweets. The distribution of the tweets shows a marked concentration around major Italian cities. Figure 1b shows epicenters of all the 212 earthquakes with $M \geq 3$ recorded by the Istituto Nazionale di Geofisica e Vulcanologia (INGV) seismic network in the same time interval. Even if we consider tweets posted 1 h after every event with $M \geq 3$ (Fig. 1c) and for every event with $M \geq 4$ (Fig. 1d) we notice that the distribution is still strongly correlated to the background tweets distribution (Fig. 1a).

As for the temporal distribution (Fig. 2) we notice that the background rate shows a clear diurnal variation, with a minimum around 2 a.m. and a maximum around 1 p.m. (Fig. 2a). In Fig. 2c, d, we consider only tweets posted respectively 1 h after a $M \geq 3$ and $M \geq 4$ events. Compared with the spatial distribution (Fig. 1c, d) the temporal pattern shows clear peaks in correspondence of important earthquakes. Four of them have been annotated as: Conero ($M = 4.9$ 21/7/2013 01:32), Garfagnana ($M = 4.8$ 25/1/201314:48 UT), Matese ($M = 4.9$ 29/12/2013 17:08 UT) and Ernici ($M = 4.8$ 16/2/2013 21:16 UT). During the considered period another important earthquake ($M = 5.2$ Northern Italy) occurred on 21/6/2013 10:33 UT,

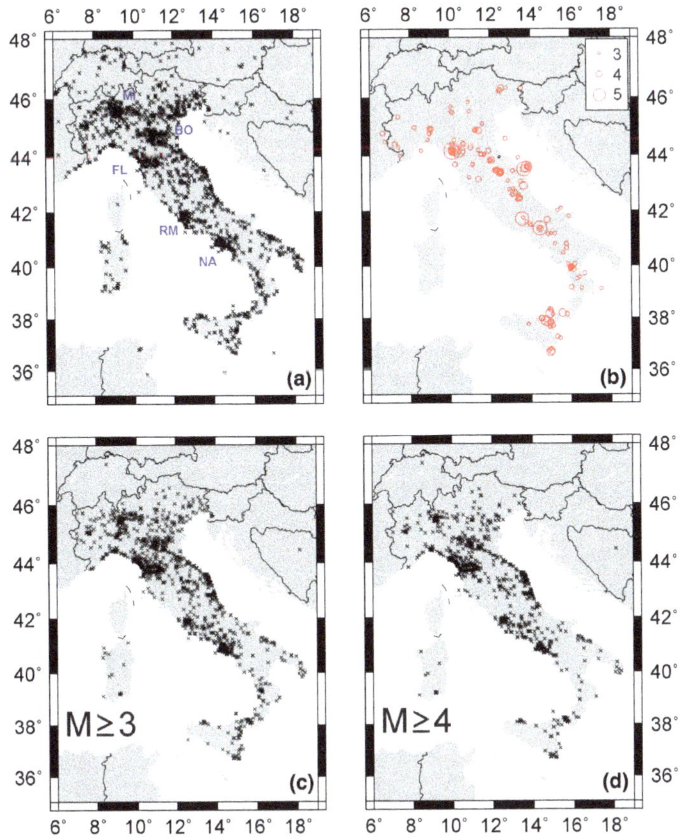

Fig. 1 Spatial distribution of the tweets. **a** Location of all the tweets in the dataset (a total number of 10,500). Labels indicate major Italian cities (*MI* Milan, *BO* Bologna, *FL* Florence, *RM* Rome, *NA* Naples). **b** Location of 212 earthquakes (M ≥ 3) occurred from Nov. 2012 to Dec. 2013. **c** Tweets posted within 1 h after an M ≥ 3 event (3400 tweets). **d** Tweets posted after events the 19 events with M ≥ 4 (2160 tweets)

but has been excluded from the analysis since TwiFelt was not fully operative in that period for technical problems.

2 Data Mining

The analysis of Twitter streams has shown that after felt earthquakes, there is an increase in the number of tweets containing at least one among the keywords "terremoto" and "scossa" (meaning respectively earthquake and shock), geolocated around the epicentral area (Earle et al. 2010; Sakaki et al. 2010). However, among the retrieved tweets there is an amount of "false positives" that means tweets

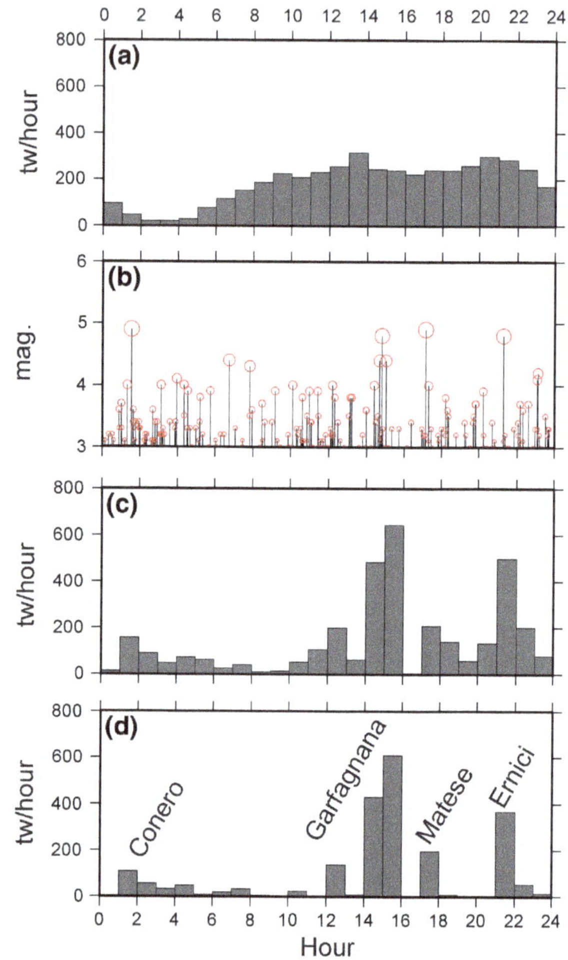

Fig. 2 Temporal distribution of tweets as function of the hour of posting. **a** Histogram showing a clear diurnal variation. **b** Event magnitudes as a function of the hour. **c** Tweets associated with events with M ≥ 3. **d** Tweets associated with events with M ≥ 4. Some of the strongest events are annotated in **d** (see text for details)

containing one of the selected keywords but not related to the actual perception of an earthquake. This can be related, for instance, to tweets using the word "earthquake" metaphorically or talking about an earthquake occurred elsewhere (Sakaki et al. 2010). Especially for moderate earthquakes (M < 5) it is important to take into account the effect of false positives on the analysis.

Different approaches to deal with this task are possible. One is a semantic analysis of the post content (Sakaki et al. 2010, 2013). This approach requires language-specific procedures and its performance can be unsuccessful in more than 10 % of cases (Sakaki et al. 2013).

In the present work we propose a data-driven approach based on a quantitative estimation of the "background" tweet rate and its use to assign a weight to each tweets. In this context estimating the "background" means defining a spatio-temporal function with the average number of tweets not directly related to the perception of an earthquake. For this purpose we have excluded all the tweets recorded in the 10 h following every earthquake with M ≥ 3 located in the Italian region. The remaining tweets (about 4400 over a total of 10,500) have been used to define a normalization function n(lat, lon, t) = s (lat, lon) t(t), which assigns a weight to each tweet on the basis of its location and time of posting. The two functions s and t are defined empirically, on the basis of the observed spatial and temporal distributions of tweets (Figs. 1a and 2a).

In Figs. 3 and 4 we show the effect of the normalization on the spatial and temporal tweets distribution. In practice, we observe that assigning a weight *w* = 1/n (lat, lon, t) greatly reduces the effect of concentration around major cities and of the diurnal variations.

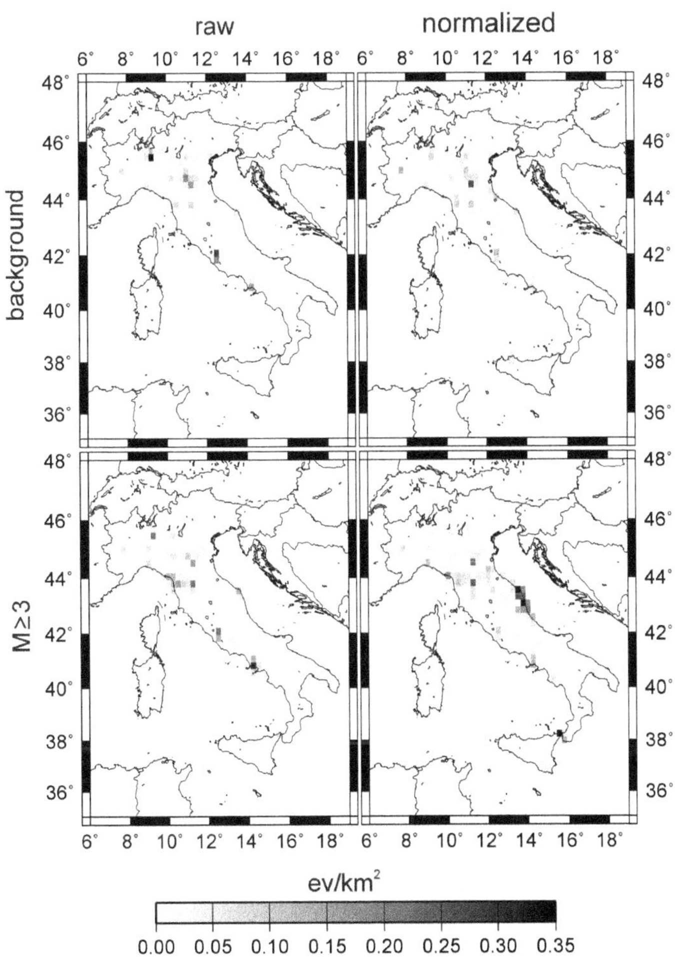

Fig. 3 Effect of the normalization on the spatial distribution of tweets. The maps on the *top* show the effect of normalization on the background tweets, while the two maps on the *bottom* represent the same but for tweets associated to events with M ≥ 3

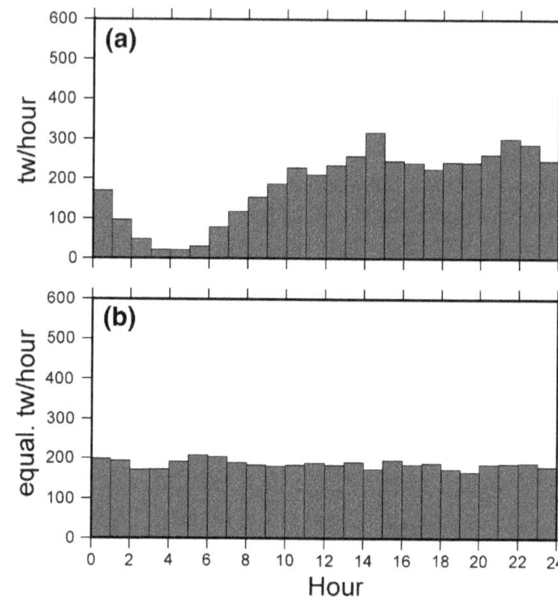

Fig. 4 Effect of the normalization on the temporal distribution of tweets. **a** Distribution before normalization. **b** Distribution after normalization

This approach allowed us to define EPIT (Earthquake Perception Index based on Tweets) as:

$$EPIT(A, D) = \frac{1}{A * D} \sum_i w_i(lat_i, lon_i, t_i),$$

where A and D are the considered area and time interval, respectively. The index i runs over all the tweets within A and D. The EPIT index allows a more reliable quantification of the earthquake perception and shows a good correlation with ground motion parameters, as we show in the next section.

3 Correlation with Ground-Motion Parameters

In Fig. 5a we represent the value of EPIT (computed on concentric rings around the epicenter) as a function of the epicentral distances for all the considered events. It shows a decay, which roughly follows a power law, as most of the GMPE (Ground Motion Prediction Equation) used in seismic hazard studies. This observation motivated us in studying the theoretical relationship between EPIT and theoretical ground motion parameters.

Actually we found a good correlation between EPIT and Peak ground acceleration (PGA) and Peak ground velocity (PGV) computed by using a GMPE valid for the Italian region (Bindi et al. 2011) (Fig. 5b, c). The linear correlation coefficients

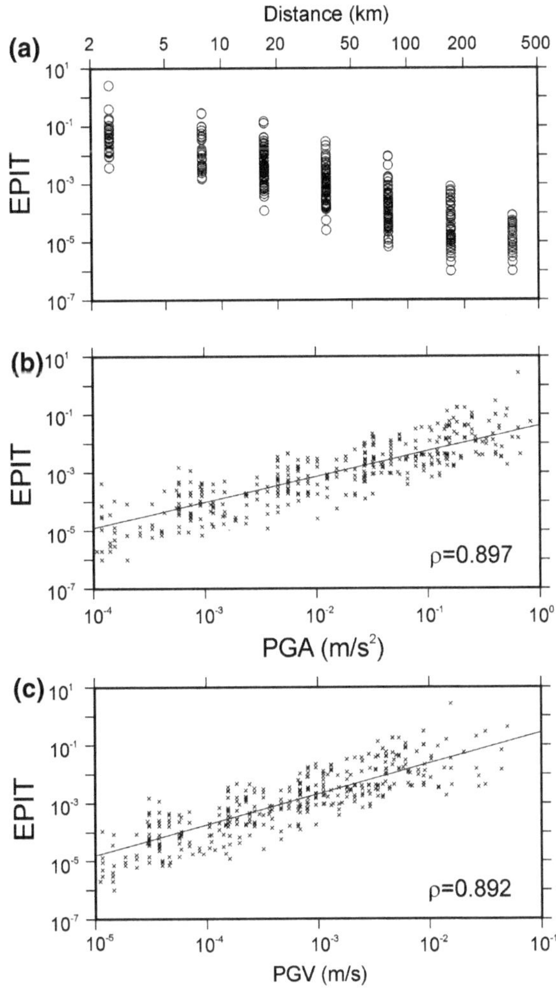

Fig. 5 Relationship of EPIT with ground motion parameters. **a** EPIT as a function of the distance for all the considered events. **b** Correlation between EPIT and PGA. **c** Correlation between EPIT and PGV. The values indicated on the *bottom* right of panels **b** and **c** are linear correlation coefficients

on a log-log scale found for the two relationships are $\rho = 0.897$ and $\rho = 0.892$ for EPIT versus PGA and EPIT versus PGV, respectively.

Nonetheless for a given PGA or PGV value the EPIT shows a variation within at least two orders of magnitudes. This suggests that using only the instrumentally recorded ground motion may be not sufficient to properly characterize the actual earthquake perception.

We maintain that our relationship holds for small and moderate magnitude earthquakes in the Italian region. Thereby, we cannot exclude that this relationship maybe no more linear for higher magnitude values and hence higher macroseismic intensities (Bossu et al. 2014; Burks et al. 2014).

4 Examples: The Garfagnana (M = 4.8) and the Matese (M = 4.9) Earthquakes

We have applied the techniques mentioned above to the whole earthquake dataset. Here we present the results from two among the most significant earthquakes occurred in Italy in 2013.

The first event occurred in the Garfagnanaregion (North-Central Italy) on Jan. 25th with M = 4.8. The earthquake occurred at 15:48 local time and was strongly felt in Florence and Bologna but also in other important cities as Milan. The distribution of tweets related to this event (Fig. 6a) is strongly concentrated around the cities of Florence and Bologna. The EPIT spatial distribution (Fig. 6b) still shows peaks localized over these two cities, but it allows delimiting fairly accurately the perception area by contouring the EPIT value of 0.005. In fact, comparing this distribution with the web based macroseismic intensities (Fig. 6c), we notice that there is a good agreement between the contour and the places where the reported MCS is higher than 2.

Another example is the Matese (Southern Italy) earthquake occurred on Dec. 29th with M = 4.9. The event occurred at 18:08 local time with an epicenter located about 60 km away from Naples. It was widely felt in the city and in all the surrounding area hosting about 5 million of people. The distribution of tweets shows a concentration on Naples, with a markedly asymmetric distribution compared to the epicenter location (Fig. 7a). However, the effect of normalization (notice the smaller symbol size on the city) makes the EPIT distribution more symmetric (Fig. 7b). Actually, also in this case the 0.005 contour delimits fairly accurately the perception area deduced from macroseismic surveys (Fig. 7c). We have verified empirically that also for other events the contour of 0.005 represents a good choice to delimit the earthquake perception area.

5 Conclusions

Compared with web-based macroseismic surveys (Wald et al. 1999; Sbarra et al. 2009) the analysis of Twitter streams provides a less detailed information. However, the response of Twitter users to an earthquake is very timely. Usually, within 5–20 min the 50 % of tweets have already been posted (Fig. 8). This feature seems to be quite irrespective of the magnitude of the earthquake. In Fig. 9 we show the times when 30, 60 and 90 % of tweets have been collected, as a function of the magnitude. It shows no significant correlation with the magnitude of the events.

We have shown that using a simple, data-driven approach, allows defining a quantitative index named EPIT that can be efficiently used to delimit the earthquake perception area. Comparison with macroseismic surveys has shown that the spatial distribution of EPIT matches with the "a posteriori" elaboration of the surveys: the contour of 0.005 of EPIT has shown to agree fairly well with points where MCS > 2

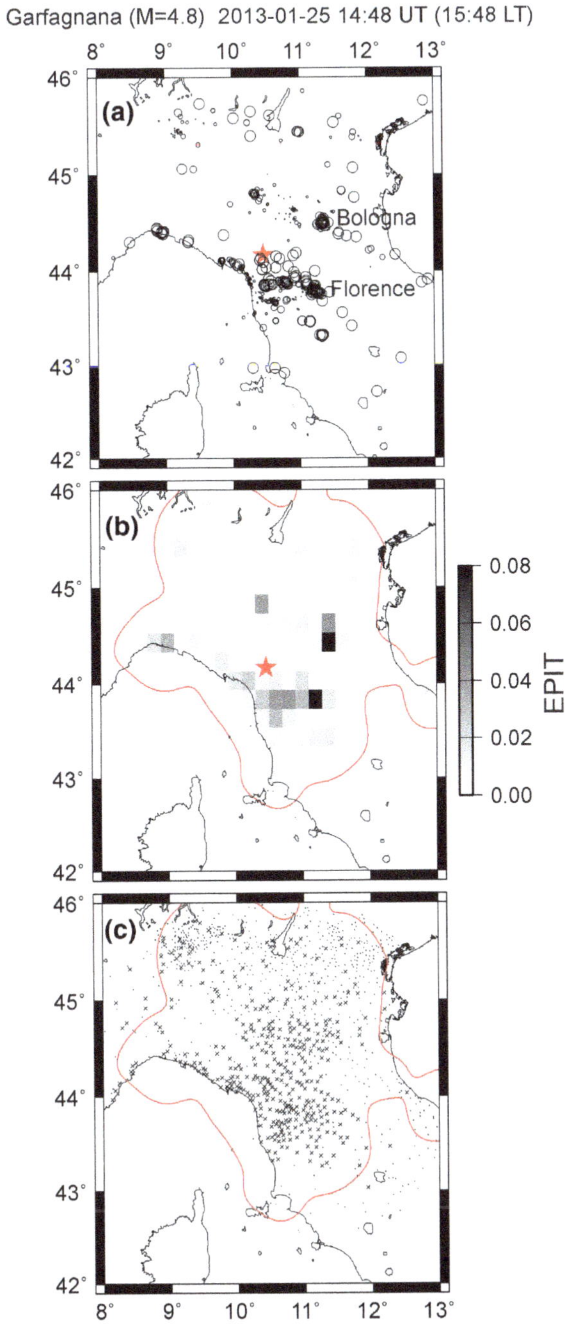

Fig. 6 Synthesis of the Garfagnana earthquake. **a** location of the 701 tweets associated to the events (*black circles*). The size of the circles is proportional to their weight. *Red star* corresponds to the epicenter. **b** Spatial distribution of EPIT. The *red line* is the contour of 0.005 for the EPIT (smoothed with a filter of 10 km radius). **c** Comparison with web-based macroseismic surveys (Sbarra et al. 2009). *Black crosses* identify places with MCS > 2, while *black points* indicate MCS ≤ 2

Fig. 7 Synthesis of the Matese earthquake. The meaning of symbols is the same as for Fig. 6

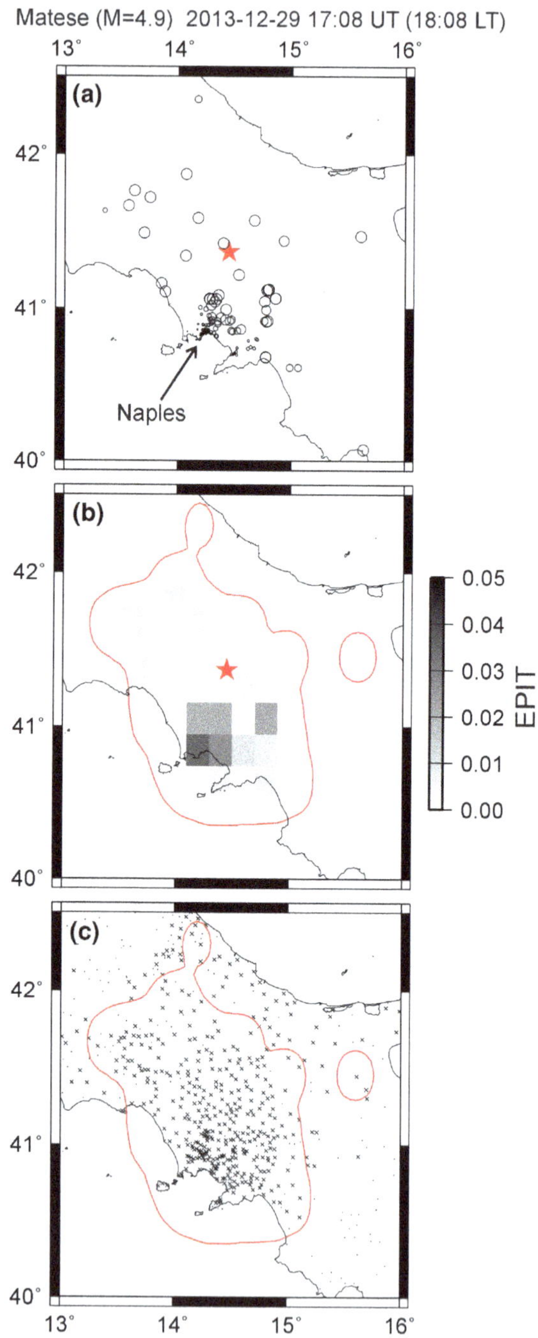

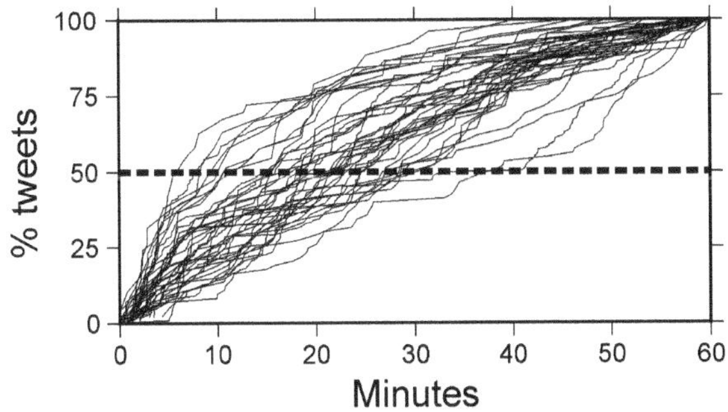

Fig. 8 Cumulative number of tweets posted within one hour since the origin time for 41 events that were associated to at least 20 posted tweets

Fig. 9 Relationship between the event magnitude and the exceedance of the 30, 60, 90 % thresholds of the cumulative number of tweets (see Fig. 8)

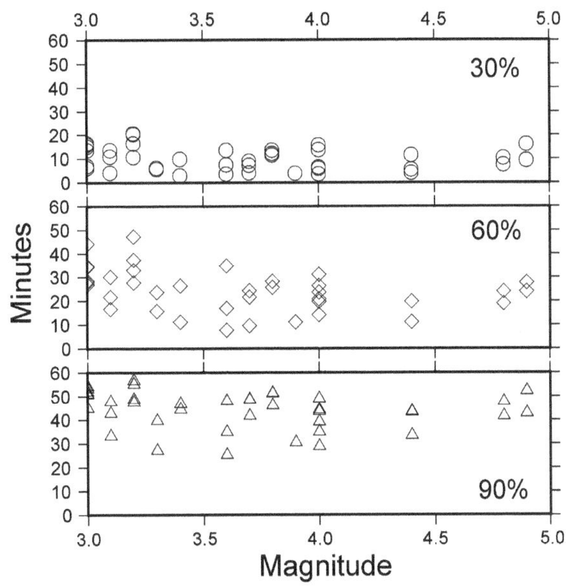

(Fig. 6c and 7c). A real-time analysis could provide an efficient and timely (less than 20 min, see Fig. 8) mapping of the earthquake perception area even for moderate earthquakes (M < 5).

This kind of analysis could have various potential applications. For instance the mobile phone communication system can be severely affected even by moderate earthquakes: studying the extension of the area where the earthquake has been perceived and the amount of tweets posted can give an indication of the potential phone traffic in near real-time (Vervaeck 2011; Moumni et al. 2013).

Future efforts could be devoted in improving earthquake detection systems, relying heavily on citizen-driven data, which provide direct information about the impact of an event on the population (Young et al. 2013).

References

Allen, R. M. (2012). Transforming earthquake detection? *Science, 335*(6066), 297–298.

Bindi, D., Pacor, F., Luzi, L., Puglia, R., Massa, M., Ameri, G., & Paolucci, R. (2011). Ground motion prediction equations derived from the Italian strong motion database. *Bulletin of Earthquake Engineering, 9*(6), 1899–1920.

Bossu, R., Mazet-Roux, G., Douet, V., Rives, S., Marin, S., & Aupetit, M. (2008). Internet users as seismic sensors for improved earthquake response. *Eos, Transactions, American Geophysical Union, 89*(25). http://www.agu.org/pubs/crossref/2008/2008EO250001.shtml.

Bossu, R., Gilles, S., Mazet-Roux, G., Roussel, F., Frobert, L., & Kamb, L. (2011). Flash sourcing, or rapid detection and characterization of earthquake effects through website traffic analysis. *Annals of Geophysics, 54*(6). doi:10.4401/ag-5265.

Bossu, R., Lefebvre, S., Cansi, Y., & Mazet-Roux, G. (2014). Characterization of the 2011 Mineral, Virginia, earthquake effects and epicenter from website traffic analysis. *Seismological Research Letters,* 85(1). doi:10.1785/0220130106.

Burks, L., Miller, M., & Zadeh, R. (2014). Rapid estimate of ground shaking intensity by combining simple earthquake characteristics with tweets. *Proceedings of the 10th National Conference in Earthquake Engineering, Earthquake Engineering Research Institute, Anchorage, AK.*

D'Auria, L., & Giudicepietro, F. (2013). TwiFelt: Real-time mapping of earthquake perception areas through the analysis of Twitter streams. Rapporti Tecnici INGV n.254 ISSN 2039-7941 (PDF available at. http://istituto.ingv.it/l-ingv/produzione-scientifica/rapporti-tecnici-ingv/rapporti-tecnici-2013/2013-05-09.2596813703).

Earle, P. S., Bowden, D. C., & Guy, M. (2011). Twitter earthquake detection: Earthquake monitoring in a social world. *Annals of Geophysics, 54*(6). doi:10.4401/ag-5364.

Earle, P., Guy, M., Buckmaster, R., Ostrum, C., Horvath, S., & Vaughan, A. (2010). OMG earthquake! Can Twitter improve earthquake response? *Seismological Research Letters, 81*(2), 246–251.

Mercalli, G. (1902). Sulle modificazioni proposte alla scala sismica De Rossi-Forel. Società tipografica modenese.

Moumni, B., Frias-Martinez, V., Frias-Martinez, E. (2013). Characterizing social response to urban earthquakes using cell-phone network data: The 2012 Oaxaca Earthquake. *PURBA 2013: Workshop on Pervasive Urban Applications*. UbiComp'13, September 8–12, 2013, Zurich, Switzerland.

Sakaki, T., Okazaki, M., & Matsuo, Y. (2010). Earthquake shakes Twitter users: Real-time event detection by social sensors. In *WWW 2010*, April 26–30, Raleigh, North Carolina.

Sakaki, T., Okazaki, M., & Matsuo, Y. (2013). Tweet analysis for real-time event detection and earthquake reporting system development. *IEEE Transactions on Knowledge and data engineering, 25*(4). doi:10.1109/TKDE.2012.29.

Sbarra, P., Tosi, P., & De Rubeis, V. (2009). Web-based macroseismic survey in Italy: Method validation and results. *Natural Hazards, 4*, 563–581. doi:10.1007/s11069-009-9488-7.

Sysomos. (2014). Inside Twitter: An In-depth look inside the Twitter World. http://www.sysomos.com/docs/Inside-Twitter-BySysomos.pdf.

Thelwall, M., Buckley, K., & Paltoglou, G. (2011). Sentiment in Twitter events. *Journal of the American Society for Information Science Technology, 62*(2), 406–418. doi:10.1002/asi.21462.

Vervaeck, A. (2011). Mobile communications and earthquakes: A very "disturbing" marriage. http://earthquake-report.com/2011/09/16/mobile-communications-and-earthquakes-a-very-disturbing-marriage.

Young, J. C., Wald, D. J, Earle, P. S., & Shanley, L. A. (2013). *Transforming earthquake detection and science through citizen seismology*. Washington, DC: Woodrow Wilson International Center for Scholars. (http://www.wilsoncenter.org/publication-series/commons-lab).

Wald, D. J., Quitoriano, V., Dengler, L. A., & Dewey, J. W. (1999). Utilization of the Internet for rapid community intensity maps. *Seismology Research Letters, 70*, 680–697.

The Easter Sunday 2011 Earthquake Swarm Offshore Malta: Analysis on Felt Reports

Matthew R. Agius, Sebastiano D'Amico and Pauline Galea

1 Introduction

Malta is a small archipelago, with a land mass covering a total area of 316 km^2 in the Sicily Channel, located between Sicily, Tunisia and Libya (Fig. 1). Local seismicity occurs offshore and is considered to be low when compared to the seismicity of other regions in the Mediterranean such as Italy or Greece (Vannucci et al. 2004). Despite the numerous local and regional earthquakes recorded by the seismograph located in the south of the islands (Boschi and Morelli 1994), very few earthquakes are actually felt. Until recently, such felt tremors were only reported in the local newspapers, giving limited qualitative and quantitative information about the shaking experience felt across the various localities. Historical records, however, indicate that Malta is susceptible to stronger shaking of higher intensity, powerful enough to damage buildings (Fig. 1, Galea 2007), such as the 1693 earthquake in south-eastern Sicily (e.g., Boschi et al. 2000). How such shaking would affect Maltese society today, taking into consideration the rapid urbanisation that has taken place on the islands in the last century, is still relatively unknown.

In order to better assess the shaking intensity of an earthquake on the Maltese islands, the Seismic Monitoring and Research Unit (SMRU) at the Department of Physics of the University of Malta has since 2007 set up a dedicated online page for the local community to report their earthquake experiences. The reported felt effects, and any damage are then manually translated into an intensity value on the European Macroseismic Scale 1998 (EMS-98, Grünthal et al. 1998). Ideally, in the case of a strong earthquake, the long established practice is of trained personnel to

M.R. Agius (✉) · S. D'Amico · P. Galea
Department of Geosciences, Faculty of Science, University of Malta, Msida, Malta
e-mail: matthew.agius@um.edu.mt

S. D'Amico
e-mail: sebastiano.damico@um.edu.mt

S. D'Amico (ed.), *Earthquakes and Their Impact on Society*,
Springer Natural Hazards, DOI 10.1007/978-3-319-21753-6_27

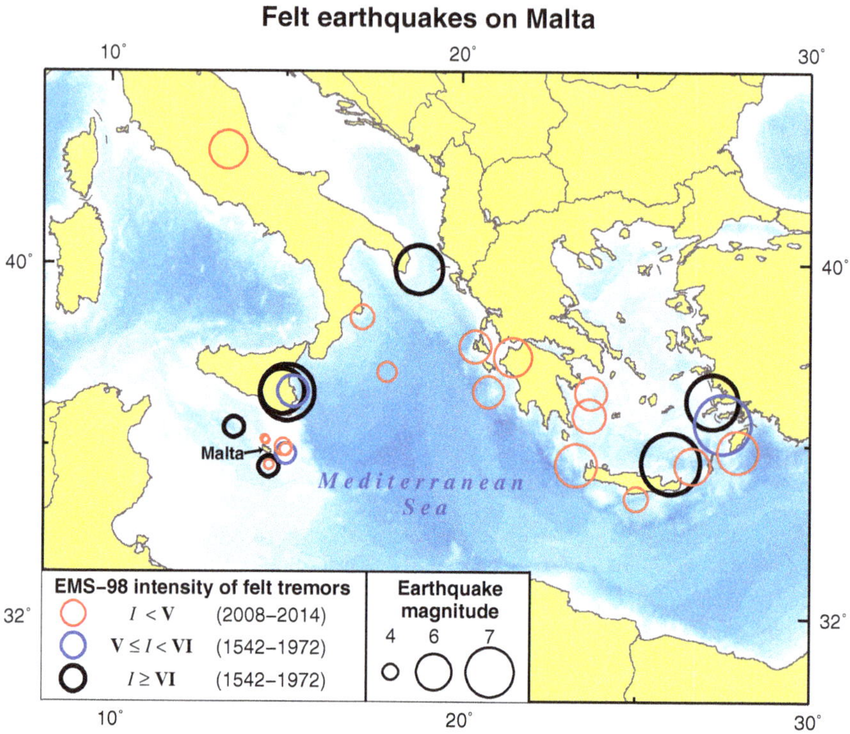

Fig. 1 Map of earthquake epicentres that were felt or produced damage on Malta and their corresponding felt intensities based on the European Macroseismic Scale 1998 (Grünthal et al. 1998). Data of felt earthquakes prior 1972 are from Galea (2007). Recent felt reports are from submitted online reports to the SMRU

make on-site inspections (e.g., Dandoulaki et al. 1998; Karababa and Pomonis 2010), however, detailed assessment of an entire locality on a macro scale is a time-consuming process. Online questionnaires are an alternative assessment on a local scale, especially when there is no structural damage. The compilation of such data reports is generally used to plot intensity maps, such as ShakeMaps (Wald et al. 2006), to better visualise the felt effects of the earthquake. Intensity maps are also used in conjunction with other studies such as ground acceleration, civil engineering, disaster management and civil protection.

Here we present a summary of felt reports for an earthquake swarm that occurred on Easter Sunday of 2011, felt widely across Malta. The compilation of the data is a first of its kind for the islands. The data reflects the demographics as well as the different types of buildings found across the archipelago.

2 Earthquake Sequence Over the Easter Weekend

An earthquake swarm is different from an aftershock sequence. The latter contains a mainshock followed by a sequence of aftershocks of ever-decreasing magnitude, whereas a swarm shows no particular pattern of magnitude variation with time, and the largest shock may occur anywhere in the sequence. The occurrence of earthquake swarms is quite common in the vicinity of the Maltese islands, especially on the offshore fault systems to the south. Historical documentation also records the occurrence of a number of earthquake swarms that were felt by the public, such as that between 14th and 21st of August, 1886. About 16 events in this sequence were large enough to be felt. The strongest shock made most of the residents run out onto the streets at night and caused general alarm among the population, to the extent that public calls for prayer and adoration were made and churches remained open throughout the night (The Malta Times, Saturday 21/08/1886). Numerous other small swarms have been recorded instrumentally in recent years but many of them were unfelt.

The swarm under consideration here started early on Easter Sunday (00:10 local time) and continued for at least three days. The largest event in the swarm occurred on Sunday 24th of April 2011 at approximately 13:02 UTC (15:02 local time). It is the largest magnitude earthquake to occur near the Maltese islands in the past decade, estimated to be of a local magnitude (M_L) 4.1 (D'Amico 2014). Figure 2 shows the recorded seismic activity throughout the day on Sunday 24th of April.

The detection and location of many of the earthquakes in the swarm were limited due to the small magnitudes and poor seismic station coverage. With only one nearby station on southern Malta (WDD, Agius et al. 2014), the epicentres of these earthquakes could only be analysed using the standard single-station technique: P-wave polarisation analysis used to establish the back azimuth from the station to the source, and the S-P time difference used to infer the distance to the earthquake. Such a technique is implemented in an automated manner at WDD through the software LESSLA (Agius and Galea 2011). All automated locations were also reviewed manually. In total the SMRU detected 15 earthquakes over four days, all located within the same source area about 38 km east of Malta (Table 1 and Fig. 3). Their local magnitudes range from 1.8 to 4.1, with most earthquakes having a magnitude of less than 3.5.

In the case of the largest event, the seismic energy reached farther stations located in the Central Mediterranean area and belonging to different institutions and/or networks (e.g., INGV, MEDNET, NOA, TT). This event has been relocated using the Computer Programs in Seismology location code *elocate* (Herrmann 2013) by applying a suitable velocity model for the region. D'Amico (2014) obtained a moment tensor solution of the earthquake applying the CAP (Cut-and-Paste) method (Zhu and Helmberger 1996; Tan et al. 2006; D'Amico et al. 2010) which is based on modelling of regional waveforms. The source depth, moment magnitude and focal mechanisms are determined using a grid search technique. For any fixed depth, the procedure attempts to find the best fit by

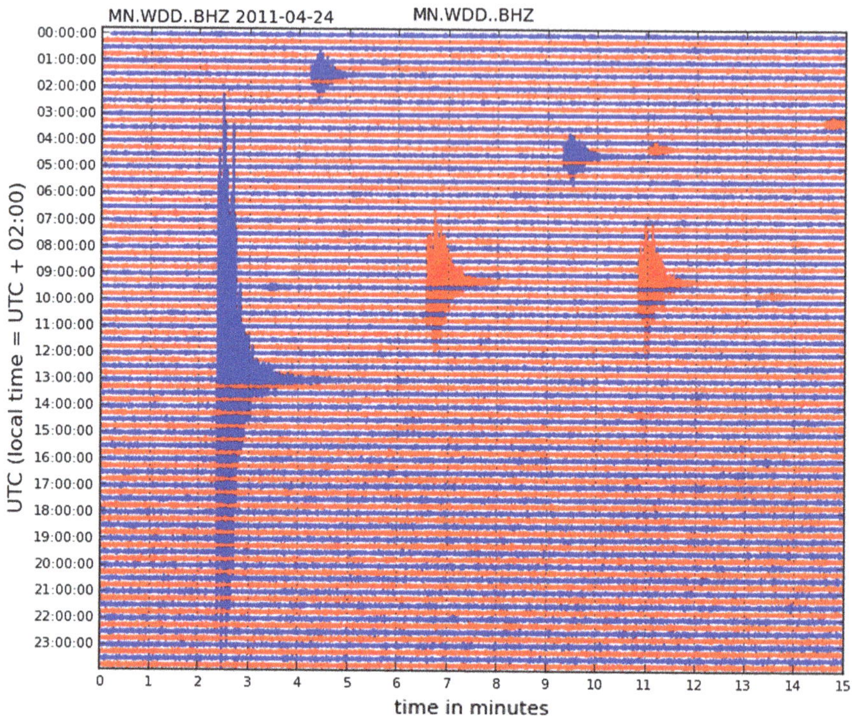

Fig. 2 The 24-hour seismic trace recorded at station WDD on Sunday 24th of April 2011. A couple of the 'stronger' earthquakes that took place during the day are clearly visible, with the strongest earthquake being registered at approximately 15:02 local time

aligning automatically the data with the synthetics. It has been shown that a good focal mechanism estimation can be obtained using a few stations (D'Amico et al. 2011) and the CAP method can be considered a stable and powerful approach to compute moment tensor solutions. For the largest event in the swarm the best fitting solutions suggests a moment magnitude (M_W) of 4.0 or M_L 4.1 and a focal depth of about 10 km. The best fitting focal mechanism shows a strike slip solution on a fault plane striking at 187° and dipping at about 71°.

The region of the 2011 swarm is close to a shallow platform, known as Hurd's Bank (Fig. 3). Published seismotectonic maps (e.g., Gallais et al. 2011) do not indicate any apparent surface feature or fault in this area. In contrast, most of the Sicily Channel is marked by a dense network of normal and strike-slip faults manifesting near surface expression. Instrumentally located seismicity in the area prior to this event shows only a few, sparse earthquakes reported in conventional seismic bulletins. However, it is likely that the 1886 swarm originated in approximately the same location as that of the Easter 2011 swarm, since it was similarly

Table 1 Parameters of all the earthquakes located by the SMRU during the 2011 swarm activity

Date	Time (UTC)	M_L	Latitude (°N)	Longitude (°E)
2011-04-23	22:10:58	2.6	35.96	14.91
2011-04-24	01:34:00	3.3	35.89	14.95
2011-04-24	02:44:34	2.5	36.01	14.91
2011-04-24	03:29:07	2.8	35.87	14.92
2011-04-24	04:25:39	2.7	35.94	14.94
2011-04-24	04:38:58	3.0	35.94	14.98
2011-04-24	09:21:19	3.1	35.88	14.94
2011-04-24	09:25:27	2.9	35.88	14.99
2011-04-24	09:33:07	1.8	35.77	14.91
2011-04-24	09:57:58	1.8	36.05	14.86
2011-04-24	13:02:12	4.1	35.94	14.92
2011-04-25	06:10:18	3.2	35.97	14.95
2011-04-26	04:10:27	3.4	35.84	15.01
2011-04-26	18:00:13	3.0	35.96	14.88
2011-04-27	05:40:38	2.9	35.81	14.95

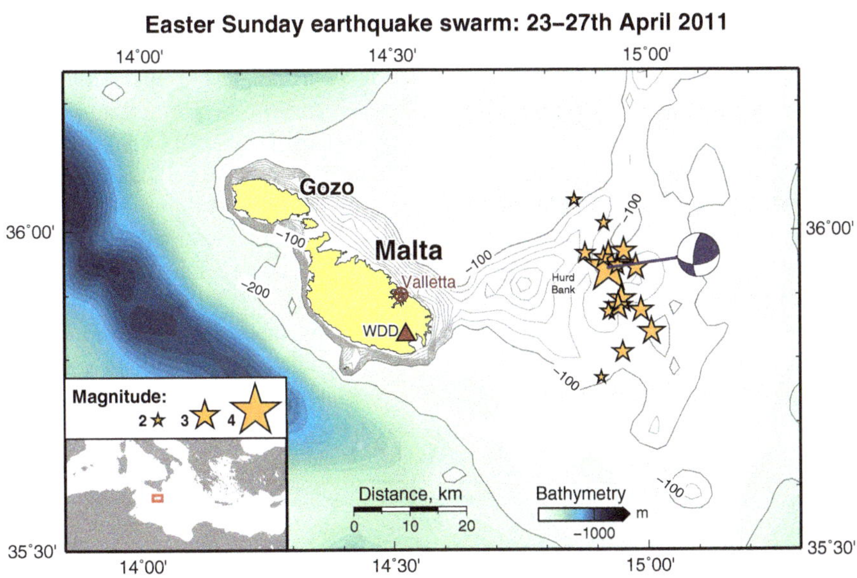

Fig. 3 The epicentres of the 15 earthquakes in the swarm located by the SMRU. *Yellow stars*: Earthquake location estimates using single-station analysis on WDD (*red triangle*) seismic data. *Blue beach ball*: The earthquake solution (location marked with the *blue line*) of the strongest earthquake M_L 4.1 (D'Amico 2014). Dense contour lines at 10 metres interval up to 100 m depth show shallow bathymetry. Valletta is the capital city of Malta, surrounded by two harbours. *Inset map*: Location of the map region within the Mediterranean Sea

felt mostly in the eastern half of Malta and hardly felt in Gozo. More long-term monitoring and interpretation of seismic activity in this area, therefore, may shed light on buried, or previously unmapped fault systems.

2.1 Media Coverage and Public Sentiment

The sequence of earthquakes and felt tremors prompted the attention of the media throughout the day. The first report was published on the Times of Malta online page on Sunday 24th of April 2011 at 00:21 local time. The news item was followed by a second article some 2 h later titled "Divorce and earthquake forecast at Borġ in-Nadur". This article featured a story of a lay man who claimed religious apparitions of the Virgin Mary. A few weeks before the earthquake, the man allegedly dictated a message conveyed to him by the holy figure to his followers: "The time has come for Malta's turn to experience the tremors and you shall see buildings shake, especially in that area that your ancestors built for defence: the area around the port". This claim came at a time when Malta was counting down the days for a national referendum on whether or not to introduce legislation for divorce. The social divide was strong particularly because of the religious sentiment. The coincidence of the forecast time and location of the earthquake (off the Grand Harbour of Valletta) fuelled the controversy with many suggesting that the prophecy was a sign of God wanting people to vote 'no' in the upcoming referendum. The article was also given full-page prominence in the print edition of The Sunday Times of Malta, on the same day (The Sunday Times 2011). Furthermore, three weeks before, another tremor was felt and also reported in the local news, this time from a distant, magnitude 6.1 earthquake in Crete on 1st of April 2011. The combination of events created a sensation that led to an increased attention on various media sources. This led to higher publicity, encouraging more people to submit the online SMRU questionnaire.

Being Easter Sunday, shortly after lunch time, families are likely to have been sitting down in a quiet environment, hence increasing their chances of experiencing shaking. At home, unlike at a workplace environment, people are more likely to have had easy access to various media sources such as television, radio and internet.

With the advent of social media, news items are easily 'shared' with friends across various online platforms, typically using smart phones. In total the online news items reported by the Times of Malta were shared more than 800 times. The extended online audience reached from such shares is hard to quantify but it is expected to grow exponentially with every 'share'—easily reaching a good percentage of the 400,000 population on the islands. This process indirectly helped with the promotion of the SMRU website resulting in hundreds of people filling in the 'Did you feel an earthquake?' questionnaire.

3 Online Questionnaire

The 'Did you feel an earthquake?' online questionnaire has 28 questions divided into four sections (Fig. 4). Section A is the only mandatory section and refers to the location and time of the felt shaking. Users are asked to specify if they were outdoor, inside a building, in a stationary or moving vehicle, or other. A Google Maps window is included for users to voluntarily give a more precise location.

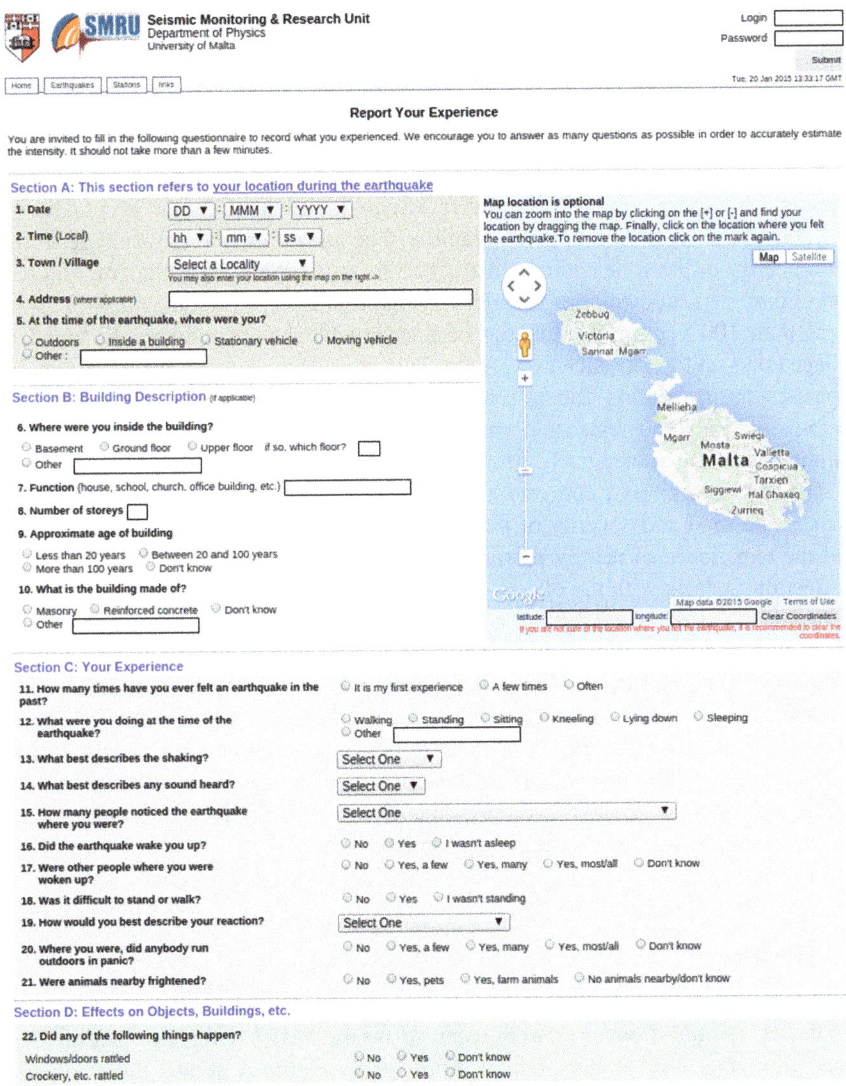

Fig. 4 Screen shot of the 'Did you feel an earthquake?' online questionnaire on the SMRU website

Fig. 5 Photos of typical dwellings found across Malta. **a** Masonry buildings older than 100 years built in historic areas such as Valletta. **b** Masonry houses built between 20 and 100 years ago outside historic towns such as Ħaż-Żabbar and Marsaskala. **c** Modern high-rise apartments built in recent years, replacing old houses, as is the case in San Ġiljan and Tas-Sliema

Section B focuses on the building description: age, use, height and construction typology (masonry, concrete, or other). Such information might eventually be useful in the study of building vulnerability. The age of a building would generally be indicative of the style and/or construction method of that particular era. Figure 5 shows three typical dwellings found across the islands. The buildings in Fig. 5a are more than 100 years old, built out of masonry blockwork, commonly found in village cores and towns such as Valletta. Two-storey houses were the more popular type of housing during the last century, involving a combination of masonry blockwork walls and reinforced concrete roofs (Fig. 5b). Nowadays, new, taller, buildings are constructed using a mix of reinforced concrete and masonry (Fig. 5c).

The third section, C, concerns the respondent's perception of the earthquake, such as the kind and severity of the shaking felt, his/her current position/activity, and the experience of nearby persons or animals.

Section D deals with the effects of the earthquake on objects, buildings and the environment, such as the rattling of windows, doors and crockery; disturbance of the motion of pendulum clocks; swaying of plants, splashing of liquids, swinging of hanging objects; shifted furniture, etc. There is also a series of questions on the damage, if any, to the building, such as cracks in plaster, fallen pieces of plaster from walls or ceilings; cracks in brick or stone walls; fallen masonry walls, etc., as well as on the effects on natural surroundings (landslips, cracks in the ground, or effects on ponds or streams). The respondent is also given the opportunity to add any other comments he/she deems appropriate.

4 Results

A total of 489 questionnaires were received by the SMRU over the span of three days, from the 24th to the 26th of April 2011. Figure 6 shows the frequency distribution of the submitted questionnaires in relation to the sequence of earthquakes. Following each earthquake is a spike of submitted felt reports. During the

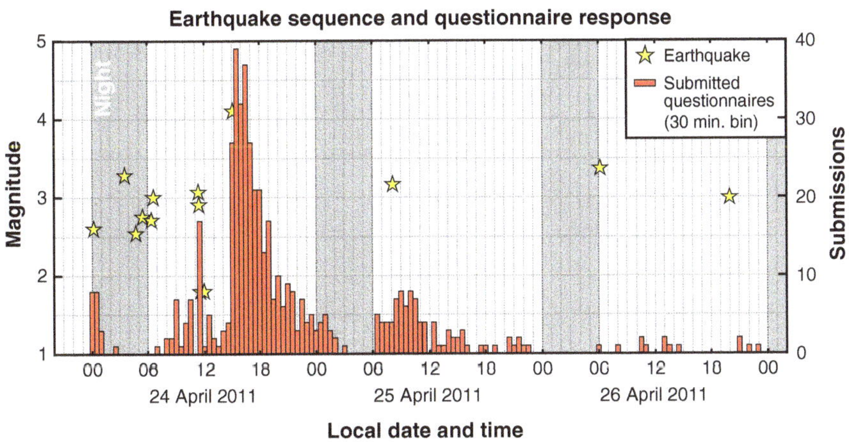

Fig. 6 Graph showing the time series of the earthquake swarm and the corresponding submitted felt reports. *Yellow stars*: The 14 earthquakes of various magnitudes that took place from the 24th to the 26th of April. *Red bars*: Submitted reports grouped in 30 min bins. *Grey shade*: Night (mid-night to 6 am)

night (grey shade) fewer or no questionnaires were registered. The majority of questionnaires were submitted following the magnitude 4.1 earthquake.

Figure 7 illustrates graphically the contents of the questionnaire reports relating specifically to the largest shock: a total of 346 questionnaires. Most reports were from people, who at the time of the earthquake, were inside a building (97 %), in an upper floor (72 %), either in a house (64 %) or in an apartment (12 %), and sitting down (61 %). The majority of the buildings had 2 or 3 storeys, were built in the last century (86 %), and made of masonry (53 %). Over 40 % claim that most people who were at the respondent's location felt the earthquake. Fifty percent of the respondents reported rattling of doors and windows whereas only 26 % reported rattling of crockery. 54 respondents (16 %) claimed that their pets were frightened; 18 reports from various localities in Malta indicated that a few people ran outdoors in panic. This contrasts with the few responses received from the western coast and the island of Gozo; no rattling of doors, and no people running outdoors were reported.

Figure 8 shows two maps; one of the population distribution across the Maltese islands, and the other showing the distribution of the submitted questionnaires related to the main shock on Sunday afternoon 24th of April. Most felt reports originated from the east–south-east parts of the main island; the more inhabited areas of Tas-Sliema, Marsaskala and Ħaż-Żabbar. Tas-Sliema has the highest number of reports (27). Only 5 reports were from Gozo.

Several fields were marked as 'unknown', while towards the end of the questionnaire, many fields were left empty, probably because the questions concerned a high level of shaking intensity and structural damage that may have not been relevant in this case.

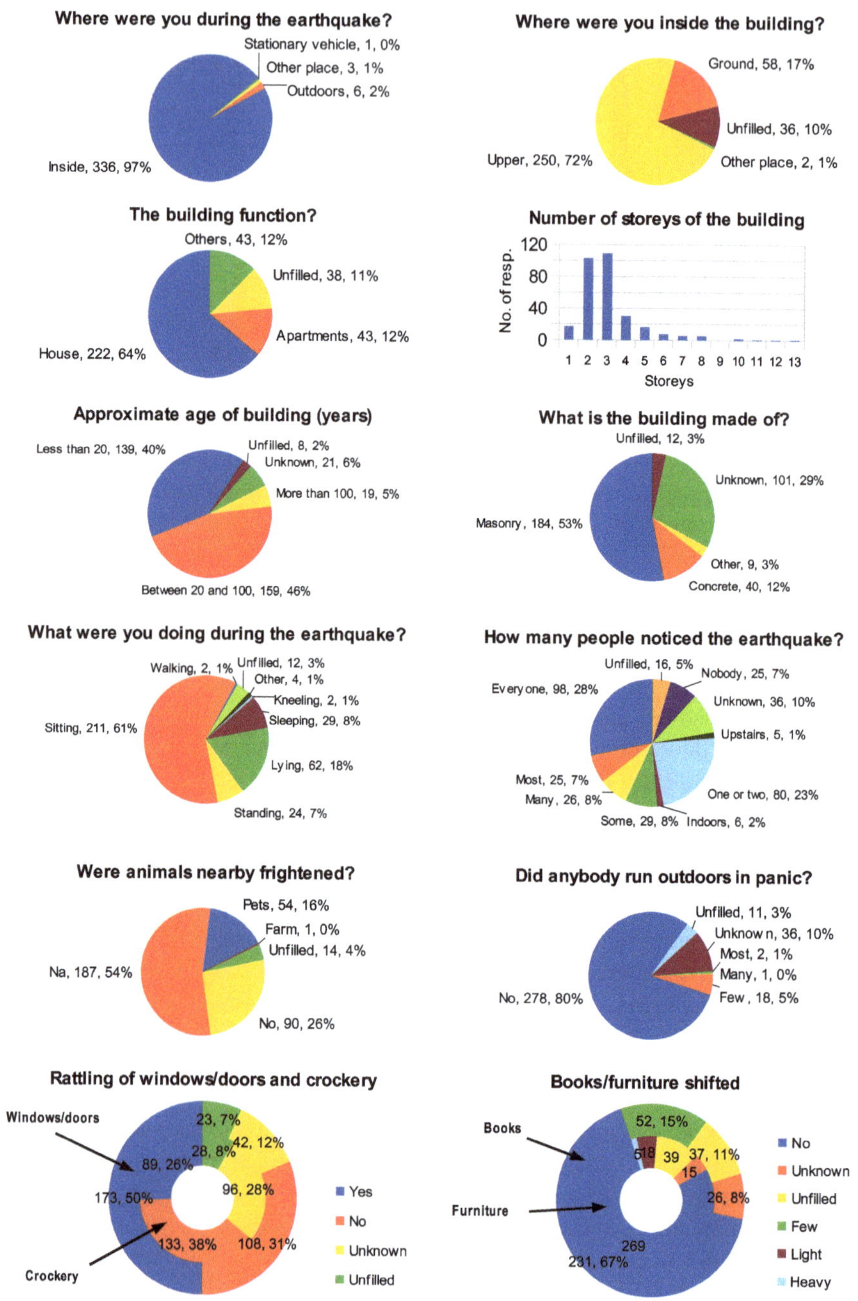

Fig. 7 Pie and bar charts showing selected statistics from the questionnaire response submitted for the M_L 4.1 earthquake on Sunday 24th of April 2011. Comma separated labels show the selected answer, the number of reports, and the percentage

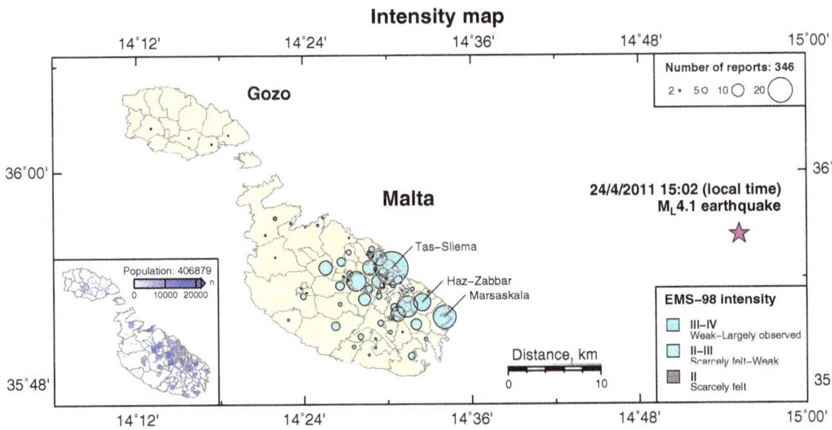

Fig. 8 Map showing local felt intensities based on the European Macroseismic Scale 1998 (Grünthal et al. 1998) intensity scale for the M_L 4.1 earthquake on Sunday 24th of April. *Colour-coded circles*: indicate the number of felt reports and the maximum intensity reported. *Purple star*: The earthquake epicentre. *Inset map*: The population distribution of Malta based on the national census (National Statistics Office 2011). *Shaded inland regions*: indicate the built-up areas and the respective local population. *Map contours*: indicate the local council boundaries

5 Discussion

The general aspects from the felt reports, with a particular focus on the intensity of the main shock in the context of the present urbanisation of the Maltese islands, is discussed hereunder.

5.1 *Seismic Intensity*

The reports, particularly the results from sections C and D, were used to assign intensity values in the various localities, according to the European Macroseismic Scale 1998 (EMS-98, Grünthal et al. 1998). The strongest reported effect is of 'shifted furniture', claimed by 23 respondents from various central–southern areas of Malta (Qawra, Rabat, Iklin, Ħamrun, Santa Venera, San Ġiljan, Tas-Sliema, Ta' Xbiex, L-Imsida, Il-Marsa, Valletta, Il-Kalkara, Ħal Tarxien, Il-Fgura, Ħaż-Żabbar, Marsaskala, Marsaxlokk, Birżebbuġa, Is-Siġġiewi, L-Imqabba, and Iż-Żurrieq).

A maximum intensity of IV has been assigned to reports claiming shifted household furniture. Because the number of such reports was only a small subset within the respective locality, an intensity range of III–IV was assigned to that area. An intensity in the range of II–III was assigned to localities that had a maximum reported level of shaking from rattling doors and crockery. An intensity II was assigned to reports that had basic felt experience that did not include effects on objects and buildings as a result of the earthquake. Figure 8 maps the shake intensity according to the reports.

The general trend of the earthquake intensity agrees with the automated report generated by the European-Mediterranean Seismological Centre (EMSC) online questionnaire. EMSC received 36 responses from 5 communities, all from the central–south-east areas of the mainland Malta. A maximum grade of intensity IV, described as largely observed, was assigned to three localities.

The intensity map shows a natural decrease in intensity with distance from the epicentre. This is also reflected by the lower number of reports towards the north-west of the archipelago. Taking into account the close proximity of both islands to the epicentre, the frequency content of the propagating seismic waves from the earthquake may have contributed to the rapid change of the felt intensity at a relatively short distance. Unlike the expected shaking felt from a regional earthquake, where the low-frequency waves would affect the islands more or less the same throughout, the energetic high-frequency seismic waves from a local earthquake—responsible for the rattling of crockery, doors and windows—are likely to attenuate quickly during propagation. Future deployment of permanent, broadband seismic stations in central Malta and Gozo will aid with investigating such characteristics.

5.2 *Citizens' Reaction on Social Media*

As new tremors were felt throughout Sunday, many took to the Internet to report their experience publicly, beneath news items and on social media soon after each earthquake. Many simply reported the locality of where they felt the tremor, saying, for example: "I felt it in Sliema too". Some were more descriptive: "There was another one… around 3.03 pm. Sounded like a fireworks factory explosion. Trembling sensation on the floor, and cabinet doors rattled. In my children's bedroom, a soft toy fell off a shelf. I'm in Qormi and I live on first floor". Others gave it a religious connotation in relation to the current national affairs: "This is a warning from above… With the divorce referendum coming up". A few comments were amusing: "I blamed my son—how typical. I thought he had moved something in the kitchen". On the other hand, a couple of comments reported otherwise: "Not a single one was felt in Gozo". Despite the numerous online posts none reported any damage. News agencies confirmed that the Civil Protection Department received no calls for assistance.

Comparatively, the overall information obtained from the snippet textual comments expressed by citizens on various online platforms is coherent with those submitted to the SMRU online questionnaire. Several new studies are showing that web crawling and data mining of online social media can be used to generate near real-time alerts for the occurrence of something phenomenal (e.g., Sakaki et al. 2010; Bossu et al. 2011; D'Auria et al. 2014). The new dataset obtained here can be used to calibrate such an Internet system were it to be adapted in the future by the SMRU.

5.3 Some Observations on Local Building Patterns

The contents of the completed questionnaires also provide some insight into construction patterns in Malta. Until recently many residential buildings in villages across Malta were constructed in of unreinforced masonry using local limestone blocks, and typically consisting of 2 storeys. In the last few decades this scenario has been changing rapidly; many low-rise houses are being demolished to be replaced with taller apartment blocks, including both masonry blockwork walls and reinforced concrete structural elements. The change in the building style is reflected in the questionnaires. Nearly 60 % of the masonry buildings were built between 20 and 100 years ago whereas about 30 % in the last 20 years. In contrast, 85 % of the buildings reported as 'reinforced concrete' were built in the last 20 years.

The seaside town of Tas-Sliema is one of the most re-developed areas that has undergone such a rapid change in a relatively short time. In fact, nearly half of the reported reinforced concrete buildings in the questionnaires are located here or in the vicinity. Today this upmarket area has many tall buildings, some exceeding 10 storeys as noted from the respondents (Fig. 5c). In the absence of real data from strong ground motion, the behaviour of these taller buildings in comparison to low-storey masonry buildings is still unclear and needs to be better understood through other techniques, especially with respect to investigation of particular construction typologies peculiar to the islands.

The largest number of questionnaires from a single locality following the main shock were from Tas-Sliema, with 27 reports (Fig. 8). Marsaskala and Ħaż-Żabbar, two localities which are approximately at equal distances to the earthquake's epicentre as Tas-Sliema, only had 19 and 14 reports, respectively. All three localities have similar population: 13,621, 11,059, and 14,916, respectively (National Statistics Office 2011), and, all three localities overlie similar geology (Pedley and Clarke 2002). The main contrasting attribute between the localities is the building height. Many respondents from Tas-Sliema were inside higher floors compared to the other localities. The average reported building height in Tas-Sliema was 5.5 storeys and the average height for Marsaskala and Ħaż-Żabbar was 2.8. It is likely that the difference in the number of submitted reports is a result of the shaking being more noticeable on higher floors, although the different response of the buildings, or the social response of the Sliema population could also be contributing factors.

6 Conclusions

Fifteen earthquakes that took place offshore Malta over the Easter of 2011 were located to be about 38 km off the east coast of Malta. These earthquakes were of various magnitudes with the largest being of M_L 4.1 on Sunday 24th of April. The latter earthquake was felt by many of the inhabitants on the islands, particularly along the south-eastern coast. SMRU personnel located the earthquakes and updated the website for immediate public information.

A total of 489 felt reports were submitted through the online 'Did you Feel an Earthquake' questionnaire run by the SMRU website. The questionnaire had been in place since 2007 but had not yet achieved its full potential until these events. The reports following the main shock were analysed and benchmarked to the EMS-98 intensity scale. The highest reported shaking was 'shifted furniture'—no structural damage was reported. Hence, a maximum intensity of IV has been assigned.

The different number of reports between localities that have similar population, geological setting, and epicentre distance may be explained, in part, by the different building types. Interestingly, the locality with the most number of reports was from an area that had a large number of high rise apartments, whereas the other localities had two-storey houses. In this regard, a detailed investigation from an engineering point of view is still desirable. The questionnaires highlight the role citizens and online social media can play when investigating a regional or large-scale area.

The earthquake swarm can help to add new constraints to the regional geodynamic model and contribute to the current investigations of seismotectonics and seismic hazard in the area. A new level of seismicity has been revealed for this part of the Sicily Channel, previously only marked by a few earthquakes in conventional seismic bulletins. Important parameters such as the depth of the earthquakes are not well constrained mainly due to the lack of station coverage. Additional seismic stations on land and future missions involving ocean-bottom seismometers and detailed sea-floor mapping could provide clues on the earthquake mechanisms in this seismically active area.

Acknowledgements We thank Alex Bezzina for insightful comments that helped us to improve the manuscript. We are grateful to Carlos Cañas Sanz for his help in generating the demographic local maps. Some figures were created using the Generic Mapping Tools (Wessel and Smith 1998) and ObsPy: A Python Toolbox for Seismology (Beyreuther et al. 2010). This study was supported by SIMIT Project part-financed by the European Union, European Regional Development Fund (ERDF) under the Italia-Malta Cross-Border Cooperation Programme, 2007–2013.

References

Agius, M. R., D'Amico, S., Galea, P., & Panzera, F. (2014). Performance evaluation of Wied Dalam (WDD) seismic station in Malta. *Xjenza, 2*(1), 78–86.

Agius, M. R., & Galea, P. (2011). A single-station automated earthquake location system at Wied Dalam station, Malta. *Seismological Research Letters, 82*(4), 545–559.

Beyreuther, M., Barsch, R., Krischer, L., Megies, T., Behr, Y., & Wassermann, J. (2010). ObsPy: A Python toolbox for seismology. *Seismological Research Letters, 81*(3), 530–533.

Boschi, E., Guidoboni, E., Ferrari, G., Mariotti, D., Valensise, G., & Gasperini, P. (2000). Catalogue of strong Italian earthquakes from 461 B.C. to 1997. *Annals of Geophysics, 43*(4).

Boschi, E., & Morelli, A. (1994). The MEDNET program. *Annals of Geophysics, 37*(5).

Bossu, R., Gilles, S., Mazet-Roux, G., & Roussel, F. (2011). Citizen seismology: How to involve the public in earthquake response. In D. S. Miller & J. D. Rivera (Eds.), *Comparative emergency management: Examining global and regional responses to disasters* (pp. 237–260). Boca Raton: CRC Press.

D'Amico, S. (2014). Source parameters related to a small earthquake swarm off-shore of Malta (Central Mediterranean). *Development in Earth Science, 2*(1), 8–13.

D'Amico, S., Orecchio, B., Presti, D., Gervasi, A., Zhu, L., Guerra, I., Neri, G., & Herrmann, R. B. (2011). Testing the stability of moment tensor solutions for small earthquakes in the calabro-peloritan arc region (Southern Italy). *Bollettino di Geofisica Teorica ed Applicata.*

D'Amico, S., Orecchio, B., Presti, D., Zhu, L., Herrmann, R. B., & Neri, G. (2010). Broadband waveform inversion of moderate earthquakes in the messina straits, Southern Italy. *Physics of the Earth and Planetary Interiors, 179*(34), 97–106.

D'Auria, L., Convertito, V., & Giudicepietro, F. (2014). Real-time mapping of earthquake perception areas in the Italian region from Twitter streams analysis. In *EGU General Assembly Conference Abstracts* (Vol. 16, p. 6236).

Dandoulaki, M., Panoutsopoulou, M., & Ioannides, K. (1998). An overview of post-earthquake building inspection practices in Greece and the introduction of a rapid building usability evaluation procedure after the 1996 Konitsa earthquake. In *Proceedings of 11th European Conference on Earthquake Engineering.*

Galea, P. (2007). Seismic history of the Maltese Islands and considerations on seismic risk. *Annals of Geophysics, 50*(6).

Gallais, F., Gutscher, M.-A., Graindorge, D., Chamot-Rooke, N., & Klaeschen, D. (2011). A miocene tectonic inversion in the Ionian Sea (Central Mediterranean): Evidence from multichannel seismic data. *Journal of Geophysical Research, 116*(B12).

Grünthal, G. E., Musson, R., Schwarz, J., & Stucchi, M. (1998). *European Macroseismic Scale 1998 (EMS-98)*, volume 15 of *Centre Européen de Géodynamique et de Séismologie*. Cahiers du Centre Européen de Géodynamique et de Séismologie, Luxembourg.

Herrmann, R. B. (2013). Computer programs in seismology: An evolving tool for instruction and research. *Seismological Research Letters, 84*(6), 1081–1088.

Karababa, F. S., & Pomonis, A. (2010). Damage data analysis and vulnerability estimation following the August 14, 2003 Lefkada Island, Greece, Earthquake. *Bulletin of Earthquake Engineering, 9*(4), 1015–1046.

National Statistics Office (2011). *Census of population and housing 2011: Final report.* Technical report, National Statistics Office, Valletta.

Pedley, M. & Clarke, M. H. (2002). *Limestone Isles in a Crystal Sea: the geology of the Maltese Islands.* Publishers Enterprises Group.

Sakaki, T., Okazaki, M., & Matsuo, Y. (2010). Earthquake shakes twitter users: Real-time event detection by social sensors. In *Proceedings of the 19th International Conference on World Wide Web*, WWW '10, pp. 851–860, New York: ACM.

Tan, Y., Zhu, L., Helmberger, D. V., & Saikia, C. K. (2006). Locating and modeling regional earthquakes with two stations. *Journal of Geophysical Research, 111*(B1).

The Sunday Times (2011, April 24). Divorce and earthquake forecast at Borg in-Nadur.

Vannucci, G., Pondrelli, S., Argnani, A., Morelli, A., Gaperini, P., & Boschi, E. (2004). An atlas of Mediterranean seismicity. *Annals of Geophysics, 47*(1).

Wald, D. J., Worden, B. C., Quitoriano, V., & Pankow, K. L. (2006). *ShakeMap manual: Technical manual, user's guide, and software guide.* U.S. Geological Survey.

Wessel, P. & Smith, W. H. (1998). New, improved version of Generic Mapping Tools released. *Eos, Transactions of the American Geophysical Union, 79*(47).

Zhu, L., & Helmberger, D. V. (1996). Advancement in source estimation techniques using broadband regional seismograms. *Bulletin of the Seismological Society of America, 86*(5), 1634–1641.

Earthquake Readiness and Recovery: An Asia-Pacific Perspective

Douglas Paton and Li-ju Jang

1 Introduction

People living in countries situated on the circum-Pacific seismic belt (the Pacific Ring of Fire), where some 90 % of the world's earthquakes occur, have to live with high levels of seismic risk. When large earthquakes occur, affected residents are abruptly faced with loss, challenges and demands that differ significantly from anything they would encounter under normal conditions and in circumstances in which normal societal functions and resources are suddenly marked by their absence. Not only are affected populations subjected to these circumstances without warning, they may have to respond to them repeatedly as they negotiate aftershock sequences over periods of many months. However, it is also evident that people are not equally affected.

While some loss and disruption is reduced through the implementation of mitigation strategies (e.g., land use planning, building design), the impact people experience and the rate at which they adapt to and recover from earthquake events is also influenced by personal and community characteristics. Recognition of the fact that some people and communities can utilize their own resources to positively influence their own recovery led to the Hyogo Framework for Action 2005–2015 (ISDR 2005) identifying a need for disaster risk reduction (DRR) readiness strategies to facilitate the development of this so called resilience and adaptive capacity (e.g., Klein et al. 2003; Norris et al. 2008; Paton 2006a, b; Pelling and High 2005).

D. Paton (✉)
School of Psychological and Clinical Sciences, Charles Darwin University, Darwin, NT, Australia
e-mail: Douglas.Paton@utas.edu.au

L. Jang
Department of Social Work, National Pingtung University of Science and Technology, 912 Pingtung, Taiwan
e-mail: ljthird@gmail.com

S. D'Amico (ed.), *Earthquakes and Their Impact on Society*,
Springer Natural Hazards, DOI 10.1007/978-3-319-21753-6_28

This chapter discusses theoretical analyses of earthquake readiness and empirical studies of earthquake response and recovery in the citizens of two countries situated on the Ring of Fire; New Zealand and Taiwan (Jang 2008; Paton 2013; Paton and Jang 2011; Paton et al. 2014).

By identifying how personal, community and cultural characteristics interact to influence earthquake readiness, response and reduction, this chapter offers insights that can inform the development of the risk communication and community outreach programs required help answer the call for the development of DRR strategies issued by the ISDR (2005). To effectively pursue this goal it is important to accommodate the fact that the impossibility of predicting when the next earthquake will occur (it could be tomorrow or it may not occur for years, decades or longer), means that much DRR readiness research and intervention must occur prior to an earthquake occurring.

A need to facilitate readiness prior to earthquakes occurring introduces several challenges for this aspect of risk management. One issue derives from the fact that readiness intervention (e.g., identifying what people should do and communicating and engaging with them in ways that facilitate comprehensive readiness) is delivered about events people may have little or no experience of and which may not occur until some indeterminate time in the future. This necessitates adopting theoretical and quasi-experimental approaches to intervention design and delivery. However, this approach leaves issues regarding the adequacy, representativeness, and effectiveness (e.g., how appropriate are recommendations, how effective will they be across the range of earthquake intensities that could occur etc.) of DRR readiness strategies open to question. In order to provide tangible evidence for the legitimacy of theoretical approaches and the evidence base required to inform the development and implementation of practical intervention strategies it is necessary to assess the readiness-resilience relationship in the context of people experiencing significant levels of disruption from earthquakes (Bruijn 2004; Carpenter et al. 2001; Gaillard 2007; Klein et al. 2003; Paton 2006b). This chapter discusses both issues. It first discusses the applicability of earthquake readiness theory in New Zealand and Taiwan. It then tests the validity of this theory by examining the degree to which its theoretical components are endorsed in people's accounts of their experience of earthquake recovery.

The value of including a comparative New Zealand-Taiwan analysis derives from it affording an opportunity to tackle an additional challenge posed by working in an area of risk management that seeks to increase readiness by influencing how people think and act. This challenge arises because earthquakes occur in many different countries. A consequence of this is a need to consider whether the associated social and cultural diversity (e.g., Hofstede 2001) affects the validity of readiness theories that focus on identifying how people think and behave.

For example, in countries with a relatively individualist cultural orientation (e.g., New Zealand), people act consistently across situations in accordance with a self-concept that is relatively independent of social situation and in which achieving personal goals is a prominent objective. If collective action occurs, it reflects personal choice regarding levels of collaboration and cooperation rather than a cultural predisposition. In contrast, in collectivist countries like Taiwan, actions are underpinned by culturally-embedded beliefs that are reflected in shared purpose and activity involving alignment with social norms, achieving collective goals, and engaging in activities related to future goals that emphasize social relations (Triandis 1995). Accordingly, because culture introduces different ways in which people might interpret risk and make decisions about readiness, it becomes important to consider whether theories developed and tested in countries that are culturally highly individualistic (e.g., the USA, Australia, New Zealand) are equally applicable in Asian countries that tend to fall at the collectivistic end of the cultural spectrum.

This chapter discusses two interrelated issues. The first focuses on identifying the degree to which an earthquake readiness theory can predict readiness in different cultures. The second discusses the extent to which people's accounts of their earthquake experiences (in culturally diverse countries) can validate, and contribute to developing, earthquake readiness theory.

Pursuing this objective requires several pieces of information. One is identifying what people need to know and be able to do (individually and collectively) to cope with, adapt to and recover from earthquake hazard consequences (i.e., what being comprehensively prepared looks like and what risk management strategies are trying to get people to adopt, know and do). The second focuses on identifying why it is that some people prepare and others less so or not at all (i.e., identifying the risk interpretation processes that risk communication and community engagement strategies need to accommodate to facilitate sustained, comprehensive readiness). Recognition of the importance of these issues has been encapsulated in theoretical efforts to identify what constitutes comprehensive readiness and to account for differences in the degree to which people prepare. The first issue addressed here concerns what is meant by comprehensive earthquake readiness.

2 Comprehensive Earthquake Readiness

Readiness DRR strategies aim to facilitate people's capability to deal with the full spectrum of the physical, personal and social demands earthquakes create for affected populations. Expert analysis of how earthquake hazards interact with people, what people value and the physical and natural environment led to defining readiness as comprising separate but related functional earthquake readiness categories.

For example, Russell et al. (1995) identified three categories. The first, structural readiness, describes activities that secure the house (e.g., secure house to foundations) and its contents (e.g., securing water heaters and tall furniture etc.) to prevent contents from injuring inhabitants (e.g., from ground shaking). The second category, survival readiness, covers activities that increase people's capacity for self-reliance during periods of disruption (e.g., ensuring a supply of water to cope with loss of utilities for several days, having a radio with spare batteries etc.).

Finally, planning readiness encompasses for example, developing household earthquake plans and attending meetings to learn about earthquakes and how to deal with their consequences. A subsequent factor analytic study (Lindell et al. 2009) proposed that readiness comprise Direct Action (e.g., learn how to shut off utilities, have a 4-day supply of canned food, strap heavy objects etc.) and Capacity Building (e.g., join an earthquake-related organization, attend meetings about earthquake hazards) factors. These studies identify how readiness strategies that facilitate the comprehensive adoption of structural, survival, planning and capacity building (social) activities increase people's ability to cope with, adapt to, and recover from the consequences of earthquake events.

For example, ensuring the physical integrity of the house and storing water and food helps people deal with direct (e.g., impact of ground shaking, liquefaction and aftershocks on buildings and infrastructure) and secondary (e.g., loss of lifelines like water, power and sewerage services and the consequent need for people to be self-reliant and able to continue functioning despite the absence of normal services) earthquake consequences. Taking steps to ensure the structural integrity of the home not only reduces the risk of injury and death to householders, it also increases the likelihood of their able to recover in situ, reduces demands on societal resources for temporary accommodation, and increases the availability of people to participate in social (mutual aid, social support), economic and environmental recovery activities. Planning and capacity building readiness can enhance the quality of the mutual aid and social support available within communities and the quality of working relationships between community members and civic agencies during and after earthquake events (Paton and McClure 2013). These activities increase people's readiness to be able to respond to, for example, challenges emanating from recovery processes (e.g., dealing with government agencies, insurance companies, builders etc.) that may persist for weeks, months or years. The use of "comprehensive" here captures the fact that people's resilience and adaptive capacity is only truly increased if all levels of readiness are adopted and sustained over time (i.e., to maintain readiness for infrequent events).

The validity of this knowledge can be increased if it can be shown that these categories are important in actual earthquake events. One way of assessing the validity of these categories involves asking people who experienced an earthquake what they had to cope with and adapt to during the recovery period. This was done by researching people's accounts of their experience of two earthquakes; the 2011 Christchurch, New Zealand earthquake and the 921 earthquake in central Taiwan in 1999.

On 22 February 2011, an MW 6.3 earthquake struck Christchurch, causing 185 deaths, over 7000 injuries and in excess of US$12 billion in damage (Bannister and Gledhill 2012). Some 100,000 homes (approximately half the housing stock) were damaged, and about 7000 homes were rendered uninhabitable. Some 3000 of the 5000 businesses in the Christchurch CBD were displaced and 1200 CBD building had to be demolished. More than half of the road network had to be replaced. The Canterbury earthquake sequence was characterized by a high level of seismicity over an extended period within close proximity to Christchurch throughout 2011 and into 2013. The second event used to provide the context for the discussion in this chapter concerns the MW 7.6 earthquake (the 921 earthquake) that struck central Taiwan on September 21, 1999. The official report from MOI (Disaster Response Report 2008) revealed that this earthquake caused 2415 deaths, 11,306 injuries, rendered some 114,000 homes uninhabitable, damaged 1500 schools, and made some 110,000 people homeless. At the township level, Tung Shih, the location for the survey and qualitative data discussed in this chapter, suffered the highest death toll (Liao 1999).

Interviews with survivors (Jang 2008; Paton et al. 2014) in each location provided the opportunity to explore how people responded to the demands and challenges of earthquake recovery. Identifying what people had to do and how well they were able to do so provides tangible insights into readiness and its predictors. The first issue to be addressed concerns people's accounts of what helped them effectively cope with and adapt to the physical, social and environmental demands encountered.

2.1 *Earthquake Survivor's Accounts of Readiness: Examples from New Zealand and Taiwan*

Paton et al. (2014) conducted interviews with residents in Christchurch affected by the 2011 earthquake and subsequent aftershocks. One set of questions focused on eliciting people's accounts of the demands they experienced and what activities and relationships helped them deal with these demands (i.e., identify what kind of readiness they would need should an earthquake occur in future).

Christchurch respondents confirmed the importance of structural, survival, and household emergency planning readiness. Their accounts also reinforced the importance of readiness strategies facilitating people's ability to work with neighbors and other community members to develop (self-help) groups to confront local demands and identify and meet local needs (e.g., collaborative efforts to remove rubble, organize local efforts to repair homes, provide mutual social support, take care of those with special needs, and establish and organize community meetings etc.). A need for psychological (particularly with regard to coping with the impact of repeat aftershocks and adapting to changes in living conditions and loss of social relationships and social support) readiness singled out as being very important. Respondents accounts introduced a need for livelihood or employment (e.g., to prepare for loss of, disruption to or changes in employment over several months) and community-agency (e.g., develop community group ability to be able to formulate and represent their needs and goals to government, businesses and recovery agencies in ways that facilitated the ability of each community to take responsibility for their recovery) readiness (Paton et al. 2014). Similar categories emerged in Taiwan.

In Taiwan (Jang 2008), the development of readiness activities was identified as an important activity. In the past, the focus of DRR in Taiwan has been on strengthening infrastructures. Planning has tended to overlook the human dimension (Jang 2008). This adds credence to the Taiwanese accounts as these reflect their experience rather than issues that may have been included in earlier public education.

Survivors' accounts reiterated the need for structural (e.g., building code reinforcement, home safety), survival (e.g., resource availability, food storage, storing water and health care products, personal safety) and planning (e.g., evacuation routes and shelters) readiness. Respondents discussed the importance of disaster education in readiness programs and its role in facilitating positive attitudes towards DRR and developing self-reliance. The latter, in conjunction with residents discussing how developing hazard knowledge helped people to be able to quickly and calmly cope with crises, reinforce the need for psychological readiness. Taiwanese accounts endorsed the importance of capacity building readiness and being able to offer mutual support and being willing and able to serve others.

Community-government relationships were identified as an important aspect of readiness, with these relationships influencing the resource availability and job security (employment or livelihood readiness). Given that these data were obtained from highly agricultural areas, planning readiness interacted with livelihood

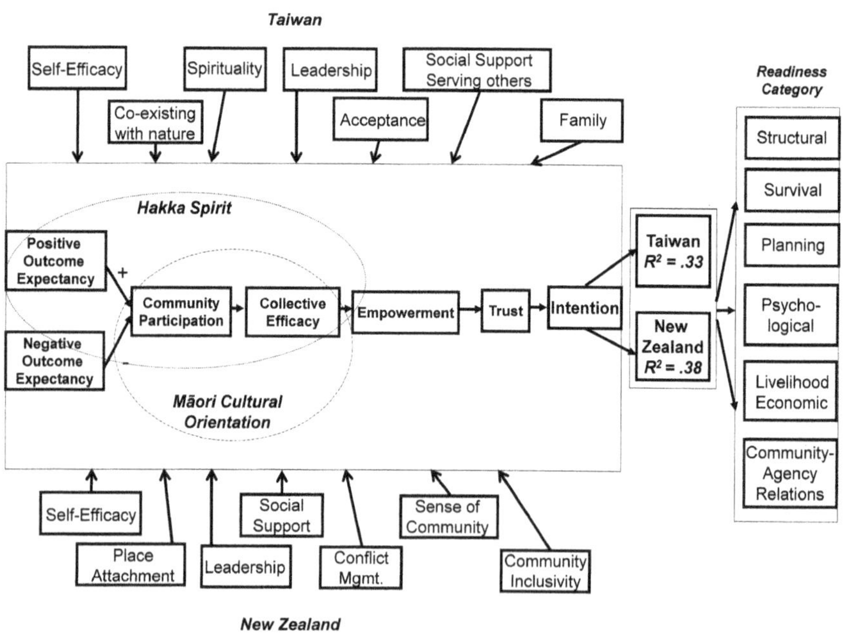

Fig. 1 Summary of the original theory (*central box*), readiness categories, and variables that inform the development of earthquake readiness and recovery theory based on recovery analysis in Taiwan (*top half*) and New Zealand (*bottom half*)

readiness to include attention being paid to safeguarding agricultural activities (e.g., identifying alternative ways of transporting goods to markets).

The analysis of survivor's accounts of readiness needs from culturally diverse countries reinforces the validity of the theoretical readiness inventories introduced above and add some additional categories. Collectively, the work discussed above identifies a need for readiness to encompass structural, survival, planning, capacity building, psychological, livelihood, and community-agency relationship readiness (Fig. 1 right hand column). However, while people experiencing earthquake events readily endorse the importance of readiness, this is not necessarily emulated when DRR strategies seek to facilitate readiness prior to earthquakes occurring.

Despite the considerable time, effort and expenditure invested in earthquake public education programs, levels of readiness for earthquakes typically remains low (Lechliter and Willis 1996; Lindell et al. 2009; Paton et al. 2014; Paton and McClure 2013). Finding considerable diversity in existing levels of readiness introduces the second issue discussed in this chapter; the need to account for why some people prepare comprehensively, some much less so and others not at all.

3 Accounting for Differences in Earthquake Readiness

Several theories (e.g., Duval and Mulilis 1995; Lindell and Hwang 2008; Paton et al. 2005) have been applied to earthquake readiness. However, the present discussion, given that one aim of this chapter is to discuss earthquake readiness from an Asia-Pacific perspective, focuses on a theory (Paton 2008) that has been tested across cultures. This theory acknowledges that an important aspect of the context in which people are called upon to prepare is uncertainty.

Large-scale earthquakes are infrequent. This means people generally have no or only limited experience of earthquakes and very few, if any, opportunities to independently test the validity of the information about risk and readiness (that come from expert sources and delivered via public education programs) for themselves. Under these circumstances, whether or not people act is a function not only of the quality of the information and advice made available to them but also of the extent to which recipients trust formal sources of information. Levels of risk acceptance and people's willingness to take responsibility for their own safety is increased, and decisions to take steps to actively manage their risk more likely, if people believe that their relationship with risk management agencies is fair and empowering (e.g., agencies are perceived as trustworthy, as acting in the interest of community members) (e.g., Paton 2008). Paton (2008) developed a community engagement theory that encapsulated how empowerment and trust, and their personal and social precursors, influenced readiness decision making under uncertainty.

At the person-level of analysis, outcome expectancy plays an important role in how people deal with uncertainty associated with earthquakes and earthquake readiness. Public education programs essentially advise people that if they adopt a particular behaviour the outcome will be a reduction of their risk or an increase in their safety. However, in the uncertain environment in which people are being asked to prepare, people interpret this information (e.g., because they cannot readily test the veracity of the information for themselves) and its recommendations based on beliefs (e.g. often derived from media coverage that typically emphasizes the catastrophic nature of earthquakes and their uncontrollability) rather than experience to estimate whether they believe or expect that outcome (i.e., increased safety, reduced damage, etc.) to occur. The 'outcome expectancy' construct describes this interpretive process.

Paton (2008) argued that if people hold negative outcome expectancy beliefs (belief that earthquakes are uncontrollable and their consequences are so catastrophic that personal action is futile) their likelihood of preparing is reduced. In contrast, holding positive outcome expectancy beliefs (beliefs that personal actions can enhance personal safety and/or mitigate earthquake consequences) motivates people to start to prepare. However, the infrequent and complex nature of earthquakes means that believing that preparing can be effective does not necessarily equate with knowing what to do or how to prepare.

If they are to act to reduce their uncertainty, the theory suggests that people look to other community members and to expert sources for information and guidance about risk and what they could do to manage their risk (Paton 2008). Consequently, outcome expectancy beliefs interact with community-level interpretive processes (community participation and collective efficacy—which serve as contexts for developing socially construct risk beliefs, readiness plans and actions) and interpretive processes at the community-agency relationship level (empowerment and trust) to predict earthquake readiness. This theory (Paton 2008) has demonstrated a capacity to act as a good predictor of earthquake readiness (McIvor et al. 2009; Paton 2013).

As introduced above, the fact that earthquakes occur in culturally diverse countries makes it important to test the validity of a theory in different cultures. A preliminary investigation of cultural equivalence was undertaken by examining whether Paton's (2008) community engagement theory (see above) of preparedness could account for differences in earthquake preparedness in Taiwan. This afforded an opportunity to test the utility of the theory in a highly collectivistic culture (Hofstede 2001). Empirical analysis of the ability of this theory to predict earthquake preparedness in Taiwan (Jang et al., in press) supports the view that this theory has a utility capable of transcending cultural boundaries.

Just as readiness research has found considerable diversity in levels of pre-earthquake readiness, research into post-earthquake response experiences has also found differences between people and communities with regard to their relative ability to deal with recovery demands (Jang 2008; Paton et al. 2014). The existence of such differences provided a foundation for exploring the community capacities and characteristics that influenced recovery (e.g., their ability to develop neighborhood groups, deal with external agencies etc.) and thus an opportunity to test the validity of the predictors in readiness theories and how they might be developed.

One way of further testing the validity of the theoretical predictors is to examine factors that account for differences in the effectiveness of people's earthquake recovery.

4 Accounting for Differences in People's Earthquake Recovery: Cross Cultural Comparison

Using the recovery research undertaken in New Zealand and Taiwan, this section outlines the factors that influenced people's recovery. Discussion starts with a summary of factors that emerged during recovery from the 2011 Christchurch earthquake.

4.1 New Zealand

4.1.1 Individual Factors

In New Zealand (Paton et al. 2014), factors contributing to effective response included confidence in self-help and self-sufficiency, flexibility and adaptability to respond to changing demands, and having a 'mind-set' of getting the job done. The concepts of self-help, flexibility and persistence echo elements of self-efficacy (Bandura 1989), thus identifying an additional variable that could be included in a revised theory. Respondents did not mention outcome expectancy. The effectiveness of people's recovery was also influenced by social or capacity building activities.

4.1.2 Community Factors

Engaging with others and maintaining relationships emerged as important predictors of effective response (Mamula-Seadon et al. 2012; Paton et al., in press), thus validating the role of community participation in recovery. Maintaining relationships contributed to people being able to develop a sense of belonging and identity that served to create feelings of connectedness to people and place and generated a sense of commitment to the recovery and well-being of the community and its members. This adds a need to consider how sense of community, place attachment and social support within a revised theory. Analysis of Christchurch recovery experiences also identified how mutual understanding and support facilitated the development of trust in others (Winstanely et al. 2011). Sense of community was also identified as arising not just from sharing the same fate as others, but from having to work with them to effectively confront and resolve local issues over time.

Groups who had to fend for themselves during the early stages of the response phase discussed how enduring (over several weeks) collaborative efforts with neighbors to meet immediate survival needs contributed to their developing problem solving and planning competencies. This reiterates the importance of collective efficacy as a predictor of adaptation and recovery, and one that plays an important role in facilitating community members' capacity to take responsibility for their own recovery.

Focus group data (Paton 2012) identified how community members' being proactive and willing to take action to deal with emergent needs even when operating under conditions of scarce or unreliable information; developing bottom-up planning and community inclusivity (e.g., conflict management, problem prioritization and collective planning) competencies; and empowering, encouraging and supporting community initiative were seen as important predictors of effective recovery. These observations are consistent with theoretical arguments that community competencies such as active community participation and collective efficacy are predictors of readiness (Paton 2008) and identify a need to develop theory to encompass issues such as facilitating inclusivity and conflict management. Similar capabilities were evident in Māori response. However, the Māori cultural value of 'aroha kitetangata' (love to all people) brought a unique and important cultural element to understanding how community participation influences readiness and recovery in this population (Paton et al. 2014).

A community resource that was pivotal to effective community response was community leaders. Emergent leaders brought people together, facilitated the development and implementation of a coherent community response, and linked the community with external agencies and specialists to secure the resources and help required to meet local needs. Effective leaders were defined as possessing: a strong sense of commitment to helping others within the community; knowledge of how the community works/how to get the community working; good local knowledge (people and place); leadership experience; willingness to take action in novel, challenging circumstances; and having good connections or knowing who/how to call on government and business to assist their community. It is clear from this discussion that community leadership needs to be incorporated into readiness theory as a predictor.

The leadership and community processes discussed above represent resources that contribute to creating empowered communities capable of taking responsibility for their own recovery. However, in complex disaster recovery environments, communities also need information and resources that cannot be met from within the community but only from the wider societal context; from NGOs, government agencies and businesses. The quality of the relationships between communities and NGOs, government and businesses that evolve in the recovery environment will thus influence the degree to which communities are empowered to enact their own recovery initiatives. This introduces a need to consider how relationships with NGOs, government and businesses influence recovery.

4.1.3 Community-Agency/Government Relationships

The relationship between government agencies and businesses and communities can facilitate or marginalize community recovery. Positive community-agency relationships in Christchurch were facilitated by government and relief agencies using community leaders as the conduit into communities. Community recovery was also facilitated by groups being able to work through dedicated local recovery organizations (e.g., CERA, CanCERN) to meet their needs (Mamula-Seadon et al. 2012).

However, evidence of agency marginalizing community response was also evident from focus group participants (Paton 2012) describing the government emergency response as often poorly coordinated, lacking local contacts, and not knowing of local distribution locations. Agencies were perceived to be working in isolation with little coordination of functions, reducing their capacity to empower community initiatives. These illustrate how community-agency relationships can manifest in low empowerment or disempowerment in ways that affect trust and future relationships.

4.2 Taiwan

This section discusses data obtained from interviews with residents in Tung Shih, Taiwan. Most of the residents of Tung Shih are Hakka. Hakka people form the second largest ethnic group in Taiwan and comprise some 20 % of the population. They are descended largely from populations in Guangdong, China. Many Hakka people live in the hills or remote mountain regions. Hakka spirit, the essence of Hakka culture, comprises characteristics such as frugality, diligence, self-reliance, shared responsibility, and persistence (Yang 2004).

Several Tung Shih residents discussed the key role that Hakka spirit played in their recovery following the 921 Earthquake. The Hakka spirit was implicated in generating a sustained sense of purpose and perseverance (which creates a kind of culture-specific self-efficacy) despite hardships and assisted people's efforts to re-build their homes and communities (Jang 2008).

4.2.1 Individual Factors

In Taiwan, some participants, after witnessing how destructive the 921 earthquake was, expressed the belief that earthquake consequences were too catastrophic for personal action to make any difference to peoples' safety (i.e., negative outcome expectancy) (Jang 2008; Paton and Jang 2011). However, a stronger tendency towards a characteristic comparable to positive outcome expectancy was also present. This may reflect an enduring aspect of Hakka people's history of dealing hazard consequences. Jang and LaMendola (2006) discussed how farming practices that evolved to limit the impact of typhoons on people's livelihoods generated enduring outcome expectancy beliefs (regarding the effectiveness of collective actions in overcoming earthquake consequences) that carried over into their beliefs about the value of readiness for earthquakes.

In Taiwan, self-reliance and perseverance (implicit facet of the Hakka Spirit) were significant predictors of readiness. Self-reliance was closely linked to job security and good health and these, in turn, were linked to strengths required to move forward and face daily challenges. Participants emphasized that being self-reliant made them feel happy and have a sense of achievement in how they responded to the earthquake consequences. Perseverance was manifest in respondents stressing that no matter how hard the tests were, life must go on. These characteristics make this facet of the Hakka Spirit comparable to self-efficacy (Bandura 1989) and perceived control of resource availability (Jang and LaMendola 2006).

Being proactive was important. In Tung Shia, survivors discussed how they could not just sit and wait for assistance from the government or NGOs. No matter what happened, community members had to rely on their own strengths to improve the conditions (i.e., empowering internal community action is important). Thus, as in New Zealand, being proactive under conditions of uncertainty is an important predictor of community recovery.

A potentially very interesting addition to readiness theory derived from Hakka beliefs regarding the importance of learning to accept disasters as part of life experience and of living and co-existing harmoniously with nature. Both characteristics had positive impacts on people's response to the 921 earthquake (Jang 2008; Jang and LaMendola 2006).

For those who accept an established fact, they understand that they have no power to change the fact, but are capable of coping with its effects. Participants likened natural disasters to ordeals from Heaven. Participants indicated that because the established fact already happened, they accepted their fate. This sense of acceptance is a kind of cultural fatalism. However, compared with fatalistic attitudes in western populations, which typically undermine readiness (Paton and McClure 2013), in Taiwan this belief manifest itself in increasing the likelihood of residents adopting a more problem-focused coping approach to assist their adaptive recovery (Jang 2008). Focus group participants also discussed how their gratitude to gods and to others within their community helped them cope in times of need and facilitated their ability to focus on the positive aspects of their experience. The inclusion of religiously-oriented gratitude introduces the prominent role spirituality played in recovery in Taiwan.

Spirituality offers explanations of death, life, loss, and natural disasters. It brings comfort to participants. Nathanson (2003, p. 63) asserts that "spirituality is an important force in recovery," because it helps the survivors identify inner strengths and find meaning in people's sufferings and behaviors. In Taiwan, participants indicated that religious beliefs helped them cope with challenges, be more optimistic, and to see the positive aspects of their experience (e.g., how everyone pulled together to promote the recovery of the whole community). Being able to impose meaning on earthquake events and accept the established fact of loss contributed to respondents adopting problem-focused coping to deal with their loss and to assist their problem solving.

Spirituality guides norms and people's values. Taiwanese people often use Poe divination to denote gods' will (Jordan, n.d.). Participants claimed that religious beliefs encouraged people to work on personal merits and practice moral culture. This reflects how spirituality in Taiwan is a way of life, and one that promotes shared meanings and understandings that facilitates resident's acceptance of earthquakes and positive interpretations and actions (Jang and LaMendola 2007). Spirituality influenced residents' acceptance of earthquake consequences, enhanced their sense of community, and strengthened their social support networks.

4.2.2 Community Factors

Participants indicated that social support networks played a key role in helping them deal with the ordeals encountered. Those who re-established their social support networks tended to be more resilient than those who did not. A related theme was the role serving others played in making people feel needed, wanted and belonging (cf. sense of community). It brings people a sense of being useful to others and helps survivors develop a better understanding of the meaning and values of life. The process of serving others both contributed to being able offer meaningful help to others and assisted survivors themselves to grow. This culturally inherent reciprocal helping contributed to survivors gaining new meaning and explanation for life after disasters (see also Farley 1998). An essential source of social support networks came from family.

Family life was significantly affected. Family change was evident in the marriage rate increasing in affected areas in the year following the 921 earthquake. An interesting cultural change dynamic, and one that linked family and livelihood processes, emerged as a result of the fact that many women from affected families became bread winners (Lo 2010). Women tended to be more resilient than men in his regard. They were willing to take whatever job was available. Men struggled with finding positions similar to those they held before the disaster. Sense of responsibility to family and community is a key trait within the Hakka spirit. It is also a key factor influencing coping style. Participants with a sense of responsibility toward their families and communities were more likely to apply a problem-focused coping style to their recovery efforts (Jang 2008).

Taiwanese respondents also endorsed the key role of community processes and community leadership as drivers of effective recovery. Community leaders were described in terms of their ability to understand local needs and priorities and for their role in local response planning and emergency operations (Jang and LaMendola 2006). As in New Zealand, this did not mean that external agencies were seen as unimportant.

Residents regarded themselves as experts with regard to knowing their community concerns and practical and coping resources needed to respond to meet local needs. External experts were described as being technical assistants, information providers, and resource brokers (this reinforces the role of empowerment in the readiness theory). At the same time, residents acknowledged the importance of being able to secure appropriate (relative to their needs) resources from sources external to their community.

4.2.3 Community-Agency/Government Relationships

Community relationships with external agencies, particularly with government sources, provided evidence of both facilitation and marginalization. In Taiwan, the Government created temporary positions for survivors. However, people who were in need seldom benefited from that policy. Community members reported how it was people with connections with local government that took those positions (Jang 2008). The nature of the community-government relationship could thus marginalize or disempower residents. However, facilitation and empowerment was also evident in relationships with other agencies. Temporary housing (prefabs) was built by religious groups (in Taiwan, religious groups are major sources of community services) and NGOs to provide shelters for survivors, with the inclusion of survivors in reconstruction projects (e.g., being inx (re)building their own houses) providing examples of empowerment and facilitation. Facilitation was also evident in the government providing free loans to assist these rebuilding efforts. Developing relationships through joint ventures with local people, local government, Farmers Associations and other occupational associations, non-profit organizations, and religious groups were identified as important for effective readiness (community-agency readiness).

5 Conclusion

The comparative New Zealand-Taiwan analyses offer some support for the view that a theory developed and initially tested in occidental contexts can be adopted for use in oriental contexts. This cross cultural equivalence has several theoretical and practical benefits. For example, it would provide a common theoretical and practical foundation for collaborative learning and research on earthquake readiness and intervention across national borders and make such findings available to countries that lack the resources to undertake this work themselves.

The discussion illustrates how everyday or mainstream community and cultural competencies and characteristics (e.g., the community participation and collective efficacy variables reflect the outcomes of people's accumulated experience of community life) influence people's ability to cope with, adapt to and recover from earthquakes. This illustrates the potential benefits that could accrue from integrating risk management and community development to develop cost-effective, sustainable readiness strategies (Paton and Jang 2011). However, the discussion also highlighted a need to accommodate cultural differences.

Cultural studies distinguish between etic (culturally universal) and emic (culture specific) processes. While the cross cultural analyses discussed above identified how the same process could operate across different countries (i.e., it describes an etic process), this does not mean that the process is enacted in the same way in different cultures. For example, community participation was important in each country. However, the origins of participatory activities and how they are sustained and enacted to influence people's risk beliefs and preparedness options differ from country to country. For example, the Hakka Spirit describes a unique and culturally implicit set of cultural beliefs and practices that underpinned the implicit application of qualities such as self-reliance, responsibility, persistence, reciprocal support and collaborative problem solving to earthquake recovery. In contrast, in New Zealand, the community participation and collective efficacy scores reflect community experiences, with their presence and use in earthquake readiness activities being more dependent on community development than might be the case in Taiwan. Thus emic processes (i.e., how and why participation occurs in a specific culture) capable of affecting earthquake preparedness need to be identified. Similar issues apply with regard to understanding how Māori cultural values can inform readiness theory and intervention.

The comparative analysis discussed here reiterated the need for readiness to encompass structural, survival readiness and highlighted the need to expand the readiness repertoire to include psychological, livelihood/economic and community agency categories (Fig. 1). Drawing on the inherent variability evident in how people and communities recovered, it was possible to confirm the validity of existing readiness theory (e.g., confirm roles for outcome expectancy, community participation, collective efficacy, empowerment and trust) and illustrate the need to expand the range of variables included. Examination of how people and communities coped with and adapted to the physical, community and societal recovery challenges they encountered identified the need to include self-efficacy, place attachment, community leadership, inclusivity, sense of community, and conflict management variables in an expanded readiness theory.

The issues canvassed above are summarized in Fig. 1. The central part of Fig. 1 summaries the components of the original theory and its role in predicting readiness in New Zealand and Taiwan. The right hand side of Fig. 1 lists the readiness categories. The top half of Fig. 1 lists the variables suggested by the Taiwan analysis that could be incorporated into theoretical conceptualizations of readiness, and the bottom represents those variables suggested by the New Zealand analysis that could contribute to the development of comprehensive earthquake readiness theory.

References

Bandura, A. (1989). Human agency in social cognitive theory. *American Psychologist, 44*, 1175–1184.

Bannister, S., & Gledhill, K. (2012). Evolution of the 2010–2012 Canterbury earthquake sequence. *New Zealand Journal of Geology and Geophysics, 55*, 295–304.

Bruijn, K. M. D. (2004). Resilience indicators for flood risk management systems of lowland rivers. *International Journal of River Basin Management, 2*, 199–210.

Carpenter, S., Walker, B., Andries, J. M., & Abel, N. (2001). From metaphor to measurement: Resilience of what to what? *Ecosystems, 8*, 941–1044.

Disaster Response Report. (2008). *Ministry of the interior department of statistics*, Taiwan. http://www.moi.gov.tw/stat/index.aspx. Accessed 28 Dec, 2008.

Duval, T. S., & Mulilis, J.-P. (1995). A person-relative-to-event (PrE) approach to negative threat appeals and earthquake readiness: A field study. *Journal of Applied Social Psychology, 29*, 495–516.

Farley, J. E. (1998). Down but not out: Earthquake awareness and readiness trends in the St. Louis metropolitan area, 1990–1997. [survey]. *International Journal of Mass Emergencies and Disasters, 16*, 303–319.

Gaillard, J.-C. (2007). Resilience of traditional societies in facing natural hazards. *Disaster Prevention and Management, 16*, 522–544.

Hofstede, G. (2001). *Culture's consequences: Comparing values, behaviors, institutions and organizations across nations*. Thousand Oaks CA: Sage.

International Strategy for Disaster Risk Reduction (ISDR). (2005). *Hyogo framework for action 2005–2015: Building the resilience of nations and communities to disasters*. Kobe: International Strategy for Disaster Reduction.

Jang, L. (2008). *Natural disasters: Effects of cultural factors on resilience*. North Charleston: VDM Verlag Dr Muller Aktiengesellschaft & Co KG and Licensors.

Jang, L., & LaMendola, W. (2006). The Hakka spirit as a predictor of resilience. In D. Paton & D. Johnston (Eds.), *Disaster resilience: An integrated approach*. Springfield: Charles C. Thomas.

Jang, L., & LaMendola, W. (2007). Social work in natural disasters: The case of spirituality and posttraumatic growth. *Advances in Social Work, 8*, 67–78.

Jordan, D. K. (n.d.) *The traditional Chinese family and lineage*. http://weber.ucsd.edu/~dkjordan/chin/hbfamilism-u.html. Accessed 14 June 2004.

Klein, R., Nicholls, R., & Thomalla, F. (2003). Resilience to natural hazards: How useful is this concept? *Environmental Hazards, 5*, 35–45.

Lechliter, G. J., & Willis, F. N. (1996). Living with earthquakes: Beliefs and information. *The Psychological Record, 46*, 391.

Lindell, M. K., Arlikatti, S., & Prater, C. S. (2009). Why people do what they do to protect against earthquakes risk: Perceptions of hazard adjustment attributes. *Risk Analysis, 29*, 1072–1088.

Lindell, M. K., & Hwang, S. N. (2008). Households' perceived personal risk and responses in a multi-hazard environment. *Risk Analysis, 28*, 539–556.

Lo, J. C. (2010). The impact of the Chi-Chi earthquake on demographic changes: An event history analysis. In S. Kurosu, T. Bengtsson, & C. Campbell (Eds.), *Demographic responses to economic and environmental crises* (pp. 193–203). China/Japan: Kashiwa.

Mamula-Seadon, L., Selway, K., & Paton, D. (2012). Exploring resilience: Learning from Christchurch communities. *Tephra, 23*, 5–7.

McIvor, D., Paton, D., & Johnston, D. M. (2009). Modelling community preparation for natural hazards: Understanding hazard cognitions. *Journal of Pacific Rim Psychology, 3*, 39–46.

Nathanson, I. (2003). Spirituality and the life cycle. In T. Tirrito & T. Cascio (Eds.), *Religious organizations in community services: A social work perspective* (pp. 63–77). New York: Springer.

Norris, F. H., Stevens, S. P., Pfefferbaum, B., Wyche, K. F., & Pfefferbaum, R. L. (2008). Community resilience as a metaphor, theory, set of capacities, and strategies for disaster readiness. *American Journal of Community Psychology, 41*, 127–150.

Paton, D. (2006a). Disaster resilience: Integrating individual, community, institutional and environmental perspectives. In D. Paton & D. Johnston (Eds.), *Disaster resilience: An integrated approach.* Springfield: Charles C. Thomas.

Paton, D. (2006b). Disaster resilience: Building capacity to co-exist with natural hazards and their consequences. In D. Paton & D. Johnston (Eds.), *Disaster resilience: An integrated approach.* Springfield: Charles C. Thomas.

Paton, D. (2008). Risk communication and natural hazard mitigation: How trust influences its effectiveness. *International Journal of Global Environmental Issues, 8*, 2–16.

Paton, D. (2012). *MCDEM Christchurch community resilience project report.* Wellington: Ministry of Civil Defence and Emergency Management.

Paton, D. (2013). Disaster resilient communities: Developing and testing an all-hazards theory. *Journal of Integrated Disaster Risk Management, 3*, 1–17.

Paton, D., & Jang, L. (2011). Disaster resilience: Exploring all-hazards and cross cultural perspectives. In D. Miller & J. Rivera (Eds.), *Community disaster recovery and resiliency: Exploring global opportunities and challenges.* London: Taylor & Francis.

Paton, D., Johnston, D., Mamula-Seadon, L., & Kenney, C. M. (2014). Recovery and development: Perspectives from New Zealand and Australia. In N. Kapucu & K. T. Liou (Eds.), *Disaster and development: Examining global issues and cases.* New York: Springer.

Paton, D., & McClure, J. (2013). *Preparing for disaster: Building household and community capacity.* Springfield: Charles C. Thomas.

Paton, D., Smith, L., & Johnston, D. (2005). When good intentions turn bad: Promoting natural hazard readiness. *Australian Journal of Emergency Management, 20*, 25–30.

Pelling, M., & High, C. (2005). Understanding adaptation: What can social capital offer assessment of adaptive capacity? *Global Environmental Change, 15*, 308–319.

Russell, L. A., Goltz, J. D., & Bourque, L. B. (1995). Readiness and hazard mitigation actions before and after two earthquakes. *Envionmental and Behavior, 27*, 744–770.

Triandis, H. C. (1995). *Individualism and collectivism.* Boulder: Westview.

Winstanely, A., Cronin, K., & Daly, M. (2011). Supporting communication around the Canterbury earthquakes and other risks. GNS Science Miscellaneous Report 2011/37. 39 p.

Yang, W. S. (2004). *A census of the Hakka population in Taiwan.* Taipei: Council for Hakka Affairs, Executive Yuan, R. O. C. (in Chinese).

Geoethics, Neogeography and Risk Perception: Myth, Natural and Human Factors in Archaic and Postmodern Society

Francesco De Pascale, Marcello Bernardo, Francesco Muto, Alessandro Ruffolo and Valeria Dattilo

1 Natural Catastrophes. Risk and Loss of Presence

Since the dawn of civilisation, in every instant of everyday life from birth to death and in all cultures, man is exposed to the risk of not being-there, that is, to the risk of catastrophe hitting him or the world around him. This may occur in connection to economic and social mutations, for example in times of war, or to the unpredictability of natural catastrophes which are out of human control, for example seaquakes.

Taking this as our starting point, we will analyse the crucial matter of the crisis or loss of presence, that is, the risk of not being-there in critical moments of historical existence, limiting ourselves to consideration of forms of defense from risk represented by natural catastrophe (for example, seaquakes and volcanic eruptions) amongst so-called primitive people, from an anthropological-physical point of view.

We will look at the historical-religious thought of Italian philosopher Ernesto de Martino (1908–1965) and in particular some of his critical lectures published posthumously in *La fine del mondo. Contributo alle analisi delle apocalissi culturali* (1977). We will treat philosophical concepts like anthropological evidence

F. De Pascale (✉) · M. Bernardo · F. Muto · A. Ruffolo · V. Dattilo
University of Calabria, 87036 Rende, CS, Italy
e-mail: francesco.depascale@unical.it

M. Bernardo
e-mail: bernardo@unical.it

F. Muto
e-mail: francesco.muto@unical.it

A. Ruffolo
e-mail: alessandro.ruffolo@gmail.com

V. Dattilo
e-mail: valeria.dattilo@gmail.com

S. D'Amico (ed.), *Earthquakes and Their Impact on Society*,
Springer Natural Hazards, DOI 10.1007/978-3-319-21753-6_29

with the aim of identifying different mechanisms of defense from the risk of not being-there, even in cultures very distant from Western ones. We will specifically consider apocalyptic representations connected with experience of natural catastrophe in traditional cultures.

One of the examples de Martino uses comes from an essay by Rudolf Lehmann, *Weltuntergang und Welterneurung im Glauben schriftloser Völker*. The author is cited by de Martino because his essay contains a description of some apocalyptic stories witnessed by Stephan Lehner amongst indigenous people, and subsequently cited by Lehmann. The natives of Namatanai in central New Meklenburg, in the Laur region, describe future cosmic catastrophes with reference to already experienced natural catastrophe:

> the sea that will swallow up the island for ever (seaquakes), the obscuring of the sun and the endless night (ashfall during volcanic eruption) (de Martino 1977).

Another example of apocalyptic representation is found amongst the Kenta or Kintak Bong' and amongst the indigenous people of the Caroline islands:

> Namoluk executes the end of the world upon order of the supreme being enraged by man's guilt, and using thunder, hurricane, lightning and two supreme beings, one of whom squeezes the earth in all its parts whilst the other throws big rocks from the sea to the earth. Total destruction then ensues (de Martino 1977).

We can see from these stories that the end is represented as destruction—the idea of the world collapsing—referring to already experienced cosmic catastrophes (seaquakes, volcanic eruptions), and alluding, as explanation of such catastrophes, to an element recurring not only in primitive traditions: punishment on the part of a supreme being, for faults deriving from having transgressed his orders. Even James George Frazer (1854–1941), one of the founding fathers of social anthropology, in his main work *The Golden Bough* (1911), shows how so-called primitives sometimes believe, as in the above-mentioned case of Namoluk, that their safety and even that of the world are connected to the life of one of these man-gods or human incarnations of deities. Through his work we can point to cases of killing of the man-god because of the gradual waning of his powers and their extinction with death. This is the case in some African tribes, where murder is necessary to avoid catastrophe or the world collapsing with the fall or ruination of the man-god:

> The people of the Congo believed, [...] that if their pontiff the Chitomé were to die a natural death, the world would perish, and the earth, which he alone sustained by his power and merit, would immediately be annihilated. Accordingly, when he fell ill and seemed likely to die, the man who was destined to be his successor entered the pontiff's house with a rope or a club and strangled him or clubbed him to death (Frazer 1911).

The tradition of sacrifice to avoid catastrophe, like that of murdering the godly kings—considered as incarnate deities upon whom the wellbeing of men and of the world depends—at the first signs of illness or old age, seems to have prevailed in this part of Africa until modern times. Individual destiny as well as the destiny of the world were linked to the destiny of the sacred or divine gods:

> To guard against these catastrophes it is necessary to put the king to death whilst he is still in the full bloom of his divine manhood, in order that his sacred life, transmitted in unabated force to his successor, may renew its youth, and thus by successive transmissions through a perpetual line of vigorous incarnations may remain eternally fresh and young, a pledge and security that men and animals shall in like manner renew their youth with a perpetual succession of generations, and that seedtime and harvest, and summer and winter, and rain and sunshine shall never fail (Frazer 1911).

Going back to de Martino, the Italian philosopher identifies in *repetition* the characteristic behaviour of so-called primitives faced with the risk of not being-there—i.e. the risk of *the end of the world* and of man. Repetition of what? Of certain critical episodes (the first catastrophe, the first hunt, the first fishing, the first giving birth) or of certain critical passages, such as that from chaos to cosmos. This series of acts or episodes have the value of *archetypes*, of models, exemplary acts—not necessarily linked with religion—that are repeated for survival in light of their perceived protective function. This enables a series of mechanisms of defense of presence at critical moments:

> ritual repetition operates a function of reenactment and reintegration relative to possible critical episodes and the darkness of current or future existence; ritual repetition operates a function of concealment or abatement of the historicity of becoming, in order to experience possible critical moments as if they had already been experienced with the desired outcome: where we are in history "as if" we were not there (de Martino 1995).

All this implies a process that de Martino defines *mythical-ritual dehistorification*. By this technique the Italian thinker refers to the suspension, the abolition of history, through the mythical-ritualistic symbol or mechanism:

> The dehistorification of a critical moment of existence is, first of all, the "myth" of such a moment. Secondly, it is the possibility of repeating the myth every time that a certain critical moment comes about; it is thus ritual (de Martino 1995).

In *La fine del mondo*, de Martino defines the use of this mechanism on the part of the primitives as a «behaviour that always re-conduces the historical "now" to a metahistorical "then", which is also "then and never again"» (de Martino 1977). The *now* is the contingent event, the *hic et nunc* represented by the critical moment (for example, the flood, the seaquake or the volcanic eruption) which is reconduced to the *then and never again*, that is to the origins of that event which was for the first time founded and resolved, not by man but by gods, reabsorbing the proliferation of critical moments into the course of an always selfsame metahistorical reality. On this subject, de Martino draws upon Helmut Preti's description of the belief in Ungud, the myth of the rainbow-serpent, amongst some native Australian tribes of the Kimberley:

> Ungud, the mythical rainbow-serpent, has its *camp* in the earth or in the depths of water caves. Creator of the world, origin of all living and growing things on earth, unitary divine essence that in the beginning of time made the world emerge from chaos, extracted living beings from the earth, or with rain sent them on earth. […] In the rainy season Ungud makes the water flow in river-beds and creeks. Then it rises to the sky as colourful rainbow-serpent to devour the rains, that is to stop them (de Martino 1977).

This example of the myth of Ungud, the rainbow-serpent of the natives of Kimberley, rests upon ritual action. If certain rites are not performed then Ungud, divine essence and force for the good, can invoke all destructive catastrophes of rain and flood, thus plunging the world into chaos and disorder. The view that the world is sustained by these exemplary behaviours is characteristic of these cultures: disorder—for example the above-mentioned risk of floods in the rainy season—is kept away through a ritual action that always reinscribes the "historical now", that is, the event, into the "then and never again", that is, into the mythical model, a meta-historical order, founded in *illo tempore* and guaranteed by a force:

> For a collective of hunter-gatherers, in a land where the means of sustenance largely depend on the rainy season and the quick springing of vegetation after a long drought, in a regime of existence in which water – both fertile rain and thirst quenching liquid – is at the center of existential possibilities, we can understand why order is represented by the rainbow-serpent: the image draws a poignant and emotional connection between the sky from which the rain falls and earth at the moment in which the rains stop and vegetation is about to explode. The image excludes […], both drought and flood: it performs this exclusion because it is a metaphysical order, founded in *illo tempore* and guaranteed by a strength that, at the same time, sits in the crystal of the sky […]. Without doubt historical order is much more problematic, there might be drought or flood, plants can dry up and animals be lacking beyond seasonal limits: but the real event becomes bearable and surpassable as long as its historicity is concealed, masked, excluded from the dominant cultural consciousness (de Martino 1977).

In *The Myth of the Eternal Return. Architypes and Repetitions*, Romanian born philosopher Mircea Eliade (1907–1986), the main historian and theorist and philosopher of rituals and religious experience as anthropological experience, also studies the mythical-ritualistic mechanism in terms of a repetition of the *arché*, the *quid* from which everything begins and derives, dominant in the primitive world. Both de Martino and Eliade accept the centrality of *repetition* in the mythical-ritualistic mechanism; both accept the idea that without repetition there exists no history, that which Mircea Eliade defines *exemplary history*, which can be repeated, finding in its repetition its very sense and value:

> The interest in the "irreversible" and the "new" in history is a recent discovery in the life of humanity. On the contrary, archaic humanity, […], defended itself, to the utmost of its powers, against all the novelty and irreversibility which history entails (Eliade 1945).

But contrary to Eliade who considers the cyclic nature and repetition as the truth of human time, de Martino goes in a different direction, considering repetition episodic—a remedy to critical moments of existence in which there is a rupture between the grammatical plane and the empirical plane, between the metahistorical and historical planes. It is at such critical moments that primitive or archaic societies feel the need to repeat certain discourses, certain facts or episodes in life that bear an important meaning for the individual and society—the first hunt, the first storm, the first volcanic eruption, the first seaquake—that constitutes a model or archetype. In linguistic terms one could say that a concrete fact of life regarding the empirical plane, that which Wittgenstein in *On Certainty* (1969) defined with the expression *movement of the waters*, becomes the norm, it is thus converted into a *rule*, into *river-bed*:

> I distinguish between the movement of the waters on the river-bed and the shift of the bed itself; though there is not a sharp division on the one from the other (Wittgenstein 1969).

A concrete fact of life thus comes to coincide with the rule, with the origin (the first storm, the first seaquake). The possibility of ritual as repetition of a mythical model allows primitive man to exert some control, to overcome a certain critical situation, not on the worldly or profane plane, but on the mythical-ritualistic plane, that is, on the plane of the already-happened in metahistory, not by the hand of men but by gods, in order to start over, once again:

> the mythical-ritualistic mechanism nonetheless presents an ambivalent character because it appears as the "undecided that is decided upon" and at the same time as the "decidable which is brought back to the already decided upon": its character is in the dynamic for which the undecided is recalled and reopened to decision, and the currently decidable is resolved in the exemplary decision of the beginnings (de Martino 1977).

But the novelty of de Martino is not that of having read the myths or having told them, rather it is that of having interpreted myths and rites *philosophically*. De Martino's interest in the intimate, *ambivalent* structure of the mythical-ritualistic mechanism, marks—according to the Italian philosopher himself, a fracture, a hiatus with Eliade's interpretation:

> The protective (and thus technical) function of the mythical-ritual mechanism is twofold: on the one hand it protects individual presence from the unrelated presence of unsurpassed situations and from the current crisis of alienation; on the other hand, it protects the presence of the historicity of the human condition. The mythical-ritual mechanism performs this twofold function thanks to the institution of a metahistorical plane that operates as a space for acceptance and reenactment of an unrelated return of the past, and at the same time as a plane of dehistorification of the proliferation of historical becoming (de Martino 1977).

The mythical-ritualistic moment thus *reestablishes* historical experience, but first of all it *dehistoricises*, suspends, puts history within parenthesis, recalling and *reestablishing* an ancient condition, a situation of departure which we have abandoned. But why do we suspend or abolish history through mythical-ritualistic repetition? Because he who repeats the archetype does not think he is imitating a certain historical moment, something that once happened. Rather, it is a simultaneity, a contemporaneous identification with that time, reactivating that *illud tempus*, very similar to the phenomenon of *déjà vu*, of the already-experienced, in which we are overcome by a sense of re-living a past moment. It is a human tendency, for which we tend to remember a historical event not as accomplished or performed by this or that person but in terms of an example of an event, in terms of exemplarity. Further, the Neapolitan thinker remembers how the possibility of making and unmaking time, assembling it and disassembling it, is a possibility which belongs also to the discipline of *Psychoanalysis*, or Depth Psychology, as a technique as well as to archaic thought as described above:

> The view that psychoses are, in general, repetitions and returns of an archaic stage of the psyche's development was put forward maybe for the first time by a Romantic doctor, Carlo Gustavo Carus, in his *Vorlesungen über Psychologie* (1831) (de Martino 1948).

Freud's thought is useful on this point. We refer mainly to his neurophilosophical essay *Beyond the Pleasure Principle* (1920):

> A drive might [...] be seen as a *powerful ten*dency in every living organism to restore a prior state, which prior state the organism was compelled to relinquish due to the disruptive influence of external forces; [...] it must aspire to an *old* state, a primordial state from which it once departed, and to which via all the circuitous byways of development it strives to return (Freud 1920).

Restoring a prior state is the key word that, in our view as well in de Martino's, characterises the *eternal return*. In both cases we repeat the primordial—childhood for the patient in analysis, cosmogony for archaic people:

> Psychoanalysis too as a technique entails a return to the past. But in the context of modern spirituality, and accordingly with the Judaeo-Christian conception of irreversible historical time, the primordial could be nothing other than early childhood, the only real individual *initium*. Psychoanalysis thus introduces historical and individual time in therapy. The psychic patient like he who suffers following a shock suffered in its temporal duration, of a personal trauma that took place in the primordial *illud tempus* of childhood. A trauma forgotten or rather, that never surfaced to consciousness. Therapy thus consists in "going backwards", in inverting time in order to bring the crisis up to the present time, in re-living the psychic trauma and reintegrating it into consciousness (de Martino 1977).

In both cases, repetition has an apotropaic value—it protects us from a psychic trauma, from the risk of presence not being-there or from a critical moment. The difference is that in Freud's theory, the phenomenon which replaces the archetype—the original that is constantly reenacted—are *shocks*, unpleasant experiences or acts which are repeated with the aim of appropriating that which is unpleasant; in the case of Eliade's theory on the other hand they are acts which have nothing intrinsically traumatic about them, if anything they are instances of complete success (for example, the first hunt). They are thus entirely positive models that are reenacted, repeated at critical moments of historical existence, when faced with a change for example, placing he who repeats in the same original time in which the prototype or model is placed. Not a simple reminder but a going back to the original place.

De Martino defines these critical moments, limit situations in which the risk of losing presence and remaining imprisoned in a *present* intended as *nunc stans*, as an a-chronous or atemporal situation, in a quantic, immutable and *eternal* world. These are situations of great suffering and deprivation, for example situations of mourning, or of war, a drought, a catastrophe. In these critical moments, the limit situation—earlier on in the article we referred for example to the experience of natural catastrophe in some native tribes of Namatanai and the Caroline islands—can be hierophanised, that is, re-conduced to its origins, *ab illo tempore*:

> The critical moment of existence where presence is faced with change, whether it be a change upon which presence has some control (for example, the killing of the beast), a

> change that announces itself without presence having any control over it (for example, the storm), or an irrevocable change with which presence is presented (for example, a death). History in critical moments reveals itself through the affective tension which imposes them on presence (de Martino 1995).

In other words, the critical moment imposes a *decision*, a quick adaptation to reality: the hunter before the beast, the farmer before the storm, and so on. In current language, we are used to associating *decision* with the generic notion of will, sometimes even using the two terms as synonymous of one another. Yet, "to decide" has nothing to do with aristocratic will intended as an elitist activity, an aristocracy of spirit or intellect—the mechanism of human decision-making is often connected to risk.[1] *To decide*, the etymology of which derives from the latin *decaedere*, "cut off", and resonates also in contemporary words such as "truncate", "interrupt", is a modest action or operation of adaptation to the environment, to the society which is necessary to our survival and has nothing to do with free will. Without "truncating", "interrupting", "cutting short", the animal who has language would not be able to survive but would be lost to himself and to the world. It is before a critical moment that this great uncertainty, by which we are overcome when we find ourselves faced with risk or in a critical moment "in which it seems like there is nothing we can do", that man develops a series of practical strategies necessary to interrupting the unlimited that unsettles our life and stops us from acting, that which the Greeks called *ápeiron* (ἄπειρον). The main strategies are the function of dehistorification of myths and that of the rituals of primitive societies, defined by de Martino as a technique that mediates reintegration into history. The mythical-ritualistic mechanism deflects risk, placing it on the metahistorical plane, yet it does not eliminate it:

> In these conditions, what is culturally relevant is the mythical-ritualistic protection of wordly affairs – the particular system of techniques that constitute the sphere of the "sacred", of "magic", of "religion" (de Martino 1995).

Amongst contemporary theoretical positions on this topic, we want to mention the work of Paolo Virno, *Ripetizione dell'antropogenesi* (Repetition of Anthropogenesis), in *Quando il verbo si fa carne. Linguaggio e natura umana* (2003) (When the verb becomes flesh. Language and human nature). Virno's merit is that of showing how belief in periodic destruction and regeneration of the cosmos, defined by de Martino as "one of the philologically most ancient of human behaviours", has to do with the biological configuration of our species which, in the absence of specialised instincts, has to deal with a partially undetermined existential context—the world we inhabit—and not an environment in which all details are predictable:

> Slightly radicalising de Martino's diagnosis, we might say: human nature consists in always having to deal with the origins of man as a particular species. Or: a distinctive feature of

[1]On risk intended as something calculable and always implicated in human decision-making, see Andrea Tagliapietra, *Filosofia. Il rischio e il limite*, 2012.

> *anthropos* is the constant *repetition of anthropogenesis*. The inaugural act does not automatically sink into an already archived "otherwhen" but always remains in the foreground, implicated in all concrete articulations of social and political praxis. Prehistory is wedged in every single historical moment (Virno 2003).

It could thus be said that through rituals, the genesis of the species is re-lived. Virno, however, takes issue with de Martino's notion of the *ethos of transcendence*. According to the Italian philosopher, the mechanism sustaining repetition cannot be the *ethos of transcendence* of to *having-to-be-there*:

> De Martino believes that the repetition of anthropogenesis is ascribable to a bizarre categorical imperative: that the being-in-the-world, in and of itself liable and subject to catastrophe, is itself supported by a prior and foundational *having-to-be-there*; an always precarious *Dasein* is protected or reaffirmed once again (when it is, to be sure) by an ethical impulse which makes of presence itself a "value" (Virno 2003).

Let us read de Martino's definition of the *ethos of presence*:

> The primordial ethos of transcendence, in its tension between situation and value, between the risk of not-being-in-the-world and reintegration-into-the-world, between passing (*passare*) and letting pass (*far* passare) – and thus overcoming (*oltrepassare*) – between immediate identification with the situation and detachment from it, is an all-encompassing and not further derivable experience: all analysis takes place within this experience, as clarification of the moments of overcoming (and thus of the dangers of the impossibility of overcoming and of the struggle to overcome) (de Martino 1995).

Contrary to de Martino, according to whom the process of human evolution or of anthropogenesis can be attributed to a categorical imperative, that is, to the *having-to-be-there* or *ethos of transcendence*, Virno goes in another direction. According to the Italian philosopher, the repetition of anthropogenesis calls in question not ethics but ontology, that is, the very biological constitution of our species:

> The renegotiation and re-edition of prerogatives characteristic to the *Homo sapiens* cannot depend on the state of health of the "ethos of transcendence". As often happens, the notion of the having-to-be-there betrays (*tradisce*) — in the double sense of the word: obliquely reveal and distort — a matter that has to do instead with *being* as such (Virno 2003).

To conclude: the structure supporting repetition is the biological configuration of our species, that is, the innate and undefined nature of the human animal, of his endless childhood which goes by the name of *neoteny* and that in some way recalls the anthropology of Herder and that of Gehlen, important and valid to understand the way in which the human animal adapts to the context of existence, and to understand the way in which man reacts before certain unpredictable events such as an earthquake. This notion is important to understand the way in which the so-called primitives reacted to such events through the development of "end of the world" representations, the evocation of which was reserved to rituals. It is also important to understand how traditional mythical-ritualistic behaviors all over the world are not exclusive to primitive societies but echo today in the everyday life of contemporary societies every time they are faced with a risk, reenacting the dialectic between risk of loss of presence and its periodical reaffirmation.

2 True Risk; Danger, Vulnerability and Exposed Value

Ugo Leone suggests that the term 'risk' is most appropriately used when we talk about natural phenomena which, when they occur, are capable of inflicting great harm upon people and property. In the case of environmental risk, according to the definition made by the Office of the United Nations Emergency Relief Coordinator, 'risk' is the possible loss of value of one or more elements (population, property, social or economic activity) due to the danger of specific natural phenomena. In other words, we are dealing with consequences such as numbers of dead and injured, social and economic damage etc. which are brought about by specific 'dangerous' natural phenomena. This is expressed as the product of three parameters: danger, vulnerability and value exposed according to the 'classical' formula $R = H \times Vu \times Va$ in which H indicates danger, Vu vulnerability and Va the value exposed. Danger is the probability that, within a given time period, an event of a given intensity will take place in a given area. For example, it is the probability that an earthquake will recurrently hit an area of the Earth's surface, or that a volcanic area will effected by lava produced by an eruption. Vulnerability is an estimation of the man-made structures that would not resist the event in question and the presumed loss of human life. The value exposed to risk is calculated both in terms of the loss of human life and predictable economic damage. As is clear, the 'danger' parameter in the formula can only be expressed in terms of probability. As a consequence, 'risk' too can only be expressed in terms of probability (Leone 2005).

2.1 Risk Perception in Italy

Over the past fifty years, population growth and an increase in productive activities has caused the expansion of cities. Urban expansion has caused soil depletion as well as having the effect of human settlements reaching areas prone to potentially harmful natural phenomena. Accordingly, there has been a steep rise in the risk level to which our societies are exposed. Positing danger, i.e. the number and intensity of natural phenomena taking place on the planet, as constant in time, risk has considerably risen because of the increase in exposed elements (people, things, activities). Vulnerability to disastrous events is also steeply rising, especially if we consider the fact that globalisation creates ever greater conditions of interdependence between countries.

This increase in risk does not correspond to a rise in our risk perception. How many of us are aware they live in a territory subject to natural phenomena of a certain seriousness, which could take on great enough intensity to constitute a real danger for our safety? How many citizens have an idea of the degree of vulnerability of their homes, or at least are aware of the most secure places in their homes, where to seek shelter in case of an earthquake? Why do we keep building in areas prone to river and stream overflow or near the crater of active volcanos, despite the normative prescriptions and dramatic images we see on the news from all over the world?

Italy lacks a culture of risk, a full awareness of the fragility and value of our territory. Our social knowledge does not currently include appropriate basic knowledge of natural phenomena that could be useful in a situation of emergency. Individual and collective prevention methods, necessary to limit the negative effects of a natural event, are still insufficient. This lack manifests itself in our consistent unpreparedness in dealing not only with the rarest and most extreme phenomena, but with more common and frequent ones too (Peppoloni 2014).

2.2 *Seismic Risk Perception in Calabria and the Importance of Historical Memory*

Earthquakes have severely marked the organisational, economic and social systems of several regions—Calabria being one of them—by causing devastation and consequences for the population that have often taken on apocalyptic dimensions and connotations. Through knowledge of seismic history, on the one hand people would have a clear picture of what an earthquake is capable of causing; on the other hand, they might also understand why its effects have often been so negative. Existing literature on some of the biggest seismic catastrophes in Calabrian history, from authors both past and present, does not fail to highlight human responsibility in the disasters, in the sense that the thousands of victims caused by the earthquake are imputable certainly to the violence of the phenomenon, but also to the inadequate way of building houses, the techniques and materials used, which were unable to offer valid resistance to the strain of the tremors (Kostner 2002).

Through our analysis of a sample of surveys conducted children and adults in the Pollino area, which over the past three years has been experiencing a seismic swarm, it emerges that with the passing of time, archaic societies' "incontestable certainty" of the earthquake as divine punishment, which we referred to in the beginning of this paper, is today unravelling and that it is mostly the youngest who tend to perceive human responsibility in the unleashing of natural catastrophe.

3 Geological Framework of Pollino, Southern Italy

From a geodynamic viewpoint, Southern Italy is divided into two regions. The southernmost one, the so-called Calabrian Arc, is the area where the Ionian lithosphere still subducts beneath the Tyrrhenian Sea; the subduction is characterized by an eastward rollback (e.g., Malinverno and Ryan 1986; Doglioni et al. 1996; Critelli et al. 2013). North of the Calabrian Arc there are the so-called Southern Apennines that constitute the accretionary prism of the Adriatic plate subduction (e.g., Doglioni et al. 1996, and references therein); These two geodynamically separated regions meet in the Pollino Chain.

The Calabrian arc is a part of the Mediterranean orogenic belt connecting the NW-SE trending Southern Apennines with the W-SW striking Maghrebian thrust belts (Monaco et al. 1996). The Aspromonte and Serre mountains (Fig. 1) are composed of a pile of several thrust sheets (granites and metamorphic rocks).

The Pollino Chain and the strictly adjacent areas represent a minimum of the seismic activity in terms of moment release along the ridge of the Italian Peninsula. They are characterized by the occurrence of a relatively large number of events, all however of moderate size. Maximum intensities, observed in 1831, 1836, 1894 never attained severity higher than degree VIII on the MCS macroseismic scale.

In the northern Calabria the last seismic sequence, before the actual, that followed an earthquake located on the western slope of the Pollino Chain (Southern Italy).

The 1998 seismic sequence followed an earthquake that occurred in the Mercure Basin, immediately to the northwest of the Pollino Chain (Guerra et al. 2005).

This seismic event occurred on 9th September 1998 and caused some damage in several towns and villages located in the area (Galli et al. 1998), attaining a maximum intensity of VII MCS. The whole sequence was confined to the upper 15 km of the crust and involved the sedimentary cover of the region where two important geological contacts exist. At the surface there is the boundary between the Apennine Chain and the Calabrian Arc; more in depth there is the limit between the Adriatic and the African plate (Guerra et al. 2005). In the four years a sequence of earthquake involved the Pollino zones. In particular the event occurred the 26/10/2012 triggered in the Mormanno village reaching a magnitude 5.0 (Figs. 1 and 2).

In the entire arc the effects of intense Quaternary tectonics are well represented by a prominent normal fault belt that extends along the inner side of the arc and offsetting the previous wrenching tectonics (Tortorici et al. 1995; Tansi et al. 2007; Tripodi et al. 2013) (Fig. 1).

4 A Questionnaire on Seismic Risk Perception in Schools in the Pollino Area

Knowledge and study of seismic history, particularly in schools, is the best way—at least in an initial stage—to create greater awareness about earthquakes and educate the public on certain sensitive issues. It is easy to see why we give such great importance to the study of past seismic events. First of all, it can invite reflection on the difficult moments experienced by the various communities affected by earthquake over the centuries. Their study in the classroom can thus help pupils acknowledge the potential consequences of such events both in terms of human losses and of damage caused to the environment.

The school system is extremely important in man's educational and growth process. Schools today have the great responsibility of educating the masses on

Fig. 1 Major seismotectonic faults of the Calabria Region and historical earthquakes; from Monaco and Tortorici (2000)

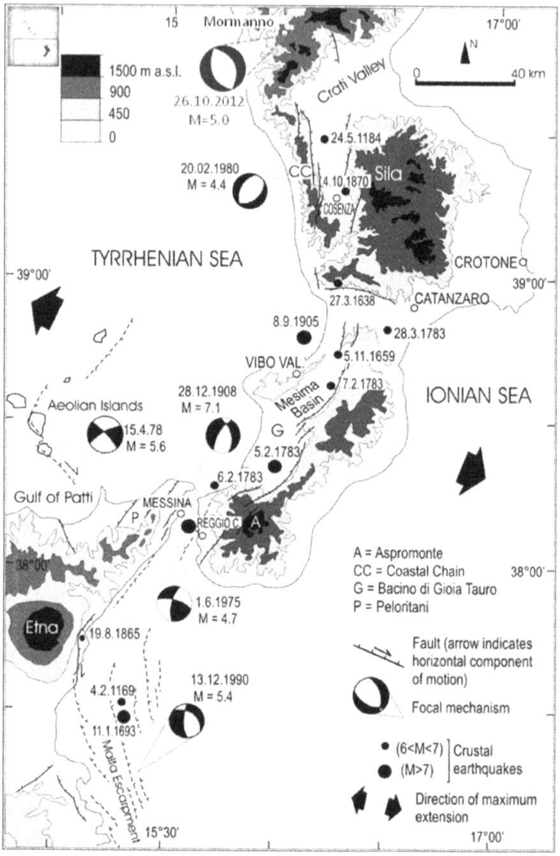

natural risk. Discussion on this topic should have a well defined space and represent one of the qualifying points of teaching. For this to happen, educational curricula need to be rethought, making space for new professional figures in school. Further, the importance of displaying informative material and relevant publications should be considered (Kostner 2002). The International Association for Promoting Geoethics led by INGV researchers Silvia Peppoloni and Giuseppe Di Capua is doing a great job in the spreading of Geoethics (Peppoloni and Di Capua 2012). This new discipline deals with the ethical, social and cultural implications of geological and geographical research, at the intersection of Geoscience, Geography, Philosophy and Sociology. Through identification of the principles that should guide our actions towards the Geosphere, Geoethics represents an opportunity for scientists to become more aware of their social responsibilities and it is a tool to orient society on matters relating to defence from natural risk, sustainable use of resources and environment conservation. Geoethics can contribute to the construction of correct social knowledge, strengthening the connection with the territory, our common heritage (Peppoloni and Pievani 2013).

Fig. 2 Location and data of the Pollino earthquake of October 26, 2012. Screenshot of a GIS project

This research involved giving a questionnaire to 641 primary and lower-secondary school students in some of the villages affected by the Pollino earthquake swarm, with the aim of discovering the effective knowledge possessed and how this related to age, experience, home area and the perceptions that the pupils had of seismic risk. Another questionnaire was also given to a sample of 40 adults (between the ages of 18 and 70) from the towns of Morano Calabro and Castrovillari in Calabria. The students' questionnaire included 35 questions, 33 of which were multiple choice while one which was "open" asked the students to describe their own first-hand earthquake experiences. The final question asked for a "mental map" to be designed with regard the actions to be taken in the eventuality of there being an earthquake while the students were in the classroom with their colleagues and teacher. The adults' questionnaire was, on the other hand, made up of 26 questions, including the one from the students' questionnaire regarding first-hand experience. Before the questionnaire was distributed in the schools, young graduate students from the University of Calabria gave a classroom lesson on earthquakes and the correct behaviour to adopt in the case of a seismic event. The sample was chosen on the basis of its being geographically representative of the Pollino area and included municipalities from both Basilicata and Calabria: Terranova del Pollino, Noepoli, Cersosimo and San Costantino Albanese (Basilicata), Mormanno, Morano Calabro, Castrovillari, Laino Borgo, Laino Castello, Saracena and Tortora (Calabria). The methodology used in analysing the knowledge of seismic risk followed the idea of a geography of perception, looking at individual and collective perceptions of space. This line of study aims at clarifying human behaviour by studying the concepts and images of the real world that men elaborate and considering the psychological and social aspects of man's behaviour within his environment. Therefore, each subject provides a personal re-elaboration of the information available and creates specific elaborations and visions. Quantitative methods are utilised through the building of a representative sample, the elaboration of a questionnaire of structured or semi-structured questions, and the codifying and analysis of the data. To this is added the elaboration and analysis of the mental maps. With regard the geographical research, the term "mental map", a subjective elaboration of an image of the environment, is associated with the name of Peter Gould (Gould and White 1993). The collected data will be of use to local institutions as it will provide a true picture of the public's level of knowledge about arguments regarding seismic risk, so aiding in making the territory more resilient through the reinforcing of prevention. On the other hand, Risk geography has developed a new approach in which greater relevance is given to the incidence and direct responsibility of the human factor (social, political and economic) in calamitous events, to the point of the coining of the term 'pseudo-natural catastrophes' (Migliorini 1981). This term refers to events where the role of anthropic action is considered along side the natural component (Tecco 2011).

4.1 Results

On the basis of the multiple choice questions, some of which required a single response while others allowed for more than one possibility, differences emerged according to pupils' home areas and age/year at school, 4th and 5th years at primary school and 1st, 2nd and 3rd years at lower secondary school. Only 39 % of pupils from Basilicata said they had personal experience of an earthquake, while all of the Calabrian pupils said that they had. Of the Basilicata children who had experienced seismic activity, 83 % said that it had occurred while they were at home and only 2 % at school, whereas 57.38 % of the Calabrian pupils had experience of earthquakes at school and 87 % at home (Fig. 3). Only 5.60 % of the Calabrians had experienced seismic activity while outdoors. Most of the pupils were sleeping at home or were studying at school. The most common reactions were fear (58.50 % in Basilicata and 69.53 % in Calabria) and confusion (24 % Basilicata, 39 % Calabria). 32.71 % of the Calabrians immediately ran outside, while only 5 % of their Basilicata counterparts did so. 19.25 % of the Calabrian children and 15 % of those in Basilicata froze as they were incapable of reacting to the earthquake (Fig. 4). Those people who were near to the pupils at the time of the event had the same reactions of fear and confusion, to which they added anxiety, panic and worry. Most pupils (93 % Basilicata, 91.96 % Calabria) gave correct answers with regard the behaviour to adopt in the event of a seismic movement, i.e. get under your desk or a door architrave (Fig. 5). Moreover, 88.50 % in Basilicata and 96.26 % in Calabria said that they should keep away from windows, cupboards and the blackboard (Fig. 5). 26 % of the Basilicata pupils said that people should stay calm, while 17 % of the Calabrians declared that one should get out of the building immediately (Fig. 5).

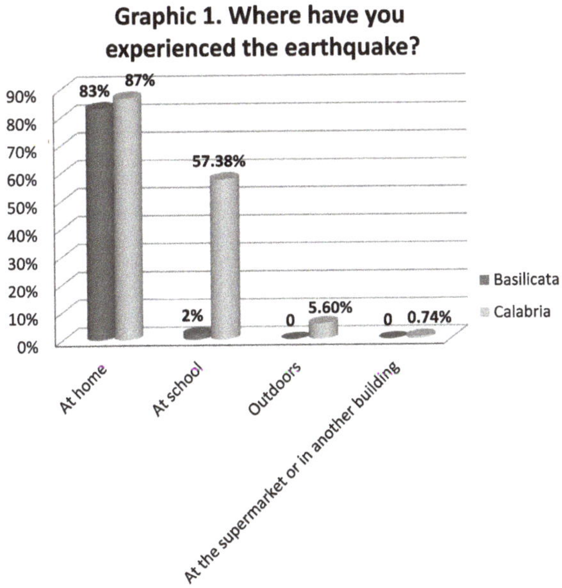

Fig. 3 Data as a percentage of the students' answers to the question: "Where have you experienced the earthquake?" (Graphic 1)

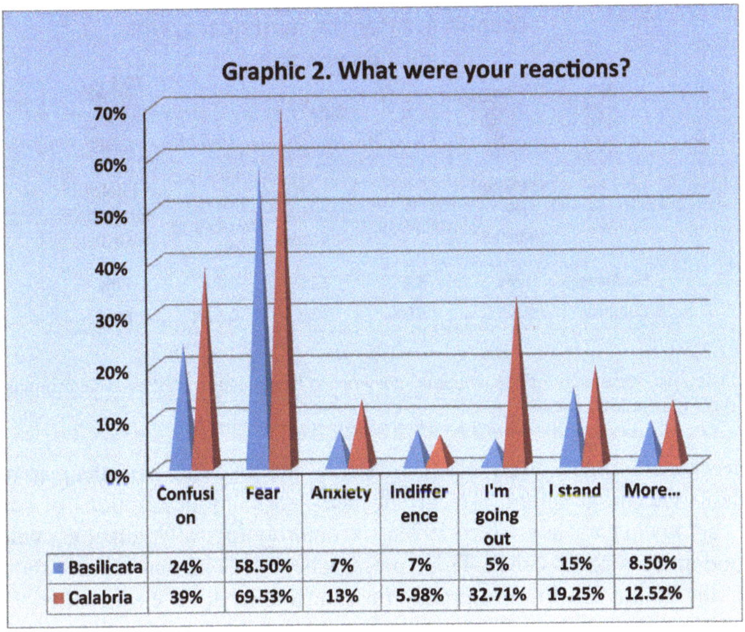

Fig. 4 Data as a percentage of the students' answers to the question: "What were your reactions?" (Graphic 2)

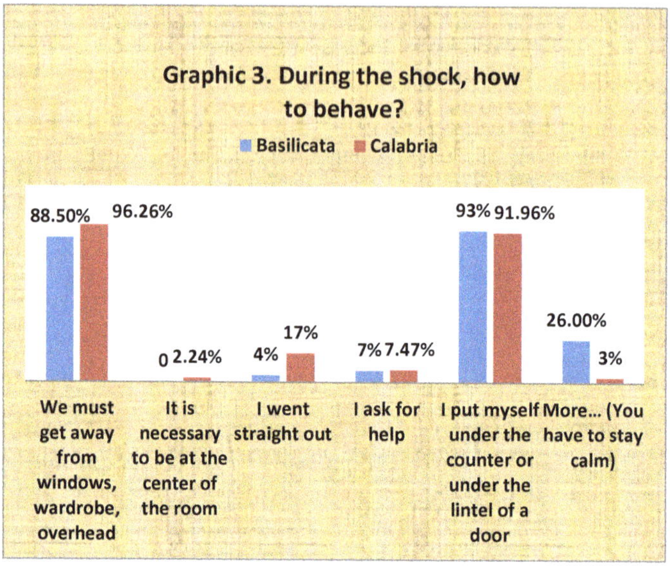

Fig. 5 Data as a percentage of the students' answers to the question: "During the shock, how to behave?" (Graphic 3)

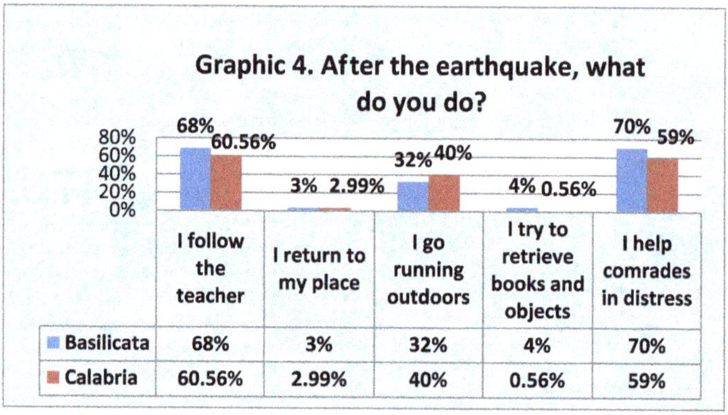

Fig. 6 Data as a percentage of the students' answers to the question: "After the earthquake what do you do?" (Graphic 4)

This answer, confirmed elsewhere, shows how the unstable, impatient, impulsive and bilious Calabrian nature, a recurrent theme in many descriptions of southern Italian regional characters (Teti 1993), is anthropologically already evident in childhood and pre-adolescence. Therefore, teachers in Calabrian schools should try to curb this inclination to immediately rush outside in the event of an earthquake because it could represent a further threat to pupils' safety. After the event, 68 % (Basilicata) and 60.56 % (Calabria) declared said they followed the teacher and 70 % (Basilicata) and 59 % (Calabria) offered help to colleagues in difficulty (Fig. 6).

In this case too, the Calabrian tendency to run outside is confirmed (40 %), while only 32 % of the Basilicata pupils had this inclination. All of the pupils had taken part in evacuation exercises, mostly for seismic risk, but some for fire risk. One alarm sign which should not be undervalued is that 49.5 % of the Basilicata pupils declared that their school building was not safe in terms of seismic risk (Fig. 7). In Calabria, on the other hand, 49.53 % of pupils considered their school to be a safe place, whereas 13.27 % said that their school was not safe (Fig. 7). A significant percentage of pupils declared that they did not know whether their school was safe or not (27 % Basilicata, 31.40 % Calabria). On the other hand, a clear majority of pupils think that their home is safe with regard seismic risk (Basilicata: 60 %; Calabria: 74.76 %), but 31 % of the Basilicata pupils did not know whether their home was safe or not (Fig. 8). One alarming statistic is the high percentage of students from Basilicata (76 %) who indicated that the emergency plan had been drawn up by the Fire Department. Only 36 % of the students from Basilicata said that the municipal authorities had prepared the plan. 58 % of the students from Basilicata did not know whether there was an emergency plan in the municipal area where they were resident. 79 % of the students in Basilicata and 75,14 % in Calabria knew that it is impossible to predict when and where an earthquake will occur, but that the

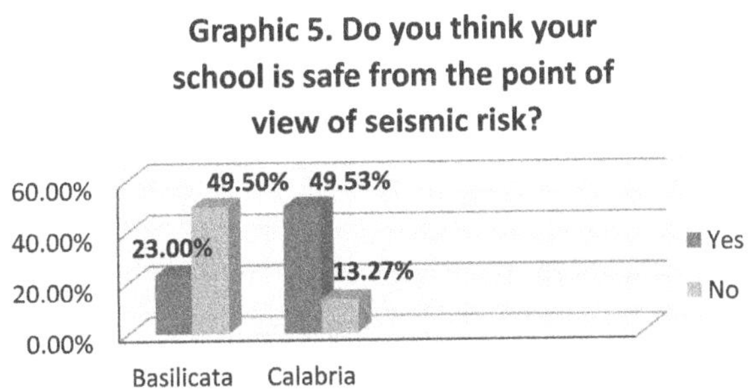

Fig. 7 Data as a percentage of the students' answers to the question: "Do you think your school is safe from the point of view of seismic risk?" (Graphic 5)

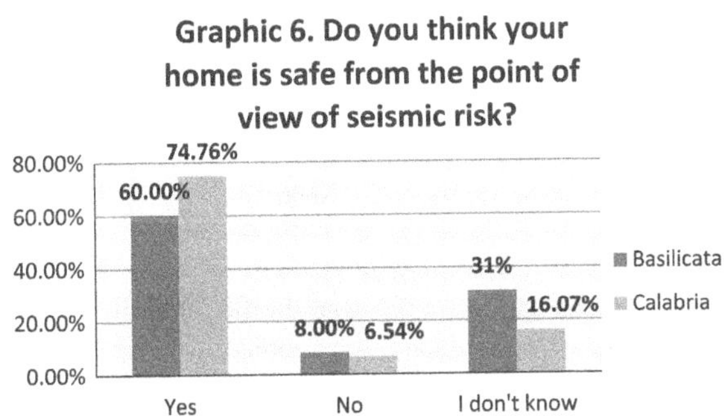

Fig. 8 Data as a percentage of the students' answers to the question: "Do you think your home is safe from the point of view of seismic risk?" (Graphic 6)

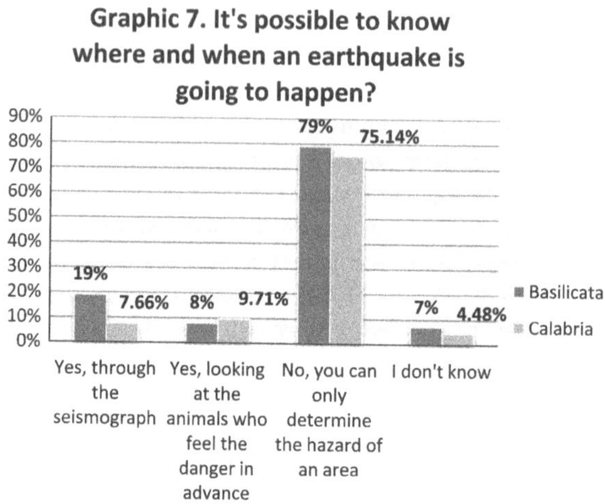

Fig. 9 Data as a percentage of the students' answers to the question: "It's possible to know where and when an earthquake is going to happen?" (Graphic 7)

dangerousness of an area can be evaluated (Fig. 9). 8 % Basilicata pupils and 9.71 % Calabria believed that some forewarning could be gained from observing the behaviour of animals that become aware of the danger in advance and 19 % of the Basilicata pupils and 7.66 % of the Calabrians thought that an earthquake could be foreseen via a seismograph (Fig. 9). The most common danger in the event of a seismic event is being hit by a falling object (57 % Basilicata and 42.41 % Calabria); then come "being caught up by a collapsing building" (52 % Basilicata and 51.05 % Calabria). Furthermore, we thought it useful to ask the pupils, in a question with multiple non-exclusive answers, how they would describe the territory where they lived with respect to an earthquake, bearing in mind certain key easily-identifiable adjectives with an immediate cognitive impact. The Calabrians considered their territory to be organised (60.32 %), inhabited (53.89 %), old (42.66 %) and looked after (26.37 %). The pupils from Basilicata were more divided with 43 % declaring that their territory was organised, while 29 % said it was disorganised, 39 % not looked after and 35 % old. This statistic confirms the feeling of insecurity which the Basilicata pupils have regarding their territory in terms of organisation in the event of an earthquake. The same methodology which was used in this question was applied to the next where pupils were asked how they perceived of earthquakes. The various possible answers were: "predictable", "unpredictable", "caused by fate", "caused by divine punishment", "natural", "caused by man" and, finally, "an event whose damage can be limited by environmental planning". The most common answers were "unpredictable" (88 % Basilicata and 81.49 % Calabria) and "natural" (83 % Basilicata and 69.15 % Calabria). However, the most significant statistic is that 8 % of the pupils in Basilicata and 13.27 % of those in Calabria believed that earthquakes are caused by man (Fig. 10). If this is added to the fact that 24.85 % in Calabria and 10 % in Basilicata were convinced that an earthquake was "an event whose damage can be limited by environmental planning", then an immediate, interesting reflection upon this is necessary (De Pascale et al. 2013). According to psychologist Felice Perussia, today there is an ongoing process of restructuring the experience of

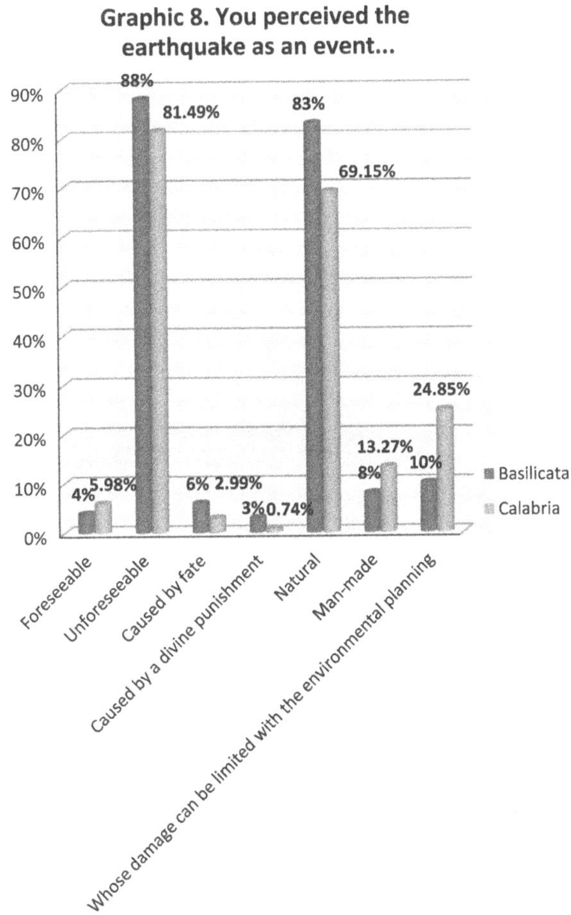

Fig. 10 Data as a percentage of the students' answers to the question: "You perceived the earthquake as an event…" (Graphic 8)

environmental risk, which accentuates the human, rather than natural, components. Therefore, there is a marked tendency towards attributing a human cause to phenomena, such as floods and even earthquakes, which would have been considered natural in the past. This does not mean that man is considered to be the original cause of the phenomena, but that he bears some responsibility regarding potential prevention or protective intervention (Perussia 1989). Therefore, it is no surprise that an increasing number of pupils consider human factors to be influential with regard catastrophic events. Regarding the "mental maps" 71 % of the drawings by Basilicata and 55 % of those by Calabrian pupils showed people under desks and door architraves. In some of the drawings, the images were simple and approximate while in others they were more complete and correct (Figs. 11 and 12). Different stages of "cognitive mapping" (Downs and Stea 1973) could be identified which evolve as a consequence of the pupil's level of spatial awareness. Therefore, some maps showed a space which was limited to the classroom, people under desks and a teacher who is asking everyone to stay calm, whereas other, more advanced, maps

Fig. 11 Advanced mental map drawn by students classroom: *watches* are represented as a current symbol of time perception. Note the Euclidean traits. From De Pascale et al. (2013)

might show more meaningful objects within the classroom. Some other "maps" showed the actions to be carried out in the event of an earthquake and their order. The most advanced drawings were produced by the 5th year primary school and the secondary school pupils. This confirms that "cognitive mapping" is a process of stages which evolves with age.

4.2 *The Questionnaires Given to Adults*

From the answers the adults gave to their questionnaires, it emerges that everyone had felt the earthquake of 26/10/12 while in their own homes. Nearly everyone had been in bed. Most had been asleep, while some had been watching television. The immediate reactions had been fear and confusion while 13 people indicated that they had chosen to "go outside". This fact, which was confirmed by the majority of respondees' choosing "I immediately go outside" as an answer to the next question

Fig. 12 Mental map that represents the students under the counter and under the lintels of the door during an earthquake (correct behavior to adopt in case of earthquake)

(on the behaviour to adopt in the eventuality of an earthquake), supported the anthropological data that had already emerged from the school questionnaire, i.e. that the Calabrians had an "impulsivity of catastrophe". Scholars had already noted this behaviour at the end of the 19th century, over a century after the 1783 earthquake, so establishing a relationship between the convulsion of nature (Lenormant 1976) and the people's temperament, and making an analogy between a land which moves and an unstable and turbulent population. Only 20 people gave the correct answer, i.e. "I get under a table or a door architrave". Even for the following question, regarding the behaviour to adopt after the earthquake, the majority replied that "You hurry outside" (23/40). 21/40 said that you should, "leave the building after putting on shoes and go to an open space, far away from buildings and high voltage electric cables". Only 8 people gave the answer, "turn off the gas, electricity and water at the mains". Instead, 19 people believed that their home was safe from the point of view of seismic risk, 8 people gave a negative answer and 13 did not know whether it was safe or not. 10 of those who said "yes" justified their answer by stating that their home had a reinforced concrete structure, 9 said, "it was built according to anti-seismic criteria", and 5 indicated, "it was built recently". With regard the question, "If your home is in a seismic area, what strategies can be adopted in order to be safer while indoors?", 21 indicated, "avoid keeping heavy objects on shelves and high furniture", while 20 chose, "secure bookshelves, wardrobes and cupboards to the wall". 33 people did not possess an emergency kit to take with them if they had to leave their homes immediately. According to 25 people, the kit should contain clothing, blankets and candles, while 13 said it should include a first aid box, a torch and a radio. 20/40 knew what the seismic classification was, i.e. the division of the territory into 4 zones on the basis of the frequency and intensity of earthquakes in the past. 4 people gave the wrong answer, indicating that they thought that the classification was the list of the most disastrous earthquakes to have hit our country and 13 did not know. 22/40 did not even know which zone their municipality was in and 21 people did not know whether their municipal authorities had an emergency plan. 29/40 knew that there were areas they should go to in the case of an emergency, i.e. the squares, car parks and open spaces not considered to be at risk and indicated in the municipal emergency plan, but 22 said that they would not know exactly where these places were in the case of an earthquake. 19 people knew that it was the municipal authorities' responsibility to prepare an emergency plan, while 18 thought it was the role of the fire brigade. 38 people thought that seismic activity was the main risk present in their municipality, while 7 thought it was the risk of fire and 6 hydrogeological risk. Only 24 people thought that it was not possible to predict where and when an earthquake would occur, but that you could determine how dangerous a zone was. 4 people indicated that seismic activity could be foreseen, "via a seismograph" and "by watching animals which are aware of the danger in advance", while 7 "do not know at all". To the question, "Is it true that earthquakes always take place in the same zones?", 6 people answered affirmatively, 10 were convinced that earthquakes always occur in different zones, while only 15 gave the correct answer, i.e. "It is a matter of chance. They can occur in new areas or in places that have already been affected". In line with the school pupils' answers, the majority of the adults indicated that the most common risks in the case of an earthquake were being hit by falling objects and being caught in a collapsing building. With regard the specific question referring to the territory in which they lived and an eventual earthquake, most of the adults were more pessimistic, although possibly more realistic, than the school pupils. Indeed, most of them considered the territory to be unorganised, neglected and ancient. The only perception that they had in common was that of the "ancient" territory, which, in effect, brings to mind the historical memory of earthquakes and the abandoning of former sites which were destroyed by natural calamities of the past, so causing the existence of "double" villages. To the final multiple choice question regarding perception of earthquakes, most answered using the adjectives "unpredictable" and "natural", but 4 people indicated "caused by man" and just one put a cross next to "the damage could be limited through environmental planning". Therefore, the awareness that man's actions can influence natural catastrophes was also apparent amongst the adults. Other than this, the adults exhibited great ignorance, especially in terms of the specific terminology relating to seismic risk.

5 Seismic Risk Perception in the Postmodern Digital Era

Growing numbers of people every day use a personal computer, smartphone, or tablet for various reasons, amongst them work, news, socialising, leisure. The Internet allows us to interact with others or gather information on whichever subject, at any time and from any place. The concept of mobility is crucial here: the use of smartphones has grown so much in the past few years that in 2013, 90 % of world mobile phone sales were smartphones with Android or iOS operating system. The performances of such devices have grown to such an extent that they allow the average user to use them instead of the more classic personal computer for the most common functions, such as managing e-mail and online browsing. In the same year, 46 % of Italian internet users (over 21,000 people) used a smartphone or a tablet.

People read and comment on the news, they exchange information and opinions on diverse topics, especially through blogs and social networks. It is no surprise, then, that all which pertains to seismic risk and earthquakes is amongst the most searched topics on the web. The internet, however, also presents a danger, as it can contain erroneous and misleading information: there are countless blogs and websites generating unverified or plainly false information, sometimes based on pseudoscience, appealing to people's fear and ignorance to spread such information for personal interest or just as a game. As an example of people's knowledge about seismology being altered by partially or completely untrue information, it suffices to look at what happened in the aftermath of the 2012 earthquake in the region of Emilia Romagna, when a group of so-called "conspiracy theorists" blamed the earthquake on fracking.

So if on the one hand people look for information on seismic risk, on the other their perception of it is compromised by the unruliness of the information available online. The most efficient way to counter this situation is to educate people first of all to be critical and verify what they read, and more in general to the science of facts. This work can and must be done in primary and secondary schools. But how to educate adults?

A fundamental tool could be participation, that is, allowing everyone, through the appropriate channels, to actively contribute to the gathering and distributing of real and verifiable information. The Istituto Nazionale di Geofisica e Vulcanologia (INGV) (National Institute for Geophysics and Volcanology), for example, has used this idea to create the website www.haisentitoilterremoto.it, allowing anyone to report on their own experience. From the homepage: "This website was born to monitor in real time the effects of Italian earthquakes and to inform the population on seismic activity. It is made possible thanks to the contribution of each of the people who describe their own experience by answering our macroseismic questionnaires. Maps of the earthquakes felt by the population are elaborated using data from the macroseismic questionnaires and they are updated every time a new questionnaire is submitted. The intensity of the phenomenon (represented by a dot on the map) is determined considering all reports from all towns. Gathered data is subject to automatic filtering of a statistical type, but is not individually verified.

Intensities greater than or equal to grade VI on the Mercalli scale require on-site verification on the part of specialised staff".

The website also provides information through the social networks and re-directs users to a specific app for Android which allows them to make the same reports and to access information via smartphone and tablet. The result is that people feel involved and are willing to participate as they receive and share truthful information.

But it is possible to take this one step further.

The quasi-totality of mobile devices is equipped with a built-in GPS (Global Positioning System) antenna which allows us to detect its position with a precision —in optimal conditions—approaching the meter, but most of them also have another set of sensors: Gyroscopic (along the three Cartesian axes), Accelerometer (along the three Cartesian Axes), Geomagnetic (along the three Cartesian axes), Barometric, Thermometric, Hydrometric, of proximity and other less common ones depending on the device's specific hardware.

For example, by using the accelerometer in conjunction with the gyroscope, it is possible to establish the direction and speed of a device.

Using such characteristics, it is possible to create a network of mobile devices which can be used as seismographic tools alongside and integrated with conventional ones. The importance of such a network is undoubtedly considerable both on the national level, given the proneness of the Italian territory to seismic events and their critical consequences, and on the global level; if it is true that the sensors present in such devices do not have the sensitivity of seismographs distributed on the national territory, it is also true that the number of the latter is extremely low. The INGV has installed just over 100 seismographic stations, forming the Rete Sismica Nazionale Centralizzata (RSNC) (Centralised National Seismic Network). The capillary presence of mobile devices, as well as their potential to provide a high quantity of useful data, cannot be ignored.

An application for mobile devices which implements such a system should preliminarily analyse the devices' hardware characteristics and sensitivity of the GPS, of the gyroscope and the accelerometer; in such a way, the level of reliability of the data collected by the device can be ascertained. Then, using data mining algorithms for the analysis of the detected movements and vibrations, via clusters and outliers analysis, a mathematical model can be created capable of identifying and classifying the different types of movements and vibrations and subsequently focus on those of interest, that is to say those deriving from seismic events. This information would then be filtered and sent out to a data centre created ad hoc, together with the GPS's detected position and a synchronised timestamp.

The data thus obtained would constitute an important source of information which could be integrated with that obtained by the RSNC, re-elaborated and used to aid the study of seismic events, thanks to potentially enormous quantities of useful data.

Further, it is worth mentioning that the fact that both mobile devices and built-in sensors are in continuous technological evolution, so the possibility of such sensors growing in sensitivity to the point of being comparable to that of traditional seismographs should not be excluded; considering the fact that, as previously shown, the diffusion of such devices is on the rise, it follows that with the passing of time such a network could grow not only in size but in precision and reliability, with close to zero economic and organisational effort.

Participation is useful, therefore, not only on the educational level but also for its pro-active potential, and this constitutes an important incentive for the population: knowing they are an active part in the creation of something useful and potentially capable of saving human lives.

Moreover, the fact that there is enormous interest in the topic from the digital point of view is witnessed by the impressive amount of applications that can be found in mobile app stores under the heading "earthquake", each of these with a number of downloads somewhere between a few hundred and several millions. The functions of such applications are varied and, at times, overlapping; some allow visualisation of historical maps of seismic events on a global scale; others limit themselves to notifying the user in real time of local or non-local seismic events; there are even several apps guiding the user on how to behave immediately after an earthquake or allowing him or her to notify friends and family through the social networks or SMS on their state of health.

Unfortunately, due to the aforementioned phenomenon of false information circulating through the internet, there also exist some applications which claim to be able to predict earthquakes through mathematical calculation based on magnetic waves, physical, and even astronomical vibrations. Communication technology can provide information on seismic risk in a effective and timely manner, following also the premises of Neogeography (Goodchild 2007; Turner 2006). This has the objective of shortening the distance between the producers of geographical knowledge and its users, thanks to the new communication technologies. Neogeography aims at participation in cartographic production to guarantee awareness of each community in the use of the cultural mechanisms that affect it. This change is made possible by technological innovations which no longer require great skill to be used. The simplification of cartographic production allows for the multiplication of agents of knowledge production, generating a multiplicity of possible interpretations of the same place, in which communities previously left out of official representation acquire space. The mechanism is a sort of collective mental map and it is a precious instrument for spatial studies that use the techniques of participatory cartography; further, it can translate into an instrument of participatory democracy or, rather, active citizenship or even participatory science. This instrument could also be useful for the perception of earthquakes and be employed to find out the reactions of young people and adults in the occurrence of a seismic event.

6 Conclusions

Most of the students involved in the survey declared that they had learnt about earthquakes from the television and, in fact, the media are one of the main vehicles of scientific communication. One of the main shortcomings of the communication system is that we do not currently have a sufficient number of specialists on the subject, this can affect the quality of information and educational efforts coming from the scientific world. This is why, when there is talk about earthquakes, it is important to provide a clear and correct message to the public. Considering the grave lexical ignorance which became apparent from the adults' answers and the lack of communication between the municipal authorities and the population with regards the emergency plan, attention should also be given to the significance of the terminology used (INGV and Protezione Civile 2013). Finally, it is also significant that many adults reported their having hurried to connect with others via a social network, particularly Facebook, immediately after the 26/10/12 earthquake so as to obtain greater information about the seismic event. In this way, this Neogeography, which involves sharing geolocalised information with friends, helps people to put what is happening into a context and transmits comprehension through local spatial knowledge, has a fundamental role to play in the communication of risk. Indeed, due to its irresistible utilisation, internet helps in the spreading and continuous implementation of geographical information and is particularly effective in processes of participation in planning. It is, therefore, important for each of us to report information about a seismic event correctly, even on social networks. At the same time, we should wait for updates from reliable earthquake monitoring sites, such as INGV, and report precise perceptions of the earthquake, "update our own situation" and use "geotagging". This would create a form of "virtual solidarity" between users and constitute an action of bottom up information sharing with the general public through new platforms of participation.

The data collected from the answers to the questionnaires on the perception of earthquakes invites a reflection. Awareness is developing amongst the youngest of the responsibilities of human actions that can favour catastrophes and transform extreme events into disasters. By contrast, the percentage of students who believe earthquakes to be divine punishment is close to null. The impact of man, perceived by the students, on natural catastrophes is symptomatic of the context relative to the new geological era in which we live: the Anthropocene (Crutzen 2005). This is the new era of men, the Age of Man and it is characterised by the fact that the human footprint on the global environment has become so big and intense that it has come into competition with certain forces of Nature in terms of their impact on the Earth system (Steffen et al. 2011; Bonneuil and Fressoz 2013). Sandro Calvani wrote that in 2013, the fact of humanity having entered Anthropocene was accepted by the whole global scientific community (www.sermig.org). The definition is now recognised by the United Nations, which have used this as a starting point for several international debates on sustainable development. There is, however, wide diversity of opinion as to the beginnings of this geological era: for some, it was eight thousand years ago, for others around two thousand and according to others still, the era would have started only with the industrial revolution. It is certain that we live for the first time in a planetary system of relations between humanity and Nature in which humanity controls all other variables and is responsible for all the consequences. The human genre is able for the first time ever to destroy or save its future.

References

Bonneuil, C., & Fressoz, J. (2013). *L'événement anthropocène. La Terre, l'histoire et nous*. Paris: Éditions du Seuil.

Critelli, S., Muto, F., Tripodi, V., & Perri, F. (2013). Link between thrust tectonics and sedimentation processes of stratigraphic sequences from the southern Apennines foreland basin system, Italy. *Rendiconti Online della Società Geologica Italiana, 25*, 21–42.

Crutzen, P. J. (2005). Benvenuti nell'Antropocene!. In A. Parlangeli (Eds.). Milano: Mondadori.

De Martino, E. (1948). *Il mondo magico*. Torino: Boringhieri.

De Martino, E. (1977). La fine del mondo. Contributo alle analisi delle apocalissi culturali. In C. Gallini (Eds.). Torino: Einaudi.

De Martino, E. (1995). *Storia e metastoria. I fondamenti di una teoria del sacro*. Lecce: Argo Editore.

De Pascale, F., Bernardo, M., & Muto, F. (2013). I terremoti dell'Irpinia e del Pollino: memoria storica, comunicazione e percezione attuale tra Geoetica e Geografia. In *Atti del 32° Convegno Nazionale del Gruppo Nazionale di Geofisica della Terra Solida, Trieste* (pp. 375–381).

Doglioni, C., Harabaglia, P., Martinelli, G., Mongelli, F., & Zito, G. (1996). A geodynamic model of the Southern Apennines. *Terra Nova, 8*, 540–547.

Downs, R. M., & Stea, D. (1973). Cognitive maps and spatial behavior: Process and products. In R. M. Downs & D. Stea (Eds.), *Image and environments* (pp. 8–26). Chicago: Aldine Publishing.

Eliade, M. (1945). Le mythe de l'éternel retour. Archétypes et répétition, Paris, Gallimard; trad. it. (1989). Il mito dell'eterno ritorno, Roma, Borla, III ed.

Frazer, J. G. (1911). The Golden Bough. A Study in Magic and Religion; trad. it (1973) Il ramo d'oro. Studio sulla magia e la religione, Torino, Editore Boringhieri.

Freud, S. (1920). Jenseit des Lustprinzips; trad. it. (1975). Al di là del principio di piacere, Torino, Bollati Boringhieri.

Galli, P., Molin, D., & Falzone, G. (1998). Terremoto Calabro-Lucano del 9 Settembre 1998: Rilievo Macro1007 sismico preliminare (on-line: http://www.serviziosismico.it/RT/RRP/980909/terremoto.html).

Goodchild, M. F. (2007). Citizens as sensors: The world of volunteered geography. *GeoJournal, 69*, 211–221.

Gould, P., & White, R. (1993). *Mental maps*. New York: Rutledge.

Guerra, I., Harabaglia, P., Gervasi, A., & Rosa, A. B. (2005). The 1998–1999 Pollino (Southern Apennines, Italy) seismic crisis: tomography of a sequence. *Annals of Geophysics, 48*, 6.

Istituto Nazionale di Geofisica e Vulcanologia and Dipartimento della Protezione Civile. (2013). Rischio sismico: nota alle redazioni giornalistiche. Comunicazione congiunta della Protezione Civile e dell'INGV. In: www.protezionecivile.gov.it/jcms/it/view_com.wp?contentId=COM40762.

Kostner, F. (2002). Terremoti in Calabria. Cronache, problemi e prevenzione, Mendicino, Klipper.

Leone, U. (2005). Il rischio ambientale, vulnerabilità, sicurezza e informazione. *Rivista della Scuola Superiore dell'Economia e delle Finanze, 3*, 262–273.

Lenormant, F. (1976). La Magna Grecia, Paesaggi e storia, 3 voll., Chiaravalle Centrale, Frama Sud.

Malinverno, A., & Ryan, W. B. F. (1986). Extension in the Tyrrhenian Sea and shortening in the Apennines as a result of arc migration driven by sinking of the lithosphere. *Tectonics, 5*, 227–245.

Migliorini, P. (1981). *Le calamità naturali*. Roma: Editori Riuniti.

Monaco, C., Tortorici, L., Nicolich, R., Cernobori, L., & Costa, M. (1996). From collisional to rifted basins: An example from the southern Calabrian arc (Italy). *Tectonophysics, 266*, 233–249.

Monaco, C., & Tortorici, L. (2000). Active faulting in the Calabrian arc and eastern Sicily. *Journal of Geodynamics, 29*, 407–424.

Peppoloni, S. (2014). Convivere con i rischi naturali. Conoscerli per difendersi, Bologna, Il Mulino.

Peppoloni, S., & Di Capua, G. (Eds.). (2012). Geoethics and geological culture. Reflections from the Geoitalia Conference 2011. *Annals of Geophysics, 55*(3). ISSN 2037-416X.

Peppoloni, S., & Pievani, T. (2013). *Le Scienze della Terra e il loro contributo al rinnovamento culturale della società*. Contributo al Festival della Scienza, Genova, 23 ottobre–3 novembre, 2013.

Perussia, F. (1989). *Pensare verde. Psicologia e critica della ragione ecologica*, Milano, Guerini e Associati.

Steffen, W., Grinevald, J., Crutzen, P., & McNeill, J. (2011). The Anthropocene: conceptual and historical perspectives. *Philosophical Transactions of the Royal Society A, 369*(1938), 842–867. doi:10.1098/rsta.2010.0327.

Tansi, C., Muto, F., Critelli, S., & Iovine, G. (2007). Neogene-Quaternary strike-slip tectonics in the central Calabrian Arc (Southern Italy). *Journal of Geodynamics, 43*(3), 393–414.

Tecco, N. (2011). Educazione geografica, resilienza e catastrofi naturali. In: C. Giorda & M. Puttilli (Eds.), *Educare al territorio, educare il territorio. Geografia per la formazione* (pp. 308–320). Roma: Carocci.

Teti, V. (1993). *La razza maledetta. Origini del pregiudizio antimeridionale*. Roma: Manifestolibri.

Tortorici, L., Monaco, C., Tansi, C., & Cocina, O. (1995). Recent and active tectonics in the Calabrian arc (Southern Italy). *Tectonophysics, 243*, 37–55.

Tripodi, V., Muto, F., & Critelli, S. (2013). Structural style and tectono-stratigraphic evolution of the Neogene-Quaternary Siderno Basin, southern Calabrian Arc, Italy. *International Geology Review, 4*, 468–481.

Turner, A. J. (2006). *Introduction to Neogeography*. Sebastopol: O'Reilly Media, Inc.

Virno, P. (2003). *Quando il verbo si fa carne. Linguaggio e natura umana*. Torino: Bollati Boringhieri.

Wittgenstein, L. (1969). *On certainty*. Blackwell publishing; trad. it. (1978). Della Certezza, Torino, Einaudi.

Sitography

http://is.pearson.it/magazine/filosofia-il-rischio-e-il-limite.
http://www.haisentitoilterremoto.it.
http://www.sermig.org/nponline/163-articoli/12418-la-nuova-era-antropocene.

Psychosocial Support to People Affected by the September 5, 2012, Costa Rica Earthquake

Mario Fernandez, Lorena Saenz, Marco Carranza, Cristina Matamoros, Oscar Duran, Marlen Brenes, Andrea Alfaro, Carolina Solis, Stephanie Macluf, Auria Zarate, Diana Montealegre, Laura Hernandez, Vanessa angulo, Daniel Chavarria, Diseiry Fernandez, Evelyn Rivera, Leonardo Umaña, Maria Fernanda Meneses, Patricia Zamora, Harold Suarez, Augusto Benavides and Edward Ruiz

1 Introduction

Psychosocial intervention in communities of the Santa Cruz County (Guanacaste province) and Cobano district (Puntarenas province) was carried out by the Brigade for Psychosocial Support of the University of Costa Rica, an interdisciplinary group integrated mainly by psychologists. The work implied to accompany the affected by the September 5, 2012 earthquake (Fig. 1), dissipate rumors and fears and contribute to the management of the various emotions caused by the earthquake. The support plan was executed between Saturday 8 September and Wednesday, 26 September. The Brigade of the UCR was accompanied by members of the Red

M. Fernandez (✉) · M. Brenes
Red Sismológica Nacional (RSN: ICE-UCR), San Jose, Costa Rica
e-mail: mario.fernandezarce@ucr.ac.cr

L. Saenz · M. Carranza · C. Matamoros · A. Alfaro · C. Solis · S. Macluf · A. Zarate · D. Montealegre · L. Hernandez · V. angulo · D. Chavarria · D. Fernandez · E. Rivera
School of Psychology, University of Costa Rica, San Jose, Costa Rica

M. Fernandez · O. Duran
Program Preventec, University of Costa Rica, San Jose, Costa Rica

L. Umaña
School of Sociology, University of Costa Rica, San Jose, Costa Rica

M.F. Meneses · P. Zamora · H. Suarez · A. Benavides · E. Ruiz
Risk and Emergency Management Postgraduate Program, University of Costa Rica, San Jose, Costa Rica

S. D'Amico (ed.), *Earthquakes and Their Impact on Society*,
Springer Natural Hazards, DOI 10.1007/978-3-319-21753-6_30

Sismologica National (RSN: ICE-UCR) and the program of the UCR for prevention and mitigation of disasters, Preventec.

The Psychosocial support is strictly necessary after the manifestation of a threat because disastrous events cause big fears, doubts, anguish, anxiety, uncertainty, misinformation and, in many cases, a feeling of loneliness in the affected. These feelings could arise when no one accompany the affected during and after the threatening event. In such condition, some people cry, others can not sleep and others experience a great fatigue. While ones cannot sleep, others fall into a deep sleep, a product of the great mental effort they made at the time of the event, perhaps helping other people who went into panic and anguish. In addition, an event of great magnitude, natural or man made, alters the mental condition of the affected and takes the peace and tranquility away from them.

The intervention was made to help those affected to control their emotions and return to peace and tranquility, which could be very difficult without professional help. Among the objectives of the activity was also to empower local actors to handle these situations and contribute to the communal organization for the same purpose.

The work was carried through sessions that began with presentations of specialist in seismology, who talked about earthquakes and tsunamis, answered questions and evacuated doubts of the people. Later, a member of Preventec gave general guidelines for handling emergencies. And finally, the group of psychologists provided psychological services to the affected.

Among the main results are the training to the Brigade of Psychologists and Sociologists of Guanacaste and the psychological attention to the affected people in the communities of Ostional, Nosara, Sámara, Playa Lagarto, San Juanillo, Marbella, Tamarindo, Playa Langosta, Cóbano, Santa Teresa de Cóbano, Mal País y Montezuma (Fig. 1).

2 Methodology

The first action of the intervention was a meeting in the National Office for Management of Risks and Emergencies of Costa Rica in which such institution decided that the psychological attention to the affected of the Santa Cruz county and the district of Cobano would be carry out by the Brigade of the University of Costa Rica. After that, members of the Brigade and Preventec decided that the work would be made in the following way:

1. A specialist in seismology would provide technical information to dispel rumors and reduce anxiety of those affected by the earthquake. The seismologist would use two different tools to transfer the information: (a) a Power Point presentation about earthquakes and tsunamis and (b) the direct answering of questions from the participants.

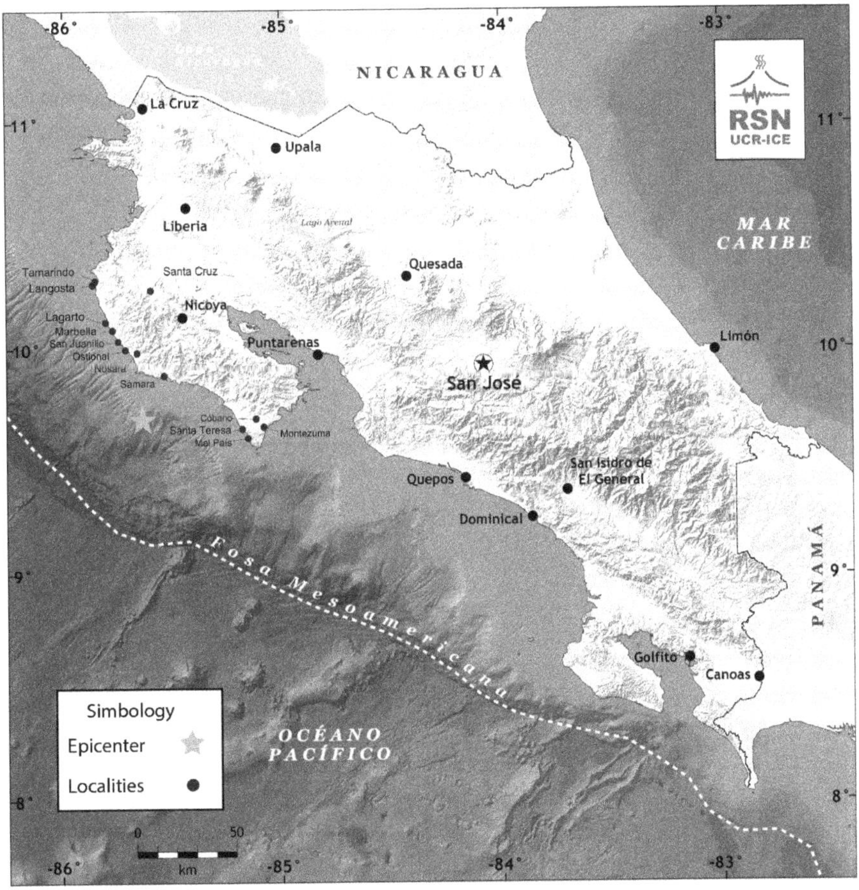

Fig. 1 Epicenter of the 7.6 Costa Rica earthquake and communities visited by the brigade for psychosocial support of the university of Costa Rica (*small circles*)

2. A trained professional would do a presentation about emergency plans and incidents command system. The professional would use several strategies to achieve his objectives, in some cases the presentation in Power Point and a dynamic to improve public participation.
3. The psychologists would provide psychological attention to individually and collectively address the traumas of the affected. They would use group dynamics, individualized contentions and ludic material, especially for working with children, to carry out their task.

In order to implement the plan, the group would work in coordination with the local committees for emergencies and local leaders. After coordinating the field work, the brigade would be allowed to attend the inhabitants of the communities. At the end of each day a self-evaluation of what has been done would be made to know the efficiency of the group and to improve the work.

3 Intervention in Communities

The first activity carried out by the group was training members of the Association of Psychologists of Guanacaste (the province where the earthquake occurred), which was held Saturday 8 in the afternoon at the headquarters of the University of Costa Rica at Liberia. At least 20 psychologists from Guanacaste received psychological attention by post-earthquake stress. We found the members of such group very sensitive, with feelings repressed due to their role as guides and leaders who attend emergencies. This training provided concrete results the next day, according to reports of the treated professionals. The importance of this activity was that it could have had a multiplier effect in the region.

In the following days, the brigade visited the communities of Santa Cruz and Cobano. In almost all the communities we talk about earthquakes and tsunamis to members of the community, which included adults, older adults, and children. The brigade helped the people to build emergency plans and form their local committee for risk and emergency management. Children were asked how the adults treated the emergency. Members of the brigade visited homes in company of a group of neighbors, previously treated, which became a very valuable and fruitful practice that reinforced the knowledge acquired and prepared the people helped to help others. They observed the work of the members of the brigade and now they are better prepared to help other community members in crisis situations. We identified the existing dangers and aspects to improve in the visited houses. Another important action was to evaluate the conditions of the hill where the people escaped and their evacuation route. In this project, the intervention of a group of masters students in risk and emergencies management of the University of Costa Rica was crucial. In Ostional, we found that the hill to escape has good condition as evacuation site but there is a telecommunications antenna located on it which represents a danger for the people.

In Nosara, the brigade worked on strengthening and unifying the local groups in charge of intervention in disasters. In this community we found a lack of communication and union of the groups involved in local risk and disaster management. A member of Preventec performed a dynamic to unify the human network of actors in risk management. The highest local authorities were in the meeting and expressed great interest in the risk management of the county of Nicoya.

In Samara, we found a group of residents (Fig. 2) waiting for explanations about the past and future seismic activity. The first thing done was to address concerns of the affected. Then, the brigade supplied psychosocial intervention and worked on the organization of the community. A great achievement of this encounter, in which the local authorities were also present, was the reactivation of local committee for emergencies.

On September 13 the brigade visited Playa Lagarto to provide psychosocial care and explain how to develop an emergency plan for the community. A diagnostic meeting was made with 22 inhabitants of the community. The group found a fishing community without social cohesion because of problems between them, lack of

Fig. 2 Meeting carried out in Sámara

resources, and a local Committee for emergencies, formed years ago, disable. In response to emergencies, this community has been evacuated several times after a quake due to the possibility of a tsunami, so they have identified people with special needs and manage information about safe areas. As part of the activities, six children with high levels of post- earthquake stress received psychological attention. We encouraged people to form a local committee for risk and emergency management.

Also on Thursday September 13, 2012, we went to the community of San Juanillo where we meet teachers and members of the community representing the local committee for emergencies, committee for sports, a religious volunteer, the fishermen's association and the committee for health. Neighbors expressed their fears about how to respond to an earthquake or tsunami, as well as complaints because they consider that they are not receiving adequate care. They also said that rumblings after earthquakes kept them very concerned.

During the visit to Marbella we helped the students and teachers of the educational center. We gave them tools to be used if another earthquake occurs. Other activities were: identifying weak areas of the center, evacuation routes and safe places. In Tamarindo and Playa Langosta we had a meeting with more than 35 emergency risk managers in the hotel Capitan Suizo. We heard their concerns and explained to participants how seismic phenomena happen and encouraged them to implement an incident command system.

On Monday 24, the Brigade arrived to the District of Cobano and was received by the Mayor. Members of the committee for emergencies suggested intervening in the high school and having a meeting with teachers of several schools from the Cobano District. In the evening we met almost 100 students of the Cobano High School. In this intervention we answered questions about earthquakes and tsunamis.

At the request of educators at the Santa Teresa School from Cobano, we visited that school Tuesday 25. The Group observed an evacuation simulation organized and directed by the members of the community. Members of the brigade gave psychological support to the school staff affected by the earthquake. In the afternoon we had a meeting with all the teachers in which we showed them useful material to implement the 'Retorno a la Alegria' (return to happiness) methodology. The group shared with them the results of the intervention and the evaluation of the activities.

Before arriving at the site of interest in Montezuma, we visited the coastal community in order to observe the degree of damage to the infrastructure and to the inhabitants. Of what was observed, it was concluded that the earthquake caused virtually no damage to the infrastructure of the town, everything was standing and in very good condition. According to the perception of the group, people were relatively calm and did not show signs of great fears. During the short tour of Montezuma we observed very good signs indicating the evacuation routes to the hills. This shows that Montezuma is adequately prepared to face the threat of tsunami and that the efforts made to manage the risk of disaster have not been in vain. This community has close high terrain and routes available to them, which makes it a place relatively safe to destructive tsunamis.

4 Psychosocial Effects of the September 5 Earthquake

In almost all the visited communities the most obvious fear of the inhabitants was the possibility of a future earthquake of large magnitude. This possibility generated much anxiety in the people and delayed the return of them to a normal life.

Closely linked to the above fear, the aftershocks associated with the main event fed fear in the mind of the people. Each felt aftershock reproduced the feeling of insecurity in the people. The word aftershock was so important to them that even children wore it with high frequency and easiness.

Since the Costa Rican population knows that the tsunami threat is real in the country and that a tsunami generated by a coastal earthquake may affect them, the occurrence of the September 5 earthquake fed the fear of the occurrence of a tsunami in the coastal residents. This fear was typical in all coastal communities. In Ostional, for example, the most, but all, the population went to a hill after the earthquake and they remained there for long time because of their fear. But once they returned to their homes, the fear followed in their minds. Also in Samara and Santa Teresa de Cóbano, there were expressions of fear about the possibility of a tsunami. The fear aroused many concerns in the people and questions such as: how long they would have to escape a tsunami? what is the minimum height of a location to be a sure place in case of tsunami?, how much they would have to move away from the coast so that no tsunami surprised them? And for the first time they questioned whether the hill where they moved was safe and this gave birth to another fear: be affected by a landslide trying to escape from the tsunami.

During the presentation at the high school of Cobano there were concerns and questions about the possibility that the Nicoya Peninsula detached from the rest of Costa Rica. We do not know the origin of this fear but it was in many people. It is likely that such a possibility was associated with the vertical bounce of the Peninsula during the earthquake, and with the release of the Caribbean plate which was being dragged down by the subduction of the Cocos plate.

An inhabitant of Mal Pais stated that this community had fear of gases emitted from a crack formed during the earthquake of September 5. They were worried about the possibility that a toxic gas kill them by inhalation. A visit to the site was made but we could not find such crack and therefore the gas emission could not be confirmed. As observed in the field and in geologic maps of Mal Pais the bad smell reported could correspond with the release of organic gases through cracks caused by the earthquake.

An inhabitant of the Montezuma expressed concern about a luminosity observed at the bottom of the Ocean near Cabo Blanco and expressed fear over the possibility of emergence of a volcano. According to him, such fear was in many people of Montezuma. At that distance from the junction of the Coco and Caribbean plates is practically unlikely the formation of a volcano and it is likely that such luminosity corresponds to a fishing bank that cause such an effect.

5 Expectations, Requests and Future Work

The psychosocial intervention in Guanacaste and Puntarenas awoke many expectations in local authorities, members of local committees of emergencies and in the population in general. From the moment we started coordination meetings with the leaders of the Municipal Committee of Emergencies of Santa Cruz (Fig. 3) the local authorities manifested their desire that the brigade makes a permanent project in the area.

The psychological attention given by the Brigade of the University of Costa Rica was very appreciated by those affected, not only because it helped them to return to a normal life and joy, but because helped them to be better prepared to deal with emergencies and disasters. On this occasion, as never before, the Brigade offered a very complete and well designed plan of intervention, which included technical talks about earthquakes, tsunamis, emergency management and psychological support. The intervention was effective. Geologists or seismologists explained the geological phenomena and reduced fears associated with them; a professional with knowledge in handling emergencies help to organize the communities and finally,

Fig. 3 Members of the local committee for emergencies of Santa Cruz and members of the brigade of the UCR in a meeting to coordinate actions

the psychologists controlled the emotions of the affected. This interdisciplinary work was very effective and helpful.

The results and requests show that the University of Costa Rica must have a permanent project on psychosocial support in crisis. The need is evident and the University of Costa Rica has a good interdisciplinary group to fill it.

6 Conclusions

The September 5, 2012 Samara, Costa Rica, earthquake did not cause neither great material damage nor loss of life but it caused psychosocial effects in people of Guanacaste and Puntarenas provinces, among which stand out: anguish, insomnia and sadness.

The main fears found were: fear of a future large earthquake, fear of aftershocks, fear of a tsunami, fear of the detachment of the Nicoya Peninsula from the rest of the Costa Rica territory, fear of poisoning gases and fear of the emergence of an underwater volcano.

The work carried out must be permanent. The University of Costa Rica has a valuable human resource that can definitely make a great contribution to the welfare of the Costa Ricans. The psychosocial intervention in Guanacaste and Puntarenas after the earthquake of September 5, 2012 showed the capacity of that group.

Acknowlegments Thanks to the authorities of the University of Costa Rica (Rectoría, Vicerrectoría de Administración and Vicerrectoría de Acción Social) for support to carry out the interventions in Guanacaste and Puntarenas. Special thanks to Wajiha Sasa Marín, Director of the Office of Communication of the University of Costa Rica, for sending a journalist and a photographer to Guanacaste and Puntarenas to document the work of the interdisciplinary group. Special recognition to Audi Paniagua for the efficient management of transport provided. Thanks also to Laura Rodríguez, José Granados, Juan Carlos Sánchez and Edgar Zeledón for their help and assistance.

The Lisbon Earthquake in the French Literature

Rosarianna Zumbo and Maria S. Casella

The great Lisbon earthquake occurred in Portugal on Saturday morning, 1 November 1755. Several thousand of people were killed and more than 10,000 building destroyed in Lisbon alone. Modern research indicates that the mainshock had a moment magnitude in the range 8.5–9.0 and it occurred in the Atlantic Ocean about 200 km west-southwest of Cape St. Vincent (Gutscher et al. 2006). The earthquake generated a tsunami that produced waves about 6 m high at Lisbon and up to 20 m in several other locations. This large and catastrophic event caused damages and losses also in Marocco and Algeria (Blanc 2009), in Europe it caused considerable damage in Spain, mainly in Madrid and Seville. Shaking was also felt in France, Switzerland, and Northern Italy. However, the worst damage occurred in the south-west of Portugal and particularly Lisbon and its inhabitants were hit. At that time Lisbon was one of the most beautiful cities in Europe and it was a city legendary for its wealth and prosperity. Eighty-five percent of Lisbon's buildings were destroyed, including famous palaces, churches, libraries, and hospitals as well as the royal palace and wonderful examples of Portugal's architecture. Several buildings were also destroyed by the subsequent fire such as the new Opera House which burned to the ground. Furthermore, archives and other precious documents were completely destroyed. The 1755 event is very important since it destroyed a major cultural centre of Europe. Its widespread physical effects aroused a wave of scientific interest and research into earthquakes and it showed that an earthquake which occurs in the eastern Atlantic Ocean can create a tsunami which can cross the Atlantic and move into the shores of eastern North American. The Lisbon earthquake, the first to be studied scientifically for its effects over a large area, can be considered as the event which led to the birth of modern seismology and earthquake engineering.

Freelancer

R. Zumbo (✉)
Msida, Malta
e-mail: rosarianna.z@gmail.com

M.S. Casella
Archivio di Stato, Messina, Italy

S. D'Amico (ed.), *Earthquakes and Their Impact on Society*,
Springer Natural Hazards, DOI 10.1007/978-3-319-21753-6_31

The earthquake surely contributed to accentuate political tensions in the Kingdom of Portugal and profoundly disrupted the country's colonial ambitions. The event was widely discussed and dwelt upon by European philosophers, and inspired major developments in theodicy and in the philosophy. On this regard Vandelli (1796) "Sometimes miracles are necessary, natural phenomena, or great disasters in order to shake, to awaken, and to open the eyes of misled nations about their interests, [nations] oppressed by others that simulate friendship, and reciprocal interest. Portugal needed the earthquake to open his eyes, and to little by little escape from slavery and total ruin" (Pereira 2009).

The 1755 Lisbon earthquake can be characterized as a "turning point in human history which moved the consideration of such physical events as supernatural signs toward a more neutral or even secular, proto-scientific causation" (Dynes 1999, 2003). It was so devastating to be considered as a topic of discussion and disputation among intellectuals involved in what has come to be known as the Enlightenment and it seemed to reaffirm the presence and the responsibility of God, that many were trying to make abstract, distant and benign. As Marques (2005) asserts in his essay, this may be because this earthquake took place in All Saint day and destroying churches full of people but the main question now was if God would prevent such disaster, why did He let it happen? And why would a benevolent being consent the deaths of so many innocent people (Marques 2005)?

This brief essay wants to proof that the 1755 Lisbon earthquake had a significant impact for changing the cultural context, that's why it is appropriate to call the 1755 Lisbon earthquake the first modern disaster, as affirmed by Dyrnes (1999). Voltaire's *Candide* as well as *Poème sur le Désastre de Lisbonne* and Rousseau's *Lettre à Monsieur de Voltaire* are some of the most famous efforts to give some kind of answers to the questions raised by this terrible earthquake. But, if Voltaire's works can still be read as two philosophical attempts for undermining the passivity inspired by Leibniz's philosophy, Rousseau's *Lettre* can be considered as a kind of writing that trigger progressive ideas defining the disaster as a "social construct" (Chester 2008).

Voltaire's *Poème sur le Désastre de Lisbonne* has a meaningful subtitle "Examen de cet axiome: tout est bien", which suggest already his future attack on Leibniz and Pope thesis that "all is well" (Marques 2005). However, what Voltaire criticizes in his *Poème* is not the fact that "all is well" but the fact that it is quite difficult to say that all is well after having seen the Lisbon earthquake and its deaths and, above all, "he condemns (…) the sort of fatalism that may come from that view, as stated in José Oscar de Almeida Marques' essay (2005) and as written by Voltaire:

> Si tout est bien, disait-on, il est donc faux que la nature humaine soit déchue. Si l'ordre général exige que tout soit comme il est, la nature humaine n'a donc pas été corrompue; elle n'a donc pas eu besoin de rédempteur. Si ce monde, tel qu'il est, est le meilleur des mondes possibles, on ne peut donc pas espérer un avenir plus heureux. Si tous les maux dont nous sommes accablés sont un bien général, toutes les nations policées ont donc eu tort de rechercher l'origine du mal physique et du mal moral. (…) Le mot "Tout est bien", pris dans un sens absolu et sans l'espérance d'un avenir, n'est qu'une insulte aux douleurs de notre vie.

But in *Candide* Voltaire's attitude seems to be more sceptical. Published in 1759, *Candide* published in 1759, became quite early an international best seller (A. Betâmio de Almeida reports 30 000 copies sold only in the first year, 2005), "which became astounding at the time for a work of fiction" (Dynes 2003). Lisbon earthquake shocked so deeply Voltaire to give him the chance to express not so much the doubt of God's existence but to affirm the existence of Evil and its coexistence with God on earth (Chester 2008).

Containing thirty chapters, *Candide* is an attack on Leibniz's philosophy, which asserts that the world, no matter how we may perceive it, is necessarily the "best of all possible worlds" as well as the meaning of "all is well" (De Almeida 2005). About that Voltaire writes, with a certain bit of sarcastic humour:

> Après le tremblement de terre qui avait détruit les trois quarts de Lisbonne, les sages du pays n'avaient pas trouvé un moyen plus efficace pour prévenir une ruine totale que de donner au peuple un bel autodafé; il était décidé par l'université de Coïmbre que le spectacle de quelques personnes brûlées à petit feu, en grande cérémonie, est un secret infaillible pour empêcher la terre de trembler.

Those lines involve a ironical meaning: Voltaire wants to underline that even with the special intervention of God and the religious institution with all its ceremonial—(la) grande cérémonie—the Providence did not prevent the earthquake disaster, the suffering and the deaths to occur because of the existence of Evil and suffering in the world (Marques 2005).

On the contrary, Jean Jacques Rousseau's *Lettre à Monsieur de Voltaire* tries the more ambitious risk of defending the Providence which had been so attacked by Voltaire's works, especially in *Candide.* Rousseau must have been shocked by Voltaire ardours attack on Providence even if he knew that Providence may be often considered as a difficult topic to defend, considering especially "the general trade towards pessimism in the second half of the eighteen century" (Marques 2005) (the war of Austrian succession and the Seven years war as well as the tensions escalated to outbreak of the French revolution in 1789). Nevertheless, Rousseau pointed out on the fact that "sufferings imposed by nature are gentle and briefer (…) than evils derived from civilization [and] (…) inflicted by other human beings" (Marques 2005): "De tant hommes écrasés sous les ruines de Lisbonne, plusieurs, sans doute, ont évité de plus grands malheurs". What Rousseau wants to remark here was the fact that so many natural catastrophes are the result of social plans and institutions, as asserted by the French writer himself in his *Lettre*: "Sans quitter [du] sujet de Lisbonne, [on] conv[ient] (..) que si les habitants de cette grande ville eussent été dispersés plus également et plus légèrement logés, le dégât eut été beaucoup moindre, et peut-être nul".

In this way, Rousseau affirms that when a earthquake occurs the main responsibility of destruction and human deaths is the actions of the man in what concerns the decision of the buildings construction. This could be identified as the first concept of what we call today "vulnerability", or how A. Betâmio de Almeida writes in his essay "the concept of the probability of damage or loss should an hazard occur" (De Almeida 2005). All this new ideas and concepts are was very different from the contemporary position of Voltaire, as we already pointed out. In Rousseau's *Lettre* there is no place for any religious belief or any philosophical scepticism against the Nature or the Providence, as in Voltaire's works. The reaction of Rousseau is the symbol of progress and change and it also introduces the idea that if men construct properly their buildings the dimension of the risk decreases.

References

Blanc, P.-L. (2009). Earthquakes and tsunami in November 1755 in Morocco: A different reading of contemporaneous documentary sources. *Natural Hazards and Earth Systems Sciences, 9*, 725–738.

Chester, D. K. (2008). *The effects of the 1755 Lisbon earthquake and tsunami on the Algarve Region* (p. 34). Southern Portugal: University of Liverpool.

De Almeida B. A. (2005). The 1755 Lisbon earthquake and the genesis of the risk management concept. In *Proceedings of the 250th anniversary of the 1755 Lisbon earthquake* (pp. 1–9).

Dynes, R. R. (1999). *The dialogue between Voltaire and Rousseau on the Lisbon earthquake: The emergence of a social science view*.

Dynes, R. R. (2003). *The Lisbon earthquake in 1755: The first modern disaster*. University of Delaware—Disaster Research Center (Preliminary paper #333).

Gutscher, M.-A., Baptista, M. A., Miranda, J. M., & Miranda, J. M. (2006). The Gibraltar Arc seismogenic zone (part 2): Constraints on a shallow east dipping fault plane source for the 1755 Lisbon earthquake provided by tsunami modeling and seismic intensity. *Tectonophysics, 426*, 153–166. doi:10.1016/j.tecto.2006.02.025.

Marques, J. O. A. (2005). The paths of providence: Voltaire and Rousseau on the Lisbon earthquake. *Cadernos de historia e Filosofia de Ciencia, 15*(1), 33–57.

Pereira, A. S. (2009). The opportunity of a disaster: The economic impact of the 1755 Lisbon earthquake. *The Journal of Economic History, 69*(02), 466–499.

Vandelli, D. (1796). Modo de evitar a ruína do reino ameaçado pelos ingleses com os contrabandos, e pelos franceses com as suas excessivas pretensões. *Aritmética Política, Economia e Financas*.

GPSR Compliance
The European Union's (EU) General Product Safety Regulation (GPSR) is a set of rules that requires consumer products to be safe and our obligations to ensure this.

If you have any concerns about our products, you can contact us on

ProductSafety@springernature.com

In case Publisher is established outside the EU, the EU authorized representative is:

Springer Nature Customer Service Center GmbH
Europaplatz 3
69115 Heidelberg, Germany

www.ingramcontent.com/pod-product-compliance
Lightning Source LLC
Chambersburg PA
CBHW071405230926
51846CB00001B/12

* 9 7 8 3 3 1 9 3 4 8 5 8 2 *